Exposed Cross-Sections of the Continental Crust

NATO ASI Series

Advanced Science Institutes Series

A Series presenting the results of activities sponsored by the NATO Science Committee, which aims at the dissemination of advanced scientific and technological knowledge, with a view to strengthening links between scientific communities.

The Series is published by an international board of publishers in conjunction with the NATO Scientific Affairs Division

A Life Sciences **B Physics**	Plenum Publishing Corporation London and New York
C Mathematical **and Physical Sciences** **D Behavioural and Social Sciences** **E Applied Sciences**	Kluwer Academic Publishers Dordrecht, Boston and London
F Computer and Systems Sciences **G Ecological Sciences** **H Cell Biology**	Springer-Verlag Berlin, Heidelberg, New York, London, Paris and Tokyo

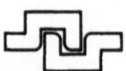

Series C: Mathematical and Physical Sciences - Vol. 317

Exposed Cross-Sections of the Continental Crust

edited by

Matthew H. Salisbury

Atlantic Geoscience Centre,
Geological Survey of Canada,
Bedford Institute of Oceanography,
Dartmouth, Nova Scotia, Canada

and

David M. Fountain

Department of Geology and Geophysics,
University of Wyoming,
Laramie, Wyoming, U.S.A.

Kluwer Academic Publishers

Dordrecht / Boston / London

Published in cooperation with NATO Scientific Affairs Division

Proceedings of the NATO Advanced Study Institute on
Exposed Cross-Sections of the Continental Crust
Killarney, Ontario, Canada
September 17–27, 1988

Library of Congress Cataloging-in-Publication Data

NATO Advanced Study Institute on Exposed Cross-Sections of the
  Continental Crust (1988 : Killarney, Ont.)
    Exposed cross-sections of the continental crust : proceedings of
  the NATO Advanced Study Institute on Exposed Cross-Sections of the
  Continental Crust, Killarney, Ontario, Canada, 17-27 September 1988
  / edited by Matthew H. Salisbury and David M. Fountain.
        p.   cm. -- (NATO ASI series. Series C, Mathematical and
  physical sciences ; vol. 317)
    Includes bibliographical references and index.
    ISBN-13:978-94-010-6788-1   e-ISBN-13:978-94-009-0675-4
    DOI: 10.1007/978-94-009-0675-4

    1. Earth--Crust--Congresses.   2. Continents--Congresses.
  3. Geology, Stratigraphic--Precambrian--Congresses.   4. Geology,
  Stratigraphic--Cambrian--Congresses.   I. Salisbury, Matthew H.
  (Matthew Harold), 1943-   .   II. Fountain, David.   III. Title.
  IV. Series: NATO ASI series.   Series C.   Mathematical and physical
  sciences ; no. 317.
  QE511.N274   1988
  551.1'3--dc20   90-44393

ISBN-13:978-94-010-6788-1

Published by Kluwer Academic Publishers,
P.O. Box 17, 3300 AA Dordrecht, The Netherlands.

Kluwer Academic Publishers incorporates the publishing programmes of
D. Reidel, Martinus Nijhoff, Dr W. Junk and MTP Press.

Sold and distributed in the U.S.A. and Canada
by Kluwer Academic Publishers,
101 Philip Drive, Norwell, MA 02061, U.S.A.

In all other countries, sold and distributed
by Kluwer Academic Publishers Group,
P.O. Box 322, 3300 AH Dordrecht, The Netherlands.

Printed on acid-free paper

TABLE OF CONTENTS

Geophysical Structure and Properties of the Crust:

Overview:

PREFACE

In the Fall of 1988, 64 geologists and geophysicists from 11 countries met in Killarney, Ontario, on the north shore of Lake Huron to examine evidence that suggests that the continental crust is exposed in cross-section at several key locations on the Earth's surface. The meeting, which was held under NATO auspices as an Advanced Study Institute, was a landmark event in that it was the first time that many of the lead scientists working on these complexes in relative isolation around the world had ever gathered together to compare results.

The present volume is a compendium of the invited lectures given on the principle sections, plus an array of supporting papers on these and other sections as well as on related topics such as crustal emplacement mechanisms, deformation and rheology. Nearly all of the best known sections are represented, including the Ivrea Zone, Calabria, the Kapuskasing Zone, Fiordland and many others. It is our hope that this Volume will serve as a reference for Earth scientists who are trying to understand levels of the crust not normally exposed to view, as well as a point of departure for new research and a teaching aid to new entrants in this relatively new field of study.

We wish to thank NATO for funding the meeting; our colleagues on the Steering Committee, John Percival from the Geological Survey of Canada, Stephan Mueller from ETH in Switzerland and Jack Oliver from COCORPS, for their assistance in organizing the meeting; John Percival and Tony Davidson of the Geological Survey of Canada for organizing and leading field trips to the Kapuskasing Zone and the Grenville Front; Jane Barrett of Dalhousie University for her invaluable services as meeting coordinator; and the proprietors of the Red Pine Lodge and Killarney Mountain Lodge for their gracious hospitality during our meetings and discussions.

Matthew H. Salisbury, David M. Fountain
May 1990.

THE IVREA CRUSTAL CROSS-SECTION (NORTHERN ITALY AND SOUTHERN SWITZERLAND)

A. ZINGG
Geologisches Institut
Bernoullistrasse 32
CH-4056 Basel
Switzerland

KEY-WORDS. Ivrea Zone / Southern Alps / lower continental crust / granulite facies / exposed crustal cross-section

ABSTRACT. The Ivrea crustal section is situated at the internal border of the Alpine arc, south and east of the Insubric Line. A large gravity anomaly and inverted seismic velocities have lead to the model of an exposed crustal section comprising the following units: 1) the Ivrea Zone with peridotites, mafic rocks and paragneisses in the amphibolite and granulite facies (lower crust and basal intermediate crust); 2) the Strona-Ceneri and Val Colla Zones with granitic gneisses, paragneisses and micaschists in the amphibolite facies (intermediate crust); 3) the Late Paleozoic to Tertiary sedimentary cover of the aforementioned units (upper crust). The crustal section is not strictly continuous, since it is truncated by several, variously aged faults.

1. Introduction

This review of the major features and history of the Ivrea crustal cross-section primarily addresses the following questions: What are the observations leading to the view of the Ivrea Zone as an exposed section of continental crust? What were the lower crustal conditions preserved in the section? To what extent were the original deep crustal features subsequently modified? What geotectonic history lead to the exposure of this lower crustal section? Can the Ivrea crustal section be used as a model for continental crust?

2. Regional framework

The European continent comprises from north to south the Precambrian shield, the Caledonian, Variscan and Alpine belts. The Alpine chain contains basement units yielding Variscan and partly also Caledonian and older radiometric ages. The Ivrea Zone is a piece of Paleozoic crust that was incorporated in Alpine orogen as a relatively rigid slice (Figs. 1 and 2).

During Mesozoic extension, the Ivrea section was located at the southern passive continental margin of the Tethyan ocean and was part of the Adriatic plate, a promontory of Africa. The polyphase Alpine collision of the Adriatic and European plates started in the Upper Cretaceous with high-pressure/low-temperature metamorphism observed in ophiolites and in some basement units (e.g.,Dora Maira, Sesia Zone). Thus, not only oceanic crust, but also continental crust was involved in Eoalpine subduction tectonics that brought low-density granitic and supracrustal rocks as deep as 70 km (Chopin, 1984). It is assumed that the basic geometry of the Ivrea density inversion (Fig. 2) was acquired at that time. This crustal configuration was subsequently modified during the Paleogene nappe thrusting episode in the Central and Western Alps and during the Oligo-Miocene episode with deformations concentrated in the external (Helvetic nappes) and internal margins (the so-called Root Zone and Southern Alps) of the Alpine

1

M. H. Salisbury and D. M. Fountain (eds.), Exposed Cross-Sections of the Continental Crust, 1–19.
© 1990 *Kluwer Academic Publishers.*

2

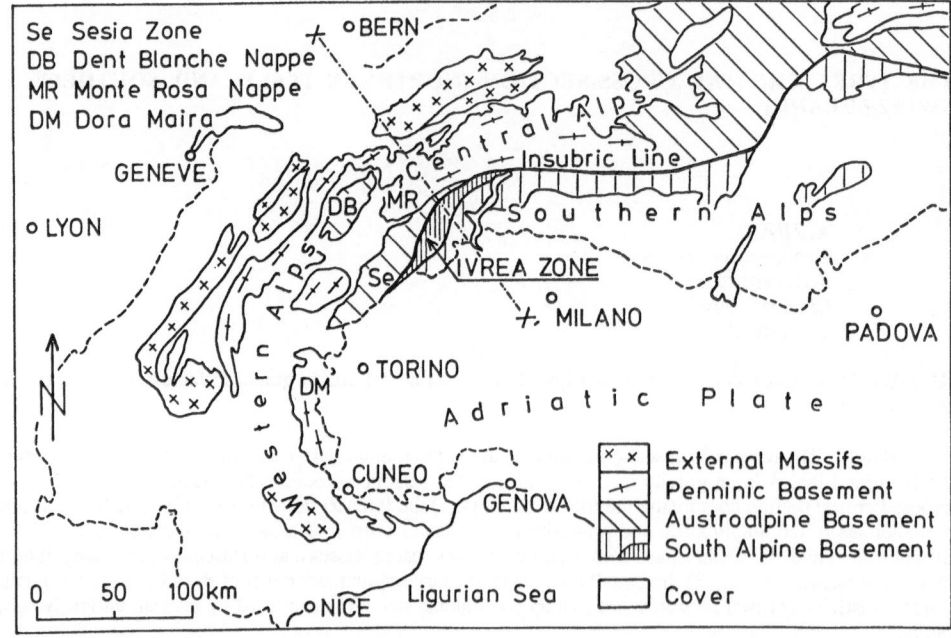

Figure 1: Location of the Ivrea Zone within the Alpine orogen. X-X': trace of profile figure 2.

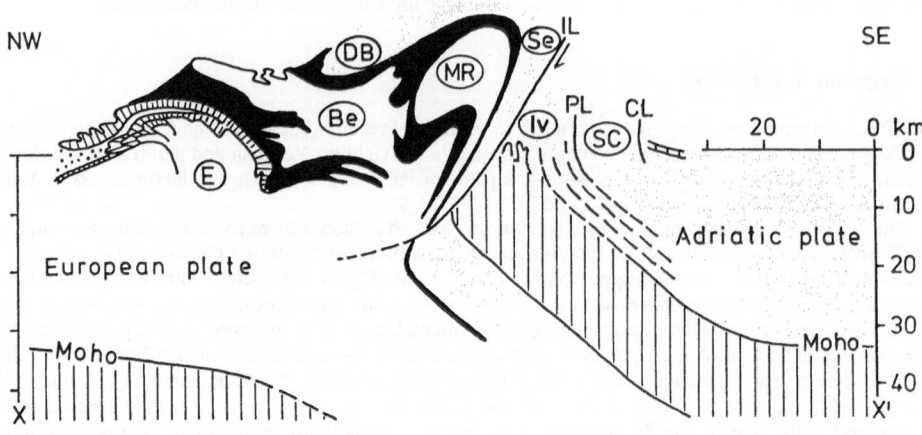

Figure 2: Simplified section across the Alps modified from Schmid et al. (1987). E: External Massifs, Be: Bernhard Nappes, MR: Monte Rosa Nappe, DB: Dent Blanche Nappe, Se: Sesia Zone, Iv: Ivrea Zone, SC: Strona-Ceneri Zone; IL: Insubric Line, PL: Pogallo Line, CL: Cremosina Line.

orogen. Dextral transpression prevailed during all these episodes with east-west shortening in the Western Alps and north-south shortening in the Central Alps.

The Ivrea Zone is situated within the internal part of the Alpine arc (Fig. 1), where the east-west striking Central Alps merge with the Western Alpine arc. It represents the deepest unit of the Southern Alps and is bounded to the north and west by the Insubric Line. This major fault zone separates the Central and Western Alps, with polyphase Alpine pervasive deformation and metamorphism, from the South Alpine basement with virtually no Alpine metamorphism and only localized faulting, thrusting and folding.

3. Geophysical data

The internal part of the Western Alpine arc is characterized by large geophysical anomalies (Fig. 3). These anomalies support a model for the crustal configuration in which a south and east dipping sliver of deep crust and mantle (geophysical Ivrea body) overrides crustal rocks of lower density (Fig. 2).

A strong positive Bouguer anomaly reaching values of + 70 mgal is observed between Cuneo and Locarno (Fig. 3A; and e.g., Vecchia, 1968; Guillaume and Guillaume, 1980). If the negative effects of the Alpine root and sediments of the Po-plain are subtracted, a Bouguer anomaly of + 170 mgal is obtained for the eastern part of the Ivrea body (Kissling, 1984). An attenuated anomaly is documented by geoid measurements east of Locarno (Bürki, this volume). However, its unequivocal geological interpretation is not yet established.

Refraction seismic studies revealed a large-scale velocity inversion at depths between 10 and 45 km. It was concluded that the ancient South Alpine Moho nearly reaches the surface in the Ivrea Zone region (Fig. 3B and e.g., Berckhemer, 1968; Mueller et al., 1980; Giese et al., 1982). First interpretations of the ECORS-CROP reflection seismic profile (Bayer et al., 1987) suggest a more complex configuration with several mantle imbricates beneath the internal part of the Western Alps. An other model is proposed by Laubscher (in press) who considers the Ivrea body as a composite structure. According to this new model the upper part of the body comprising the Ivrea section is geometrically unconnected with the Adriatic lower crust and upper mantle, wedged during the Neogene beneath internal parts of the Central Alps.

The total intensity of the geomagnetic field ranges from 45000-48000 nT (nanotesla) in the Ivrea Zone region (Wagner et al., 1984). After subtraction of the regional field, an anomaly with a maximum of + 650 nT west of Torino remains (Fig. 3C). Several submaxima are observed whose trend deviates slightly from the average strike of the Ivrea Zone. Variable depths of the top of the Ivrea body (Guillaume and Guillaume, 1980) as well as variations in its lithological composition and structural discontinuities of the arc-shaped Ivrea body may produce this effect.

4. Geology of the Ivrea crustal section

4.1. GEOLOGICAL UNITS

The surficial expression of the anomalous crustal configuration revealed by geophysics is the exposure of different levels of the South Alpine crust (Fig. 4). From the northwest to the southeast, i.e. from the originally deepest to shallowest levels, the following basement units are observed: (1) The Ivrea Zone (= zona Ivrea-Verbano or zona diorito-kinzigitica of the Italian literature) which represents lower crust and basal intermediate crust; (2) the Strona-Ceneri Zone; and (3) the Val Colla Zone. These latter two zones comprise the "Formazione dei Laghi" and represent intermediate crust. Late Paleozoic to Tertiary sediments are exposed further to the south. Several variously aged faults separate the different units, so the Ivrea crustal cross-section is not strictly continuous.

4.2. FAULTS AND CONTACTS BETWEEN THE UNITS

The Insubric Line forms the northern and western limit of the Ivrea section (Fig. 4) and consists of a mylonite belt that accommodated both uplift and backthrust of the Central Alps onto the Ivrea Zone and slightly younger dextral strike slip movement during the Oligo-Miocene (S.M. Schmid et al., 1987, 1989).

4

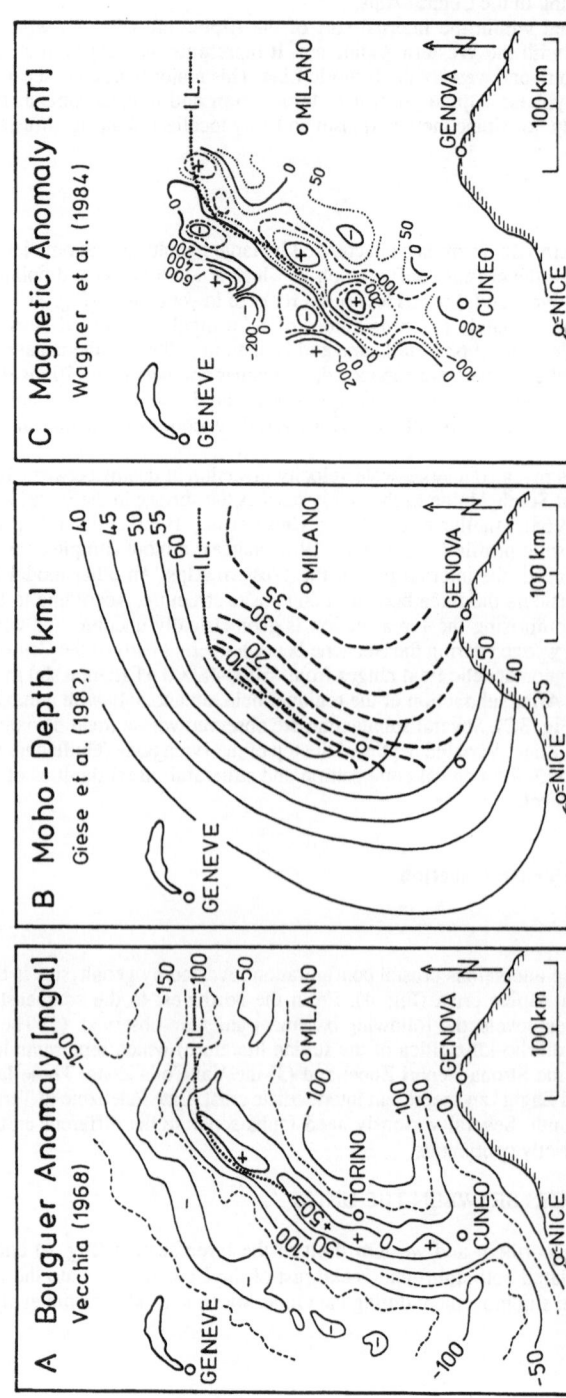

Figure 3: Geophysical anomalies in the Ivrea Zone region. A: Bouguer anomalies according to Vecchia (1968). B: Moho depth according to Giese et al. (1982) based on refraction seismics. The superposition of the European and Adriatic Moho beneath the Insubric Line is confirmed by reflection seismic profiling (Bernoulli et al., in press). C: Magnetic anomalies according to Wagner et al. (1984)

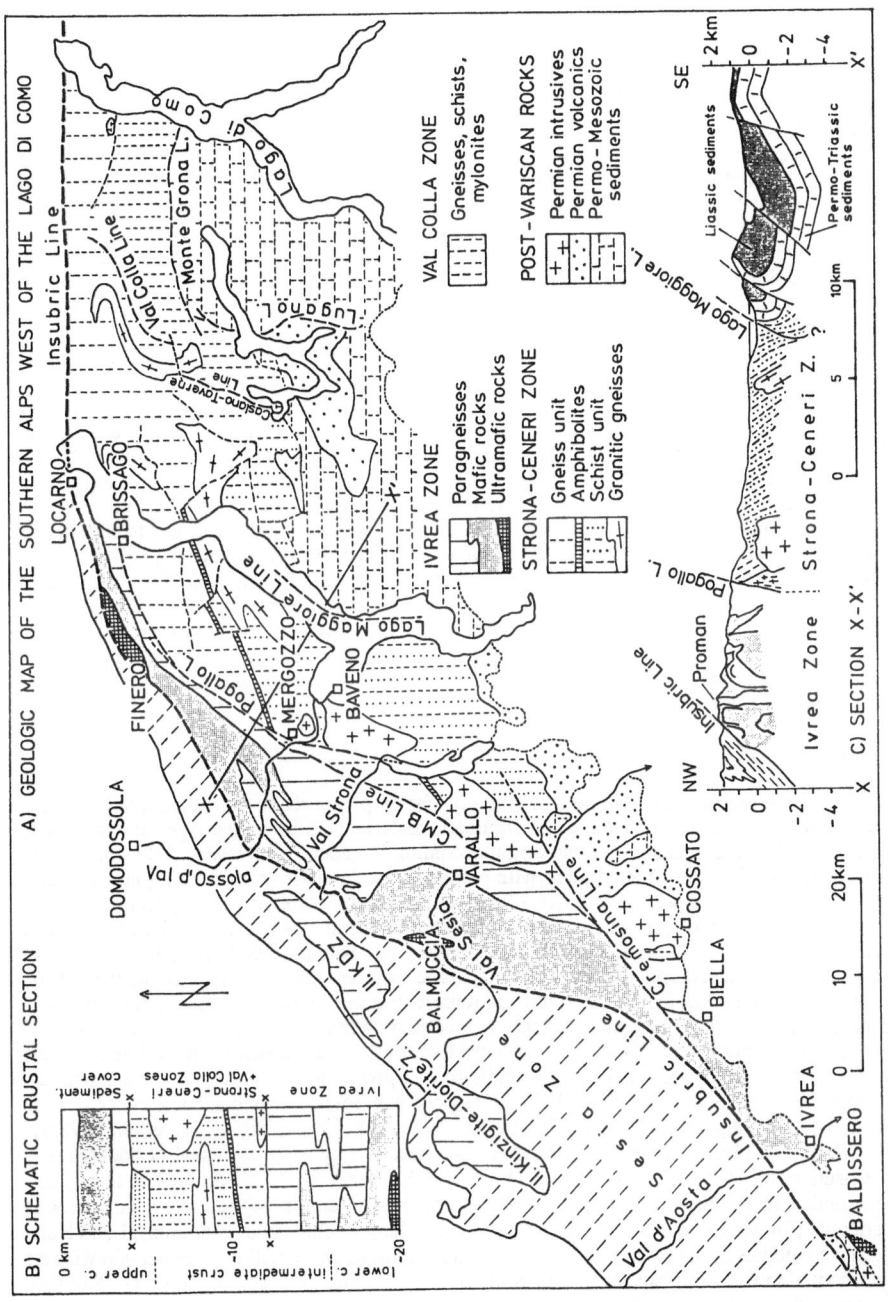

Figure 4: A: Geological sketch map of the Southern Alps. B: Restored crustal section for the Late Jurassic. Alpine tectonics, especially dextral strike-slip movements make such a reconstruction approximate. C: Profile across the Southern Alps (simplified from Zingg et al., in press).

The contact between the Ivrea and Strona-Ceneri Zones is characterized by several features: 1) mostly concordant compositional banding and foliation; 2) localized faults; 3) small Late Paleozoic mafic intrusives and dykes. Locally, the lithologic contacts are gradational and it is difficult to place a boundary between the two zones. According to Boriani and Sacchi (1973) and Boriani et al. (1977, in press) the whole contact is tectonic and formed by the so-called Cossato-Mergozzo-Brissago Line (CMB Line), a purported Late Paleozoic transcurrent fault. The contact between the Ivrea and Strona-Ceneri Zones bears a basic, yet unsolved question for the Pre-Mesozoic evolution of the South Alpine crustal section: do these two zones represent a continuous crustal section already in the Paleozoic or were they formed at different places and sealed together in the Late Carboniferous to Early Permian for example (Handy, 1986, 1987; Borghi, 1989; Boriani et al., in press).

In the Ossola region and to the east, the Pogallo Line (Boriani, 1970) coincides with part of the contact between the Ivrea and Strona-Ceneri Zones. In addition, the southeastern rim of the Ivrea Zone consists of a km-wide shear zone (Pogallo Ductile Fault Zone; Handy, 1987). According to Boriani and co-workers (1973, 1977, in press) the Pogallo Line is a Late Paleozoic transcurrent fault. For Hodges and Fountain (1984), Handy (1986, 1987) and S.M. Schmid et al. (1987), the Pogallo Line represents a rotated deep-seated, low angle fault that was active during Early Mesozoic rifting of the Tethyan continental margin.

The border between the Strona-Ceneri and the Val Colla Zone is a 1 km thick mylonite zone (Val Colla Line; Reinhard, 1964) formed under greenschist facies conditions. A Late Paleozoic age of shearing is inferred from the synkinematic metamorphic grade and the radiometric mineral ages (Mc Dowell, 1970). Similar tectonites are found to the southwest along the Caslano-Taverne Line. The latter fault must have been reactivated because unmetamorphosed Carboniferous sediments are pinched in along this line.

Permian volcanics and intrusives are juxtaposed along the Alpine aged Cremosina Line. In addition to a vertical component, a dextral displacement of about 12 km is postulated (Boriani and Sacchi, 1974).

Facies and thicknesses of the Early Mesozoic sediments differ strongly on both side of the Lago Maggiore and Lugano Lines leading to their interpretation as Liassic synsedimentary normal faults (Van Houten, 1929; Bernoulli, 1964; Kälin and Trümpy, 1977). These faults were extensively reactivated during the Alpine orogeny. The north-south striking Lugano Line merges with the east-west striking Monte Grona Line. This line may represent the basal part of the Lugano normal fault exposed during the Alpine steepening of the whole section (Bertotti, in press).

4.3. LITHOLOGIES

The Ivrea Zone is composed of amphibolite and granulite facies paragneisses (so-called kinzigites and stronalites, respectively) associated with mafic and ultramafic rocks and marbles. Four types of mafic rocks can be distinguished: 1) mafic rocks of oceanic origin according to geochemical trends (Sills and Tarney 1984; Mazzucchelli and Siena 1986) alternating with paragneisses; 2) mafic rocks forming larger bodies like the "Anzola Gabbro"; 3) banded mafic rocks, often associated with ultramafic rocks, occurring in the granulite facies part of the Ivrea Zone. These three occurrences of mafic rocks have experienced the same regional metamorphism and deformation as the paragneisses (R. Schmid, 1967; Bertolani, 1968). In contrast, the fourth type of mafic rocks shows relict magmatic features. This latter type forms the dioritic rim of the Mafic Formation in the Val Sesia region and represents a deeply seated intrusion presumably of Late Paleozoic age. The Mafic Formation commonly represented on maps of the Ivrea Zone (e.g., Fig. 4) includes the mafic rock types 2-4 and thus is a composite of variously aged intrusives. In the Val Sesia, the Mafic Formation consists of about 2 km of layered mafic rocks showing locally intrusive contacts with the Balmuccia peridotite, about 3.5 km of relatively homogeneous gabbroic rocks and about 2 km of dioritic rocks at the rim of the paragneisses (e.g., Rivalenti et al., 1975, 1981, 1984; Bigioggero et al., 1978/79; Pin and Sills, 1986). The ultramafic rocks of the Ivrea Zone comprise pyroxenites, spinel-peridotites (e.g. Balmuccia; Garuti et al., 1978/79; Rivalenti et al., 1978/79; Shervais, 1979; Sinigoi et al., 1983), phlogopite-peridotites (e.g. Finero; Lensch, 1968, 1971; Coltori and Siena, 1984) and two occurrences of kelyphite-peridotite (former garnet-peridotites reequilibrated under granulite facies conditions, Lensch and Rost, 1972). The peridotites occur as isolated, variously sized bodies in the granulite facies domain with the exception of one peridotite body associated with eclogitic amphibolites (Boriani and Peyronel Pagliani, 1968) at the southeastern border of the Ivrea Zone.

The Strona-Ceneri Zone consists of amphibolite facies granitic gneisses, various paragneisses and micaschists (Boriani et al., 1977; in press). A northern gneissic unit ("gneiss dei Laghi") is separated by a small band of amphibolites and hornblende-bearing gneisses from a schistose unit ("scisti dei Laghi") in the southeast. Both series contain bodies of granitic gneiss (Boriani et al., 1977, 1982/83;). Postmetamorphic Permian granites of the Baveno suite intrude this zone. The Val Colla Zone is made of micaschists, and subordinate amphibole-bearing gneisses and granitic gneisses metamorphosed under amphibolite and (retrograde?) greenschist facies conditions (Reinhard, 1964). Parts of this zone are mylonitized.

In the cover of the South Alpine basement sedimentation initiated with terrigenic clastics in the Late Carboniferous, followed by Permian volcanics and clastics. A marine environment was restored in the Early to Middle Triassic with the deposition of shallow water carbonates. During the Liassic, platforms and basins with siliceous limestones document the continental margin position of the Southern Alps with respect to the Tethyan ocean (Bernoulli, 1964; Kälin and Trümpy, 1977; Winterer and Bosellini, 1981). In the Late Jurassic, the former platform and basin domains contain deep water sediments, indicating the subsidence of the entire continental margin. The Alpine orogenic cycle begins with Cenomanian to Campanian flysch sediments, which lie concordantly on the pelagic sediments. A molasse-type sequence with detritus originating from the Central Alps was deposited during the Late Oligocene and Early Miocene.

4.4 METAMORPHISM

Regional metamorphism is the dominant feature of the Ivrea and Strona-Ceneri Zones and only few relics of an older high-P/low-T metamorphism are known. After the climax of regional metamorphism localized shear zones formed under variable metamorphic conditions.

In the Ivrea Zone the metamorphic grade increases from the amphibolite to the granulite facies towards the northwest (Fig. 5 and R. Schmid, 1967; Bertolani, 1968; Zingg, 1980). Most of the textures related to regional metamorphism are well equilibrated and annealed and postdate two or three phases of deformation (Steck and Tièche, 1976; Kruhl and Voll, 1976). Textures of incomplete prograde reactions of the type described by Carmichael (1969) are observed in the paragneisses. Retrograde reaction rims are occasionally found in unsheared rocks, especially in the impure marbles. Migmatitic rocks occur frequently in the upper amphibolite facies part of the Ivrea Zone. Partial melting is not only postulated from local field observations but also from bulk geochemistry. R. Schmid (1978/79) and Sighinolfi and Gorgoni (1978)

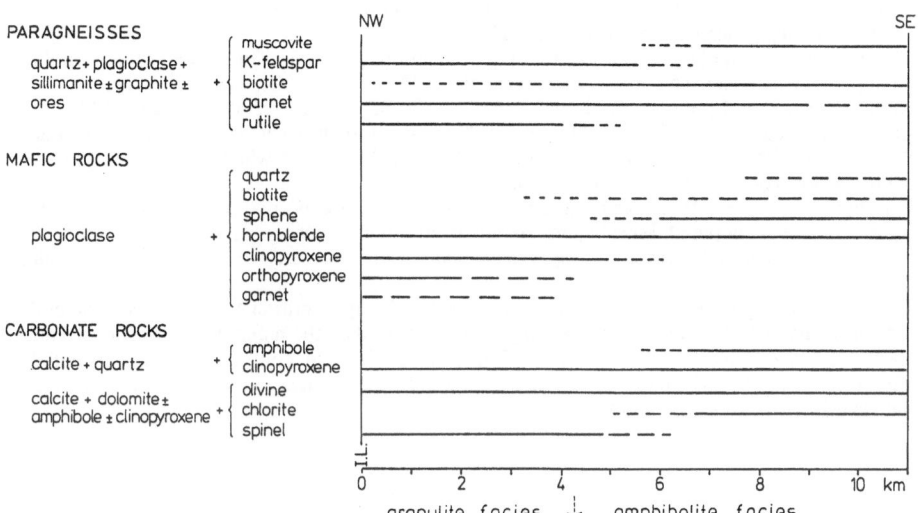

Figure 5: Changes of the mineral assemblages with increasing metamorphic grade in the central part of the Ivrea Zone. Cordierite is frequently found in the Val Sesia region and to the southwest. See R. Schmid and Wood (1976) and Zingg (1980) for location of mineral isograds.

consider the paragneisses to be restites resulting from degranitization of the Ivrea paragneisses, because their compositions are not found in a sedimentary environment. The Nd-systematics of the paragneisses are consistent with such an interpretation (Voshage et al., 1987).

The pressure and temperature estimates of regional metamorphism scatter enormously. For the granulite facies part of the Ivrea Zone temperatures between 500 and 940 °C and pressures between 5 and 11 kb are reported (R. Schmid and Wood, 1976; Hunziker and Zingg, 1980; Sills, 1984). These estimates depend strongly on the applied thermometer and barometer as well as on the calibration, modes of measuring and retrograde cation exchange. R. Schmid et al. (1988) have found a marked jump in PT estimates between the upper amphibolite facies (5.5 kb / 730 °C) and the granulite facies (8.3 kb / 940 °C) in the Val d'Ossola region. This jump coincide with shear zones at the amphibolite-granulite boundary (Brodie and Rutter, 1987). In other places the transition from the amphibolite to the granulite facies seems continuous.

In the Strona-Ceneri Zone, the metamorphic grade increases weakly and diffusely towards the northwest (Boriani et al., 1977) and thus contrasts with the strong increase in the Ivrea Zone. No jump of the regional metamorphic grade is observed at the contact between the two zones. Paragneiss assemblages with garnet, kyanite and staurolite and a later growth of fibrolitic sillimanite are common (Boriani et al., 1977, in press; Borghi, 1988, 1989). The latter author obtained pressures of 6.5-7 kb and temperatures of 650-660 °C for the peak of the regional metamorphism in the Strona-Ceneri Zone. During an initial stage of retrograde metamorphism, coronas of margarite were formed around kyanite according to the same author.

Little is known about metamorphism in the Val Colla Zone. Reinhard (1964) reports garnet-kyanite-staurolite assemblages and also andalusite-bearing assemblages from the southeastern part of the zone. Thus the Val Colla Zone was positioned at about the same level as the Strona-Ceneri Zone during regional metamorphism. In the Val Colla fault zone certain mylonitic and retrograde series show staurolite and kyanite relics in a fine-grained biotite-bearing matrix.

5. Radiometric age determinations

The radiometric ages from the Ivrea and Strona-Ceneri Zones compiled in figure 6 show a small Early Paleozoic cluster (Caledonian orogeny) and a large group reaching from the Late Paleozoic (Variscan orogeny) to the Early Mesozoic (Tethyan rifting). There is little consensus in the interpretation of several of these data, as discussed by Zingg (1983), Boriani et al. (1985; in press), Zingg et al. (in press). However, the following general comments can be made:

1) The different systems (mineral, whole rock) give a normal age succession as expected from their closing temperatures: Zircon and total rock, monazite U-Pb (ca. 600 °C), muscovite (ca. 500 °C), muscovite K-Ar (ca. 350 °C), biotite Rb-Sr and K-Ar (ca. 300 °C), zircon fission track (ca. 240 °C) and apatite fission track (ca. 120 °C). Remarkably, the Sm-Nd mineral isochrons yield ages in the same range as the Rb-Sr muscovites. These isochron ages are largely defined by garnet which show extended cation mobility during slow cooling, at least for main elements.

2) For the majority of the Mesozoic mica ages there is no indication of mixing between Variscan and Alpine ages (Zingg et al., in press). Most biotites are unaltered and yield concordant Rb-Sr and K-Ar ages if measured with both methods. However, mixing ages are obtained if chloritized biotites are measured (Bürgi and Klötzli, in press).

3) The radiometric ages document discontinuities in the South Alpine crustal section and the role of the faults separating the different units. Most mica ages are younger than the major Variscan unconformity (about 305 Ma). The Ivrea and Strona-Ceneri Zones have different cooling histories, except the northwestern rim of the Strona-Ceneri Zone which shows locally mica ages similar to those in the Ivrea Zone (Mc Dowell, 1970; Hunziker, 1974).

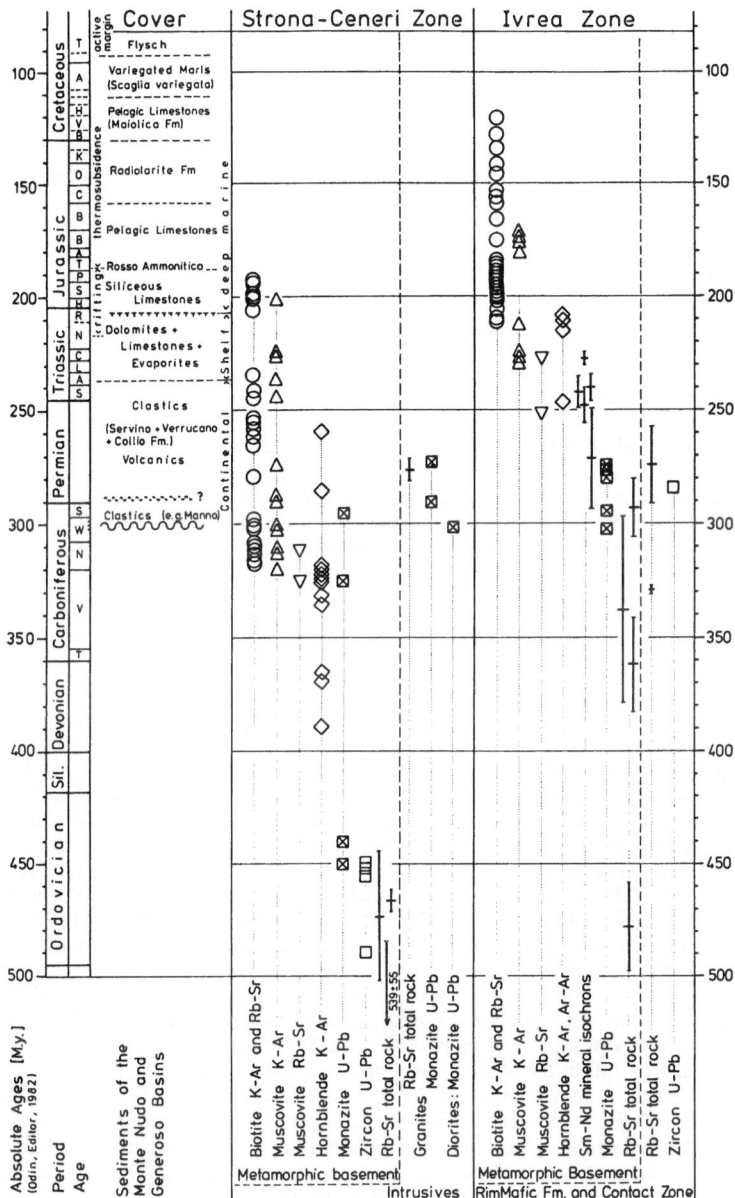

Figure 6: Synoptic view of stratigraphy and radiometric ages of the South Alpine basement (from Zingg et al., in press).

6. History of the Ivrea section

6.1. EVOLUTION PRIOR TO THE REGIONAL METAMORPHISM

The knowledge of the early evolution of the Ivrea section is very fragmentary because regional metamorphism has erased most traces of the previous history. Discordant paragneiss zircon suites from the Ivrea and Strona-Ceneri Zones have an upper intercept with the concordia of 1900-2500 Ma (Köppel and Grünenfelder, 1971; Köppel, 1974) indicating an Early Proterozoic source for the sedimentary material. Crustal residence model ages of 1200 to 1800 Ma are obtained from the Nd isotopes of the Ivrea paragneisses (Voshage et al., 1987). From Sr model calculations the paragneiss material was deposited between 480 and 700 Ma (Hunziker and Zingg, 1980). In the eastern part of the South Alpine basement, pollen of Cambrian to Silurian age were found in phyllites (Sassi et al., 1984; Gansser and Pantic, 1988). Thus material from a Proterozoic source was deposited during the lower Paleozoic. Sills and Tarney (1984) propose an accretionary wedge setting to account for the association of these continental sediments with the mafic rocks of oceanic origin (type 1 mafic rocks). Such a tectonic setting also explains the incorporation of series with high pressure metamorphism, like the kelyphite-peridotites (Lensch and Rost, 1972) and eclogitic amphibolites (Boriani and Peyronel Pagliani, 1968; Buletti, 1983).

Sm-Nd isochrons of about 600 Ma were obtained on peridotites and on the layered series of the Mafic Formation (Pin and Sills, 1986; Voshage et al., 1987). They were tentatively interpreted by the latter authors as dating a crustal formation process involving the differentiation of undepleted mantle into depleted mantle (residual peridotites) and gabbroic crust (lower part of the Mafic Formation). Such an interpretation is not easily reconcilable with other isotopic dating results and is not substantiated by detail investigations of the Balmuccia peridotite and associated mafic dykes (Voshage et al., 1988).

6.2. REGIONAL METAMORPHISM

An Ordovician, Carboniferous and Permian age was proposed for the peak of the regional metamorphism in the Southern Alps based on different interpretations of the Rb-Sr whole rock isochrons and the U-Pb mineral ages. For Allègre et al. (1974) multistage and multiepisodic models for the U-Pb data lead to the interpretation of a Cadomian (520-580 Ma) event overprinted by Variscan amphibolite to granulite facies metamorphism. Hunziker and Zingg (1980) proposed an Ordovician age for the regional metamorphism and anatexis followed by a very slow cooling. Their interpretation is based on the 478 Ma whole rock isochron obtained on 20-40 kg paragneiss samples and the paragneiss zircons from the Strona-Ceneri Zone. These discordant zircon suites have a lower intercept with the concordia between 450 and 500 Ma which corresponds approximately with two concordant monazite ages (Köppel and Grünenfelder, 1971, 1978/79; Köppel, 1974). For Boriani et al. (1985, 1988, in press), these zircons were rejuvenated during granitic intrusions into very low grade metamorphic sediments. These intrusions are dated by a Rb-Sr whole rock isochron on orthogneisses of 466 Ma (Boriani et al., 1982/83) and by zircons with U-Pb ages ranging from 429 to 487 Ma (Köppel and Grünenfelder, 1971). The regional metamorphism is Variscan for Boriani and co-workers (e.g.1982/83, 1988, in press) and is dated by Rb-Sr muscovites with ages around 325 Ma in the Strona-Ceneri Zone and by the Rb-Sr isochrons of 338 Ma (Graeser and Hunziker ,1968; age recalculated with lambda = 1.42) on small paragneiss samples and bands in the Ivrea Zone. The Permian age for the climax of the regional metamorphism proposed by Pin (1986) and Bürgi and Klötzli (in press) is based on studies of the dioritic rim of the Mafic Formation and contact zone in the Val Sesia region. Pin (1983) obtained a concordant U-Pb zircon age of 285 Ma on a dioritic sample. He assumes a synmetamorphic intrusion of the Mafic Formation and concludes that the metamorphism must be Permian (see also Pin and Vielzeuf, 1983). Bürgi and Klötzli (in press) obtained a Rb-Sr whole rock isochron yielding 274±17 Ma on migmatitic paragneisses at the contact with the Mafic Formation. They argue that partial melting dated by the isochron occurred during the climax of the thermal evolution.

It would be helpful for future discussions on the age of regional metamorphism to consider the following points: 1) The interpretations of the Rb-Sr whole rock isochrons are based on circumstantial arguments and not on precise knowledge on Sr-isotope homogenization processes in metamorphic rocks; 2) There are indications of a metamorphism prior to the dominant regional metamorphism (high-P relics, polymetamorphism in the scisti dei Laghi; Borghi, 1988, 1989; Boriani et al., in press); 3) The Ivrea and Strona-Ceneri Zones may have been juxtaposed after the regional metamorphism (CMB Line, Pogallo

Line). Thus the two zones may have evolved separately in different tectonic settings prior to Early Mesozoic time. Moreover the thermal climax may not be synchronous in all units. In the Strona-Ceneri Zone the regional metamorphism is obviously older than the Permian granitic intrusions, i.e. older than 270 Ma. The Ivrea Zone in the Varallo region may well have attained the highest temperatures during the intrusion of the rim of the Mafic Formation (type 4 mafic rocks). However, this is a local and not a regional metamorphic feature. The Permian ages of Pin (1983) and Bürgi and Klötzli (in press) are likely to date the intrusion of the rim of the Mafic Formation and not the regional thermal peak.

Although none of the proposed ages for the regional metamorphism give a satisfactory explanation to all available data, all authors agree that high temperatures prevailed until Early Permian time in the Ivrea Zone. This is documented by a Rb-Sr isochron of 293 Ma obtained on the phlogopite-peridotites (Voshage et al., 1987) and by the U-Pb ages. The monazite ages of about 270 Ma coincide with the lower intercept of discordant paragneiss zircon suites with the concordia (Köppel, 1974). The overall cooling of the Ivrea Zone started in the Early Permian.

The views on the post Variscan evolution are also controversial. For Vai and Cocozza (1986) and Boriani et al. (1988, in press), the Ivrea section acquired its present-day steep orientation already in the Late Paleozoic, a view previously also proposed by Hunziker and Zingg (1980). Boriani et al. (1988, in press) argue that the mafic dykes along the Ivrea/Strona-Ceneri boundary intruded the already steeply dipping series and that the Permian Baveno granite was only moderately tilted after its intrusion in a shallow crustal level. In this model, the Ivrea Zone represents the central part of the Variscan orogen, and the Permian cooling is related to its emplacement into a shallow crustal level. Consequently, the Ivrea Zone is, like many other high-grade terrains, the exposed deep portion of an orogen and has few common features with the in situ unexposed lower crust at depth. A further consequence of a Late Paleozoic emplacement and steepening of the Ivrea cross-section is that most Permian and Mesozoic radiometric ages of the Southern Alps would have to be rejuvenated ages. Alternatively, rotation and steepening of the Ivrea section and its emplacement into surficial crustal levels is largely due to Alpine orogeny (e.g. Laubscher, 1971a, b; S.M. Schmid et al., 1987; Zingg et al., in press). In this case, Late Paleozoic and Early Mesozoic cooling are due to various stages of extensional uplift (Handy and Zingg, in press). This view is considered in the present paper.

6.3. LATE PALEOZOIC MAGMATISM

Late Carboniferous and Early Permian time are characterized by magmatic activity in all crustal levels of the Southern Alps: Basalts and rhyolites in the cover, granites of the Baveno suite (Boriani et al., 1988) in Strona-Ceneri Zone, slightly older mafic intrusives at the contact between the Ivrea and Strona-Ceneri Zone (Giobbi Origoni et al., 1988) and the dioritic rim of the Mafic Formation in the Ivrea Zone (Bigioggero et al., 1978/79). According to the Sr and Nd isotope systems, all magmatic rocks show a pronounced mantle component indicating magma mixing and assimilation (Hunziker and Zingg, 1980; Pin and Sills, 1986; Voshage et al., 1987; Stille and Buletti, 1987; Pinarelli et al., 1988). To explain the geochemical trend and mantle signature of this Late Paleozoic magmatism the subduction of oceanic crust of a "Proto-Tethys" is proposed by Stille and Buletti (1987).

The granites within the Strona-Ceneri Zone and the small mafic intrusions at the contact with the Ivrea Zone cut discordantly the foliation related to regional metamorphism and locally contain inclusions of metamorphic basement. The dioritic rim of the Mafic Formation in Val Sesia is concordant to the foliation of the adjacent paragneiss series and the contact between the two units is not clear-cut. Anatectic melting of the paragneisses is observed near the contact (Bürgi and Klötzli, in press). Thus the postmetamorphic character of the intrusions is obvious at shallow and intermediate crustal levels but not in deep crustal levels where higher temperatures prevailed.

6.4. HIGH-TEMPERATURE SHEARING IN THE IVREA ZONE

Localized shear zones formed under high temperature conditions are found in many places of the Ivrea Zone. These zones are characterized in the regional metamorphic granulite facies domain by the fine-grained recrystallization of plagioclase, K-feldspar, quartz and biotite in the paragneisses, plagioclase, amphibole and clinopyroxene in the mafic rocks and olivine, amphibole and clinopyroxene in the peridotites. All these fine-grained minerals in the mylonitic matrix have compositions similar to the large porphyroclasts and to the minerals from adjacent samples unaffected by mylonitization. The metamorphic conditions of this shearing are not precisely established by geothermometry and -barometry. However, the dynamic recrystallization of olivine suggests that shearing occurred at temperatures of at least 600-650 °C (Carter, 1976; Kirby, 1983).

Brodie and Rutter (1987) have mapped these shear zones in the Val d'Ossola and attribute the high-temperature shearing to crustal extension. The lower age limit is given by Ar-Ar ages of 210 and 215 Ma obtained on newly formed amphiboles within the mylonite matrix of metabasites (Brodie et al., 1989). As these ages are grain-size dependant, they can be interpreted as cooling ages that yield a minimum age for the shearing. The minimum 600 to 650 °C temperatures suggested by the dynamic recrystallization of olivine in sheared peridotites can be loosely correlated with the closing temperature of monazite, which yields about 270 Ma (Köppel, 1974). Despite the crude dating of this shearing, this age is distinctly older than age estimates for the late stages of the Pogallo shearing (see below). On the other hand both the high-temperature and Pogallo shear zones have about the same structural orientation and sense of shear. Basically two interpretations can be proposed: 1) Dating of the shear zones based on the correlation of the synkinematic mineral assemblages with the radiometrically established cooling curves is erroneous because of the high and rapidly changing geothermal gradient. In this case, the shear zones can have about the same age and belong to the same tectonic setting. 2) In spite of the same kinematics high-temperature and Pogallo shearing belong to two independent geotectonic cycles (Handy and Zingg, in press). The high-temperature shear zones are related to the Late Paleozoic evolution which is characterized not only by magmatic activity but also by the formation of basins filled up with volcanoclastic rocks (Verrucano and Collio formations) and the Pogallo shearing is related to the Early Mesozoic Tethyan rifting. The second interpretation is favoured.

6.5. POGALLO SHEARING

The transition from retrograde amphibolite facies to greenschist facies mylonitization is observed in the km-wide Pogallo ductile fault zone, the southeastern border of which is the Pogallo Line (Handy, 1986, 1987). Along and across strike of the Pogallo ductile fault zone the same orientation of the foliation (subvertical with SW-NE strike) and northeast plunging stretching lineations with sinistral sense of shear are observed. Parallel to the strike of the fault zone, increasing temperatures towards northeast are inferred from the quartz microstructures, synkinematic mineral assemblages and the disappearance of the discordance between the compositional banding of the Ivrea and Strona-Ceneri Zones. This suggest that different levels of this ductile fault are exposed.

The Pogallo Line forms the southeast boundary of the Pogallo ductile shear zone coincides in the Val d'Ossola with a jump of the radiometric biotite ages from about 270 to 180 Ma (Hunziker, 1974; Handy, 1987; Zingg et al., in press). According to the cooling path derived from radiometric age determinations, the transition from retrograde amphibolite to greenschist facies occurred between the Late Paleozoic and Early Mesozoic in the Ivrea Zone. For the late stage of faulting under lower greenschist facies conditions and close to the ductile-brittle transition an Early Jurassic age can be inferred. If the whole Ivrea section is rotated into its presumed pre-Alpine orientation, a ESE-dipping low angle extensional fault results (Hodges and Fountain, 1984; Handy, 1986, 1987). Its original orientation coincides with the orientation of Early Mesozoic synrift-basins in the Southern Alps. Therefore, the Pogallo Line is interpreted as a deep low-angle fault active during the Early Mesozoic rifting. For this stage, a crustal thickness of 10-20 km is estimated (Laubscher and Bernoulli, 1983; Handy, 1987).

6.6. INSUBRIC SHEARING AND STEEPENING OF THE SECTION

Greenschist facies shearing is observed along the northwestern rim of the Ivrea Zone, i.e. along the Insubric Line, (Kruhl and Voll 1976; Steck and Tièche, 1976; S.M. Schmid et al., 1987, 1989). Mylonitization and concomitant retrogade reactions develop preferentially in the paragneisses which are less flow resistant than the mafic rocks. Garnet is replaced by chlorite, biotite by chlorite and sphene, sillimanite and K-feldspar by white mica, and plagioclase by clinozoisite and albite. The mylonitized rim of the Ivrea Zone shows horizontal to northeast dipping lineations with a dextral sense of shear.

This mylonitization of the Ivrea rim is locally slightly younger than the Insubric mylonites derived from the adjacent Sesia Zone which formed during Late Oligocene to Miocene uplift and backthrust of the Central Alps onto the Southern Alps (S.M. Schmid et al., 1987). The huge Proman antiform (Fig. 4C and

R. Schmid, 1968; Brodie and Rutter, 1987) formed under lowermost greenschist facies or even lower conditions probably as a response of the Ivrea Zone to backthrusting (S.M. Schmid et al., 1987; Zingg et al., in press). The steeply dipping series of the Ivrea section are regarded as the southeastern limb of the Proman antiform. A post-Oligocene rotation of the Ivrea Zone and adjacent part of the Sesia Zone is also suggested by a tilted Oligocene volcanoclastic series and paleomagnetic investigations in the Biella region (Heller and R. Schmid, 1974; Lanza, 1979; S.M. Schmid et al., 1989).

Alpine deformation of the South Alpine basement is not restricted to the Ivrea section. The shortening of the Permo-Mesozoic sediment cover (Laubscher 1985; Roeder, 1989) and the strong variations of the basement thickness (Bernoulli et al., in press) suggest large-scale Alpine basement tectonics in the Southern Alps.

6.7. FORMATION OF THE GEOPHYSICAL IVREA BODY

Alpine tectonics in the South Alpine basement and the position of the Ivrea Zone at the front of the Adriatic plate and within the Western Alpine arc strongly suggest an Alpine formation of the geophysical Ivrea body. Laubscher (e.g. 1971a, b) relates the formation of the geophysical Ivrea body to Neogene westward motion of the Adriatic plate. S.M. Schmid et al. (1987) propose the following multistage history for the formation of the Ivrea body: The Early Mesozoic rifting caused an inflection of the Moho in the Ivrea Zone region. During the Upper Cretaceous, continental crust (e.g. the Sesia Zone) was subducted and generated the density inversion at the base of the Ivrea body. A Cretaceous age of this inversion is also suggested by the evolution of the II. Kinzigite-Diorite Zone. This unit on top of the Sesia Zone sensu stricto (Fig. 4A) shows the same lithological association, age and prealpine tectonometamorphic evolution as the Ivrea Zone (Dal Piaz et al., 1971). Its basal part, however, was locally obliterated by high-P/low-T metamorphism (Lardeaux et al., 1982). Thus, segments of the Adriatic deep crust were incorporated during Cretaceous subduction tectonics into the Alpine metamorphosed edifice. The final geometry of the geophysical Ivrea body was established during the Neogene. A large portion of this body now underlies the Penninic and Austroalpine units, which appear now to be rootless (Fig. 2 and Bayer et al., 1987; Laubscher, 1988).

7. The Ivrea Zone, an example of lower continental crust

Circular argumentation is common in discussions of exposed crustal sections and on the nature of lower continental crust. Criteria (e.g. geophysical anomalies, lithological constitution and metamorphic conditions) are needed to decide if a specific terrain represents exposed lower crust or not. At the same time exposed sections are used to substantiate what is defined as lower crust, e.g. on seismic sections.

The large geophysical anomalies indicate an unusual crustal configuration, with lower crust exposed in the Ivrea Zone. The lithological constitution is somewhat surprising: Besides peridotite bodies and variously aged mafic rocks large amounts of paragneiss, (i.e. metasedimentary material) is found. Granulite facies conditions prevailed in the northwestern part of the Ivrea Zone with pressures of about 8 kb. This seems low for the base of continental crust. However, it is not precisely known what stage of the history is recorded by the geobarometer used and how much of the section is cut out by the Insubric Line. The sum of all information suggests that the granulite facies part of the Ivrea Zone really corresponds to lower continental crust as proposed by Mehnert (1975), Fountain (1976), Fountain and Salisbury (1981).

Synthetic seismograms based on the Ivrea Zone (Hale and Thompson, 1982; Fountain 1986; Hurich and Smithson, 1987) show layered patterns similar those revealed by reflection seismic for the lower crust in many places. The amphibolite/granulite facies transition of the central part of the Ivrea Zone may correspond to the Conrad discontinuity (Fountain, 1976). Laterally and at about the same level, the limit between the paragneiss series and the Mafic Formation may represent another discontinuity. On seismic records both compositional and metamorphic boundaries would appear at the same depth, i.e. two structures of different origins would be imaged as a unique and continuous feature.

Each exposed lower crustal section is disturbed to some extent during emplacement. The South Alpine section is steepened and transected by several late- and post-Variscan faults. In addition some considerations suggest that the CMB Line could seal two crustal portions which evolved at different places during the

Paleozoic. Thus models of crustal differentiation processes based on the immediate superposition of the Ivrea and Strona-Ceneri crustal segments may be erroneous.

Various stages of localized shearing are recognized. However, within the Ivrea and Strona-Ceneri Zones large portions have retained their original tectonometamorphic features, with minerals only altered by retrograde cation exchange and unmixing during cooling. The preservation of prealpine features in the Ivrea Zone is partly due to rheological contrasts of the Ivrea rocks with the adjacent Austroalpine units during emplacement within the Alpine orogenic belt. Alpine deformation is accomodated within the quartz-bearing lithologies of the adjacent Sesia Zone. These behaved ductilely under low grade conditions, in contrast to the mafic rocks and peridotites whose behavior is governed by feldspar, amphibole, pyroxene and olivine (S.M. Schmid et al., 1987, 1989).

During its history, the Ivrea Zone has repeatedly changed its regional tectonic setting. In an early stage it was presumably situated at an active continental margin, at the end of the Paleozoic within the Variscan belt, during the Early Mesozoic at a passive continental margin and finally during the Tertiary within the Alpine belt. Thus the Ivrea section has a very specific tectonometamorphic evolution and is not representative for the Variscan crust in general. Rather the Ivrea cross-section (Fig. 4B) is an excellent example of Early Mesozoic attenuated continental crust.

ACKNOWLEDGEMENTS

The stimulating discussions on exposed crustal cross-sections during the Killarney conference and field-trips organized by J. Barrett, M. Salisbury, D.M. Fountain and collaborators are highly appreciated. Comments on an early draft by T. Noack, G. Rivalenti and S.M. Schmid and the constructive reviews by D.M. Fountain and M.R. Handy significantly improved the present contribution. The support of the Swiss National Research Foundation (NFP-20 grant 4.903-0.85.20 and NF grant 2.991-0.88) is acknowledged.

REFERENCES

Allègre, C.J., Albarède, F., Grünenfelder, M. and Köppel, V. (1974): 238U/206Pb - 235 U/207Pb - 232Th/208Pb zircon geochronology in Alpine and non-Alpine environment. *Contr. Mineral. Petrol.*, **43**, 163-194.

Bayer, R., Cazes, M., Dal Piaz, G.V. and 20 co-authors (1987): Premiers résultats de la traversée des Alpes occidentales par sismique reflexion verticale (Programme ECORS-CROP). *C.R. Acad. Sci. Paris*, **305**, Ser.II, 1461-1470.

Berckhemer, H. (1968): Topographie des "Ivrea-Körpers", abgeleitet aus seismischen und gravimetrischen Daten. *Schweiz. mineral. petrogr. Mitt.*, **48**, 235-246.

Bernoulli, D. (1964): Zur Geologie des Monte Generoso (Lombardische Alpen). *Beitr. geol. Karte Schweiz*, NF **118**, 134 p.

Bernoulli, D., Heitzmann, P. and Zingg, A. (in press): Central and Southern Alps in Southern Switzerland: tectonic evolution and first results of reflection seismics. In: Roure, F., Heitzmann, P. and Polino, R. (Eds.): Deep structure of the Alps. *Mém. Soc. géol. France*, **155**, *Mém. Soc. géol. Suisse*, **1**, *Soc. geol. Ital.*, vol. spec. **1**.

Bertolani, M. (1968): La petrografia della Valle Strona (Alpi Occidentali Italiane). *Schweiz. mineral. petrogr. Mitt.*, **48**, 695-732.

Bertotti, G. (in press): The deep structure of the Monte Generoso basin: an extensional basin in the South-Alpine Mesozoic passive continental margin. In: Roure, F., Heitzmann, P. and Polino, R. (Eds.): Deep structure of the Alps. *Mém. Soc. géol. France*, **155**, *Mém. Soc. géol. Suisse*, **1**, *Soc. geol. Ital.*, vol. spec. **1**.

Bigioggero, B., Boriani, A., Colombo, A. and Gregnanin, A. (1978/79): The "diorites" of the Ivrea Basic Complex (Central Alps, Italy). *Mem. Sci. geol. (Padova)*, **33**, 71-85.

Borghi, A. (1988): Evoluzione metamorfica del settore nord-est della serie dei laghi (alpi meridionali - cantone Ticino). *Rend. Soc. Geol. Ital.*, **11**, 165-170.

-        (1989): L'evoluzione metamorfico-strutturale del settore nord-orientale della serie dei laghi (alpi meridionali). *Diss. Consorio Univ. Torino - Genova - Cagliari*, 187 p.

Boriani, A. (1970): The "Pogallo-Line" and its connection with the metamorphic and anatectic phases of "Massicio dei Laghi" between the Ossola valley and Lake Maggiore (Northern Italy). *Boll. Soc. Geol. Ital.*, **89**, 415-433.

Boriani, A. and Peyronel Pagliani, G. (1968): Rapporti fra le plutoniti erciniche e le metamorfiti del "Massiccio dei Laghi" nella zona del M. Cerano (bassa Val d'Ossola). *Rend. Soc. Ital. Mineral. Petrol.*, **24**, 111-142.

Boriani, A. and Sacchi, R. (1973): Geology of the junction between the Ivrea-Verbano and Strona-Ceneri Zones. *Mem. Ist. Geol. Mineral. (Padova)*, **28**, 35p.

-        (1974): The "Insubric" and other tectonic lines in the Southern Alps (NW Italy). *Mem. Soc. Geol. Ital.*, **13**, supl., 1-11.

Boriani, A., Bigioggero, B. and Origoni Giobbi, E. (1977): Metamorphism, tectonic evolution and tentative stratigraphy of the "Serie dei Laghi" - geological map of the Verbania area (Northern Italy). *Mem. Sci. geol. (Padova)*, **32**, 25p.

Boriani, A., Burlini, L. and Sacchi, R. (in press): The Cossato-Mergozzo-Brissago line and the Pogallo line (Southern Alps, N-Italy) and their relationships with the Late-Hercynian magmatic and metamorphic events. *Tectonophysics*.

Boriani, A., Colombo, A. and Macera, P. (1985): Radiometric geochronology of Central Alps. *Rend. Soc. Ital. Mineral. Petrol.*, **40**, 139-186.

Boriani, A., Origoni Giobbi, E. and Del Moro, A. (1982/83): Composition, level of intrusion and age of the "Serie dei Laghi" orthogneisses (Northern Italy - Ticino, Switzerland). *Rend. Soc. Ital. Mineral. Petrol.*, **38**, 191-205.

Boriani, A., Giobbi Origoni, E., Borghi, A. and Caironi, V. (in press): The evolution of "Serie dei Laghi" (Strona Ceneri and Scisti dei Laghi): the upper component of the Ivrea-Verbano crustal section; Southern Alps, N-Italy and Ticino, Switzerland. *Tectonophysics*.

Boriani, A., Burlini, L., Caironi, V., Giobbi Origoni, E., Sassi, A. and Sesana E. (1988): Geological and petrological studies on the hercynian plutonism of Serie dei Laghi - geological map of its occurrence between Valsesia and Lago Maggiore (N-Italy). *Rend. Soc. Ital. Mineral. Petrol.*, **43**, 367-384.

Brodie, K.H. (1987):Deep crustal extensional faulting in the Ivrea Zone of Northern Italy.*Tectonophysics*, **140**, 193-212.

Brodie, K.H., Rex, D. and Rutter, E.H. (1989): On the age of deep crustal extensional faulting in the Ivrea Zone, northern Italy. In: Coward, M.P., Dietrich, D. and Park, R.G. (Eds.): *Alpine Tectonics. Geol. Soc. London, Spec. Publ.*, **45**, 203-210.

Buletti, M. (1983): Zur Geochemie und Entstehungsgeschichte der Granat-Amphibolite des Gambarognogebietes, Ticino, Südalpen. *Schweiz. mineral. petrogr. Mitt.*, **63**, 233-247.

Bürgi, A. and Klötzli, U. (in press): New data on the evolutionary history of the Ivrea Zone (Northern Italy). *Tectonophysics*.

Bürki, B. (this volume): Geophysical interpretation of astrogravimetric data in the Ivrea Zone.

Carmichael, D.M. (1969): On the mechanism of prograde metamorphic reactions in quartz-bearing pelitic rocks. *Contr. Mineral. Petrol.*, **20**, 244-267.

Carter, N.L. (1976): Steady state flow of rocks. *Rev. Geophys. Space Phys.*, **14**, 301-360.

Chopin, C. (1984): Coesite and pure pyrope in high-grade pelitic blueschists of the Western Alps: a first record and some consequences. *Contr. Mineral. Petrol.*, **86**, 107-118

Coltori, M. and Siena, F. (1984): Mantle tectonite and fractionate peridotite at Finero (Italian Western Alps). *N. Jb. Mineral., Abh.*, **149**, 225-244.

Dal Piaz, G.V., Gosso, G. and Martinotti, G. (1971): La II zona diorito-kinzigitica tra la Valsesia e la valle d'Ayas (alpi occidentali). *Mem. Soc. Geol Ital.*, **10**, 257-276.

Fountain, D.M. (1976): The Ivrea-Verbano and Strona-Ceneri Zones, Northern Italy: a cross-section of the continental crust - new evidence from seismic velocities of rock samples. *Tectonophysics*, **33**, 145-165.

-        (1986): Implications of deep crustal evolution for seismic reflection interpretation. In: Barazangi, M. and Brown, L. (Eds.): *Reflection Seismology: The Continental Crust.. Amer. Geophys. Union, Geodyn. Series*, **14**, 1-7.

16

Fountain, D.M. and Salisbury, M.H. (1981): Exposed cross-sections through the continental crust: implications for crustal structure, petrology and evolution. *Earth Planet. Sci. Lett.*, **56**, 263-277.

Gansser, A. and Pantic, N. (1988): Prealpine events along the eastern Insubric Line (Tonale Line, northern Italy). *Eclogae geol. Helv.*, **81**, 567-577.

Garuti, G., Rivalenti, G., Rossi, A. and Sinigoi, S. (1978/79): Mineral equilibria as geotectonic indicators in the ultramafics and related rocks of the Ivrea-Verbano basic complex (Italian Western Alps): pyroxenes and olivine. *Mem. Sci. geol. (Padova)*, **33**, 147-160.

Giese, P., Reutter, K.-J., Jacobshagen, V. and Nicolich, R. (1982): Explosion seismic crustal studies in the Alpine Mediterranean region and their implications to tectonic processes. In: Berckhemer, H. and Hsü, K. (Eds.). *Alpine-Mediterranean Geodynamics. Amer. Geophys. Union, Geodyn. Ser.*, **7**, 39-73.

Giobbi Origoni, E., Bocchio, R., Boriani, A., Carmine, M. and De Capitani, L. (1988): Late-Hercynian mafic and intermediate intrusives of Serie dei Laghi (N-Italy). *Rend. Soc. Ital. Mineral. Petrol.*, **43**, 395-410.

Graeser, S. and Hunziker, J.C. (1968): Rb-Sr- und Pb-Isotopen-Bestimmungen an Gesteinen und Mineralien der Ivrea-Zone. *Schweiz. mineral. petrogr. Mitt.*, **48**, 189-204.

Guillaume, A. and Guillaume, S. (1980): Nouvelles cartes des anomalies de la pesanteur dans les Alpes occidentales. *C. R. Acad. Sci. Paris*, **290**, ser. D, 163-166.

Hale, L.D. and Thompson, G.A. (1982): The seismic reflection character of the continental Mohorovicic Discontinuity. *J. geophys. Res.*, **87**, 4625-4635.

Handy, M.R. (1986): The structure and rheological evolution of the Pogallo fault zone, a deep crustal dislocation in the Southern Alps of northwestern Italy (Prov. Novara). *Diss. Univ. Basel*, 327p.

- (1987): The structure, age and kinematics of the Pogallo Fault Zone; Southern Alps, northwestern Italy. *Eclogae geol. Helv.*, **80**, 593-632.

Handy, M.R. and Zingg, A. (in press): The rheological evolution of the Ivrea crustal cross-section (Southern Alps of northwestern Italy and southern Switzerland). *Bull. Geol. Soc. Amer.*

Heller, F. and Schmid, R. (1974): Paläomagnetische Untersuchungen in der Zone Ivrea-Verbano (Prov. Novara, Norditalien): Vorläufige Ergebnisse. *Schweiz. mineral. petrogr. Mitt.*, **54**, 229-242.

Hodges, K.V. and Fountain, D.M. (1984): Pogallo Line, South Alps, northern Italy: An intermediate crustal level, low-angle normal fault? *Geology*, **12**, 151-155.

Hunziker, J.C. (1974): Rb-Sr and K-Ar age determination and the Alpine tectonic history of the Western Alps. *Mem. Ist. Geol. Mineral. Univ. Padova*, **31**, 54p.

Hunziker, J.C. and Zingg, A. (1980): Lower Palaeozoic amphibolite to granulite facies metamorphism in the Ivrea Zone (Southern Alps, Northern Italy). *Schweiz. mineral. petrogr. Mitt.*, **60**, 181-213.

Hurich, C.A. and Smithson, S.B. (1987): Compositional variation and the origin of deep crustal reflections. *Earth planet. Sci. Lett.*, **85**, 416-426.

Kälin, O. and Trümpy, D.M. (1977): Sedimentation und Paläotektonik in den westlichen Südalpen: Zur triasisch-jurassischen Geschichte des Monte Nudo-Beckens. *Eclogae geol. Helv.*, **70**, 295-350.

Kirby, S.H. (1983): Rheology of the lithosphere. *Rew. Geophys. Space Phys.*, **21**, 1458-1487.

Kissling, E. (1984): Three-Dimensional Gravity Model of the northern Ivrea-Verbano Zone. *Matér. Géol. Suisse, Géophys.*, **21**, 53-61.

Köppel, V. (1974): Isotopic U-Pb ages of monazites and zircons from the crust-mantle transition and adjacent units of the Ivrea and Ceneri Zones (Southern Alps, Italy). *Contr. Mineral. Petrol.*, **43**, 55-70.

Köppel, V. and Grünenfelder, M. (1971): A study of inherited and newly formed zircons from paragneisses and granitised sediments of the Strona-Ceneri-Zone (Southern Alps). *Schweiz. mineral. petrogr. Mitt.*, **51**, 385-409.

- (1978/79): Monazite and zircon U-Pb ages from the Ivrea and Ceneri Zones. Abstract, 2nd Symposium Ivrea-Verbano, Varallo. *Mem. Sci. geol.(Padova)*, **33**, p. 257.

Kruhl, J.H. and Voll, G. (1976): Fabrics and metamorphism from the Monte Rosa Root Zone into the Ivrea Zone near Finero, Southern Margin of the Alps. *Schweiz. mineral. petrogr. Mitt.*, **56**, 627-633

Lanza, R. (1979): Palaeomagnetic data from the andesitic cover of the Sesia-Lanzo Zone (Western Alps). *Geol. Rundschau*, **68**, 83-92.

Lardeaux, J.-M., Gosso, G., Kienast, J.-R. and Lombardo, B. (1982): Relations entre le métamorphisme et la déformation dans la zone Sésia-Lanzo (Alpes occidentales) et le problème de l'éclogitisation de la croûte continentale. *Bull. Soc. géol. France*, (7), **24**, 793-800.

Laubscher, H.P. (1971a): Das Alpen-Dinariden-Problem und die Palinspastik der südlichen Tethys. *Geol. Rundsch.*, **60**, 813-833.

- (1971b): The large-scale kinematics of the western Alps and the northern Apennines and its palinspastic implications. *Amer. J. Sci.*, **271**, 193-226.

- (1985): Large-scale, thin-skinned thrusting in the Southern Alps: Kinematic models. *Geol. Soc. Amer. Bull.*, **96**, 710-718.

- (1988): Material balance in Alpine orogeny. *Geol. Soc. Amer. Bull.*, **100**, 1313-1328.

- (in press): Deep seismic data from the central Alps: mass distributions and their kinematics. In: Roure, F., Heitzmann, P. and Polino, R. (Eds.): Deep structure of the Alps. *Mém. Soc. géol. France*, 155, *Mém. Soc. géol. Suisse*, 1, *Soc. geol. Ital.*, vol. spec. 1.

Laubscher, H.P. and Bernoulli, D. (1982): History and Deformation of the Alps. In: Hsü, K.J. (Ed.): *Mountain building processes*. Academic Press, London, 169-180.

Lensch, G. (1968): Die Ultramafitite der Zone von Ivrea und ihre geologische Interpretation. *Schweiz. mineral. petrogr. Mitt.*, **48**, 91-102.

- (1971): Die Ultramafitite der Zone von Ivrea. *Ann. Univ. sarav.*, Heft 9, 146 p.

Lensch, G. and Rost, F. (1972): Kelyphitperidotite in der mittleren Ivreazone zwischen Val d'Ossola und Val Strona. Ein Beitrag zur Herkunftstiefe der Ultramafitite der Ivrea Zone. *Schweiz. mineral. petrogr. Mitt.*, **52**, 237-250.

Mazzucchelli, M. and Siena, F.(1986): Geotectonic significance of the metabasites of the kinzigitic series, Ivrea-Verbano zone (Western Italian Alps). *Tschermaks mineral. petrogr. Mitt.*, **35**, 99-116

McDowell, F.W. (1970): Potassium-Argon ages from the Ceneri Zone, Southern Swiss Alps. *Contr. Mineral. Petrol.*, **28**, 165-182.

Mehnert, K.R. (1975): The Ivrea Zone, a model of the deep crust. *N. Jb. Mineral. Abh.*, **125**, 156-199.

Mueller, St., Ansorge, J., Egloff, R. and Kissling, E. (1980): A crustal cross section along the Swiss Geotraverse from the Rhinegraben to the Po Plain. *Eclogae geol. Helv.*, **73**, 463-483.

Pin, C. (1986): Datation U-Pb sur zircons à 285 Ma du complexe gabbro-dioritique du Val Sesia - Val Mastallone et âge tardi-hercynien du métamorphisme granulitique de la zone Ivrea-Verbano (Italie).*C. R. Acad. Sci. Paris*, Ser. II, **303**, 827-830.

Pin, C. and Sills, J.D. (1986): Petrogenesis of layered gabbros and ultramafic rocks from Val Sesia, the Ivrea Zone, NW Italy: Trace element and isotope geochemistry. In: Dawson, J.B., Carswell, D.A., Hall, J. and Wedepohl, K.H. (Eds): *The nature of lower continental crust. Geol. Soc. London, Spec. Publ.*, **24**, 231-249.

Pin, C. and Vielzeuf, D. (1983): Granulites and related rocks in Variscan median Europe: a dualistic interpretation. *Tectonophysics*, **93**, 47-74.

Pinarelli, L., Del Moro, A. and Boriani, A. (1988): Rb-Sr geochronology of Lower Permian plutonism in Massiccio dei Laghi, Southern Alps (NW Italy). *Rend. Soc. Ital. Mineral. Petrol.*, **43**, 411-428.

Reinhard, M. (1964): Über das Grundgebirge des Sottoceneri im Süd-Tessin und die darin auftretenden Ganggesteine. *Beitr. geol. Karte Schweiz*, N.F., **117**, 89p.

Rivalenti, G., Garuti, G. and Rossi, A. (1975): The origin of the Ivrea-Verbano basic formation (Western Italian Alps) - whole rock geochemistry. *Boll. Soc. Geol. Ital.*, **94**, 1149-1186.

Rivalenti, G., Garuti, G., Rossi, A. and Sinigoi, S. (1978/79): Spinels as petrogenetic indicators in the Ivrea-Verbano Basic Complex (Italian Western Alps). *Mem. Sci. geol. (Padova)*, **33**, 161-171.

Rivalenti, G., Garuti, G., Rossi, A., Siena, F. and Sinigoi, S. (1981): Existence of different peridotite types and of a layered ingneous complex in the Ivrea Zone of the Western Alps. *J. Petrol.*, **22**, 127-153.

Rivalenti, G., Rossi, A., Siena, F. and Sinigoi, S. (1984): The layered series of the Ivrea-Verbano igneous complex, Western Alps, Italy. *Tschermaks mineral. petrogr. Mitt.*, **33**, 77-99.

Roeder, D. (1989): South-Alpine thrusting and trans-Alpine convergence. In: Coward, M.P., Dietrich, D. and Park, R.G. (Eds.): *Alpine Tectonics. Geol. Soc. London, Spec. Publ.*, **45**, 211-227.

Sassi, F.P., Kalvacheva, R. and Zanferrari, A. (1984): New data on the age of deposition of the South-Alpine phyllitic basement in the Eastern Alps. *N. Jb. Geol. Paläont., Mh.*, 741-751.

Schmid, R. (1967): Zur Petrographie und Struktur der Zone Ivrea-Verbano zwischen Valle d'Ossola und Val Grande (Prov. Novara, Italien). *Schweiz. mineral. petrogr. Mitt.*, **47**, 935-1117.

- (1978/79): Are the metapelites of the Ivrea-Verbano Zone restites? *Mem. Sci. geol.(Padova)*, **33**, 67-69.

Schmid, R. and Wood, B.J. (1976): Phase relationships in granulitic metapelites from the Ivrea-Verbano Zone (Northern Italy). *Contr. Mineral. Petrol.*, **54**, 255-279.

Schmid, R., Dietrich, V., Komatsu, M., Newton, R.C., Ragettli, R., Rushmer, T., Tuchschmid, M. and Vogler, R. (1988): Metamorphic evolution and origin of the Ivrea Zone. In: Boriani, A. and Burlini, L. (Eds.): *Italy-U.S. workshop on the nature of the lower continental crust, Verbania (Italy) May 23-27. Programme-Abstracts-Field Trip Guide. Univ. Milano, Ricerca Sci. Educazione permanente*, Suppl. **65a**, 60-61.

Schmid, S.M., Zingg, A . and Handy, M. (1987): The kinematics of movements along the Insubric Line and the emplacement of the Ivrea Zone. *Tectonophysics*, **135**, 47-66.

Schmid, S.M., Aebli, H.R., Heller, F. and Zingg, A. (1989): The role of the Periadriatic Line in the tectonic evolution of the Alps. In: Coward, M.P., Dietrich, D. and Park, R.G. (Eds.): *Alpine Tectonics. Geol. Soc. London, Spec. Publ.*, **45**, 153-171.

Shervais, J.W. (1979): Thermal emplacement model for the Alpine Lherzolite massif at Balmuccia, Italy. *J. Petrol.*, **20**, 795-820.

Sighinolfi, G.P. and Gorgoni, C. (1978): Chemical evolution of high-grade metamorphic rocks - anatexis and remotion of material from granulite terrains. *Chem. Geol.*, **22**, 157-176.

Sills, J.D. (1984): Granulite facies metamorphism in the Ivrea Zone, N.W. Italy. *Schweiz. mineral. petrogr. Mitt.*, **64**, 169-191.

Sills, J.D. and Tarney, J. (1984): Petrogenesis and tectonic significance of amphibolites interlayered with metasedimentary gneisses in the Ivrea Zone, Southern Alps, Northwestern Italy. *Tectonophysics*, **107**, 187-206.

Sinigoi, S., Comin-Chiaramonti, P., Demarchi, G. and Siena, F. (1983): Differentiation of partial melts in the mantle: evidence from the Balmuccia peridotite, Italy. *Contr. Mineral. Petrol.*, **82**, 351-359.

Steck, A. and Tièche, J.-C. (1976): Carte géologique de l'antiforme péridotitique de Finero avec des observations sur les phases de déformation et de recristallisation. *Schweiz. mineral. petrogr. Mitt.*, **56**, 501-512.

Stille, P. and Buletti, M. (1987): Nd-Sr isotopic characteristics of the Lugano volcanic rocks and constraints on the continental crust formation in the South Alpine domain (N-Italy-Switzerland). *Contr. Mineral. Petrol.*, **96**, 140-150.

Vai, J.B. and Cocozza, T. (1986): Tentative schematic zonation of the Hercynian chain in Italy. *Bull. Soc. géol. France*, Ser. 8, **2**, 95-114.

Van Houten, J. (1929): Geologie der Kalkalpen am Ostufer des Lago Maggiore. *Eclogea Geol. Helv.*, **22**, 1-40.

Vecchia, O. (1968): La zone Cuneo-Ivrea-Locarno, élément fondamental des Alpes. Géophysique et géologie. *Schweiz. mineral. petrogr. Mitt.*, **48**, 215-225.

Voshage, H., Hunziker, J.C., Hofmann, A.W. and Zingg, A. (1987): A Nd and Sr isotopic study of the Ivrea Zone, Southern Alps, N-Italy. *Contr. Mineral. Petrol.*, **97**, 31-42.

Voshage, H., Sinigoi, S, Mazzucchelli, M., Demarchi, G., Rivalenti, G. and Hofmann, A.W. (1988): Isotopic constraints on the origin of ultramafic and mafic dikes in the Balmuccia peridotite (Ivrea Zone). *Contr. Mineral. Petrol.*, **100**, 261-267.

Wagner, J.-J., Klingelé, E. and Mage, R. (1984): Regional geomagnetic study of the southern border of the Western Alps - the Ivrea body. *Matér. Géol. Suisse, Géophys.*, **21**, 21-29.

Winterer, E.L. and Bosellini, A. (1981): Subsidence and sedimentation on Jurassic passive continental margin, Southern Alps, Italy. *Bull. Amer. Assoc. Petroleum Geol.*, **65**, 394-421.

Zingg, A. (1980): Regional metamorphism in the Ivrea Zone (Southern Alps, N-Italy): Field and microscopic investigations. *Schweiz. mineral. petrogr. Mitt.*, **60**, 153-179.

- (1983): The Ivrea and Strona-Ceneri Zones (Southern Alps, Ticino and N-Italy) - A review. *Schweiz. mineral. petrogr. Mitt.*, **63**, 361-392.

Zingg, A., Handy, M.R., Hunziker, J.C. and Schmid, S.M. (in press): Tectonometamorphic history of the Ivrea Zone and its relationship to the crustal evolution of the Southern Alps. *Tectonophysics*.

GEOLOGICAL MAPS:

Carta geologica d'Italia, 1:100'000: Domodossola (15), Cannobio (16), Varallo (30), Varese (31), Ivrea (42), Biella (43).

Geologische Karte der Schweiz, 1:25'000: Tesserete (39), Bellinzona (66), Lugano (69).

Boriani, A., Bigioggero, B. and Origoni Giobbi, E. (1977): Carta geologica della zona di Verbania, 1:50'000. *Mem. Sci. geol. (Padova)*, **32**.

Boriani, A., Burlini, L., Caironi, V., Giobbi Origoni, E., Sassi, A. and Sesane, E. (1988): Carta geologica dei graniti dei Laghi, 1:50'000. *Rend. Soc. ital. Mineral. Petrol.*, **43**.

Hermann (1937): Carta geologica delle Alpi Occidentali, 1:200'000.

Reinhard (1953): Geologische Kartenskizze des Grundgebirges des Sottoceneri, des westlich angrenzenden italienischen Gebietes und der Morcote Halbinsel, 1:100'000. *Eclogae geol. Helv.*, **46**.

THE EXPOSED CRUSTAL CROSS SECTION OF SOUTHERN CALABRIA, ITALY: STRUCTURE AND EVOLUTION OF A SEGMENT OF HERCYNIAN CRUST

VOLKER SCHENK
Mineralogisch-petrographisches Institut
Universität Kiel
Olshausenstr. 40
D-2300 Kiel 1
FRG

ABSTRACT. The Calabrian crustal section, situated at the border of the Adriatic microplate, has a distinct layered structure: Over a distance of about 50 km one can trace a rock sequence from granulite-grade (from 7.5 to 5.5 kbar) through granitoids and amphibolite-facies gneisses into unmetamorphosed Paleozoic and younger sediments. The upper and the lower crustal sections are believed to have been part of the same crustal sequence since the late Hercynian because intrusive contacts of this age are found in both sections.

The *lower crustal section* is younger than 1.3 b.y. and, at least in part, as young as 0.55 b.y. It experienced its main metamorphism, crustal differentiation, and first uplift event during the Hercynian orogeny (~ 300 m.y.), but was brought to the surface (~ 10 m.y.) in the course of movements along the Adriatic plate boundary during Alpine time. The lower crust consists of an upper metapelite unit (5 - 6 km) resting on a mainly metabasic unit (2 km) in which zircons point to a magmatic formation age of about 550 m.y. The high proportion (75 %) of granulite-facies, mainly pelitic metasediments characterize the lower crustal section. The pelitic rocks are highly restitic and a 40 % peraluminous granite component is estimated to have been extracted.

The *upper crust* consists of a variety of metasediments (amphibolite- to greenschist-facies) and a high proportion of mostly peraluminous orthogneisses (500 - 520 m.y.), which possibly represent the earliest partial melts of the now exposed lower crustal metapelites. Hercynian granitoids (300 - 280 m.y.) range (bottom to top) from calc-alkaline diorites and tonalites to granodiorites and peraluminous S-type granites. Only the latter display Nd model ages (1.3 b.y.) identical to the lower crustal metapelites, which seem to be the source of these granites.

The *thermal and baric evolution* of the lower crust is characterised by a clockwise P-T-time path (kyanite - sillimanite - andalusite - kyanite). Almost all garnets in metapelites grew in the stability field of sillimanite under kinematic conditions, but the latest and highest metamorphic stage occurred after any penetrative deformation. The end of the granulite-facies metamorphism was induced by a tectonic event during which the lower crust was uplifted (5 - 10 km) into middle crustal levels, where it slowly cooled (2 °/m.y.) isobarically during Mesozoic time. Final uplift was rapid and started some 25 m.y. ago. The oldest sediments on top of lower crustal rocks are 10 m.y. old.

The most likely geotectonic setting of the granulite-facies metamorphism in Calabria is a continental margin above a subduction zone.

M. H. Salisbury and D. M. Fountain (eds.), Exposed Cross-Sections of the Continental Crust, 21–42.
© 1990 *Kluwer Academic Publishers.*

1. Introduction

In the Serre of Southern Calabria, lithological units representing all levels of the continental crust are exposed. From north to south, i.e. from the bottom to the top, the following crustal levels are outcropping, as shown in the geological map (Fig. 1) and in the profile (Fig. 2) for the area.

The lower crustal section (7 - 8 km thick) begins in the north at an Alpine fault zone (Curinga-Girifalco line) and consists of both the lower, mainly metabasic "granulite-pyriclasite unit" (about 2 km thick) and the overlaying "metapelite unit" (about 6 km thick). The highest levels of the metapelite unit have been intruded by small gabbro bodies and by schistose blastomylonitic diorites, which are themselves overlain by schistose tonalites, as well as by massive granodiorites and S-type granites. The granodiorites and S-type granites intruded discordantly into amphibolite to greenschist-facies schists of the upper crust and into unmetamorphosed Paleozoic sediments. Carbonate rocks of Mesozoic age lie discordantly on top of the Paleozoic schists. They are overlain by Tertiary and Pleistocene clastic sediments.

Petrological data and isotopic age data point to two important tectonic events during which the lower crustal section was uplifted and tilted. The first uplift occurred during the Hercynian, the last uplift and tilting during the Apennine orogenic cycle (Schenk, 1980, 1989).

The present paper reviews the *lithological composition* of the Calabrian crust, the *physical properties of its rocks*, and estimates the *bulk chemical composition* of the lower crust. The *formation ages*, the *pressure-temperature-time path* of the lower crust, and the *differentiation process of the crust* are also discussed.

2. Southern Calabria - a true crustal cross section?

There is no general consensus in the literature about the interpretation of the Calabrian section described above. The main disagreement concerns the nature of the crustal section: Is it a single Hercynian crustal section or does it represent a composite crustal section (upper and lower parts) assembled during a later Alpine tectonic event?

Italian authors (e.g. Amodio Morelli et al., 1976; Borsi et al., 1976; Del Moro et al., 1986) have stated that the upper and lower crust have different histories and are separated by a major tectonic contact of Alpine age (s.l.). Until 1986, a nappe thrust has been assumed to lie between the Stilo granodiorite (an "Africa derived tectonic unit") and the underlying Cardinale tonalite (an "European tectonic unit") which are both of late Hercynian age. On the basis of Rb-Sr biotite ages between 130 and 140 m.y. determined in rocks from the lower crustal unit and in the overlying tonalite as well, Del Moro et al. (1986) shifted the assumed tectonic contact from the top to the base of the Cardinale tonalite (Fig. 1). Transcurrent movements are now considered to be responsible for the tectonic contact.

The present author's opinion is that southern Calabria represents a continuous crustal profile (Schenk 1980, 1981, 1984, 1989) and that the tectonic movement along the contacts of both calc-alkaline plutons, the tonalite and the underlying diorite, and the lower crustal gneisses are of minor importance. Both, the tonalite and the underlying diorites are of late Hercynian age (Schenk, 1980). The similar Rb-Sr biotite ages in the tonalite and the metapelite unit suggest a common cooling history, and are not due to tectonic movement along their contact 130 - 140 m.y. ago, as suggested by Del Moro et al. (1986). Nevertheless, a fault with cataclastic deformation occurs between both calc-alkaline plutons, which might be due to differential movements during the Apennine uplift and tilting of the lower crust.

In view of the conflicting interpretations of the Calabrian section, the main arguments which are consistent with the author's interpretation (i e., a single, exposed cross section through a Hercynian crust) are summarised below. A more detailed discussion of these points follows in later sections.

1) The continuous metamorphic zonation in the granulite facies section, which parallels the lithological stratification, indicates a thickness of 7 - 8 km for the exposed lower crust.

2) The intrusion of calc-alkaline diorites and tonalites into the middle crust was contemporaneous with the lower crustal metamorphism (300 ± 10 m.y.). There is a magmatic contact between these diorites and the underlying lower crustal rocks.

3) Upper crustal S-type granites are only slightly younger than the end of the lower crustal metamorphism (295 - 290 m.y.). They have nearly identical Nd-isotopic composition, Nd-model ages, oxygen isotopic composition and complementary chemistry to the restitic lower crustal metapelites. They are therefore best interpreted as partial melts of the lower crustal metapelites.

4) The more-or-less *continuous change* of biotite cooling ages from Hercynian ones (295 - 280 m.y.) at the top to younger ones (110 m.y.) at the base of the exposed crustal section, points to a common post-Hercynian cooling history of the whole section.

5) Two tectonic events, a Hercynian (~ 295 m.y.) and an Apennine one (~ 25 - 10 m.y.), have been recognised in the lower crustal rocks by petrological and isotopic methods. These events brought the lower crust to the surface in two stages. Both lower crustal events have their synchroneous upper crustal counterparts: Hercynian (295 - 280 m.y.) biotite cooling ages in amphibolite-facies upper crustal rocks reflect the rapid erosion of the uppermost crustal levels during the lower crustal uplift. The increased tilting of Mesozoic and Tertiary sediments in the southeast records the Apennine uplift. Biotite ages of 25 m.y. in upper crustal shear zones (Bonardi et al., 1987) may also point to this younger uplift event.

3. Geological setting of southern Calabria

Southern Calabria is a part of the Calabrian massif, which is situated at the boundary of the so-called "Adriatic microplate" or the "Adriatic promontory of Africa", and like the Ivrea zone, its position is in the sharp bend of the Alpine-Apennine mountain system (Fig. 1). Seismic refraction studies revealed a doubling of the crustal thickness in Calabria and along the whole plate boundary with respect to the adjoining Tyrrhenian and Adriatic crusts (Fig. 1, 3) (Schütte, 1978; Giese & Reutter, 1978). The Calabrian massif is regarded as a piece of pre-Alpine crust, which was involved in the Alpine orogeny. Later the Alpine mountain system was dismembered due to movements of microplates in the Western Mediterranean area (e.g. Alvarez, 1976).

In southern Calabria a major cataclastic and mylonitic fault zone of Alpine age separates the granulite-facies rocks of the crustal section from the underlying phyllonitic rocks (Fig. 1, 2). In this phyllonitic unit, in which Hercynian muscovite and Alpine biotite ages (43 m.y.) have been found, lenses of fresh granulite-facies gneisses occur locally (Schenk, 1980). According to Paglionico & Piccarreta (1976), Paleozoic phyllites form a further, and deeper tectonic unit, the so-called Pomo river unit, at the northeastermost edge of the Serre. Yet, relics of garnet and muscovite porphyroblasts in an extremely fine-grained recrystallized quartz-feldspar matrix point to a mylonitic origin of these phyllites, which once reached at least uppermost greenschist-facies conditions. Thus, they may represent an ultra-mylonitic equivalent of the overlying retrograded gneisses.

24

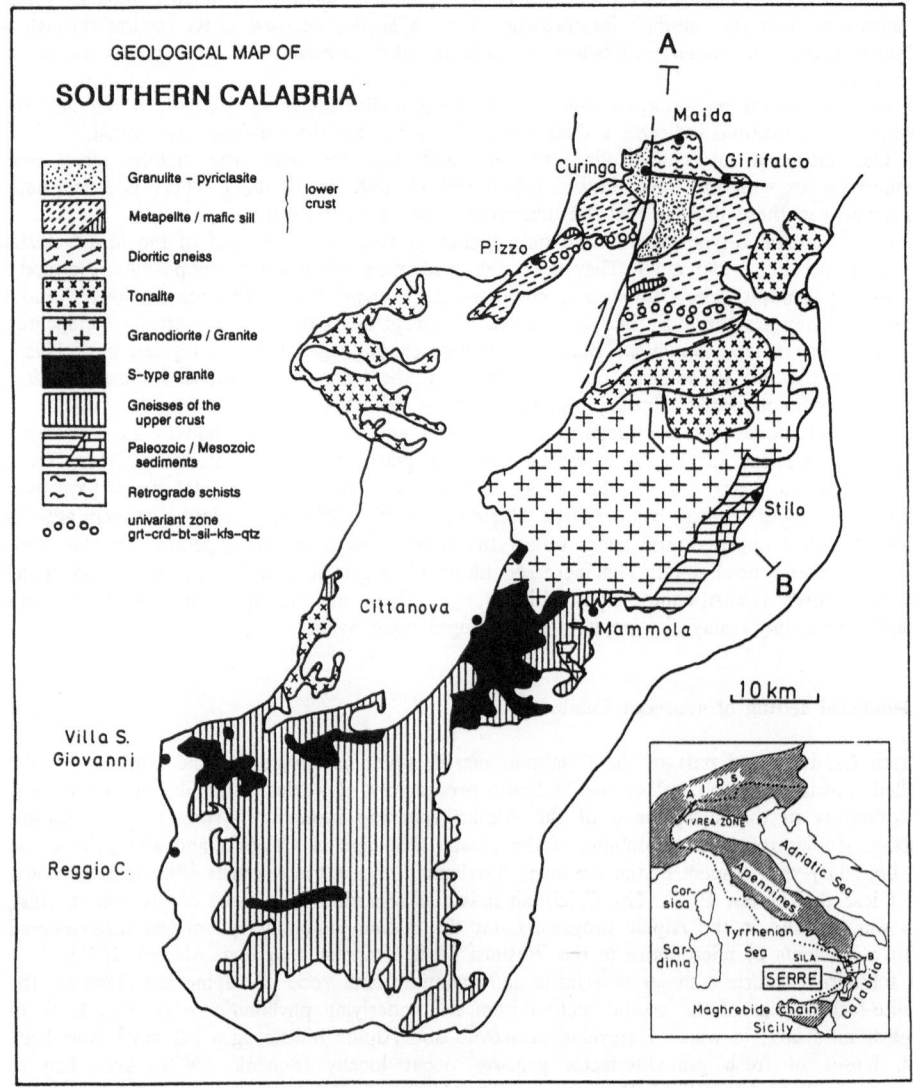

Fig. 1: Simplified geologic map of southern Calabria. The southern part (granitoids and upper crust) is modified from Atzori et al. (1984), Borsi et al. (1976), Amodio-Morelli et al. (1976). The "univariant" garnet-cordierite zone represents a metamorphic bathozone at about 5.5 to 6.0 kbar (~ 20 km depth).

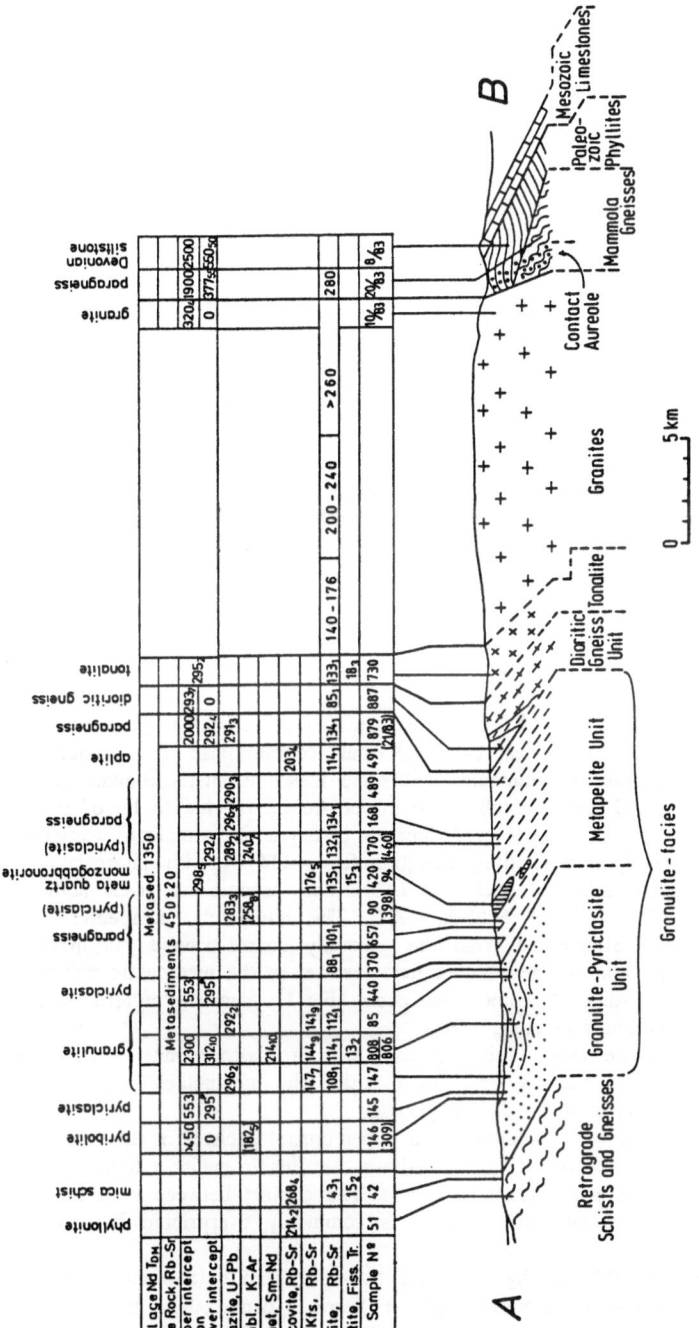

Fig. 2: Geologic section through the Serre of southern Calabria with isotopic dates (updated from Schenk, 1980). Biotite cooling ages of Serre granitoids and upper crustal gneiss after Del Moro et al. (1986).

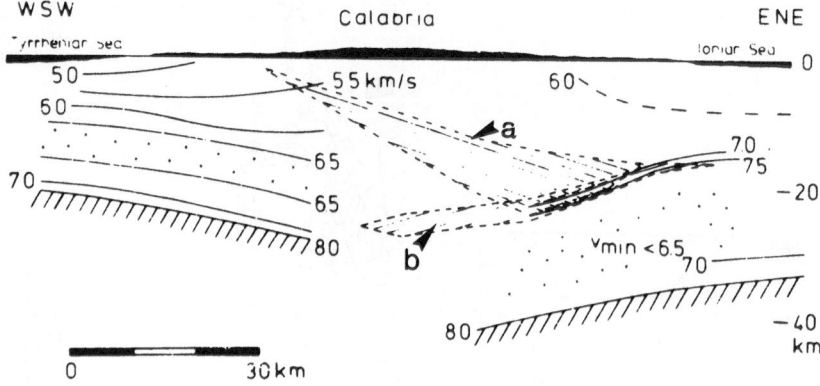

Fig. 3: Model structure of the crust in northern Calabria, according to Schütte (1978) and Schwarz (1978). Both models (a. and b.) are compatible with the magnetic and seismic data.

4. Lithology of the crustal section

The *granulite pyriclasite unit* at the base of the exposed crustal section consists predominantly of metabasic rocks, i.e. foliated pyriclasites ($\sim 80\%$), but contain also fine-grained felsic orthopyroxene-granulites, fine-grained metapelites and some lenses of metacarbonate rocks of clearly sedimentary origin. Small lenses of ultramafic rocks are interpreted to be of cumulate origin. The overlying *metapelite unit* consists predominantly of migmatitic aluminous paragneisses (76 %) and semipelitic orthopyroxene-biotite-garnet-granofelses (14 %) with intercalations of metabasic ($\sim 10\%$) and metacarbonate rocks ($< 1\%$). The proportions given above are estimated from the map (Fig. 4, after Schenk, 1984). The aluminous metapelites of the granulite-pyriclasite unit are distinguished from the generally coarse-grained porphyroblastic paragneisses of the metapelite unit by their fine-grained texture. In the metapelitic unit sills of quartz-monzogabbroic to quartz-dioritic compositions occur (Fig. 1); from the eastern sill zircon dates of 298 ± 5 m.y. were obtained (Schenk, 1980).

At the top of the metapelite unit, dioritic gneisses with a blastomylonitic ("perl-gneiss") texture occur. Lenses of dioritic gneisses were found inside the metapelites and metapelitic lenses up to 200 m long also occur in the dioritic gneisses.

Granitic augen gneisses of the type common in the upper crustal gneisses of the Aspromonte were found at several places in the dioritic gneiss. Above the dioritic gneiss a foliated tonalite (Cardinale) containing numerous mafic xenoliths occurs. The contact between both Hercynian calc-alkaline plutons, the diorite and the tonalite, is commonly overprinted by cataclastic deformation.

The tonalite is bordered in the east by a massive two-mica granite (Satriano) and in the south by a very large mesaluminous granodioritic pluton (Serra S. Bruno). The granodiorite has intrusive contacts with both Paleozoic sediments (at Stilo) and amphibolite-facies gneisses (at Mammola). According to Colonna et al. (1973) there is a (pre-intrusion) tectonic contact between the upper crustal gneisses and the Paleozoic sediments. Peraluminous S-type granites

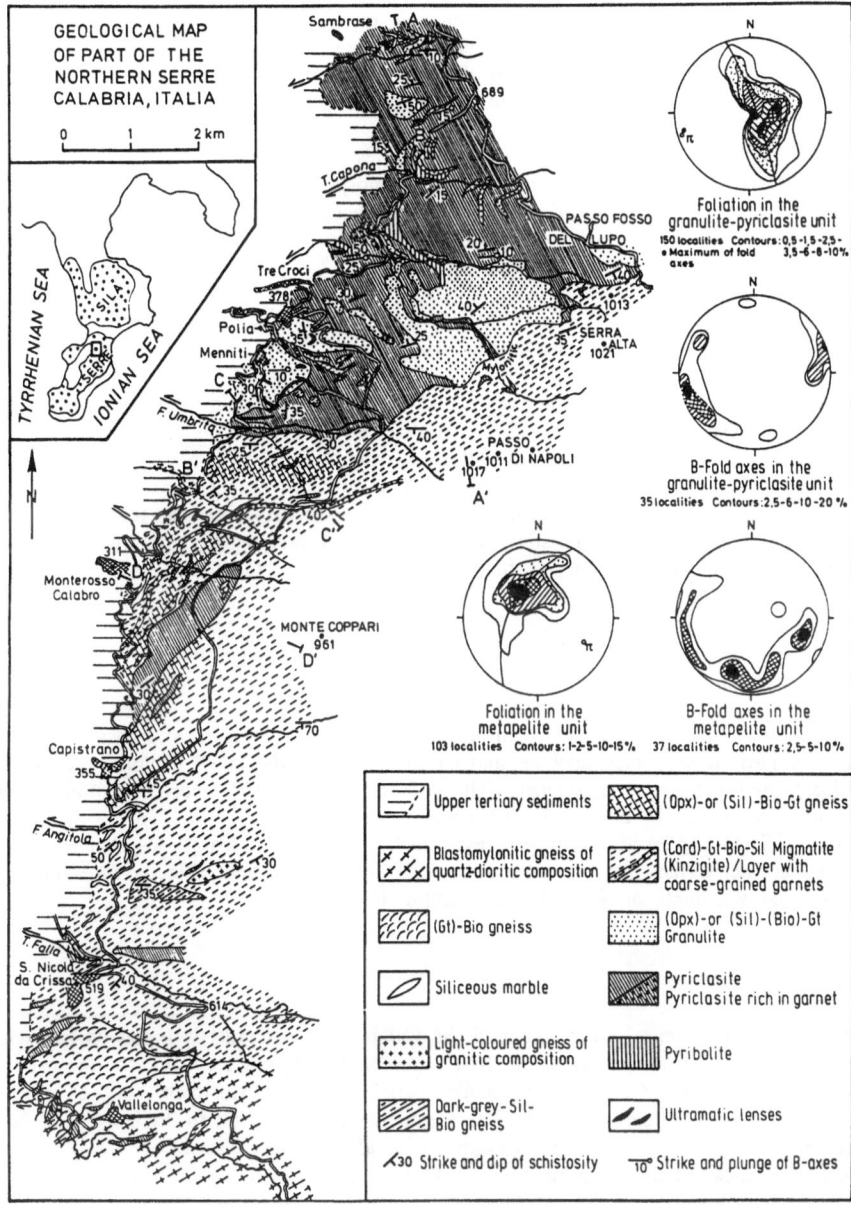

Fig. 4: Detailed geological map of part of the northern Serre that emphasizes the granulite facies areas (after Schenk, 1984).

(with cordierite, sillimanite, muscovite, biotite) intruded about 280 to 290 m.y. ago (Del Moro et al., 1983) into the upper crustal gneisses and the granodiorite pluton. The upper crustal paragneisses are of amphibolite- to greenschist-facies grade (staurolite, chloritoid; kyanite, sillimanite and andalusite, corresponding to different metamorphic stages) and are associated with widely distributed peraluminous granitic augen gneisses of Pan-African age (500 - 520 m.y., Schenk & Todt, 1989).

The upper crust area, especially in the Aspromonte, is still not completely mapped (Bonardi et al., 1984; Atzori et al., 1977) and no regional metamorphic subdivision has been recognised. The Hercynian structures were disturbed by Alpine tectogenesis which gave rise to a complex nappe structure, in which many yet unresolved problems exist (D'Amico et al., 1983; Lorenzoni et al., 1980; Bonardi et al., 1984).

5. Physical properties of the Calabrian crustal rocks

As discussed in the introduction, the lower and upper crustal units of southern Calabria constitute a single Hercynian crustal section. Rotating this presently tilted section back to the vertical, one obtains a pile of horizontal lithostratigraphic units (Fig. 5) which represent a hypothetical crustal profile through the Adriatic crust of southern Calabria (Schenk, 1981).

Laboratory measurement of compressional and shear wave velocities at high temperatures and high confining pressures in representative samples from all lithological units resulted in the construction of a hypothetical velocity-depth profile (mean V_P) of the Calabrian crust (Kern & Schenk, 1985, 1988). The salient features of this profile are (Fig. 5):

- a velocity inversion in the upper crust, caused by the S-type granites (V_P = 6.2 km/s)
- a high velocity layer in the middle crust (V_P = 7.2 km/s), corresponding to the garnet- and sillimanite-rich restitic metapelites, which, interestingly, have higher velocities than the underlying metabasites.

However, some caution must be paid regarding the high velocities measured in the metapelitic samples. Leucosomes may be under-represented in the small samples (cubes with 43 mm edges) used for the measurements. The presence of 10 - 20 % more leucosome would lower the mean V_P values by about 0.1 - 0.2 km/s.

A seismic reflection program is in progress which will test the hypothesis that the lower crustal rocks at the surface are an extension of an intracrustal high velocity layer detected beneath Calabria at a depth of about 20 km (Schütte, 1978).

6. Chemistry of the lower crust

The estimation of the chemical composition of the Calabrian lower crust is based on geological mapping (1/10000) by the author along the best exposed strip through all units of the exposed lower crust section (see Fig. 4).The lower granulite-pyriclasite unit (2 km thick) consists of about 80 % metabasite and 20 % felsic granulite. The metapelite unit (5 - 6 km thick) consists of 76 % aluminous paragneisses, 14 % granofelses (semipelitic) and 10 % metabasites.

The mean chemical composition of each rock type is given in Tab. 1, together with the calculated bulk compositions of the whole granulite-pyriclasite unit and the whole metapelite unit. The estimated bulk composition of the lower crust differs significantly from that presented by Maccarrone et al. (1983). This difference is related to two discrepancies:

- The distribution of the different rock types within the lower crust, as given by Maccarrone et al. (1983), is not representative because it is based only on mapping in poorly exposed areas, mainly along roads.

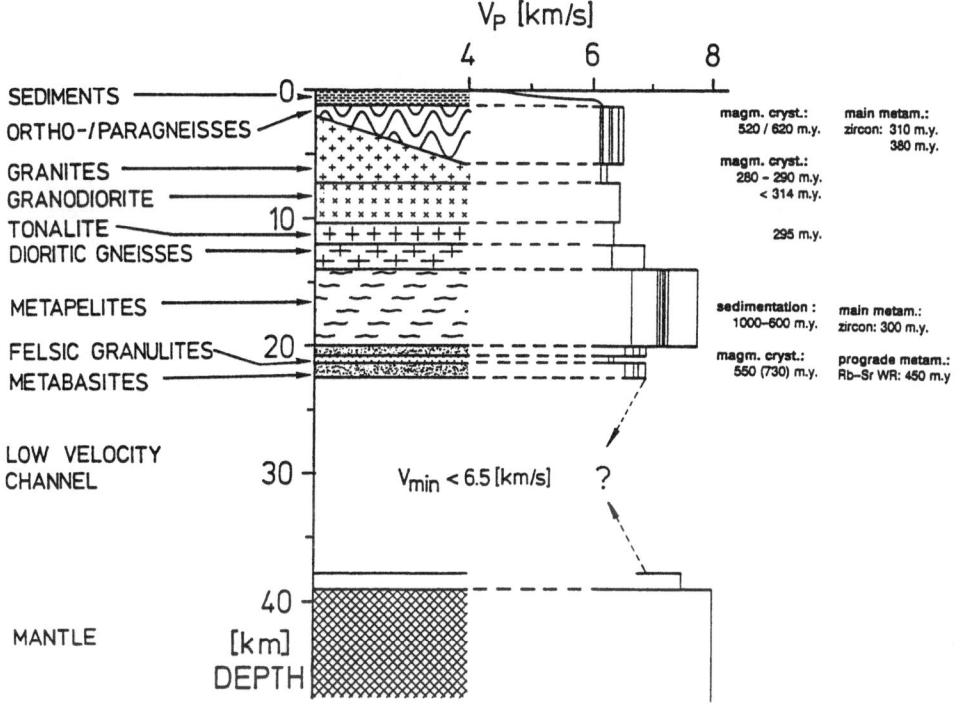

Fig. 5: Conjectural lithological section through the Calabrian crust, modified from Schenk (1981). Wave velocity data from Kern & Schenk (1985, 1988). Evolution model according to isotopic dates given in Fig. 2.

- Maccarrone et al. (1983) have under-estimated the proportion of leucosome within the metapelites, which, in fact, amounts to about 20 %.

Indeed, the estimated 20 % for leucosome in the lower crustal metapelites has been confirmed by the analysis of a mega-sample, a 2.4 m long and 2 cm thick slice perpendicular to the layering. This representative mega-sample includes both leucosomes and peraluminous melanosomes but is, nevertheless, highly restitic (Table 1). The addition of 40 % peraluminous granite would be necessary to shift its composition towards that of a normal pelite (cf. Shaw, 1956). This is in general agreement with the conclusions of Maccarrone et al. (1983).

A more detailed chemical study of the lower crustal rocks, including trace element determination in several mega-samples is in progress (in cooperation with R. Emmermann, Gießen). This study has already confirmed the clear restitic character of the lower crustal metapelites. In contrast, the upper crustal paragneisses in Calabria have normal pelitic compositions (Table 1).

Table 1:

I. Chemical compositions of the main rock types of the Calabrian crust based on data from the author and R. Emmermann and from Moresi et al. (1979) (column 1).

	1 Meta-basite Average	2 Granu-lite Average	3 Grano-fels Average	4 Al-para-gneiss. Average small samples	5 Leuco-some Average	6 Calculated metapelite 80 % of no 4 20% of no 5	7 Meta-pelite mega-sample
	N = 110	N = 13	N = 6	N = 16	N = 17		
SiO2	49.9	67.9	56.0	51.1	76.2	56.1	56.2
TiO2	1.5	0.7	1.3	1.4	0.13	1.15	0.95
Al2O3	17.3	14.9	18.4	24.0	13.6	21.9	21.6
FeOtot	9.8	5.2	9.1	11.2	0.86	9.1	9.3
MgO	6.9	2.3	4.5	4.7	0.44	3.85	4.03
CaO	9.5	2.6	4.6	2.0	0.48	1.7	1.90
Na2O	2.4	2.7	2.7	1.5	1.9	1.6	1.78
K2O	0.7	2.7	1.7	1.8	4.9	2.4	2.82

II. Estimated bulk composition of the Calabrian lower crust (normalized) compared with an estimate of Taylor & McLennan (1985).

	A Metabasite unit 80% of no 1 20% of no 2	B Metapelite unit 76% of no 7 14% of no 3 10% of no 1	C Whole lower crust 25% of A 75% of B	D Estimated lower crust (Taylor & McLennan)
SiO2	54.5	56.4	56.0	54.4
TiO2	1.2	1.1	1.1	1.0
Al2O3	17.1	21.0	20.0	16.0
FeOtot	9.1	9.4	9.3	10.6
MgO	6.1	4.5	4.9	6.3
CaO	8.3	3.0	4.4	8.5
Na2O	2.5	2.0	2.1	2.8
K2O	1.1	2.5	2.1	0.34

Table 1 (cont.):

III. Estimation of pelite composition prior to partial melt extraction. The composition of upper crustal metapelites (gneisses) from Calabria and an average pelite are given for comparison.

	8 Metapelite megasample	9 Peraluminous granite Cittanova	10 Calculated pelite comp.: 60% of no 8 40% of no 9	11 Upper crustal gneiss N=6	12 Average pelite (Shaw, 1956)
SiO_2	56.2	72.5	62.7	62.4	61.54
TiO_2	0.95	0.25	0.67	0.9	0.82
Al_2O_3	21.6	14.8	18.9	18.5	16.95
FeO_{tot}	9.3	1.83	6.3	6.7	6.21
MgO	4.03	0.61	2.8	2.45	2.52
CaO	1.90	1.61	1.8	0.9	1.76
Na_2O	1.78	3.49	2.5	1.8	1.84
K_2O	2.82	3.93	3.3	3.3	3.45

7. Formation ages of the Calabrian crust

Indications about the formation ages of the Calabrian crust have been gained from Sr and Nd model ages of metasediments and from U-Pb zircon dates of metamagmatic rocks.

Sr model ages of lower crustal metapelites are around 950 m.y. (Fig. 6) and thus somewhat younger than the Nd model ages (T_{DM}) of 1.3 b.y. obtained from mega-samples (Köhler., pers. communication). Assuming that the Sr isotopic composition of the pelitic sedimentary material was equilibrated with ocean water at the time of its deposition, a probable sedimentation age between 1000 and 600 m.y. can be advanced (Fig. 6). This is similar to the results reported from the Ivrea zone (Hunziker & Zingg, 1980).

U-Pb zircon analyses, which give information on the formation age of the Calabrian crust, have been performed on two metabasites from the lower crust as well as on two peraluminous orthogneisses and one hornblende-bearing I-type orthogneiss from the upper crust. The zircons of an unmetamorphosed Paleozoic siltstone (Devonian?, at Stilo, see Fig. 1) have also been analysed (Schenk & Todt, 1989).

The zircons of both lower crustal metabasites are highly discordant and plot along a common discordia near the lower intercept at about 295 m.y. The discordia defined by only one of the metabasite samples (no 145) intercepts the concordia at 312 and 730 m.y. Combining data from both samples and using a lower intercept of 295 m.y., which is the age of metamorphism obtained from zircon and monazite dates for the overlaying paragneisses, the corresponding upper intersection of the metabasite discordia is then at 553 ± 27 m.y. (Fig. 7). This 553 m.y. age is interpreted as the probable intrusion age of the metabasites, although the upper intercept at 730 m.y. for sample 145 could also represent the intrusion age. In any case, the upper intercept of the metabasite discordia is totally different from the corresponding intercept of the paragneiss discordia, which is 2000 - 2300 m.y. This excludes the interpretation that the metabasite zircons represent material incorporated from the surrounding metasediments during intrusion.

32

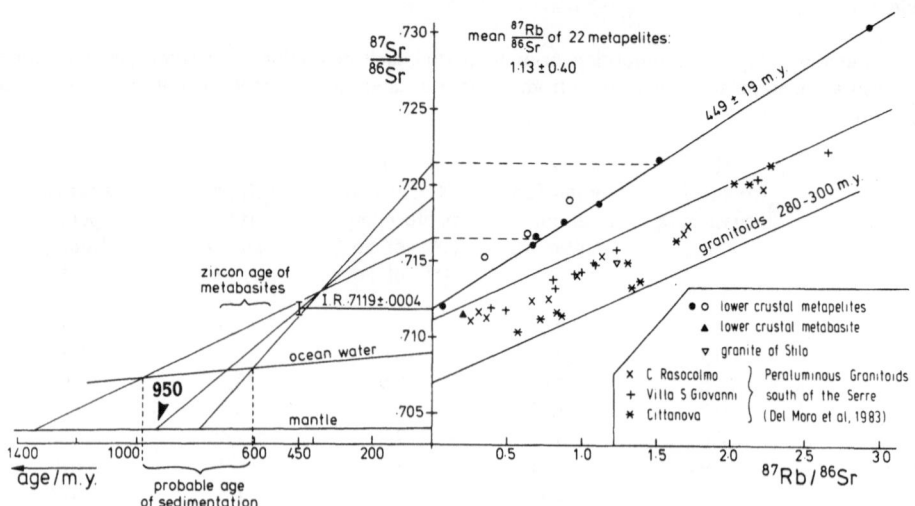

Fig. 6: Rb-Sr evolution diagram for lower crustal metapelites. The isochron points to a (metamorphic?) homogenization 450 m.y. ago. Sr model age is at about 950 m.y. and the probable sedimentation age lies between 1000 and 600 m.y.

The zircons of the hornblende-bearing orthogneiss of the upper crust are less discordant and their discordia intersects the concordia at 311 and 622 m.y. The zircons of both peraluminous orthogneisses point to an intrusion age of 516 ± 25 m.y. (Schenk & Todt, 1989), an age which is known from orthogneisses of the Kabylides of Northern Africa (Bossière & Peucat, 1985).

The detrital zircons of the Devonian (?) siltstone from Stilo are made up of different populations, including red grains. The analytical results from all fractions fall along a common discordia intersecting the concordia at 550 ± 50 m.y. and at 2500 m.y. With the exception of the red grain fractions, all points plot near the lower intersection point. Obviously, the detrital zircons represent a two-component mixture of an Archean and a Pan-African component, with the latter being much more common. The predominance of Pan-African zircons in this Devonian siltstone indicates that the Pan-African orthogneisses of southern Calabria might be part of the source rocks.

In summary, all formation ages obtained from Calabrian rocks so far point to a relatively young crust, formed later than 1.3 b.y. (Nd model ages for the metasediments of the lower crust) but in part as late as 550 m.y. (most probable formation age of the metabasites which underlay the metapelite unit). A continental margin seems to have been the most probable place for the sedimentation and crustal formation of southern Calabria. This is indicated by the high proportion of metamorphosed clastic sediments in the lower crust and the presence of numerous intercalations of metacarbonates, which suggests shallow water sedimentation.

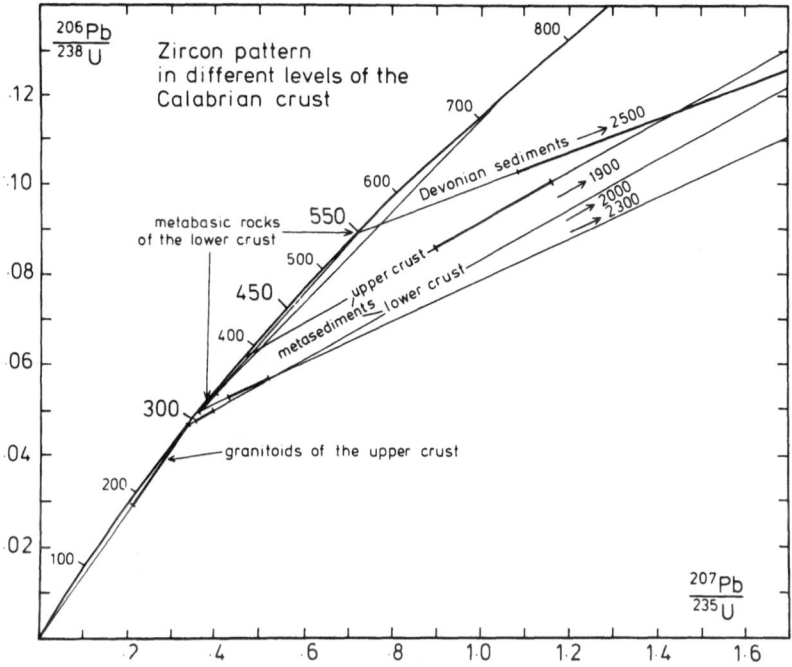

Fig. 7: U-Pb zircon pattern for metamagmatic and metasedimentary rocks from different levels of the Calabrian crust. The spread of zircon data in each rock is indicated by heavy lines. The upper intersection points of the discordia from metasedimentary rocks are at about 2 b.y.; those from metabasic rocks lie between 550 and 730 m.y.

8. Metamorphic zonation in the lower crustal section

The presence of a metamorphic zonation in the granulite-facies unit in the Serre was first recognised by Schenk (1984) and represents an important argument for the presence of a continuous lower crustal section. This is further supported by geological mapping, resulting in the recognition of two distinct lithostratigraphic units (the granulite-pyriclasite unit and the overlying metapelite unit) and a well preserved lithostratigraphy in the metapelite unit.

The position of the metapelite garnet-cordierite "isograde" (garnet-cordierite-biotite-sillimanite-K-feldspar-quartz) is now approximately mapped over the whole exposed lower crustal area (Fig. 1). Since coexisting garnets (X_{Mg} = 0.30) and cordierites (X_{Mg} = 0.77 - 0.80) are very similar in composition along the strike in the "univariant" zone, the latter probably represents a true bathozone of 5.5 to 6.0 kbar (~ 20 km depth) in the crust.

In the western area the univariant zone is displaced about 7 km towards the north due to a fault in the Tertiary basin. Similar displacements are seen at the boundary between the granulite-pyriclasite unit and the metapelite unit, as well as at the Alpine Curinga-Girifalco line

(Fig. 1). A "staurolite zone" is restricted to lower grades, i.e. to south of the "garnet-cordierite isograde". In the deeper, northern part, staurolite stability was exceeded and the high-temperature breakdown products of staurolite (garnet-sillimanite-spinel-corundum and garnet-sillimanite-spinel-cordierite) are found. Pseudomorphs of sillimanite-K-feldspar-biotite after muscovite occur in metapelitic leucosomes near the contact with the overlying dioritic gneisses. Thus, at the top of the granulite-facies section, muscovite + quartz stability was overstepped only during the late stages.

Further evidence for a temperature and pressure gradient in the lower crustal segment during the highest stage of metamorphism is the systematic Mg --- Fe shift of garnet, biotite and cordierite in pressure- or temperature-dependent divariant assemblages from the base towards the top of the section (Schenk, 1984, 1989). In the cordierite-garnet-sillimanite-quartz assemblage cordierites shift from $X_{Mg} = 0.87$ (base) towards 0.72 (top) in the metapelite unit, whereas the coexisting garnets shift from $X_{Mg} = 0.38$ to 0.19. Using Holdaway & Lee's (1977) calibration of this divariant equilibrium, a pressure difference of 1.5 to 2.0 kbar between the top and base of the metapelite unit has been obtained.

In metabasic rocks, which form intercalated layers and lenses in the metapelites, ortho-pyroxene was stable even in the uppermost levels of the metapelite unit.

Estimates of peak temperatures in the deeper, granulite-pyriclasite unit were done using the garnet-orthopyroxene- (Harley & Green, 1982) and the garnet-clinopyroxene-thermometers (Ellis & Green, 1979) on both felsic granulites and metabasic rocks. The results fall in the range 770 - 820 °C at a pressure of 7.5 kbar. Pressure estimates using the orthopyroxene-garnet-plagioclase-quartz assemblage (Newton & Perkins, 1982) give results of 6.5 - 7.0 kbar for felsic granulites, but 7 - 8 kbar for metabasic rocks.

A *temperature gradient* of about 100 - 120 °C has been deduced for the lower crustal section based on temperature estimates of 800 °C in the granulite-pyriclasite unit, the stability of staurolite + quartz at the top of the section, and the systematic shift towards Fe-richer compositions in the garnet-biotite-sillimanite assemblages in the upper part of the section.

According to the P-T data presented above, the mean *geothermal gradient* during the highest metamorphic stage was 30 °/km with the whole lower crustal segment in the stability field of sillimanite (Fig. 8).

At shallower levels in the upper crustal phyllites of the Aspromonte andalusite that probably formed during the Hercynian regional metamorphism occurs as the stable alumosilicate. This suggests that the late Hercynian geotherm passed through the andalusite stability field.

9. P-T path of the lower crust

Information about the P-T conditions *prior to the peak metamorphism* is scarce. However, the common presence of relict kyanite in rocks of the upper metapelite unit as well as of one kyanite inclusion in garnet from a rock of the granulite-pyriclasite unit indicate that, during an earlier metamorphic stage, the geotherm of the lower crust passed through the stability field of kyanite (Fig. 8). On the basis of textural evidence, the prograde evolution can be subdivided into an earlier kinematic stage and a later static one, during which the highest temperatures were reached (Schenk, 1984). Granulite-facies conditions prevailed, however, during both stages, as shown by the formation of orthopyroxene. Large late-stage 'netlike' orthopyroxene porphyroblasts postdate any penetrative deformation. This leads to the important conclusion that the main heating stage during the granulite-facies metamorphism took place under static conditions and survived the penetrative deformation of the lower crust.

The *end of the high-temperature static metamorphic stage* was induced by a tectonic (erosional uplift) event. This has been deduced from mineral reaction textures involving garnet in SiO_2-undersaturated metabasic and metapelitic rocks (Fig. 9). These remarkable textures show that the

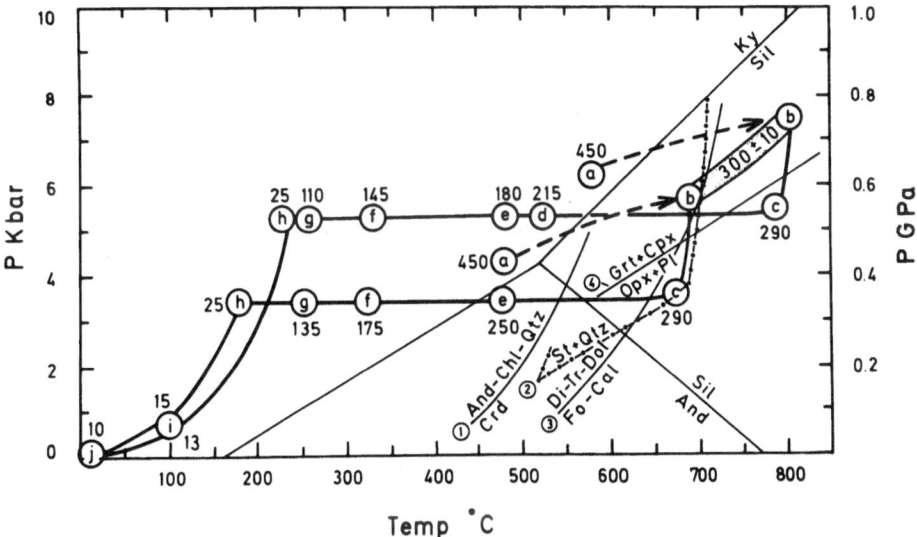

Fig. 8: P-T-t paths of the lower and the upper part of the lower crustal section of southern Calabria. *Stippled field:* P-T conditions over the lower crustal section during peak metamorphism. *Reaction curves* according to: (1) Seifert & Schreyer (1970), (2) Richardson (1978), (3) Käse & Metz (1980), (4) Green & Ringwood (1967); Al_2SiO_5 triple point after Salje (1986). *Age data* (in m.y.): (a) Rb-Sr WR metasediments, (b) and (c) U-Pb zircon and monazite, (d) Sm-Nd garnet and plagioclase, (e) K-Ar hornblende, (f) Rb-Sr K-feldspar and plagioclase, (g) Rb-Sr biotite, (h) intersection of the post-Hercynian and Apennine temperature-time-curves, (i) fission track apatite, (j) oldest overlying sediments. Rb-Sr closing-temperatures of biotite are taken at 250 °C and feldspars are taken at 315 °C according to Dodson's (1979) calculations for cooling rates of 3 °C/m.y. (Schenk, 1989).

breakdown of the dense, garnet-bearing parageneses was followed by new growth of these parageneses (Fig. 9). The garnet breakdown is related to the uplift, whereas the regrowth stage took place during subsequent cooling (Schenk, 1984, 1989). The reactions involved are:

garnet + clinopyroxene = orthopyroxene + plagioclase
garnet + sillimanite = cordierite + spinel

which have flat dP/dT slopes (Fig. 8). Since reaction curves must have been crossed a second time after the pressure decrease, the cooling path of the lower crust must have been nearly isobaric (Fig. 8).

During this *nearly isobaric cooling*, the pressure within the granulite-pyriclasite unit was around 5.6 kbar (at 650 °C), using the garnet-orthopyroxene-plagioclase-quartz barometer (Newton & Perkins, 1982) and the garnet-clinopyroxene thermometer (Ellis & Green, 1979) on late stage parageneses in quartz-bearing metabasites. These pressure values are about 2 kbar lower than the peak metamorphic pressures.

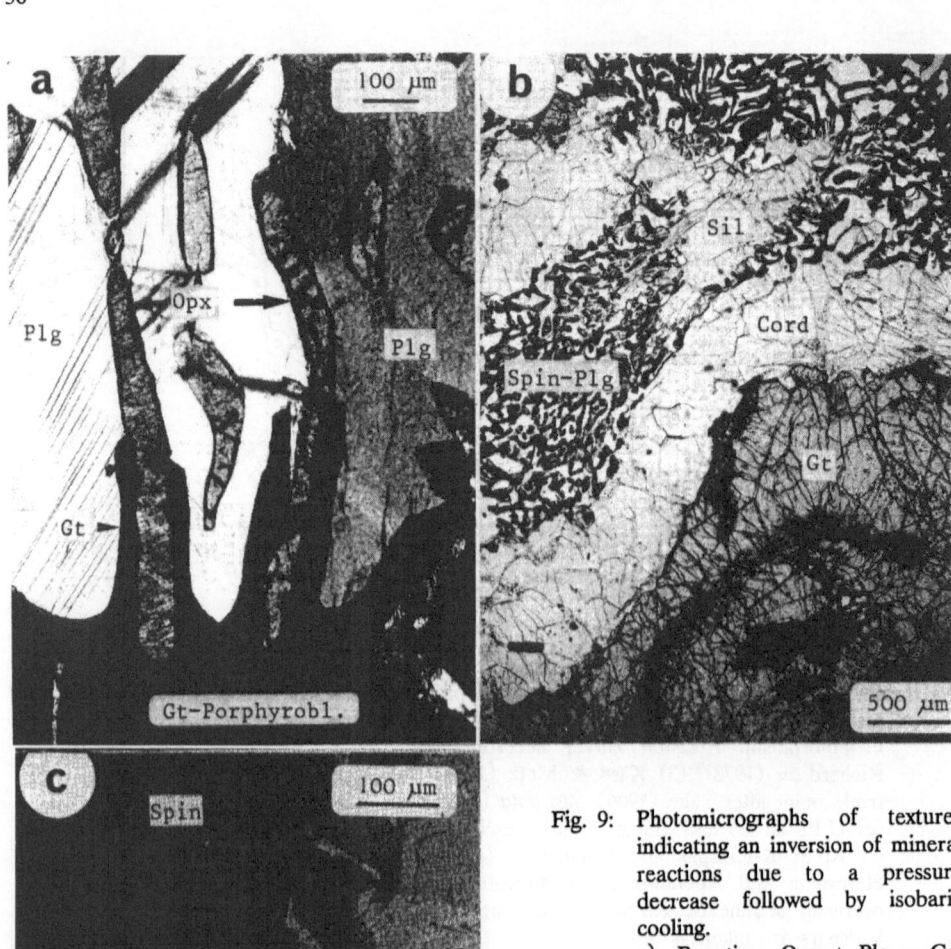

Fig. 9: Photomicrographs of textures indicating an inversion of mineral reactions due to a pressure decrease followed by isobaric cooling.

a) Reaction Opx + Plg = Grt + Cpx; late-stage reaction rims of Grt and Cpx grown between worm-like Opx and Plg of a corona around a garnet porphyroblast.

b), c) Reaction
Grt + Sil = Crd + Spin.

b) Grt is rimmed by Crd and Sil is surrounded by a Spin-Pl corona.

c) Regrowth of a Sil rim around Spin within the corona.

In the upper part of the metapelite unit, i.e. in the "univariant zone" and south of it, growth of andalusite at the expense of cordierite indicates that the pressure during the cooling stage was below 4 kbar. The most common reaction leading to the partial breakdown of cordierite is:

$$cordierite + CO_2 = magnesite + andalusite + quartz$$

Yet, in several samples sillimanite and/or kyanite also occur with andalusite as a reaction product, giving further support for nearly isobaric cooling after the first tectonic uplift: The cordierite began breaking down within the stability field of sillimanite, then cooling proceeded through the andalusite field and entered the kyanite field at temperatures still high enough to allow the growth of 0.5 cm large kyanite blasts. In some samples chlorite occurs with magnesite in the breakdown products of cordierite indicating the presence of some water in the fluid phase. But the occurrence of magnesite as the common breakdown product points to a very high XCO_2 (> 0.7) in the fluid phase of the metapelites, even during the retrograde metamorphism.

The main results of the petrological work are (1) the recognition of a metamorphic zonation in the 7 - 8 km thick granulite-facies lower crustal section and (2) the evidence for the clockwise P-T path. The earliest part of the path was in the kyanite field; most of the subsequent mineral growth occurred syntectonically in the stability field of sillimanite, but peak temperatures were reached after any penetrative deformation. The end of the static heating was caused by a tectonic event, after which the uplifted lower crust cooled isobarically in middle crustal levels.

10. Isotopic dating of the P-T path

In order to date the prograde and the peak metamorphism, two isotopic methods have been applied to the Calabrian lower crustal rocks.

(I) Rb-Sr whole-rock analyses of large (about 30 kg) metapelitic samples, taken across the whole lower crustal section, and

(II) U-Pb analyses of zircons and monazites from felsic granulites and metapelites (Schenk, 1980).

The Rb-Sr data show some scatter but define an isochron of 450 ± 20 m.y. (Fig. 6). The zircons are highly discordant and define a discordia intersecting the concordia at 300 ± 10 m.y. (Fig. 7). Six out of seven monazites are concordant between 296 and 289 m.y. and lie near the lower intersection point of the zircon discordia.

The meaning of the poorly defined Rb-Sr isochron of the metapelites is not clear, since the scale of Sr homogenization during regional metamorphism depends on several factors (temperature, kinematic conditions, fluid phase) whose extent are not precisely known. Possibly, in Calabria Sr homogenization was enhanced by the presence of a water-rich fluid phase and the penetrative deformation during prograde metamorphism (Schenk, 1989). The U-Pb system of zircons and monazites presumably dates the peak temperatures reached by the dry granulite-facies rocks during the static metamorphism of Hercynian age, about 300 m.y. ago. In Calabria, monazite ages obviously do not reflect cooling below a closing temperature, since all monazites yield very similar ages despite their different stratigraphic position and the different peak metamorphic temperatures of their host rocks. In contrast, K-Ar ages of hornblende, Rb-Sr ages of feldspar (plagioclase + K-feldspar) and of biotite, and a Sm-Nd age of garnet are much younger than the Hercynian granulite-facies event and are therefore interpreted as cooling ages. Using any of these mineral systems, younger cooling ages are obtained from the hotter, deeper granulite-pyriclasite unit than from the overlying metapelite unit (Schenk, 1980, 1989).

Interestingly, the Sm-Nd garnet age of 214 ± 10 m.y. from a felsic granulite is much younger than the peak metamorphism and points to a "closing temperature" of about 500 °C (Fig. 8).

The systematic sequence of mineral ages in both lower crustal units points to very slow cooling (~ 2 °/m.y.), undisturbed by any tectonic event between 290 and 110 m.y. This interpretation is supported by the fact that undisturbed carbonate sedimentation occurred at the surface above the Calabrian crust during the Mesozoic (Görler, 1978).

According to the tectono-thermal reconstruction proposed here, the Calabrian crust cooled slowly through the Mesozoic and by the Tertiary attained temperatures of about 200 °C at a depth of about 15 km. This indicates a very low geothermal gradient, which is supported by recent heat flow measurements in Calabria (El Ali & Giese, 1978).

The age of the final uplift of the former lower crust section from intermediate levels up to the surface can be estimated from (1) the age of the oldest sediments on top of the lower crustal rocks which is Tortonian (about 10 m.y.) and (2) the apatite fission track ages which are only slightly older (13 - 15 m.y.). Consequently, the last phase of uplift must have occurred rather rapidly during the Apennine orogeny, starting about 25 m.y. ago (Fig. 8).

Parts of the closing stages of uplift may be oversimplified, since we cannot document the geological record in the lower crust between the biotite cooling ages at 110 m.y. and the apatite fission track ages at 15 m.y. Some uplift between 110 m.y. and 15 m.y. during Alpine time cannot be ruled out. On the other hand, the change in inclination of Mesozoic and Tertiary sedimentary strata in southeastern Calabria (at Stilo) points to an uplift event starting about 30 m.y. ago (Görler, 1978).

As in the case of the Apennine uplift, the Hercynian uplift event seen in the lower crust is also recorded in upper crustal rocks; Rb-Sr biotite ages in both S-type granites of the Aspromonte and the amphibolite-facies country rocks are of late Hercynian age (280 - 295 m.y., Del Moro et al., 1983). These cooling ages are interpreted as the result of rapid erosion of the uppermost crustal levels occurring synchroneously with the lower crustal uplift 295 - 290 m.y. ago. Interesting, inside the large area of granitoids in the Serra S. Bruno, Del Moro et al. (1986) found biotite ages transitional between the Hercynian upper crustal ones and the 135 to 110 m.y. lower crustal ones (Fig. 2). Thus, there is a nearly continuous decrease in biotite ages towards the bottom of the Serre granitoids where the cooling ages are very similar to those of the underlying tonalite of Cardinale (140 compared with 135 m.y.). In addition, this continuous change of biotite ages strongly supports the author's interpretation of a continuous crustal profile. It should be noted that the above is a re-interpretation of the Del Moro et al. (1986) data.

11. Crustal differentiation

The presence of a 5 - 6 km thick unit in the lower crust, which consists mainly of highly restitic metapelites, and the presence of large masses of peraluminous granites, granodiorites, and orthogneisses in the upper crust shows the probable genetic interrelation of the whole section. Because of their composition and age, the peraluminous, cordierite and sillimanite bearing two-mica granites from Cittanova, Villa S. Giovanni and M. S. Demetrio (Fig. 1) are the most likely melt products of the lower crustal metapelites.

The intrusion ages of the granites have been determined through internal Rb-Sr mineral isochrons between 280 and 290 m.y. (Del Moro et al., 1983). Thus, their ages coincide with or are only slightly younger than the end of the granulite-facies metamorphism and the tectonic uplift of the lower crust 295 to 290 m.y. ago (Fig. 8). These granites have nearly the same Sm/Nd ratios and Nd model ages (1.3 b.y.) as the lower crustal metapelites (Köhler, pers. communication). However the Sr initial ratios of these Hercynian granites are lower than the initial ratio of the 450 m.y. isochron obtained for the lower crustal metapelites (Fig. 6). Such low Sr initial ratios of granites are a common, but a not yet well understood phenomenon, which can be explained by the addition of low $^{87}Sr/^{86}Sr$ mantle material to crustal melts.

The biotite-bearing Serre granodiorite is mesaluminous to peraluminous and slightly older (< 314 m.y.) than the peraluminous granites (D'Amico et al., 1982). Its Nd model age and its oxygen isotopic composition are intermediate to those of the S-type granites and the calc-alkaline tonalite (pers. communication of H. Köhler and St. Hoernes) possibly indicating that these rocks are a hybrid of the S-type melts generated in the metapelites and the calc-alkaline tonalites. The peraluminous orthogneisses, which occur as inclusions in the Hercynian dioritic gneiss and in the paragneisses of the Aspromonte are much older (500 - 520 m.y.) than the granulite-facies metamorphism in the lower crust (300 m.y.) and are even older than the process leading to the Sr-homogenization (450 m.y.) in the lower crustal metapelites. Nevertheless, the orthogneisses could represent the oldest differentiation products of the Calabrian lower crust. Alternatively, these "Cadomian" or "Pan-African" gneisses could have been formed elsewhere as part of a different crustal section and, subsequently, became part of the Calabrian crust through a tectonic event in early or pre-Hercynian time. In this case, upper and lower crust would have been "welded" by the Hercynian intrusions.

12. Summary and conclusion

The exposure of the Calabrian lower and intermediate crustal layers is due to tectonic movements at the Adriatic plate boundary during the upper Tertiary. Nevertheless, the overall lithologic structure and evolution of the section was complete by the late Hercynian (290 - 280 m.y. ago). The structure is the result of a long history and of different geological processes:

1) The intrusion of basaltic magmas (granulite-pyriclasite unit) occurred about 550 or (less probable) 730 m.y. ago. Hornblende-bearing orthogneisses in the upper crust point to a similar formation age (620 m.y.).

2) The sedimentation of large amounts of clastic sediments (metapelite unit) is younger than 1.3 b.y. (Nd model ages) and probably occurred between 1000 and 600 m.y. (Sr model ages).

3) The Hercynian granulite-facies metamorphism in the lower crust (300 ± 10 m.y.) was connected with an exceptionally steep geotherm and lead to a crustal differentiation process resulting in the formation of restitic metapelites in the lower crust and the intrusion of peraluminous S-type granites (290 - 280 m.y.) in the upper crust.

4) I. An earlier differentiation process of the lower crust may have resulted in the formation of peraluminous melts 520 to 500 m.y. ago which are now orthogneisses in the upper crust. II. Alternatively, the upper crust including these orthogneisses evolved elsewhere from a different crust and became part of the Calabrian crust only in early or pre-Hercynian time through a tectonic event.

5) The extensive calc-alkaline magmatism (< 314 - 295 m.y.), which is represented by diorites and tonalites, coincides with the granulite-facies metamorphism during the Hercynian orogeny.

6) The end of the granulite-facies metamorphism is marked by a tectonic uplift, which also initiated a general cooling process of the whole crust.

7) After the Hercynian uplift, the former lower crust remained at an intermediate crustal level (13 - 17 km). During this time it slowly cooled (2 °/m.y.) from 800 - 700 °C down to about 200 °C; it was then rapidly uplifted to the surface between 25 and 10 m.y. ago.

The remarkably high proportion of metapelites in the lower crust might be due to its paleogeographic position during the late Proterozoic sedimentation, possibly at the margin of an unusual large continental mass. A similar high proportion of sediments in the deep crust are found in the Ivrea zone and in the Moldanubian zone of central Europe.

Pressure determinations of 7.5 kbar indicate that in Calabria the deepest exposed unit came from a depth of about 25 to 30 km. Since the Hercynian crust in central Europe and in the Adriatic plate generally is only around 30 km thick (e.g. Mueller, 1977), it seems possible that in Calabria the deepest parts of a Hercynian type crust are exposed.

The most likely geotectonic setting of the Hercynian granulite-facies metamorphism and crustal differentiation in Calabria is a continental margin above a subduction zone. This model is based on the occurrence of granulite-facies metamorphism that was accompanied by an extensive calc-alkaline magmatism and the subsequent tectonic uplift, which may have been due to a continent-continent collision (Schenk, 1984, 1989). Heating by mantle magmas intruding at the crust mantle boundary could explain, how the highest late-stage metamorphism survived the penetrative deformation and why the S-type granites are somewhat younger than the calc-alkaline magmas.

Acknowledgements

This work was partly supported by the Deutsche Forschungsgemeinschaft (SPP "Kontinentale Unterkruste"). Some of the results reported here stem from work in collaboration with R. Emmermann (Gießen), St. Hoernes (Bonn), H. Köhler (München), and W. Todt (Mainz). I thank S. Harley (Edinburgh), D. Lattard (Berlin), J. Schumacher (Kiel) and two anonymous reviewers for critical reading and discussion.

References

Amodio-Morelli, L., Bonardi, G., Colonna, V., Dietrich, D., Giunta, G., Ippolito, F., Liguori, V., Lorenzoni, S., Paglionico, A., Perrone, V., Piccarreta, G., Russo, M., Scandone, P., Zanettin Lorenzoni, E., Zupetta, A. (1976): L'arco Calabro-Peloritano nell' orogene Apenninico-Maghrebide. Mem. Soc. Geol. It. 16, 1-60

Alvarez, W. (1976): A former continuation of the Alps. Bull. geol. Soc. Am. 87, 891-896

Atzori, P., Pezzino, A., Rottura, A. (1977): La massa granitica di Cittanova (Calabria Meridionale): Relazioni con le rocce granitoidi del massiccio delle Serre e con le metamorfiti di Canolo, S. Nicodemo e Molochio (nota preliminare). Boll. Soc. Geol. It. 96, 387-391

Bonardi, G., Messina, A., Perrone, V., Russo, S., Zuppetta, A. (1984): L'unità di stilo nel settore meridionale dell' arco Calabro-Peloritano. Boll. Soc. Geol. It. 103, 279-309

Bonardi, G., Compagnoni, R., Del Moro, A., Messina, A., Perrone, V. (1987): Riequilibrazioni tettono-metamorfiche alpine nell' Unità dell' Aspromonte, Calabria meridionale. Rend. Soc. It. Miner. Petrol. 42/2, 301

Borsi, S., Hieke, M., Lorenzoni, O., Paglionico, A., Zanettin Lorenzoni, E. (1976): Stilo Unit and "dioritic kinzigitic" Unit in Le Serre (Calabria, Italy). Geological petrological and geochronological characters. Boll. Soc. Geol. It. 95, 219-245

Bossière, G., Peucat, J. J. (1985): New geochronological information by Rb-Sr and U-Pb investigations from the pre-Alpine basement of Grande Kabylie (Algeria). Can. J. Earth Sci. 22, 675-685

Colonna, V., Lorenzoni, S., Zanettin Lorenzoni, E. (1973): Sull'esistenza di due complessi metamorfici lungo il bordo sud-orientale del massicio "granitico" delle Serre (Calabria). Boll. Soc. Geol. It. 92, 801-830

D'Amico, C., Rottura, A., Maccarrone, E., Puglisi, G. (1982): Peraluminous granitic suite of Calabria-Peloritani Arc (southern Italy). Rend. Soc. It. Mineral. Petrol. 38, 35-52

Del Moro, A., Pardini, G., Maccarrone, E., Rottura, A. (1983): Studio radiometrico Rb-Sr di granitoidi peraluminosi dell'arco Calabro-Peloritano. Rend. Soc. It. Mineral. Petrol. 38, 1015-1026

Del Moro, A., Paglionico, A., Piccaretta, G., Rottura, A. (1986): Tectonic structure and post-Hercynian evolution of the Serre, Calabrian arc, southern Italy: Geological, Petrological and radiometric evidences. *Tectonophysics* **124**, 223-238

Dodson, M. H. (1979): Theory of cooling ages. In: Jäger, E., Hunziker, J. C. (eds), *Lectures in isotope geology*, Springer, Berlin, 194-214

El Ali, H., Giese, P. (1978): A geothermal profile between the Adriatic and the Tyrrhenian Sea. In: Closs, H., Roeder, D., Schmidt, K. (eds), *Alps, Apennines, Hellenides*. Schweizerbart, Stuttgart, 324-327

Ellis, D. J., Green, D. H. (1979): An experimental study of the effect of Ca upon garnet-clinopyroxene Fe-Mg exchange equilibria. *Contrib. Mineral. Petrol.* **71**, 13-22

Giese, P., Reutter, K.-J. (1978): Crustal and structural features of the margins of the Adria microplate. In: Closs, H., Roeder, D., Schmidt, K. (eds), *Alps, Apennines, Hellenides*. Schweizerbart, Stuttgart, 565-588

Görler, K. (1978): Critical review of postulated nappe structures in Southern Calabria. In: Closs, H., Roeder, D., Schmidt, K. (eds), *Alps, Apennines, Hellenides*. Schweizerbart, Stuttgart, 349-354

Green, D. H., Ringwood, A. E. (1967): An experimental investigation of the gabbro-garnet granulite-eclogite transformation and its petrological applications. *Geochimica Cosmochimica Acta* **31**, 767-833

Harley, S. L., Green, D. H. (1982): Garnet-orthopyroxene barometry for granulites and peridotites. *Nature* **300**, 697-701

Holdaway, M. J., Lee, S. M. (1977): Fe-Mg cordierite stability in high-grade pelitic rocks based on experimental, theoretical, and natural observations. *Contrib. Mineral. Petrol.* **63**, 175-198

Hunziker, J. C., Zingg, A. (1980): Lower paleozoic amphibolite to granulite facies metamorphism in the Ivrea Zone (Southern Alps, Northern Italy). *Schweiz. mineral. petrogr. Mitt.* **60**, 181-213

Käse, H. R., Metz, P. (1980): Experimental investigation of the metamorphism of siliceous dolomites. IV. Equilibrium data for the reaction: 1 Diopside + 3 Dolomite = 2 Forsterite + 4 Calcite + 2 CO_2. *Contrib. Mineral. Petrol.* **73**, 151-159

Kern, H., Schenk, V. (1985): Elastic wave velocities in rocks from a lower crustal section in Southern Calabria (Italy). *Phys. Earth Planet Int.* **40**, 147-160

Kern, H., Schenk, V. (1988): A model of velocity structure beneath Calabria, southern Italy, based on laboratory data. *Earth and Planetary Science Letters* **87**, 325-337

Lorenzoni, S., Orsi, G., Zanettin Lorenzoni, E. (1980): The Hercynian range in southeastern Aspromonte (Italy). Its relationship with the Alpine Stilo Unit. *N. Jb. Geol. Paläont. Mh.* **1980**, 404-416

Maccarrone, E., Paglionico, A., Piccarreta, G., Rottura, A. (1983): Granulite-amphibolite facies metasediments from the Serre (Calabria, Southern Italy): Their protoliths and the processes controlling their chemistry. *Lithos* **16**, 95-111

Moresi, M., Paglionico, A., Piccareta, G., Rottura, A. (1979): The deep crust in Calabria (Polia-Copanello unit): A comparison with the Ivrea-Verbano zone. *Mem. Sci. Geol. Padova* **33**, 233-242

Mueller, S. A. (1977): A new model of the continental crust. In: Heacock, J. G. (ed.), *The earth's crust: Its nature and physical properties*. Geophys. Monogr. Ser. **20**, AGU, Washington D.C., 289-317

Newton, R. C., Perkins, D. (1982): Thermodynamic calibration of geobarometers based on the assemblages garnet-plagioclase-orthopyroxene(-clinopyroxene)-quartz. *Am. Mineral.* **67**, 203-222

Paglionico, A., Piccarreta, G. (1976): Le Unita del fiume Pomo e di Castagna nelle Serre settentrionalii (Calabria). *Boll. Soc. geol. Ital.* **95**, 1-11

Richardson, S. W. (1968): Staurolite stability in a part of the system Fe-Al-Si-O-H. *Journal of Petrology* **9**, 467-499

Salje, E. (1986): Heat capacities and entropies of andalusite and sillimanite: The influence of fibrolitization on the phase diagram of the Al_2SiO_5 polymorphs. *American Mineralogist* **71**, 1366-1371

Schenk, V. (1980): U-Pb and Rb-Sr radiometric dates and their correlation with metamorphic events in the granulite-facies basement of the Serre, Southern Calabria (Italy). *Contrib. Mineral. Petrol.* **73**, 23-38

Schenk, V. (1981): Synchronous uplift of the lower crust of the Ivrea Zone and of Southern Calabria and its possible consequences for the Hercynian orogeny in Southern Europe. *Earth and Planetary Science Letters* **56**, 305-320

Schenk, V. (1984): Petrology of felsic granulites, metapelites, metabasics, ultramafics, and metacarbonates from Southern Calabria (Italy): Prograde metamorphism, uplift and cooling of a former Lower Crust. *J. Petrol.* **25**, 255-298

Schenk, V. (1989): P-T-t path of the lower crust in the Hercynian fold belt of southern Calabria. In: Daly, J. S., Cliff, R. A., Yardley, B. W. D. (eds), *Evolution of metamorphic belts*, Geol. Soc. Spec. Publ. **43**, 337-342

Schenk, V., Todt, W. (1989): The age of the Adriatic crust in Calabria (Southern Italy): constraints from U-Pb zircon data. *Terra Abstracts* **1**, 350

Schmid, R. (1979): Are the metapelites of the Ivrea-Verbano zone restites? *Mem. Sci. Geol. Padova* **33**, 67-69

Schütte, K. G. (1978): Crustal structure of Southern Italy. In: Closs, H., Roeder, D., Schmidt, K. (eds), *Alps, Apennines, Hellenides*. Schweizerbart, Stuttgart, 321-323

Schwarz, G. (1978): Magnetic measurements in Northern Calabria. In: Closs, H., Roeder, D., Schmidt, K. (eds), *Alps, Apennines, Hellenides*. Schweizerbart, Stuttgart, 321-323

Seifert, F., Schreyer, W. (1970): Lower temperature stability limit of Mg-cordierite in the range 1 - 7 kb water pressure: A redetermination. *Contrib. Mineral. Petrol.* **27**, 225-238

Shaw, D. M. (1956): Geochemistry of pelitic rocks. Part III: Major elements and general geochemistry. *Geol. Soc. Am. Bull.* **67**, 919-934

Taylor, S. R., McLennan, S. M. (1985): *The continental crust: its composition and evolution*. Blackwell, Oxford, 312 p.

AN EXPOSED CROSS-SECTION OF CONTINENTAL CRUST, DOUBTFUL SOUND
FIORDLAND, NEW ZEALAND; GEOPHYSICAL & GEOLOGICAL SETTING.

G.J.H. Oliver
Geology Division
The University
St Andrews
Fife
Scotland KY16 9ST

ABSTRACT. Doubtful Sound apparently has the largest positive Bouguer
anomaly in the World. The anomaly lies adjacent to the Alpine Fault,
an active plate collision contact between the Pacific and Indian
Plates. The exceptional setting has led to the uplift and exposure of
25km of a 35km thick continental crustal section of the Pacific Plate
margin. The base of the section is composed of >12kb granulite
orthogneisses. They are in tectonic contact with 5-9kb ortho- and
para-amphibolites of the middle crust. The contact is an extensional
ductile shear zone called the Doubtful Sound Décollement. Structures
dip away from a central dome such that in eastern Fiordland, the upper
crust of basic and acid plutons, volcanics and sediments is exposed.
Radiometric dating suggests the following senario: late Mesozoic
continental extension associated with incipient Tasman Sea spreading,
uplifted unusually deep structural levels of a Mid-Cretaceous
metamorphic core complex. Granulite orthogneisses of the basement-core
have protoliths of gabbroic calc-alkaline arc plutonics dated at 119-
130 Ma, similar in age to diorites and granites of the upper crustal
level in eastern Fiordland and equivalents in NW Nelson. The covering
carapace of early to mid-Palaeozoic Tuhuan ortho- and para-schists
(386-428 Ma igneous protoliths) were separated from the basement-core
granulites by <93 Ma mylonites. The dynamic metamorphism and uplift
associated with the extension tectonics reset the Tuhuan mineral dates
to give mica cooling ages of >80 Ma. Late Eocene-early Miocene
transcurrent and more recent transpressive faulting, has brought the
Fiordland metamorphic core complex to the surface along the Pacific-
Indian Plate boundary - the Alpine Fault.

1.   INTRODUCTION

1.1.   Aim

The aim of this review paper is to demonstrate that Fiordland, through
its unique geological history and present day setting at an active
plate collision, represents a tilted crustal cross-section exposing
some 25km of a 35km thickness. The paper is divided into sections on
the regional geophysics and regional geology; this is then interpreted
in general terms of a metamorphic core complex in which a crustal
cross-section is presented. Finally, Fiordland's part in the
development of the south west margin of Gondwanaland is displayed in
palaeogeographic maps and sections. In Gibson's paper (this volume),

43

M. H. Salisbury and D. M. Fountain (eds.), Exposed Cross-Sections of the Continental Crust, 43–69.
© 1990 Kluwer Academic Publishers.

the structural details of the metamorphic core complex model are discussed.

1.2. Previous Work

The first geological map of Fiordland was published by Hector in 1864. His map displays a string of diorites and syenites in the western fiords intruding "gneisses and crystalline metamorphic rock". In the SW he shows "plastic clay and brown coal series" over "contorted and foliated schist". Park (1921, 1924) set up a stratigraphy assuming a Precambrian age for gneisses and discovered Ordovician graptolites in the SW. Bensen and Bartrum (1936) recognised the progressive regional metamorphism of the Ordovician slates towards the northern schists and gneisses. Turner (1937, 1939) reported on these high grade gneisses between Lake Manapouri and Doubtful Sound, dividing them into provinces, and assigning them to Eskola's amphibolite facies. He recognised the Doubtful Sound garnet-diopside and hypersthene-diopside assemblages as characteristic of the highest grades of regional metamorphism. Turner (1948, 1958, 1968) used these Doubtful Sound rocks as examples of his various granulite facies schemes. Wood (1960, 1962, 1966) produced NZ Geological Survey 1:250,000 geology maps of Fiordland correlating various formations across the region.

Landis and Combs (1967) and Aronson (1968) recognised paired metamorphic belts in the South Island on the basis of radiometric dates. Fiordland was seen as a foreland or western province of rocks first metamorphosed during the 350 Ma Tuhuan Orogeny but recycled for a second time during the 100 Ma Rangitata Orogeny.

Detailed mapping in Doubtful Sound allowed Oliver (1976, 1980) and Oliver and Coggon (1979) to extrapolate a basement-cover relationship over the greater part of western Fiordland (see Fig. 7) with a regional décollement separating granulite orthogneiss basement of western Fiordland from an amphibolite facies cover of Tuhuan rocks of central Fiordland. In recognising that the mylonitonised granulites of the décollement were retrogressed to amphibolites, Oliver (1980) proposed that the basement must be older than the cover and probably old Precambrian. Thus he called the décollement the Doubtful Sound Thrust. However, radiometric dating has recently shown the granulites are actually Cretaceous (Mattinson et al. 1986, McCulloch et al. 1987, Gibson et al. 1988) and were nearly contemporaneous with unmetamorphosed calc-alkaline arc plutonics of eastern Fiordland. Further mapping by Gibson (1982a, b, 1988) has shown that the mylonites of the Doubtful Sound décollement have the extensional fabrics of a ductile shear zone. This contradiction of Oliver's basement-cover concept has been rationalised by Gibson et al. (1988) who have proposed a model of a deeply eroded metamorphic core complex. This model will be elaborated by Gibson (this volume).

2. REGIONAL GEOPHYSICAL SETTING

2.1. Gravity

Reilly (1965) showed that a very large positive Bouguer gravity anomaly exists under Fiordland, centred on Doubtful Sound. Woodward (1972) measured a corresponding gravity low offshore and presented gravity models showing mantle rocks 10km below the surface at Doubtful Sound. Oliver (1973)and Oliver and Cogon (1979), using additional land based stations, (see Fig. 1), explained the anomalies in terms of subduction of Tasman Sea lithosphere under the New Zealand lithosphere such that Fiordland has been underthrust and uplifted 20km (see Fig. 2) to expose the lower crust. Thus successively higher levels of the middle and upper continental crust would be progressively exposed in a west to east traverse. This model is strikingly similar to that proposed for the Ivrea Zone (Berkhemer 1969). Davey and Smith (1983) used the same gravity data but different rock densities to propose a subduction model whereby the Moho was placed 25km below Doubtful Sound. Lower crustal rock densities were placed at less than 10km below the surface. The size of the positive Bouguer anomaly, +186m.gal, probably makes it the largest in the world.

2.2. Active Seismicity

Smith (1971) and Scholz et al. (1973) located medium and micro-earthquakes under Fiordland, confined in a 150km long by 30km wide zone orientated parallel to the coast. The configuration of the seismic activity is shown in Fig. 3: it defines a narrow steeply dipping Benioff Zone down to 160km depth. The dip and strike of this zone was used to constrain the gravity model in Fig. 2. On a regional scale, the seismicity is part of the plate boundary between the Indian and Pacific plates. Composite focal mechanism solutions by Scholz et al. (1973) indicate thrust faulting for Fiordland, operating in a NW-SE compressional regime, compatible with overthrusting of the Fiordland lithosphere over the Tasman lithosphere. This current activity probably explains why a huge gravity anomaly is being supported at a shallow depth.

2.3. Seismic Refraction and Reflection.

Davey and Broadbent (1980) obtained reversed seismic refraction measurements for Fiordland. One seismic line passed down Doubtful Sound across Fiordland (see Fig. 4). Rocks with compressional wave velocities of 6.8 km/s occur within 3.5km of the surface whilst those with a velocity of 7.3 km/s lie at about 8km. Broadbent and Davey (1978) recorded an unreversed 8.6 km/s velocity which was roughly modelled by introducing 8.4 km/s material at 18km depth. However, they commented on the possibility of combined reflected and refracted waves generating the 8.6 km/s arrival and did not discuss this in their 1980 paper. Clearly, further seismic refraction research is required.

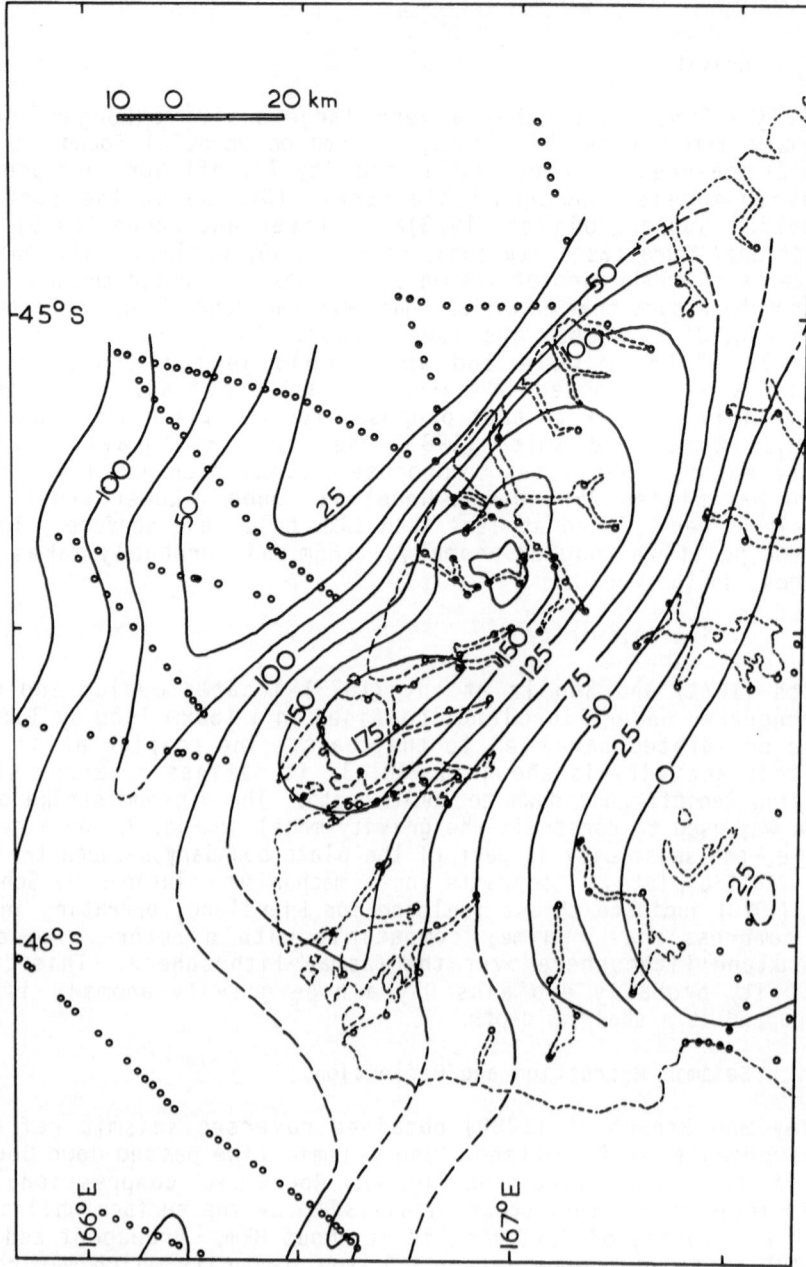

Figure 1. Bouguer gravity map of Fiordland (after Oliver and Coggon, 1979).

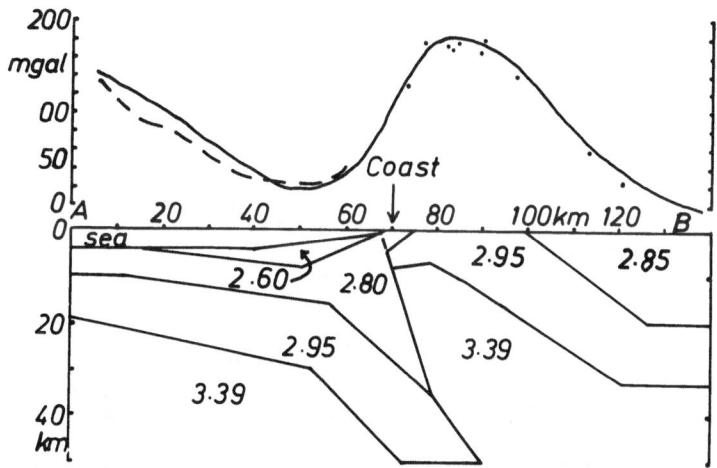

Figure 2. Gravity model for Fiordland from Oliver and Coggon (1979). Numbers indicate density; dots are values on land, broken line on sea. Solid line is the model anomaly assuming infinite strike length.

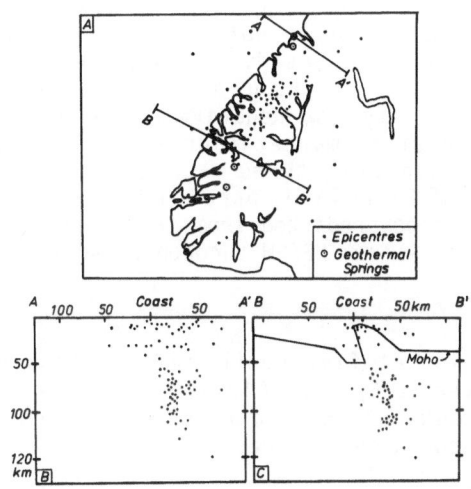

Figure 3. Active seismicity in Fiordland after Smith (1971) and Scholz et al. (1973). A is plan view, B and C are cross-sections. Crust mantle boundary from gravity model.

48

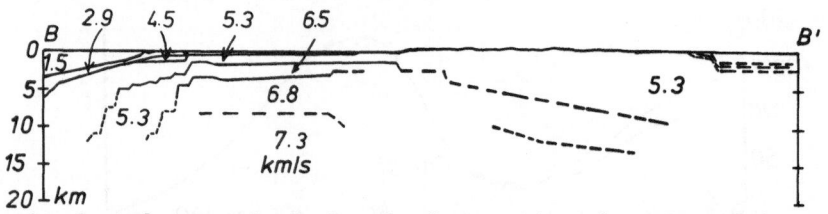

Figure 4. Postulated seismic P-velocity section along Doubtful Sound, (from Davey and Broadbent 1980).

Velocities of 6.8 and 7.3 km/s are appropriate for lower continental crustal levels (Priestley and Davey 1983) and mimic laboratory results for granulites at the appropriate pressures (Christensen and Fountain 1975). Since 7.3 km/s lower crustal rocks are ~8km below the surface, considerable (>20km) uplifts are suggested.

Davey and Smith (1983) displayed a seismic reflection section across the offshore gravity low. The profile shows a sedimentary wedge with thrust fault features compatible with accretion against Fiordland crust.

2.4.   Aeromagnetics

The aeromagnetic map of Fiordland (Fig. 5) shows that the region can be divided into magnetic provinces based on differences of magnetic relief (Oliver and Coggon, 1979). The western magnetic province corresponds with the positive Bouguer gravity anomaly and has moderate magnetic relief. The relief does not correspond with areas of metamorphic retrogression, (see Section 3.1.1). The relief declines east of Doubtful Sound on the flanks of the gravity anomaly into a central province of relatively low negative anomalies. There is a sharp contrast with an eastern magnetic province of high magnetic relief. There appears then to be three magnetic provinces running NE/SW along the length of Fiordland, parallel to the strike of the gravity anomaly. It might be predicted from this that at least three contrasting elongate terranes run NE/SW through Fiordland.

2.5.   Geothermal Activity

Fig. 3 shows the location of five geothermal springs reported by hikers and hunters (Oliver 1976). The Alpine Fault is associated with geothermal springs throughout the South Island and presumably the Fiordland activity is associated with recent faulting.

2.6.   Geophysical Model

Without even discussing surface geology, the geophysical data available for Fiordland allows some important predictions to be made: gravity, earthquake, reflection and refraction results allow a crustal model such as that illustrated in Fig. 6 to be constructed. The Pacific plate with Fiordland continental crust is actively overriding

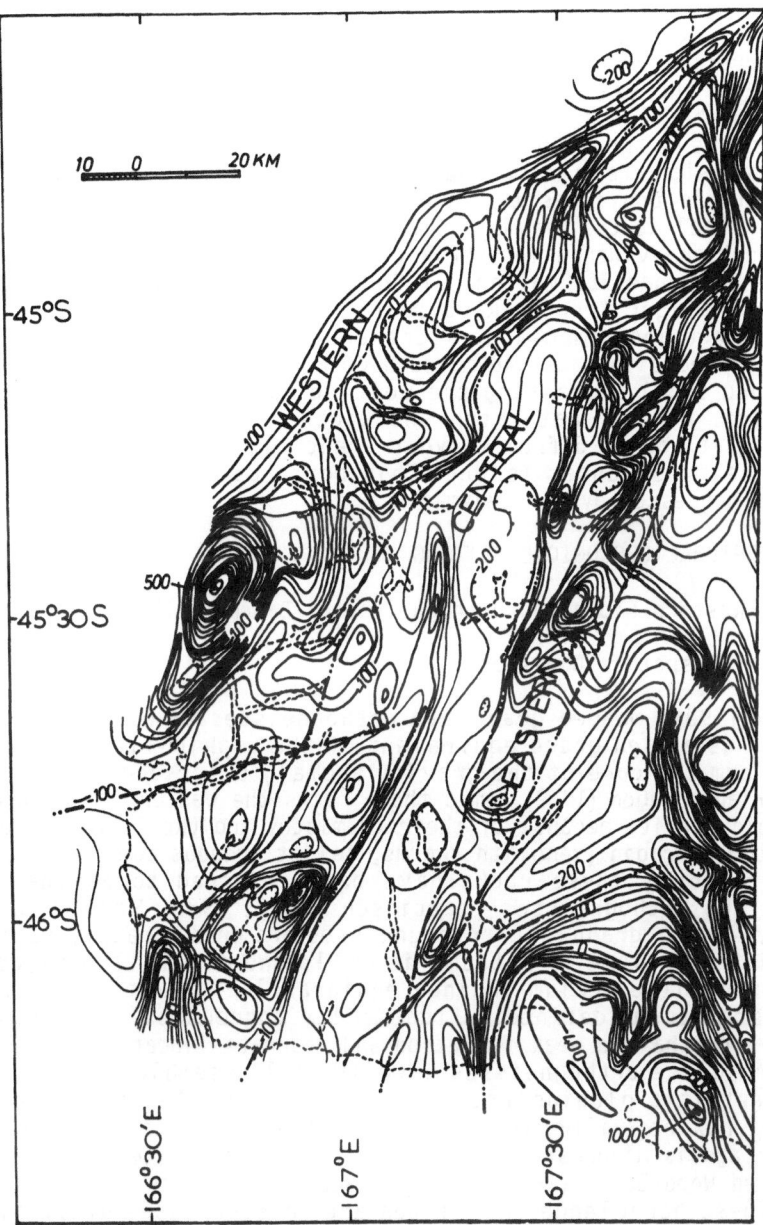

Figure 5. Aeromagnetic map of Fiordland showing magnetic provinces and anomalies, after Oliver and Coggon 1979.

subducting Indian crust along an extension of the Alpine Fault. Lower crustal rocks, presumably granulites, have been brought to the surface in a crustal dome: the crust is anomalously thin here, and there is a hint from the seismics that the MOHO lies at 18 km depth. To the east, on the flanks of the crustal dome, the crust is tilted such that successively the middle and then the upper crust is exposed. These three crustal levels may have been defined by the three different aeromagnetic provinces that run parallel to the length of Fiordland and parallel to the gravity anomaly.

Further evidence for the active subduction model is given by the Quarternary andesite volcano of Solander Island on the continental side of Pusyger Trench (Reed 1951) and by submarine "volcanoes" of unknown age and composition further south (Summerhayes 1969). The Solander andesite, just 40km south of Fiordland, has 61% SiO_2 and 2.3% K_2O (Reed 1969) suggesting the volcano may be 200km above a Benioff zone (cf. Fig. 1, Hatherton and Dickenson 1968). A Benioff zone dipping at about 75 from Pusyger Trench would pass through a point 200km below Solander Island. This compares well with the $70° - 80°$ values suggested by the seismicity.

3. REGIONAL GEOLOGICAL SETTING

3.1. Western Fiordland Orthogneisses: Basement-Core

3.1.1 Feldspathic Gneisses

Turner (1939) mapped granulite orthogneisses of western Fiordland (Fig.7) as a high grade equivalent of the amphibolites to the east. They form the core of the gravity anomaly described in Section 2.1. Oliver and Coggon (1979) named these rocks the Western Fiordland Meta-Gabbroic Diorite because of their chemistry and the field evidence for a plutonic origin (xenolith swarms, relict igneous cumulate layering). More recently these rocks have been referred to as the Western Fiordland Orthogneisses (Mattinson et al. 1986, McCulloch et al. 1987). Fresh granulite orthogneisses are predominately feldspathic with 15% by volume of ultramafic sills and dikes. The feldspathic varieties have medium grained equigranular, granoblastic, elongate microstructures defined by lenticular aggregates of mafic grains. These rocks have a weak foliation and moderate lineation. The bulk of the granulite orthogneisses have mineral assemblage consisting of antiperthitic plagioclase, olive-brown hornblende, clinopyroxene, orthopyroxene, ilmenite and magnetite, apatite, zircon. This assemblage is diagnostic of medium pressure granulites. Oliver (1976) measured Wood-Banno (1973) two pyroxene temperatures of ~750°C.

These hornblende granulites are criss-crossed by plagioclase pegmatites and veins studded with garnet and/or hornblende crystals, the latter are up to 10 cm long. Often there is a garnetiferous reaction zone between the pegmatite veins and the hornblende granulites. The new mineral assemblage contains antiperthitic plagioclase, garnet, Na-rich clinopyroxene, quartz, K-feldspar, rutile, apatite, zircon. These are referred to as garnet granulites. The garnet and quartz form wormy intergrowths with clinopyroxene

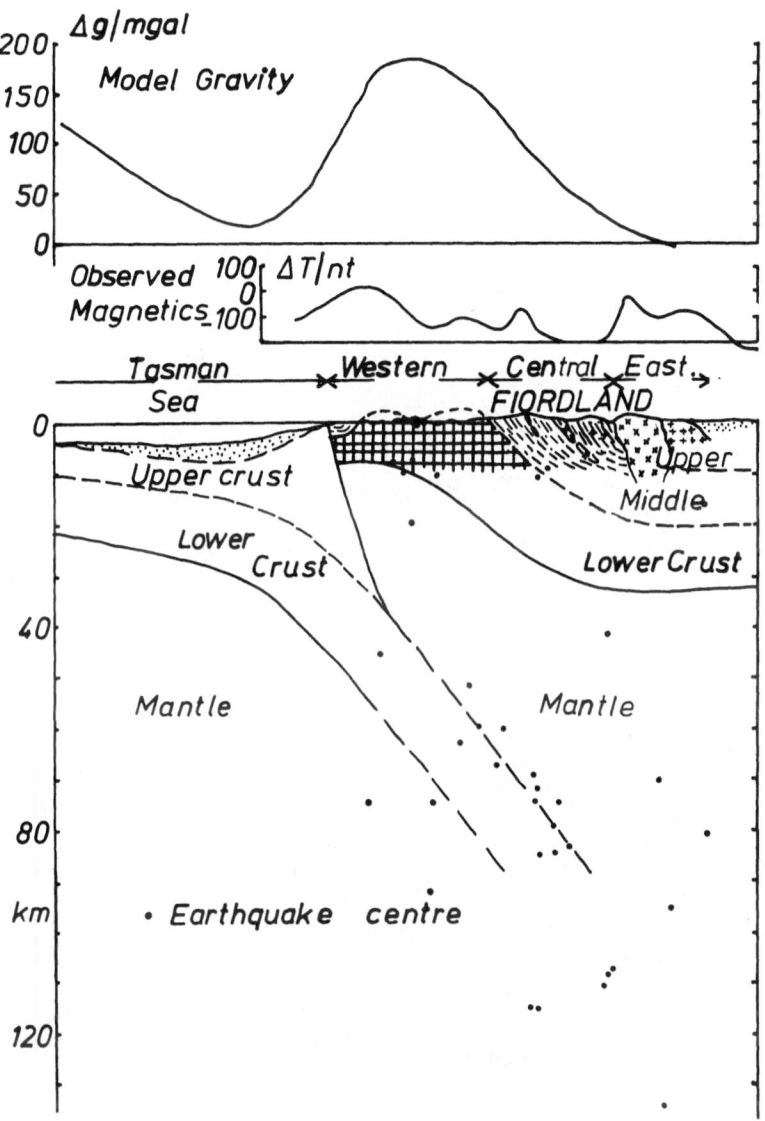

Figure 6. Structure of Fiordland. Line of section XY of Fig. 7. The gravity profile is that calculated for the model of Fig. 2; the magnetic profile is taken from the aeromagnetic map, Fig. 5; and data on seismicity are taken from Scholz et al. (1973). Key to symbols in Fig. 8.

52

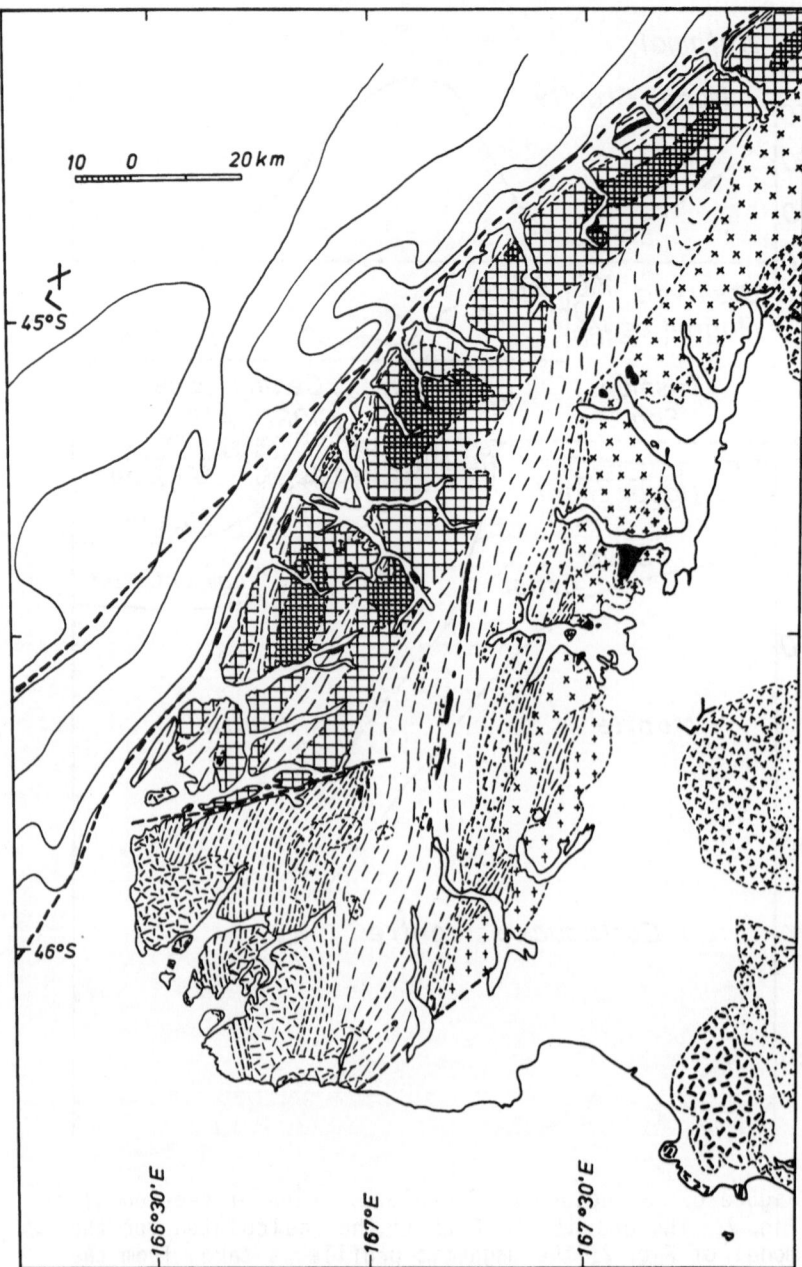

Figure 7. Geology Map of Fiordland (after Oliver and Coggon 1979).

Sediments: calcareous mudstones, arkosic sandstones, minor terrestrial deposits

Unconformity

PRESERVATION
GRANITES
(in SW Fiordland)

EASTERN FIORDLAND GRANODIORITES
AND GRANITES Unfoliated post-
metamorphic intrusions.

EASTERN FIORDLAND GABBROS AND
DIORITES Major foliated diorite
and gabbro masses, minor basic
intrusions. Darran Complex.

SOUTHWESTERN
FIORDLAND
METASEDIMENTS
AND GRANODIORITES
Ordovician slates,
quartzose greywackes,
schists (greenschist-
amphibolite facies);
and foliated grano-
diorite intrusions.

CENTRAL FIORDLAND GRANODIORITES
AND GRANITES Large foliated
bodies, and injection complexes.

CENTRAL FIORDLAND METASEDIMENTS
Quartzofeldspathic and micaceous
gneisses, schists, minor metabasites, marbles and calc-silicate
gneisses, Amphibolite facies
metamorphism.

Doubtful Sound Decollement separates basement core from cover

WESTERN FIORDLAND
ORTHOGNEISSES

Gneisses partly or completely
retrogressively metamorphosed
to amphibolite facies, from
granulite facies.

Uniform feldspathic gneisses of
granulite facies.

Sediments east of
Takitimu Mts.
(Permian, Triassic).

Ultramafic-mafic rocks
of various ages.

Volcanics and sediments
of Takitimu Mts.
(Permian-Carboniferous).

Faults and fault zones.

Gabbros, norites, granites
etc. of Longwood Range
(Permian-Carboniferous)

Cretaceous-Recent

Mesozoic: Rangitata orogeny.

Upper Palaeo-zoic-Mesozoic.

Sedimentation Lower-Mid Palaeozoic; metamorphism and plutonism Devonian: Tuhua orogeny. Rangitata reactivation.

COVER

Gabbroic diorite emplacement and granulite facies metamorphism, uplift and retrogressive metamorphism in Rangitata orogeny.

BASEMENT-CORE

Figure 8.   Geological legend to Figure 7.

around orthopyroxene relicts (see Fig. 1, Oliver 1977). Oliver (1977) proposed the following dehydration reaction to account for the change in mineralogy:

Hbl + Plag + Ilm + Mag + Opx = Gnt + Cpx + Qtz + K-fel + Rut + H_2O

similar to garnet producing reactions in charnockites described by McLelland and Whitney (1977). Oliver (1977) argued that although garnet granulites are normally regarded as high pressure equivalents of hornblende varieties, bulk-rock compositional changes associated with partial melting and dehydration of the hornblende (producing pegmatites and veins) would allow both varieties to coexist. Blattner (1976) had previously argued that CO_2 infiltration could trigger the reaction by lowering the fH_2O for hornblende breakdown. The occurrence of scapolite in pegmatites supports this. Mattinson et al. (1986) maintained that the two types of granulites indicate two phases of metamorphism: first, synkinematic autometamorphic recrystallisation of magmatic plagioclase, orthopyroxene, clinopyroxene and hornblende at ~6Kb and second, static crystallisation of garnet bearing assemblages at ~12 kb/ 650-700°C. However, no details of how these 6kb and 12kb pressures were calculated have been published by McCulloch et al. (1987) or Mattinson et al (1986). They stated that zoned clinopyroxenes coexisting with plagioclase and quartz in granulite show a core to rim increase in jadeite. However, Oliver (1976) found the opposite: typical core compositions are 20-22% Jd whilst rims are 18-19% Jd. Some eclogites (see below) have up to 28% Jd cores with 24% Jd rims. These compositions indicate pressures of 13-14kb if Holland's (1983) geobarometer is applied at 750-800°C. Oliver (1977) calculated Raheim-Green (1975) garnet-clinopyroxene temperatures of 760°C. Newton and Perkins (1982) used Oliver's (1977) data to calculate 9-12.1kb using garnet, orthopyroxene plagioclase geobarometers. Using the same analyses but varying thermodynamic assumptions, Bohlen et al. (1983) calculated a maximum of 11.2kb; Perkins and Chipera (1985) calculated a maximum of 12.7kb; Moecher et al. (1988) calculated maxima of 23, 16.1 and 14kb. Assuming 12kb means that the garnet-clinopyroxene geothermometer should be revised to 800°C, still within error of the two pyroxene geothermometer results. Some caution is required in accepting these PT values ..... they could represent conditions frozen in at a particular point along the PT trajectory and might not necessarily be the maximum P.T.'s (McLellan, 1990).

Both Oliver (1980) and Gibson et al. (1988) would argue against Mattinson et al. (1986) and McCulloch et al. (1987) who state that the granulites were originally intrusive into the present day cover (the Tuhua sequence, see below). The Doubtful Sound Décollement (DSD) always intervenes between the basement-core granulites and the cover amphibolites. Oliver (1980) described a zone of progressive retrogression underneath the DSD: 1.5km structurally below the DSD, pyroxenes and garnets begin to hydrate to simplectic intergrowths of blue-green hornblende and quartz or biotite and quartz. By 1km below the DSD, hypersthene, garnet and olive-green hornblende have been replaced; by 0.75km below the DSD, all clinopyroxene has been replaced and a new assemblage of plagioclase, green hornblende, biotite, epidote, sphene, oxide, apatite and zircon has been established. In this state, the rocks are now very strongly foliated and lineated compared to the granulite protolith. Although Oliver (1976) described flaser and augen structures in these streaky gneisses, he recognised the textures as thrust tectonites, not as extensional shear fabrics as Gibson et al. (1988) did. Gibson et al. (1988) described these same rocks as L-S tectonites, with an exaggerated stretching lineation trending NE/SW on gently dipping mylonitic foliations. S-C bands, asymmetric porphyroclast tails and fold vergence indicate that the amphibolite cover has been dragged off the granulite basement-core towards the NE. Figures 6 and 7 show the DSD is folded such that inliers of cover occur north of Dusky Sound.

3.1.2. Dating and geochemistry of feldspathic orthogneisses

Mattinson et al. (1986) measured U-Pb isotopes in zircons from the gneisses which give the age of synkinematic magma emplacement of between ~120 and ~130Ma age. These ages have been corroborated by a Rb/Sr whole rock 120Ma isochron (McCulloch et al. 1987) and ion probe sigle zircon 126Ma spot ages (Gibson et al. 1988). The U/Pb, Rb/Sr and Nd/Sm isotope systematcs are consistent with an arc subduction related origin with little or no cotamination from ancient continental crust. Unlike Precambrian granulites, these Fiordland examples have primitive eNd values overlapping present day mantle compositions. McCulloch et al. (1987) considered that the Fiordland data could be explained by a variety of mechanisms including either direct derivation from the mantle, variable contamination of mantle magmas with country rocks, or derivation from low Rb/Sr, LREE enriched protolith having a 100 to 300Ma prehistory. McCulloch et al. (1987) prefer the third mechanism involving the 370Ma (Oliver, 1980) or older cover rocks such as found in Doubtful Sound. However, the cover rocks analysed by Oliver (1976) have neither the appropriate Rb/Sr ratios nor the appropriate LREE patterns. Further isotope studies of xenoliths within the granulites orthogneisses and ortho-and para-gneisses from the cover rocks are required.

Mattinson et al. (1986) give zircon ages from a retrogressed (and presumably mylonitised) granulite boulder that are younger (116Ma) than the 120 to 130 Ma emplacement age. They point out that this may indicate a slightly younger magmatic age or the age of the amphibolite facies resetting. These ages are in agreement with the ion probe zircon spot 119Ma ages for an amphibolitised mylonite measured by Gibson et al. (1988). Gibson et al. (1988) also presented K/Ar hornblende (93Ma) and K/Ar biotite (77Ma) cooling ages for the same mylonite whilst Mattinson et al. (1986) have apatite U/Pb ages of 78Ma for their sample. These age dates allow a thermal history of the basement-core granulites to be postulated, see Fig. 9. Plutonism at 130 to 120Ma was quickly followed by granulite facies metamorphism at 750-800°C and ~12Kb (say 45km depth assuming a 2.8 g/cm density). Mylonitisation under amphibolite conditions occurred almost immediately at 119Ma and rapid uplift and cooling through various mineral blocking temperatures subsequently took place, i.e. hornblende K/Ar blocking (550°C) at 93Ma (say 22km depth, assuming a geothermal gradient of 35°C/km), apatite U/Pb blocking (400°C) at 78Ma (16km)

56

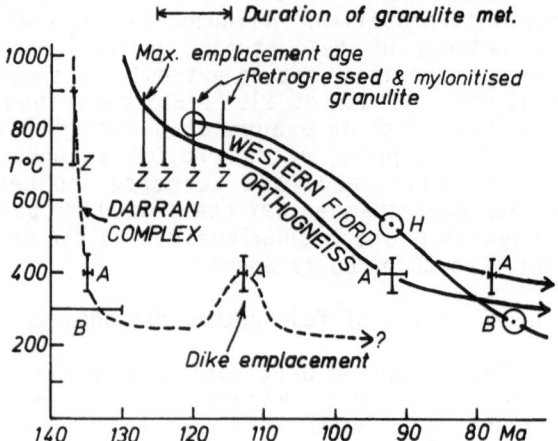

Figure 9. Thermal histories of Western Fiordland Orthogneisses and Darran Complex. Blocking temperatures given in Mattinson et al. (1986). Age dates from Mattinson et al. (1986), Gibson et al. (1988) and Williams and Harper (1978).
    (A = apatite, B = biotite, H = hornblende, Z = zircon).

and biotite K/Ar blocking (300°C) at 77Ma (~12km). This means that the granulite basement-core was uplifted ~33km within 43Ma at an average rate of 0.8mm/yr for the Doubtful Sound area. McCulloch et al. (1987) using 120Ma for the age of magmatism and granulite metamorphism and a 101Ma Rb/Sr biotite age (blocking at $300\text{-}350^{\circ}$C) calculated uplift rates of between 2 and 3.5mm/yr for northern Fiordland. Recent uplift in the New Zealand Southern Alps has been estimated as up to 17mm/yr (Wellman, 1979).

3.1.3. Mafic Orthogneisses

Hornblendite sheets make up a significant volume of the basement-core. In Crooked Arm, at the centre of the positive Bouguer gravity anomaly, they make up 15% of the outcrop although certain mountainsides appear to have as much as 60% of hornblendite sheets. A small proportion of the sheets contain olivine and pyroxene: there is one example of a hornblende-bearing harzburgite sheet in Crooked Arm (Oliver, 1980). These sheets have experienced the granulite facies metamorphism along with their country rocks, producing garnet and pyroxene eclogites (Oliver, 1980). They have also suffered the retrograde metamorphism and deformation associated with the extension along the Doubtful Sound Décollement. Therefore, most hornblendite sheets lie parallel to the foliation in the stretched felsic orthogneisses and appear to be sills. In the unaltered granulites, they are more cross-cutting.
    Gibson et al. (1988) describe one dike which cross-cuts the stretching lineation at high angle yet has only amphibolite facies minerals co-facial with the altered country rocks: i.e. intrusion of hornblendite continued whilst the décollement was operating (119 to

93Ma). Attempts to radiometrically date the hornblendite intrusion with accuracy have not been sucessful (McCulloch et al. 1987) but the geological constraints listed above restrict the intrusion to very soon after the feldspathic orthogneisses crystallised from magma (130 to 120Ma) but before or during the granulite facies episode. McCulloch et al. (1987) showed that the mafic orthogneisses and their minerals had different initial strontium isotope ratios at 120Ma from the feldspathic orthogneisses, indicating that they were not cogenetic. However, a mantle origin is suggested by their 143Nd/144Nd and 86Sr/87Sr ratios falling along the mantle array. The average chemistry is that of a basanite. They are strongly tholeiitic compared with the calcalkalic feldspathic orthogneisses. The REE pattern is nearly flat at 10 times chondrite values (Oliver, 1976). On basalt discrimination diagrams such as TiO2 versus FeO /MgO, FeO versus FeO /MgO and Ti versus Zr, they plot in ocean floor basalt fields. On Ti/100 : Zr : Yx3 and Ti/100 : Zr : Sr/2 triangles, they plot in or near the low potassium tholeiite field. There are chemical similarities with Jurassic North Sea basanites formed, it was suggested, during the stretching of the North Sea lithosphere (Foerseth et al. 1976, Latin et al. in press).

3.2.   Central Fiordland Cover

3.2.1   Lithologies

The cover is made up of a predominantly metasedimentary sequence with subsidiary volcanics and various gabbro and anorthosite sills, overlying and in tectonic contact with the Fiordland Orthogneiss basement-core. Further to the east around Lake Manapouri, foliated granodiorite and granites predominate. Gibson (1982) has subdivided the cover into the Doubtful Sound Province and the Western Manapouri Province, following Turner (1937a, b). The cover constitutes the Tuhua Sequence which was regionally metamorphosed and intruded by granites during the Upper Devonian-Lower Carboniferous. Oliver (1976, 1980) mapped Kellard Point and Deep Cove Gneisses in contact with the basement along the Doubtful Sound Décollement. These consist of amphibolite facies marbles (Kellard Point Gneisses), migmatised quartzo- feldspathic metasediments and interbanded hornblende plagioclase gneisses (Deep Cove Gneisses). Oliver (1976) suggested that these thick marbles might have correlatives with the Ordovician fossiliferous Mt. Arthur marbles of NW Nelson (Benson 1934). In south western Fiordland low grade pelites have Lower Ordovician graptolites (Benson et al. 1934, 1935). If the Fiordland cover is comparable with NW Nelson then a Cambrian through to Silurian age of sedimentation is predicted. Oliver (1976) used REE pattern normalised to North American Shale (Haskin et al. 1966) and Misra (1971) discrimination diagrams (TiO$_2$/MnO/FeO/MgO) to show that representative samples of Deep Cove hornblende plagioclase gneiss plot in both sedimentary and igneous fields. Thus tuffs or lavas make up significant volumes of this miogeosynclinal sequence. Foliated granite bodies (orthogneisses), peridotite, anorthosite and hornblendite sills occur. In places metamorphosed anorthosite can be mapped as the Black Giants

Anorthosite (Gibson, 1982a), whilst the Mount George (layered) Gabbro extends for 50km and was syn-or late-metamorphism in age. Psammite, schist, paragneiss, marble and calc-silicate formations make up the rest of the Doubtful Sound Province. In Doubtful Sound itself, Oliver (1980) estimated maximum metamorphic conditions to be ~7kb and ~675°C whilst Gibson et al. (1988) quote 5-9kb from further east. Aronson (1968) dated discordant zircon from a gneissic granite boulder from Deep Cove at 320-360 m.y. Oliver (1980) dated the Deep Cove Gneiss migmatite event at 388±23 m.y. using a whole rock Rb-Sr isochron. A foliated granite from the west coast had a whole rock Rb-Sr age of 372 ±12 ma (Oliver, 1980). Aronson (1968) named this Devonian to Carboniferous orogenic event the Tuhuan Orogeny. It appears to be contemporaneous with the Lachlan Orogeny in SE Australia.

Gibson (1982) noted that the Western Manapouri Province, equivalent to Oliver and Coggon's (1979) Central Fiordland Granodiorites and Granites, could be clearly separated from his Doubtful Sound Province on the basis of lack of lithological correlation and the predominance of granitoids in the former. Uniform hornblende and biotite gneisses predominate and might have originated as eugeosynclinal greywackes (Gibson 1982). Sheets and dikes of trondhjemite and pegmatite form an injection complex (Turner, 1937b). Granitic to quartz monzonite and hornblende and biotite orthogneisses are also characteristic and perhaps related to 372 Ma foliated granites from the west coast at Deas Cove (Oliver, 1980). Non-foliated, post-tectonic leucogranite with xenoliths of granitic orthogneiss make up a pluton on the south side of the West Arm of Lake Manapouri. Although undated, this pluton presumably represents a Cretaceous intrusion, more characteristic of the Eastern Fiordland Plutonics.

3.2.2. Timing of Deformation

On the basis of geological mapping (Oliver 1976, Gibson 1979) and radiometric dating (Aronson 1968, Oliver 1980, Gibson et al. 1988) the following sequence of events can be suggested:

1. Deposition of a thick pile of Cambrian to Silurian sediments and subsidiary volcanics on a shallow marine shelf.

2. D1 recumbent folding, kyanite grade regional metamorphism up to 9 kb and up to 675 C. K/Ar hornblende ages (428±16, 346±4, 322±3, 316±3 Ma Gibson et al. 1988) are somewhat disturbed but indicate Silurian through to Carboniferous events. Formation of S1.

3. Synmetamorphic emplacement of granitoids and migmatisation during Upper Devonian-Lower Carboniferous times (360-388 Ma WR Rb/Sr isochrons, Oliver 1980; 386 Ma zircon spot age, Gibson et al. 1988). Emplacement of Mt. George Gabbro with sillimanite grade aureole.

4. D2 recumbent folding and formation of S2 crenulation
    cleavage. Minimum ages for D2 hornblende orientated in the
    S2 foliation of a 386 Ma (zircon) tonalite is given by a
    K/Ar mineral age of 242 Ma whilst the enclosing country
    rock gives a K/Ar hornblende age of 265 Ma (Gibson et al.
    1988). D1 and D2 together make up the Tuhuan Orogeny.

5. D3 major extensional faulting and mylonite formation in
    shear zones. Doubtful Sound Décollement between basement-
    core and cover formed during amphibolite facies
    metamorphism. Zircon ages of 116 and 119 Ma (Mattinson
    et al. 1986, Gibson et al. 1988) from retrogressed
    granulite in the Doubtful Sound Décollement zone give
    maximum ages for the shearing. Hornblende from the same
    rocks gives a K/Ar cooling age of 93 Ma, the minimum age of
    shearing (Gibson et al. 1988).

6. Regional uplift during and following D3 crustal extension:
    hornblende and biotite K/Ar ages from the cover get
    progressively younger from E to W and towards the Doubtful
    Sound Décollement, hornblende ages decrease from 428 to 99
    Ma; biotite ages decrease from 245 to 79 Ma; (Gibson et al.
    1988). Gibson et al. (1980) relate this resetting to
    dynamothermal metamorphism within the decollement zone.
    However, a similar pattern of K/Ar mineral ages decreasing
    westwards has been documented in the Alpine Schists
    (Sheppard et al. 1975) and interpreted as an artifact of Ar
    loss associated with uplift and nearness to the Alpine
    Fault. As yet ages are not available from Fiordland
    adjacent to the Alpine Fault. D3 crustal extension and
    uplift is correlated with the Ragitatan Orogeny, a
    contemporary of the New England Orogen in E Australia
    (Cawood, 1984).

7. D4 open folds and reverse faulting, presumably Tertiary.
    It is suggested that the folding of the Doubtful Sound
    Décollement (Figs. 6 and 7) might be a combination of
    post-D3 uplift and D4 folding.

This sequence of events, 1-7, is discussed in more detail by
Gibson (this volume).

3.3   East Fiordland Plutonics

3.3.1. Darran Complex

These rocks make up the Eastern Manapouri Province of Turner (1937a,
b, c) or the Eastern Belt of Mattinson et al. (1986). Oliver and
Coggon (1979) recognised a more westerly gabbro-diorite batholithic
region (e.g. Darran Complex) and an easterly granite-granodiorite
batholithic region. The Darran Complex consists of unmetamorphosed
gabbronorite with lesser amounts of leucogranite, diorite, gabbro and

ultramafics. The western contact is a 300m wide shear zone. In the east, Darran gabbronorites intrude Permo-Triasic Eglington Volcanics. Mattinson et al. (1986) give concordant U/Pb zircon ages of 137±1 Ma for the age of intrusion. This is older than the 120 Ma Western Fiordland Orthogneiss protolith (see section 3.2.2). U/Pb apatite (136 Ma, Mattinson et al. 1986) and K/Ar biotite (136-130 Ma, Williams and Harper 1978) ages confirm that the batholith was intruded at a relatively high level in the crust and cooled quickly. Aronson (1968) gives mineral and whole rock Rb/Sr dates of 105-120 Ma for the Pomona Island Granite from Lake Manapouri. Concordant 100 Ma muscovite and biotite Rb/Sr mineral ages for the unmetamorphosed Preservation Granite in SW Fiordland shows that Cretaceous granite intrusion was not confined to Eastern Fiordland (Aronson, 1968).

3.3.2   Geochemistry

The Darran Complex gabbros have major and trace element chemistry suggesting an island arc setting similar to the Western Fiordland Orthogneisses (Mattinson et al. 1986). However, consideration of their substantially different isotope compositions (WFO e Nd = 0.1 to 3.0; Darran = 3.9 to 4.6; WFO 87Sr/86Sr = 0.70391; Darran 87Sr/86Sr = 0.70373 to 0.70386) rule out their origin from a common source. As noted above (section 3.3.1.) the Darran Complex is somewhat older i.e. 137 Ma versus 120-130 Ma.

4.   CRETACEOUS-TERTIARY SEDIMENTS

Upper Cretaceous to Lower Tertiary coal measures in the Waiau Basin to the east of Fiordland (Brothers, 1959) have detrital kyanite and garnet indicating that uplift and erosion did in fact accompany the extension tectonics described in section 3.3.2. Hypersthene and clinopyroxene have not been described perhaps suggesting that Western Fiordland Orthogneisses were not exposed at this time. Similar Cretaceous to lower Tertiary sediments occur in SW Fiordland at Puysegur Point. Clasts in basal conglomerates consist of local granite, schists and hornfelses but not granulites (Wood 1960).
    Wellman (1954) reports vertically dipping sandstones of lower Pliocene age on the east coast of Fiordland. The contact with gneisses is puggy and could either be a sedimentary unconformity or a fault. Proximal sandstones have granite fragments together with detrital biotite, muscovite, plagioclase, hornblende, epidote, clinozoisite, garnet, zircon, sphene, and lutile. Interestingly, minerals characteristic of the granulite orthogneisses are not listed and the inference is that the granulites were not exposed at 5Ma.

5.   CRUSTAL CROSS-SECTION FOR FIORDLAND

From the foregoing it is apparent that Pre-Mesozoic Fiordland continental crust was made up of Lower Palaeozoic sediments, volcanics and plutonics which had been cycled through the Tuhuan Orogeny. In Nelson and Westland there is evidence for Proterozoic basement e.g. Aronson (1968) dated 1170 to 1480Ma detrital zircons in L. Palaeozoic

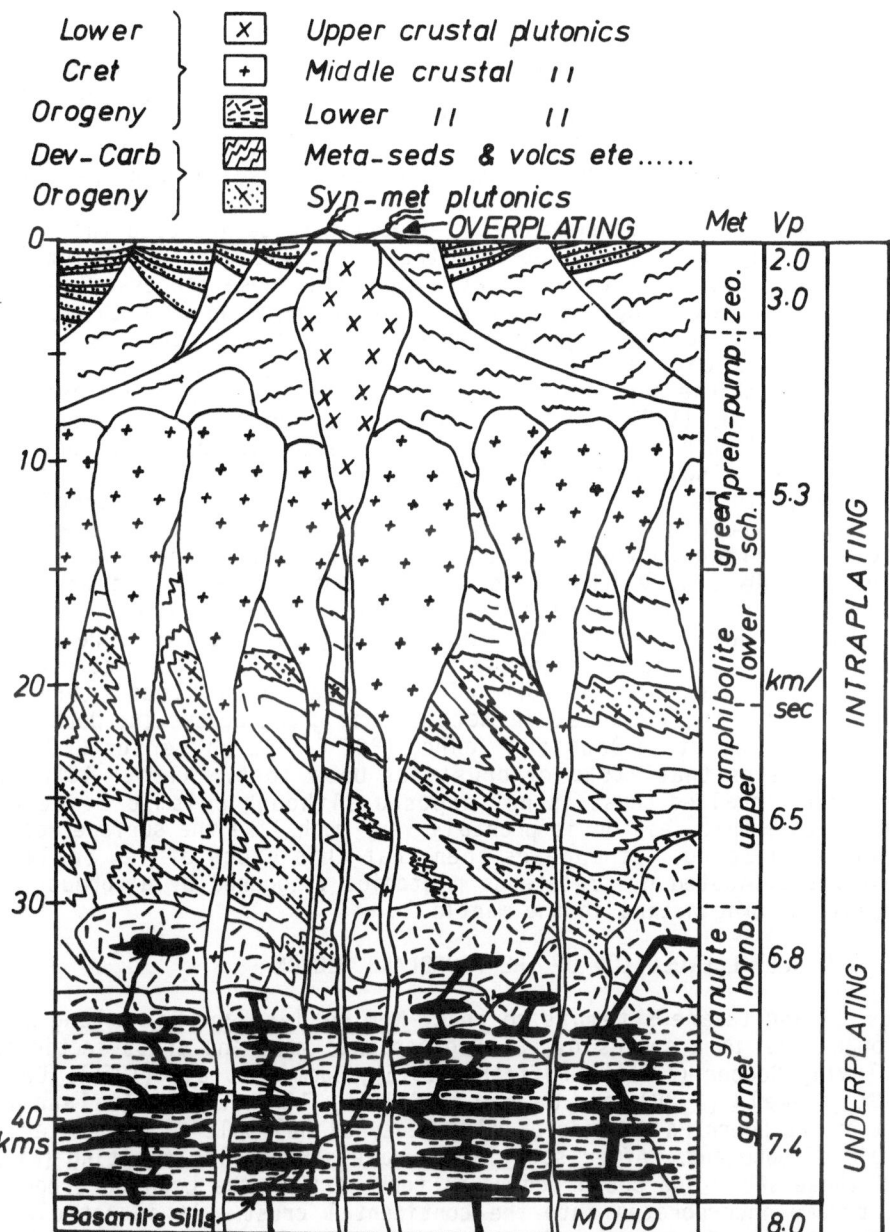

Figure 10. Crustal model for western South Island, N. Z., during Lower Cretaceous times.

Greenland Series sediments, which lie unconformably on Proterozoic Charlestown Gneisses. The Lower Cretaceous calcalkaline arc magmatism, associated with a "westerly" dipping subduction zone appears to have occurred at two levels in the Proterozoic-Palaeozoic crust. In western Fiordland, intrusion was confined to the lower crust, i.e. magma from the mantle underplated the existing crust. Possibly, this gabbroic magma was too dense to rise through the lower crust. In eastern Fiordland and Nelson/Westland a near contemporaneous less dense dioritic plutonic suite intruded the upper crust on a regional scale. Some of the Nelson/Westland magmatism was manifested in surface volcanic activity. All these processes served to thicken the existing continental crust to maybe 45-50km thick. It is feasible that the high pressure episode in western Fiordland granulite was caused by this plutonic thickening and not by collisional thickening involving crustal overthrusting as proposed by Mattinson et al. (1986) and McCulloch et al. (1987). As yet there is no direct evidence that the Hokonui or Torlesse Terranes overthrusted Fiordland or Nelson/Westland.

Figure 10 is a cartoon that illustrates the form of Fiordland continental crust in the period 130 to 120Ma when the crust was at its thickest. Although the upper and middle crustal regions are well exposed in Nelson/Westland and eastern and central Fiordland, it should be noted the contact between the middle crustal (Tuhuan) amphibolites and lower crustal (Rangitata) granulites is a regional decollement. Experience from the Western Cordilleran core complexes suggest that maybe 5 or 10km of crust has been "lost" through attenuation. The contrast in metamorphic grade across the Doubtful Sound Décollement in Doubtful Sound, (800°C & 12 kb versus 675°C & 7kb), suggests that ~15km of crust has been lost. Also, ~15km of lowermost Fiordland crust is not exposed although seismic evidence suggest that this section is probably very much like the surface rocks in Doubtful Sound. It appears then that this piece of New Zealand crust was dominated by Cretaceous plutonics triggered by a subduction process and originating in the mantle.

6.   CRUSTAL EVOLUTION OF THE FIORDLAND REGION

Figs. 11 and 12 are attempts to trace New Zealand and Fiordland back through time in a series of palaeogeographic maps and cross-sections. Following Cooper (1975), Griffiths (1975), Coombs et al. (1976), Crook (1981), Howell (1980), and Cawood (1984) Fig. 11 shows New Zealand as part of the Lower Palaeozoic Lachlan Geosynclinal Belt on the eastern margin of Gondwana. After some large scale strike slip faulting (Fig. 11B) these shelf and basin sediments took part in the Lachlan Orogeny and became incorporated into the continental crust. It is not clear if the Lachlan Orogeny involved continental collision and accretion. From the Carboniferous through to the Lower Cretaceous (Fig. 11. D, E, F, G, H after Griffiths, 1975) the New Zealand section of the Gondwana margin became active again. A subduction zone persisted in the region as Pacific ocean crust was consumed and marginal basins and island arcs were accreted. Sedimentation in the New Zealand Geosyncline occurred in the volcanic arc dominated Hokonui Terrane to the 'west'

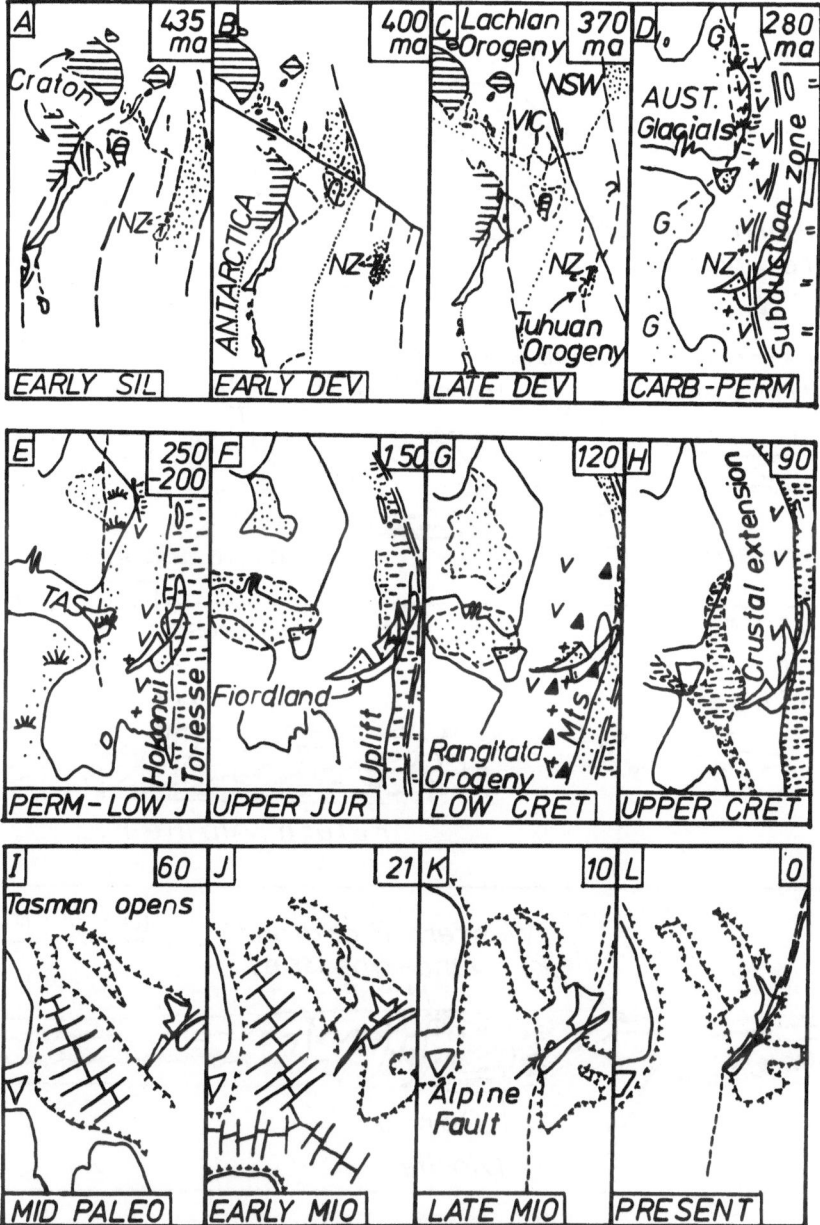

Figure 11. Plate tectonic evolution of the SW Pacific margin of Gondwanaland; see text for details.

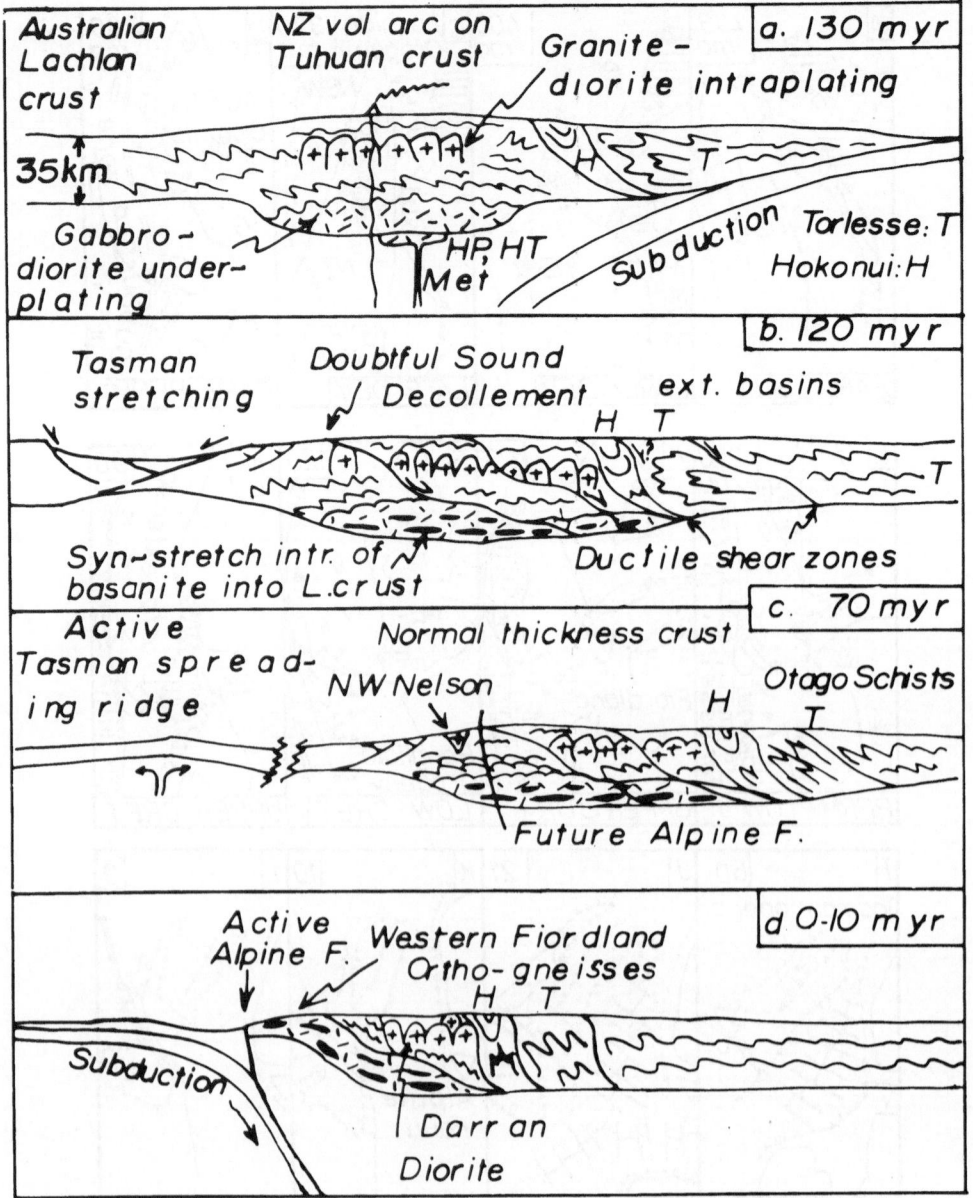

Figure 12. Schematic crustal cross-section through S. New Zealand showing a) L. Cret. crustal thickening in the volcanic arc; b) and c) crustal thinning during Tasman stretching; d) final exhumation of lower crust along the Alpine Fault. See text for details.

and in the deep sea fan-trench dominated Torlesse Terrane to the 'east'. Subduction ended at about 130Ma (Fig. 11G) and the huge piles of accreted sediments and volcanics were subjected to regional tectonism and metamorphism of the Rangitata Orogeny. In the 'east' the Torlesse Terrane experience a HP, LT metamorphic regime without any plutonic activity. In the 'west' the Fiordland foreland crust was suddenly invaded by copious amounts of calc-alkaline magma, perhaps doubling the original thickness of crust (Fig. 12a). Almost immediately, this crust found itself under an extensional regime as the incipient Tasman Sea was formed (Figs. 11H and 12b). During the Upper Cretaceous and Lower Palaeocene, the Fiordland crust became so stretched that the lowermost granulite grade crust became detached from its amphibolite grade cover along a ductile shear zone system called the Doubtful Sound Décollement. The crust continued to thin as the cover was dragged off towards the NE (see Fig. 12b). At higher levels, e.g. Nelson, these extensional tectonics were displayed in brittle graben structures. Mineral cooling ages suggest that western Fiordland lower crust was uplifted to within 12km of the surface by 77Ma. Uplift rates varied between 0.8 and 3.5mm/y. With the initiation of the Kermadec/Maquarie subduction zones in the later Eocene (Fig. 11J), the Alpine Fault changed from a mainly dextral strike-dip transcurrent fault in Fiordland into a Benioff zone dipping to the SE (Figs. 11K and 12d). Fiordland and the rest of eastern South Island was transported to the SW and as it moved, it overthrust the leading edge of the Indian Plate. In this way the middle and lower crust of Fiordland were finally brought to the surface. The present day situation of active easterly subduction and overthrusting maintains the lower crust/mantle boundary at a very unusual high level.

7.   CONCLUSIONS

Fiordland is perhaps unique in that the effects of lower crustal stretching, with the production of a metamorphic core complex, can now be seen at the surface. The décollement between basement-core and cover is a 1-2km thick ductile extensional shear zone which operated in the plastic regime of the middle-lower crust. This structure would make a strong reflection for seismic surveys and would have been at an appropriate depth prior to the Tertiary uplift, to have been labelled as the Conrad Discontinuity.

Fiordland should be a high priority for on going geophysical and geological research. The seismic refraction data needs to be extended to penetrate and define the Moho. Present knowledge places the Moho between 10 and 25km depth. Perhaps seismic reflection techniques could be applied. Once the 'field' geophysics are constrained, it would then be possible to measure the same criteria in the laboratory, actually knowing the crustal position. This data could then be extrapolated to other continental areas where there are abundant geophysical features but little outcrop to constrain them.

8.   REFERENCES

Aronson, J.L. (1965) 'Reconnaissance Rb/Sr geochronology of New
    Zealand plutonic and metamorphic rocks', N.Z.J. Geol. Geophys.,
    8, 401-24.
Aronson, J.L. (1968) 'Regional geochronology of New Zealand', Geochim.
    Cosmochim. Acta., 32, 669-97.
Bensen, W.N. and Bartrum, J. (1936) 'The geology of the region about
    Preservation and Chalky Inlets, South-West Fiordland, New
    Zealand. Part III', Trans. N.Z. Inst., 65, 108-52.
Blattner, P.(1976) 'Replacement of hornblende by garnet in granulite
    facies assemblages near Milford Sound, New Zealand', Contr. Min.
    Pet., 55, 181-90.
Bohlen, S.R., Wall, V.J., and Boettcher, A.L. (1983) 'Experimental
    investigation of model garnet granulite equilibria', Contr. Min.
    Pet., 83, 52-6.
Broadbent, M. and Davey, F.J. (1978) 'The Fiordland Seismic Refraction
    Survey, 1974-5' Report No. 124. Geophysics Div., D.S.I.R.,
    Wellington.
Brothers, R.N. (1959) 'Heavy Minerals from Southland, part II: Upper
    Cretaceous-lower Tertiary coal Measures', N.Z.J. Geol. Geophys.,
    2, 164-6.
Cawood, P.A. (1974) 'The development of the SW Pacific margin of
    Gondwana: Correlations between the Rangitata and New England
    orogens', Tectonics, 3, 539-53.
Christensen, N.I. and Fountain, D.M. (1975) 'Construction of the lower
    continental crust based on experimental studies of seismic
    velocities in granulite', Geol. Soc. Am. Bull., 86, 227-236.
Coombs, D. S., Landis, C.A., Norris, R.J., Sinton, J.M., Borns, D. and
    Craw, D. (1976) 'The Dun Mountain Ophiolite Belt, New Zealand:
    Its tectonic setting, constitution and origin, with special
    reference to the southern portion', Am. J. Sci., 267, 561-603.
Cooper, R.A. (1975) 'New Zealand and south-east Australia in the early
    Palaeozoic', N.Z.J. Geol. Geophys., 18, 1-20.
Crook, K.A.W. (1981) 'The break up of the Australian-Antarctic segment
    of Gondwanaland', In Keast, A. (ed) Ecological Biogeography of
    Australia. Dr. W. Junk bv Publishers, The Hague.
Davey, F.J. and Smith, G.C. (1983) 'The tectonic setting of the
    Fiordland region, South West New Zealand', Geophys. J. Roy.
    Astro. Soc., 71, 23-38.
Davey, F.J. and Broadbent, M.O. (1980) 'Seismic measurements in
    Fiordland, Southwest New Zealand', N.Z.J. Geol. Geophys., 23,
    395-409.
Faerseth, R.B., R.M. MacIntyre and Naterstad, J. (1976) 'Mesozoic
    alkaline dykes in the Sunnhordland region, western Norway: ages,
    geochemistry, and regional significance', Lithos, 9, 331-45.
Gibson, G.M. (1979) 'Metamorphites at Wilmot Pass, central Fiordland,
    New Zealand', Unpublished Ph.D. thesis lodged in the library,
    Univ. Otago, Dunedin, New Zealand.
Gibson, G.M. (1982a) 'Stratigraphy and petrography of some
    metasediments and associated intrusive rocks from central
    Fiordland, New Zealand', N.Z.J. Geol. Geophys., 25, 21-43.

Gibson, G.M. (1982b) 'Polyphase deformation and its relationship to metamorphic crystallisation in rocks at Wilmot Pass, Fiordland, New Zealand', N.Z.J. Geol. Geophys., 25, 45-65.

Gibson, G.M., McDougall, and T.R. Ireland, (1988) 'Age constraints of metamorphism and the development of a metamorphic core complex in Fiordland southern New Zealand', Geology, 16, 405-8.

Haskin, L.A., Frey, F.A. and Schmitt, R.H. (1966) 'Meteorite, solar and terrestial rare earth distributions', Phys. Chem. Earth., 7, 167-177.

Hatherton, T. and Dickenson, W.R. (1968) 'Andesitic volcanism and seismicity in New Zealand', J. Geophys. Res., 73, 4615-19.

Hector, J. (1864) 'Geological Map of the province of Otago, New Zealand', Displayed in the Geological Museum, Otago University, New Zealand.

Holland, T.J.B. (1983) 'The experimental determination of activities in disordered and short-range ordered jadeitic pyroxenes', Contr. Min. Pet., 82, 214-220.

Howell, D.F.G. (1980) 'Mesozoic accretion of exotic terranes along the New Zealand segment of Gondwanaland', Geology, 8, 487-91.

Landis, C.A. and Coombs, D.S. (1967) 'Metamorphic belts and orogenesis in Southern New Zealand', Tectonophys., 4, 501-18.

Latin, D.M., Dixon, J.E, and Fitton, J.G. (in press) 'Rift related magmatism in the North Sea Basin', In: Tectonic evolution of the North Sea Rifts: Eds. Blundell, D.J. and A. Gibbs. Oxford Univ. Press.

Mattinson, J.M., Kimbrough, D.L. and Bradshaw, J.Y. (1986) 'Western Fiordland Orthogneiss: Early Cretaceous arc magmatism and granulite facies metamorphism, New Zealand', Contr. Min., Pet., 92, 383-9.

McLellan, E. (1990) 'Metamorphic Studies, Geotimes', 35, 51.

McCulloch, M.T., Bradshaw, J.Y. and Taylor, S.R. (1987) 'Sm-Nd and Rb-Sr isotope and geochemical systematics in Phanerozoic granulites from Fiordland, Southwest New Zealand,' Contr. Min. Pet., 97, 183-95.

McLelland, J.M. and Whitney, P.R. (1977) 'The origin of garnet in the anorthosite charockite suite of the Adirondaks', Contr. min. pet., 60, 161-81.

Misra, S.N. (1971) 'Chemical distinction of high grade ortho and para metabasites' Norsk. Geol. Tidsskr., 51, 311-16.

Moecher, D.P., Essene, E.J. and Anovitz. (1988) 'Calculation and application of clinopyroxene-garnet-plagioclase-quartz geobarometers', Contr. Min. Pet., 100, 92-106.

Newton, R.C. and Perkins, D. III. (1982) 'Thermodynamic calibrations of geobarometers for charnockites and basic granulites based on the assemblages garnet-plagioclase-orthopyroxene-(clinopyroxene)-quartz, with applications to high grade metamorphism', Am. Mineral., 67, 203-22.

Oliver, G.J.H. (1973) 'Gravity anomalies and the structure of Fiordland:', Comment. N.Z.J. Geol. Geophys., 16, 307-10.

Oliver, G.J.H. (1976) 'The High Grade Metamorphic rocks of Doubtful Sound, Fiordland, New Zealand: A Study of the Lower Crust', Thesis, lodged in the Library, University of Otago, Dunedin, New

Zealand.

Oliver, G.J.H. and Coggon, J.H. (1979) 'Crustal structure of Fiordland New Zealand', Tectonophys., 54, 253-192.

Oliver, G.J.H. (1980) 'Geology of the granulite and amphibolite facies gneisses of Doubtful Sound, Fiordland, New Zealand', N.Z.J. Geol. Geophys., 23, 27-41.

Park, J. (1921) 'The geology and mineral resources of western Southland', N.Z. Geol. Surv. Bull., 23.

Park, J. (1924) 'The Pre-Cambrian complex and pyrrohotite bands, Dusky Sound, New Zealand', Econ. Geol., 19, 750-55.

Perkins, D. III and Chipera, S.J. (1985) 'Garnet-plagioclase-quartz barometry: refinement and application to the English River Subprovince and the Minnesota River Valley', Contr. Min. Pet., 89, 69-80.

Priestly, K. and Davey, F.J. (1983) 'Crustal structure of Fiordland, southwestern New Zealand, from seismic-refraction measurements', Geology, 11, 660-662.

Raheim, A. and Green, D.H. (1975) 'Experimental determination of the temperature and pressure dependance of Fe-Mg partition between coexisting garnet and clinopyroxene', Contr. Min. Pet., 48, 179- 203.

Reed, J.J. (1951) 'Hornblende andesites from Solander Island', Trans. Roy. Soc. N.Z. 79, 119-25.

Reilly, W.I. (1965) 'Gravity Map of New Zealand 1:4,000,000 1st ed.', D.S.I.R., Wellington.

Scholtz, C.H., Ryann, J.M.W., Weed, R.W. and Frohlich, C. (1973) 'Detailed seismicity of the Alpine Fault Zone and Fiordland region, New Zealand', Geol. Soc. Am. Bull., 84, 3297-316.

Sheppard, D.S., Adams, C.J., and Bird, G.W. (1975) 'Age of metamorphism and Uplift in the Alpine Schist Belt, New Zealand', Geol. Soc. Am. Bull., 86, 1147-53.

Smith, W.D. (1971) 'Earthquates at shallow and intermediate depths in Fiordland, New Zealand', J. Geophys. Res., 16, 4901-7.

Summerhayes, C.P. (1969) 'Marine geology of the New Zealand Sub-Antarctic sea floor', N.Z. D.S.I.R. Bull., 190. N.Z. Oceanographic Inst. Mem. 50, 94pp.

Turner, F.J. (1937a,b,c) 'The metamorphic and plutonic rocks of Lake Manapouri', Parts I, II, III. Trans. Roy. Soc. N.Z., 67, 83-100, 227-49, 68, 122-40.

Turner, F.J. (1939) 'Hornblende Gneisses, marbles, and associated rocks from Doubtful Sound, Fiordland', Trans. Roy. Soc. N.Z., 69, 5170-98.

Turner, F.J. (1948) 'Mineralogical and structural evolution of the metamorphic rocks', Geol. Soc. Am. Mem., 30, 342 pp.

Turner, F.J. (1958) 'Mineral assemblages of individual facies', In Fyfe, W.S., Turner, F.J. and Verhoogen, J. Metamorphic reactions and metamorphic facies. Geol. Soc. Am. Mem., 73, 199-239.

Turner, F.J. (1968) 'Metamorphic petrology, mineralogical and field aspects', McGraw-Hill Inc., 403 pp.

Wellman, H.W. (1954) 'Marine Pliocene at Resolution Island, Dusky Sound, Fiordland', N.Z.J. Sci. Tech., 35, 378-89.

Wellman, H.W. (1979) 'A uplift map for the South Island of New
    Zealand, and a model for the uplift of the Southern Alps', In:
    Walcott, R.I., Cresswell, M.M. (eds). The origin of the Southern
    Alps. Roy. Soc. New Zealand.
Williams, J.G. and Harper, C.T. (1978) 'Age and status of the MacKay
    Intrusives in the Eglington-upper Hollyford area', N.Z.J. Geol.
    Geophys. 21, 733-42.
Wood, B.J. and Banno, S. (1973) 'Garnet - pyroxene and orthopyroxene-
    clinopyroxene relationships in simple and complex systems',
    Contr. Min. Pet., 42, 109-17.
Wood, B.L. (1962) 'Sheet 22, Wakatipu Geological Map of New Zealand',
    1:250,000. D.S.I.R., Wellington, New Zealand.
Wood, B.L. (1966) 'Sheet 24, Ivercargill, Geological Map of New
    Zealand, 1:250,000', D.S.I.R., Wellington, New Zealand.
Woodward, D.J. (1972) 'Gravity Anomalies in Fiordland, south west New
    Zealand', N.Z.J. Geol. Geophys., 15, 22-32.

UPLIFT AND EXHUMATION OF MIDDLE AND LOWER CRUSTAL ROCKS IN AN EXTENSIONAL TECTONIC SETTING, FIORDLAND, NEW ZEALAND

G.M. GIBSON
School of Applied Science
University College of Southern Queensland
Toowoomba
Australia   4350

ABSTRACT.   Late Mesozoic continental extension combined with more recent transpression along the mid-Tertiary Alpine Fault have combined in Fiordland to uplift and expose unusually deep structural levels of a mid-Cretaceous metamorphic core complex. High pressure Early Cretaceous granulite facies orthogneisses occupy the axial region of this complex and were emplaced from deeper levels beneath a thinned and ductilely deformed cover sequence comprising mid-Palaeozoic granites, metasediments and metavolcanic rocks. Emplacement of the core complex was associated with widespread resetting of mineral isotopic systems in the (amphibolite facies) cover sequence whilst at higher crustal levels in the once-contiguous regions of Westland and Nelson crustal extension was accompanied by high-level silicic magmatism, normal faulting and nonmarine sedimentation in rapidly-subsiding fault-bounded asymmetric basins. Although the original boundary between the orthogneisses and cover sequence may have been intrusive, the present contact is a ductile shear zone in which the granulites have been mylonitized and retrogressed to the amphibolite facies. Strongly lineated mylonites in this zone developed before 93 Ma consistent with emplacement of the core complex during continental rifting and breakup of the Pacific margin of Gondwana, and prior to the onset of seafloor spreading in the Tasman Sea.

1.   INTRODUCTION

Although granulite facies rocks have been described at the base of many exposed crustal sections (Fountain and Salisbury, 1981; Percival and Card, 1983; Brodie and Rutter, 1987) there remains the problem of how such deep crustal rocks get exhumed and uplifted to the surface. Continental obduction and/or thrust faulting offer one possible solution to the problem but are clearly not applicable in all cases. In Doubtful Sound, New Zealand (Gibson et al., 1988; Gibson, 1988) and the Ivrea Zone, northern Italy (Hodges and Fountain, 1984; Brodie and Rutter, 1987; Handy, 1987) there is clear evidence that continental extension played an equally important, if

71

M. H. Salisbury and D. M. Fountain (eds.), Exposed Cross-Sections of the Continental Crust, 71–101.
© 1990 *Kluwer Academic Publishers.*

not dominant, role in the exhumation process. This paper describes the structural and metamorphic evidence for such a mechanism in the Doubtful Sound section and compares the tectonic evolution of this region to that of the metamorphic core complexes in the North American Cordillera (eg Crittenden et al., 1980; Davis and Lister, 1988). In a companion paper, Oliver (this volume) describes the regional geology and geophysical setting of the Doubtful Sound crustal section.

2. REGIONAL GEOLOGY AND TECTONIC SETTING

South Island, New Zealand has been subdivided into nine tectonostratigraphic terranes (Bishop et al., 1985), all but two of which occur within the eastern province (Fig. 1). This province encompasses several late Palaeozoic-Mesozoic magmatic arc assemblages and accretionary complexes (Coombs et al., 1976; Bishop et al., 1985) and is separated from the western province by a major fault, the Median Tectonic Line (Landis and Coombs, 1967). The western province comprises two terranes made up of variably metamorphosed Late Proterozoic-Lower Palaeozoic rocks (Suggate et al., 1978; Cooper, 1979).

At least two major tectonothermal events have affected the western province (Fig. 2): the first in mid-Palaeozoic times at 350-380 Ma and the second during the late Mesozoic including a strong thermal pulse at 100-120 Ma (Landis and Coombs, 1967; Aronson, 1968). These events have traditionally been identified (eg Suggate et al., 1978) as the Tuhua and Rangitata Orogenies and are primarily based on radiometric age data obtained from magmatic rocks associated with each event. Late Mesozoic magmatic rocks are particularly well represented in Fiordland and southeast Nelson and include both granitoid and intermediate-basic plutonic complexes (Tulloch, 1983a), most of which are virtually undeformed and unmetamorphosed. The most notable exception is the western Fiordland orthogneiss (Mattinson et al., 1986) which comprises Early Cretaceous plutonic rocks metamorphosed to the granulite facies. This unit (Fig. 1), together with many other late Mesozoic plutonic complexes in western New Zealand, exhibit calc-alkaline compositional trends consistent with magmatism in a subduction-related volcanic arc environment (Mattinson et al., 1986; McCulloch et al., 1987; Tulloch, 1983b; Williams, 1983) and were intruded during an episode of Late Jurassic-Early Cretaceous plate convergence preceding juxtaposition and/or collision between the eastern and western provinces. This collision represents the climactic phase of the Rangitata Orogeny (Bradshaw et al., 1981) and was the last in a series of such events whereby successive terranes in the eastern province were amalgamated and accreted onto the Pacific margin of Gondwana (Howell, 1980; MacKinnon, 1983). The western province had been part of Gondwana since mid-Palaeozoic time. In the absence of any formal name, the Late Proterozoic-Lower Palaeozoic rocks of Fiordland are simply referred to here as the cover sequence (cf. Oliver and Coggon, 1979).

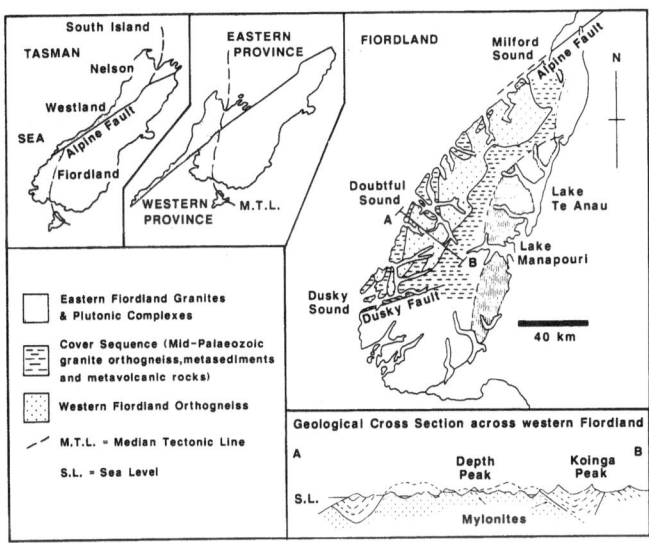

Figure 1. Simplified geological map and cross-section of Fiordland
(Oliver, 1980; Mattinson et al., 1986) showing tectonic window in the
amphibolite facies cover sequence through which the Early Cretaceous
granulite facies orthogneisses (western Fiordland orthogneiss) are
exposed. Insets show configuration of the western province (South
Island) for the present day and prior to the inception of the mid-
Tertiary Alpine Fault.

Amphibolite facies metasediments and subordinate metavolcanic
rocks make up the bulk of the cover sequence in Fiordland where they
have been deformed into a broad anticlinorium and subsequently eroded
to form a tectonic window (Fig. 1) through which the underlying Early
Cretaceous granulite facies orthogneisses are exposed. These
orthogneisses constitute an entirely different structural entity to
the cover sequence and with the possible exception of northern
Fiordland where an original intrusive contact is apparently preserved
(Mattinson et al., 1986), are everywhere separated from the latter by
a major structural discontinuity or decollement. This discontinuity
was originally defined (Oliver and Coggon, 1979) as the Doubtful Sound
Thrust but has since been interpreted as a ductile shear zone
separating upper and lower crustal plates of a metamorphic core
complex emplaced during late Mesozoic continental rifting and breakup
of the Pacific margin of Gondwana (Gibson et al., 1988; see also Kamp
and Hegarty, 1989).

74

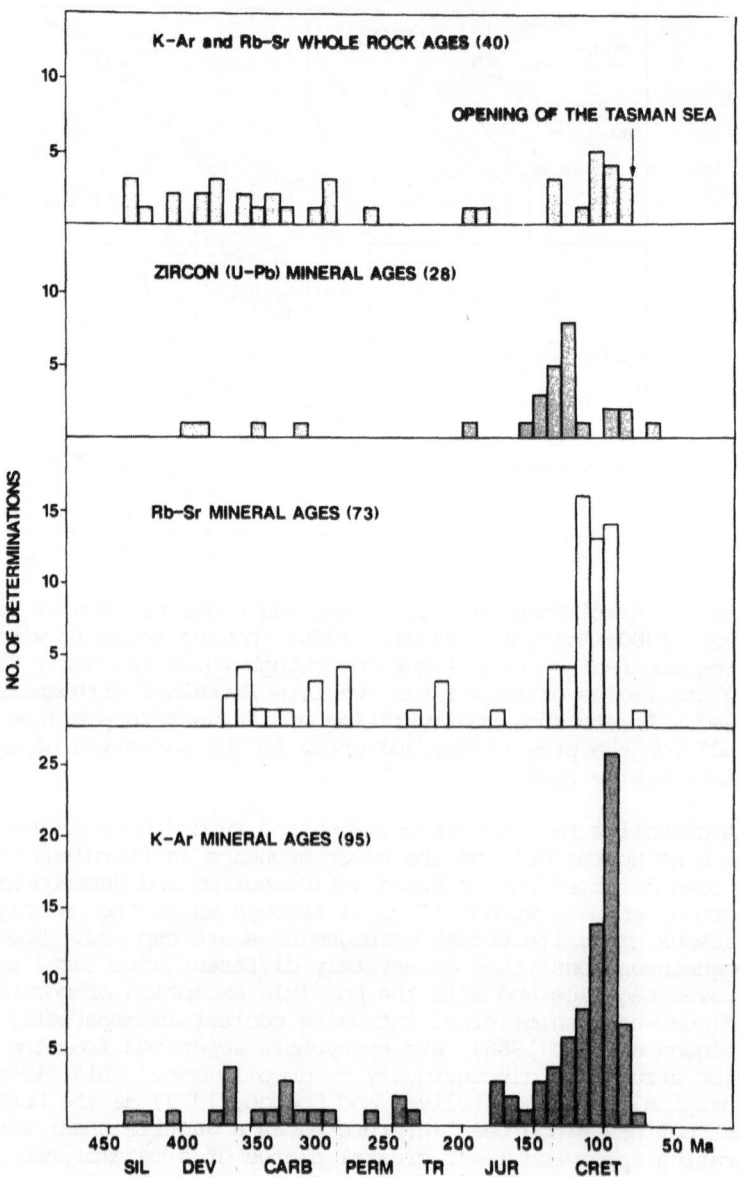

Figure 2. Histogram of published geochronological data for western province. Data from Adams and Nathan (1978); Aronson (1965; 1968); Gibson et al. (1988); Grindley (1980); Harrison and McDougall (1980); Mattinson et al. (1986); Oliver (1980); Williams and Harper (1978).

3. FIORDLAND STRATIGRAPHY

3.1. Cover Sequence

Oliver and Coggon (1979; see also Oliver, this volume) subdivided
Fiordland north of Dusky Sound into three northeast-trending regions
on the basis of the observed rock associations: (1) a western region
underlain by granulite facies orthogneisses and their associated cover
rocks; (2) a central region comprising mainly amphibolite facies
metasediments and granitic orthogneisses; and (3) an eastern region
composed predominantly of late Mesozoic granite rocks and post-
tectonic intrusive complexes. However, whilst the case for
recognising an eastern region (Fig. 1) composed mainly of late
Mesozoic plutonic rocks may be perfectly justified, there is little to
recommend retention of the western and central regions as separate
entities (cf. Gibson, 1982a; Mattinson et al., 1986). These two
regions are not only characterised by identical metasedimentary
sequences but these metasedimentary sequences have shared a common
structural and metamorphic history; they are all part of the cover
sequence and together with various granitic orthogneisses and
gabbroic rocks (Fig. 3) form an eastward-thickening crustal wedge
which dips eastwards off the underlying western Fiordland
orthogneiss (cf. Oliver and Coggon, 1979). If there is a fundamental
structural break in the Doubtful Sound area of Fiordland, it is
between this cover sequence and the underlying granulites.

The cover sequence in the Doubtful Sound area is subdivided here
(Figs. 3 and 4) into three major lithostratigraphic units: Deep Cove
Gneiss, Black Giants Anorthosite and related rocks, and the central
Fiordland metasediments. Of these three units, only Deep Cove Gneiss
and the Black Giants Anorthosite have been afforded formational
status (Oliver, 1980; Gibson, 1982a); the other unit is a 10 km-thick
sequence of thin-bedded and isoclinally folded metasedimentary rocks
which have been intruded to varying degrees by syntectonic mid-
Palaeozoic granitic orthogneisses. These metasediments make up the
bulk of the cover sequence in central Fiordland and are all grouped
together here as the central Fiordland metasediments (cf. Oliver and
Coggon, 1979). They are particularly well represented in the Merrie
Range, Mackenzie Pass and Mt Barber areas (Fig. 3) and include calc-
silicate rocks, pyritic metaquartzites, metapsammites, pelitic
schists and feldspathic biotite gneisses. Subsidiary amounts of
amphibolite and hornblende $\pm$ biotite gneiss interstratified with
these rocks probably represent metamorphosed mafic lava flows or minor
intrusions. Pelitic rocks in this sequence are migmatized and
typically contain kyanite or sillimanite but rarely both (Gibson,
1982a).

Underlying the central Fiordland metasediments is Deep Cove
Gneiss (Fig. 4); it makes up the remainder of the cover sequence in
central Fiordland and is just over 2000 m thick. It is in low-angle
fault contact with the central Fiordland sediments (Fig. 3) and
consists mainly of biotite, hornblende and quartzofeldspathic
gneisses which are everywhere interstratified with subordinate amounts

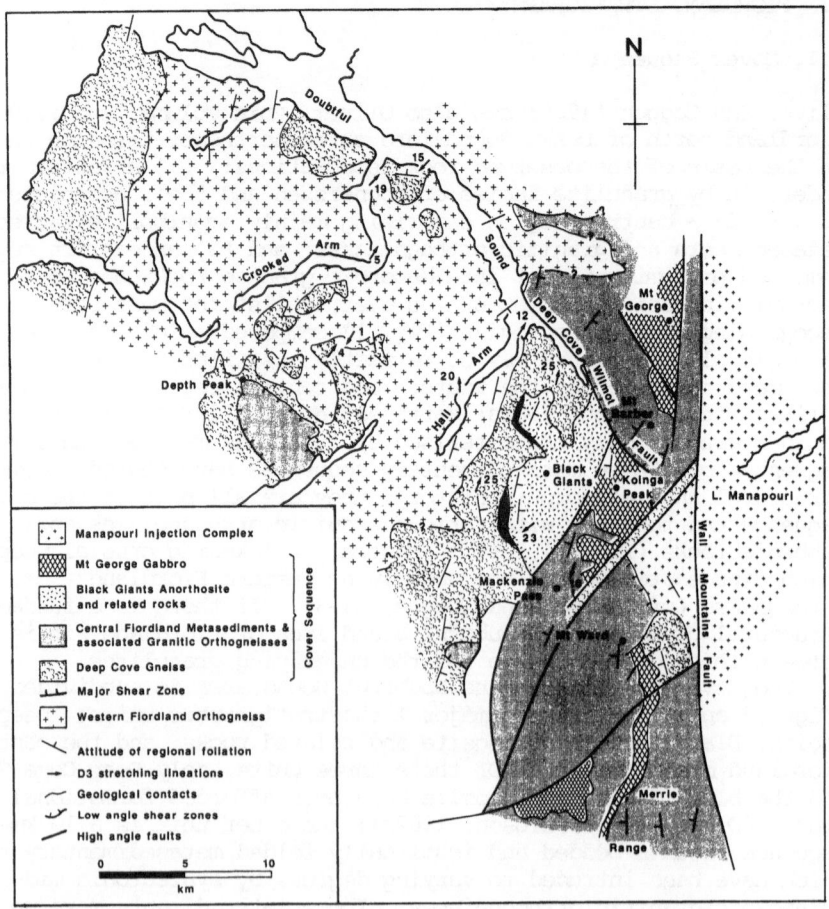

Figure 3. Simplified lithological and structural map of the Doubtful Sound area (after Oliver, 1980; Gibson, 1982a and Gibson unpubl. data).

of amphibolite, marble and calc-silicate rocks (Gibson, 1982a; see also Oliver, 1980). Gneisses in Deep Cove Gneiss are characteristically thick-bedded although banded varieties with strongly developed mylonitic fabrics also occur (Fig. 5a); they coincide with areas of high shear strain and are particularly common towards the base of the unit. Similarly banded tonalitic gneisses occur in the type locality at Deep Cove (Fig. 3); they are layered on a scale of 1-15 mm, possess a fine to medium grain size and contain highly attenuated intrafolial folds. Such strongly foliated gneisses are unusual in Deep Cove Gneiss and reflect greatly increased amounts of strain towards the contact with the underlying western Fiordland orthogneiss.

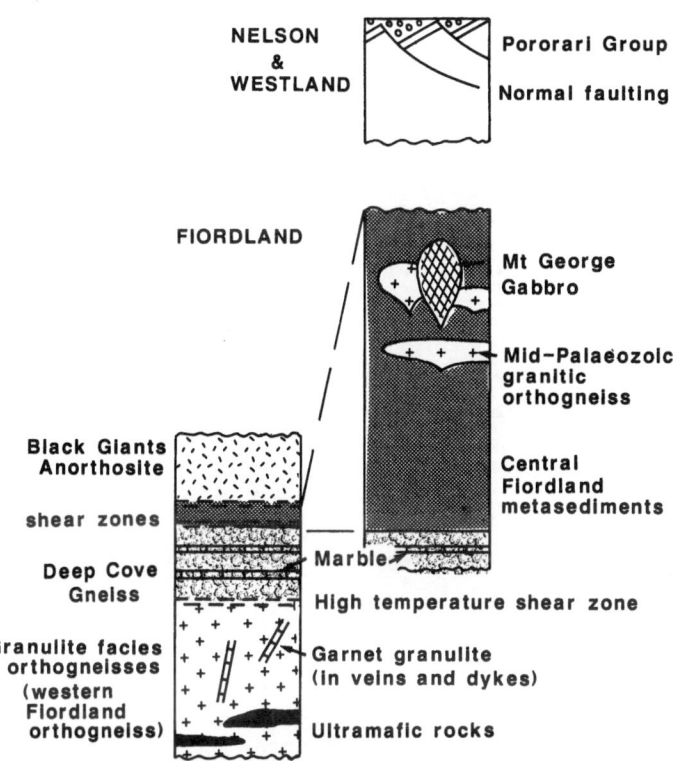

Figure 4. Schematic vertical sections through the crustal sequences in the Doubtful Sound area. Upper part of section is for the once-contiguous regions of Nelson and Westland which preserve much higher crustal levels than Fiordland.

The third unit in the cover sequence comprises a dismembered and regionally metamorphosed layered intrusion (Black Giants Anorthosite) and a conformably overlying sequence of massive amphibolites, hornblende ± garnet ± biotite gneisses. This sequence represents the country rocks into which the former was intruded and for this reason is not shown separately in Figure 3; it is 1000 m thick and was probably derived from a pile of volcanic and/or minor intrusive rocks. Black Giants Anorthosite is marginally thinner at 600-800 m and consists mainly of amphibolites and metamorphosed gabbroic anorthosite (An_{80-95}) together with

lesser amounts of hornblende schist, meta-anorthosite and metaperidotite (Gibson, 1982a). These rocks are in low-angle fault contact with the underlying central Fiordland metasediments (Fig. 3) although an original intrusive relationship is indicated by the presence of rare screens of calc-silicate gneiss within the Black

78

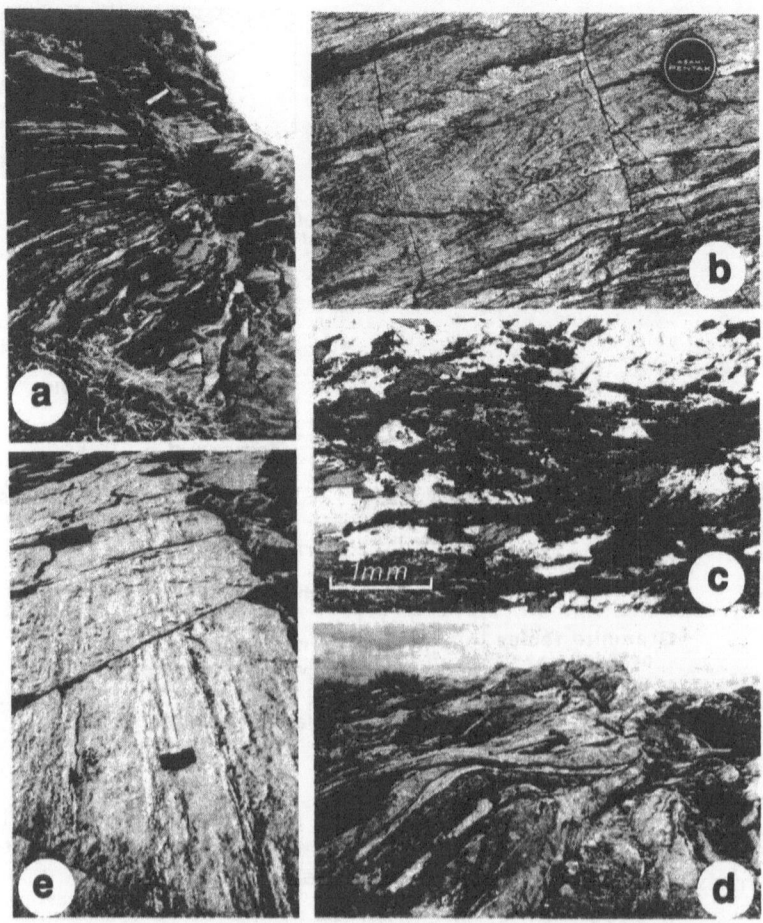

Figure 5. (a) Laminated, mylonitic marbles (light) and hornblende
gneisses in Deep Cove Gneiss 200 m above contact with western Fiordland
orthogneiss, Crooked Arm. Hammer is 80 cm long. (b) Tight, asymmetric
D2 folds in migmatized pelitic schists in Merrie Range. Note granitic
material injected parallel to axial-surface of folds. (c) Oriented
biotite and sillimanite define the S2 foliation in pelitic schist from
the Merrie Range. An earlier S1 fabric also defined by biotite and
sillimanite is deformed into a tight isocline (centre of photograph).
(d) Ductile normal fault in hornblende and quartzofeldspathic gneisses,
Deep Cove Gneiss. Section is oriented SW-NE parallel to D3 stretching
with hangingwall downthrown to northeast. (e) D3 stretching lineation
defined by quartz ribbons in biotite schists, Depth Peak area,
Crooked Arm.

Giants Anorthosite.

Also included in Figure 3 is Mt George Gabbro, a linear but
discontinuous belt of Early Cretaceous gabbroic plutons, some of
which were emplaced at very shallow crustal levels; they largely
post-date regional metamorphism and have narrow contact aureoles in
which low pressure sillimanite + biotite + potash feldspar +
cordierite assemblages or more rarely (eg west of Merrie Range)
hypersthene + biotite + cordierite assemblages have developed.
Other plutons were emplaced at slightly deeper structural levels and
have undergone limited amounts of recrystallization to the
amphibolite facies. Individual plutons in Mt George Gabbro have been
dated at 120-126 Ma (Kimbrough, pers. comm., 1984; Gibson and
McDougall, unpubl. data).

3.2. Western Fiordland Orthogneiss

This unit comprises mainly Early Cretaceous plutonic rocks
metamorphosed to the granulite facies. Two-pyroxene feldspathic
orthogneisses predominate and consist largely of hypersthene,
clinopyroxene, antiperthitic plagioclase and olive-green hornblende.
These orthogneisses possess a streaky to weakly developed gneissic
fabric and are locally cut by subplanar zones of garnet granulite
developed in and around plagioclase-rich veins and dykes (Oliver,
1977; Blattner, 1976; 1978). Garnet granulite is typically coarser
grained than its two-pyroxene host rocks and has been attributed to
a second, higher pressure phase of granulite facies metamorphism
(Bradshaw, 1989; Mattinson et al., 1986). Metamorphic pressures
calculated for these garnet granulites range from 9-14 kbar (Newton
and Perkins, 1982; Mattinson et al., 1986; Moecher et al., 1988;
Bradshaw, 1989). In contrast the earlier two-pyroxene granulites
are inferred to have developed during low- to medium-pressure
metamorphism (Bradshaw, 1989).

In addition to these garnet zones, the two-pyroxene
orthogneisses are also host to various ultramafic rocks some of
which have been transformed into eclogite. A more complete
description of these and other lithologies in the western Fiordland
orthogneiss is presented elsewhere in this volume by Oliver.

3.3. Manapouri Injection Complex

The cover sequence is bounded to the east by the Manapouri
Injection Complex (Fig. 3). This complex was originally included in
the central Fiordland region by Oliver and Coggon (1979) but is
treated here as a quite separate entity (cf. Turner, 1937; Gibson,
1982a); it is in fault contact with the cover sequence and consists
mainly of coarse grained granitic orthogneisses, post-tectonic
leucogranites and minor paragneiss, all of which have been
extensively intruded by sheets of trondjhemite and pegmatite
(Turner, 1937; Bishop, 1967; Gibson, 1982a). These rocks have
yielded only Cretaceous K-Ar and Rb-Sr mineral ages (Aronson, 1965;

Gibson and McDougall, unpubl. data) and in this respect would appear
to have more in common with the late Mesozoic plutonic complexes of
eastern Fiordland. They are also of comparable age to the western
Fiordland orthogneiss although whether there is any genetic
relationship between this unit and the Manapouri Injection Complex is
uncertain. At best these two units may represent different crustal
levels within the same Early Cretaceous magmatic arc (cf. Oliver,
this volume).

4. STRUCTURE AND METAMORPHISM

4.1. Cover Sequence

The cover sequence in Fiordland has undergone at least four phases of
deformation (D1-D4) and metamorphism up to the amphibolite facies.
Peak metamorphic conditions occurred during the Tuhua Orogeny and
coincided with migmatization, granite emplacement and the widespread
development of sillimanite + biotite $\pm$ potash feldspar assemblages in
pelitic schists throughout the Merrie Range and Mt Barber areas.
More rarely these rocks also contain minor amounts of almandine garnet
and cordierite. Metamorphic temperatures for these assemblages are
estimated at 640-680°C for pressures of 3.5-5 kbars. Rare
occurrences of kyanite in the Mt Barber area are attributed to
subsequent and possibly higher pressure and/or lower temperature
metamorphism.
     First generation folds have not been observed in the Merrie Range
and Mt Barber areas and the dominant structures are tight to
isoclinal, overturned to recumbent D2 folds (Fig. 5b) with a well
developed axial-surface fabric (S2) which trends northeast-southwest.
This fabric defines the regional schistosity and in pelitic rocks
varies from a crenulation cleavage to a transposition foliation; it
overprints an earlier schistosity (S1) defined by oriented
sillimanite and biotite but is itself defined by the same mineralogy
(Fig. 5c). This suggests that the D1 and D2 deformations formed under
similar metamorphic conditions and thus cannot have been entirely
independent of each other. Rather they form a continuum and developed
during a single progressive tectonothermal event. On the basis of
316-342 Ma hornblende dates obtained from sillimanite grade
amphibolites and hornblende gneisses in the Merrie Range and Mt Barber
areas (Gibson et al., 1988), this event has a mid-Palaeozoic age
(see section 5 for further details).
     Further west where deeper structural levels and hence higher
pressure (5-9 kbar) assemblages in the cover sequence are preserved,
pelitic rocks contain kyanite, biotite, almandine garnet and
occasionally potash feldspar. The D2 foliation in these rocks is a
penetrative schistosity defined by oriented kyanite and biotite.
Sillimanite, if it is present at all, usually occurs as inclusions
within garnet porphyroblasts and in some cases may even define an
earlier fabric oriented at high angles to the external schistosity.

Biotite inclusions in garnet may be similarly oriented. These relations constitute evidence for at least two phases of deformation in these rocks and demonstrate that the second phase of deformation occurred under increased pressures. Deformation under increased metamorphic pressures is also supported by compositional profiles through garnet porphyroblasts showing grossular enrichment towards garnet rims (Gibson, 1979; Bradshaw, 1985) and by the observation (Ward, 1984) that kyanite is occasionally included in the rims of these same porphyroblasts. The second phase of deformation was evidently associated with significant crustal thickening in order that this part of the cover sequence be carried to sufficient depth to stabilize kyanite at the expense of sillimanite. Less certain is whether this second phase of deformation is equivalent to the D2 event in sillimanite grade rocks at higher structural levels in the Merrie Range. Both are assumed to be of comparable age here, largely because hornblende oriented in the D2 foliation of both kyanite and sillimanite grade amphibolites and hornblende gneisses gives Palaeozoic ages (section 5).

Second generation fold axes are usually subhorizontal or else pitch at shallow angles within their axial surfaces (Fig. 6). However, in some areas the D2 fold axes become steeper and more variably plunging along strike even though their axial surfaces show no corresponding deviation from the northeast regional trend (Fig. 6). This suggests that the D2 structures are non-cylindrical on a regional scale although this could either be an original feature of the second generation folds or the result of subsequent reorientation of the D2 fold axes during later deformation. In this context it is perhaps worth noting that in parts of the cover sequence most affected by the subsequent D3 deformation (eg marble horizons in Deep Cove Gneiss), D2 fold axes are more variably oriented and in some cases may be aligned parallel or subparallel to the D3 stretching direction (cf. Figs. 7d and 7e). At higher structural levels in the Merrie Range where the effects of the D3 deformation are less pronounced, the D2 fold axes usually trend northwest-southeast (Fig. 6a).

The third phase of deformation (D3) was associated with subvertical flattening, boudinage, ductile normal faulting (Fig. 5d) and the development of a conspicuous northeast-southwest-trending stretching lineation (Fig. 7e). This lineation is most spectacularly developed in rocks immediately overlying the western Fiordland orthogneiss and takes the form of mineral streaks, quartz rods and ribbons (Fig. 5e), straie and slickensides and elongated mineral aggregates. Further away from this contact, this lineation diminishes in intensity and at higher structural levels in the cover sequence may be absent altogether. It has not been identified in the Merrie Range and Mt Barber areas where the D3 deformation mainly finds expression in boudinage and retrogressive metamorphism related to reactivation of the D2 foliation. Ptygmatically or isoclinally folded post-D2 quartz veins and pegmatite in Deep Cove Gneiss are attributed to significant

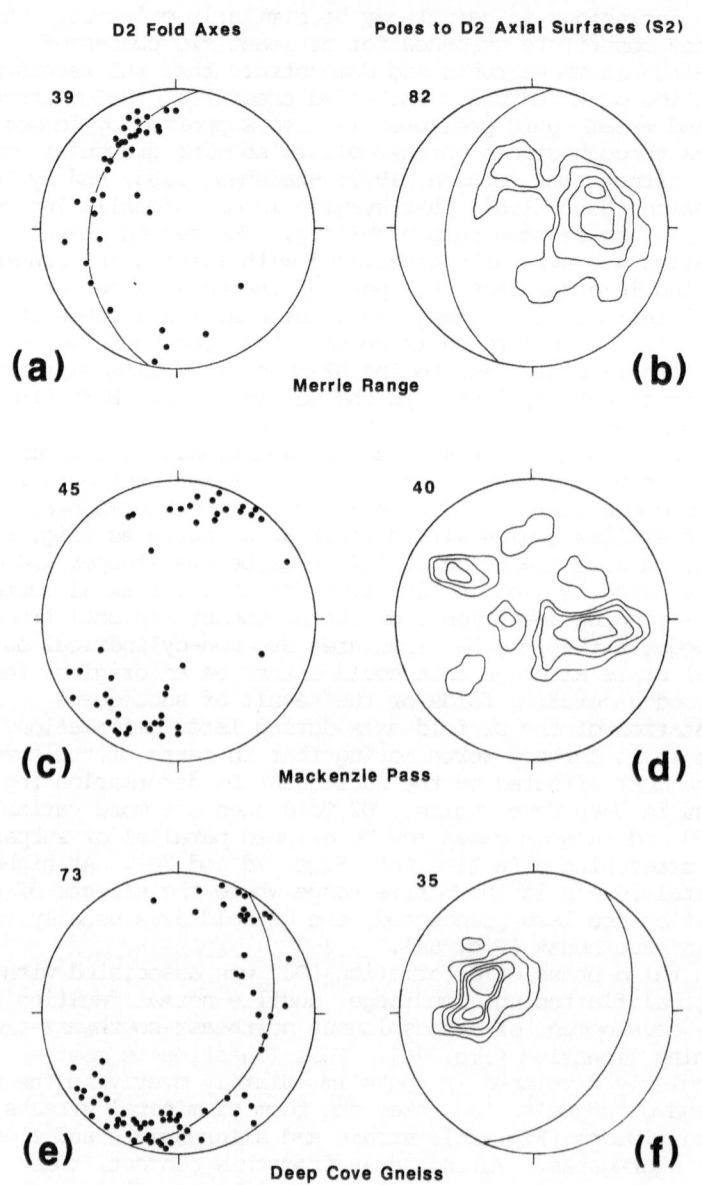

D2 Fold Axes   Poles to D2 Axial Surfaces (S2)

(a)   (b)

Merrie Range

(c)   (d)

Mackenzie Pass

(e)   (f)

Deep Cove Gneiss

Figure 6. Lower hemisphere equal area projections of D2 structural elements for central Fiordland. Change of attitude in regional S2 foliation (shown as single great circle) is due to effects of D4 post-metamorphic folding. Contour interval is 1%, 3%, 6%, 9% and 12% per 1% area.

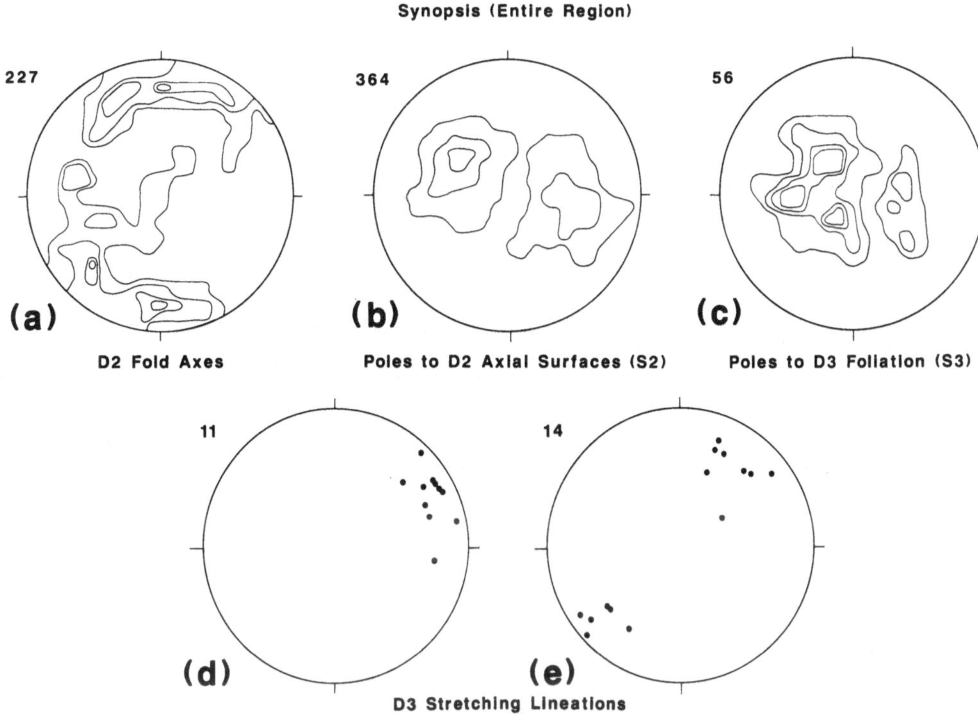

Synopsis (Entire Region)

(a) D2 Fold Axes

(b) Poles to D2 Axial Surfaces (S2)

(c) Poles to D3 Foliation (S3)

(d) (e) D3 Stretching Lineations

Figure 7. Lower hemisphere equal area projections for D2 and D3 structural elements for central Fiordland. (a) All D2 fold axes and intersection lineations. (b) Poles to regional foliation (S2). The two focii represent opposing limbs of the post-metamorphic D4 folds. (c) Poles to D3 foliation; note that it is subparallel to S2 in (b); (d) and (e) D3 stretching lineations for western Fiordland orthogneiss and cover sequence respectively. Contour interval as in Fig. 6.

subvertical flattening (shortening) during the D3 deformation.
Foliation produced during the development of D3 deformation is
parallel or subparallel to the D2 regional schistosity (cf. Figs. 7b
& 7c).

In Deep Cove Gneiss the D3 deformation also gave rise to several
high strain shear zones characterized by banded or thinly laminated
mylonitic gneisses. Foliation in these rocks is a transposed
compositional layering which parallels the regional D2 foliation and
occasionally contains rootless intrafolial isoclines. Mineral
assemblages in these gneisses typically contain abundant epidote
consistent with crystallization under retrograde (lower amphibolite
facies) metamorphic conditions. Marble horizons near the base of
Deep Cove Gneiss are similarly mylonitized and retrogressed.

A fourth phase of post-metamorphic deformation (D4) rotated all
earlier-formed structures about subhorizontal northeast- to northwest-
trending axes giving rise to broad, open folds in the regional
foliation (Figs. 1 and 7). These folds developed during Tertiary times
in response to transpressive motion along the offshore extension of
the Alpine Fault.

4.2. Western Fiordland Orthogneiss

The D3 deformation also extends downward into the underlying granulite
facies orthogneisses which become increasingly strongly foliated and/or
lineated as the contact with the cover sequence is approached.
Deformation was accompanied by retrograde metamorphism of the
granulites to the amphibolite facies and a marked reduction in grain
size so that rocks immediately below the contact are thoroughly
reconstituted mylonitic gneisses consisting mainly of hornblende,
biotite, plagioclase and quartz with subsidiary amounts of epidote and
sphene; grain size is 0.5-2 mm. Contrary to earlier interpretations
(Turner, 1968; see also Mattinson et al., 1986), there is no simple
prograde amphibolite-granulite facies transition in this part of
Fiordland. Rather the amphibolite facies orthogneisses constitute a
high-temperature ductile shear zone which was superimposed upon rocks
with a pre-existing plutonic and/or granulite facies fabric (Gibson
et al., 1988). In Doubtful Sound this shear zone is several hundred
metres thick and everywhere concordantly underlies the cover sequence.
Enclaves of less strongly mylonitized and/or retrogressed rocks
within this zone (cf. Blattner, 1976) indicate deformation was
heterogeneous and probably accommodated on a series of anastomosing
shear zones rather than on any single discontinuity.

Lineation in D3 tectonites is subhorizontal (Fig. 8a) and trends
northeast-southwest parallel to the D3 stretching direction in the
overlying cover sequence (cf. Figs. 7d & 7e). It is most
spectacularly developed in amphibolite facies gneisses near the
entrance to Hall Arm but also occurs in partially retrogressed
granulites at several localities throughout Crooked Arm. In
amphibolite facies rocks it is typically defined by elongated
lenticular aggregates of dynamically recrystallized quartz and

feldspar bounded by thin foliae of hornblende and biotite but may
also be expressed as mineral streaks and striae on foliation
surfaces. At deeper structural levels lineation is more commonly a
coarse rodding defined by elongated mafic aggregates (Fig. 8a) in
which original pyroxene and garnet have undergone varying degrees
of comminution and replacement by hornblende-quartz and biotite-
quartz symplectites. Plagioclase in these same rocks may be
similarly comminuted and/or fractured.

Foliation (Fig. 8b) in the D3 mylonites is usually flat-lying or
gently folded and everywhere parallelled by a penetrative closely
spaced jointing. It is the most conspicuously developed structure in
the western Fiordland orthogneiss and has been superimposed on all
pre-existing lithologies including many once vertical or steeply
dipping veins and dykes of garnet granulite which have since been
deformed into recumbent isoclines following subvertical D3 flattening.
These folds have S3 as the axial-surface foliation; it is
subhorizontal and continuous with the D3 mylonitic fabric in the
adjacent host rocks. Undeformed veins and dykes of garnet granulite
were either emplaced after deformation ceased or else reflect the
heterogeneous nature of deformation during the D3 event.

Amphibolite facies mylonites just below the contact with the cover
sequence are often cut by secondary shear zones attributed to
continued deformation under increasingly brittle conditions. They
contain cataclasites and lower grade metamorphic assemblages in which
original biotite and hornblende have been replaced by chlorite and/or
epidote. Additional retrogression and/or shearing accompanied
development of the post-metamorphic D4 folds.

4.3. Origin of D3 Fabrics

Data presented in the previous two sections clearly show that the D3
fabrics are of regional significance and affect the western Fiordland
orthogneiss and cover sequence alike. It is equally clear that
fabric development in both units intensifies towards their mutual
contact. This suggests a genetic relationship between this contact
and the D3 fabrics. However, whereas previous workers (Oliver and
Coggon, 1979; Oliver, 1980) interpreted these fabrics to be the
result of regional overthrusting, an extensional origin is favoured
here (cf. Gibson et al., 1988). Particularly critical in this regard
is the occurrence of ultramafic dykes emplaced into the western
Fiordland orthogneiss during the course of the D3 deformation. One
such dyke is illustrated in Figure 8a; it cuts the D3 stretching
lineation at right angles and is itself metamorphosed and deformed.
A synkinematic emplacement age is indicated by the fact that the dyke
possesses only amphibolite facies mineralogy and thus is isofacial
with its retrogressed host rocks. Other ultramafic dykes post-date
the D3 fabrics and indicate that dyke emplacement continued after the
cessation of deformation.

Mafic dykes also cut the D3 fabrics in the cover sequence. They
too are oriented at high angles to the regional D3 stretching

lineation and trend northwest-southeast. However, unlike older, and much more strongly deformed amphibolite dykes of inferred mid-Palaeozoic age, these younger intrusions are more typically composed of massive to weakly foliated diorite or coarse grained amphibolite and were evidently emplaced rather late in the deformational cycle.

4.4. Kinematic Indicators

Shear sense indicators in the D3 mylonites from Doubtful Sound include S-C and C-C' fabrics (Figs. 8b and 8c) as well as asymmetric tails on porphyroclasts, most of which give a consistent top-to-the-northeast sense of shear (Gibson et al., 1988). They indicate that the western Fiordland orthogneiss has been dragged upward from beneath the overlying cover sequence.

Although S-C fabrics are widely distributed in amphibolite facies mylonites throughout Hall Arm, the S-surfaces are not always readily discernible and it is the C-surfaces which define the dominant foliation. However, as both surfaces are defined by the same amphibolite facies mineralogy (Fig. 8d), they are interpreted to have formed penecontemporaneously under the same metamorphic conditions. According to the criteria developed by Lister and Snoke (1984), these rocks would be classified as type II S-C mylonites.

4.5. Low-angle shear zones in the cover sequence

A northeast sense of shear is also shown by the Wilmot Fault (Fig. 3). This structure strikes northwest at high angles to the regional D3 stretching lineation and is one of the few low-angle shear zones in the cover sequence for which a normal (top-to-northeast) sense of shear can be unequivocally demonstrated. It dips northeast and juxtaposes sillimanite grade schists (Mt Barber) against deeper structural levels (kyanite grade rocks) to the south. Unfortunately neither the age of this fault nor its relationship to other low-angle shear zones in the cover sequence is well known. However, the Wilmot Fault cannot be older than the Early Cretaceous Mt George Gabbro which it truncates south of Mt Barber (Fig. 3). Nor can it be younger than the high-angle faults of inferred mid-Tertiary age (Norris et al., 1978) which displace both it and other low-angle shear zones in the Doubtful Sound area (Fig. 3). A mid-Cretaceous to Early Tertiary age for the Wilmot Fault is therefore indicated.

Other low-angle shear zones in the Doubtful Sound area have been interpreted as thrust faults developed after the peak of regional metamorphism and its associated D2 deformation (Gibson, 1982b; see also Oliver, 1980). They have traces marked by fault breccias and cataclasites and at the very least must have been reactivated during the D3 deformation. However, in view of their post-D2 age, they too may include structures of extensional origin.

5. AGE OF METAMORPHISM AND DEFORMATION

Recently published geochronological data (Oliver, 1980; Mattinson et al.,

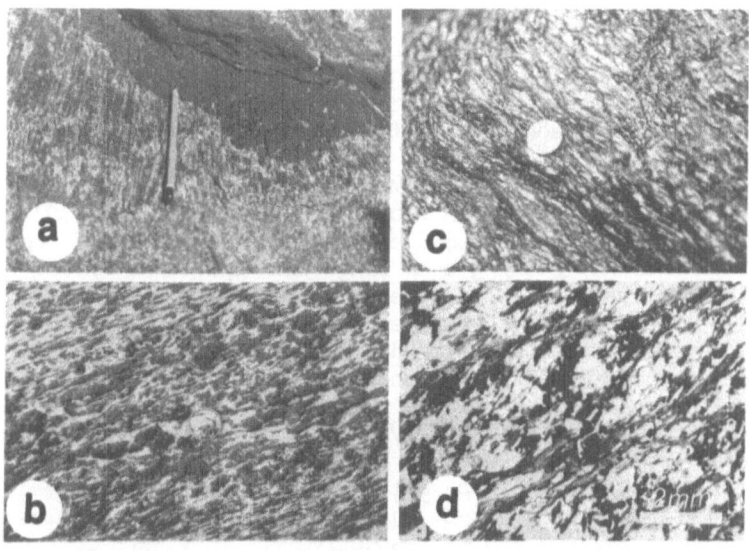

Figure 8. (a) Metamorphosed ultramafic dyke oriented at high angles to subhorizontal D3 stretching fabric developed in partially retrogressed granulites, Crooked Arm. Mineral assemblage in dyke records only amphibolite facies metamorphism. (b) Streaky, flat-lying mylonitic foliation in partially retrogressed garnet granulite, Crooked Arm. Note secondary shear zone with sinistral (top-to-the-northeast) sense of shear (northeast to left in the photograph). (c) S-C fabrics in amphibolite facies mylonite, Hall Arm. Sense of shear is dextral (top-to-the-northeast). (d) Hornblende (dark) and biotite (light) oriented parallel to S- and C-surfaces (less steeply dipping) in D3 mylonite, Hall Arm.

1986) have firmly established metamorphism and deformation in Fiordland as polyphase (see also Wood, 1972). These data not only indicate a high temperature Cretaceous event but a significant component dating from the Tuhua Orogeny or earlier. Thus migmatized hornblende gneisses from Deep Cove have yielded a Rb/Sr whole rock age of 388 + 23 Ma (Oliver, 1980) whilst a synmetamorphic tonalitic orthogneiss from the same locality has been dated at 386 + 10 Ma (U/Pb zircon age; Gibson et al., 1988). Hornblende crystals in these gneisses and the immediately adjacent tonalitic orthogneiss are oriented parallel to the D2 regional foliation and give K-Ar ages of 265 Ma and 242 Ma respectively (Gibson et al., 1988). However, neither date gives more than a minimum age for the D2 deformation and its associated metamorphism because hornblende in these rocks incurred significant loss of radiogenic argon during a subsequent mid-Cretaceous thermal event (Gibson et al., 1988). The least disturbed hornblende samples derive from amphibolites interstratified with sillimanite + potash feldspar-bearing pelitic schists in the Merrie Range and Mt Barber areas at higher structural levels in the cover sequence.

They give (Fig. 9) cooling ages of 316-342 Ma (Gibson et al., 1988) and demonstrate that low-pressure sillimanite grade regional metamorphism in central Fiordland is primarily a manifestation of the mid-Palaeozoic Tuhua Orogeny rather than the late Mesozoic Rangitata Orogeny (cf. Mattinson et al., 1986; McCulloch et al., 1987; Bradshaw, 1989).

Progressively younger K-Ar mineral ages are encountered towards the contact with the western Fiordland orthogneiss (Fig. 9). In the Black Giants Anorthosite and deeper structural levels of Deep Cove Gneiss, hornblende cooling ages range from 160-98 Ma whilst the corresponding values for biotite are 125-80 Ma (Gibson et al., 1988; see also Aronson, 1968). However, because rocks from these deeper levels have undergone the same deformational history and share the same deformational fabrics as the rest of the cover sequence, it follows that this trend towards younger ages must be a secondary effect. The mineral isotopic systems at these structural levels were disturbed to varying degrees by a late mesozoic thermal event and in some cases lost most, if not all, of their original radiogenic argon. The two most important exceptions are the hornblende samples from Deep Cove dated at 242 and 265 Ma respectively. They were only partially degassed and thus retain some vestige of their original pre-Mesozoic thermal history.

Granulite facies orthogneisses in the underlying western Fiordland orthogneiss have given a Rb/Sr whole rock isochron age of 120 Ma (McCulloch et al., 1987) and U/Pb zircon ages of 120-126 Ma (Mattinson et al., 1986; Gibson et al., 1988). These have been interpreted as the age of the granulite facies protolith and compare with a 116 Ma zircon age inferred to be the maximum age for the cessation of granulite facies metamorphism (Mattinson et al., 1986). This 116 Ma age was derived from an amphibolitized granulite from southern Fiordland and is in good agreement with a 119 Ma age obtained from a similarly retrogressed (but mylonitized) granulite from Crooked Arm. However, because neither this zircon sample nor the 116 Ma zircons analysed by Mattinson et al. (1986) showed any evidence for inherited components, these data could also represent the age of the granulite protolith. Alternatively the U/Pb zircon system was completely reset during granulite and/or amphibolite facies metamorphism. In either case these data give an upper age limit of about 116-119 Ma for the D3 deformation. A hornblende K-Ar cooling age of 93 Ma obtained from a D3 mylonite gives a minimum age for this same event (Gibson et al., 1988).

6. CRUSTAL EVOLUTION IN FIORDLAND

6.1. Previous Tectonic Models

Although geological maps of Fiordland have been available since the early 1960s (Wood, 1960; 1962), it is only within the last decade or so that Oliver and Coggon (1979) published the first tectonic model for the region. They suggested Fiordland was made up of two components (Table 1): a Precambrian granulite facies basement and a

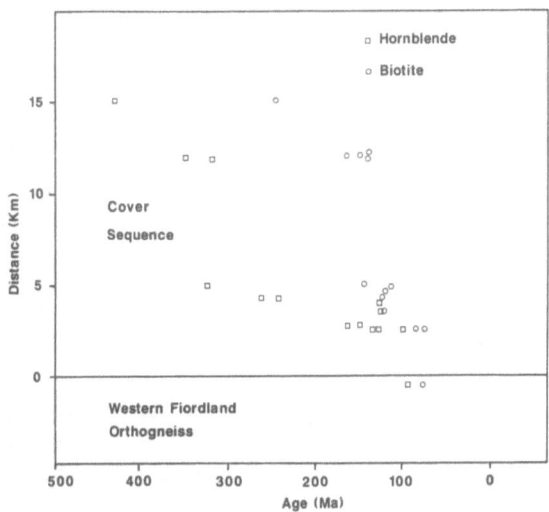

Figure 9. K-Ar mineral ages for rocks of cover sequence plotted against distance from contact with western Fiordland orthogneiss. Data from Gibson et al. (1988).

structurally overlying mid-Palaeozoic plutonic-metasedimentary cover sequence which was separated from the former by a major structural discontinuity or decollement. This discontinuity was identified as the Doubtful Sound Thrust (cf. Oliver, this volume) and on this basis Oliver and Coggon (1979) proposed that the cover sequence had been thrust westward over the granulite facies basement during the Tuhua Orogeny. However, it is now clear that both this shear zone and the underlying granulites are of Early Cretaceous age. Kinematic indicators throughout this shear zone are also consistent with an eastward rather than westward sense of tectonic transport for the cover sequence.

Mattinson et al. (1986; see also Bradshaw, 1989; McCulloch et al., 1987; Bradshaw and Kimbrough, 1989) proposed a very different tectonic model for Fiordland (Table 1). They argued for an intrusive relationship between the cover sequence and western Fiordland orthogneiss and attributed virtually all regional metamorphism in Fiordland north of Dusky Sound to magmatic heating accompanying synkinematic intrusion of the granulite protolith. Metamorphism of the granulite protolith and its host rocks initially occurred at mid-crustal levels involving low- to medium-pressure metamorphism with the development of two-pyroxene feldspathic orthogneisses in the western

TABLE 1.   Recently proposed tectonic models for Fiordland

Model 1:   Cover sequence has been thrust westwards over a granulite
           facies basement of inferred Precambrian age.   Contact
           between basement and cover is identified as the Doubtful
           Sound Thrust (Oliver and Coggon, 1979; Oliver, 1980).

Model 2:   Granulite protolith intruded into the cover sequence during
           the Early Cretaceous; no major structural break between
           cover and the underlying granulites (western Fiordland
           orthogneiss) (Mattinson et al., 1986; McCulloch et al.,
           1987; Bradshaw, 1989).

Model 3:   Western Fiordland orthogneiss and the cover sequence are
           separated by a ductile shear zone of extensional origin
           related to late Mesozoic continental rifting and breakup of
           the Pacific margin of Gondwana ("metamorphic core-complex
           model") (Gibson et al., 1988; this paper).

Fiordland orthogneiss and sillimanite-bearing mineral assemblages in
the cover sequence.   Thereafter rocks of the cover sequence and
western Fiordland orthogneiss were subjected to high pressure
metamorphism induced by regional overthrusting and concomitant
crustal thickening brought about by an island arc-continent or
continent-continent collision (see especially McCulloch et al., 1987).
This led to overprinting of the earlier assemblages by garnet
granulite in the western Fiordland orthogneiss and kyanite grade
metamorphism in the cover sequence.   Thus according to Mattinson
et al. (1986) there have been two episodes of Early Cretaceous
metamorphism in Fiordland.

A major difficulty for this model is that low- to medium-pressure
metamorphism in the cover sequence is now known to have occurred during
the mid-Palaeozoic Tuhua Orogeny (Gibson et al., 1988; 1989) and
therefore cannot be related to any metamorphic event in the underlying,
and much younger, western Fiordland orthogneiss.   Moreover, if the
sillimanite grade rocks of central Fiordland were ever subjected to
increased lithostatic pressures following an episode of regional
overthrusting there is little or no evidence for this preserved within
their mineral assemblages, at least not within samples collected in
the Merrie Range (Fig. 3).   Sillimanite and sillimanite + potash
feldspar + cordierite mineral assemblages are still widely distributed
in the Merrie Range and occur in both regionally metamorphosed rocks
and the contact aureoles of Early Cretaceous gabbros (Mt George
Gabbro).   Neither these assemblages nor the gabbros show any sign of
having re-equilibrated during higher pressure metamorphism.   Kyanite

is unknown in this region although it does occur occasionally in sillimanite grade rocks on Mt Barber. Either the rocks of the Merrie Range were never sufficiently deeply buried during overthrusting for such re-equilibration to take place or else these rocks are part of the overthrust sheet itself. Alternatively there are inherent problems in the Mattinson et al. (1986) model and some other explanation must be sought to account for the geology of the Doubtful Sound area.

A third model (Table 1) was put forward by Gibson et al. (1988; see also Kamp and Hegarty, 1989). They confirmed earlier suggestions (Oliver and Coggon, 1979; Oliver, 1980) that the western Fiordland orthogneiss and cover sequence in Doubtful Sound were everywhere separated by a major shear zone but in contrast to Oliver and Coggon (1979) they proposed that this structure was of extensional, rather than compressional, origin. In their model (Gibson et al., 1988) the western Fiordland orthogneiss and its overlying cover sequence represented lower and upper crustal plates of a mid-Cretaceous metamorphic core complex whose emplacement was linked to continental rifting and the late Mesozoic breakup of Gondwana. This model is developed further here but first it is necessary to review the evidence for continental extension in other parts of western New Zealand, particularly in the once-contiguous regions of Nelson and Westland (Fig. 1).

6.2. Mid-Cretaceous crustal extension in Nelson and Westland

Late Mesozoic sedimentary basins in the Nelson-Westland region have long been attributed (Laird, 1980) to continental rifting and breakup of the Pacific margin of Gondwana. These basins are typically asymmetric and take the form of fault-angle depressions or half-grabens (Fig. 4) bounded by NNE-NNW-trending normal faults whose hanging walls are commonly downthrown to the east or northeast (Laird, 1980; Nathan et al., 1986; Tulloch and Kimbrough, in press). Sedimentation in these basins was nonmarine (Pororari Group) and coeval with high-level silicic magmatism (Fig. 10) as evidenced by the presence of dacitic tuffs near the base of the Pororari Group (Nathan et al., 1986). These tuffs have been interpreted (Adams and Nathan, 1978) as the eruptive phase of the nearby 110 Ma Berlins Porphyry which together with various other high-level intrusions, their associated Lower Palaeozoic host rocks and the Pororari Group comprise the upper plate of a mid-Cretaceous metamorphic core complex (Tulloch and Kimbrough, in press). Upper plate rocks are in low-angle normal fault contact with a mylonitized and ductiley deformed lower plate in which late Precambrian paragneiss and granites of both mid-Palaeozoic and Early Cretaceous age predominate. Among other Early Cretaceous granites in the Lower plate is the 96-99 Ma Buckland Granite (Fig. 10), clasts of which occur towards the top of the sedimentary sequence in the Pororari Group (Nathan et al., 1986). Thus normal faulting, sedimentation and emplacement of the metamorphic core complex were all contemporaneous and occurred during the period

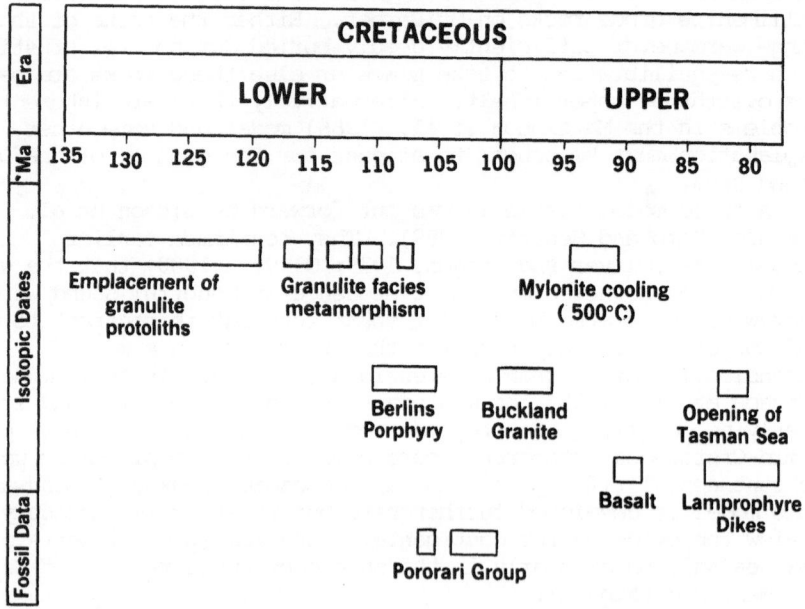

Figure 10. Summary of chronological data for western province (modified from Nathan et al., 1986).

from about 110 Ma through to at least 95 Ma and probably beyond (cf. Tulloch and Kimbrough, in press).

Superimposed upon lower plate mylonites in the Nelson-Westland region is a conspicuous northeast- to southwest-trending stretching lineation (Tulloch and Kimbrough, in press). It is of Early Cretaceous age and has the same trend and character as the D3 stretching lineations (Fig. 7) in Fiordland. It therefore provides an important link with the mylonitic fabrics of D3 age in the western Fiordland orthogneiss. These D3 fabrics give 93 Ma K-Ar cooling ages and cannot be older than their plutonic and/or granulite facies protoliths dated at 116-126 Ma (Mattinson et al., 1986; Gibson et al., 1988). They evidently formed during the same time interval as the mylonites in Nelson and Westland, indicating that if the core complexes of Fiordland and Nelson-Westland are not part of one and the same structure (Fig. 11), then they at least developed within the same late Mesozoic tectonic regime (cf. Gibson et al., 1989).

6.3. Core-Complex Model for Fiordland

It is evident from the previous sections, especially sections 4 and 5, that the western Fiordland orthogneiss and its overlying cover sequence comprise rocks with very different structural and metamorphic histories. They constitute separate structural entities and with the possible exception of northern Fiordland where an original intrusive relationship is apparently preserved (Bradshaw, 1989; Bradshaw and Kimbrough, 1989) are everywhere separated by a major shear zone in which Early Cretaceous mylonites have developed. Whilst several workers (eg Bradshaw and Kimbrough, 1989) have argued that this shear zone is little more than a tectonised intrusive contact, several lines of evidence militate against such a conclusion and point instead to it being a much more fundamental structure related to continental extension: (1) this shear zone juxtaposes rocks metamorphosed under very different pressures (12 kbar for the western Fiordland orthogneiss as opposed to 7-9 kbar for the immediately overlying cover sequence) indicating some 10-12 km of crust has been excised during deformation; (2) the mylonites are of extensional origin and developed at a time when continental extension/rifting was already well advanced in western New Zealand; (3) shear sense in the mylonites (hangingwall downthrown to the northeast) matches the sense of asymmetry (Nathan et al., 1986; Tulloch and Kimbrough, in press) developed in many of the late Mesozoic sedimentary basins of Nelson and Westland, regions once contiguous with Fiordland. Such relations are unlikely to be coincidental unless there was a genetic link between these different aspects of the late Mesozoic geology; (4) the mylonites are only one manifestation of a much more pervasive D3 deformation which affected virtually all crustal levels in Fiordland and was associated with significant thinning and attenuation of the cover sequence; (5) mylonites of comparable age and trend occur in Nelson and Westland and have been similarly linked to the emplacement of a metamorphic core complex. Given all these factors, the case for uplift and emplacement of the granulites in an extensional environment is difficult to refute. Fiordland, together with the once-contiguous regions of Nelson and Westland, exhibit striking similarities with the metamorphic core complexes of the North American Cordillera and have an infrastructure which probably developed in much the same way.

One such model to explain this infrastructure is illustrated in Figure 11 and has the western Fiordland orthogneiss as part of the exhumed lower plate. The granulites were dragged upward from beneath the hangingwall along a shallow-dipping ductile shear zone in much the same manner as postulated for metamorphic core complexes elsewhere (Wernicke, 1981; 1985; Davis and Lister, 1988; Lister and Davis, 1989). This model not only provides a ready explanation for the mylonites and their associated D3 structures but it also accounts for the prevalence of S-C fabrics and other asymmetric microstructures (eg asymmetric tails on garnet porphyroclasts) throughout these rocks. Such asymmetries are to be expected in a shear zone where

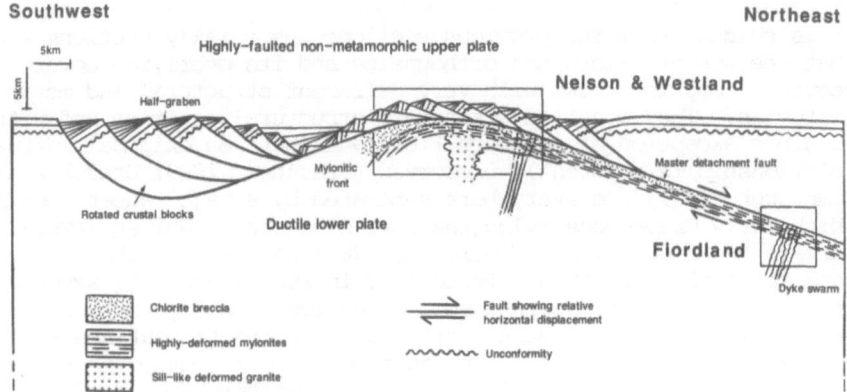

Figure 11. Section through western New Zealand in mid-Cretaceous time showing idealized metamorphic core complex (after Lister and Davis, 1989) and the different crustal levels represented by Fiordland and the Nelson-Westland region.

non-coaxial laminar flow (simple shear) was probably the dominant deformational process (Lister and Snoke, 1984).

Mylonitization of lower plate rocks during emplacement of the western Fiordland orthogneiss was accompanied by hydration and retrogressive metamorphism of the granulites to the amphibolite facies. Similar retrograde effects have been observed in other metamorphic core complexes (Crittenden et al., 1980; Rehrig, 1986; Reynolds and Lister, 1986) and are not altogether unexpected because mylonitization inevitably leads to greater fluid mobility by maintaining a dynamically-induced permeability (Etheridge et al., 1983). By such means water was able to penetrate to considerable depths in the western Fiordland orthogneiss thereby promoting both increased dynamic recrystallization and a change towards greater ductility in the lower plate. This in turn may have facilitated rapid uplift of the core complex. Uplift rates as high as 2-3.5 mm yr^{-1} have been calculated for northern Fiordland (McCulloch et al., 1987) although similar calculations for Doubtful Sound indicate uplift was no higher than 1-1.5 mm yr^{-1} (cf. Oliver, this volume). This value is based on the assumption that (1) high pressure metamorphism took place at depths of 45 km between 116-110 Ma (Mattinson et al., 1986; McCulloch et al., 1987) and (2) amphibolite facies retrograde metamorphism and cooling of the lower plate mylonites below 500°C (equivalent to a depth of 17 km assuming a geothermal gradient of 30°C/km) occurred within 17-23 Ma as evidenced by a hornblende K-Ar cooling age of 93 Ma (Gibson et al., 1988).

Emplacement of the core complex brought high temperature lower plate tectonites into contact with a cooler amphibolite facies cover sequence which had undergone significant thinning following subvertical D3 flattening. However, whilst this may in part explain the trend towards younger mineral ages in the cover sequence (Fig. 9), it is doubtful whether tectonic emplacement of the western Fiordland orthogneiss in itself could have raised isotherms in the upper plate to the extent needed for widespread resetting of the mineral isotopic systems. It is much more likely that some other process acting in concert with continental extension was responsible for this pattern. One obvious possibility is heating accompanying intrusion of the granulite protolith but this is seemingly precluded by metamorphic and isotopic data (Mattinson et al., 1986; Bradshaw, 1989) indicating that granulite facies metamorphism in Fiordland was linked to deformation and events accompanying the earlier Rangitata Orogeny (section 6.1).

It is, nevertheless, interesting to note that garnet granulite in Fiordland is almost entirely confined to veins and dykes that bear a striking resemblance to tensional joint systems (Blattner, 1976; Oliver, 1977). In fact Blattner and Black (1980) have already likened these veins and dykes to hydraulic fractures initiated by decompression. It is tempting to suggest that these high pressure granulites owe little to the Rangitata Orogeny but instead developed during the incipient stages of continental rifting.

Another possibility to explain resetting of the mineral isotopic systems is heating following magmatic underplating and/or the upwelling of hotter mantle material at depth. Wood (1972) reported a large body of mylonitized peridotite in northern Fiordland and minor dykes and sills of ultramafic rocks are common throughout the Doubtful Sound area (Fig. 4; Oliver, 1980). Moreover, some of these dykes and sills were emplaced during the course of D3 deformation (Fig. 8a). These observations lend support to the idea that crustal extension was accompanied by deep-seated magmatic activity although at this stage it is still uncertain whether the ultramafic rocks of northern Fiordland represent part of a dismembered intrusive body or a slice of tectonised upper mantle.

7. UPLIFT AND EXHUMATION OF THE LOWER CRUSTAL SECTION IN DOUBTFUL SOUND

K-Ar and Rb-Sr biotite ages for the deepest levels of the cover sequence and immediately underlying western Fiordland orthogneiss range from 78-95 Ma (Gibson et al., 1988; Aronson, 1968). Mattinson et al. (1986) reported comparable U/Pb apatite ages of 78-93 Ma for the western Fiordland orthogneiss. Assuming an average crustal gradient of $30^{\circ}C/km$ and blocking temperatures of $300^{\circ}C$ and $400^{\circ}C$ for the biotite and apatite isotopic systems, these data indicate that the western Fiordland orthogneiss was uplifted to within 10-12 km of the surface by 80 Ma. The greater part of uplift and cooling of the western Fiordland orthogneiss must therefore have occurred during the interval of continental rifting immediately preceding (Fig. 10) opening of the Tasman Sea at about 82 Ma (Weissel et al., 1977). This is consistent with the proposition (Gibson et al., 1988; 1989) that continental rifting and emplacement of the western Fiordland

orthogneiss were concomitant processes; both resulted from
continental extension associated with late Mesozoic breakup of the
Pacific margin of Gondwana (cf. Tulloch and Kimbrough, in press).

Figure 11 pertains only to the situation that existed in mid-
Cretaceous time before development of the mid-Tertiary Alpine Fault
(Fig. 1). Prior to development of this structure the western
Fiordland orthogneiss remained buried as evidenced by the fact that
no Fiordland-derived detritus with the appropriate high grade
metamorphic mineralogy has been reported from the late Mesozoic-early
Tertiary sedimentary sequences that border the region (Wood, 1960;
Carter and Lindquist, 1977; Suggate et al., 1978).

In fact final uplift and exhumation of the western Fiordland
orthogneiss is not likely to have commenced much before 10 Ma when
there was a marked acceleration in compressive motion across the
fault (Walcott, 1979; Norris and Carter, 1980; Allis, 1986).
Throughout the early stages of its development from late Oligocene
through Miocene times (Carter and Norris, 1976; Kamp, 1986),
displacement on the Alpine Fault was predominantly strike-slip
thereby leading mainly to fragmentation and dispersion of the core
complex along the plate boundary (see Fig. 1). However, within the
last 10 Ma Fiordland has been subjected to a much greater rate of
uplift. Whilst some of this uplift may be related to crustal
shortening across the region (as evidenced by D4 post-metamorphic
folding of the core complex and its associated detachment surface),
most was probably generated during thrusting of the Pacific plate
(of which Fiordland is part) over and above oceanic crust of the
Tasman Sea. This process commenced at 7 Ma (Davey and Smith, 1983)
and continues to the present day.

8. CONCLUSIONS

High pressure Early Cretaceous granulite facies orthogneisses in
Fiordland, New Zealand represent the deeper levels of a mid-Cretaceous
core complex which was emplaced from beneath a cover sequence
comprising mainly multiply-deformed, Late Proterozoic-Lower Palaeozoic
sediments and volcanic rocks metamorphosed to the amphibolite facies.
Collectively these various lithologies represent one of the deepest
exposed crustal sections in the world. However, compared to most
other regionally metamorphosed granulite terrains, Fiordland
possesses neither great antiquity nor has it had a protracted uplift
history involving crustal residence times of several hundred million
years. Rather uplift has been extremely rapid, for a time exceeding
1 mm yr^{-1}, and involved two separate stages. The first stage was
associated with attenuation and tectonic denudation of the cover
sequence accompanying emplacement of the core complex whilst the
second stage occurred mainly within the last 10 Ma in response to
transpression across the mid-Tertiary Alpine Fault and/or the
thrusting of Fiordland crust over oceanic crust of the Tasman Sea.

9. ACKNOWLEGEMENTS

This work was supported by the Australian Research Grants Scheme and the Darling Downs Institute Research Services and has benefitted from comments and discussions offered by Drs. J.D. Bradshaw, J.Y. Bradshaw, C.M. Ward and R.J. Norris. For incisive and constructive reviews of the manuscript I thank Dr B. Wernicke and an anonymous reviewer. My sincere thanks also to the organisers of the NATO conference on Exposed Crustal Sections, and in particular Professors M. Salisbury and D. Fountain for their invitation and support to attend the meeting in Ontario, Canada. Attendance costs were defrayed by a grant from the Natural Sciences and Engineering Research Council of Canada. For support in the field I thank Mr D. Gleeson. Miss Marie Schulz typed the manuscript.

10. REFERENCES

Adams, C.J.D. and Nathan, S. 1978. Cretaceous chronology of the Lower Buller Valley, South Island, New Zealand. N.Z. J. Geol. Geophys., 21: 455-462.

Allis, R.G. 1986. Mode of crustal shortening adjacent to the Alpine Fault, New Zealand. Tectonics, 5: 15-32.

Aronson, J.L. 1965. Reconnaissance rubidium-strontium geochronology of New Zealand plutonic and metamorphic rocks. N.Z. J. Geol. Geophys., 1: 401-423.

Aronson, J.L. 1968. Regional geochronology of New Zealand. Geoch. et Cosmoch. Acta, 32: 669-697.

Bishop, D.G. 1967. The geology of the powerhouse access tunnell, West Arm, Manapouri. N.Z. J. Geol. Geophys., 10: 831-38.

Bishop, D.G., Bradshaw, J.D. and Landis, C.A. 1985. Provisional terrane map of South Island, New Zealand: in Howell, D.G. (ed.): Tectonostratigraphic Terranes of the Circum-Pacific Region: 515-521. Circum-Pacific Council for Minerals and Energy, Houston, Texas.

Blattner, P. 1976. Replacement of hornblende by garnet in granulite facies assemblages near Milford Sound, New Zealand. Contrib. Mineral. Petrol., 55: 181-90.

Blattner, P. 1978. Geology of crystalline basement between Milford Sound and the Hollyford Valley, New Zealand. N.Z. J. Geol. Geophys., 21: 33-47.

Blattner, P. and Black, M.B. 1980. Apatite and scapolite as petrogenetic indicators in granulites of Milford Sound, New Zealand. Contrib. Mineral. Petrol., 74: 339-48.

Bradshaw, J.D., Andrews, P.B. and Adams, J.G. 1981. Carboniferous to Cretaceous on the Pacific margin of Gondwana: The Rangitata phase of New Zealand: in Cresswell, M.M. and Vella, P. (eds.): Gondwana Five: 217-221. Balkema, Rotterdam.

Bradshaw, J.Y. 1989. Origin and metamorphic history of an Early Cretaceous polybaric granulite terrain, Fiordland, southwest New Zealand. Contrib. Mineral. Petrol., 103: 346-360.

Bradshaw, J.Y. and Kimbrough, D.L. 1989. Comment on age constraints on metamorphism and the development of a metamorphic core complex in Fiordland, southern New Zealand. Geology, 17: 381-82.

Brodie, K.H. and Rutter, E.H. 1987. Deep crustal extensional faulting in the Ivrea Zone of Northern Italy. Tectonophys., 140: 193-212.

Carter, R.M. and Lindquist, J.K. 1977. Balleny Group, Chalky Island, southern New Zealand: an inferred Oligocene submarine Canyon and fan complex. Pacific Geology, 12: 1-46.

Carter, R.M. and Norris, R.J. 1976. Cainozoic history of southern New Zealand: an accord between geological observations and plate-tectonic predictions. Earth Planet. Sci. Lett., 31: 85-94.

Coombs, D.S., Landis, C.A., Norris, R.J., Sinton, J.M., Borns, D.J. and Craw, D. 1976. The Dun Mountain Ophiolite Belt, New Zealand, its tectonic setting constitution and origin, with special reference to the southern section: Am. J. Sci., 276: 561-603.

Cooper, R.A. 1979. Lower Palaeozoic rocks of New Zealand. J. R. Soc. N.Z., 9: 29-84.

Crittenden, M.D. Jr., Coney, P.J. and Davis, G.H. (eds.). 1980. Cordilleran metamorphic core complexes. Geol. Soc. Amer. Mem., 115: 490 pp.

Davey, F.J. and Smith, E.G.C. 1983. The tectonic setting of the Fiordland region, southwest New Zealand. Geophys. J. Roy. Astro. Soc., 72: 23-38.

Davis, G.A. and Lister, G.S. 1988. Detachment faulting in continental extension; perspectives from the southwestern U.S. Cordillera. Geol. Soc. Am. Special Paper, 218: 133-59.

Etheridge, M.A., Wall, V.J. and Vernon, R.H. 1983. The role of the fluid phase during regional metamorphism and deformation. J. Metamorphic Geol., 1: 205-226.

Fountain, D.M. and Salisbury, M.H. 1981. Exposed cross-sections through the continental crust: implications for crustal structure, petrology and evolution. Earth Planet. Sci. Lett., 56: 263-277.

Gibson,. G.M. 1979. Metamorphites at Wilmot Pass, Fiordland, New Zealand. Unpubl. Ph.D. Thesis, University of Otago, 315 p.

Gibson, G.M. 1982a. Stratigraphy and petrography of some metasediments and associated intrusive rocks from central Fiordland, New Zealand. N.Z. J. Geol. Geophys., 25: 21-43.

Gibson, G.M. 1982b. Polyphase deformation and its relationship to metamorphic crystallization in rocks at Wilmot Pass, Fiordland, New Zealand. N.Z. J. Geol. Geophys., 25: 45-65.

Gibson, G.M. 1988. Exposed cross-section of continental crust: Doubtful Sound, Fiordland, New Zealand: Part 2 - interpretation as a deeply eroded Mesozoic metamorphic core complex. (Abstract). NATO Advanced Studies Programme on Exposed Crustal Sections of the Continental Crust: p. 10. Killarney, Ontario, Canada.

Gibson, G.M., McDougall, I. and Ireland, T.R. 1988. Age constraints on metamorphism and the development of a metamorphic core complex in Fiordland, southern New Zealand. Geology, 16: 405-408.

Gibson, G.M., McDougall, I. and Ireland, T.R. 1989. Reply to comment on age constraints on metamorphism and the development of a metamorphic core complex in Fiordland, southern New Zealand. Geology, 17: 381-382.

Grindley, G.W. 1978. West Nelson: in Suggate, R.P., Stevens, G.R. and Te Punga, M.T. (eds.): The Geology of New Zealand. Government Printer, 2 volumes, 820 p.

Grindley, G.W. 1980. Sheet S13 Cobb (1st ed.). Geological Map of New Zealand. 1:63 360. Map (1 sheet) and notes (48 p.). N.Z. Dept. Sci. Industrial Research, Wellington.

Handy, M.R. 1987. The structure, age and kinematics of the Pogallo Fault Zone; southern Alps, northwestern Italy. Eclog. Geolog. Helv., 80: 593-632.

Harrison, T.M. and McDougall, I. 1980. Investigations of an intrusive contact, northwest Nelson, New Zealand - I. Thermal, chronological and isotopic constraints. Geochim. Cosmochim. Acta, 44: 1985-2003.

Hodges, K.V. and Fountain, D.M. 1984. Pogallo Line, South Alps, northern Italy: An intermediate crustal level, long-angle normal fault? Geology, 12: 151-155.

Howell, D.G. 1980. Mesozoic accretion of exotic terranes along the New Zealand segment of Gondwanaland. Geology, 8: 487-491.

Kamp, P.J.J. and Hegarty, K.A. 1989. Multigenic gravity couple across a modern convergent margin: inheritance from Cretaceous asymmetric extension. Geophys. Jl., 96: 33-41.

Kamp, P.J.J. 1986. The mid-Cenozoic Challenger Rift System of Western New Zealand and its implications for the age of Alpine Fault inception: Bull. Geol. Soc. Am., 97: 255-81.

Laird, M.G. 1980. The late Mesozoic fragmentation of the New Zealand segment of Gondwana. Fifth International Gondwana Symposium, Wellington, New Zealand: 311-318.

Landis, C.A. and Coombs, D.S. 1967. Metamorphic belts and orogenesis in southern New Zealand. Tectonophys., 4: 501-518.

Lister, G.S. and Davis, G.A. 1989. The origin of metamorphic core complexes and detachment faults formed during Tertiary continental extension in the northern Colorado River region, U.S.A. J. Struct. Geol., 11: 65-94.

Lister, G.S., Etheridge, M.A. and Symonds, P.A. 1986. Detachment faulting and the evolution of passive continental margins. Geology, 14: 246-50.

Lister, G.S. and Snoke, A.W. 1984. S-C mylonites. J. Struct. Geol., 6: 617-38.

McCulloch, M.T., Bradshaw, J.Y. and Taylor, S.R. 1987. Sm-Nd and Rb-Sr isotopic and geochemical systematics in Phanerozoic granulites from Fiordland, southwest New Zealand. Contrib. Mineral. Petrol., 97: 183-195.

Mackinnon, T.C. 1983. Origin of the Torlesse terrane and coeval rocks, South Island, New Zealand, Geol. Soc. Amer. Bull., 94: 967-985.

Mattinson, J.M., Kimbrough, D.L. and Bradshaw, J.Y. 1986. Western Fiordland orthogneiss: Early Cretaceous arc magmatism and granulite facies metamorphism, New Zealand. Contrib. Mineral. Petrol., 92: 383-392.

Moecher, D.P., Essene, E.J. and Anovitz, L.M. 1988. Calculation and application of clinopyroxene-garnet-plagioclase-quartz geobarometers. Contrib. Mineral. Petrol., 100: 92-106.

Nathan, S., Anderson, H.J., Cook, R.A., Herzer, R.H., Hoskins, R.H., Raine, J.I. and Smale, D. 1986. Cretaceous and Cenozoic sedimentary basins of the West Coast region, South Island, New Zealand. Wellington, New Zealand Geological Survey Basin Studies 1, 90 pp.

Newton, R.C. and Perkins, D. 1982. Thermodynamic calibration of geobarometers based on the assemblages garnet-plagioclase-orthopyroxene (clinopyroxene)-quartz. Am. Mineral., 67: 203-22.

Norris, R.J. and Carter, R.M. 1980. Offshore sedimentary basins at the southern end of the Alpine Fault, New Zealand. Spec. Publ. Int. Ass. Sediment., 4: 237-65.

Norris, R.J., Carter, R.M. and Turnbull, I.M. 1978. Cainozoic sedimentation in basins adjacent to a major continental transform boundary in southern New Zealand. Jl. Geol. Soc. Lond., 135: 191-205.

Oliver, G.J.H. 1977. Feldspathic hornblende and garnet granulites and associated anorthosite pegmatites from Doubtful Sound, Fiordland, New Zealand. Contrib. Mineral. Petrol., 65: 111-21.

Oliver, G.J.H. 1980. Geology of the granulite and amphibolite gneisses of Doubtful Sound, Fiordland, New Zealand. N.Z. J. Geol. Geophys., 23: 27-41.

Oliver, G.J.H. and Coggon, J.H. 1979. Crustal structure of Fiordland. Tectonophys., 54: 253-292.

Percival, J.A. and Card, K.D. 1983. Archean crust as revealed in the Kapuskasing uplift, Superior Province, Canada. Geology, 11: 323-26.

Rehrig, W.A. 1986. Processes of regional Tertiary extension in the western Cordillera: insights from the metamorphic core complexes, in Mayer, L. (ed.): Extension tectonics of the southwestern United States: a perspective on processes and kinematics. Geol. Soc. Am. Special Paper 208: 97-122.

Reynolds, S.J. and Lister, G.S. 1987. Structural aspects of fluid-rock interactions in detachment zones. Geology, 15: 362-66.

Suggate, R.P., Stevens, G.R. and Te Punga, M.T. 1978. The Geology of New Zealand. Government Printer, Wellington, 2 volumes, 820 pp.

Tulloch, A.J. 1983a. Granitoid rocks of New Zealand - A Review: in Roddick, J.A. (ed.), Circum-Pacific plutonic terranes: Geol. Soc. Am. Memoir, 159: 5-20.

Tulloch, A.J. 1983b. Tectonic settings and source rocks for Palaeozoic and Mesozoic granitoid rocks of New Zealand. Geol. Soc. N.Z. Program with Abstracts, Auckland, New Zealand.

Tulloch, A.J. and Kimbrough, D.L. In press. The Paparoa metamorphic core complex, Westland-Nelson, New Zealand: Cretaceous extension associated with fragmentation of the Pacific margin of Gondwana. Tectonics.

Turner, F.J. 1937. The metamorphic and plutonic rocks of Lake Manapouri, Parts I, II, III. Trans. Roy. Soc. N.Z., 67: 83-100, 227-49; 68: 122-40.

Turner, F.J. 1968. Metamorphic petrology, mineralogical and field aspects. McGraw-Hill Inc., 403 pp.

Walcott, R.I. 1979. Plate motions and shear strains in the vicinity of the Southern Alps. Bull. R. Soc. N.Z., 18: 5-12.

Ward, C.M. 1984. Geology of the Dusky Sound area, Fiordland, with emphasis on the structural-metamorphic development of some porphyroblastic staurolite pelites. Unpubl. Ph.D. Thesis, University of Otago, Dunedin, New Zealand.

Weissel, J.K., Hayes, D.E. and Herron, E.M. 1977. Plate tectonic synthesis: The displacements between Australia, New Zealand and Antarctica since the late Cretaceous. Marine Geol., 25: 231-277.

Wernicke, B. 1981. Low-angle normal faults in the Basin and Range province; nappe tectonics in an extending orogen. Nature, 291: 645-648.

Wernicke, B. 1985. Uniform-sense normal simple shear of the continental lithosphere. Can. J. Earth Sci., 22: 108-25.

Williams, J.G. 1983. The Hollyford Gabbronorite - a calc-alkaline cumulate. N.Z. Jl. Geol. Geophys., 26: 345-357.

Williams, J.G. and Harper, C.T. 1978. Age and status of the Mackay Intrusives in the Eglinton-Upper Hollyford area. N.Z. J. Geol. Geophys., 21: 733-42.

Wood, B.L. 1960. Sheet 27, Fiord Geological Map of New Zealand, 1:250,000. D.S.I.R., Wellington, New Zealand.

Wood, B.L. 1962. Sheet 22, Wakatipu Geological Map of New Zealand, 1:250,000. D.S.I.R., Wellington, New Zealand.

Wood, B.L. 1972. Metamorphosed ultramafites and associated deformation near Milford Sound, New Zealand. N.Z. J. Geol. Geophys., 16: 88-127.

A CRUSTAL CROSS-SECTION FOR A TERRAIN OF SUPERIMPOSED SHORTENING AND
EXTENSION: RUBY MOUNTAINS-EAST HUMBOLDT RANGE METAMORPHIC CORE
COMPLEX, NEVADA

A.W. Snoke, A.J. McGrew, P.A. Valasek, and S.B. Smithson
Department of Geology and Geophysics
Program for Crustal Studies
University of Wyoming
Laramie, WY 82071-3006
U.S.A.

ABSTRACT. Geologic mapping coupled with geochronological,
thermobarometric, and seismic reflection studies provide the data for
constructing a crustal cross-section through a Tertiary extensional
orogen in the eastern Great Basin, western U.S. Cordillera.
Oligocene-Miocene sedimentary and volcanic rocks were deposited
during brittle, upper-level crustal extension dominated by high-angle
normal faults, rotation of the strata and the faults themselves, and
the progressive evolution of a low-angle detachment fault system.
Together with nonmetamorphosed to very low-grade rocks of the
Cordilleran miogeocline, the synextensional strata comprise an upper
crustal suprastructure that was attenuated during Tertiary crustal
extension. Structurally below the suprastructure and commonly
separated from it by a regional detachment fault is a transitional
metasedimentary/granitoid zone which preserves a principally Mesozoic
magmatic, metamorphic, and deformational history. In turn, this zone
grades downward abruptly into a 1.5- to 2.0-km-thick, upper
amphibolite facies Tertiary shear zone that forms the top of a
mylonitic to nonmylonitic migmatitic infrastructure of variable
age. This infrastructural zone clearly records a complex Mesozoic
history, but is in part characterized by a Tertiary magmatic-
metamorphic-deformational history that appears to increase in
intensity with depth.
     Seismic reflection data collected across the Ruby-East Humboldt
complex indicate a highly reflective character throughout the
crust. Nonetheless, distinct zones of strong seismic reflectivity
occur at 4 and 6 s and are interpreted to be zones of intense plastic
deformation that probably acted as middle crustal decoupling zones.
Middle and lower crustal velocities derived from wide-angle seismic
reflection data suggest the presence of mafic rocks interpreted as
additions during crustal extension; and these underplated mafic
intrusions together with their wallrocks probably experienced
granulite facies metamorphism during Cenozoic crustal extension.
Taken as a whole, this crustal cross-section suggests that noncoaxial

M. H. Salisbury and D. M. Fountain (eds.), Exposed Cross-Sections of the Continental Crust, 103–135.
© 1990 *Kluwer Academic Publishers.*

strain was principally partitioned along distinct km-scale
extensional shear zones, but when viewed on the scale of the entire
middle and lower crust this deformation probably constituted a bulk
coaxial strain regime that generated a distinctive crustal
extensional metamorphic fabric.

1. INTRODUCTION

In the hinterland of the Sevier orogenic belt of the western U. S.
Cordillera, metamorphic complexes commonly occur as domiform
footwalls to regionally extensive plastic-to-brittle, normal-sense
shear zones. The rocks of the hanging wall, shear zone, and footwall
comprise a distinct petrotectonic assemblage traditionally referred
to as a "Cordilleran metamorphic core complex". An important aspect
of these complexes is that various structural levels of the
hinterland have been juxtaposed, and consequently constitute a
natural, albeit much attenuated, cross-section from brittlely
deformed upper crust (suprastructure) to a plastically deformed
middle crust (infrastructure). Nonetheless, a simple suprastructure-
infrastructure model for these complexes is quite inadequate, for the
various tectonic levels almost invariably display a polyphase
magmatic, metamorphic, and deformational history that reflects a
complex evolution involving Mesozoic contraction and superimposed
Cenozoic extension.

In the Ruby Mountains-East Humboldt Range in northeastern Nevada
(Fig. 1) the combined effects of doming, late Cenozoic Basin and
Range block faulting, and Pleistocene glaciation have produced a
superbly exposed terrain of contrasting structural levels (Fig. 2).
These disparate crustal levels were juxtaposed along regionally
extensive plastic-to-brittle, normal-sense fault zones during
Oligocene to Miocene crustal extension. These large-scale structural
characteristics qualify the Ruby Mountains-East Humboldt Range as a
classic example of a Cordilleran metamorphic core complex.

The purposes of this paper are: 1) to describe the various parts
of an exposed partial crustal cross-section as manifested in the
northern Ruby Mountains and East Humboldt Range; and 2) to evaluate
the implications of this exposed crustal cross-section for the
crustal structure of the eastern Great Basin. The achievement of
these objectives is further facilitated by the availability of
vertical incidence and wide-angle seismic reflection data collected
from this area (Valasek et al., 1987, 1989).

2. WHAT IS A CORDILLERAN METAMORPHIC CORE COMPLEX?

The concept of a Cordilleran metamorphic core complex has evolved
greatly during the 1970s and 1980s as increasing emphasis was placed
on detailed studies of these enigmatic terranes. The now-classic
core complexes of the western North American Cordillera appear to
represent the first well-studied examples of a newly recognized

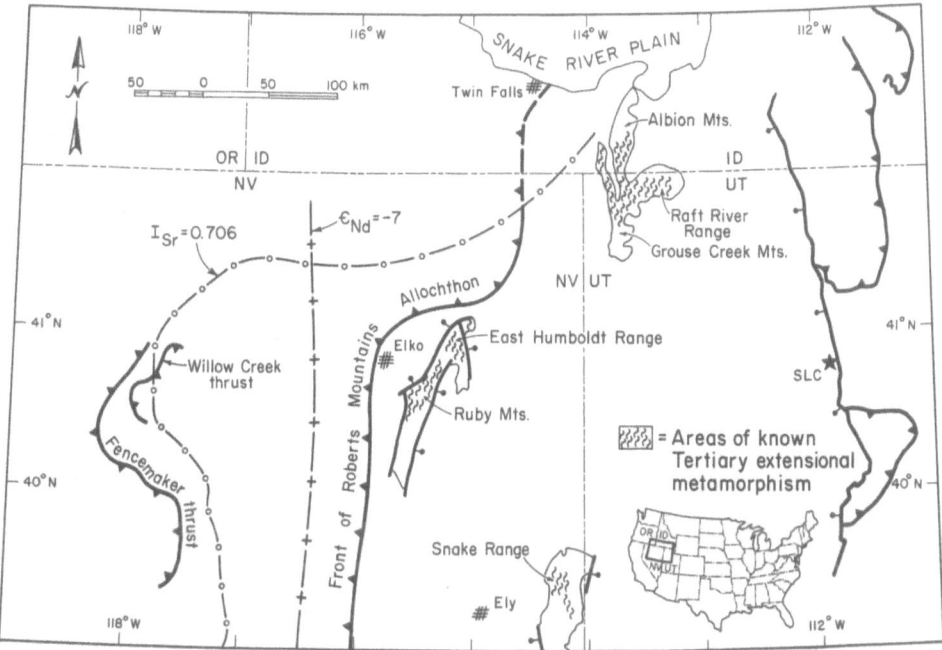

FIGURE 1. Generalized geologic map of the northern Basin and Range
Province, western U. S. Cordillera, showing the distribution of
exposed Tertiary extensional metamorphic terrains including the Ruby
Mountains-East Humboldt Range, northeastern Nevada. Some important
geochemical and structural boundaries are plotted to provide a
regional reference frame. Base map was King and Beikman (1974).
Boundaries from Farmer (1988), R. W. Kistler (personal comm., 1987),
Stewart (1980), and Speed et al. (1988).

tectonic phenomenon. Their evolution clearly involves important
crustal extension, invariably associated with syntectonic magmatism
and metamorphism. A manifestation of this extreme crustal extension
is the division of the crustal column into three distinct zones: a
footwall of igneous and metamorphic rocks; a plastic-to-brittle, low-
angle normal-sense fault zone; and a brittlely attenuated hanging
wall. The footwall can consist of Precambrian crystalline basement
rocks, metamorphosed supracrustal rocks, or even Tertiary plutonic
rocks (Coney, 1979). The Ruby-East Humboldt(R-EH) complex is
particularly interesting in that all three of these elements are
present and comprise a composite, deep-seated footwall that provides
a "window" into the deepest exposed crust of the eastern Great
Basin. Perhaps the greatest challenge in deciphering the tectonic
evolution of core complexes is to distinguish features that developed
during the pre-extensional history from features that developed
during extension.

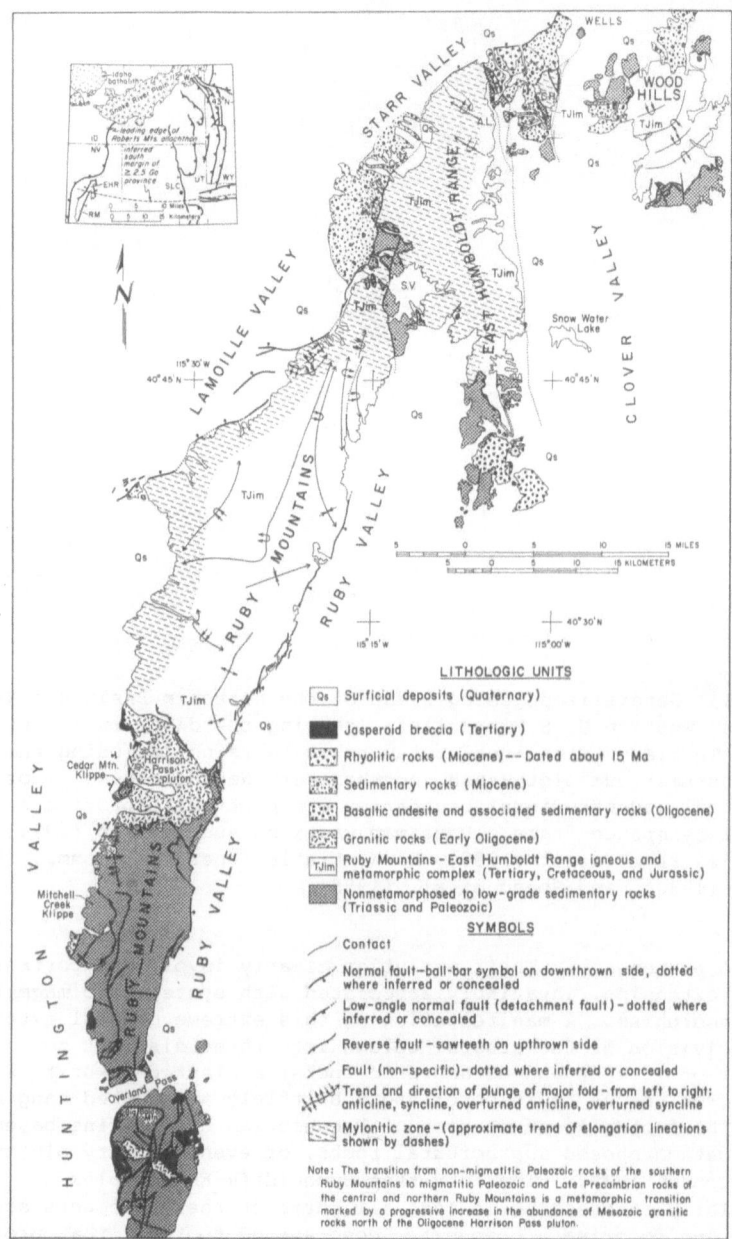

FIGURE 2. Generalized geologic map of the East Humboldt Range, Wood
Hills, Ruby Mountains, and Big Bald Mountain, Nevada (after Sharp,
1939; Snelson, 1957; Thorman, 1970; Hose and Blake, 1976; Howard et
al., 1979; Coats, 1987). A.L.=Angel Lake, C.H.=Clover Hill,
S.V.=Secret Valley.

3. STUDY OF AN ATTENUATED CRUSTAL CROSS-SECTION

3.1. Structural observations

The most fundamental structural features in the R-EH complex are the
R-EH mylonitic shear zone (Fig. 3) and the shallower and brittle R-EH
detachment fault (Fig. 4). The Tertiary R-EH shear zone is a
kilometer-scale zone (1.5 to 2.0 km thick in the northern Ruby
Mountains – East Humboldt Range) of WNW-directed, normal-sense noncoaxial strain superimposed on rocks that range in age from Archean
to Oligocene and in metamorphic grade from upper amphibolite facies
in the north to lower greenschist facies in the south. As the R-EH
shear zone is traced southward along strike, the character of the
footwall changes drastically from a gneiss/metasedimentary complex in
the East Humboldt Range to low-grade to nonmetamorphosed lower
Paleozoic rocks in the southern Ruby Mountains (Snoke et al.,
1990). The hanging-wall rocks, however, are relatively similar along
strike consisting chiefly of unmetamorphosed, brittlely deformed
rocks ranging in age from Carboniferous to Tertiary. In the
southeastern East Humboldt Range, an interlayered sequence of
metasedimentary and granitic rocks forms an extensive terrain
structurally above the mylonitic shear zone (Figs. 4 and 5). This
metasedimentary/granitoid terrain is structurally overlain and

FIGURE 3. Exposures of flaggy, mylonitic and migmatitic Cambrian
Prospect Mountain Quartzite, Echo Canyon, northern Ruby Mountains.

108

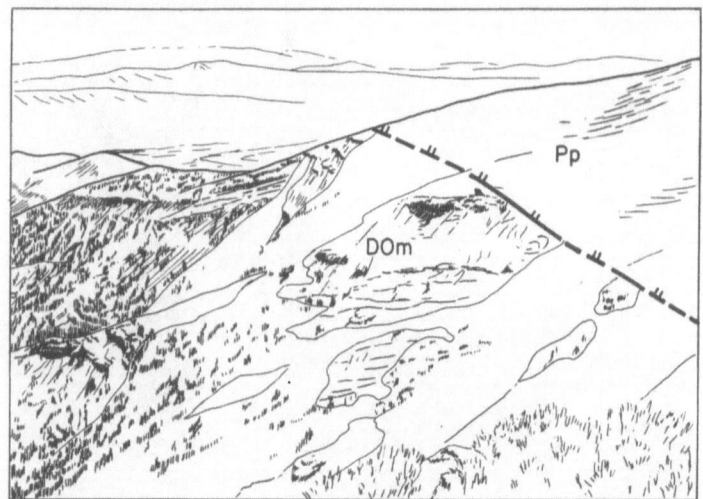

FIGURE 4. Detachment fault in the southeastern East Humboldt Range separating a hanging-wall plate of unmetamorphosed Lower Permian rocks (Pp=Pequop Formation) from a footwall of lower amphibolite-facies, middle to lower Paleozoic metasedimentary rocks (DOm) of the Cordilleran miogeoclinal sequence (i.e., structural top of the transitional metasedimentary/granitoid zone).

truncated by the R-EH detachment fault with a hanging wall consisting
of very low-grade to nonmetamorphosed rocks ranging in age from
Carboniferous to Oligocene (Fig. 5). The metasedimentary/granitoid
terrain is an extensive tract of primarily Mesozoic magmatic,
metamorphic, and deformational features; an analogous Mesozoic
terrain forms the lower metamorphosed plate in the Wood Hills (Fig.
2; Thorman, 1970).

   In many localities, however, this transitional
metasedimentary/granitoid terrain has been severely overprinted and
attenuated by a complex system of Tertiary plastic-to-brittle
anastomosing fault zones. In these localities, the transitional zone
essentially constitutes an extensional duplex structure of fault-
bounded lithic slices with the brittle R-EH detachment forming the
roof fault (Gibbs, 1984). Overlying this detachment fault is a
complexly faulted zone of low-grade to nonmetamorphosed
suprastructure. At these highest structural levels a thick sequence
of Upper Paleozoic to Lower Triassic miogeoclinal rocks and overlying
syntectonic Tertiary sedimentary and volcanic rocks has been severely
attenuated along several generations of closely spaced, shallowly to
steeply inclined normal faults.

3.2. Geochronometric observations

Numerous geochronometric studies (Kistler et al., 1981; Dallmeyer et
al., 1986; Dokka et al., 1986; Wright and Snoke, 1986; Lush et al.,
1988) throughout the Ruby Mountains-East Humboldt Range clearly
indicate a polyphase tectonothermal evolution for the igneous-
metamorphic complex. The Mesozoic magmatic-metamorphic history is

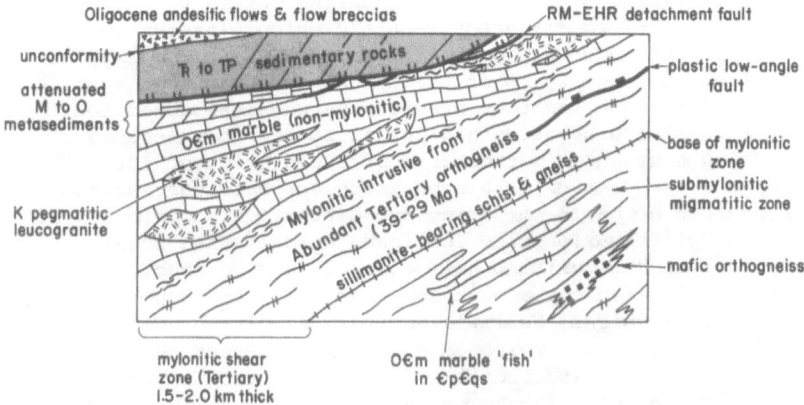

FIGURE 5. Diagrammatic interpretative cross-section for the
southeastern East Humboldt Range as derived from the unpublished
geologic mapping of A. W. Snoke and G. K. Taylor and U-Th-Pb
geochronometric studies of J. E. Wright.

still incompletely known, but Late Jurassic (ca 155 Ma) and Late
Cretaceous (ca 80 Ma) igneous events are now well documented by
zircon and monazite U-Th-Pb dates (J. E. Wright and coworkers, work
in progress). Amphibolite-facies regional metamorphism and
deformation clearly accompanied the Jurassic magmatism (Kistler et
al., 1981; Hudec and Wright, 1990). The details of the Cretaceous
history are still very sketchy in the northern Ruby Mountains and
East Humboldt Range, but regional relations suggest that important
Cretaceous metamorphism and deformation were probably synchronous
with magmatism (Miller and Gans, 1989). The sum of this poorly-
constrained history is that during the mid- to late Mesozoic the R-EH
terrain was part of an overthickened crustal section that formed a
vast tract in the hinterland of the Sevier orogenic belt (Coney and
Harms, 1984).

The Tertiary magmatic-metamorphic-deformational history has been
documented by a combination of U-Th-Pb, ^{40}Ar/^{39}Ar, and fission-track
geochronometric techniques. Tertiary igneous rocks consist of
andesitic to rhyolitic volcanic rocks in the synextensional
supracrustal sequence, scattered basalt dikes, a locally deformed
composite granitoid pluton (Harrison Pass pluton, Fig. 2), and quartz
dioritic to granitic plutonic orthogneisses in the igneous-
metamorphic complex.

3.3. Thermobarometric observations

Thermobarometric studies are still in an embryonic stage in the R-EH
complex and adjacent terrains (e.g., Clover Hill and the Wood Hills,
Fig. 2). Nonetheless, phase petrology plus the presently available
data allow some preliminary statements to be made. The most detailed
studies are by Hurlow (1987, 1988) who applied quantitative
thermobarometry to estimate the P-T regime during Tertiary
mylonitization. His P-T estimate for Tertiary mylonitization is ca
600±50°C and 400±50 MPa (Hurlow, personal comm., 1990). The samples
were obtained from the lower part of the mylonitic shear zone exposed
in the East Humboldt Range and thus provide the physical conditions
of a stage of a complex exhumation history along the normal-sense
shear zone. Using a standard lithostatic pressure gradient (26
MPa/km, Carter and Tsenn, 1987), this pressure estimate corresponds
to an inferred depth of ca 15 km during mylonitization. Obviously as
the footwall was uplifted along the inclined, normal-sense shear
zone, the physical conditions must have changed, and no doubt,
substantial mylonitization could have occurred at lower temperatures
and pressures. Nevertheless, Hurlow's estimate serves as a reference
point in the framework of our restored crustal cross-section.

A migmatitic, amphibolite-facies terrain forms much of the
footwall of the core complex in the northern Ruby Mountains and East
Humboldt Range. Sillimanite is widespread in metapelitic rocks
throughout this terrain, and locally this sillimanite appears to
postdate an earlier kyanite-bearing assemblage. We interpret much of
the sillimanite growth to be the product of Tertiary high-grade
metamorphism during crustal extension, but the kyanite-bearing

assemblage probably records metamorphic conditions that developed during Mesozoic crustal shortening and magmatism. This Mesozoic history is best preserved in two areas east of the East Humboldt Range -- the Wood Hills (Thorman, 1970; Thorman and Snee, 1988) and Clover Hill (Snoke, 1989). Metamorphic mineral assemblages from the Wood Hills and Clover Hill both contain kyanite + garnet + biotite; however, in the Wood Hills embayed staurolite porphyroblasts are present, whereas at Clover Hill sillimanite is intergrown with biotite. The reaction Q + Mu + St = Ky + Gt + Bt + vapor was probably critical in petrogenesis at both localities, but at Clover Hill this reaction has proceeded to completion, whereas in the Wood Hills it has not. Thus Clover Hill may have experienced somewhat higher temperatures than the Wood Hills during Mesozoic metamorphism, but both assemblages indicate pressures in excess of bathograd 5 (P > 4.8 kb; Carmichael, 1978). Furthermore, accessory rutile forms inclusions in garnet at both localities (Snoke, 1989; observation by B. Ronald Frost, 1989) indicating that the GRAIL equilibrium applies. GRAIL is a well-calibrated geobarometer that normally yields pressure estimates in excess of 6 kb for typical mineral assemblages (Bohlen et al., 1983).

3.4. Geophysical observations

Seismic reflection data collected along a traverse from North Ruby Valley through Secret Valley and along Secret Creek gorge (Fig. 2) indicate a very reflective crust in the northern Ruby Mountains-East Humboldt Range (Figs. 6 and 7; Valasek et al., 1989). The reflectivity is especially pronounced in the middle to lower crust but is also manifested in the upper crust by an anastomosing reflection fabric that dips west or east. Geologic mapping near the seismic reflection traverse suggests that the upper crustal reflections are related to the presently exposed mid-Tertiary mylonitic shear zone and recumbently-folded migmatitic infrastructure (Valasek et al., 1989). The fold-nappes of the infrastructure are of uncertain age but clearly have been much intruded, flattened, and reworked during mid-Tertiary crustal extension.

Insight into the nature of the deeper reflections is provided by wide-angle reflection data (Fig. 8) that were obtained to supplement the vertical incidence reflection traverse (Valasek et al., 1987). The wide-angle data has been used to develop a crustal velocity model for the R-EH complex (Table 1).

4.   COMPONENTS OF A CRUSTAL CROSS-SECTION FOR THE RUBY-EAST HUMBOLDT COMPLEX

4.1. Synextensional and older rocks of the suprastructure

An Oligocene-Miocene sequence of volcanic and sedimentary rocks is an important component of the suprastructure. Originally these rocks were deposited unconformably on unmetamorphosed Upper Paleozoic to

112

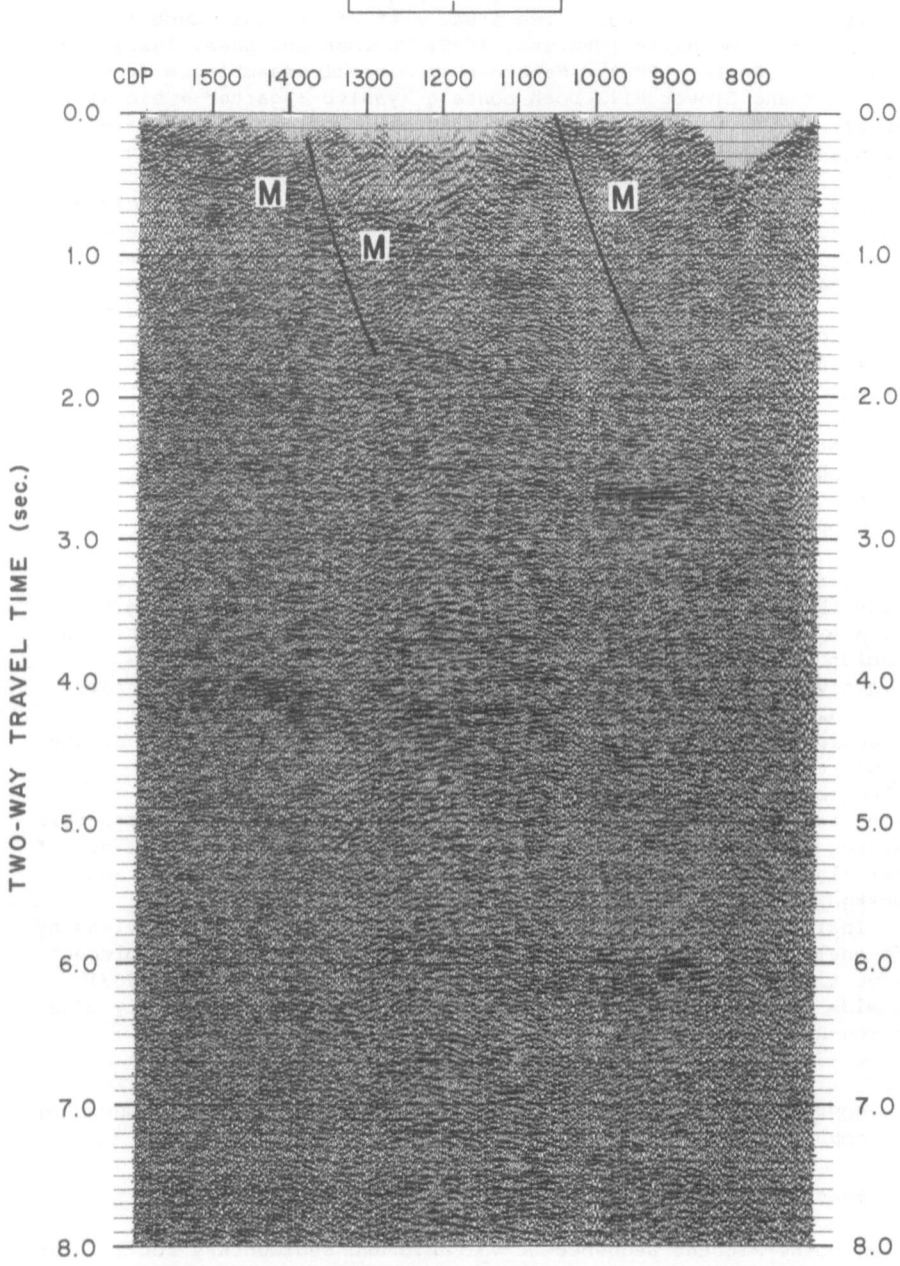

FIGURE 6. CDP 48-fold seismic section from the northern Ruby Mountains-southwestern East Humboldt Range (see Valasek et al., 1989). M is the mylonite zone reflection set.

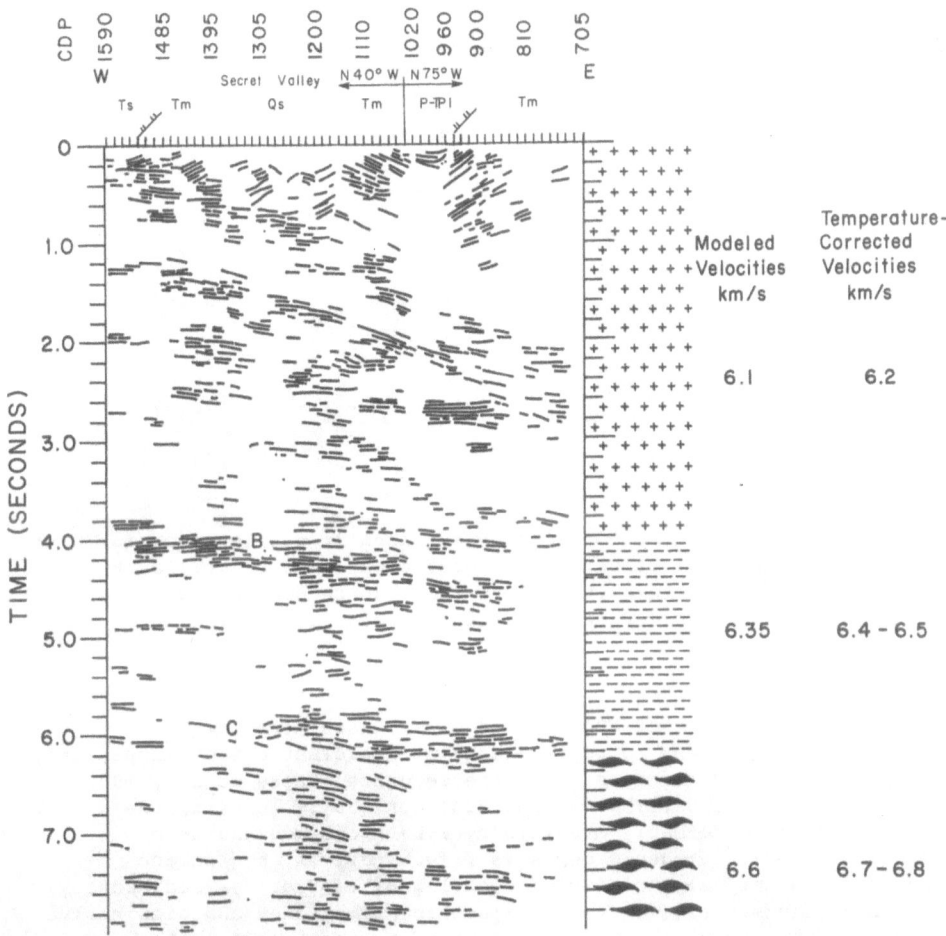

FIGURE 7. Line drawing of seismic section (after Valasek et al., 1989) from Figure 6. Reflections are discussed in the text.

TABLE 1. VELOCITY STRUCTURE, NORTHERN RUBY MOUNTAINS-EAST
HUMBOLDT RANGE, NEVADA

Discontinuity	Depth(km)	Velocity (km/s)	T-corrected velocity*
	0.0		
		3.80	
	1.5		
		5.50	
A	3.0		
		6.10	6.20
B	12.5		
		6.35	6.4-6.5
C	19.5		
		6.6	6.7-6.8
X	28.0		
		7.2-7.5	7.4-7.8
M	31.0		
		7.8	7.9-8.1

*Temperature-corrected velocity is based on estimates of the
temperature derivative and the geothermal gradient (Valasek et
al., 1987) to obtain velocities that can be compared with room-
temperature velocity measurements on rock samples in the
laboratory.

Lower Mesozoic rocks of the eastern Great Basin miogeoclinal
succession. Oligocene to Upper Eocene volcaniclastic rocks apparently
comprise the initial deposits of the sequence (Snoke et al., 1983;
Snoke and Lush, 1984; Thorman and Brooks, 1988). In turn, these
rocks are disconformably overlain by a heterogeneous suite of
synextensional tectogenic deposits (lower part of the Miocene
Humboldt Formation) that include conglomerate (Fig. 9), sedimentary
breccia, sandstone, siltstone, lacustrine limestone, and distinctive
lenses of megabreccia. This sedimentary succession was interrupted
by an outpouring of 15 Ma rhyolite lavas, and both the older
sedimentary rocks and the lavas are intruded by masses of rhyolitic
quartz porphyry (ca 13.8 Ma -- K-Ar, sanidine, Snoke et al., 1983;
Snoke and Lush, 1984). This suite is in turn overlain by the upper
part of the Humboldt Formation which includes conglomerate,
sandstone, siltstone, tuff, and sparse limestone. Metamorphic rock
fragments (especially metamorphosed Ordovician Eureka Quartzite) are
a conspicuous component of the coarse detritus in the conglomeratic
deposits of the younger (post-15 Ma) Humboldt Formation. Mylonitic
rock fragments, however, do not become a conspicuous detrital
component until the Quaternary, forming ubiquitous flaggy clasts in
uplifted and dissected alluvial fan deposits.
     The other important component of the suprastructure is
nonmetamorphosed to very low-grade units that range in stratigraphic
age from Carboniferous to early Triassic and constitute part of the

eastern Great Basin miogeoclinal sequence. These units, like the
Oligocene-Miocene sequence, are cut by steeply- and shallowly-dipping
normal faults, but locally older-on-younger faulting is present in
the pre-Tertiary rocks of the suprastructure, suggesting important
pre-extensional imbrication of the miogeoclinal sequence. This
imbrication (i.e., thrust faulting) probably occurred during the
middle to late Mesozoic when lateral compression is inferred to be
the dominant regional stress system (Misch, 1960; Armstrong, 1968;
Snoke and Miller, 1988).

Nevertheless, brittle attenuation as a result of Cenozoic
crustal extension is the fundamental structural style of the
suprastructure. Normal faults of variable dip and inclination to
bedding have partitioned the suprastructure into a mosaic of fault-
bounded slices structurally separated from subjacent metamorphic
rocks by the R-EH detachment fault. The geometry of the fault
systems of this composite extensional allochthon indicate that
brittle attenuation involved polyphase rotation of originally
steeply-inclined normal faults to shallower orientations during the
progressive evolution of a plastic-to-brittle regional normal fault
system (e.g., Wernicke and Axen, 1988; Buck, 1988).

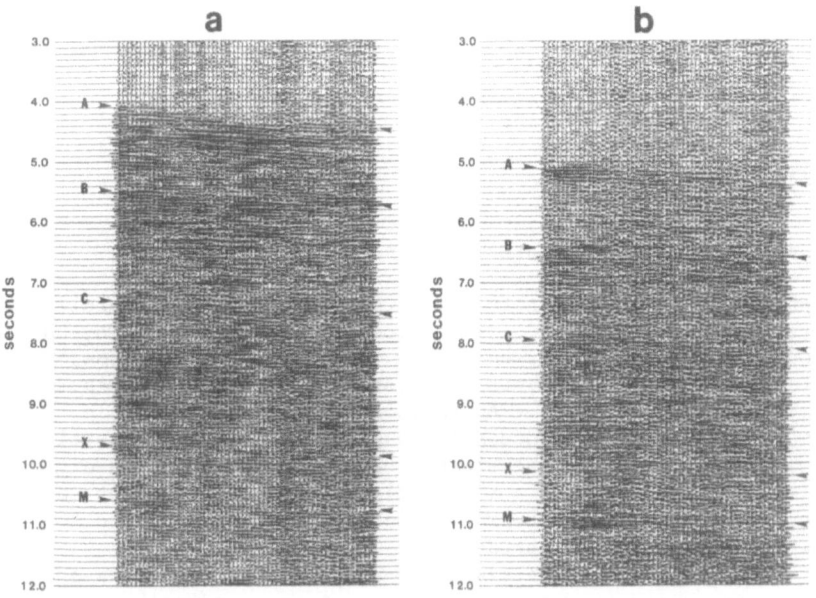

FIGURE 8. Examples of "wide-angle" shot gathers used to determine
crustal velocity in the R-EH core complex. (a) is an offset of 20 km
and (b) is offset of 26 km. Events A through M were used to
determine interval velocities in Table 1. M is Moho. (From Valasek
et al., 1989.)

FIGURE 9. Poorly-sorted fanglomerate deposit from Miocene
synextensional deposits, northeastern East Humboldt Range. Clasts
are exclusively middle to upper Paleozoic rocks of the Cordilleran
miogeoclinal sequence.

4.2. Transitional metasedimentary/granitoid zone

Metamorphosed Middle Cambrian to Lower Carboniferous(?) miogeoclinal
carbonate, siliciclastic, and pelitic rocks locally intruded by Upper
Cretaceous (J. E. Wright, personal comm., 1988) leucocratic muscovite
± garnet granite form a distinct structural zone in the southeastern
East Humboldt Range (Fig. 5). An analogous structural level is
extensively exposed in the Wood Hills and northern Pequop Mountains
(Thorman, 1970). In all these areas, the metamorphosed Paleozoic
rocks lie structurally below a regionally extensive detachment fault
which separates them from a suprastructure of nonmetamorphosed to
very low-grade miogeoclinal strata.

In the upper part of the transitional zone granitic rocks are
virtually absent; and they gradually become more conspicuous with
structural depth. Individual, mappable granitoid bodies occur as
intrusive sheets (Fig. 10) with complex, interdigitated wallrock
contacts, locally marked by massive skarns. Compositional layering
(i.e., relict bedding features parallel to metamorphic foliation, S_1)
in the metasedimentary rocks is commonly truncated by the intrusive
granitic masses. With increasing structural depth, distinct mappable
leucogranite bodies become progressively smaller in size and

eventually grade continuously into a deep-seated magmatic injection complex near the base of the transitional zone.

FIGURE 10. Muscovite leucogranite (on the left) intrusive contact with overlying Cambrian impure metacarbonate rocks, transitional metasedimentary/granitoid zone, southeastern East Humboldt Range.

4.3. Migmatitic mylonitic to nonmylonitic infrastructure

The migmatitic infrastructure of the Ruby-East Humboldt complex can be subdivided into four structural domains, each characterized by a distinctive association of structural features developed over a range of scales. In order of increasing structural depth these domains are:

1) Mylonitic, WNW-lineated Horse Creek assemblage;
2) Mylonitic, WNW-lineated quartzite-schist-gneiss unit;
3) Submylonitic, WNW-lineated metasedimentary/gneiss domain;
4) Nonmylonitic, variably lineated metasedimentary/gneiss domain.

It is important to emphasize that this domainal sequence represents a composite of structural features exposed in both the northern Ruby Mountains and East Humboldt Range. Domain 3, for instance, appears to be omitted or poorly developed in the northern Ruby Mountains, whereas domain 4 is imperfectly preserved only at the deepest structural levels observed in the East Humboldt Range. It should also be noted that in both the northern Ruby Mountains and in the East Humboldt Range, the structural organization delineated above is superimposed upon well-developed, large-scale recumbent fold-nappe complexes. We will first discuss the individual domains and then the large-scale fold relations upon which this structural organization was superimposed.

4.3.1. Mylonitic, WNW-lineated Horse Creek assemblage.
The contact between the transitional metasedimentary/granitoid zone and the mylonitic Horse Creek assemblage is a deformed intrusive contact between mylonitic, Tertiary plutonic orthogneiss and older metasedimentary and/or granitic wallrock (Fig. 5). The mylonitic intrusive front is gradational but rapid and mappable in the southeastern East Humboldt Range. At other localities, however, it is commonly overprinted by late brittle low-angle faulting, thereby juxtaposing slices of metasedimentary rocks against a mylonitic footwall of Horse Creek assemblage rocks (e.g., Secret Creek gorge area, see Snoke and Howard, 1984, STOP 12).

The Horse Creek assemblage consists of metasedimentary rocks and abundant intrusive igneous rocks ranging from pegmatitic leucogranite to various felsic and mafic orthogneisses. All the rocks of the Horse Creek assemblage have experienced a pervasive mylonitic deformation (Fig. 11) subsequent to an earlier history that included either amphibolite facies metamorphism and polyphase folding or igneous crystallization under mid-crustal conditions. The metasedimentary paleosome that served as wallrock for the intrusive igneous rocks consists chiefly of impure metacarbonate rocks considered to be metamorphosed equivalents of Cambrian and Ordovician miogeoclinal rocks of the eastern Great Basin.

4.3.2. Mylonitic, WNW-lineated quartzite-schist-gneiss unit.
The base of the Horse Creek assemblage is marked by an abrupt, sometimes faulted, transition into flaggy, upper amphibolite facies migmatitic micaceous feldspathic quartzite and pelitic schist that form the

FIGURE 11. Mylonitic foliation surface with strong WNW-trending
elongation lineation in granitic gneiss. The white part of the
foliation surface is a mylonitic quartz vein within the gneiss. Note
conjugate joint system associated with the elongation lineation. The
granitic gneiss is part of the Horse Creek assemblage, southwestern
East Humboldt Range.

upper part of a deeper level mylonitic strain facies in the R-EH
shear zone. The mylonitic rocks, including impure quartzite, pelitic
schist, marble, calc-silicate rock, and a variety of quartzofeldspathic gneisses, are generally characterized by strong,
moderately- to gently-dipping grain shape foliations and pervasive,
well-developed mineral elongation lineations with an azimuth of

approximately 290°. Abundant WNW-trending isoclinal folds of variable vergence, but commonly southward, characterize this domain. The mylonitization is superimposed on several major fold-nappe complexes, including the Winchell Lake fold-nappe in the northern East Humboldt Range and the Lamoille Canyon and Soldier Creek fold-nappes in the northern Ruby Mountains. A suite of 29 Ma biotite monzogranite sheets becomes increasingly abundant with structural depth in this domain, and these sheets are fully involved in mylonitization and partially involved in isoclinal folding (Wright and Snoke, 1986; McGrew and Snoke, 1988).

Grain-size reduction in these rocks has been moderate to intense, and extensive dynamic recrystallization is commonly evident in thin section. Composite fabrics, including both type I and type II S-C mylonites (Lister and Snoke, 1984), are widely developed and in approximately 90% of the cases indicate top-to-the-WNW normal-sense shear. Asymmetric mica fish (Lister and Snoke, 1984) and sigma-type feldspar porphyroclasts (Passchier and Simpson, 1986) are also generally compatible with top-to-the-WNW normal-sense shear. In addition, the impure quartzites exhibit well-developed crystallographic fabrics, and preliminary quartz c-axis microfabric determinations indicate top-to-the-WNW shear in consonance with all other kinematic indicators (A. J. McGrew, work in progress).

4.3.3. Submylonitic, WNW-lineated metasedimentary/gneiss domain. With increasing structural depth in the East Humboldt Range, the strongly mylonitic rocks described above grade continuously into a domain of coarse-grained gneisses which are commonly slabby to massive and often show relatively wavy or anastomosing foliation (Figs. 12A,B). The WNW-trending mineral elongation lineation continues to be an important fabric element but is less pervasive than at higher structural levels. WNW-trending isoclinal folds are abundant. The quantity of Tertiary orthogneiss appears to increase with structural depth, and the 29 Ma biotite monzogranite sheets show well-developed grain shape foliations and, as in the strongly mylonitic domain, are partially involved in isoclinal folding (Fig. 13).

Despite the fact that these rocks lack many of the features considered to be most characteristic of mylonitic deformation, there continues to be evidence that the deformational history involved a significant component of non-coaxial strain even at structural depths in excess of a kilometer beneath the gradation with the strongly mylonitic zone. Consequently, we consider that these rocks constitute a "submylonitic" strain facies developed at deep structural levels within the R-EH shear zone. Composite fabrics are uncommon, and preliminary quartz c-axis results suggest a weaker component of non-coaxial strain than in the overlying mylonitic zone (McGrew and Snoke, 1990). Nevertheless, shallowly inclined, normal-sense biotitic shear surfaces and high-grade plastic shear bands are extensively developed and commonly yield a reliable sense of displacement. Surprisingly, approximately 70% of all kinematic indicators in this domain yield an ESE-directed sense-of-shear, antithetic to the ambient shear sense in the superjacent mylonites.

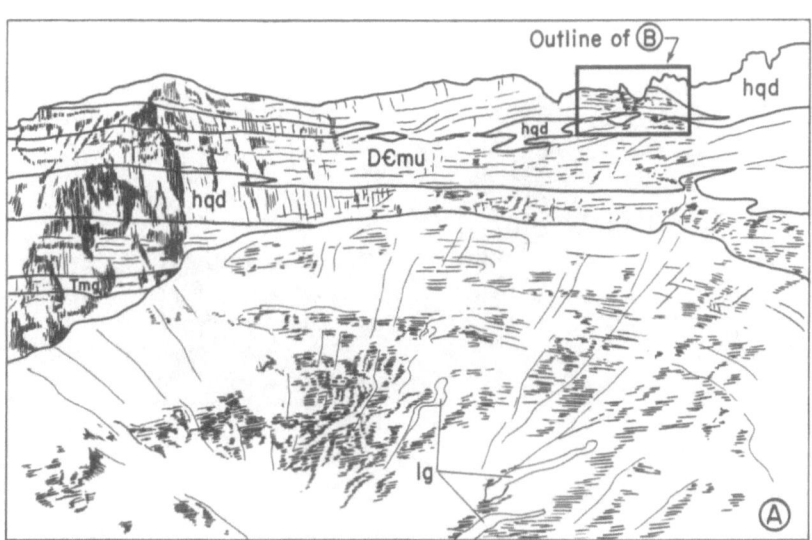

FIGURE 12A. Submylonitic, WNW-lineated metasedimentary/gneiss
domain, west wall of Lizzie's Basin cirque, central East Humboldt
Range. hqd = hornblende-biotite quartz diorite orthogneiss, lg =
pegmatitic leucogranite, DЄmu = lower Paleozoic metacarbonate rocks,
Tmg = Tertiary biotite monzogranitic orthogneiss.

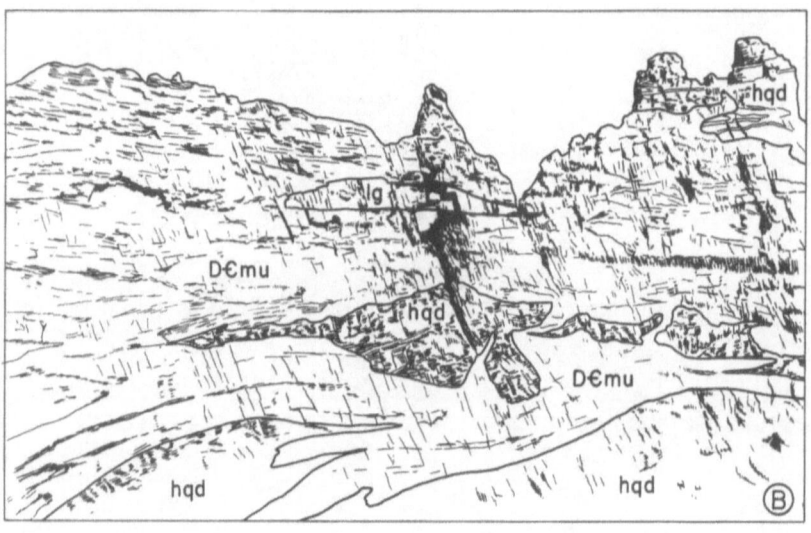

Figure 12B. Closeup of Hole-in-the-Mountain (elevation = 11,127 ft) and surrounding area showing strongly deformed mafic and felsic orthogneiss layers within migmatitic metasedimentary rocks. hqd = hornblende-biotite quartz dioritic orthogneiss, lg = pegmatitic leucogranite; DЄmu = lower Paleozoic metacarbonate rocks.

FIGURE 13. Sketch from photograph of folded 29 Ma biotite monzogranite sheet intruded into a strongly layered, leucosome-rich migmatite sequence, Angel Lake area, northern East Humboldt Range.

4.3.4. <u>Nonmylonitic, variably lineated metasedimentary/gneiss domain</u>. At the very deepest levels in the northern Ruby Mountains, the rocks are nonmylonitic but commonly contain an elongation lineation in pelitic horizons defined by oriented bundles of fibrolitic sillimanite (Howard, 1966). This lineation shows considerable variation in trend from approximately ENE to NS but a general maximum lies between NNE to NE with a shallow plunge (Howard, 1968; Stryhas, 1988). In the Soldier Creek area where the elongation lineation is approximately NS, Malavieille (1987) reported a southward sense-of-shear. On the northwall of Lamoille Canyon where the lineation is strongly localized in the NE quadrant, asymmetric sigma-type feldspar porphyroblasts indicate a southwestern sense-of-shear (Stryhas, 1988).

At the deepest structural levels exposed in the East Humboldt Range the proportion of leucogranite increases rather abruptly, commonly comprising in excess of 75% of the exposed rock, and the WNW-trending lineation becomes increasingly less systematically developed.

4.3.5. <u>Fold-nappe complexes</u>. In both the East Humboldt Range and northern Ruby Mountains the above described sequence of structural domains is superimposed on large-scale fold-nappe complexes,

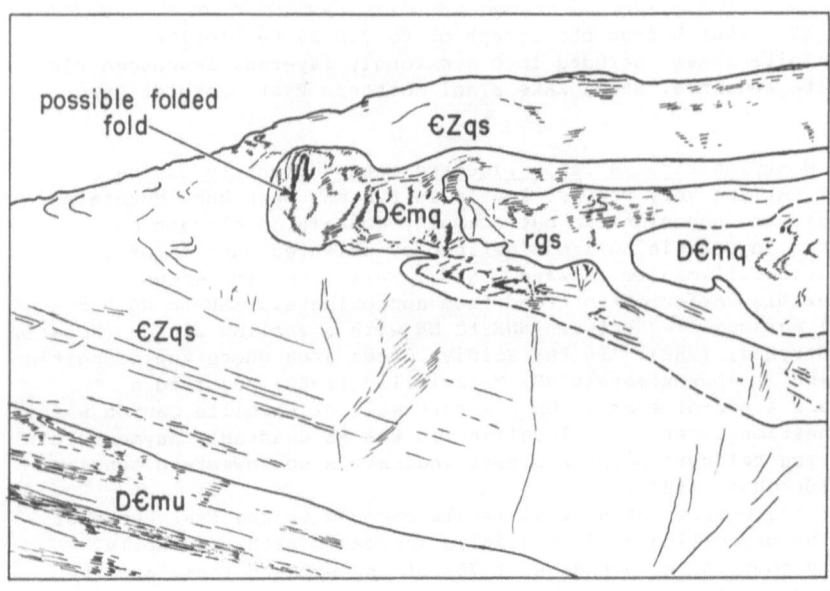

including the Winchell Lake fold-nappe in the northern East Humboldt
Range (Fig. 14) and the Lamoille Canyon (Fig. 15) and Soldier Creek
fold-nappes in the northern Ruby Mountains. In addition, in several
localities large "tectonic fish" of recognizable stratigraphic units
entirely encased within older stratigraphic units can be recognized
within the mylonitic shear zone. These relations suggest that some
large recumbent folds may have been profoundly attenuated and
entirely detached from their roots during Tertiary mylonitic
deformation. In the northern Ruby Mountains, some of the large-scale
folds can be traced into the mylonitic shear zone where they appear
to have been re-worked causing an overall flattening and attenuation
of the folds as well as a rotation of fold-nappe hingelines into
parallelism with the pervasive mylonitic stretching lineation. In
the northern East Humboldt Range, however, the hingeline of the
Winchell Lake fold-nappe trends approximately parallel to the
pervasive WNW-trending stretching lineation characteristic of the
mylonitic shear zone, and it is impossible to discern whether the
fold developed in or rotated into this orientation.

The Winchell Lake fold-nappe is cored by a thick Precambrian
basement gneiss complex consisting of Archean biotite monzogranitic
orthogneiss (Lush et al., 1988) and inferred Early Proterozoic
pelitic to quartzitic paragneisses. Numerous small amphibolitic
bodies interpreted as metamorphosed mafic intrusions are common
throughout this gneiss complex but are virtually absent in the
surrounding rocks which consist of inferred Upper Precambrian and
Lower Cambrian quartzites and schists and Middle Cambrian to
Devonian(?) calcite and dolomite marbles. The Upper Precambrian to
Devonian(?) sequence is structurally inverted with respect to the
Precambrian basement gneiss complex enclosed in the core; and
consequently, a pre-folding, pre-metamorphic low-angle tectonic
contact must intervene (Lush et al., 1988). Beneath the fold-nappe

---

FIGURE 14. Hinge zone of the Winchell Lake fold-nappe as exposed
along the back wall of Winchell Lake cirque on the eastern face of
the northern East Humboldt Range. The core of the fold at this
locality consists of dark outcrops of rusty-weathering graphitic
paragneiss surrounded by white-weathering Ordovician-Cambrian marble
and Ordovician metaquartzite. Enveloping the fold is a thick
sequence of Lower Cambrian and Upper Precambrian flaggy quartzite and
schist. These metasedimentary rocks, particularly the metaclastic
units, all contain abundant sheet-like bodies of leucogranitic
orthogneiss. The hingeline of this major structure trends west-
northwest, approximately parallel to mineral elongation lineations in
the deformed rocks. Closure is to the south. To the north, the fold
is cored by Archean and Lower Proterozoic gneisses, and a
premetamorphic, pre-folding thrust is inferred to separate the
metasedimentary rocks shown in this photo from the basement complex
enclosed in the core of the fold. €Zqs = Cambrian and Upper
Precambrian quartzite and schist, rgs = rusty-weathering graphitic
schist, D€mq = lower Paleozoic metacarbonate rocks and quartzite,
D€mu = lower Paleozoic metacarbonate rocks.

complex the metasedimentary sequence is repeated at least twice more suggesting major pre-extensional and pre-metamorphic duplication of section.

At present, timing relations between major isoclinal folding events are best constrained in the northern East Humboldt Range. Here, 29 Ma biotite monzogranitic sheets are partially involved in isoclinal folding that appears to be parasitic to the Winchell Lake fold-nappe. In one locality a large biotite monzogranite sheet appears to wrap around the nose of the Winchell Lake fold-nappe itself. These relations raise the exciting possibility that the Winchell Lake fold-nappe may have formed and evolved in the context of large-scale plastic flow in the deeper reaches of an extensional orogen. Alternatively, the Winchell Lake fold-nappe may be a composite feature that initiated at deep crustal levels during Mesozoic compressional orogenesis and was subsequently modified during large-scale Tertiary extensional deformation. However, regardless of the age of the Winchell Lake fold-nappe, the East Humboldt Range preserves unambiguous evidence for pre-extensional (presumably Mesozoic) isoclinal folding, migmatization, and low-angle thrust faulting. In fact, it is clear that the compositional layering that is folded around the Winchell Lake fold-nappe was itself transposed during an earlier phase of isoclinal folding, and at numerous localities examples of well-developed "hook structure" are preserved. Migmatitic layering is sometimes involved in these coaxially refolded folds, but folded 29 Ma biotite monzogranite sheets are never observed to be involved in hook structure (McGrew and Snoke, 1988). It should be emphasized that because multiple phases of recumbent isoclinal folding have been documented in the East Humboldt Range, it is by no means clear whether the Winchell Lake fold-nappe belongs to the same fold generation as the fold-nappes exposed in the northern Ruby Mountains.

4.4. Deep crustal rocks

The strong layering demonstrated by the seismic reflection section suggests that the crust is strongly anisotropic. The nature of this anisotropy depends on modal mineral content and overall fabric of the rock units. Furthermore, the strong multicyclic nature of the

FIGURE 15. The north wall of Lamoille Canyon, northern Ruby Mountains, which provides an excellent cross-sectional view of the east-vergent Lamoille Canyon nappe (Howard, 1966, 1980, 1987). The hinge of the nappe is outlined by brown-weathering cliffs of upper amphibolite facies Prospect Mountain Quartzite (Cambrian-Late Precambrian, €Zpm). The quartzite nose is surrounded by marble and calc-silicate rock of the marble of Verdi Peak (Ordovician and Cambrian, O€m) pervasively intruded by numerous sills of white-weathering pegmatitic leucogranite. Ordovician-Cambrian marble and calc-silicate rocks also occur in the core of the fold and have been interpreted by Howard (1966) as the lower plate of the pre-metamorphic and pre-folding Ogilvie thrust fault.

reflections indicates that the interval velocities must be averages for interlayered zones of high and low velocities.

Laboratory velocity determinations (Christensen, 1982) suggest that the "temperature-corrected" velocities of 6.1 - 6.2 km/s (Table 1, Fig. 7) may correspond to quartzo-feldspathic rocks in agreement with surface geologic observations in the northern Ruby Mountains-East Humboldt Range indicating that the upper crust consists chiefly of quartzite, pelitic schist, marble, and granitic orthogneiss. Velocities of 6.4 to 6.5 km/s correspond to quartzo-feldspathic rocks (deformed granitic and metasedimentary rocks) combined with about one-third mafic material, possibly intrusive. The next zone, delineated between 20 to 28 km, is characterized by velocities of 6.7 to 6.8 km/s. These velocities indicate a predominantly mafic composition, but more felsic rocks (i.e., immediate to silicic) must also be present to provide lithologic layering, an important factor in reflectivity. Although additions of mafic rocks recrystallized under granulite facies conditions are inferred to be common in this zone, a wide range of rocks such as garnet-sillimanite gneiss, dolomite, and anorthosite may have the same velocity. Indeed, some of the layering implied by the multi-cyclic reflections could be restite with some unusual high-velocity mineral combinations such as garnet-sillimanite and garnet-pyroxene.

Actual evidence for intrusions is not provided by the character of the seismic reflection data (i.e., subhorizontal, parallel reflection bands). Rather this character suggests lithologic layering and plastic flow. Any original intrusive geometry, whether it be pre- or syn-extensional, would be rotated subparallel to the horizontal flow pattern of the deep-seated extensional deformation. The seismic reflection fabric of the lower crust therefore suggests extensional metamorphism and plastic deformation. The role of magmatic underplating can only be inferred from the estimated interval velocity data and the widespread distribution of crustally-derived, anatectic Tertiary granitic rocks in the presently exposed upper crust (Wright and Snoke, 1986).

The lowermost 3 km in the crust (the zone between the "X" reflection and the Moho) has a relatively high velocity of 7.4 to 7.8 km/s (Table 1) that lies between normal velocities for lower crust and upper mantle. These interval velocity data indicate a composition intermediate between mafic and ultramafic such as pyroxene-rich and possibly garnet-rich rocks. This zone, which contains numerous discontinuous reflectors, could consist of residuum from partial melting and/or layered ultramafic-mafic cumulate rocks generated during local differentiation of batches of the inferred underplated magmas.

5. A PROPOSED CRUSTAL CROSS-SECTION AND ITS IMPLICATIONS FOR EASTERN GREAT BASIN CRUSTAL STRUCTURE

The crust in the eastern Great Basin is clearly the product of a long and complex tectonic evolution. In a nutshell, the highlights of this polyphase evolution include the petrogenesis and accretion of Archean and Proterozoic basement complexes; the development of the late Precambrian to early Mesozoic Cordilleran miogeocline; Mesozoic magmatism, metamorphism, and polyphase deformation; and a protracted Cenozoic extensional history that began in the mid—Tertiary and continues to the present. Our proposed crustal cross-section (Fig. 16) for the eastern Great Basin reflects elements of this complex history at various structural levels.

The crustal cross-section consists of two essential parts: an exposed part based on geologic mapping and structural studies and an unexposed part based on vertical incidence and wide-angle seismic reflection studies (Valasek et al., 1987, 1989). A basic tripartite subdivision of the exposed part of the cross-section is proposed: brittlely attenuated suprastructure, transitional metasedimentary/ granitoid zone, and mylonitic to nonmylonitic migmatitic infrastructure (e.g., see Fig. 5). An important feature of this basic crustal architecture is that the Tertiary magmatic, metamorphic, and deformational history apparently becomes increasingly pervasive with structural depth. This aspect is clearly manifested by the abundance of highly deformed Tertiary gneissic rocks that comprise the deeper structural levels of the migmatitic infrastructure (Wright and Snoke, 1986; McGrew and Snoke, 1988) and by mid—Tertiary cooling ages on high-temperature metamorphic minerals (e.g., hornblende) from such structurally deep exposures (Dallmeyer et al., 1986). Consequently, we believe that the mid—Tertiary deformed granitic rocks characteristic of the migmatitic infrastructure might be even more pervasive at still deeper levels of the crust. The migmatitic infrastructure would pass downward into a broad zone consisting chiefly of granitic rocks of various ages (Precambrian, Jurassic, Cretaceous, and/or mid—Tertiary), but with a conspicuous Tertiary granitoid component derived from crustal anatexis, with all rocks overprinted by extensional strain.

The present-day middle crust is characterized by two prominent sets of subhorizontal seismic reflections at 4 and 6 s (Fig. 6 and 7). These strongly reflective zones have been interpreted as zones of intense, possibly mylonitic, plastic strain (Valasek et al., 1989). Major changes in crustal interval velocity appear to coincide with these prominent sets of subhorizontal reflections suggesting that they may represent fundamental crustal boundaries. Furthermore, similar reflective zones have also been recognized on a seismic reflection profile that traversed part of a core complex in the Albion Mountains, Idaho (S. B. Smithson, unpublished data, 1989). In this light, these reflection sets probably represent mid-crustal decoupling zones that have served to partition crustal strain during Cenozoic crustal extension (Valasek et al., 1989).

Crustal interval velocities of 6.7-6.8 km/s from a depth of 20 to 28 km (Table 1) indicate an important mafic component, and therefore, we propose that this zone is the primary locus of magmatic underplating. These underplated mafic magmas and their intruded wallrocks would have recrystallized in the granulite facies during crustal extension (Sandiford and Powell, 1986). The deformed

130

Cover rocks

Detachment fault

Mylonite zone (mid-Tertiary)

Migmatitic core (includes
Jurassic, Cretaceous and
Tertiary deformed igneous
rocks as well as p€ to
mid-Paleozoic paleosome)

Mid-crustal decoupling zone
(mylonitic shear zone?)

Large-scale pure shear
accommodated by simple
shear concentrated along
mid- & lower crustal
shear zones

Mid-crustal decoupling zone
(mylonitic shear zone?)

Possible gabbroic intrusions,
largely sheared and rotated,
intruded into highly deformed
& metamorphosed p€ rocks

Mafic cumulate or residuum

Moho consisting of inter-
layered mantle & crustal
rocks—possible magma

AWS & SBS'88

FIGURE 16. Interpretative crustal cross-section for the northern
Ruby Mountains-East Humboldt Range, eastern Great Basin. Upper plate
suprastructural rocks overlie a detachment fault system and a
mylonitic shear zone. Recumbent fold nappes of the migmatitic
infrastructure are underlain by granitic intrusions and a midcrustal
decoupling zone (mylonitic shear zone?). This zone passes downward
into deformed Precambrian basement rocks intruded by gabbroic bodies
that have been sheared and rotated toward horizontality during middle
to lower crustal extensional deformation, resulting in a granulite
facies metamorphism. Crustal-scale pure shear strain is accommodated
by simple shear strain concentrated along midcrustal and lower
crustal shear zones. The layered Moho is accentuated by shear
accompanying extension and/or intrusion. Magma may be ponded at the
Moho. Although this diagrammatic model is not drawn to scale, it
contains the essential elements of a crustal cross-section
approximately 30 km thick, as envisioned for the Ruby-East Humboldt
core complex. (After Valasek et al., 1989.)

Tertiary granitoids of the migmatitic infrastructure have a crustal isotopic signature (Wright and Snoke, 1986; J. E. Wright, per. comm., 1989), and were consequently generated in the deep crust during lithospheric thinning and attendant lower crustal heating. Therefore, some of the material in the 6.7-6.8 km/s zone is inferred to be restite generated by partial melting of wallrock as the intrusive underplating process proceeded during crustal extension. The lowermost 3 km in the crust (the zone between the "X" reflection and the Moho) has a relatively high velocity of 7.4 to 7.8 km/s (Table 1). This zone may consist of mafic residuum or layered cumulates. These very mafic to ultramafic rocks probably would be highly deformed and characterized by tectonitic fabrics as described by Moore (1973) from the deep-seated Giles complex of central Australia.

The maximum amount of intruded mafic material during Cenozoic extension has been estimated as 6-8 km (Valasek et al., 1987, 1989). However, this estimate assumes only a limited amount of mafic rock in the lower crust prior to extension. If substantial mafic material was added during Mesozoic magmatism, our estimate of underplated rock would be much less (e.g., 3-4 km). Also, anatexis in the deep crust will differentiate the crust into a more mafic lower crust and a more felsic upper crust.

Reflections from the Moho show no significant relief on the base of the crust beneath the R-EH complex, compatible with regional findings of very little structural relief on the Moho across the northern Basin and Range province (Klemperer et al., 1986). This uniformity of crustal thickness across the northern Basin and Range province suggests that heterogeneous supracrustal thinning and extension may be compensated in part by fluid-like flow at deeper crustal levels (Wernicke, 1989). The net effect of Cenozoic extension, which began at least by early Oligocene time, was to produce a regional-scale re-equilibration of the Moho from inferred depths in excess of 50 km at the end of Mesozoic contraction to its present depth of approximately 31 km.

ACKNOWLEDGEMENTS

This research was supported by National Science Foundation research grants EAR87-07435 to A.W. Snoke and EAR85-19153 to S.B. Smithson, A.W. Snoke, and R.A. Johnson. Continual discussion and collaboration with J.E. Wright on the geochronological evolution of the Ruby-East Humboldt complex provided many temporal constraints. During an informal field trip in 1981, Warren Hamilton suggested that the first author consider the evolution of the Ruby-East Humboldt complex in whole-crust scale dimensions. This paper is our present best answer to the provocative challenge provided by Hamilton's insightful questions concerning the evolution of this Cordilleran metamorphic core complex. Numerous comments on an earlier version of this manuscript by M.R. Hudec, P.A. Camilleri, K.A. Howard, and an anonymous reviewer were very helpful in the final preparation of this

article. Finally, we especially appreciate the superb help of
Phyllis A. Ranz in the preparation of the illustrations for the
article.

REFERENCES

Armstrong, R.L. (1968) 'Sevier orogenic belt in Nevada and Utah',
    Geological Society of America Bulletin, 79, 429-458.
Bohlen, S.R., Wall, V.J., and Boettcher, A.L. (1983) 'Experimental
    investigations and geological applications of equilibria in the
    system $FeO-TiO_2-Al_2O_3-SiO_2-H_2O$', American Mineralogist, 68, 1049-
    1058.
Buck, W.R. (1988) 'Flexural rotation of normal faults', Tectonics, 7,
    959-973.
Carmichael, D.M. (1978) 'Metamorphic bathozones and bathograds: A
    measure of the depth of post-metamorphic uplift and erosion on the
    regional scale', American Journal of Science, 278, 769-797.
Carter, N.L., and Tsenn, M.C. (1987) 'Flow properties of continental
    lithosphere', Tectonophysics, 136, 27-63.
Christensen, N.I. (1982) 'Seismic velocities', in R.S. Carmichael
    (ed.), Handbook of Physical Properties of Rocks, vol. 2, CRC
    Press, Boca Raton, Florida, pp. 2-228.
Coats, R.R. (1987) 'Geology of Elko County, Nevada', Nevada Bureau of
    Mines and Geology Bulletin 101, 112 pp.
Coney, P.J. (1979) 'Tertiary evolution of Cordilleran metamorphic
    core complexes', in J.M. Armentrout, M.R. Cole, and H. TerBest,
    Jr. (eds.), Cenozoic Paleogeography of the Western United States:
    Pacific Coast Paleogeography Symposium 3, The Pacific Section,
    Society of Economic Paleontologists and Mineralogists, Los
    Angeles, Calif., pp. 15-28.
Coney, P.J., and Harms, T.A. (1984) 'Cordilleran metamorphic core
    complexes: Cenozoic extensional relics of Mesozoic compression',
    Geology, 12, 550-555.
Dallmeyer, R.D., Snoke, A.W., and McKee, E.H. (1986) 'The Mesozoic-
    Cenozoic tectonothermal evolution of the Ruby Mountains, East
    Humboldt Range, Nevada: A Cordilleran metamorphic core complex',
    Tectonics, 5, 931-945.
Dokka, R.K., Mahaffie, M.J., and Snoke, A.W. (1986)
    'Thermochronologic evidence of major tectonic denudation
    associated with detachment faulting, northern Ruby Mountains-East
    Humboldt Range, Nevada', Tectonics, 5, 995-1006.
Farmer, G.L. (1988) 'Isotope geochemistry of Mesozoic and Tertiary
    igneous rocks in the western United States and implications for
    the structure and composition of the deep continental
    lithosphere', in W.G. Ernst (ed.), Metamorphic and crustal
    evolution of the Western United States, Prentice-Hall, Englewood
    Cliffs, N.J., pp. 87-109.
Gibbs, A.D. (1984) 'Structural evolution of extensional basin
    margins', Journal of Geological Society of London, 141, 609-620.
Hose, R.K., and Blake, M.C., Jr. (1976) 'Geologic map of White Pine

County, Nevada', in Geology and mineral resources of White Pine County, Nevada, Nevada Bureau of Mines and Geology Bulletin 85, scale 1:250,000.

Howard, K.A. (1966) 'Structure of the metamorphic rocks of the northern Ruby Mountains, Nevada', Ph.D. thesis, Yale University, New Haven, Conn., 170 pp.

Howard, K.A. (1980) 'Metamorphic infrastructure in the northern Ruby Mountains, Nevada', in M.D. Crittenden, Jr., P.J. Coney, and G.H. Davis (eds.), Cordilleran Metamorphic Core Complexes, Geological Society of America Memoir 153, Boulder, Colo., pp. 335-347.

Howard, K.A. (1968) 'Flow direction in triclinic folded rocks', American Journal of Science, **266**, 758-765.

Howard, K.A. (1987) 'Lamoille Canyon nappe in the Ruby Mountains metamorphic core complex, Nevada', in Geological Society of America Centennial Field Guide - Cordilleran Section, 1987, pp. 95-100.

Howard, K.A., Kistler, R.W., Snoke, A.W., and Willden, R. (1979) 'Geologic map of the Ruby Mountains, Nevada, scale 1:125,000', U.S. Geological Survey Miscellaneous Geological Investigation Map I-1136.

Hudec, M.R., and Wright, J.E. (1990) 'Mesozoic history of the central part of the Ruby Mountains - East Humboldt Range metamorphic core complex, Nevada', Geological Society of America Abstracts with Programs, **22**, no. 3, 30.

Hurlow, H.A. (1987) 'Structural geometry, fabric, and chronology of a Tertiary extensional shear zone - detachment system', M.S. thesis, University of Wyoming, Laramie, Wyo., 141 pp.

Hurlow, H.A. (1988) 'P-T conditions of mylonitization in a Tertiary extensional shear zone, Ruby Mountains-East Humboldt Range, Nevada', Geological Society of America Abstracts with Programs, **20**, 170.

King, P.B., and Beikman, H.M. (1974) 'Geologic map of the United States (exclusive of Alaska and Hawaii)', U.S. Geological Survey, scale 1:2,500,000.

Kistler, R.W., Ghent, E.D., and O'Neill, J.R. (1981) 'Petrogenesis of garnet two-mica granites in the Ruby Mountains, Nevada', Journal of Geophysical Research, **86**, 10,591-10,606.

Klemperer, S.L., Hauge, T.A., Hauser, E.C., Oliver, J.E., and Potter, C.J. (1986) 'The Moho in the northern Basin and Range province, Nevada, along the COCORP 40°N seismic-reflection transect', Geological Society of America Bulletin, **97**, 603-618.

Lister, G.S., and Snoke, A.W. (1984) 'S-C mylonites', Journal of Structural Geology, **6**, 617-638.

Lush, A.P., McGrew, A.J., Snoke, A.W., and Wright, J.E. (1988) 'Allochthonous Archean basement in the northern East Humboldt Range, Nevada', Geology, **16**, 349-353.

Malavieille, J. (1987) 'Les mechanismes d'amincissement d'une croute epaissie: Les "Metamorphic core complexes" du Basin and Range (USA)', These, Docteur d'Etat, Université des Sciences et Techniques du Languedoc, Montpellier, 332 pp.

McGrew, A.J., and Snoke, A.W. (1988) 'Cenozoic and Mesozoic folding

134

and migmatization in the East Humboldt Range metamorphic core complex, Nevada', Geological Society of America Abstracts with Programs, **20**, A393–A394.

McGrew, A.J., and Snoke, A.W. (1990) 'Styles of deep-seated flow in an extending orogen: the record from the East Humboldt Range, Nevada', Geological Society of America Abstracts with Programs, **22**, no. 3, 66.

Miller, E.L., and Gans, P.B. (1989) 'Cretaceous crustal structure and metamorphism in the hinterland of the Sevier thrust belt, western U.S. Cordillera', Geology, **17**, 59–62.

Misch, P. (1960) 'Regional structural reconnaissance in central northeast Nevada and some adjacent areas –– Observations and interpretations', in Geology of East-Central Nevada, Intermountain Association of Petroleum Geologists 11th Annual Field Conference Guidebook, pp. 17–42.

Moore, A.C. (1973) 'Studies of igneous and tectonic layering in the rocks of the Gosse Pile Intrusion, Central Australia', Journal of Petrology, **14**, 49–79.

Passchier, C.W., and Simpson, C. (1986) 'Porphyroclast systems as kinematic indicators', Journal of Structural Geology, **8**, 831–843.

Sandiford, M., and Powell, R. (1986) 'Deep crustal metamorphism during continental extension: Modern and ancient examples', Earth and Planetary Science Letters, **79**, 151–159.

Sharp, R.P. (1939) 'Basin-range structure of the Ruby-East Humboldt Range, northeastern Nevada', Geological Society of America Bulletin, **50**, 881–920.

Snelson, S. (1957) 'The geology of the northern Ruby Mountains and the East Humboldt Range, Elko County, Nevada', Ph.D. thesis, University of Washington, Seattle, Wash., 268 pp.

Snoke, A.W. (1989) 'Clover Hill, Nevada: Structural link between the Wood Hills and East Humboldt Range', Geological Society of America Abstracts with Programs, **21**, no. 5., 146.

Snoke, A.W., McCall, R.G., and McKee, E.H. (1983) 'Tertiary stratigraphy and structure in the northern East Humboldt Range, Nevada –– A clue to Cenozoic regional tectonic patterns', Geological Society of America Abstracts with Programs, **15**, 403.

Snoke, A.W., and Howard, K.A. (1984) 'Geology of the Ruby Mountains-East Humboldt Range, Nevada –– A Cordilleran metamorphic core complex', in J. Lintz, Jr. (ed.), Western Geological Excursions, vol. **4**, Geological Society of America Annual Meeting, MacKay School of Mines, Reno, Nev., pp. 260–303.

Snoke, A.W., and Lush, A.P. (1984) 'Polyphase Mesozoic-Cenozoic deformational history of the northern Ruby Mountains-East Humboldt Range, Nevada', in J. Lintz, Jr. (ed.), Western Geological Excursions, vol. **4**, Geological Society of America Annual Meeting, MacKay School of Mines, Reno, Nev., pp. 232–260.

Snoke, A.W., and Miller, D.M. (1988) 'Metamorphic and tectonic history of the northeastern Great Basin', in W.G. Ernst (ed.), Metamorphic and crustal evolution of the Western United States, Prentice-Hall, Englewood Cliffs, N.J., pp. 606–648.

Snoke, A.W., Hudec, M.R., Hurlow, H.A., and McGrew, A.J. (1990) 'The anatomy of a Tertiary extensional shear zone, Ruby Mountains-East Humboldt Range, Nevada', Geological Society of America Abstracts with Programs, **22**, no. 3, 85.

Speed, R., Elison, M.W., and Heck, F.R. (1988) 'Phanerozoic tectonic evolution of the Great Basin', in W.G. Ernst (ed.), Metamorphic and crustal evolution of the Western United States, Prentice-Hall, Englewood Cliffs, N.J., pp. 572-605.

Stewart, J.H. (1980) 'Geology of Nevada -- A discussion to accompany the Geologic Map of Nevada', Nevada Bureau of Mines and Geology Special Publication 4, 136 pp.

Stryhas, B.A. (1988) 'Progressive refolding in high strain regimes: An application to the Maggia nappe, Ticino, Switzerland and the Lamoille Canyon nappe, Ruby Mountains, Nevada', Ph.D. thesis, Washington State University, Pullman, Wash., 188 pp.

Thorman, C.H. (1970) 'Metamorphosed and nonmetamorphosed Paleozoic rocks in the Wood Hills and Pequop Mountains, northeast Nevada', Geological Society of America Bulletin, **81**, 2417-2448.

Thorman, C.H., and Brooks, W.E. (1988) 'Preliminary geologic map of the Oxley Peak quadrangle, Elko County, Nevada', U.S. Geological Survey Open-file Report 88-755.

Thorman, C.H., and Snee, L.W. (1988) 'Thermochronology of metamorphic rocks in the Wood Hills and Pequop Mountains, northeastern Nevada', Geological Society of America Abstracts with Programs, **20**, A18.

Valasek, P.A., Hawman, R.B., Johnson, R.A., and Smithson, S.B. (1987) 'Nature of the lower crust and Moho in eastern Nevada from "wide-angle" reflection measurements', Geophysical Research Letters, **14**, 1111-1114.

Valasek, P.A., Snoke, A.W., Hurich, C.A., and Smithson, S.B. (1989) 'Nature and origin of seismic reflection fabric, Ruby-East Humboldt metamorphic core complex, Nevada', Tectonics, **8**, 391-415.

Wernicke, B. (1989) 'The fluid crustal layer and its implications for continental dynamics', Geological Society of America Abstracts with Programs, **21**, A81.

Wernicke, B., and Axen, G.J. (1988) 'On the role of isostasy in the evolution of normal fault systems', Geology, **16**, 848-851.

Wright, J.E., and Snoke, A.W. (1986) 'Mid-Tertiary mylonitization in the Ruby Mountain-East Humboldt Range metamorphic core complex, Nevada', Geological Society of America Abstracts with Programs, **18**, 795.

PROGRESS IN TECTONIC AND PETROGENETIC STUDIES IN AN EXPOSED CROSS-SECTION OF YOUNG (~100 Ma) CONTINENTAL CRUST, SOUTHERN SIERRA NEVADA, CALIFORNIA

Jason B. Saleeby
Division of Geological and Planetary Sciences 170-25
California Institute of Technology
Pasadena, CA 91125

ABSTRACT. The southern Sierra Nevada offers an oblique section through young Cordilleran-type batholith generated crust spanning surface (volcanic) to deep (granulitic) levels. Regional mapping and Pb/U zircon geochronology reveal structural continuity through this crustal section for volcanic, plutonic and metamorphic assemblages developed at ~100 Ma, making it one of the youngest sections in the world. Construction of a synthetic cross-section is well-constrained by the oblique section map pattern, abundant age data, and a published crustal structure section that was based on geophysical and lower crustal xenolith data from across the shallow levels of the batholith. The synthetic cross-section depicts the state of the sialic crustal section during its ~100 Ma petrogenesis. Critical features of the section are as follows: (1) Much of the pre-Cretaceous crust was completely reconstituted by batholith generation; (2) Major influxes of mantle-derived tonalitic to gabbroic magmas drove crustal-level melting and magma mixing; (3) Pre-existing sialic components within the batholithic magmas were contributed primarily from partial to complete melts of craton derived metasedimentary material; (4) Silicic magmas emplaced at shallow crustal levels represent moderately to well-mixed and fractionated systems, while those frozen at deeper levels appear to be more heterogeneous both lithologically and geochemically; (5) Granulite facies metamorphic assemblages in the deep crustal rocks developed primarily in a retrograde regime descending from gabbro and tonalite solidus conditions; (6) Upward rise of silicic magmas was accompanied by downward return flow of metamorphic host rocks, and locally ignimbrite sections of only slightly older age than enclosing plutons were transported to considerable depth as well; (7) Much of the upper crust responded to the intrusion of silicic magmas by large-scale extension which also promoted major ignimbrite eruptions; and (8) Within the deeper levels of the composite batholith and off its flanks, at moderate to shallow crustal levels, low-angle detachments may have developed as a primary structural feature. Each of these points as well as a number of other interesting problems form the basis for a broad spectrum of current and future research efforts.

M. H. Salisbury and D. M. Fountain (eds.), Exposed Cross-Sections of the Continental Crust, 137–158.
© 1990 *Kluwer Academic Publishers.*

1.   INTRODUCTION

The Sierra Nevada mountain range of California has long been known for its spectacular exposures of a Cordilleran-type composite batholith.   Great interest in the ultramafic-slate-greenstone belt along the western foothills of the range arose as early as the mid-eighteen hundreds due to its hosting of a major auriferous vein system. With the advent of plate tectonic theory, major interest was focused on the Sierra Nevada from two new perspectives. (1) Metamorphic rocks along the western margin of the range were suspected and subsequently shown to represent a major pre-batholithic suture zone; and (2) The composite batholith and the structurally complex ensimatic assemblages of the California Coast Ranges to the west were together considered the type example of an exhumed continent edge subduction-magmatic arc system.

The new interest in Sierra Nevada geology stimulated by plate tectonic theory spawned vigorous field and geochronological studies in the early 1970's which have continued to the present day.   Such studies have focused equally on metamorphic framework and batholithic rocks.   Geophysical studies including work in gravity, magnetism, heat flow, and seismology have progressed over the past two decades, but not nearly at the rate of the surface geological studies. Nevertheless, the modest amount of geophysical data that is available has proven to be critical in influencing thought on crustal structure, tectonics and petrogenesis.

Synthesis of the large body of geological and geophysical data for the Sierra Nevada leads to the realization that the southern half of the range offers a structurally continuous oblique section through consanguinous sialic crust from surface to deep batholithic levels. Regional and detailed geological mapping and geochronological control along the oblique section in conjunction with a well-constrained crustal section model across the high-level exposures that is based on geophysical and lower crustal xenolith data facilitates the construction of a synthetic cross-section which depicts the batholith at the time of its petrogenesis. In this paper key elements of the geological and geophysical data are reviewed as a basis for the presentation of the synthetic cross-section. The cross-section is offered as a working model for the primary structure and petrogenesis of the relatively young Sierran sialic crustal section. A number of the significant implications of the model are discussed and some of the critical questions to be addressed in the future are noted. The geological context of the southern Sierra section can be understood more clearly with a brief introduction to the regional tectonics.

2.   TECTONIC SETTING

The Sierra Nevada batholith is a composite Cordilleran-type batholithic belt consisting of numerous gabbroic to granitic, but mainly tonalitic to granodiorite plutons. Over 90 percent of the batholith was emplaced during the Cretaceous, more specifically between ~120 and ~85 Ma (Evernden and Kistler, 1972; Stern and others, 1981; Chen and Moore, 1982; Saleeby and contributors, 1986). Jurassic, Triassic and rare Paleozoic plutons occur in the western metamorphic belt and along the eastern border of the range as well as in smaller ranges to the east.

The Cretaceous batholith is genetically related, in a tectonic sense, to east-dipping subduction represented by the Franciscan Complex of the California Coast Ranges to the west. Between the batholith and the Franciscan Complex lies the Great Valley which formed a major forearc basin during Franciscan subduction and Sierran magmatic arc activity (Dickinson, 1981). The Great Valley is underlain by oceanic crust and upper mantle. Such rocks surface along the western metamorphic belt of the Sierra Nevada as a complex ophiolitic terrane (Saleeby, 1981; 1982). Within the western metamorphic belt the ophiolitic terrane is bounded tectonically to the east by a thick sequence of continent-derived metasedimentary rocks. The boundary is a major plate tectonic suture which pre-dated the batholith (Saleeby, 1981).

The contiguous western metamorphic belt is truncated in the central part of the range by a major reentrant of Cretaceous batholithic rocks. On Figure 1 this is shown as the Fresno reentrant. To the south rocks of the western metamorhic belt occur as septa scattered within the batholith. The suture between the oceanic crustal and craton-derived metasedimentary rocks is expressed in the batholith as a steep geochemical gradient represented in Figure 1 by the initial $^{87}Sr/^{86}Sr(Sr_i) = 0.706$ contour. Initial ratios as low as ~0.703 are measured in plutons to the west or oceanic side of the 0.706 line, and initial ratios as high as ~0.708 occur to the east (Kistler and Peterman, 1973; 1978). As discussed below the transverse geochemical gradients across the suture reflect differing degrees of older sialic crustal components within the batholith. Metamorphic septa along and east of the 0.706 line consist primarily of high-grade craton-derived metasedimentary rocks.

The current structural setting of the Sierra Nevada is a regional west-tilted fault block with a steep spectacular eastern escarpment (Bateman and Wahrhaftig, 1966). Uplift along the eastern escarpment is related to Basin and Range extensional faulting which commenced in the eastern Sierra region during the Pliocene. Earlier uplift and erosion along the western Sierra is poorly understood. In particular between the Late Cretaceous and Early Eocene the southwestern extremity of the range underwent major uplift which resulted in the exposure of the deep-level rocks.

As discussed below igneous and metamorphic rocks at the southern end of the range were at ~8 kb pressures at ~100 Ma. These rocks were uplifted, eroded, and unconformably overlapped by marine strata by mid-Eocene time (Nilsen and Clark, 1975). The plate tectonic setting of this rapid uplift event is not understood. Subsequent to the uplift event, during the Miocene, the southern end of the range was rotated ~45 degrees clockwise (Plescia and Calderone, 1986). Rotation may have occurred in conjunction with regional extensional tectonism that is widespread in regions to the south of the Sierra Nevada (Goodman and others, in press). The Neotectonic regime has subsequently uplifted the southern end of the range again along the Garlock left-slip fault as well as additional moderate to high-angle faults that occur primarily in the southernmost Great Valley (Goodman and others, in press). As discussed below, the polyphase uplift history of the southernmost Sierra Nevada has not disrupted the structural continuity between the deep level batholithic rocks and the shallower level batholith to the north. Thus, the southern Sierra offers an ideal sampling of deep crust whereby the sampling process has not strongly modified the primary state of the oblique crustal section.

3. THE SOUTHERN SIERRA NEVADA AS A DEEP CRUSTAL SECTION

3.1. Historical Sketch

The major batholithic and metamorphic units of the southern Sierra Nevada are shown in Figure 1. Much of the rationale for the deep crustal section model can be derived from the regional map relations. The internal structure of the batholith between latitudes 36°30' and 38°00' was well-established by over three decades of U.S.G.S. quadrangle mapping. With the addition of broad based geochronological studies (Evernden and Kistler, 1970; Stern and others, 1981; Chen and Moore, 1982), the critical structural and temporal relations were established across the central part of the batholith. Such a basis was essential for the deep crustal model.

Formulation of the deep crustal model for the southern Sierra Nevada arose from a decade of research of both regional and detailed nature carried out by the author and his students, and through collaborative efforts with a number of U.S.G.S. and academic workers. The author's introduction into the regional structure of the Sierra Nevada batholith came through systematic studies of metamorphic wallrocks of the southwestern quadrant of the range. The oceanic lithosphere framework which the batholith was generated and emplaced into was established by these studies (Saleeby, 1978; 1979; 1982; Saleeby and others, 1978). In conjunction with these studies, detailed field, geochronological and petrologic investigations were conducted on a major belt of tonalitic to gabbroic rocks which characterize much of the southwestern part of the batholith (Saleeby, 1976; Mack and others, 1979; Saleeby and Sharp, 1980; Saleeby and others, 1987). As discussed below such batholithic rocks are thought to constitute a major constituent of the lower crust beneath the entire Sierra Nevada. Such a view began to evolve through regional observations whereby the belt of tonalitic to gabbroic rocks were seen to change in character towards the southern termination of the range where they pass into the Tehachapi gneiss complex - the deepest exposed level of the batholith (Saleeby and others, 1987).

Patterns in primary and superposed structures and metamorphic textures of the Tehachapi gneiss complex are in tremendous contrast to patterns observed across the central part of the batholith. In particular igneous and metamorphic phenomena cannot be clearly distinguished in the gneiss complex. As detailed structural and geochronological work progressed on the gneiss complex the author was also active as a collaborator in a series of studies on silicic metavolcanic remnants exposed in the central Sierra (Tobisch

---

Figure 1. Generalized map of southern Sierra Nevada batholith showing major age-compositional belts, generalized metamorphic framework assemblages, $Sr_i = 0.706$ isopleth, areas for which age/paleodepth constraints exist, and broad igneous barometric contours (Sources: Kistler and Peterman, 1973; 1978; Saleeby and Sharp, 1980; Saleeby, 1981, unpub. data; Stern and others, 1981; Chen and Moore, 1982; Elan, 1986; Ligget, 1986; Saleeby and contributors, 1986; Tobisch and others, 1986; Saleeby and others, 1987; Ague and Brimhall, 1988; Pickett and Saleeby, 1989). Approximate location of Figure 2 Transect cross-section line shown at northern end of map. Extreme southern end of contiguous western Foothills metamorphic belt shown in northwest corner of map area. Batholithic rocks which truncate metamorphic belt are termed Fresno reentrant. Volcanic xenolith localities after Domenick and others (1983) and Dodge and others (1986; 1988).

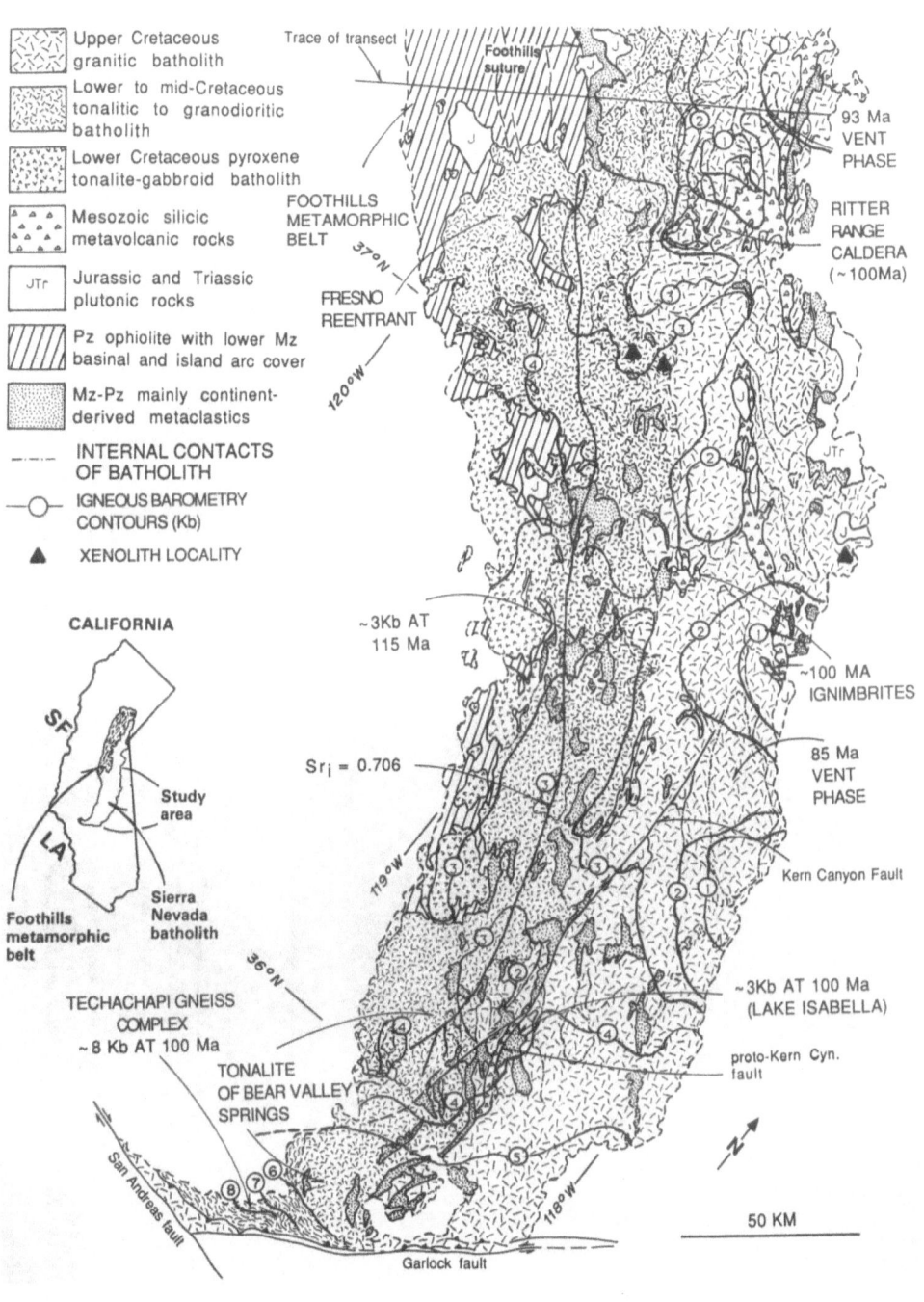

Upper Cretaceous granitic batholith

Lower to mid-Cretaceous tonalitic to granodioritic batholith

Lower Cretaceous pyroxene tonalite-gabbroid batholith

Mesozoic silicic metavolcanic rocks

JTr — Jurassic and Triassic plutonic rocks

Pz ophiolite with lower Mz basinal and island arc cover

Mz-Pz mainly continent-derived metaclastics

INTERNAL CONTACTS OF BATHOLITH

IGNEOUS BAROMETRY CONTOURS (Kb)

XENOLITH LOCALITY

CALIFORNIA

SF

LA

Study area

Sierra Nevada batholith

Foothills metamorphic belt

TECHACHAPI GNEISS COMPLEX ~8 Kb AT 100 Ma

TONALITE OF BEAR VALLEY SPRINGS

San Andreas fault

Garlock fault

Trace of transect

Foothills suture

FOOTHILLS METAMORPHIC BELT

FRESNO REENTRANT

37°N

120°W

~3Kb AT 115 Ma

$Sr_i = 0.706$

119°W

36°N

118°W

50 KM

93 Ma VENT PHASE

RITTER RANGE CALDERA (~100Ma)

JTr

~100 MA IGNIMBRITES

85 Ma VENT PHASE

Kern Canyon Fault

~3Kb AT 100 Ma (LAKE ISABELLA)

proto-Kern Cyn. fault

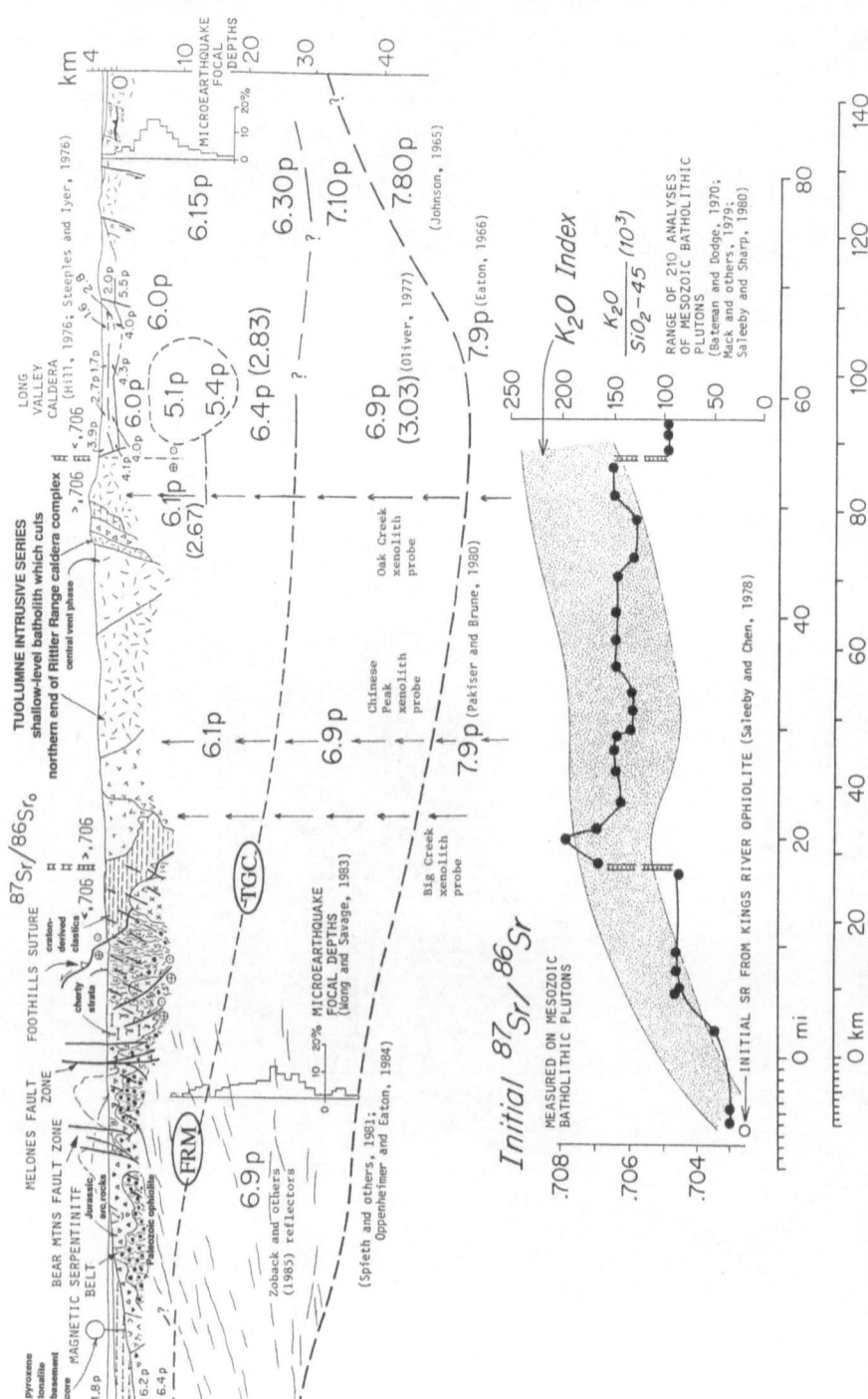

Figure 2. Crustal structure section across central Sierra Nevada ~25 km north of Fresno reentrant (from Saleeby and contributors, 1986). P-wave velocity is given with "p" and modeled density in parentheses. Vertical = horizontal scale. Hypothetical crustal positions of Fresno reentrant migmatite zone and Techachapi gneiss complex are shown by FRM and TGC symbols. Deep probe traces from volcanic xenolith studies are projected northward onto section after Domenick and others (1983) and Dodge and others (1986; 1988).

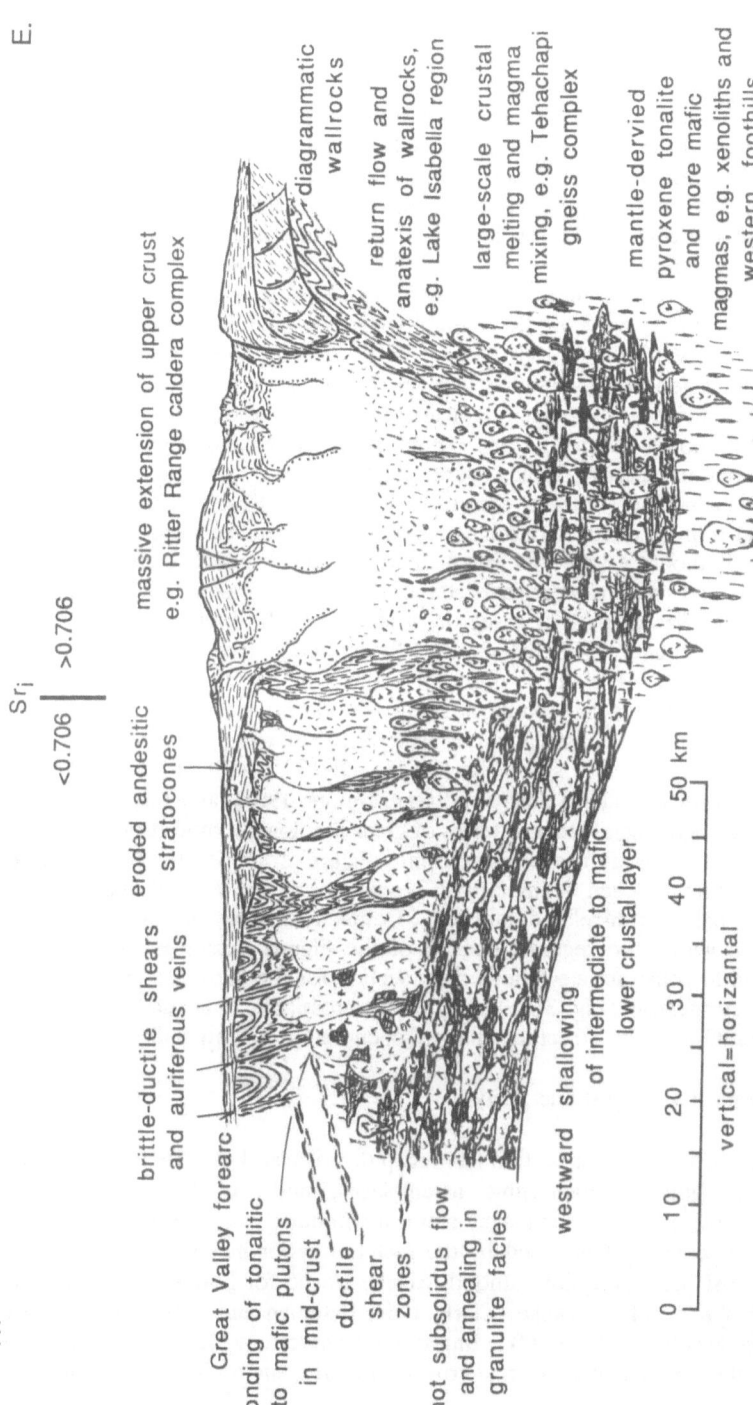

Figure 3. Synthetic cross section constructed from oblique section model of southern Sierra Nevada region (Fig. 1). The section depicts the batholithic terrane at ~100 Ma with the current axial region undergoing its petrogenesis. Western batholithic belt has already completed its petrogenesis. Areas used to constrain structure and section after Tobisch and others (1986). Wallrock structure is highly diagrammatic with pre-caldera extensional faulting along eastern margin of section. Symbols of batholithic rocks and magmas are dark cross-hatch = gabbroids; v's = pyroxene tonalite-diorite; stipples = tonalite-granodiorite; clear = granitic. Surface deposits consist primarily of silicic ignimbrites.

and others, 1986; Saleeby and others, 1988, in press). Strong similarities in zircon ages of silicic ignimbrites and related high-level plutons of this region and zircon ages of the Tehachapi gneiss complex suggested the existence of an oblique section through the composite batholith. Such a suggestion prompted regional-scale field and geochronological studies in areas between the shallow and deep-level exposures which were not covered by the U.S.G.S. studies noted above. Such on-going work is more firmly establishing structural and temporal continuity between the shallow and deep level exposures.

Throughout the course of these investigations several additional major data sets accumulated for the southern Sierra Nevada which have proven to be critical for formulation of the deep section model. (1) Large bodies of Sr, Nd and O isotopic data were published for various parts of the batholith (Kistler and Peterman, 1973; 1978; DePaolo, 1981; Masi and others, 1981; Ross, 1983; Saleeby and others, 1987). (2) Petrologic and limited geochemical studies were performed on upper mantle and crustal xenoliths that were carried in Quaternary volcanic eruptions through the central part of the Sierra Nevada batholith (Domenick and others, 1983; Dodge and others, 1986; 1988). (3) Scattered thermobarometric studies were performed on batholithic and contact metamorphic rocks from the southern part of the batholith (Sharry, 1981; Elan, 1986; Liggett, 1986), and more recently broad igneous barometric studies were completed for major parts of the batholith (Ague and Brimhall, 1988; Pickett and Saleeby, 1989). (4) Detailed mapping and geochronological studies were performed on metamorphic framework rocks and adjacent plutons along the axial part of the southern Sierra Nevada batholith (Saleeby and others, 1978; Saleeby and Busby-Spera, 1986).

The integration of all of these different sources of data indicates that for the general time frame of 100 Ma the southern Sierra Nevada offers an oblique section, or down-plunge view, of broadly consanguineous sialic crust spanning volcanic to granulitic (~8 kb) levels. Geophysical data synthesized in the DNAG Continent-Ocean Transect across the central Sierra Nevada (Saleeby and contributors, 1986), and the xenolith data referenced above strongly suggest that in terms of gross lithologic composition the oblique section is representative of a true cross-section. Figure 2 shows the Sierra Nevada segment of the Continent-Ocean Transect with projections of the vertical and transverse locations of critical paleodepth-related data sets. Detailed mapping and mesoscopic field observations placed within the regional context of the Figure 1 map and the Figure 2 section can be used to construct a synthetic cross-section for the Sierran crust at the time of its ~100 Ma petrogenesis (Fig. 3). The synthetic cross-section is used as a working model for consideration of the structure and petrogenesis of Sierran sialic crust.

3.2   Construction of the Synthetic Section

Construction of the synthetic cross-section from map data is based on structural continuity in broadly coeval igneous-metamorphic assemblages, and on the recognition of depth-related changes in these assemblages along a southerly paleodepth gradient. An additional geometric constraint is posed by the fact the major metamorphic framework assemblages that host the composite batholith can be traced for hundreds of kilometers along the structural trend of the range (Clark, 1964; Bateman and Clark, 1974; Saleeby and others, 1978; Saleeby, 1981; 1982). Important transverse variations in framework assemblages, batholith petrochemistry, gross crustal structure, and timing of magmatism are well documented across the range (Saleeby and contributors, 1986). These variations are critical in the petrogenetic and tectonic analysis of the crustal section. Of first order importance is the eastward migration of major magmatism between ~115 and 85 Ma. The synthetic section freezes the process at ~100 Ma, and thereby offers a general dynamic and petrogenetic model for sialic crust generation at that frame in time. Three well-studied areas were chosen as templates for detailed constructions to depths of ~25 km in the petrogenetically active part of the section.

Area 1 - The Ritter Range caldera complex is a 100 ± 2 Ma large volume resurgent structure perched in the high central Sierra Nevada (Fiske and others, 1977; Tobisch and others, 1986). The high country setting offers virtually 100% exposure and tremendous structural relief through intra-caldera and outflow ignimbrites, resurgent pluton and cupolas, pre-caldera substrate and caldera collapse and resurgent structures. The caldera structure clearly records the highest structural levels of the 100 Ma state of the batholithic terrane as also suggested by the <1 kb igneous barometry published for the area by Ague and Brimhall (1988). Additional exposures of ~100 Ma ignimbrite sequences occur up to 100 km south of the Ritter Range as steep infolds within slightly younger batholithic rocks which yield igneous barometers of less than 1 to as much as 3 kb (Ague and Brimhall, 1988). Pelites carry only andalusite in this region. The infold structures are typified by steep constrictional shape fabrics which developed during batholith-related dynamothermal metamorphism. Such structures are present within the Ritter range caldera fill sequence adjacent to the principal resurgent pluton. As discussed below these and other structures suggest downward return flow of host materials during upward rise of silicic magma. The return flow structures increase in intensity southward as deeper levels of the batholith are exposed.

Area 2 - The Lake Isabella region offers excellent high desert and moderate to high relief exposures of ~100 Ma batholithic plutons and high grade metamorphic wallrock septa. The plutons, wallrocks and structural features are typical of a large tract of the mesozonal batholith. Throughout this region, and more so towards the south, sillimanite predominates or is exclusive over andalusite in pelites. This tract of the batholith yields igneous barometric data of between 2 and 4 kb (Ague and Brimhall, 1988). A detailed thermobarometric study within the dynamothermal aureole of a 100 ± 2 Ma large composite pluton at Lake Isabella yields pressures of 3 ± 0.5 kb (Elan, 1986). The metamorphic wallrocks of this region exhibit extreme synbatholithic transposition into steep constrictional shape fabrics and map to mesoscale sheath folds (Saleeby and Busby-Spera, 1986; work in progress). Early dikes originating from the rising magma bodies are transposed with the wallrocks as are scattered leucosomes within partially melted pelitic and psammitic wallrocks. Traversing south from the Lake Isabella area a short distance (10-15 km) higher grade metamorphic conditions are exhibited by widespread partial to complete melting of pelitic and psammitic metamorphic units. Within a distance of ~25 km such units are for the most part completely melted and mixed into the batholith. Igneous barometric data in this area reaches 5 to 6 kb. Within this transition zone and in regions to the south, the steep constrictional fabrics take on a southerly plunging statistical maxima consistent with a significant northward tilt and a steep southward paleodepth gradient. Progressing southward along this gradient the entire batholith takes on a different character as it becomes highly foliated and ultimately gneissic.

Area 3 - The Tehachapi gneiss complex constitutes the southernmost exposures of the Sierra Nevada batholith. Systematic mapping and Pb/U geochronological studies indicate that the complex is strongly dominated by orthogneisses of mid-Cretaceous igneous age with a 100 ± 2 Ma thermal-magmatic maxima (Saleeby and others, 1987; Sams and Saleeby, 1988). Mineral assemblages typical of granulite facies metamorphism are scattered throughout the complex, although as discussed below, in most instances such assemblages are best interpreted as igneous in origin. A major tonalite batholith complex (tonalite of Bear Valley Springs - BVS batholith), whose upper mesozonal level has been traced into the area west of Lake Isabella, extends into the northeastern zone of the gneiss complex. Distinct metamorphic wallrock assemblages that can be traced for over 200 km north of the gneiss complex are stripped of all lower temperature melting fractions midway along the BVS batholith. Within the southern reaches of the batholith and within the gneiss complex, only lenses of pure metaquartzite, marble and scapolite-bearing calc-silicate rock remain as substantial refractories. Thermobarometric studies conducted within the complex (Sharry, 1981; Pickett and Saleeby, 1989) suggest a pressure of ~8 kb in mineral assemblages developed at ~100 Ma (Saleeby and others, 1987). Such pressures are generally consistent with the presence of magmatic epidote in tonalites, disappearance of olivine as an early liquidus phase in gabbroic cumulates, local solidus to hot-subsolidus breakdown of igneous hornblende to garnet in 100 Ma gabbroic and dioritic intrusives, and vestiges of kyanite in rare pelitic restite lenses (Saleeby and others, 1987; Sams and Saleeby, 1988; Pickett and Saleeby, 1989). Within the southern half of the BVS batholith, and within enclaves of metamorphic framework rock, steep synplutonic and high grade metamorphic deformation fabrics are pervasive. Southwestward into the gneiss complex, the fabrics intensify and are involved in ductile flow folding and high temperature shear. The high temperature deformation is related directly to the ~100 Ma thermal-magmatic maxima (Saleeby and others, 1987). The BVS batholith and the gneiss complex are locally cut by Cenozoic high-angle faults. These do not juxtapose unrelated units and therefore are not fundamental tectonic breaks. The Tehachapi gneiss complex represents the deepest exposed levels of the mid-Cretaceous Sierra Nevada batholith.

Transverse variations across the batholith are used in conjunction with deep crustal xenolith and geophysical data to model the deepest levels of the 100 Ma magmatic system in the synthetic section. Most notable of the transverse variations are the initial $^{87}Sr/^{86}Sr$ ratios (Sr_i) and the K_2O index of the batholithic plutons (Fig. 2). Steep gradients in these geochemical parameters coincide with a major boundary in the metamorphic wallrock assemblages where highly depleted Paleozoic ophiolite with overlying oceanic island arc and deep water strata to the west are juxtaposed against a thick accumulation of craton-derived Paleozoic to early Mesozoic marine strata. The boundary in general coincides with the $Sr_i = 0.706$ contour (Fig. 1). Strontium, Nd, O and Pb/U zircon isotopic studies show a substantial involvement of the craton derived material in batholithic plutons and volcanic derivatives east of the $Sr_i = 0.706$ line (Kistler and Peterman, 1973; 1978; DePaolo, 1981; Masi and others, 1981; Ross, 1983; Saleeby and others, 1987; Pickett and Saleeby, 1989). The contribution of such materials to the batholith drops off significantly to the west. In particular, south of the Fresno reentrant (Fig. 1) there is a belt of pyroxene tonalite to gabbroidal batholithic plutons which intrude exclusively ophiolitic wallrocks along the very western edge of the batholith (Saleeby, 1976; Mack and others, 1979; Saleeby and Sharp, 1980). On-going isotopic and petrochemical studies strongly suggest that this plutonic belt represents a little-or-non-contaminated mantle-derived suite. Such rocks are abundant in the Tehachapi gneiss complex and are the principle carriers of "granulite facies" mineral assemblages. Rocks of very similar character also occur in the mafic igneous and feldspathic granulite members of the lower crustal xenolith suites from the central part of the batholith (Dodge and others, 1986; 1988).

Seismic P-wave velocity structure across the central part of the batholith shows a distinct westward shallowing of the deep high-velocity layer (Fig. 2). This deep layer may actually be exposed as the pyroxene tonalite-gabbro belt along the western edge of the range. Bouguer gravity along the Figure 2 segment of the Transect displays a tremendous gradient from -260 mgals in the east to -30 mgals in the west indicating an eastward deepening crustal "root" for the batholith (Oliver, 1977; Saleeby and contributors, 1986). Consideration of P-wave velocity laboratory experiments (Christensen and Fountain, 1975; Fountain and Christensen, in press) and specific gravity measurements suggest that the "root" is not mafic, but of mafic-intermediate composition. The western pyroxene tonalite-gabbro belt possesses such a bulk composition, and if modified with minor subsolidus garnet to the same extent as the Tehachapi complex, such rocks would yield the geophysical signatures of the sub-batholithic layer. As discussed below, such a view for the origin of the sub-batholithic layer carries important petrogenetic implications.

Regional and detailed map patterns are used to model the pluton shapes with depth. Many of the large plutons map out as large sill-like bodies with vertical or steeply-east dipping margins. The high-level plutons and late phases of composite bodies commonly have more irregular shapes. There is little or no evidence for the plutons flattening out at depth as is commonly assumed. The BVS batholith apparently offers the greatest structural relief of any individual large pluton. The BVS batholith becomes more heterogeneous and steeply foliated to its base where it appears to be transposed by low-dipping ductile shear that occurred during solidus to hot-subsolidus conditions. Such deformation may reflect broad lateral spreading in the mid-crust, perhaps due to subsequent voluminous magmatism immediately to the east. The ductile shear, lateral flow and granulitic annealing shown at the deep to sub-batholithic levels is modeled after the base of the BVS batholith and its transition into the Tehachapi gneiss complex.

Surface expression of the composite batholith appears to have varied with the transverse zonation pattern. First, all remnants of silicic ignimbrites lie east of the $Sr_i = 0.706$ line as do igneous barometric determinations of <2 kb. Furthermore, substantial remnants of pre-batholithic metamorphic framework rocks are rare east of the $Sr_i = 0.706$ line. In contrast a major tract of pre-batholithic rocks (Foothills metamorphic belt) lies on strike to the north of the batholith west of the $Sr_i = 0.706$ line. Numerous tonalitic to gabbroic plutons within this tract of metamorphic rocks and in basement cores recovered from the eastern Great Valley suggest that the western part of the batholith plunges northward beneath the Foothills metamorphic belt (Saleeby, 1981; Saleeby and contributors, 1986). The oblique section interpretation for the southern Sierra carries as a corollary that the more tonalitic western zone of the batholith did not form substantial surface breakthroughs as did the eastern granitic zone of the batholith. The synthetic section depicts scattered andesitic stratocones as the surface expression of the western batholith. The underlying tonalitic plutons are accordingly shown in the model as ponding in the mid-crust in contrast to the granitic plutons which form major caldera structures and which vent voluminous ignimbrites.

3.3   Major Implications of the Synthetic Section

The first order implication of the Figure 3 model is that nearly the entire crust of the Sierra Nevada has been reconstituted by Cretaceous batholith generation. This view is unquestionable to depths corresponding to the Tehachapi gneiss complex as shown by basic mapping and geochronological data. An alternative view is that the lower crust is the remnants of older (Precambrian) basement and that the 6.4 to 6.0 km/sec batholithic layers have spread over the top of this layer. This is not consistent with the meager amount of isotopic data published on the xenolith suites (Domenick and others, 1983; Dodge and others, 1986), although much more work can be done on the problem. The original investigations of the Tehachapi gneiss complex were based on seeking out remnants of such Precambrian basement rock (Saleeby and others, 1987). However, every major gneiss unit of the complex has been dated by Pb/U zircon techniques, and in conjunction with Sr, Nd and O isotopic data a mid-Cretaceous petrogenetic age is demonstrated for the complex. Older continental components of typically 10% are detected within the orthogneisses, but are traced to the melt products of craton-derived metasedimentary enclaves (Sams, 1986; Saleeby and others, 1987; Pickett and Saleeby, 1989).

Reconstitution of the crust and related batholith generation are considered to have been driven primarily by mantle-derived basaltic magma influxes. The pyroxene tonalite to gabbroic rocks of the southwestern Sierra and Tehachapi complex are direct representatives of such magmas. On-going petrochemical work in the southwestern Sierra belt suggests that two different derivatives of the same or very similar mantle magma system(s) are present. Hornblende-rich calcic gabbro suites represent basaltic magmas which underwent their main crystallization and fractionation histories within the crust. Two pyroxene gabbro-diorite-tonalite suites appear to have fractionated significantly prior to entering the crust. Comingling of the two magma types is not uncommon, but in most instances layered calcic hornblende gabbro intrusions are disrupted and suspended as large xenoliths within the two pyroxene intrusions. There is strong overlap in major, minor and trace element patterns in the two suites suggesting a very similar source. However, the two pyroxene rocks clearly underwent much more extended fractionation histories.

The two mantle igneous suites are a critical constituent of the Tehachapi gneiss complex. By minor hot subsolidus deformation and/or annealing the two pyroxene rocks are commonly converted to pyroxene granulite. The principal phases of these "granulites" are igneous in origin, however. The hornblende gabbro suite underwent the most notable change as compared to shallower analogues to the north. Olivine which is a ubiquitous early liquidus phase in the north ceases to be an important phase presumably due to the pressure effect on its liquidus relations (Wyllie, 1971). At the solidus and under hot subsolidus conditions, hornblende locally reacted to garnet. This reaction is concentrated along veins or fractures and commonly results in bleached or anorthositic halos around the larger garnets. Pyroxenes within such rocks are yet to be clearly identified as metamorphic versus igneous. The textural and structural relations in the Tehachapi complex in conjunction with the geochronological constraints indicate that highly nonpervasive granulite facies metamorphism occurred under retrogressive conditions from tonalite and gabbro solidus conditions. Additional but subordinate disequilibrium reactions under amphibolite and upper greenschist facies conditions locally record further thermal decay along the retrogressive path. The lower crust on Figure 3 is accordingly modeled as a retrogressive granulite terrane of relatively young mean age. Additional tests of this model will be performed on the lower crustal xenolith suite.

One of the longstanding debates on granulite terranes concerns the relations between their depletion, granulite facies metamorphism and possible partial melting. Approximately 100 major, minor and trace element analyses on the western Sierra pyroxene tonalite-gabbro suite indicate that much of the suite originated as igneous rocks already relatively depleted compared to most upper crustal rocks. A typical analysis of a two pyroxene tonalite is quite similar to the mean of 254 Scourian granulites published by Weaver and Tarney (1981) (Table 1). These data are not presented as an explanation for the origin of the Scourian granulites, but rather to make the point that general granulitic geochemical patterns are a primary feature of the Sierran pyroxene tonalite-gabbro

	A	B
Major elements (wt.%)		
SiO_2	61.2	60.21
TiO_2	0.54	0.70
Al_2O_3	15.6	16.63
Fe_2O_3	5.9	6.10
MnO	0.08	0.11
MgO	3.4	3.54
CaO	5.6	6.82
Na_2	4.4	3.43
K_2O	1.0	0.96
P_2O_5	0.18	0.11
Trace elements (ppm)		
Rb	11	24.5
Sr	569	247
Ba	757	260
Th	0.42	1.8
U	--	0.6
Element ratios		
K/Rb	755	392
Rb/Sr	0.02	0.10
Ba/Sr	1.3	1.1

Table 1. Comparison of major and selected trace element patterns between Archean Lewisian gneisses and typical pyroxene tonalite from the southwest Sierra Nevada batholith. A: Mean of 254 Scourian granulites (Weaver and Tarney, 1981); B: Sierran two-pyroxene tonalite (B.W. Chappell and Saleeby, unpub. data).

suite. Additional depletion, particularly of mobile trace elements, is possible under extended deep crustal residence and/or subsequent thermal-magmatic events.

The major features of the synthetic section also portray constraints posed by Nd, Sr, O and Pb/U zircon isotopic data. As noted above high Sr_i values in batholithic plutons (>0.706) are generally interpreted as having resulted from various components of Precambrian sialic crust (Kistler and Peterman, 1973; 1978). Neodymium, Sr and O isotopic data on high-level plutons of the central Sierra suggest that the sialic components were ultimately derived primarily from melted craton-derived sediments that mixed with mantle-derived magmas (DePaolo, 1981). The BVS batholith and the Tehachapi gneiss complex offer insights to such mixing phenomena. In these deep-level rocks various degrees of mixing between pyroxene tonalite to gabbroic magmas, heterogeneous hornblende-biotite tonalites and partial to complete metasedimentary melts can be observed and sampled in the field. DePaolo's (1981) oxygen contoured ϵ_{Nd}-ϵ_{Sr} plot is in general reproduced in such sample suites, and wallrock detrital zircon is traced as a contaminant within the hybrid rocks (Saleeby and others, 1987; Pickett and Saleeby, 1989). The southern BVS batholith in particular represents a major mixing horizon. Here a myriad of mafic tonalite pillow and dike form inclusions are mingled into lighter tonalite and local granodiorite with isotopic and restitic vestiges of melted metasedimentary rock. The dark tonalite inclusions are for the most part mineralogically re-equilibrated pyroxene tonalites (hydrated to hornblende ± biotite). Both chilling of the inclusions resulting in their isolation and magma mixing relations are preserved with chill features more prevalent for darker dioritic to gabbroic inclusions. Northward along the batholith, at higher structural levels, such mixing phenomena are less pervasive and more uniform fractionated phases of the pluton are present. Such a relation is portrayed for the 100 Ma magma system in the synthetic section. The higher well-mixed and fractionated levels of the magma system are more typical of structural levels sampled and modeled by DePaolo (1981). At the deeper structural levels incomplete mixing and fractionation patterns are frozen into the batholith. This is reflected in Figure 1 by the southward dispersion of the Sr_i = 0.706 contour line.

The synthetic section carries a number of major tectonic implications.

(1)   Along the granitic batholith belt the upper crust underwent such major extension that the principal roof rocks that remain are batholith related volcanic ejecta. Pre-batholith framework rocks are preserved primarily as steeply dipping septa. Remnants of deformed extensional faults within high-level septa of the west-central Sierra have been mapped by Tobisch and others (1986) and are shown diagramatically in Figure 3 beneath the eastern outflow ignimbrites. Such pre-caldera extensional faults may represent the initial inflation of the crust during the early phases of silicic magmatism.

(2)   Downward return flow of host rocks along rising pluton margins appears to be a major transport pattern from near-surface to at least mid-crustal levels. The regionally pervasive steep constrictional fabrics in the host rocks are shown to have resulted in net downward transport by the fact that local infolds of silicic volcanic rocks that are only slightly older than the adjacent plutons are present at depths corresponding to ~3 kb pressure (Saleeby and Busby-Spera, 1986; Saleeby and others, in press). The fact that such constrictional fabrics can be traced southward to high-grade septa within the southern BVS batholith suggests that the return flow mechanism can transport fertile upper-crustal material to major melting and mixing horizons in the deeper crust.

(3)   Deep seismic reflection surveys commonly reveal numerous horizontal or near horizontal reflectors within sialic crystalline terranes (Smithson, 1979; Klemperer and others, 1985). The synthetic section predicts that the lower crust attains a primary layered state by a combination of igneous layering and hot subsolidus flow. Such layering could control the manner in which sialic crust behaves during subsequent tectonic events be they extensional or compressional. It is apparent that most sialic crustal sections "bottom out" at depths corresponding to 10 ± 2 kb, which suggests a common delamination mechanism

or control on delamination depth (Oliver, 1988; Percival, 1988; Schenk, 1988; this report). Recent deep seismic reflection and refraction lines across the Tehachapi gneiss complex (CALCRUST, unpubl. data) suggest that the complex could be detached from its original lower crustal substrate even though it is still attached to rooted higher-level rocks to the north. Recent seismic reflection data across the Fresno reentrant reveal low west-dipping reflectors in mid- to upper-crustal levels extending westward beneath the Great Valley (Zoback and others, 1985) (Fig. 2). Synbatholithic ductile shear zones along the southern edge of the reentrant currently being mapped may represent exposed equivalents of the reflectors. Possible relations between such structures and magma tectonics are currently under investigation.

(4)   The high velocity lower crustal layer maintains a constant thickness as it shallows to the west (Fig. 2). Total thickness of the crust is greatest east of the $Sr_i = 0.706$ line where substantial components of Precambrian-derived sialic material are mixed into the batholith (Kistler and Peterman, 1973; 1978; DePaolo, 1981; Saleeby and others, 1987). Components of such sialic crustal material are minimal or nil to the west of the 0.706 line, and structural-stratigraphic studies in the western Foothills metamorphic belt suggest that the oceanic crustal sequence and its superjacent strata formed a relatively thin crustal section prior to batholith emplacement. These relations suggest that the current crustal thickness pattern mimics the pre-batholith relative thickness pattern.

4.   DISCUSSION AND CONCLUSIONS

The southern Sierra Nevada constitutes one of the more recently recognized continuous sialic crustal sections in the world. Unique aspects of this section are its relatively young age (~100 Ma), broad expanse of excellent exposure, and continuity with a major intact virtually non-deformed volcanic edifice. The large body of existing field, geochronological and isotopic data render the region ideal for the posing of topical problems in the structure and petrogenesis of sialic crust generated within Phanerozoic Cordilleran batholithic belts.

Other exposures of deep crustal sections from the Cordilleran batholithic belt in the Idaho batholith (Hyndman, 1988), northwest Washington (Kriens and Wernickie, 1988) and Prince Rupert area of British Columbia (Hollister, 1988) differ by a lack of detailed age control and/or structural continuity between strictly coeveal deep-level batholithic rocks and volcanic levels. Furthermore, these exposures show abundant evidence for major crustal shortening and even nappe tectonics in intimate association with batholithic magmatism. In marked contrast the southern Sierra Nevada shows clear evidence for major crustal extension at high crustal levels and vertical transport at medial to deep levels. Horizontal compression may have been significant at the deepest exposed levels, but with a lack of clear kinematic indicators horizontal spreading as an internal kinematic pattern of the composite batholith itself is just as likely.

The contrast in batholith kinematic patterns between the Sierran section and those to the north may be related to the degree of upper and lower plate coupling during subduction (Kanamori, 1986). High degrees of coupling in the north may have induced a regional state of compressive stress across the batholith belt during magma emplacement resulting in synplutonic crustal shortening. In contrast, lower degrees of coupling at Sierra Nevada latitudes may have induced a state of regional tensile stress in the upper plate promoting crustal spreading. Such a contrast in plate dynamic patterns is consistent with the preservation of an intact linear forearc basin at Sierra Nevada latitudes, and the apparent lack of such to the north. More importantly though, is the existence of a major west-vergent thrust belt along the western edge of the northern batholithic belt which developed during major phases of magmatism (Rubin and others, in press), and the lack of such at Sierra Nevada latitudes.

The lack of major tectonic breaks and crustal shortening structures within the southern Sierra add to its value as a natural laboratory for the study of structural and petrogenetic relations in a consanguineous sialic crustal section. A number of critical problems are currently under investigation which pertain both to better documentation of the depth of exposure and structural continuity of the section, and to the tectonic and petrogenetic phenomena of its construction.

1.   A better resolution of the depth of exposure offered by the Tehachapi gneiss complex is needed, even though a number of independent lines of evidence each suggest that pressures of ~8 kb are represented in rocks and mineral assemblages which formed at 100 ± 2 Ma. A corollary to this problem is the hot-subsolidus breakdown of hornblende in mafic intrusive rocks - can the dehydration occur under extended hot subsolidus conditions in the granulite facies conditions or is an external mechanism such as CO_2 infiltration required (Newton and others, 1980). Fluid inclusion studies on the garnet reaction products should help clarify this problem. Additionally, an important problem with these rocks is whether or not clinopyroxene formed with the garnets. Thus far, all pyroxenes identified within the garnet-bearing rocks appear to be relict igneous phases.

2.   The possibility of major constituents involved in the mixing of batholithic materials in addition to mantle-derived gabbroic to tonalitic magmas and continent-derived sedimentary materials needs close investigation. Detailed isotopic and petrochemical studies of volcanic hosted lower crustal xenoliths should aid considerably in this problem, particularly in regard to the possibility of Proterozoic sialic basement material having contributed to the batholithic magma systems. Additionally, along the southwestern margin of the Fresno reentrant oceanic crustal host rocks for the batholith are partially melted into an amphibolite-trondhjemite migmatite complex. Locally the leucosome material was mobilized and comingled into the same tonalitic and gabbroic plutons which helped promote the partial melting event. Ague and Brimhall's (1988) igneous barometric data suggest that a relatively high-pressure welt may exist in this area, and thus detailed petrogenetic studies of the migmatite complex and adjacent plutons may shed light on the possibility of crustal-level melting of oceanic crust as an added batholithic component.

3.   Several aspects of crustal geophysics are of major interest. Of fundamental importance is the strong west-dipping seismic reflection fabric that is evident off the west flank of the Fresno reentrant. Detailed field studies are currently in progress in order to test the possibility that some of the reflectors surface along the western margin of the batholith. Additional seismic work is also needed across the Tehachapi gneiss complex in order to test whether or not it is still attached to its primary crustal root, and also to test whether or not the Neotectonic regime involves an active lower crust beneath the deep-level rocks. Finally, the Tehachapi gneiss complex as well as the western belt of tonalitic to gabbroic intrusives and deep crustal xenoliths are prime targets for seismic velocity laboratory measurements. In conjunction with the deep crustal velocity structure

displayed in the Figure 2 transect such velocity measurements could provide critical constraints on the composition and petrogenesis of the lower crust.

As finer details on the structure, composition and mixing relations of the batholith accrue we will approach a state of understanding where we may be able to reasonably constrain questions of broad scale petrogenesis, and material budget and transfer. We may also be in position to more seriously pose the problem of whether or not tectonic and petrogenetic phenomena which give rise to Cordilleran-type batholithic belts might have played a major role in sialic crustal genesis throughout earth history.

ACKNOWLEDGEMENTS

Field, geochronological and petrologic studies in the southern Sierra Nevada region were supported by N.S.F. grants EAR-8018811, EAR-8218460, EAR-841973 and EAR-8708266, and a Fellowship awarded by the Alfred P. Sloan Foundation. Fieldtrips, written communications and discussions with R.W. Kistler, Warren Hamilton, D.C. Ross, F.C.W. Dodge, D.A. Pickett, O.T. Tobisch, D.B. Sams, and J.G. Moore were very helpful in the recognition of the oblique crustal section and the formulation of the synthetic cross-section. Special thanks to the Tejon Ranch Company for providing access to areas which encompass most of the Tehachapi gneiss complex. This is Contribution No. 4744, California Institute of Technology.

REFERENCES CITED

Ague, J.J. and Brimhall, G.H., 1988. Magmatic arc asymmetry and distribution of anomalous plutonic belts in the batholiths of California: Effects of assimilation, crustal thickness, and depth of crystallization: Geol. Soc. Am. Bull., 100, 912-927.

Bateman, P.C. and Clark, L.D., 1974. Stratigraphic and structural setting of the Sierra Nevada batholith: Pacific Geol., 8, 78-89.

Bateman, P.C. and Wahrhaftig, C., 1966. Geology of the Sierra Nevada : in: Geology of Northern California (Ed. E.A. Bailey) Calif. Div. Mines and Geology Bull., 190, 107-172.

Chen, J.H. and Moore, J.G., 1982. Uranium-lead isotopic ages from the Sierra Nevada batholith, California: J. Geophys. Res., 87, 4761-4784.

Christensen, N.I. and Fountain, D.M., 1975. Constitution of the Lower continental crust based on experimental studies of seismic velocities in granulite: Geol. Soc. Am. Bull., 86, 227-236.

Clark, L.D., 1964. Stratigraphy and structure of the Sierra Nevada metamorphic belt, California: U.S. Geol. Surv. Prof. Paper 410, 70 pp.

DePaolo, D.J., 1981. A neodymium and strontium isotopic study of the Mesozoic calc-alkaline granitic batholiths of the Sierra Nevada and Peninsular Ranges, California: J. Geophys. Res., 86, 10470-10485.

Dickinson, W.R., 1981. Plate tectonics and the continental margin of California: in: The Geotectonic Development of California (Ed. W.G. Ernst), 1, 1-28.

Dodge, F.C.W., Calk, L.C. and Kistler, R.W., 1986. Lower crustal xenoliths, Chinese Peak lava flow, central Sierra Nevada: Jour. Pet., 27, 1277-1304.

Dodge, F.C.W., Lockwood, J.P. and Calk, L.C., 1988. Fragments of the mantle and crust from beneath the Sierra Nevada batholith: Xenoliths in a volcanic pipe near Big Creek, California: Geol. Soc. Am. Bull., 100, 938-947.

Domenick, M.A., Kistler, R.W., Dodge, F.C.W. and Tatsumoto, M., 1983. Nd and Sr isotopic study of crustal and mantle inclusions from the Sierra Nevada and implications for batholith petrogenesis: Geol. Soc. Am. Bull., 94, 713-719.

Elan, R., 1986. High grade contact metamorphism at the Lake Isabella North Shore roof pendant, Southern Sierra Nevada, California: Ph.D. thesis, University of Southern California, 202 pp.

Evernden, J.F. and Kistler, R.W., 1970. Chronology of emplacement of Mesozoic batholithic complexes in California and western Nevada: U.S. Geol. Surv. Prof. Paper 623, 67 pp.

Fiske, R.S., Tobisch, O.T., Kistler, R.W., Stern, T.W. and Tatsumoto, M., 1977. Minarets caldera: A Cretaceous volcanic center in the Ritter Range pendant, central Sierra Nevada, California (abstract): Geol. Soc. Am. Abstr. with Prgms., 9, no. 7, 975.

Fountain, D.M. and Christensen, N.I., 1988. Composition of the continental crust and upper mantle: A review: Geol. Soc. Am. Memoir, Geophysical Framework of the United States, (Ed. L. Pakiser and W. Mooney), in press.

Goodman, E.D., Malin, P.E., Ambos, E.L. and Crowell, J.C., 1989. The southern San Joaquin Valley as an example of Cenozoic basin evolution in California: in: The Origin and Evolution of Sedimentary Basins and Their Energy and Mineral Resources (Ed. R.A. Price), in press.

Hollister, L.S., 1988. The central gneiss complex near Prince Rupert, British Columbia: in: NATO Advanced Study Institutes Programme - Exposed Cross Sections of the Continental Crust (Abstract), 12.

Hyndman, D.W., 1988. Idaho batholith magmas and tectonics: Implications for the lower continental crust of a converging margin: in: NATO Advanced Study Institutes Programme - Exposed Cross Sections of the Continental Crust (Abstract), 27.

Kanamori, H., 1986. Rupture process of subduction-zone earthquakes: Ann. Rev. Earth Planet. Sci., **14**, 293-322.

Kistler, R.W. and Peterman, Z.E., 1973. Variations in Sr, Rb, K, Na and initial $^{87}Sr/^{86}Sr$ in Mesozoic granitic rocks and intruded wall rocks in central California: Geol. Soc. Am. Bull., **84**, 3489-3512.

Kistler, R.W. and Peterman, Z.E., 1978. Reconstruction of crustal blocks of California on the basis of initial strontium isotopic compositions of Mesozoic granitic rocks: U.S. Geol. Surv. Prof. Paper 1071, 17 pp.

Klemperer, S.L., Brown, L.D., Oliver, J.E., Ando, C.J., Czuchra, B.L. and Kaufman, S., 1985. Some results of COCORP seismic reflection profiling in the Grenville-age Adirondack Mountains, New York State: Can. J. Earth Sci., **22**, 141-153.

Kriens B. and Wernicke, B., 1988. Characteristics of a continental margin magmatic arc with depth: The Skagit-Methow crustal section: in: NATO Advanced Study Institutes Programme - Exposed Cross Sections of the Continental Crust (Abstract), 27.

Liggett, D.L., 1986. Geology and geochemistry of a garnet-bearing granitoid in the southwestern Sierra Nevada, Tulare County, California: Calif. State University, Northridge, M.S. thesis, 143 pp.

Mack, S., Saleeby, J.B. and Farrell, J.E., 1979. Origin and emplacement of the Academy pluton, Fresno County, California: Geol. Soc. Am. Bull., **90**, 321-323 (Part I), 633-694 (Part II).

Masi, U., O'Neil, J.R. and Kistler, R.W., 1981. Stable isotope systematics in Mesozoic granites of Central and Northern California and Southwestern Oregon: Contrib. Mineral. Petrol., **76**, 116-126.

Newton, R.C., Smith, J.V. and Windley, B.F., 1980. Carbonic metamorphism, granulites and crustal growth: Nature, **288**, 45-50.

Nilsen, T.H. and Clarke, S.H., Jr., 1975. Sedimentation and tectonics in the early Tertiary continental borderland of central California: U.S. Geol. Surv. Prof. Paper 925, 64 pp.

Oliver, G., 1988. Exposed cross-section of continental crust: Doubtful Sound Fiordland, New Zealand: Part One - Geophysical and geological setting: in: NATO Advanced Study Institutes Programme, International Advanced Course on Exposed Cross Sections of the Continental Crust (Abstract), 14.

Oliver, H.W., 1977. Gravity and magnetic investigations of the Sierra Nevada batholith, California: Geol. Soc. Am. Bull., **88**, no. 3, 445-461.

Percival, J.A., 1988. Kapuskasing: Update and Prospectus: in: NATO Advanced Study Institutes Programme, International Advanced Course on Exposed Cross Sections of the Continental Crust (Abstract), 15.

Pickett, D.A. and Saleeby, J.B., 1989, Petrogenetic and isotopic studies of mid-crustal batholithic rocks of the Tehachapi Mountains, Sierra Nevada, California: Geol. Soc. Am. Abst. with Prgms., v. 21.

Plescia, J.B. and Calderone, G.J., 1986. Paleomagnetic constraints on the timing and extent of rotation of the Tehachapi Mountains, California: Geol. Soc. Am. Abstracts with Programs, **18**, 171.

Ross, D.C., 1983. Generalized geologic map of the southern Sierra Nevada, California, showing the location of basement samples for which whole rock ^{18}O has been determined: U.S. Geol. Surv. Open-File Report, **83**, 904.

Rubin, C.M., Saleeby, J.B., Cowan, D.S., Brandon, M.T. and McGroder, M.F., 1989. Regionally extensive mid-Cretaceous west-vergent thrust system in the northwestern Cordillera: Implications for continent-margin tectonism: Geology, in press.

Saleeby, J.B., 1976. Onset of Cretaceous magmatism, southern Sierra Nevada batholith, California: Geol. Soc. Am. Abst. with Prgms., **9**, no. 7, 1084.

Saleeby, J.B., 1978. Kings River Ophiolite, Southwest Sierra Nevada Foothills, California: Geol. Soc. Am. Bull., **89**, 617-636.

Saleeby, J.B., 1979. Kaweah Serpentinite Melange, Southwest Sierra Nevada Foothills, California: Geol. Soc. Am. Bull., **90**, 29-46.

Saleeby, J.B., 1981. Ocean floor accretion and volcano-plutonic arc evolution of the Mesozoic Sierra Nevada: in: The Geotectonic Development of California, Rubey Volume 1 (Ed. W.E. Ernst), Englewood Cliffs, N.J.: Prentice-Hall, 132-181.

Saleeby, J.B., 1982. Polygenetic ophiolite belt of the California Sierra Nevada - geochronological and tectonostratigraphic development: J. Geophys. Res., **87**, 1803-1824.

Saleeby, J.B. and Sharp, W.D., 1980. Chronology of the structural and petrologic development of the southwest Sierra Nevada Foothills, California: Summary: Geol. Soc. Am. Bull., Part I, **91**, 317-320.

Saleeby, J.B. and Busby-Spera, C.V. 1986. Fieldtrip Guide to the metamorphic framework rocks of the Lake Isabella area, southern Sierra Nevada, California: Mesozoic and

Cenozoic structural evolution of selected areas, east-central California: Geol. Soc. Amer. Cordilleran Section Guidebook Volume, 81-94.

Saleeby, J.B., Goodwin, S.E., Sharp, W.D. and Busby, C.J., 1978. Early Mesozoic paleotectonic-paleogeographic reconstruction of the southern Sierra Nevada region: in: Mesozoic Paleogeography of Western United States, Pacific Coast Paleogeography Symp. 2 (Eds. D.G. Howell and K.A. McDougall), Pac. Sect., Soc. Econ. Paleontologists and Mineralogists, 311-336.

Saleeby, J.B., Sams, D.B. and Kistler, R.W., 1987. U/Pb zircon, strontium, and oxygen isotopic and geochronological study of the southernmost Sierra Nevada batholith, California: J. Geophys. Res., **92**, 10,443-10,466.

Saleeby, J.B. and 12 contributors, 1986. Ocean-continent transect, Corridor C2, Monterey Bay offshore to the Colorado Plateau: Geol. Soc. Am. Map and Chart Series TRA C2, 2 sheets Scale 1:500,000, 87 pp.

Saleeby, J.B., Kistler, R.W., Longiaru, S., Moore, J.G. and Nokelberg, W.J. Middle Cretaceous metavolcanic rocks in the Kings Canyon area, central Sierra Nevada, California: in: Geological Society of America Symposia Volume, Cordilleran Magmatism (Ed. J.L. Anderson), in press.

Sams, D.B., 1986. U/Pb zircon geochronology, petrology, and structural geology of the crystalline rocks of the southernmost Sierra Nevada and Tehachapi Mountains, Kern County, California: Calif. Inst. Tech. Ph.D. Thesis, 315 pp.

Sams, D.B. and Saleeby, J.B., 1988. Geology and petrotectonic significance of crystalline rocks of the southernmost Sierra Nevada, California: in: Rubey Volume on Cordilleran Metamorphism (Ed. W.G. Ernst) Prentice-Hall, Inc., Englewood Cliffs, N.J., 866-893.

Schenk, V., 1988. The crustal cross section of Southern Calabria (Italy): Lithology, thermal evolution and differentiation of a continental crust in the Hercynian fold belt: in: NATO Advanced Study Institutes Programme, International Advanced Course on Exposed Cross Sections of the Continental Crust (Abstract), 16.

Sharry, J., 1981. Geology of the western Tehachapi Mountains, California: Ph.D. dissertation, Massachusetts Institute of Technology, 215 pp.

Smithson, S.B., 1979. Aspects of continental crustal structure and growth: targets for scientific deep drilling: Contrib. to Geol., **17**, 65-75.

Stern, T.W., Bateman, P.C., Morgan, B.A., Newell, M.F. and Peck, D.L., 1981. Isotopic U-Pb ages of zircon from granitoids of the central Sierra Nevada, California: U.S. Geol. Surv. Prof. Paper 1185, 17 pp.

Tobisch, O.T., Saleeby, J.B. and Fiske, R.S., 1986. Structural history of continental volcanic arc rocks along part of the eastern Sierra Nevada, California: A case for extensional tectonics: Tectonics, 5, 65-94.

Weaver, B.L. and Tarney, J., 1981. Lewisian gneiss geochemistry and Archaean crustal development models: Earth Planet. Sci. Lett., 55, 171-180.

Wyllie, P.J., 1971. The Dynamic Earth: Textbook in Geosciences: (Eds. J. Wiley and Sons, Inc.), 416 pp.

Zoback, M.D., Westworth, C.M. and Moore, J.G., 1985. Central California seismic reflection transect: II - The eastern Great Valley and western Sierra Nevada: Amer. Geophys. Union, EOS, 65, no. 45, 1102.

CHARACTERISTICS OF A CONTINENTAL MARGIN MAGMATIC ARC AS A FUNCTION OF DEPTH: THE SKAGIT-METHOW CRUSTAL SECTION

Bryan Kriens[1] and Brian Wernicke
Department of Earth and Planetary Sciences
Harvard University
Cambridge, MA 02138

ABSTRACT. Calc-alkaline arc magmatism has been documented throughout NW Washington and SW British Columbia at around 95-85 Ma. Migmatites, orthogneisses, directionless plutons, and volcanic flows of mid-Cretaceous age are exposed in the region, and represent the major constituents of the arc. In the Ross Lake area of northern Washington, upper to middle crustal levels of the arc have been uplifted and exposed by large-scale Paleogene folding as a relatively intact vertical crustal section. Thus, the vertical variation in deformation, metamorphism, and magma behavior in a magmatic arc may be directly observed. Based on stratigraphic, structural, and petrologic data, the arc-section consists of a brittlely deformed upper crust containing a few plutons, and a ductilely deformed and migmatized middle crust containing numerous large plutons. Ductility in the largely quartzose country rocks and plutons is observed only below the uppermost limit of voluminous middle crustal plutons at about 12-14 km depth in the crustal section. Assuming the onset of ductility in quartz represents a high strength gradient in the crust, the high strength "lid" at the brittle-ductile transition is suggested to have caused large-scale ponding of magma. Ambient country rock viscosity thus appears to be a major control on the mechanics of magma ascent in this region.

1. INTRODUCTION

Arc magmatism is a first-order process in the lithosphere, and a major mechanism of crustal growth. While much is known about the map-view characteristics of arcs, little is known about their properties in the vertical dimension. Cross-sectional processes are usually inferred from magma chemistry, earthquakes, seismic velocity models, xenoliths, and other properties observable from uppermost crustal levels. Exposures of deeply eroded arcs are usually not

---

[1] Now at Department of Earth Sciences, California State University Dominguez Hills, Carson, CA 90747.

159

M. H. Salisbury and D. M. Fountain (eds.), Exposed Cross-Sections of the Continental Crust, 159–173.
© 1990 Kluwer Academic Publishers.

simply "connected" to the surface; thus, even if it is assumed that a given deep-crustal tract represents a portion of an arc, upper levels are either not exposed or are found far away from the deep exposures, making it difficult to prove that the various levels of exposure constitute a representative section through a single arc. Under such circumstances, processes observed at one level of exposure may not necessarily be related to those in another.

Below we discuss the characteristics of an intact, near-vertical, ca. 25 km crustal section through a mid-Cretaceous magmatic arc, located in the North Cascades Mountains of Washington. The data which support the existence of the crustal section, and the interpretation of the geologic setting in which it has evolved, have been presented and discussed in detail elsewhere (Kriens and Wernicke, 1986; Kriens and Wernicke, 1987; Kriens, 1988; Kriens and Wernicke, 1990; see also Whitney and McGroder, 1989 for an alternative interpretation). In this paper, we summarize the geologic setting of the crustal section and describe the arc from shallow to deep levels. We then examine the implications of these observations for models of arc evolution and magma ascent.

2. SUMMARY OF OBSERVATIONS

2.1. Geologic Setting of the Skagit-Methow Crustal Section

The Skagit-Methow crustal section is located in the Ross Lake area of northern Washington (Fig. 1). We briefly describe here the Cretaceous-Tertiary tectonic evolution of this area. For a detailed account, see Kriens (1988) and Kriens and Wernicke (1990).

A study of the Ross Lake area was begun in 1984 to test a dextral-slip hypothesis proposed by Misch (1966) for the NNW-trending contact between two large continental fragments, the crystalline Skagit Metamorphic Suite (Misch, 1966, 1988) and the Jurassic(?) to Cretaceous volcano-sedimentary Methow sequence (Barksdale, 1975). Detailed mapping of this contact (named the Ross Lake fault by Misch, 1966) in the Ross Lake area shows that the brittle faults and mylonites thought by Misch to express dextral shearing along the contact are not continuous, throughgoing structures, and that the contact between Skagit orthogneisses (quartz diorite orthogneiss of Fig. 1) and Methow sequence strata is primarily intrusive. In addition, the mylonites throughout the contact zone show both right- and left-lateral shear. The Skagit-Methow contact is therefore not a major terrane boundary but rather a tectonized intrusive contact that has not accommodated significant motion between the two regions. Based on these observations and regional synthesis of timing relations, deformation in the contact zone is interpreted to express regional ENE-WSW shortening, not major dextral-slip, in the early Tertiary. The Paleogene dextral strike-slip faults and mylonite zones mapped to the north (Haugerud, 1985) and south (Miller et al., 1985; Miller and Walker, 1987; Miller, 1988; Miller et al., 1988) of the Ross Lake area can be interpreted in light of the evidence for regional early Tertiary shortening as relatively minor components of the regional fault system that records tectonic escape (Kriens et al., 1987; Kriens, 1988). Alternatively, they may record a shift

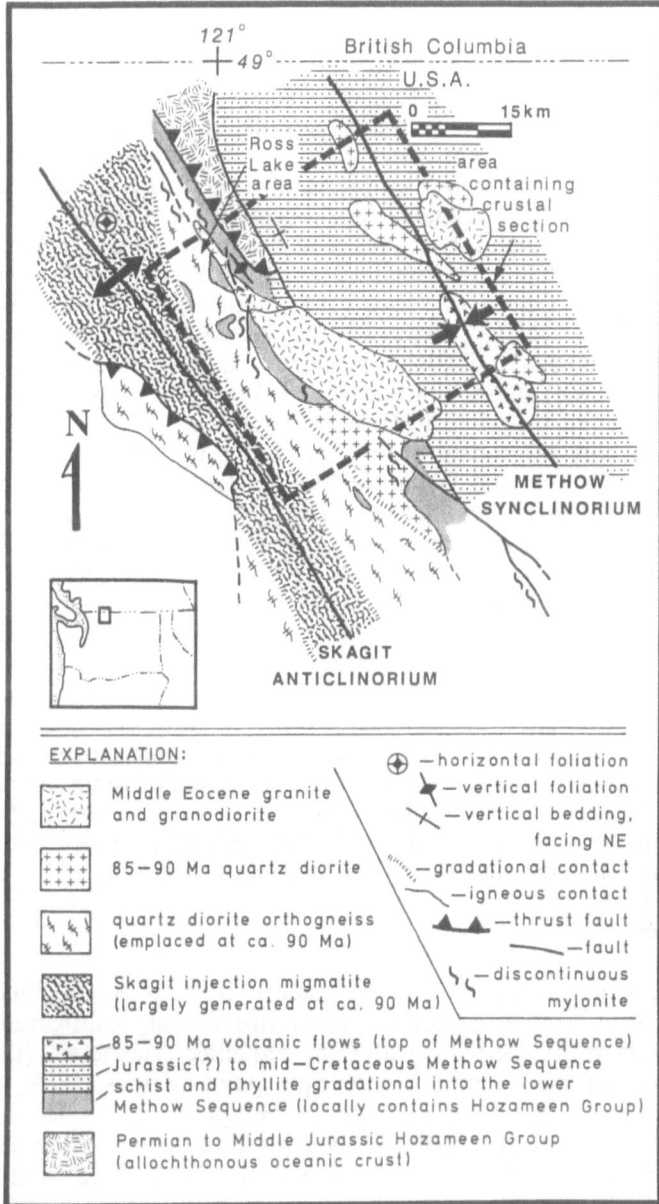

Fig. 1. Geologic map of the Ross Lake area, compiled from Misch (1966), Barksdale (1975), Roddick, et al. (1979), and Kriens (1988). Rock units older than the Eocene plutons and lying within the dashed square have been projected onto an ENE-WSW cross section, and subsequently restored to an original vertical crustal section (see Fig. 2).

from ENE-WSW shortening to NW-SE dextral shear in Eocene time.

Since no terrane-bounding fault exists in the Ross Lake area, the Cretaceous-Tertiary history of the area can be interpreted as a sequence of primarily four events: 1) Jurassic(?) to Early Cretaceous deposition of the lower Methow sequence (Barksdale, 1975) in a forearc-rift basin setting; 2) 110-95 Ma, east-vergent thrusting of late Paleozoic to Mesozoic Hozameen Group ocean floor basalts over these sedimentary rocks, with concomitant development of a west-derived foredeep (the Virginian Ridge Formation of Barksdale, 1975); 3) intrusion of the thrust/sedimentary sequence at ca. 90 Ma by arc magmas, which in addition caused deformation, regional metamorphism, and formation of some, if not all, of the Skagit Gneiss of Misch (1966); and 4) 65-50 Ma, ENE-WSW shortening, which caused folding, mylonitization, brittle faulting, and uplift of the crustal section (Kriens, 1987, 1988). Undeformed ca. 50-45 Ma plutons (Tabor et al., 1968; Hoppe, 1984) subsequently intruded the region; they are shown only in Figure 1 and are not discussed further.

The early Tertiary shortening event produced a 60 km wavelength/12 km amplitude fold pair (Kriens and Wernicke, 1987), here referred to as the Skagit anticlinorium and Methow synclinorium. The Skagit-Methow crustal section occupies the steeply-dipping common limb of the fold pair in the Ross Lake area (Fig. 1). Only one throughgoing fault, the Hozameen fault (Misch, 1966), is found to lie within the region occupied by the crustal section. It has been interpreted as a high angle reverse fault of modest displacement (5-7 km, west-side-up) and is constrained to have formed during or closely following the waning stages of early Tertiary shortening (Kriens, 1988). Because unmetamorphosed lower Methow sediments are found on both sides of the fault (Misch, 1966; Kriens, 1988; Kriens and Wernicke, in press), and Hozameen Group clast/source rock ties are also documented across the fault (Tennyson and Cole, 1978; Trexler and Bourgeois, 1985), we believe that it is not a terrane boundary. The presence of unmetamorphosed lower Methow sediments on both sides of the fault implies that little omission of crustal levels has occurred.

2.2. Spatial Observations

Using structural, stratigraphic, and geobarometric constraints, the limb region containing the crustal section is restored to its pre-Tertiary configuration (Fig. 2). The depth scale is assigned using Methow sequence thicknesses (Barksdale, 1975), the location of the aluminosilicate triple point (Kriens, 1988), and P-T estimates from mineral rim compositions in metapelitic rocks found in and adjacent to the Skagit migmatitic core. Pressures are ca. 6.4 kb in the migmatitic core, and ca. 4.5 kb near the aluminosilicate triple point (D.L. Whitney, oral communication, 1988). Pressures of ca. 9 kb from garnet cores in both localities, (Whitney and McGroder, 1989) are considered by us in light of field relations to record an earlier, pre-90 Ma high-P metamorphic event (Kriens and Wernicke, 1990).

From top to bottom, the principle features of the crustal section are: 0-3 km, andesite, pyroxene andesite, and dacite flows interbedded with fluviatile

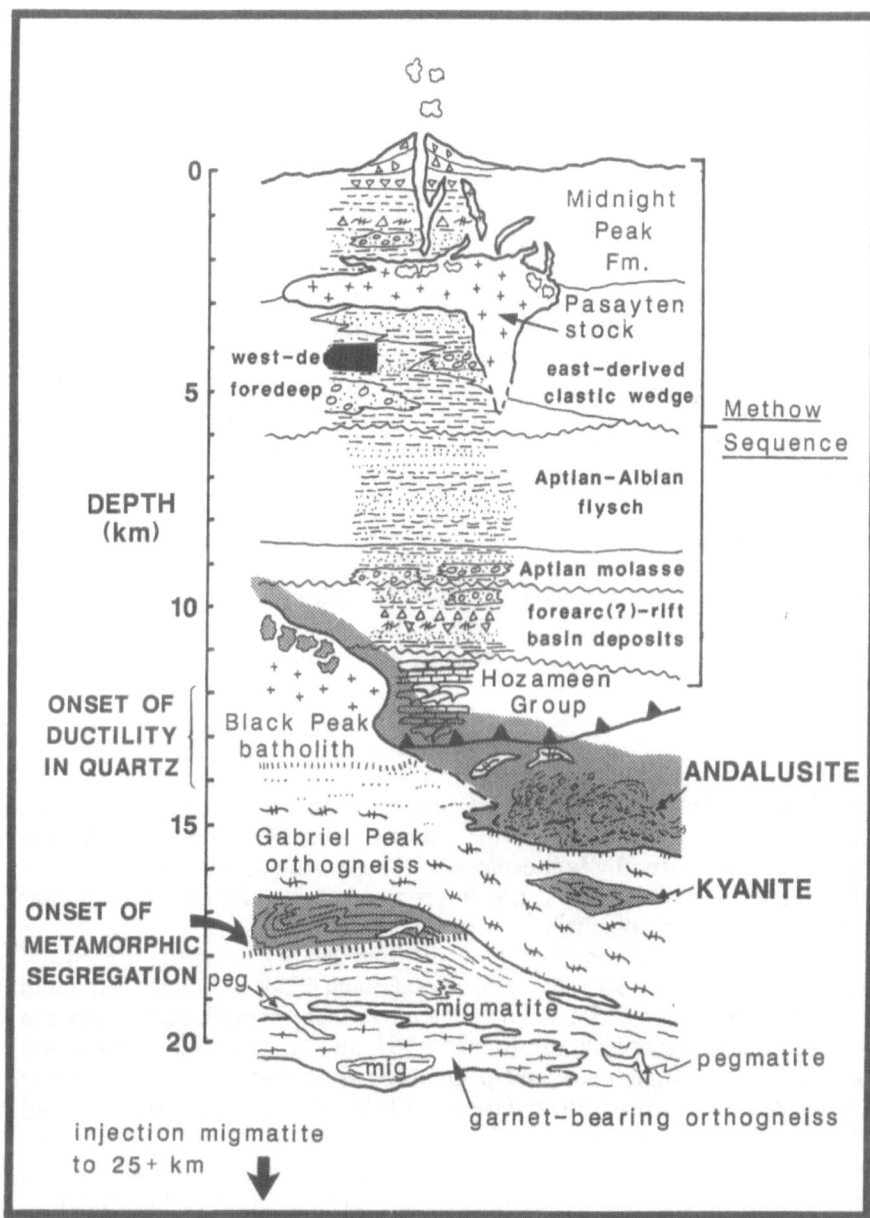

Fig. 2. The Skagit-Methow crustal section, showing the principle features of a continental margin magmatic arc (see text for description). Arc intrudes crust soon after thickening by thrusting. Shaded area denotes region of low to medium grade metamorphic rocks (protolith: Methow Sequence and locally imbricated slivers of Hozameen Group).

sediments (the Midnight Peak Formation of Barksdale, 1975); 2-10 km, patchy, discordant, tabular, directionless quartz diorite stocks locally intruding their volcanic roof via stoping and inflation (the ca. 88 Ma Pasayten and Rock Creek stocks; Tabor et al., 1968; Barksdale, 1975; ages recalculated by R.A. Haugerud, written communication, 1989); 10-13 km, relatively continuously exposed quartz diorite plutons with discordant contacts and numerous stoped blocks of little-deformed country rock in roof regions, passing gradationally downward into foliated, concordant orthogneisses containing complexly deformed xenoliths and septa of metamorphosed country rock (the 90 Ma Black Peak Batholith/Gabriel Peak Orthogneiss of Misch, 1966, and Adams, 1964); 12-14 km, onset of ductile behavior in quartz-bearing rocks; 14-16 km, andalusite to kyanite/sillimanite transition; 13-18 km, continuously exposed orthogneiss; 18-25(?) km, gradation downward from metasediments and dioritic orthogneiss into concordant garnet-bearing orthogneiss and metamorphically segregated paragneiss (migmatite) pervaded by heterogeneously deformed pegmatites (Skagit Gneiss of Misch, 1966; 90-60 Ma based on U/Pb ages of Mattinson, 1972).

2.3. Temporal Observations

At around 90 Ma the Skagit-Methow area was intruded by quartz diorite, with lesser diorite and granodiorite (Misch, 1966; Tabor et al., 1968; Mattinson, 1972; Misch, 1988). In the Methow region, andesitic to dacitic flows of the Turonian Midnight Peak Formation (Barksdale, 1975) were extruded. These flows and the underlying Cenomanian foredeep are intruded by ca. 88 Ma (K/Ar, biotite, hornblende) quartz diorites (e.g. Pasayten and Rock Creek Stocks; Tabor et al., 1968; ages recalculated by R.A. Haugerud, written communication, 1989). The stocks were more-or-less quenched at the time of their emplacement, since the K/Ar ages are not much younger than the country rocks which they intrude, and contact aureoles are virtually undeveloped. These stocks, because they are roughly the same age as the flows, are interpreted to be slightly later plutons from the same magmatic episode which gave rise to the flows.

In the Skagit region, some of the intrusive bodies fringing the Skagit migmatite yield concordant 90 Ma U/Pb ages (Black Peak batholith, Eldorado orthogneiss), yet the Gabriel Peak Orthogneiss (which is gradational into the Black Peak Batholith according to Adams, 1964 and Misch, 1966, 1988), and Skagit paragneisses, orthogneisses, and pegmatites yield discordant U/Pb zircon ages of 90-60 Ma (Mattinson, 1972; Hoppe, 1984). These rocks are suggested to have been emplaced at the same time as the 90 Ma igneous rocks in the Skagit border and Methow regions (Kriens, 1988). Some, if not all, of the migmatization may also be ca. 90 Ma old, on the basis of cross-cutting relations with Skagit orthogneisses and pegmatites. As an explanation for the spread in ages, the discordant migmatites and igneous rocks are hypothesized to have exhibited varying degrees of post-90 Ma diffusion of Pb as a result of occupying deep crustal positions prior to folding and uplift. Deep crustal positions and high temperatures are apparent from P-T studies of paragneissic pelitic rocks from the central portion of the Skagit migmatite. Mineral core P-T conditions are 720°C,

9-10 kb, and mineral rim conditions are 570-600°C, 6.4 kb (Whitney and McGroder, 1989; Whitney, oral communication, 1988). The mineral rim conditions yield a minimum depth of 23 km at 28 MPa/km geobaric gradient. An alternative interpretation is that there was a major ca. 65 Ma igneous crystallization event (Miller et al., 1988). However, the observation of analogous behavior in K/Ar systems in the area (ca. 85-90 Ma in the Methow area and Black Peak batholith, ca. 55-45 Ma in Skagit Gneiss; Tabor et al., 1968; Misch, 1964) supports the hypothesis of high temperatures in the Skagit core, and therefore the possibility of Pb diffusion. Isotopic data from the Gabriel Peak/ Black Peak unit also supports the hypothesis; Pb206/U238 and Pb207/U235 ages of zircons from this composite unit decrease as the Skagit migmatitic core is approached from the east. Although we do not discount the possibility of some deep crustal intrusions at about 65 Ma ago, it is worth noting that much of the northwestern Cordillera is characterized by a Late Cretaceous magmatic lull (Armstrong, 1988). Thus, it is suggested that prolonged high temperatures, not renewed intrusion, are the principle control of post-90 Ma U/Pb ages from the Skagit Complex.

Isotopic and stratigraphic data summarized above (and detailed in Kriens, 1988) point to a mid-Cretaceous (85-95 Ma) emplacement age for all major components of the arc. Cross-cutting relations (Kriens, 1988; Kriens and Wernicke, 1990) between the rock units yield the following evolution of the vertical structure of the arc (Fig. 3). First, lobate bodies and apophyses of garnet-bearing quartz diorite were emplaced at 18-25(?) km, and comprise most of the Skagit injection migmatite. Multiple cross-cutting relations within these rocks indicate that ductile flow and development of a subhorizontal foliation occurred during and after emplacement (Kriens, 1988). Next, trondjhemitic pegmatites cut the garnet-bearing orthogneiss foliation. Thus, these dikes are interpreted to have been intruded during the waning stages of ductile deformation which followed intrusion of the garnet-bearing orthogneisses. Ductile deformation and substantial magmatic intrusion occurred prior to and after pegmatite emplacement. These pegmatites are not found above about 16 km depth. Finally, large volumes of quartz diorite intruded as sill-like bodies immediately above the migmatite complex. Within the 12-18 km level, intrusive rocks make up about 75-80% of the outcrop. They exhibit foliation and contacts concordant with metasedimentary country rocks. Above 12 km, intrusive rocks make up only about 15-20% of the outcrop, and are largely unfoliated bodies containing numerous stoped blocks near their margins (Kriens, 1988).

The crustal section is thus divisible into two zones of contrasting intrusive and deformational style. Below 12 km depth, migmatization, pegmatite emplacement, voluminous intrusion, syn-intrusive metamorphism and ductile deformation are observed. Above 12 km, brittle stoping and intrusion, concentrated at the base of the volcanic pile, is volumetrically minor. Contact metamorphism around the plutons is negligible, little foliation is developed in the plutons, and country rocks are not ductilely deformed.

166

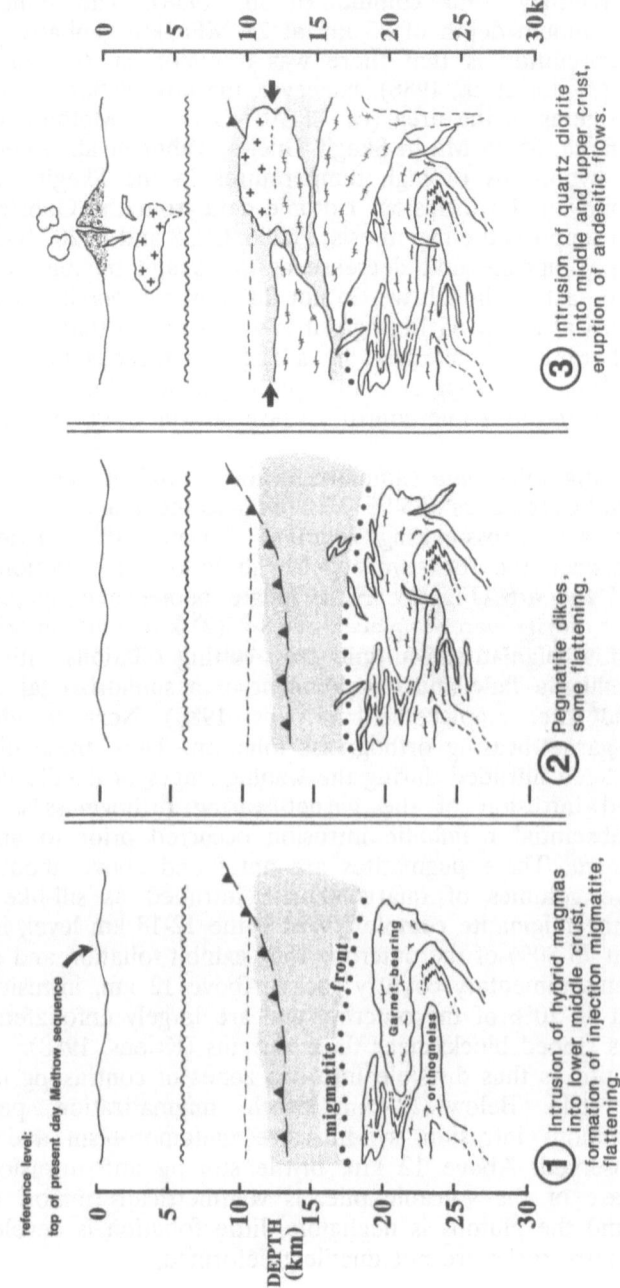

SKAGIT–METHOW ARC EVOLUTION

reference level:
top of present day Methow sequence

DEPTH (km)

migmatite front

Garnet-bearing

orthogneiss

① Intrusion of hybrid magmas into lower middle crust, formation of injection migmatite, flattening.

② Pegmatite dikes, some flattening.

③ Intrusion of quartz diorite into middle and upper crust, eruption of andesitic flows.

Fig. 3. Simplified evolution of the mid-Cretaceous Skagit-Methow arc, based on cross-cutting relations and information from Fig. 2. Shaded area denotes region of low to medium grade metamorphic rocks. Arrows in stage 3 refer to the approximate location of the proposed crustal strength maximum. Wavy line indicates the unconformity above the Aptian-Albian flysch of Fig. 2. Dashed line indicates the inferred unconformity above the Hozameen Group shown in Fig. 2. Solid line at 3 km in step 3 indicates the base of the Midnight Peak Formation. Thrust fault shown is same as that in Fig. 2.

3. VISCOUS ENTRAPMENT OF MAGMAS

An explanation for the contrasting intrusive behavior, deformational style, and marked reduction in magma volume above 12 km is afforded by considering the field observations in light of laboratory measurements. In the 12-14 km interval, brittlely deformed, well-bedded quartzose country rocks become ductilely deformed metamorphic tectonites. In plutons at this level, stoped margins give way downward to ductilely deformed, foliated, concordant contacts. These observations suggest that the brittle-ductile transition for quartzose rocks lay at about 12-14 km during intrusion. Laboratory experiments have shown that a crustal strength maximum may occur at the transition from brittle to ductile behavior (Brace and Kohlstedt, 1980). A simple explanation of these observations may be that this high strength "lid" caused ponding of large volumes of quartz diorite magma in the middle crust.

We interpret the field relations to indicate that in order to move past the high strength lid, the plutons had to stope the roof rock or move through brittle cracks, while below this lid they ascended by ductile shouldering aside of country rock, probably by Stokes-type flow (e.g. Marsh, 1982). We suggest that as the magma reached the vicinity of the high strength lid, Stokes flow became ineffective as a result of increasing country rock viscosity. Most of the plutons were unable to meet the physical requirements for stoping, as evidenced by the large volume of magma which solidified at the uppermost middle crustal level.

Field observations suggest that Skagit magmas have ascended in the middle crust by diapiric rise in a near-solidus or above-solidus state (apophyses are locally present, country rock is ductilely deformed, and sub-solidus pluton foliation occurred after emplacement, during cooling). Magma ascent by diapirism can be modeled via the familiar Stokes equation (e.g. Turcotte and Schubert, 1982):

$$U = a^2 g (p_{cr} - p_m) / 3\mu_{cr} \tag{1}$$

(where a = radius of magma body, μ_{cr} = country rock viscosity, U = velocity of ascent, and ($p_{cr} - p_m$) = density difference between country rock and magma), or by the modified Stokes equation of Marsh (1982):

$$U = 2d^2 g (p_{cr} - p_m) / 3A\mu'_{cr} \tag{2}$$

(where d = thickness of the softened aureole - in part a function of magma body radius, μ'_{cr} = smallest value of country rock viscosity in the aureole, and A = coefficient reflecting the variation in viscosity across the aureole). For plutons substantially hotter than their country rock, μ'_{cr} is 5-15 orders of magnitude < μ_{cr} and A may range from 10-35 (values are based on approximating the temperature profile in the aureole as a simple exponential function; Marsh, 1982). In general, the environment of ascent of the large orthogneiss bodies in the Ross Lake area was through country rock composed of young migmatite, whose ambient temperature was probably near its melting temperature. Thus, μ_{cr} and μ'_{cr} of (1) and (2) are probably not markedly different, corresponding to a value of A = 1, with the exception of the highest levels of intrusion, where temperature contrasts are permissibly large, in which case Equation (2) with a larger value of A would more closely approximate magmatic ascent. Therefore, as a first approximation, we investigate the effects of parameters which appear in the traditional Stokes equation in the context of magma rise. Specifically, we are concerned with the relative importance of each parameter on the cessation of magma rise in the middle crust.

Two variables, viscosity and density contrast, are of primary importance. Magma body size is not considered to be important because plutons probably do not decrease appreciably in size during ascent prior to cooling. Magma thus ceases to rise as a result of either increased country rock viscosity or decreased density differences between magma and country rock. In the Skagit-Methow crustal section, the close correspondence between the level of the crust in which country rock viscosity increases rapidly (i.e. the brittle-ductile transition; Strehlau and Meissner, 1987) and the level at which voluminous plutons occur suggests that the viscosity term is important. Holding density differences constant, and using viscosity vs. depth data (see Fig. 8 of Strehlau and Meissner, 1987), the magma velocity may decrease two to three orders of magnitude in 10 km of ascent in the vicinity of the brittle-ductile transition. Holding viscosity constant, and assuming a seismically defined, maximum variation in country rock density from 2.9 to 2.6 g/cc within this same depth interval (e.g. Fuis et al., 1984), and a constant magma density of 2.5 g/cc (the high end of values for felsic to intermediate melts; Rivers and Carmichael, 1987), the velocity of ascent may be reduced to as little as 25 % of its initial value, a substantial amount but notably less than the viscosity effect. Ascent velocity of silicic plutons is thus influenced more by viscosity contrasts than density contrasts near the brittle-ductile transition. Although a small fractional change in density may stop rise if a new, lighter country rock is encountered, densities of felsic to intermediate melts (e.g. Rivers and Carmichael, 1987) and deep-sedimentary to metasedimentary rocks (Daly et al., 1966) show no overlap (2.4-2.5 and 2.7-2.8 g/cc, respectively). Thus country rock viscosity alone may account for the ponding of the Skagit magmas, assuming that the percentage of crystals in the magmas was low enough to keep the density of the magma close to the density range measured by Rivers and Carmichael (1987) (i.e. significant density increase resulting from crystallization of melt has not occurred). The control of density contrast on magma ascent at the brittle-ductile transition in continental crust may be important only for systems involving higher density basaltic melts (e.g. Glazner and Ussler, 1988) or higher density crystal-rich (near-solidus) felsic magmas.

Because few magma bodies were able to ascend into the upper crust by brittle stoping, it appears that for the most part the magmatic system was unable to generate deviatoric stresses necessary to cross the high strength lid, even though the density contrast driving ascent did not appreciably change. Ascent by softening of a thermal boundary layer around the pluton (Equation 2) appears not to have taken place, i.e. most of the magmas stopped at precisely the depth where the mechanism may have become important, and the few plutons that continued to ascend did so without developing contact aureoles.

It is worth noting, however, that arcs such as the southern Sierra Nevada (e.g. Sams and Saleeby, 1988) contain voluminous upper crustal magma accumulations, often with ductilely deformed contact aureoles. Thus, although the "viscous entrapment" mechanism proposed here may account for the Skagit-Methow arc, it may not account for ascent in arcs in which widespread shallow crustal plutons are observed. In such arcs, thermal erosion of the high-strength lid (Wernicke, 1986) and Marsh's (1982) modified Stokes flow may be the important factors. Thus, there may be two basic types of arc: those in which the total magmatic volume supplies enough thermal energy to thermally erode the high-strength lid, and those in which it does not.

4. CONCLUSION

The migmatites, orthogneisses, plutons, and volcanic flows in the Skagit-Methow crustal section appear to be coeval igneous components of the mid-Cretaceous continental margin magmatic arc of NW Washington-SW British Columbia. The most striking feature of the section is the cospatial onset of ductile deformation and appearance of voluminous orthogneiss at about 12-14 km depth. Deformational style in both plutons and country rocks indicates that the brittle-ductile transition in quartzose rocks was located at this depth during intrusion. Apparently, the brittle-ductile transition, or viscosity maximum, was responsible for trapping magma in the middle crust of the Skagit-Methow arc. Although heat supplied by the ascending magmas may have elevated the geotherm, the temperatures and depths indicated by petrologic thermobarometry do not suggest an abnormally high geotherm. The Skagit plutons likely ascended by Stokes-type flow and subsequently spread out laterally under the high strength lid. The close association between the region of voluminous pluton accumulation and the level in the crust at which ambient country rock viscosity rapidly increases, and the probability that the magmas were less dense than the country rock on either side of the transition, suggests independently that emplacement (final entrapment) was not controlled by density differences between magma and wall rock.

The Skagit-Methow crustal section may represent one of principally two kinds of magmatic system, one in which magma ascent is primarily limited to the middle crust, as opposed to one in which magma is widespread in both upper and middle crust (e.g. the Sierra Nevada arc). The type of system obtained apparently depends on whether the strength maximum can be thermally eroded (Wernicke, 1986) or mechanically exceeded by a combination of regional and pluton-related heat flow and deviatoric stress. The heat of early migmatization and pegmatite emplacement appears to have softened the ascent path for the orthogneisses, but the plutons were never able to reach a regime where thermal aureoles around individual plutons developed (e.g. Marsh, 1982). Most of the ascent occurred through rock at or very near its melting temperature. By volume, we suggest that ascent of plutons through the brittle crust may be a less common process than Stokes flow through sufficiently hot middle crustal country rock, primarily because of ambient country rock viscosity.

Acknowledgements

We are extremely grateful to Warren Hamilton, John Bartley, and several anonymous reviewers for their helpful comments on the manuscript.
We also wish to thank Donna Whitney for her permission to discuss her unpublished thermobarometric data. This project was funded by NSF grant EAR-84-51181 awarded to Brian Wernicke.

References

Adams, J.B., 1964, Origin of the Black Peak quartz diorite, Northern Cascades, Washington: American Journal of Science, v. 262, p. 290-306.

Armstrong, R.L., 1988, Mesozoic and Early Cenozoic magmatic evolution of the Canadian Cordillera, in Clark, S.P., B.C. Burchfiel, and J. Suppe, eds., Processes in Continental Lithospheric Deformation: Geological Society of America Special Paper 218, p. 55-91.

Barksdale, J.D., 1975, Geology of the Methow Valley, Okanogan County, Washington: State of Washington, Department of Natural Resources, Division of Geology and Earth Resources Bulletin 68, 72 p.

Brace, W.F., and Kohlstedt, D.L., 1980, Limits on lithospheric stress imposed by laboratory experiments: Journal of Geophysical Research, v. 85, n. B11, p. 6248-6252.

Daly, R.A., G.E. Manger, and S.P. Clark, 1966, Density of rocks: Geological Society of America Memoir 97, p. 20-26.

Fuis, G.S., W.D. Mooney, J.H. Healy, G.A. McMechan, and W.J. Lutter, 1984, A seismic refraction survey of the Imperial Valley region, California: Journal of Geophysical Research, v. 89, n. B2, p. 1165-1189.

Glazner, A.F., and Ussler III, W., 1988, Trapping of magma at midcrustal density discontinuities: Geophysical Research Letters, v. 15, n. 7, p. 673-675.

Haugerud, R.A., 1985, Geology of the Hozameen Group and Ross Lake shear zone, Maselpanik area, North Cascades, southwest British Columbia [Ph.D. thesis]: Seattle, Washington, University of Washington, 269 p.

Hoppe, W.J., 1984, Origin and age of the Gabriel Peak Orthogneiss, North Cascades, Washington [M.Sc. thesis]: Lawrence, Kansas, University of Kansas, 79 p.

Kriens, B., 1987, Cretaceous-Tertiary tectonic evolution of the North Cascades, Washington: new findings from the Ross Lake fault zone and vicinity: Geological Society of America Abstracts with Programs, v. 19, n. 6, p. 396.

Kriens, B., 1988, Tectonic Evolution of the Ross Lake Area, Northwest Washington-Southwest British Columbia [Ph.D. thesis]: Cambridge, Massachussetts, Harvard University, 214 p.

Kriens, B., E. Aliberti, and B. Wernicke, 1987, Early Eocene convergent deformation in the North Cascades and SW British Columbia - occlusion tectonics vs. dextral slip: International Union of Geodesy and Geophysics Abstracts, 19th General Assembly, v. 1, p. 111.

Kriens, B., and Wernicke B., 1986, Crustal sections and arc magmatism: new findings from the Ross Lake fault zone, North Cascades, Washington (abstract): EOS, v. 67, n. 44, p. 1189.

Kriens, B., and Wernicke, B., 1987, Characteristics of a continental margin magmatic arc with depth: the Skagit-Methow crustal section: Geological Society of America Abstracts with Programs, v. 19, n. 7, p. 733.

Kriens, B., and Wernicke, B., 1990, Nature of the contact zone between the North Cascades crystalline core and the Methow sequence in the Ross Lake area, Washington; implications for Cordilleran tectonics: Tectonics, v. 9 (in press).

Marsh, B.D., 1982, On the mechanics of igneous diapirism, stoping, and zone melting: American Journal of Science, v. 282, p. 808-855.

Mattinson, J.M., 1972, Ages of zircons from the Northern Cascade Mountains, Washington: Geological Society of America Bulletin, v. 83, p. 3769-3784.

Miller, R.B., 1988, Possible role of ductile shear zones in linking overstepping strike-slip faults, Ross Lake fault zone, North Cascades, Washington: Geological Society of America Abstracts with Programs, v. 20, n. 7, p. A108.

Miller, R.B., S.A. Bowring, and W.J. Hoppe, 1988, New evidence for extensive Paleogene plutonism and metamorphism in the crystalline core of the North Cascades: Geological Society of America Abstracts with Programs, v. 20, n. 5, p. 432-433.

Miller, R.B., P. Misch, and W.J. Hoppe, 1985, New observations on the central and southern segments of the Ross Lake fault zone, North Cascades,

Washington: Geological Society of America Abstracts with Programs, v. 17, n. 6, p. 370.

Miller, R.B., and N.W. Walker, 1987, Structure and age of the Oval Peak Batholith: implications for the Ross Lake Fault Zone, Washington: Geological Society of America Abstracts with Programs, v. 19, n. 6, p. 433.

Misch, P., 1964, Age determinations on crystalline rocks of Northern Cascade Mountains, in Kulp, J.L., Senior Investigator, Investigations in Isotopic Geochemistry: U.S. Atomic Energy Commission Publication NYO-7243, Appendix D, Columbia University, Lamont Geological Observatory, Palisades, New York, Appendix D, pp.1-15.

Misch, P., 1966, Tectonic evolution of the Northern Cascades of Washington State: Canadian Institute of Mining and Metallurgy Special Volume 8, p. 101-148.

Misch, P., 1988, Tectonic and metamorphic evolution of the North Cascades: an overview, in Ernst, W.G., ed., Metamorphism and Crustal Evolution of the Western United States: Englewood Cliffs, New Jersey, Prentice-Hall, p. 180-195.

Rivers, M.L., and Carmichael, I.S.E., 1987, Ultrasonic studies of silicate melts: Journal of Geophysical Research, v. 92, n. B9, p. 9247-9270.

Roddick, J.A., J.E. Muller, and A.V. Okulitch, 1979, Fraser River 1:1,000,000 geological map: Geological Survey of Canada, sheet 92, map 1386A.

Sams, D.B., and Saleeby, J.B., 1988, Geology and petrotectonic significance of crystalline rocks of the southernmost Sierra Nevada, California, in Ernst, W.G., ed., Metamorphism and Crustal Evolution of the Western United States: Englewood Cliffs, New Jersey, Prentice-Hall, p. 865-893.

Strehlau, J., and Meissner, R., 1987, Estimation of crustal viscosities and shear stresses from an extrapolation of experimental steady state flow data, in Fuchs, K., and Froidevaux, C., eds., Composition, Structure, and Dynamics of the Lithosphere - Asthenosphere System: American Geophysical Union Geodynamics Series, v. 16, p. 69-87.

Tabor, R.W., J.C. Engels, and M.H. Staatz, 1968, Quartz diorite-quartz monzonite and granite plutons of the Pasayten River area, Washington - petrology, age, and emplacement, in Geological Survey Research 1968: U.S. Geological Survey Professional Paper 600-C, p. C45-C52.

Tennyson, M.E., and Cole, M.R., 1978, Tectonic significance of Upper Mesozoic

Methow-Pasayten Sequence, northeastern Cascade Range, Washington and British Columbia, in Howell, D.G. and K.A. McDougall, eds., Mesozoic Paleogeography of the Western United States, Society of Economic Paleontologists and Mineralogists, Pacific Coast Paleogeography Symposium, p. 499-508.

Trexler, J.H., Jr., and Bourgeois, J., 1985, Evidence for mid-Cretaceous wrench faulting in the Methow basin, Washington: tectonostratigraphic setting of the Virginian Ridge Formation: Tectonics, v. 4, p. 379-394.

Turcotte, D.L., and Schubert, G., 1982, Geodynamics: Applications of Continuum Physics to Geological Problems: New York, New York, John Wiley and Sons, 450 p.

Wernicke, B., 1986, The Basin and Range moho, 5-10 second reflectivity, and a simple physical model for continental rift magmatism (abstract): EOS, v. 67, p. 1184.

Whitney, D.L., and McGroder, M.F., 1989, Cretaceous crustal section through the proposed Insular-Intermontane suture, North Cascades, Washington: Geology, v. 17, p. 555-558.

THE EVOLUTION OF THE KAMILA SHEAR ZONE, KOHISTAN, PAKISTAN.

P. J. Treloar[1], K. H. Brodie[1*], M. P. Coward[1], M. Q. Jan[2],
M. A. Khan[2], R. J. Knipe[3], D.C. Rex[2] and M. P. Williams[1].

[1]Department of Geology, Imperial College, London, SW7 2BP, UK
[2]National Centre of Excellence in Geology, University of
 Peshawar, Peshawar, North West Frontier Province, Pakistan.
[3]Department of Earth Sciences, Leeds University, Leeds, LS2
 9JT, UK
[*]now at: Department of Geology, The University, Manchester, MI3
 9PL, UK

ABSTRACT. The Kamila Shear Zone is a deep to mid crustal structure
developed within the Kohistan island arc complex of North Pakistan prior
to Himalayan collision between Kohistan and India. Meta-gabbros of the
Chilas complex were transported southwards across the shear zone onto a
stack of high pressure rocks that had been assembled in the hanging wall
of the Tethyan subduction zone. The shear zone is constituted by an
anastomosing array of amphibolite facies ductile high strain zones within
which fabric intensity varies although mylonitic zones are common. Shear
criteria and kinematic indicators have a consistent SW-vergence. Little
microstructural evidence for the high temperature ductile shearing is
preserved, fabrics having been over-printed by post-deformational
processes of recrystallisation, annealing and grain growth. A subsequent
history of exhumation during decreasing temperature is documented by a
progressive sequence of down temperature retrogression and deformation in
superimposed shear zones which predominantly affected coarse grained
rocks unaffected by the earlier crystal plastic deformation. The shearing
involved dominantly cataclastic amphibolite to greenschist facies
deformation culminating in lower greenschist facies shearing and
mylonitisation, and the development of a distributed network of minor
cataclastic and gouge-filled fault rocks.

1. INTRODUCTION

In NW Pakistan (Fig. 1) the Indian and Asian Plates, which to the east
are juxtaposed across the Indus-Tsangpo Suture Zone (Searle et al.,
1987), are separated by the dominantly basic to intermediate rocks of the
Kohistan complex. This was the last fragment to be accreted to the Asian
Plate before Himalayan collision forming a suture, the "Northern Suture"

M. H. Salisbury and D. M. Fountain (eds.), Exposed Cross-Sections of the Continental Crust, 175–214.
© 1990 Kluwer Academic Publishers.

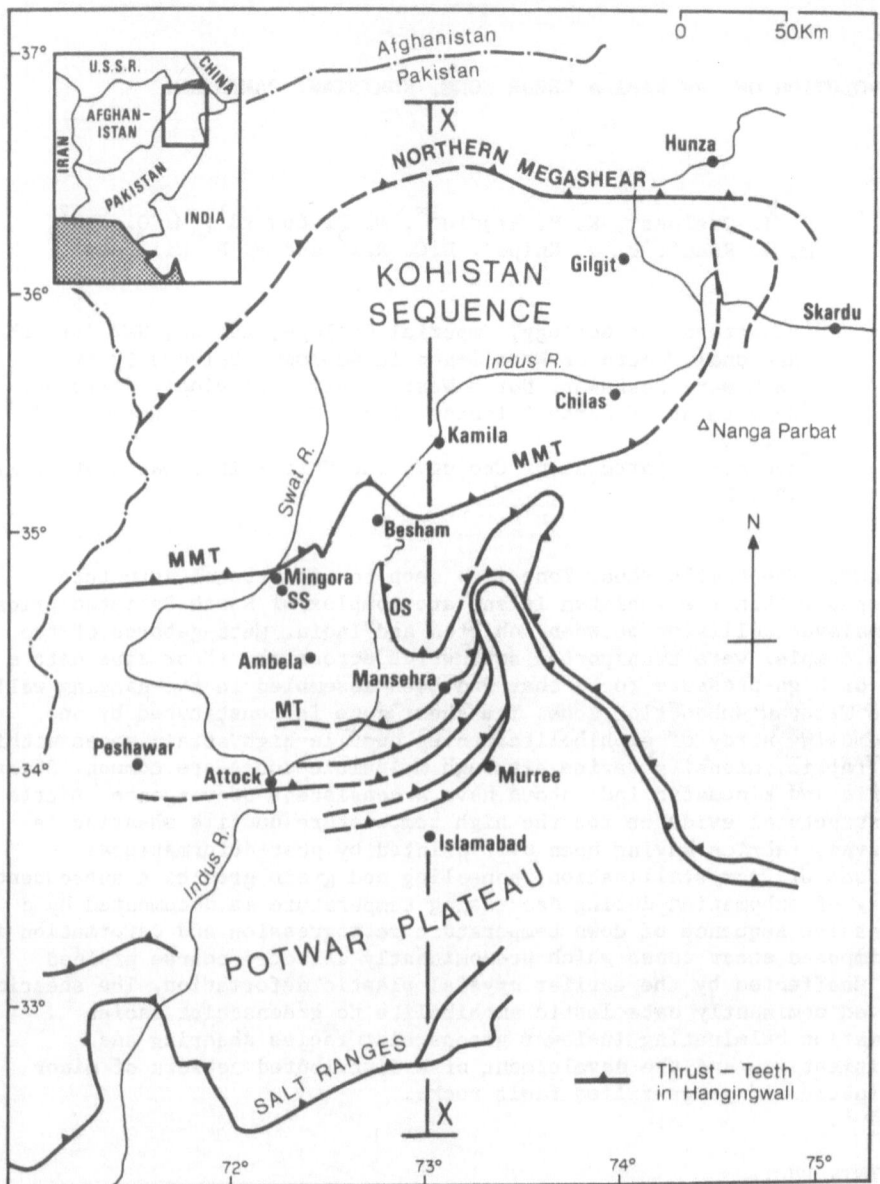

Figure 1. Outline geological map of North Pakistan to show the position of the Kohistan island arc between the Indian and Asian plates and the important thrusts developed in the Indian plate. MMT: Main Mantle Thrust. MT: Mansehra Thrust. OS: Oghi Shear. X-X: line of section in Figure 4.

(or "Shyok Suture") at between 102 and 85 Ma (Coward et al, 1986; Pudsey et al., 1986; Treloar et al., 1989a). In NW Pakistan the subsequent Himalayan collision was between Kohistan and India, with Kohistan thrust southwards over the Indian Plate along the Main Mantle Thrust (MMT). The MMT is a north-dipping crustal-scale structure (Gansser, 1964) which is the true westward extension of the Indus-Tsangpo Suture.

The Kohistan complex developed above a north-dipping subduction zone (Bard, 1983a, b; Searle et al., 1987) within which Tethyan oceanic rocks, which formed the leading edge of the Indian Plate, were subducted. Geochronological (Treloar et al., 1989a) and structural (Coward et al., 1987) evidence shows that much of the deformation within the island arc predated collision with India and is a combination of shearing and folding relating to both suturing and subduction processes. Within southern Kohistan, the Kamila Shear Zone is a particularly well defined zone of deformation up to 40 km wide, that is characterised by intense shear fabrics developed synchronously with a phase of major hydration and amphibolitisation. In this article we summarise the tectonic development of the Kamila Shear Zone, its relationship to other structures within Kohistan and the way that microstructures developed within individual shear zones document changing conditions of deformation with time.

2. GEOLOGY AND TECTONIC SETTING OF THE KOHISTAN ISLAND ARC

2.1 Geology of the Kohistan complex

The geology of Kohistan has been described elsewhere (Tahirkelli, 1979, 1982; Bard et al., 1980; Jan 1980, 1988; Jan and Howie, 1981; Coward et al., 1982, 1986, 1987; Bard, 1983a, b; Petterson and Windley, 1985; Khan et al., 1989) and only a brief summary is given here. The rocks within the Kohistan complex have been tilted and folded so that they are now steep to north dipping. As such, a north to south profile forms a section that passes stratigraphically and structurally downward through a tectonically thickened island arc sequence (Tahirkelli, 1979).

The highest level rocks, exposed just south of the Northern Suture (Fig. 2) are the deep water slates, turbidites, volcaniclastic sediments and limestones of the Albian to Aptian aged Yasin Group (Pudsey, 1986) which may have been deposited within an intra-arc basin. The Yasin Group overlies the submarine to subaerial basic to intermediate arc-type tholeiitic volcanics and pillow lavas and associated volcaniclastic sediments of the Chalt Group. These low metamorphic grade rocks were intruded by the gabbroic to granitic plutons of the Kohistan Batholith. The first batholithic phase, dated at ca. 102 Ma predates the Northern Suture. Later intrusives, which postdate the suture and mark a second stage of plutonism, were intruded into an Andean-type margin. These second stage plutons became progressively more acidic with time, early gabbros being intruded by tonalites, granodiorites and granites and finally by a dense network of aplitic-pegmatites as young as 28 Ma (Petterson and Windley, 1985). An Eocene age for the second stage plutonism is indicated both by isotope geochronology (Petterson & Windley, 1985; Debon et al., 1987; Treloar et al., 1989a) and by field

178

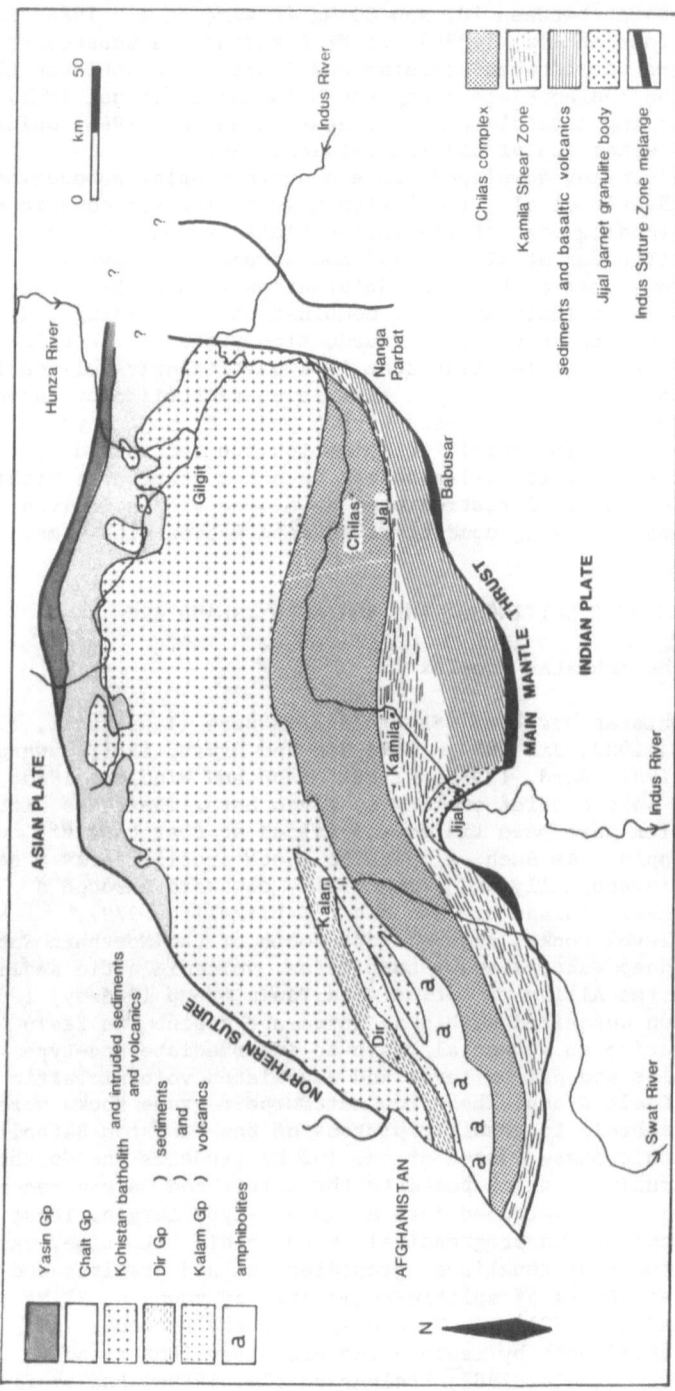

Figure 2. Simplified geological map of Kohistan. The Kohistan complex is separated from the Asian Plate to the north by the Northern Suture and from the Indian Plate to the south by the Main Mantle Thrust. The Kamila Shear Zone is a major SW-verging crustal-scale shear zone. The Chilas complex of two-pyroxene meta-gabbros outcrops in the shear zone hanging wall. Footwall lithologies include both the high pressure rocks of the Jijal block, and basaltic volcanic sequences, intruded by basic bodies, to the east of Jijal.

relations near Dir where volcanics of the Dir-Utror Group, dated by Eocene fossils, unconformably overlie some intrusions but are intruded by others.

To the south of the Kohistan Batholith, the arc volcanics and sediments pass downwards into a major basic to ultramafic complex, the "Chilas Complex" (Khan et al., 1989). This is a large (>10 km thick) stratiform body, dominated by two-pyroxene meta-gabbros and norites, that may represent a sub-arc magma chamber (Coward et al., 1986; Khan et al., 1989). Shortly after emplacement the original norites and gabbros of the Chilas complex underwent a granulite facies re-equilibration at ca. 800°C at 7 to 8 kb (Jan and Howie, 1980), during which the primary igneous textures were recrystallised to a polygonal, granoblastic metamorphic fabric. Parts of the Chilas body retain a relic igneous layering that enables the way-up and fold structure to be established (Coward et al., 1987). The Chilas complex and the overlying Dir, Chalt and Yasin sequences can be interpreted as the root zone and overlying volcanic pile of a single island arc developed above the subducting Tethyan plate.

South of the Chilas complex is a zone of amphibolite facies rocks, many of them highly sheared. This belt, known as the "Kamila Amphibolite Belt" contains variably deformed coarse-grained amphibole-bearing rocks, many of which are amphibolitised Chilas complex meta-gabbros and norites, fine-grained amphibolites with a calc-alkaline chemistry (Jan, 1988) and intrusive granitic sheets. Some of the fine-grained amphibolites are probably of volcanic origin. Within the Indus Valley, the main part of the amphibolite belt provides a section through a 38 km wide shear zone, the "Kamila Shear Zone", which is comprised of a series of anastomosing high strain zones. This shear zone, which narrows eastward until south of Jal it is no more than 3.5 km wide, developed at mid to lower crustal levels and played an important role in the assembly of the southern part of the Kohistan complex. The origin of the metabasic rocks of the shear zone is somewhat uncertain. Bard et al. (1980) argued that they represent the oceanic crust on which the island arc was built, whereas Coward et al. (1986) suggest that they are a highly deformed sequence of arc-type plutons and volcanics although with possible ocean floor relics. Basaltic volcanics and associated sediments outcrop south of Jal between the shear zone and the MMT, although such rocks are not seen south of the shear zone in the Indus Valley section.

The amphibolites and sheared rocks of the Kamila Shear Zone separate the island arc rocks of the Chilas complex from a lithologically diverse suite of rocks developed along the southern margin of the Kohistan complex. These include both greenschist and blueschist facies (glaucophane-epidote-albite-phengite bearing) lithologies of the MMT suture zone melange (Jan, 1980), and the ultramafic rocks and garnet-pyroxene-plagioclase granulites of the Jijal complex (Jan and Howie, 1981; Bard, 1983a). The garnet-pyroxene granulites, which form the northern part of the Jijal complex, were metamorphosed at 670-690°C at 12-14 kb (Jan and Howie, 1981). They contain meta-norite relics petrogenetically similar to those of the Chilas complex. Both the Chilas complex meta-gabbros and norites and the Jijal garnet-granulites are strongly calc-alkaline (Jan, 1988) with similar mantle normalised trace element spidergrams (Jan and Windley, under review) that are characteristic of plutonic rocks developed in the root of an island arc. Although it has been suggested that the Jijal garnet-granulites may be a folded limb of the high pressure basal part of the Chilas complex (Jan, 1980; Coward et al., 1982), it is as likely that the two are separate, albeit petrotectonically similar, island arc complexes.

The ultramafic rocks which form the southern part of the Jijal complex include garnet and plagioclase free dunites, peridotites, diopsidites, websterites and serpentinites. Similar peridotitic rocks outcrop in the hanging wall of the MMT at Babusar Pass, east of Jijal. Jan and Howie (1981) suggested a possible ophiolitic origin for the peridotitic rocks. In that their constituent minerals are more magnesian and the spinel more enriched in Cr, there are sufficient distinct chemical and mineralogical differences between the pyroxenite members of the Jijal peridotites and the pyroxenite layers within the Chilas body to suggest that there is no magmatic connection between the two (Jan and Windley, under review; Khan et al., 1989). Although the high bulk Mg/Fe ratios and low Al contents (Jan, 1980), the lithological variety, compositions of Cr-spinel, clinopyroxene and olivine (Jan and Howie, 1981) and the internal igneous layerings of the Jijal peridotites are typical of many peridotitic complexes (c.f. Moores, 1982), detailed geochemistry has, as yet, failed to establish firmly their tectonic setting except that they are high pressure (10 ± 2 kb) cumulates probably derived from a primitive tholeiitic or boninitic magma, probably in an arc or fore-arc region (Jan and Windley, under review).

Figure 3 shows a schematic vertical profile through the Kohistan complex. In this profile the arc rocks of the Chilas complex and overlying Chalt and Yasin sequences are shown as lying structurally above the amphibolites of the Kamila Shear Zone, with the Jijal complex structurally below. This simple representation of the relationship between the Chilas and Jijal complexes, both of which have island arc affinities, would be consistent with their having been originally either a single island arc, subsequently deformed and disrupted, or two different arc sequences. A detailed study of the Kamila Shear Zone and the arc rocks to north and south of it, as outlined in this article, may go some way to establishing which of these interpretations is correct.

2.2. Himalayan tectonic setting of the Kohistan complex

Firstly, however, we wish to summarise briefly the relationship between Kohistan and the Asian and Indian plates. Figure 4 shows a section through the Himalayan collision zone of North Pakistan. Within Kohistan dips are generally steep to northerly. In the northern part of the complex dips of bedding and cleavage are clearly related to a number of large scale folds, such as the Jaglot syncline, which predate the second stage Eocene plutons of the Kohistan batholith. Internal deformation within Kohistan postdated suturing with Asia along the Northern Suture, but predated collision with India along the MMT. The northern suture, originally steep and marked by a tectonic melange with sub-vertical fabrics (Coward et al., 1986), was further deformed during the main Himalayan collision by a series of south-verging shears and thrusts developed in the Asian plate. These structures, the Hunza Shear Zone and

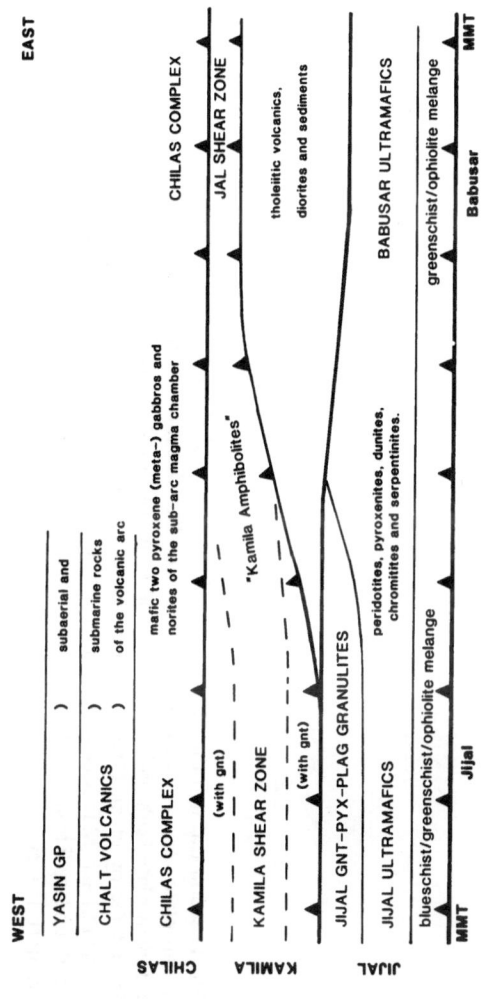

Figure 3. A schematic vertical profile (after Coward et al., 1986) drawn through the steep to northward dipping rocks of Kohistan.

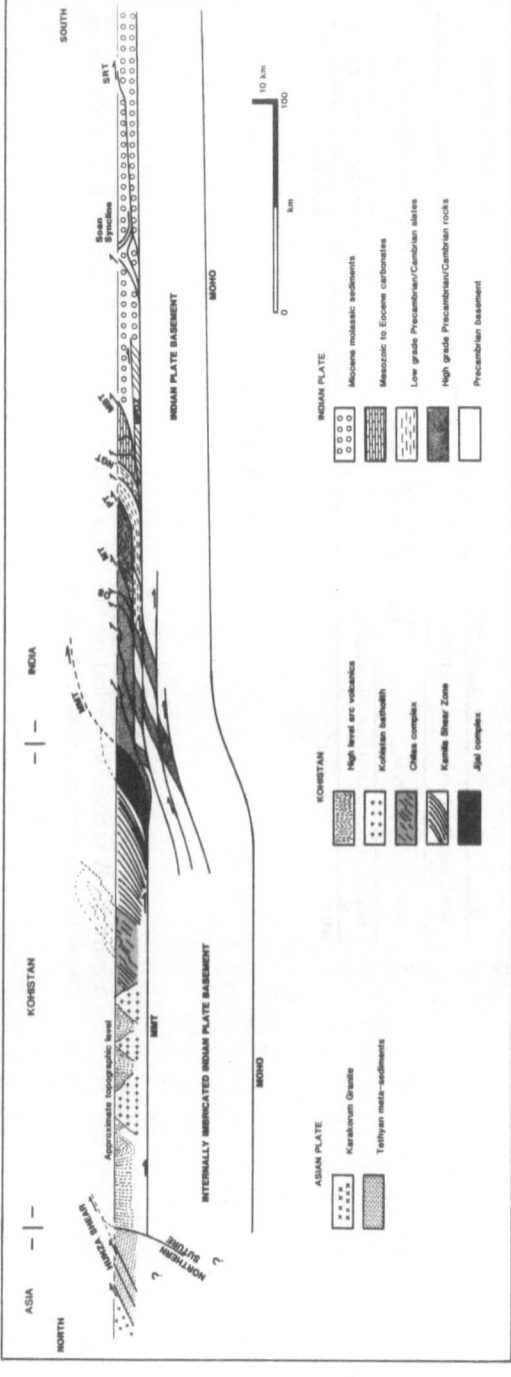

Figure 4. A simplified section through the Himalayan collision zone of north Pakistan from the Asian to the Indian plate across the Kohistan island arc. The thickness of the Kohistan complex and of the Indian plate both under Kohistan and south of the MMT is derived from Malinconico (1986) and Duroy et al. (1989). Abbreviations, from north, are: MMT: Main Mantle Thrust. OS: Oghi Shear. MT: Mansehra Thrust. PT: Panjal Thrust. NGT: Nathia Gali Thrust. MBT: Main Boundary Thrust. NPDZ: North Potwar Deformation Zone. SRT: Salt Range (or Main Frontal) Thrust.

Main Karakorum Thrust (Rex et al., 1988) are probably late Himalayan break back, or out-of sequence, thrusts developed in the hanging wall of the MMT.

After collision with Kohistan at about 50 to 55 Ma, the Indian plate was deformed and metamorphosed in the footwall of the MMT as Kohistan was transported southward across the leading edge of continental India. Restorations across the imbricated sequences yield a minimum shortening estimate of 470 km (Coward & Butler, 1985). Within the Indian plate, the most northerly exposed rocks are those of the internal crystalline zones, within which basement gneisses and overlying late Precambrian to Cambrian sediments were metamorphosed at up to sillimanite grade synchronously with early stages of the Himalayan deformation (Coward et al., 1988; Treloar et al., 1989b, c). Cooling histories demonstrate that post-metamorphic cooling had been initiated by 40 million years ago, when the pile cooled back through the hornblende blocking temperature (Treloar et al., 1989a; Treloar & Rex, under review). The metamorphic pile was later deformed within an imbricate south-verging thrust stack within which metamorphic zones were tectonically inverted (Treloar et al., 1989b, c) and which was subsequently largely unroofed by the early Miocene (Treloar & Rex, under review). This stack of cooled metamorphic rock was transported south onto low grade Precambrian to Cambrian slates along the Panjal Thrust. These slates were themselves thrust onto carbonate-rich Mesozoic to Eocene shelf sediments of the Tethyan margin along the Nathia Gali Thrust (Coward & Butler, 1985). The Mesozoic and Eocene sequences were then thrust south on to Miocene molasse along the Main Boundary Thrust.

3. AMPHIBOLE BEARING ROCKS OF THE SOUTHERN CHILAS COMPLEX

The terms "Kamila Amphibolite Belt" and "Kamila Shear Zone" have tended to be used synonymously (Coward et al, 1982, 1987). This is somewhat misleading as, to the north of the shear zone, meta-gabbros and norites of the Chilas complex are extensively amphibolitised and, sometimes, garnet-bearing. The garnet growth, and much of the amphibolitisation, reflect amphibolite facies metamorphism and deformation synchronous with the main deformation within the shear zone. The dominant lithologies within the southern part of the Chilas complex (Fig. 5a) are a series of weakly deformed, metamorphosed basic through to ultramafic hornblende-bearing coarse-grained rocks. These are dominated by a hornblende-plagioclase assemblage in which hornblende constitutes from 30% to, in hornblendites, greater than 90% of the rock. Plagioclase constitutes much of the remainder, and epidote, which is widespread, is a locally significant component. These rocks are the extensively hydrated and amphibolitised products of the plagioclase and two pyroxene-bearing Chilas meta-gabbros and norites, and the igneous layering widely present in the Chilas complex (Khan et al., 1989) is preserved within them (Fig. 6a). Coarse-grained hornblende-plagioclase rocks (which represent hydrated meta-gabbros) grade into nearly pure hornblendite layers, up to 100 m thick, and characterised by acicular hornblende crystals up to 10 cms long which represent hydrated cumulus layers.

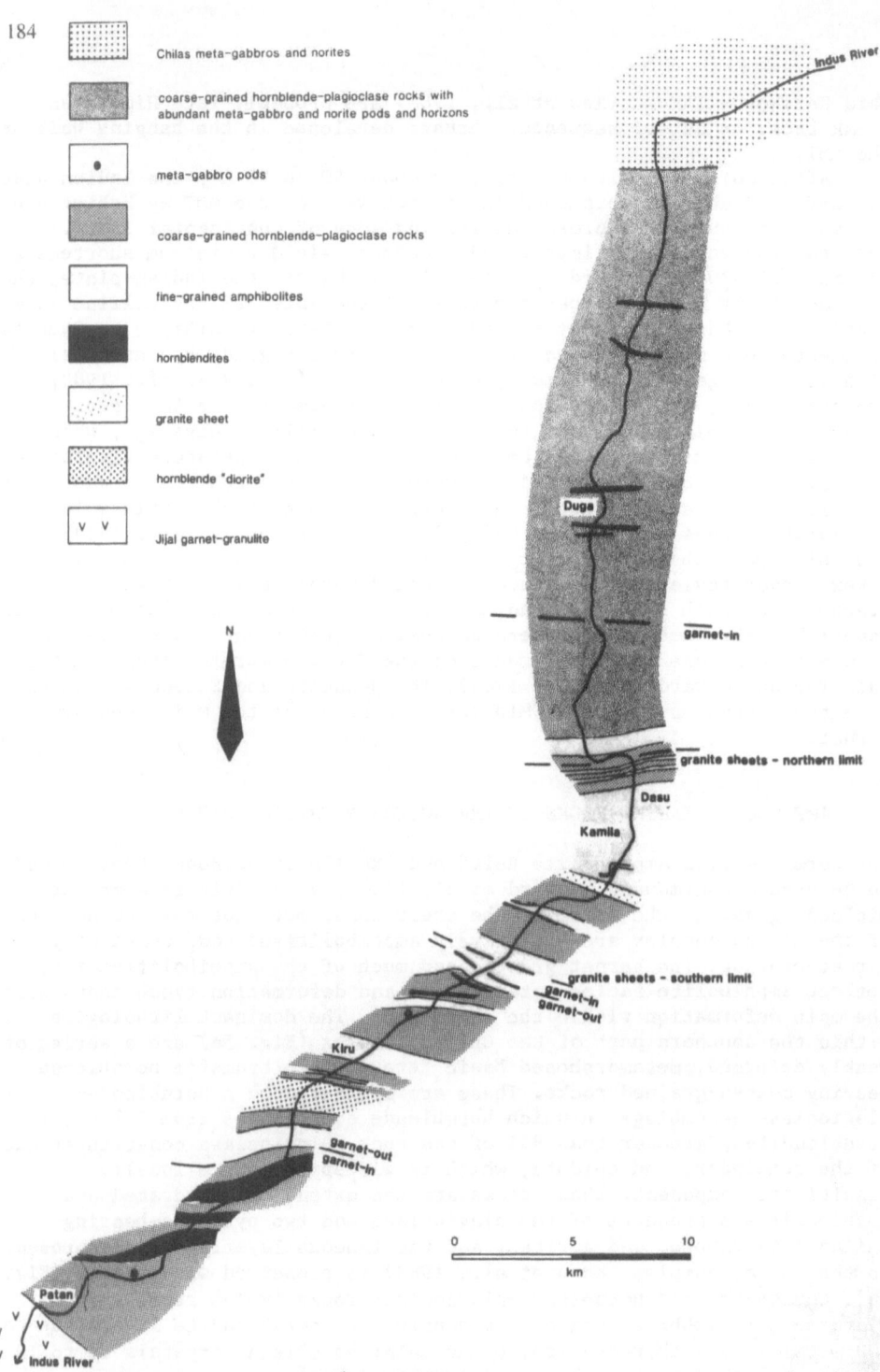

184

Chilas meta-gabbros and norites

coarse-grained hornblende-plagioclase rocks with
abundant meta-gabbro and norite pods and horizons

meta-gabbro pods

coarse-grained hornblende-plagioclase rocks

fine-grained amphibolites

hornblendites

granite sheet

hornblende "diorite"

Jijal garnet-granulite

Indus River

N

Duga

garnet-in

granite sheets - northern limit

Dasu

Kamila

granite sheets - southern limit
garnet-in
garnet-out

Kiru

garnet-out
garnet-in

Patan

Indus River

0   5   10
              km

A)

Figure 5a. Geological map along the Indus Valley traverse through the southern part of the
Chilas complex and the Kamila Shear Zone. The Kamila Shear Zone forms that part of the
traverse southward from about 2 kms north of Dasu.

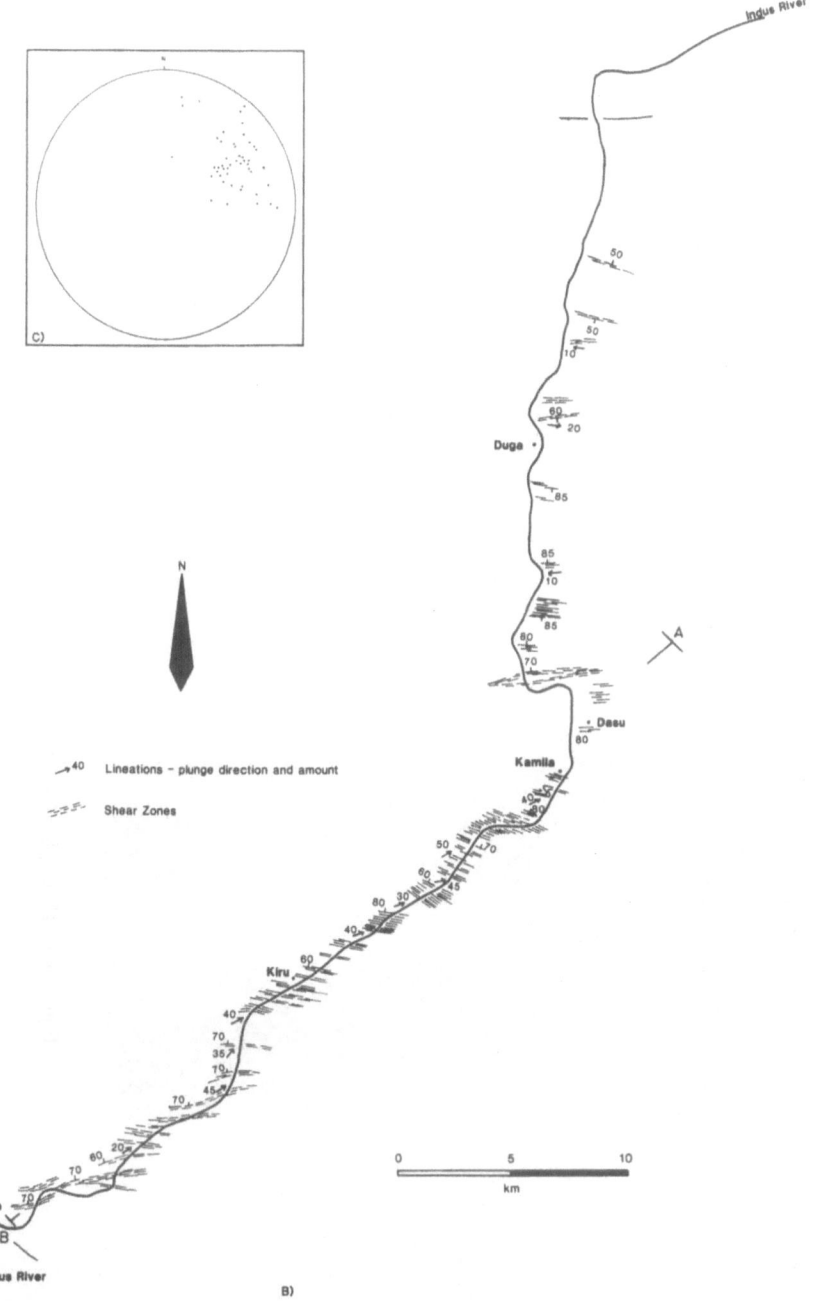

Figure 5b. A structural map drawn along the same Indus Valley traverse as Fig. 5a to show the locations of the main shear zones. A and B mark the ends of the section in Fig. 7. Figure 5c. Equal area lower hemisphere projection of NE-plunging stretching and mineral lineations within the Kamila Shear Zone.

Figure 6a. Field photograph of igneous layering within hydrated two-pyroxene meta-gabbros.

Figure 6b. Field photograph of narrow band of amphibolitised rock centred on a small brittle shear within meta-gabbro.

188

Figure 6c. Field photograph of sheared amphibolitised meta-gabbros showing a low strain pod, and the development of both fine-grained mylonitic bands and incipient flaser gneisses.

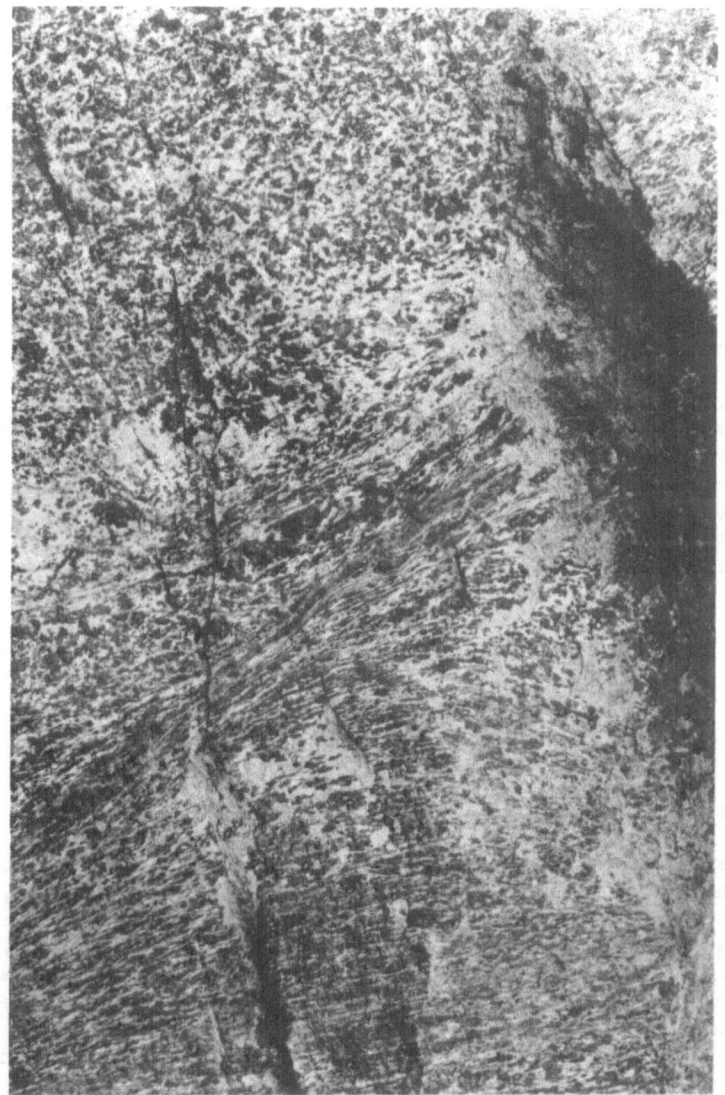

Figure 6d. Field photograph of shear zone cutting an undeformed coarse grained hornblende-plagioclase rock. Two phases of shearing can be identified here: an early ductile shear zone cut by a later narrow mylonitic shear zone.

Primary two pyroxene-plagioclase-bearing meta-gabbros and norites of the Chilas complex are preserved throughout this tract, although their frequency decreases southward. The northward limit of extensive hydration is about 13 km north of Dasu (Fig 5a). Two forms of hydration are preserved. Near to, and east of Sazin, there is a high concentration of coarse-grained amphibolite dykes and amphibole-plagioclase pegmatites which represent the late phases of magmatic activity in the Chilas complex and some of the hydration of the meta-gabbros and norites is due to this intrusion event. Southwards from here the role of shearing in controlling hydration is especially clear. The initial hydration was largely along narrow shear zones which enhanced permeation of water into the rocks. Bands of amphibolite, up to 200 m wide, are formed within unaltered meta-gabbro and norite (Fig. 6b).

Deformation within the southern part of the Chilas complex is confined to a series of folds with generally northward vergence and a number of widely spaced narrow, steeply dipping, shear zones of limited displacement. These shears include some with ductile, east-west striking, hornblende-bearing shear fabrics. In many of these, shear bands and fabric curvature indicate north-side up dip-slip displacement, whereas in others, sub-horizontal to gently east-plunging stretching lineations indicate a strike slip component of movement. A series of lower temperature shear zones, some with greenschist facies asemblages containing actinolite and/or chlorite, and others with cataclastic gouge infills, post-date the ductile shears. Although some of these have a south-verging thrust sense of movement, most are extensional north-side down faults.

4.   THE KAMILA SHEAR ZONE.

4.1. Lithologies within the Kamila Shear Zone

Within the Indus Valley section through southern Kohistan, southward of about 2 km north of Dasu (Fig. 5), the character of the rocks changes. Instead of a low overall state of deformation, with strain concentrated into a few, relatively narrow shear zones, deformation intensity is substantially higher than to the north with shear strain heterogeneously distributed across a 38 km wide zone through shear along a large number of broad high strain zones. Further east, near Jal, this belt narrows considerably to a 3.5 km wide belt of mylonites in direct contact with Chilas complex meta-gabbros. Lithological diversity is far greater than to the north. Within the Indus Valley, the Kamila Shear Zone is characterised by the presence of both fine-grained amphibolites and granite sheets. The granites are confined to a strip 8 km wide around Kamila (Fig. 5a) and include garnet-granites, hornblende-granites, trondhjemitic granites and micaceous pegmatitic granites. They occur as narrow sheets which, to the south of Kamila, constitute up to 50% of the outcrop. As some of the sheets have intense shear fabrics, and others cut the regional fabrics within the host rocks, a largely syntectonic timing for granite emplacement is inferred.

The fine-grained amphibolites can be divided into unbanded and banded types. Contacts between the two are invariably strongly sheared. The banded or striped amphibolites are characterised by alternating hornblende-rich and plagioclase-rich layers commonly one or two cm wide. Fine-grained unbanded amphibolites are locally intercalated with siliceous and calcareous meta-sediments which Jan (1988) interpreted as demonstrating a volcanic origin for the unbanded amphibolites. Support for this argument is provided by the outcrop both of basaltic to andesitic volcanics and sediments (Khan, 1988) south of the Jal high strain zone to the east of the Indus Valley (Fig. 2) and of meta-pillows in the Swat Valley. Elsewhere, however, a volcanic origin cannot be firmly demonstrated. The striped amphibolites, at least, are likely to be the products of intense shearing, of both originally fine- and coarse-grained basic lithologies.

The coarse-grained rocks are dominantly hornblende-plagioclase bearing and are similar to those of the southern part of the Chilas complex, although with a greater proportion of hornblendite layers, especially near Patan (Fig. 5a). Within the Indus Valley, near Kiru as well as about 1 km north of Patan, the amphibolites contain pods of relatively undeformed pyroxene-bearing meta-gabbro and norite, similar to those in the Chilas complex. Similar pods occur within amphibolites at Khwazakhela in the Swat Valley to the west (Jan, 1988). Pyroxenes within these pods are variably replaced by amphibole. The presence of such pods, as well as the similarity in calc-alkaline chemistries between them and Chilas gabbros and norites, led Jan (1988) to suggest that the Kamila Shear Zone included deformed Chilas complex rocks as well as basaltic volcanics.

4.2. Deformation within the Kamila Shear Zone

Within the Kamila Shear Zone strain is heterogeneously distributed, with localised zones of intense grain size reduction anastomosing around areas that show little evidence of strain, although there is a complete gradation between high and low strain areas. The distribution of the main high strain zones is shown in Figure 5b and the cross section (Fig. 7) illustrates the anatomosing nature of the shear zones. Within individual zones shear strain is also variable with varying degrees of grain size reduction and annealing. Sheared gneisses, mylonites, blastomylonites and ultramylonites may all be present (Fig. 6c, d). In general the main shear fabrics trend at $090\pm20^{\circ}$ and dip steeply ($70\pm10^{\circ}$) towards the north.

The dominant shear fabric is ductile, with the development of a compositional segregation of leucocratic material, and the alignment of acicular and platy minerals and of elongate lenses of deformed mineral aggregates. Stretching and mineral lineations plunge consistently towards 060° (Fig 5c). Shear bands, fabric curvature into shear zones, and rotation of garnet and plagioclase porphyroclasts can all be used to indicate the sense of shear. The majority of these shear criteria have a southwest-verging thrust sense with movement parallel to the stretching lineations. Some though, indicate extension in the opposite direction, which may represent some limited spreading of the hot, thickened pile shortly after the main phase of shortening.

192

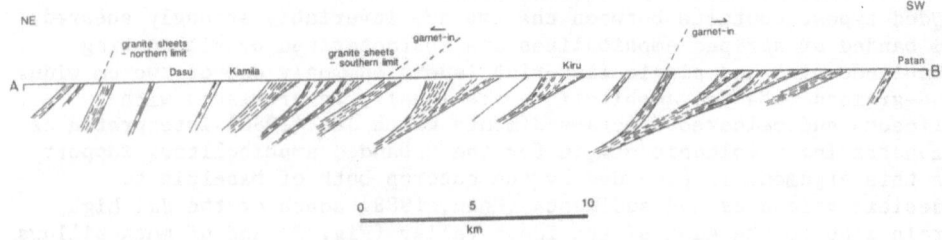

Figure 7. NE-SW section (A-B), parallel to the main stretching lineation trend, through the Kamila Shear Zone to show the anastamosing nature of the major ductile SW-verging shears that constitute the shear zone.

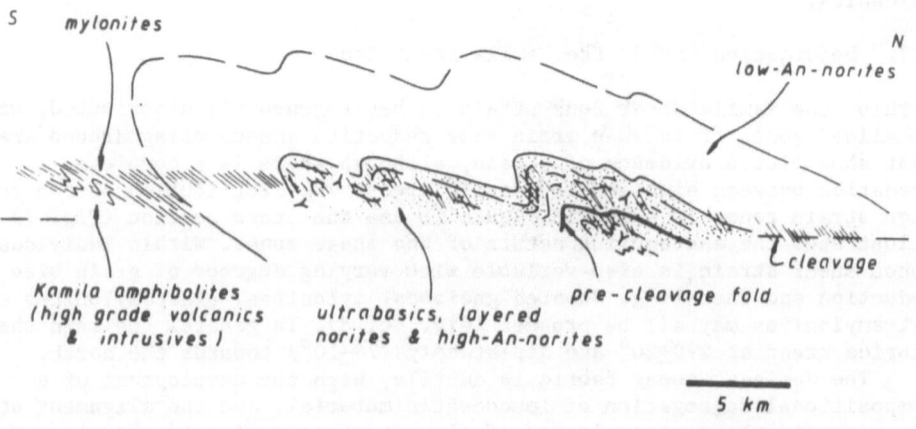

Figure 8. Cross-section through the Jal Shear Zone (the eastern continuation of the Kamila Shear Zone) south of Chilas. The section shows the folding of the Chilas Complex in the hanging wall of the Jal shear zone, a narrow south-verging mylonitic high strain zone (from Coward et al., 1987).

To the east the Kamila Shear Zone narrows and the shear strain heterogeneously distributed across 38 km within the Indus Valley section is concentrated into 3.5 km of mylonites known as the Jal Shear Zone. Here, mylonites strike east-west with steep northerly dips and a consistent sense of southward thrusting (Khan, 1988). Hydrated Chilas meta-gabbros crop out in the immediate hanging wall of the shear zone. Facing directions within the gabbros indicate the presence of an antiform to the north, interpreted (Coward et al., 1987; Khan, 1988) as an asymmetric hanging wall antiform, with an upright to steeply north-dipping axial planar cleavage (Fig. 8). The hanging wall antiform, indicated also by the north-verging folds in the northern part of the Indus Valley section, is one of a number of large folds each with more intense deformation on their southern, overturned, limbs (Coward et al., 1987).

To the south of the Jal Shear Zone, and between it and the peridotites of the MMT zone at Babusar Pass (Fig. 2), is a wide zone of basaltic to andesitic volcanics and associated sediments, intruded by dioritic bodies (Khan, 1988). These volcanics, which we equate with the sheared volcanics in the Indus Valley section of the shear zone, may be evidence for an original geographic separation of the Chilas and Jijal complexes.

The main phase ductile shearing within the Kamila Shear Zone was succeeded by continuing southwest-verging deformation during unroofing and cooling of the shear zone. The main phase ductile shear fabrics are postdated by more brittle east-west striking, steeply dipping thrust surfaces coated by fibrous actinolite and/or chlorite, themselves postdated by gouge-filled thrust systems which often form little duplex structures. It is uncertain as to how many of these late structures are related to deformation prior to collision with India during exhumation of the shear zone, although some are probably break-back structures developed during the Kohistan-India collision and which breached the MMT locally thickening the Kohistan sequence (Searle et al., 1987, Fig. 8). The youngest structures preserved in Kohistan are brittle, generally north-dipping, extensional faults associated with thinning and extension of Kohistan following the thrusting of the island arc sequence onto the Indian Plate.

4.3. P-T conditions and age of shearing within the Kamila Shear Zone

Metamorphism and deformation within the Kamila Shear Zone appears to have occurred under amphibolite facies conditions, with aligned hornblendes defining the shear fabrics. Amphibolites within the shear zone contain a main phase assemblage of hornblende and plagioclase with a variable development of garnet depending on bulk composition. The majority of the hornblendes are ferroan pargasite to tshermakite in composition (our data; Jan & Howie, 1982; Bard, 1983a). Garnet occurs in two areas: one an 8 km wide belt along the southern margin of the shear zone just to the north of Patan; and the second a 14 km wide belt, largely coincident with the granite belt, along the northern margin of the shear zone. Within both areas garnet is sporadically present, probably in response to a strong compositional control. In many cases garnets occur within, and

along the margins of, plagioclase-quartz bearing segregated veins and
layers. Garnet compositions are typically in the range Alm_{51-63} Pyr_{15-22}
Gro_{16-28} with only minor (3 - 8 mol%) spessartine.

There are few reliable barometers or thermometers available within
the garnet-hornblende-plagioclase system. Garnet-amphibole thermometry
(Graham and Powell, 1984) gives temperatures of about 550-650°C for the
garnet amphibolites, and preliminary garnet-rutile-ilmenite-plagioclase
barometry (Bohlen & Liotta, 1986) suggests maximum metamorphic pressures
of 9 to 10 kb. This is similar to a pressure estimate of 10 $\pm$ 1 kb
derived by Bard (1983a), for kyanite and staurolite bearing assemblages
among the amphibolites just north of Patan. The widespread late-stage
amphibole alteration, the common occurrence of garnet in plagioclase-rich
segregations and potential retrograde changes in plagioclase composition
suggest that these estimates should be treated with caution, although we
do infer that the highest metamorphic temperatures were in the two
garnet-bearing zones.

A minimum age for the ductile shearing and hornblende growth within
the Kamila Shear Zone is given by $^{40}Ar-^{39}Ar$ cooling ages derived from
hornblendes within the amphibolite belt. A number of samples have been
analysed from within the shear zone, but only one has a flat $^{40}Ar-^{39}Ar$
spectrum with an age of 83 $\pm$ 1 Ma (Fig. 9a). A second spectrum is U-
shaped, with a plateau defined by 73.6% of the gas that also indicates an
age of 83 $\pm$ 1 Ma (Fig. 9b). Other spectra are also U-shaped, indicative
of excess argon, but with minima that approach 83 Ma. This is taken to be
the age at which the Kamila Amphibolite Belt cooled through 500°C, and is
a minimum age for shearing within the Kamila Shear Zone. A K-Ar muscovite
age of 66 Ma from a pegmatite that cross-cuts the main shear fabric
indicates that substantial post-shearing uplift and cooling had occurred
within Kohistan by 60 Ma. (Treloar et al., 1989a).

5. DEFORMATION PROCESSES AND MECHANISMS

A series of deformation textures within both basic and granitic
lithologies can be recognised which document a range of processes from
high temperature crystal plastic deformation through a sequence of
variably brittle and plastic deformation processes, to late-stage high-
level low temperature deformation that included the development of fault
gouge. Deformation microstructures preserved in granitic lithologies
record mainly greenschist facies events, and will be only briefly
described.

5.1. High temperature metamorphism and deformation

The Chilas two-pyroxene meta-gabbros and norites that were deformed
within the Kamila Shear Zone had previously undergone a post-emplacement
granulite facies metamorphism that resulted in the development of a
granoblastic polygonal fabric involving recrystallisation of both
pyroxene and plagioclase. Pyroxenes from undeformed meta-gabbro and
norite pods within the shear zone include both clino- and orthopyroxenes.
Orthopyroxenes are En_{60-66} with up to 2.7% Al_2O_3. Clinopyroxenes are

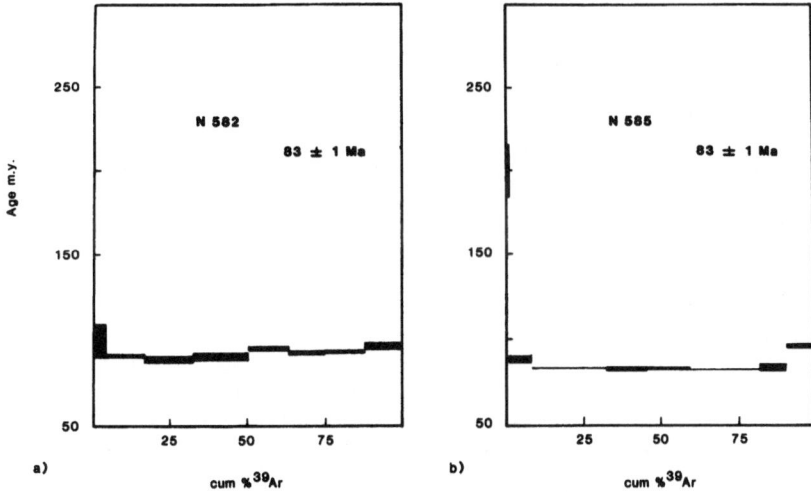

Figure 9a. $^{40}Ar^{-39}Ar$ spectrum of a hornblende from sample N582 from near Patan. A flat spectrum with a plateau defined by 90% of the gas implies an age of 83 ± 1 Ma for cooling through ca. 500°C.

Figure 9b. A U-shaped Ar-Ar spectra for sample N585 from near Patan. A plateau, characterised by 73.6% of the gas, indicates a maximum age for the sample of 83 ± 1 Ma.

Di_{67-73} with up to 5% Al_2O_3.

The earliest process that can be identified following the granulite facies re-equilibration, is the high temperature amphibolite facies static hydration of the meta-gabbros and norites (Fig. 10), with pyroxene replaced by a pargasitic hornblende. This hydration is variably developed and locally pyroxene cores are preserved within amphibole. Although this hydration is essentially static, with fluid being supplied through associated shears, some evidence is seen for high temperature deformation of the orthopyroxene, possibly prior to, or synchronous with, the hydration (see below). These variably hydrated, but relatively undeformed, meta-gabbros and norites are those preserved as pods and lenses within the shear zone.

The highest temperature shear zones preserved seem to have developed under amphibolite facies conditions. However, little microstructural evidence of high temperature crystal plasticity is preserved, and in most cases the rocks contain microstructures indicative of post-deformation annealing under high temperatures (upper amphibolite facies). In the field, grain size variations can be mapped as an indicator of strain variations, but generally annealing and grain growth has obscured evidence of earlier high strains.

The strong foliation, obvious in hand specimen, is produced by the alternation of segregated amphibole-rich and plagioclase-rich layers (Fig. 11). In some samples a strongly anisotropic fabric is defined by aligned amphibole crystals, which also define the regional stretching lineation (Fig 5). In others, however, little mineral shape fabric is evident in either the amphibole or plagioclase, which generally show equigranular granoblastic microstructures. Rare amphibole porphyroclasts are present together with, in places, smaller porphyroclasts of orthopyroxene. Where present, orthopyroxene porphyroclasts are oriented parallel to the foliation but do not seem to be stable and are partially altered to amphibole in pressure shadow regions (Fig. 11). In contrast, amphibole porphyroclasts have recrystallised, forming a matrix of similar and uniform composition. Blebs and lamellae of Fe-Ti oxides (ilmenite and rutile) contained within the amphibole porphyroclasts occur as separate phases intergrown with the matrix minerals. Quartz occurs locally as fine-grained aggregate stripes parallel to the foliation. The matrix plagioclase is An_{55-60} in composition. Amphibole compositions range from tschermakitic hornblende to ferroan pargasitic hornblende, suggesting that the deformation and recrystallisation occurred under at least upper amphibolite conditions.

The deformation mechanisms responsible for accommodating the strain in these shear zones are difficult to infer. Crystal plasticity of plagioclase would be expected at temperatures >550°C and rather higher temperatures would be required for plastic deformation of amphibole. The elongation of the orthopyroxene porphyroclasts suggests a possible, still earlier, higher temperature granulite facies deformation.

The pronounced mineral segregation into amphibole-rich and plagioclase-rich layers is a common feature in deformed amphibolite facies basic rocks. The formation of these layers as a result of the deformation of initial larger grains would require extremely high strains through either (a) cyclic dynamic recrystallisation or (b) diffusion-

Figure 10. Optical micrograph (under crossed polars of coarse grained Chilas meta-gabbro showing tabular igneous plagioclase grains. The original pyroxenes have been statically pseudomorphed by fine-grained amphibole.

198

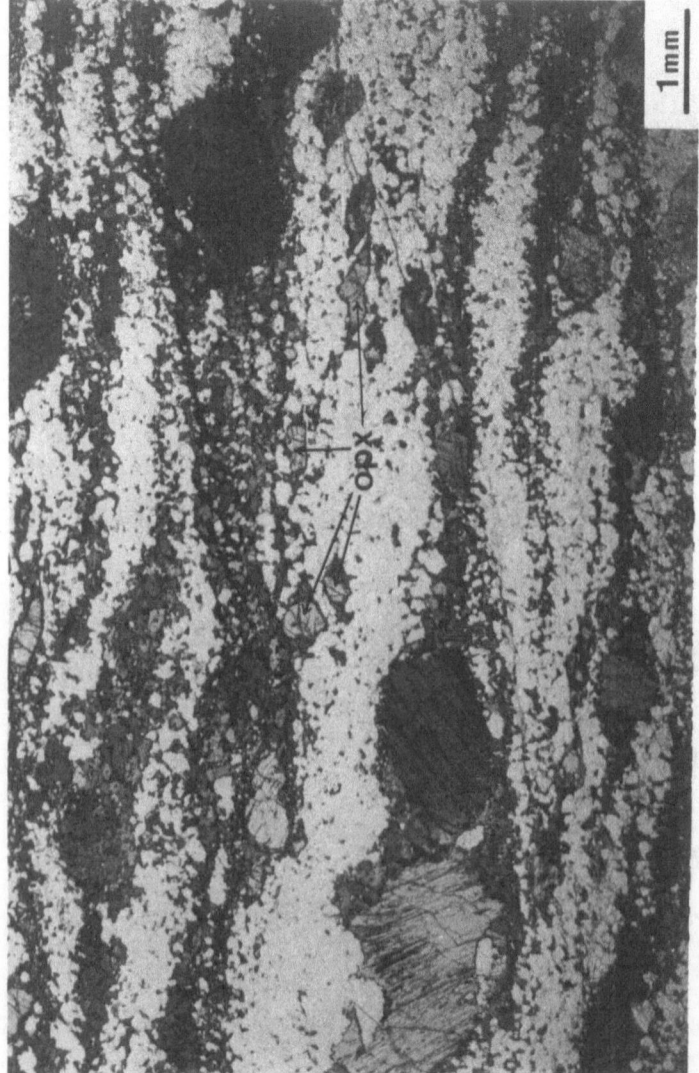

Figure 11a. Optical micrograph of a high temperature metabasic shear zone showing alternation of plagioclase and amphibole-rich layers. Porphyroclasts of orthopyroxene (light) and brown amphibole are present, both with tails of brown amphibole. The plagioclase shows a recrystallised granoblastic texture. Plane polarised light.

Figure 11b. As in Figure 11a, but under crossed polars.

accommodated grain boundary sliding, if grain size was sufficiently reduced. The grain size reduction for (b) could be achieved by either dynamic recrystallisation or through the formation of transiently fine-grained reaction products (Rubie, 1983; Brodie and Rutter, 1987a; Rutter & Brodie, in press), for example by pyroxene breakdown to fine-grained amphibole and quartz. Cyclic dynamic recrystallisation of both plagioclase and amphibole could also lead to a switch to grain size sensitive flow. Within the Chilas complex itself, a combination of reaction and grain sensitive flow controls the development, in narrow shear zones within homogeneous meta-gabbros and norites, of banded amphibolite (Khan, 1988).

The high temperature microstructures document deformation within the shear zone during the main crustal thickening and imbrication of the island arc. Microstructural evidence for the deformation mechanisms involved in this early deformation is limited owing to extensive post deformational thermal effects which have produced annealing and static recovery. This is common in high grade terranes (e.g. Myers, 1984; Brodie and Rutter, 1987b).

5.2. Deformation during uplift and cooling

The subsequent uplift and cooling history is well documented by microstructural and mineral assemblage changes accompanying lower temperature deformation in superimposed shear zones. The pargasitic hornblende-Ca-plagioclase+garnet+quartz assemblage is locally retrogressed thrugh a magnesio hornblende-sodic plagioclase-clinozoisite assemblage to a typically greenschist facies actinolitic amphibole-epidote-Na-plagioclase-quartz+epidote assemblage. The deformation of the plagioclase and amphibole becomes dominantly cataclastic. A number of stages of deformation during decreasing temperature can be recognised.

1) Plagioclase (ca. An_{60-65}) shows extensive evidence of cracking and rotation of grains (Fig. 12a) with alteration along the cracks to a more sodic composition (ca. An_{40}), and the development of very fine grained laths of epidote along the cracks (Fig. 12b). Amphibole has recrystallised to random aggregates of lath shaped grains, often rimmed by fine-grained clinozoisite aggregates. In some rocks early growth of epidote at the expense of Ca-plagioclase is more advanced. Although earlier stages of this deformation may be plastic, brittle deformation seems to become increasingly dominant, with static crystallisation of the amphibole and alteration and growth of new epidote along fractures in the plagioclase. In addition, metamorphic garnets which date from the main phase amphibolite-facies metamorphism become brecciated and rotated.

2) These fractured rocks are cut by local, highly foliated mylonitic zones containing relict porphyroclasts of both green magnesio-hornblende and calcic plagioclase (An_{64}) in a very fine-grained matrix of actinolite-clinozoisite-biotite-chlorite-plagioclase-quartz. The porphyroclasts of amphibole and plagioclase are rounded and the plagioclase shows marginal alteration in pressure shadows to a more sodic composition (Fig. 13). These foliated zones vary from narrow anastomosing networks that cut the cataclastic plagioclase - magnesio-hornblende rocks, to discrete mylonite zones up to 50 cms thick. Figure 12 shows the

Figure 12a. Optical micrograph (under plane polarised light) of a metabasic shear zone showing brittle fragmentation of plagioclase (An_{65}), and static recrystallisation of magnesio-hornblende. At the bottom of the field of view a narrow, highly foliated mylonite cuts the rock, producing further grain size reduction and retrogression, and rounding of the relict plagioclase and amphibole grains and clasts.

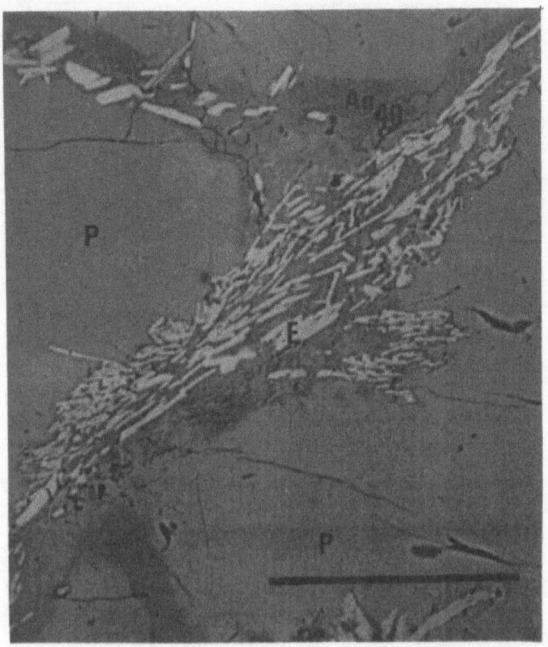

Figure 12b. Back scattered image of area outlined in (a) showing alteration along the fractures in the plagioclase (P) of composition An_{65}. The plagioclase is altered to a more sodic composition (An_{40}) shown by the darker atomic number contrast at the edges. Small laths of epidote (E, bright contrast) are also developed along the fracture surfaces. Scale bar = 100 μ.

203

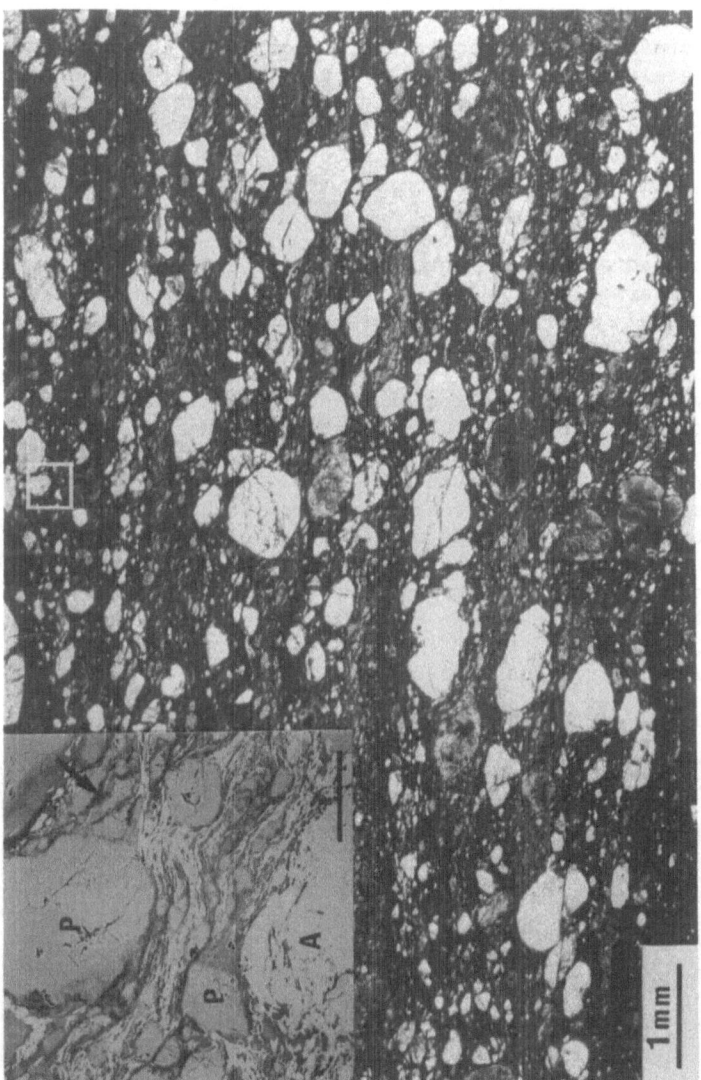

Figure 13. Optical micrograph, in plane polarised light, of a low temperature metabasic mylonite showing rounded porphyroclasts of plagioclase and rare amphibole in a very fine grained, foliated matrix of actinolite, chlorite, plagioclase, epidote and quartz. Inset is a backscattered image of part of the matrix indicated showing laths of epidote (bright contrast) and chlorite (dark grey, arrowed) surrounding relict porphyroclasts of calcic plagioclase (P) and amphibole (A). A more sodic plagioclase is also present in the prophyroclast tails and in the matrix (darker contrast). Scale bar = 100 μ.

progression from the cataclastically deformed rock to the highly deformed mylonite.

This latter deformation is characterised, in part, by the development of fine-grained fibrous chlorite-actinolite-plagioclase-quartz aggregates that define S-C fabrics and, locally, ultramylonitic quartz-chlorite aggregates (Fig. 14). Much of the strain within these rocks seems to be accommodated by deformation of chlorite which suffers plastic deformation, slip on (001), as commonly observed in phyllosilicates at temperatures as low as $250-350^{\circ}C$ (Bons, 1988), together with plastic deformation of quartz.

It is to be expected that cataclastic deformation will dominantly affect the initially coarser grained rocks rather than those which have previously undergone grain size reduction through crystal plastic processes. This appears to be the situation here, where high temperature shear zones are preserved adjacent to low temperature greenschist facies mylonites. Similarly the low temperature greenschist facies hydration and deformation which produced the fine-grained highly foliated mylonites tends to be localised within the rocks which earlier suffered cataclastic deformation and where permeability would thereby be enhanced.

Microstructures within granitic rocks also document a progressive down temperature retrogression and deformation. The original mineral assemblage of quartz-plagioclase-alkali feldspar with minor muscovite and $\pm$ biotite, garnet and amphibole, is replaced by a hydrous greenschist facies assemblage of white mica-epidote-quartz$\pm$chlorite. The fault rocks consist of strongly foliated layers of fine-grained quartz, that represent recrystallised early quartz ribbons, interbanded with fine-grained to very fine-grained aggregates of white mica and epidote (Fig. 15a). Little evidence of high temperature deformation is preserved. The relict quartz ribbons provide limited intragranular evidence for crystal plasticity of quartz, although these are usually recrystallised to fine-grained aggregates. Relict porphyroclasts of white mica, usually kinked and fractured, are locally preserved. Clusters of coarse-grained epidote porphyroblasts, that date from an earlier hydration, show evidence of brittle deformation and rotation (Fig. 15b), as do relict igneous garnets. The fine-grained epidote-white mica aggregates presumably reflect feldspar breakdown, that also pre-dated the lower temperature shearing. Chlorite and/or biotite occur within those fault rocks which are derived from granites with higher initial mafic content. Strongly developed shear bands, mica fish and rotation senses of epidote porphyroclasts consistently indicate a SW-vergance. Finally a distributed network of minor cataclastic zones, related to regional uplift, was developed. These zones, which are dominated by fine-grained chlorite with or without biotite define strong S-C fabrics and grade into a series of low temperature gouge zones.

6.   REGIONAL IMPLICATIONS.

Although both the Jijal and Chilas complexes have been studied in detail (Jan and Howie, 1981; Khan et al., 1989 and references therein), and Jan and Windley (under review) have demonstrated a similar (i.e. island arc)

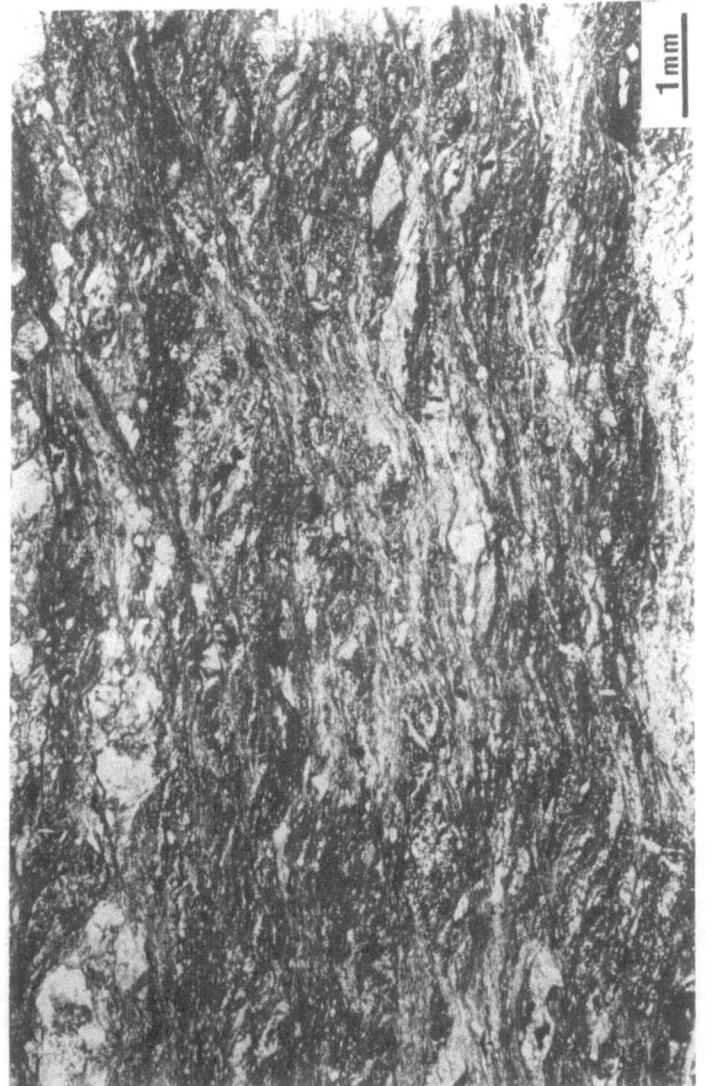

Figure 14. Optical micrograph, in plane polarised light, of a low temperature metabasic mylonite rich in chlorite-actinolite-epidote, showing development of shear bands. Few relict porphyroclasts remain.

206

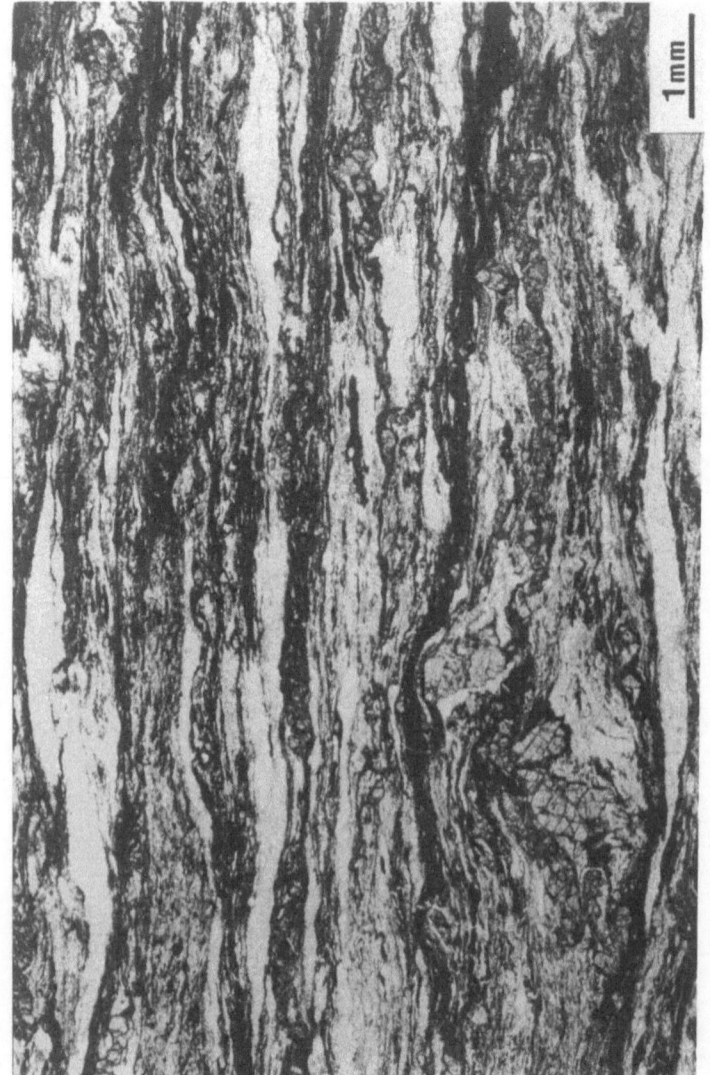

Figure 15a. Optical micrograph, in plane polarised light, of granitic mylonites. Highly foliated mylonite containing alternating elongate lenses of fine-grained recrystallised quartz and bands rich in white mica. Fine-grained epidote is present in the matrix, and coarser-grained epidote is present as porphyroclasts, often showing fracturing and granulation.

Figure 15b. Optical micrograph, in plane polarised light, of granitic mylonites. Broken and rotated epidote porphyroclast, which indicates the sense of shear.

origin for the both the Jijal garnet granulites and the Chilas meta-
gabbros and norites, there is still uncertainty as to their tectonic
relationships. In particular, it is uncertain as to whether the Jijal
garnet granulites were originally the higher pressure base of the Chilas
complex, tectonically detached and folded (Jan, 1980; Coward et al.,
1982, 1986), or whether, although they form a suite that is chemically
similar to the Chilas body and which was probably derived in a similar
tectonic setting, they form a second island arc that subsequently
collided with the Chilas arc with the suture between the two marked by
the Kamila Shear Zone (see discussion in Coward et al., 1987).

In this context we would stress the division of southern Kohistan
into three components, all structurally and lithologically distinct. The
northernmost is the Chilas complex, deformed into a series of overturned
south-facing folds developed in the hanging wall of the Kamila Shear
Zone. The Kamila Shear Zone itself incorporates both Chilas-type
lithologies (primary and hydrated) and at least some sediments and calc-
alkaline arc or fore-arc volcanics. To the south of the shear zone, in
the immediate hanging wall of the MMT, is a lithologically diverse suite
of rocks that includes both blueschists of the suture zone melange,
ultramafic rocks from Babusar, and the garnet-granulites and ultramafic
peridotites of the Jijal complex, most of which carry a demonstrably high
pressure mineralogy. We interpret this latter complex as a distinct
tectonic unit, assembled above the subduction zone, and within which the
garnet-clinopyroxene bearing granulites were thrust onto the Jijal
peridotites, which were themselves thrust south onto the blueschist
melange. Due to the similarity, at 75-80 Ma, of $^{40}Ar-^{39}Ar$ cooling ages
derived from phengites in the blueschist sequence (Maluski and Matte,
1984) to cooling ages within the Kohistan complex, we include the
blueschist-bearing rocks within this tectonic sequence, rather than as
part of a discrete subduction type melange zone. These ages give a
minimum age for the assembly of the stack, although Sm-Nd garnet ages of
103 Ma from the garnet granulites (Coward et al., 1986) suggest that
different components of the thrust stack probably crystallised prior to
its assembly. Paragonite-bearing assemblages, stable at 8-9 kb and 500-
550°C have been described from shear zones within the Jijal garnet
granulites by Jan et al. (1982). The presence of such high pressure shear
zone assemblages implies that the Jijal stack may have been asssembled at
depths approaching 30 km.

If we can recognise the Jijal thrust stack as a discrete tectonic
unit, distinct from the rest of Kohistan, the tectonics of the Kamila
Shear Zone are more easily understood. Essentially the shear zone
reflects the southward thrusting of the main island arc (Chilas Complex)
over its fore-arc region. This thrusting involved intense shortening of
the fore-arc region, and the complex interthrusting and imbrication of
the arc gabbros and volcanics with fore-arc volcanics and sediments.
Whether the Jijal stack was assembled synchronously with shearing within
the Kamila Shear Zone, essentially below the shear zone but above the
subduction zone, or was a discrete, already assembled block, is
uncertain. A schematic section across the shear zone from Chilas to
Jijal is shown in Figure 16.
$^{40}Ar-^{39}Ar$ geochronology shows that shearing within the Kamila Shear

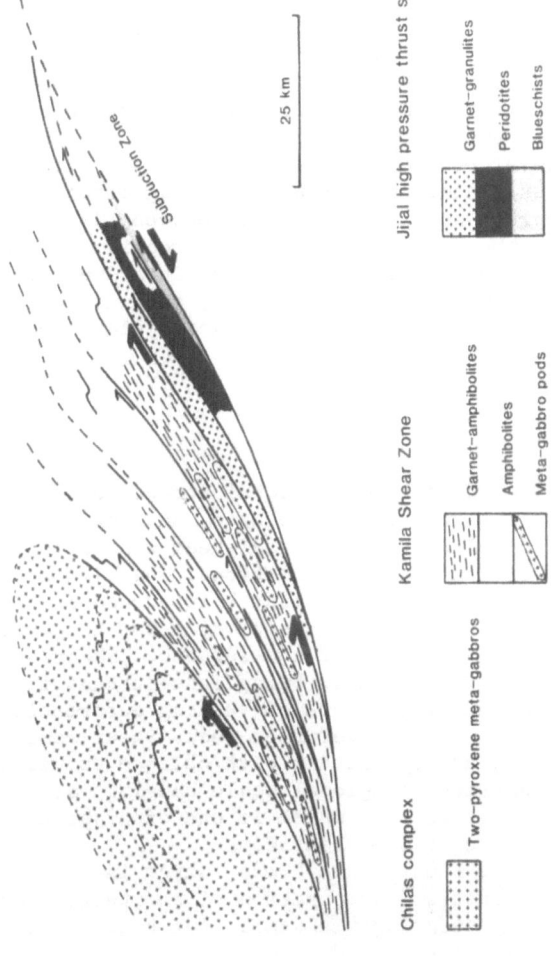

Figure 16. A N-S schematic section to show the assembly of the southern part of the Kohistan complex in the hanging wall of the N-dipping subduction zone. The Chilas complex (main part of the island arc) was thrust southwest-wards over the rocks of the Kamila Shear Zone which separate it from the Jijal thrust stack of high pressure rocks. Note that the presence of high pressure rocks within the Jijal thrust stack does not necessarily imply that the stack was assembled at high pressure.

Zone pre-dated 85 Ma. As the time window for the suturing of Kohistan to Asia along the Northern Suture is 102–85 Ma (Treloar et al., 1989a), it is possible that deformation within the shear zone was a direct consequence of changes in plate motion due to that suturing. In particular the shortening across the Kamila Shear Zone, largely accommodated by the subduction of the fore-arc sequences under the Chilas complex, would have resulted from the post-suturing cessation or slowing down of the northward movement of the Island Arc.

The consistent SW-vergence of kinematic indicators within the Kamila Shear Zone would suggest a dominant NE-SW shortening direction across the Asia-Kohistan-Tethyan plate collision zone at between 100 and 85 Ma. This direction is oblique to the southerly movement direction documented in north Pakistan for the main Himalayan collision (Coward et al., 1982, 1986, 1987, 1988; Treloar et al., 1989b, 1989c), and implies a change in relative motions of the converging Indian and Asian plates between 85 and 55 Ma.

7.   CONCLUSIONS.

The Kamila Shear Zone is a deep to mid crustal structure that separates the high temperature Chilas and Jijal complexes, each of which are part of separate island arc complexes. Within the shear zone deformation occurred under a range of temperature conditions. The highest temperature deformation seems to have been at temperatures in excess of $500^\circ C$, possibly as high as $650^\circ C$, and various stages of lower temperature deformation chart progressive uplift of the sequence. Shear strain within the shear zone is variably distributed, being concentrated into a number of relatively narrow anastomosing shears of varying deformation intensity which separate pods of relatively undeformed rock. Shear senses, derived from kinematic indicators, are consistent with oblique thrusting towards the SW. The shear zone carries a regional scale hanging wall antiform overturned towards the SW (Fig. 16), which is one of a series of such folds developed to the north of the shear zone.

The development of the Kamila Shear Zone, which is interpreted here as being related to the southward propagation of shortening across the Kohistan complex following suturing with Asia, but prior to collision with India, says little about the kind of deep structure that may normally be expected to develop within island arc regions. However, the concentration of shearing along the crustal-scale mechanical heterogeneity marked by the apparent arc-forearc boundary, however diffuse, may be a logical corollary of any substantial change in rates or vectors of relative plate motions in a still active arc environment.

Although a mid- to lower crustal high temperature ductile shear zone, such as the Kamila Shear Zone, is readily recognised in the field by fabric curvatures and variations in fabric intensity, it is to be expected that the high temperature shear fabrics will be annealed or recrystallised out as soon as local stress fields relax. Potentially, many high temperature high strain zones are not preserved owing to subsequent annealing and grain growth where the high temperatures outlast the deformation. Relatively rapid uplift associated with major faults as observed in the Kamila Shear Zone may be required to preserve these textures.

   Subsequent deformation of metabasic rocks at lower temperatures ($T<450^{\circ}C$) is likely to be dominated initially by cataclastic processes and these will tend to concentrate in coarser grained rocks which have not undergone grain size reduction by crystal plastic processes. Following cataclasis, the enhanced permeability of the rock will enable fluid influx and the development of a hydrous greenschist facies assemblage. The presence of minerals such as chlorite and quartz will allow the rock to deform plastically, producing highly foliated mylonite zones. Certainly, some of the more spectacular deformation microstructures are those which postdate the high temperature shearing and which developed sequentially during uplift and cooling.

ACKNOWLEDGEMENTS.

We acknowledge receipt of a (U.K.) Natural Environment Research Council Grant GR3/6113, and MPW a research studentship from the same body. We acknowledge discussions with various colleagues, in particular Brian Windley, Mike Petterson and Rob Butler.

REFERENCES.

Bard, J-P., Maluski, H., Matte, P. and Proust, F. 1980. The Kohistan sequence: crust and mantle of an obducted island arc. Geol. Bull. Univ. Peshawar. 11. 87-94.

Bard, J-P. 1983a. Metamorphic evolution of an obducted island arc: example of the Kohistan sequence (Pakistan) in the Himalayan collided range. Geol. Bull. Univ. Peshawar. 16. 105-184.

Bard, J-P. 1983b. Metamorphism of an obducted island arc: example of the Kohistan sequence (Pakistan) in the Himalayan collided range. Earth Planet. Sci. Lett. 65. 133-144.

Bohlen, S.R. and Liotta, J.J. 1986. A barometer for garnet amphibolites and garnet granulites. J. Petrol. 27. 1025-1034.

Bons, A-J. 1988. Deformation of chlorite in naturally deformed low-grade rocks. Tectonophysics. 154. 149-165.

Brodie, K.H. and Rutter, E.H. 1987a. The role of transiently fine-grained reaction products in syntectonic metamorphism: natural and experimental examples. Can. J. Earth Sci. 24. 554-564.

Brodie, K.H. and Rutter, E.H. 1987b. Deep crustal extensional faulting in the Ivrea Zone of northern Italy. Tectonophysics. 140. 193-21.

Coward, M.P. and Butler, R.W.H. 1985. Thrust tectonics and the deep structure of the Pakistan Himalaya. Geology. 13. 417-420.

212

Coward, M.P., Jan, M.Q., Rex. D.C., Tarney, J., Thirlwall, M.F. and Windley, B.F. 1982. Geotectonic framework of the Himalaya of north Pakistan. <u>J. Geol. Soc. London</u>. **139**. 299-308.

Coward, M.P., Windley, B.F., Broughton, R.D., Luff, I.W., Petterson. M.G., Pudsey, C.J., Rex, D.C. and Khan, M.A. 1986. Collision tectonics in the NW Himalaya. <u>Geol. Soc. London Sp. Publ</u>. **19**. 203-219.

Coward, M.P., Butler, R.W.H., Khan, M.A. and Knipe. R.J. 1987. The tectonic history of Kohistan and its implications for Himalayan structure. <u>J. Geol. Soc. London</u>. **144**. 377-391.

Coward, M.P., Butler, R.W.H., Chambers, A.F., Graham, R.H., Izatt. C.N.. Khan, M.A., Knipe, R.J., Prior, D.J., Treloar, P.J. and Williams, M.P. 1988. Folding and imbrication of the Indian crust during Himalayan collision. <u>Phil. Trans. Roy. Soc. London. ser. A</u>. **326**. 89-116.

Debon, F., Le Fort, P., Dautel, D., Sonet, J. and Zimmermann, J.-L. 1987. Granites of western Karakorum and northern Kohistan (Pakistan): a composite mid-Cretaceaous to upper Cenozoic magmatism. <u>Lithos</u>. **20**. 19-40.

Duroy, Y., Farah, A., Malinconico, L.L. and Lillie, R.J. 1989. Subsurface densities and lithospheric flexure of the Himalayan foreland in Pakistan. <u>Geol. Soc. Amer. Sp. Paper</u>. **232**. In press.

Gansser, A. 1964. <u>The geology of the Himalayas</u>. Wiley Interscience. London. 289pp.

Graham, C.M. and Powell, R. 1985. A garnet-hornblende geothermometer: calibration, testing and application to the Pelona schist, southern California. <u>J. Metamorphic Geology</u>. **2**. 13-31.

Jan, M.Q. 1980. Petrology of the obducted mafic and ultramafic metamorphics from the southern part of the Kohistan Island Arc sequence. <u>Geol. Bull. Univ. Peshawar</u>. **13**. 95-108.

Jan, M.Q. 1988. Geology of amphibolites from the southern part of the Kohistan arc, N Pakistan. <u>Min. Mag</u>. **52**. 147-159.

Jan, M.Q. and Howie, R.A. 1980. Ortho- and clino-pyroxene from pyroxene granulites of Swat, Kohistan, northern Pakistan. <u>Min. Mag</u>. **43**. 715-728.

Jan, M.Q. and Howie, R.A. 1981. The mineralogy and geochemistry of the metamorphosed basic and ultrabasic rocks of the Jijal complex, Kohistan, NW Pakistan. <u>J. Petrol</u>. **22**. 85-126.

Jan, M.Q. and Howie, R.A. 1982. Hornblendic amphiboles from basic and intermediate rocks of Swat, Kohistan, northwest Pakistan. <u>Amer. Mineral</u>. **43**. 715-728.

Jan, M.Q., Wilson, R.N. and Windley, B.F. 1982. Paragonite parageneses from the garnet granulites of the Jijal complex, Kohistan, N. Pakistan. Min. Mag. 45. 73-77.

Jan. M.Q. and Windley, B.F. Chromian spinel and silicate chemistry in ultramafic rocks of the Jijal complex, NW Pakistan. J. Petrol. Under Review.

Khan, M.A. 1988. Petrology and structure of the Chilas mafic-ultramafic complex, Kohistan, NW Himalayas, Pakistan. Unpubl. PhD thesis. Univ of London.

Khan, M.A., Jan, M.Q., Windley, B.F., Tarney, J. and Thirlwall, M.F. 1989. The Chilas mafic igneous complex: the root of the Kohistan Island Arc in the Himalayas of N Pakistan. Geol. Soc. Amer. Sp. Paper. 232. In Press.

Malinconico, L.L. 1986. The structure of the Kohistan arc terrane in northern Pakistan as inferred from gravity data. Tectonophysics. 124. 297-307.

Maluski, H. and Matte, P. 1984. Ages of Alpine tectonometamorphic events in the northwestern Himalayas. Tectonics. 3. 1-18

Moores, E.M. 1982. Origin and emplacement of ophiolites. Rev. Geophys. and Space Phys. 20. 735-760.

Myers. J.S. 1984. Archaean tectonics in the Fiskenaesset region of southwest Greenland. In: Precambrian Tectonics Ilustrated. A Kroner and R Greiling (eds). 95-112.

Petterson, M.G. and Windley, B.F. 1985. Rb-Sr dating of the Kohistan arc batholith in the Trans-Himalaya of north Pakistan and its tectonic implications. Earth. Planet. Sci. Lett. 74. 45-57

Pudsey, C.J. 1986. The northern suture, Pakistan: margin of a Cretaceous island arc. Geol. Mag. 123. 405-423.

Pudsey, C.J., Coward, M.P., Luff, I.W., Shackleton, R.M., Windley, B.F. and Jan, M.Q. 1986. The collision zone between the Kohistan arc and the Asian Plate in NW Pakistan. Trans. Roy. Soc. Edinburgh, Earth Sci. 76. 463-479.

Rex, A.J., Searle, M.P., Tirrul, R., Crawford, M.B., Prior, D.J., Rex, D.C. and Barnicoat, A.C. 1988. The geochemical and tectonic evolution of the central Karakorum, North Pakistan. Phil. Trans. Roy. Soc. London. Ser. A. 326. 229-255.

Rubie. D.C. 1983. Reaction enhanced ductility: the role of solid-solid univariant reactions in the deformation of the crust and mantle. Tectonophysics. 96. 331-352.

Rutter, E.H. and Brodie, K.H. Rheology of the lower crust. In: The Lower Continental Crust. D. Fountain, R. Kay and R. Arculus (eds). In Press.

Searle, M.P., Windley, B.F., Coward, M.P., Cooper, D.J.W., Rex. A.J., Rex, D.C., Li Tindong, Xiao Xuchang, Jan, M.Q., Thakur, V.C. and Kumar, S. 1987. The closing of Tethys and the tectonics of the Himalaya. Bull.Geol. Soc. Amer. 98. 678-701.

Tahirkelli, R.A.K. 1979. Geology of Kohistan and adjoining Eurasia and Indo-Pakistan continents. Geol. Bull. Univ. Peshawar. 11. 1-30

Tahirkelli, R.A.K. 1982. Geology of the Himalaya, Karakorum and Hindu Kush in Pakistan. Geol. Bull. Univ. Peshawar. 15. 51pp.

Treloar, P.J., Rex, D.C., Guise, P.G., Coward, M.P., Searle. M.P., Petterson, M.G., Windley, B.F., Jan, M.Q. and Luff, I.W. 1989a. K-Ar and Ar-Ar geochronology of the Himalayan collision in NW Pakistan: constraints on the timing of suturing, deformation, metamorphsm and uplift. Tectonics. 8. 881-909.

Treloar, P.J., Broughton, R.D., Williams, M.P., Coward, M.P. and Windley, B.F. 1989b. Deformation, metamorphism and imbrication of the Indian Plate, south of the Main Mantle Thrust, North Pakistan. J. Metamorphic Geology. 7. 111-125.

Treloar, P.J., Williams, M.P. and Coward, M.P. 1989c. Metamorphism and crustal stacking in the north Indian Plate, North Pakistan. Tectonophysics. 165. 167-184..

Treloar, P.J. and Rex, D.C. Cooling and uplift histories of the crystalline thrust stack of the Indian plate internal zones west of Nanga Parbat, Pakistan Himalaya. Tectonophysics. Under Review.

CRUSTAL FORMATION AT DEPTH DURING CONTINENTAL COLLISION

L. S. Hollister
Dept. of Geological and Geophysical Sciences
Princeton University
Princeton, NJ 08544

M. L. Crawford
Department of Geology
Bryn Mawr College
Bryn Mawr, PA 19010

ABSTRACT. The rapid (1-2 mm/yr) and substantial (over 10 km) Tertiary exhumation of the Coast Crystalline Complex near Prince Rupert, British Columbia led to the exposure of rocks formed at crustal depths of 15-20 km and greater. Abundant sills, mostly tonalite in composition, delivered a thermal pulse which locally raised temperatures to the low-pressure granulite facies within the intruded section. The thermal pulse and exhumation are recorded by the metamorphic mineral assemblages and textures, and by isotopic cooling ages. The net result of the thermal and deformation events is a sequence more than 10 km thick of interlayered migmatite, tonalite sills, ductile high strain zones, sillimanite- and garnet-rich rocks (stronalite), other metasedimentary rocks, and amphibolite. Foliations in all units dip gently except where cut by near vertical ductile shear zones, within which foliations are sub-parallel to the shear zone boundaries. During uplift, strain was concentrated in the ductile shear zones. If the section had not been exhumed, seismic reflection profiling would show shallow dipping reflectors in the deep crust due to the contrast in density and fabric of the layers.

Introduction

The Central Gneiss Complex near Prince Rupert, British Columbia is a portion of a deep crustal section formed during the Late Cretaceous to Early Tertiary and brought to the surface by an episode of rapid uplift in the Early Tertiary. This paper reviews the basis for concluding that deep crustal rocks are exposed, gives a general description of those rocks, and discusses how they were formed.

Construction of P-T Path for Central Gneiss Complex

The arguments for deep crust being exposed in the Central Gneiss Complex (Fig. 1, from Crawford et al., 1987) are based on

M. H. Salisbury and D. M. Fountain (eds.), Exposed Cross-Sections of the Continental Crust, 215–225.
© 1990 *Kluwer Academic Publishers.*

reconstruction of the P-T-time history of the area. This is done by
relating pressure and temperature data obtained on metamorphic rocks
to dates of intrusion of plutons, and from cooling age
determinations. The data show that the section originated at depths
greater than 20 km and was rapidly exhumed during the Early Tertiary.
This rapid exhumation resulted in exposure and preservation of the
rock associations and fabrics formed in the deep crust during
continental collision which began prior to 98 Ma.

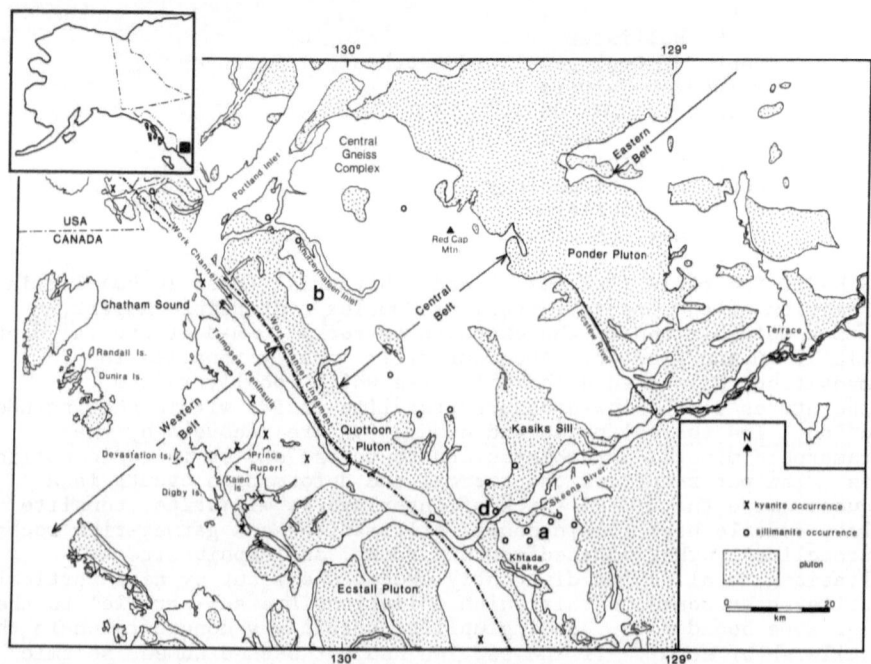

Figure 1. Map of the Prince Rupert-Terrace area showing the
distribution of the western, central (Central Gneiss Complex), and
eastern belts and of the large intrusive bodies (patterned). The Work
Channel lineament (dot-dash line) separates the western and central
belts; the boundary between the central and eastern belts lies within
the area mapped as Ponder pluton. Kyanite and sillimanite-bearing
localities are shown to emphasize the contrast in metamorphic mineral
assemblages between the western and central belts. From Crawford, et
al. (1987). Locations a, b, and d are referred to in the text and in
Figure 2.

Crawford et al. (1987) give the detailed reconstruction across the
Coast Plutonic Complex near Prince Rupert. The exhumed section
consists of arc-derived sedimentary and volcanic rocks which,
following underthrusting, were metamorphosed to kyanite-staurolite
grade. Subsequent injection of tonalite sills heated the supracrustal
rocks to over 700°C, hot enough to induce partial melting, which
resulted in the formation of migmatites (Lappin and Hollister, 1979;

Kenah and Hollister, 1983; McLellan, 1988). For some rock types, the combination of deformation and partial melting, resulted in melt removal and the formation of high density restitic layers which are similar to the stronalites characteristic of the lower crustal section exposed in the Ivrea zone in the western Alps of Italy (Mehnert, 1975).

Figure 2 shows the P-T path of the highest grade portion of the Central Gneiss Complex. The invariant point and univariant curves for kyanite, andalusite, and sillimanite are shown for reference. The temperature for point (a) is based on the presence of mineral assemblages which constrain a minimum temperature at which the observed assemblage could have formed (Hollister, 1982). Geothermometry based on cation exchange reactions was not used because, as discussed by Frost and Chacko (1989), such geothermometers are not reliable for temperatures above 600-650°C; they generally involve exchange of equal valent cations between minerals which have different blocking temperatures for cation diffusion. Pressure determinations based on mineral compositions, on the other hand, typically involve mass-transfer reactions and coupled exchange of cations of more than one valence. The pressure of point (a), Fig. 2, is based on compositions of minerals involved in the following mass-transfer reactions (Hollister, 1982):

garnet + sillimanite + quartz = cordierite
grossular + sillimanite + quartz = anorthite
garnet + quartz = hypersthene + anorthite

Samples used to determine point (a) were collected from the overturned limb of the large recumbent fold in the area near Khtada Lake (Figs. 1, 3, and 4). The pressure and temperature at this locality result from a significant input of heat from the 5 km thick Kasiks sill (Fig. 1). This pluton is approximately 54 Ma (U-Pb method, Van der Heyden, pers. comm.). Also, at this locality the orientation of minor fold axes and lineations, including aligned sillimanite needles, changes from east plunging and parallel to the hinge of the recumbent fold to NNE trending as the Kasiks sill is approached from the south. Thus, earlier structures associated with the recumbent fold were reoriented due to intrusion of the Kasiks sill while the rocks were at P-T conditions within the sillimanite stability field.

Further north, in the area to the west of Khutzeymateen Inlet (Fig. 1) the overall structure is a stacked section of nappes which has not been as pervasively injected by magma as the section in the Khtada Lake area. Consequently, the extreme metamorphic conditions recorded near the Kasiks sill were not achieved in the Khutzeymateen Inlet area, even adjacent to the Quottoon pluton (Douglas, 1986), which is dated at 60 Ma by the U-Pb method (Armstrong and Runkle, 1979). In the Khutzeymateen Inlet area, metamorphic temperatures were 600-700°C. The determinations are based on biotite-garnet exchange thermometry which gives temperatures consistent with the mineral assemblages, and pressures of ~6 kbar are based on the mass-transfer reactions, Garnet + Sillimanite + Quartz = Cordierite, and Grossularite + Sillimanite + Quartz = Anorthite (Douglas, 1986). Point b (Fig. 1) on the P-T path (Fig. 2) represents these conditions.

The estimate for conditions of metamorphism prior to those

inferred for point (b) are based in part on occurrences of kyanite and
staurolite inclusions in garnet (Hollister, 1977; 1982) and other
relict metamorphic textures (Selverstone and Hollister, 1980; Hill,
1985; Douglas, 1986; Sisson, 1985), and on geologic arguments.
Kyanite and staurolite are absent from the schist matrix everywhere in
the central belt, with the exception of one occurrence of staurolite
at Redcap Mountain (Hill, 1985). However, sillimanite pseudomorphs of
kyanite are widespread. In the Khutzeymateen area, sillimanite
pseudomorphs of staurolite in the pelitic parts of graded beds
preserve the unmistakable habit of staurolite. Crawford et al. (1987)
concluded that the kyanite and staurolite crystallized at the
pressure-temperature conditions represented by point c, Fig. 2, prior
to 98 Ma, within the tectonically thickened crust generated during the
collision with Stikinia of the crustal fragment referred to as the
Alexander terrane.

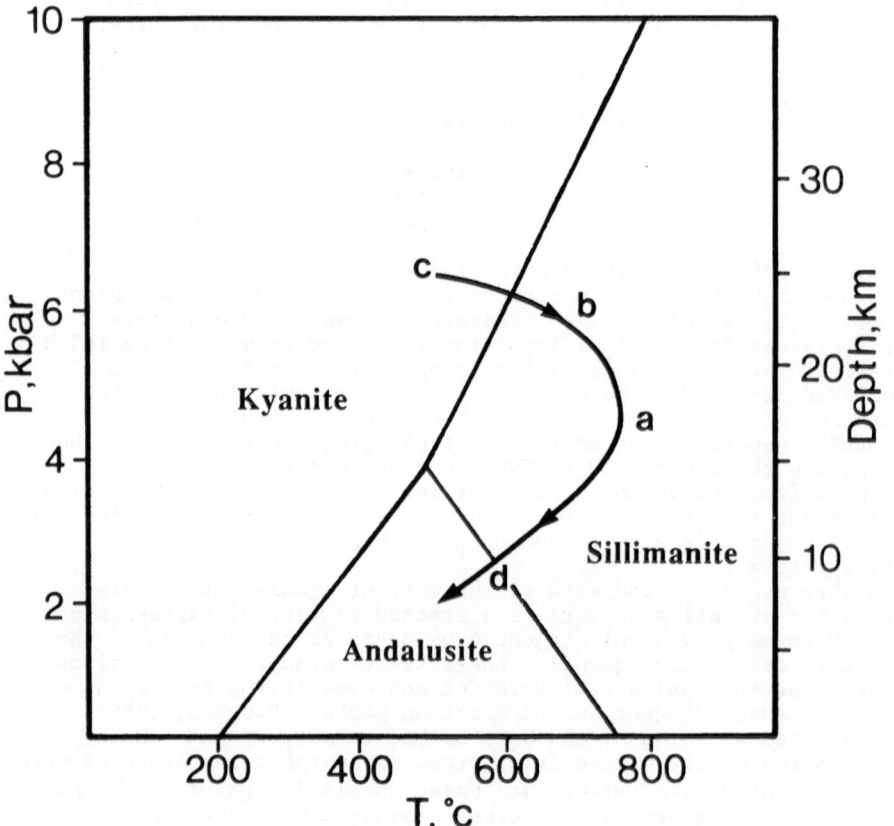

Figure 2. Pressure-temperature diagram summarizing the P-T history of
the CGC. The stability fields of kyanite, sillimanite, and andalusite
are shown for reference. The points a, b, c, d on the P-T path are
discussed in the text.

The pressure of point c is based on the argument that the Central Gneiss Complex was uplifted from a depth at least equivalent to that represented by the high pressure (8 ± 1 kbar, Crawford et al., 1987) rocks west of the Work Channel lineament (Fig. 1). The absence of the garnet + hornblende assemblage in rocks of suitable composition within the Central Gneiss Complex east of the lineament and the presence of this assemblage west of the lineament suggests the level of metamorphic origin of the Central Gneiss Complex was less than about 8 kbar. However, prior to heating to the recorded metamorphic conditions, the Central Gneiss Complex could have been at greater depth. The lowest pressure from which the Central Gneiss Complex was uplifted is constrained by the observation that the metamorphic conditions at b were set during decompression which means the rocks had to have come from greater depth.

The petrologic data in support of the lowest pressure point (d) include late, brittle veins containing andalusite, orthoclase, sillimanite, and cordierite (Hollister, 1982). These observations imply the P-T path crossed the sillimanite-andalusite univariant curve near point (d). Mineral reaction textures in gneisses of the central belt (Hollister, 1982) show that when metamorphic conditions at point (a), Fig. 2, were attained, the P-T path of these rocks was in transit from relatively high-pressure and moderate temperature to lower pressures as temperature increased. Because the highest temperature metamorphic assemblages occur in the vicinity of early Tertiary sills, such as the Kasiks sill, it is reasonable to attribute the added heat input to the injection of the sills. Subsequent exhumation without significant input of heat from plutons controlled the P-T path to point d, Fig. 2.

The rates of exhumation are constrained by U-Pb dates on zircons from several tonalite plutons, and K-Ar cooling dates on hornblende and biotite from the plutons and the Central Gneiss Complex. The earliest near-concordant date on a tonalite pluton is 85 Ma (Woodsworth, et al, 1983), which is close to the age of the first sediment derived from the Coast Mountains, the late Cretaceous west-derived Kasalka Group in the Bowser basin to the east (Woodsworth, 1979). The Quottoon pluton is 60 Ma (Armstrong and Runkle, 1979) and the Kasiks sill is approximately 54 Ma (Van der Heyden, personal communication). The latter intruded contemporaneously with metamorphism at (a), Fig. 2. Uplift was finished by about 49 Ma, the prevailing K-Ar cooling age of biotite in gneiss and tonalite throughout the Prince Rupert portion of the Central Gneiss Complex (Sutter, written communication, 1986). Uplift thus occurred concurrently with continuing injection of tonalite. The exhumation rate following intrusion of the Kasiks sill is reasonably constrained to be greater than 1 mm/yr (detailed arguments for this estimate are in Hollister, 1982, and Crawford et al., 1987).

Structures Associated with Uplift

In order to characterize the nature of the deep crust prior to exhumation, thermal effects and deformation superimposed during uplift must be removed. Detailed study in the Khtada Lake area (Fig. 3) and observations in selected areas as far north as Portland Inlet (Fig. 1) suggest that uplift was accompanied by the formation of vertical to steeply dipping ductile shear zones, tens to hundreds of meters wide. The Khtada Lake area (Fig. 3) is cut be several such shear zones. In outcrop scale, minor folds in the gneisses tighten toward and are truncated by ductile shear zones that parallel the axial planes of these minor folds. Such outcrop scale shear zones range in width from a few centimeters to tens of meters or more. The foliation in the shear zones consists of gneissic layering with leucocratic segregations up to a few centimeters thick. Within the shear zones are small, isolated isoclinal fold noses, most commonly of calcsilicate lithologies, and isolated slabs or blocks with foliation discordant to that of the shear zone. These appear to be dismembered relics of unsheared rocks. Close to the larger shear zones, this steeply dipping intense flow foliation becomes more pervasive. In the Khtada Lake area, flow foliation also increases markedly toward the lower contact of the Kasiks pluton. Within the shear zones, no traces of the pre-existing fabric remain. The large, steeply dipping ductile shear zones, hundreds of meters across and mappable, in some cases, for up to 15 km along strike, are parallel to the axial surfaces of upright, map scale open to isoclinal folds. The steeply dipping shear zones cut the Kasiks and Quottoon plutons (Fig. 3) and therefore in part post-date these intrusions.

The Work Channel lineament (Fig. 1) is the westernmost of the large, upright shear zones. The Work Channel lineament zone contains vertically foliated augen gneiss and, locally, mylonite. Lineations everywhere within this zone are nearly vertical. As with the other major shear zones, structures in the schists and gneisses adjacent to the Work Channel lineament are steep and trend in a northwesterly direction, parallel to the Work Channel lineament; lineations and fold axes plunge steeply (60-90°) down the dip of the foliation. These structures, and the marked contrast of metamorphic conditions on either side of the Work Channel lineament, suggest it is reasonable to attribute much of the later stages of uplift of the Central Gneiss Complex to vertical displacement across the Work Channel lineament and the other major northwest trending shear zones within the CGC. Consequently, the pre-exhumation crustal section would be represented by those portions of the CGC which were unaffected by the steeply dipping ductile shear zones.

Structural Development Prior to Uplift

Figure 4, a cross section of the Central Gneiss Complex between two major shear zones south of the Kasiks pluton, illustrates many of the features inferred to characterize the crust prior to development of the steep ductile shear zones. Foliation in domains between the shear zones, almost everywhere parallel to lithologic boundaries, is defined by local concentrations of biotite or sillimanite, by oriented biotite, and by banding in thin leucocratic layers. South of the Kasiks pluton, in the highest grade part of the area, the metasedimentary and metavolcanic units consist of stronalite (a garnet- and sillimanite-rich felsic rock), hypersthene- and biotite-bearing felsic gneiss, concordant well-foliated granitoid layers, and migmatite. These layers range from tens to hundreds of

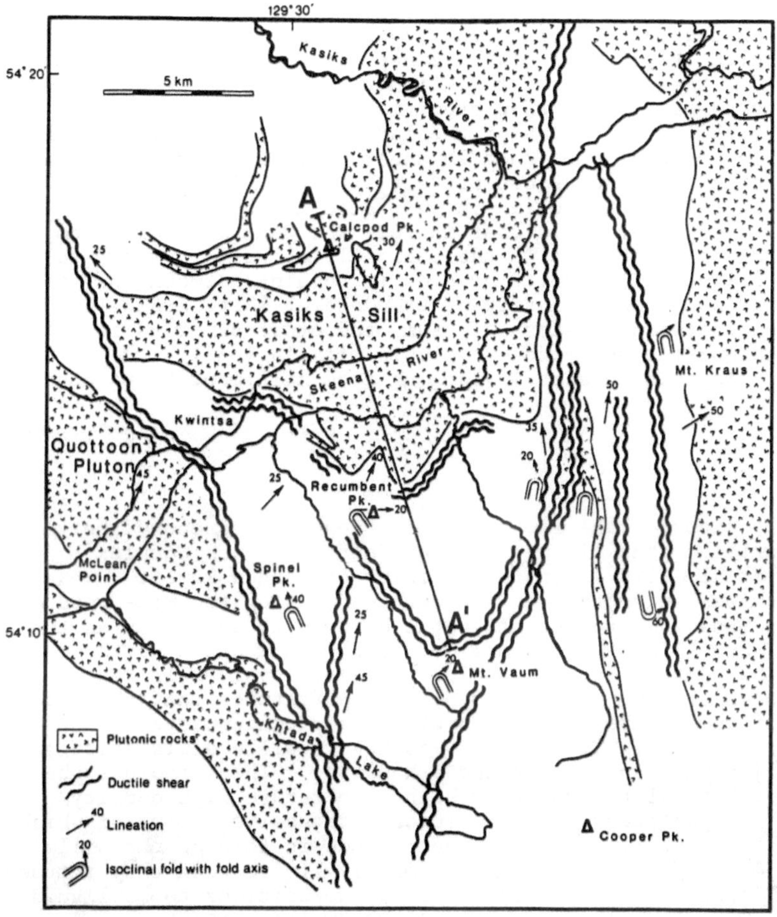

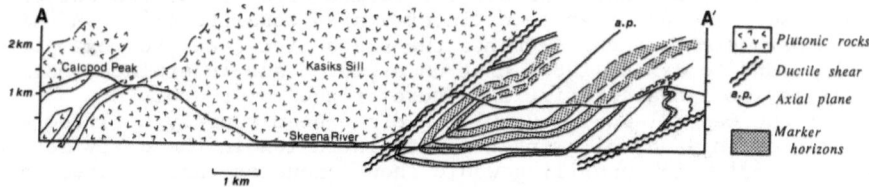

Figure 3. Map of a portion of the central belt, near Khtada Lake (Figure 1), showing the location of the cross section AA' (Figure 4). Ductile shear zones are marked by wavy lines; the patterned areas are major plutonic bodies. From Crawford, et al. (1987).

Figure 4. Cross section AA'. Symbols used are as in Figure 3; the dotted pattern marks mappable stratigraphic units; a.p., fold axial plane. No vertical exaggeration.

meters in thickness and have lateral continuity for at least 5 km.
The foliation and the compositional layering are folded by large
isoclinal folds (Fig. 3, Fig. 4). Except at the hinges of these and
similar folds, the dip of the foliation throughout the area, both
south and north of the Kasiks pluton, is gentle to the north
(Hutchison, 1982). The limbs of the large isoclinal folds are sheared
off by gently north dipping ductile shear zones which resemble the
steeply dipping shear zones but are cut by them. The gentle
north-dipping foliation thus is in part also a flow foliation. The
Kasiks sill, and other sills that intrude the Central Gneiss Complex
to the north are concordant to the regional foliation. Minor fold
hinges and aligned sillimanite needles in domains between the steeply
dipping shear zones have consistent trends over large areas, but can
vary in trend between domains. In sum, the crust prior to formation
of the steep ductile shear zones is both lithologically and
structurally layered on a scale ranging from millimenters to hundreds
of meters.

Discussion

As presently exposed, the rocks in this orogen that record the
greatest burial depths lie west of the Central Gneiss Complex
(Crawford, et al., 1987), in the vicinity of Prince Rupert. This
crust, which was metamorphosed and deformed during the mid-Cretaceous
orogenic episode, was formed at depths between 20 and 30 km as the
Alexander/Wrangellia superterrane collided with western North
America. The Central Gneiss Complex was also involved in the
mid-Cretaceous orogen; however, many of the features that characterize
the crust of the Central Gneiss Complex are a consequence of the
intrusion of the early Tertiary (60-50 Ma) tonalite sills associated
with the major phase of uplift and exhumation. These include the
formation of stronalite and other high temperature high density
gneisses, of migmatites, and the plutons themselves, many of which
have the form of sills. Cumulate hornblende in some of the sills
suggests these magmas must have been mostly liquid when they intruded
(Hollister, 1982). For the present rock compositions to be liguid
requires temperatures in excess of 900 ^{o}C (Wyllie, 1977). A
contribution of melt from the mantle into the crust therefore seems
likely. From geochemical arguments, Arth, et al (1988) concluded that
plutons similar in composition located within the extension of the
Central Gneiss Complex into southeast Alaska had a component of
subduction related magma. The addition of these mantle-derived magmas
substantially thickened the earlier formed crust at depths between 15
and 20 km. The association of interlayered tonalite sills and high
grade metamorphic rocks supports the suggestion (see, for example,
Furlong and Fountain, 1986) that underplated and interplated magmatic
rocks may be a characteristic feature of the lower crust.

If the Central Gneiss Complex had not been rapidly uplifted and
exhumed during and following tonalite intrusion in the early Tertiary,
seismic reflection profiling would show a zone at those crustal depths
characterized by pronounced near horizontal or low angle reflectors.
These reflectors would represent the interlayered gneisses with large
density constrasts between layers meters to hundreds of meters thick,

ductile shear zones meters to hundreds of meters thick, and tonalite sills and tabular plutons meters to kilometers thick. This section is thus very similar to the description given by McCarthy and Thompson (1988) of the mid to lower crustal reflecting horizons in the extended crust of the western United States.

Rapid uplift of this package of rocks along shear zones which bound the Central Gneiss Complex on both sides quenched the melt and metamorphic equilibria when the rocks were at depths at 5-10 km. Abrupt exhumation from a depth of 20 km or more to less than 10 km froze the mineralogic and structural features formed at depth. Much later, Miocene uplift (Parrish, 1983) and erosion brought the crystalline rocks to the surface.

Had rapid uplift not occurred, the deep crustal features described above might not have been preserved in their present form. Slow cooling would have set in following emplacement of the plutons during the early Tertiary, and the mineral assemblages of the high pressure granulite facies of metamorphism would have developed. Such assemblages would give the section a higher average density, and hence also higher seismic velocities, than the presently exposed section. High pressure granulite facies rocks probably now underlie the Central Gneiss Complex.

The association of the ductile shear zones with migmatites and magmatic intrusions, the scarcity of stretching lineations in the ductile shear zones, and the high temperatures recorded by the metamorphic assemblages suggest that melt may have played an important role in the deformation mechanism (Crawford et al., 1987; Hollister and Crawford, 1986). This may be an important factor in the deformation dynamics at collisional plate boundaries. The rock associations of the Central Gneiss Complex are similar to those reported elsewhere for the mid to lower crust. They are here interpreted to have formed as part of the dynamic processes which operated in the collision zone between a microplate and North America.

Acknowledgments

The research reported here was supported by NSF grants EAR81-00398 and EAR84-07014 to Crawford and EAR83-19249 and EAR87-19368 to Hollister, and the Geological Survey of Canada. We are grateful to the organizers of the 1988 NATO Workshop at Killarney, Ontario, for giving one of us (LSH) the opportunity to present orally our conclusions to an international audience. The manuscript benefitted substantially from the two thoughtful anonymous journal reviewers.

References

Armstrong, R.L., and Runkle, D. (1979) 'Rb-Sr geochronometry of the
    Ecstall, Kitkiata, and Quottoon plutons and their country rocks,
    Prince Rupert region, Coast Plutonic Complex, British Columbia.'
    Canadian Journal of Earth Science **16**, 387-399.
Arth, J.G., Barker, F. and Stern, T.W. (1988) 'Coast Batholith and
    Taku plutons near Ketchikan, Alaska: Petrography, Geochronology,
    Geochemistry, and Isotopic Character.' *American Journal of Science*
    288-A, 461-489.
Crawford, J.L., Hollister, L.S., Woodsworth, G.J. (1987) 'Crustal
    deformation and regional metamorphism across a terrane boundary:
    Coast Plutonic Complex, British Columbia.' *Tectonics* **6**, 343-361.
Douglas, B.J. (1986) 'Deformational history of an outlier of
    metasedimentary rocks, Coast Plutonic Complex, British Columbia,
    Canada.' *Canadian J. Earth Sci.* **23**, 813-826.
Frost, B.R., and Chacko, T. (1989) 'The granulite uncertainty
    principle: limitations on thermobarometry in granulites.' *Journal of
    Geology* **97**, 435-450.
Furlong, K.P. and Fountain, D.M. (1986) 'Continental crustal
    underplating: Thermal considerations and seismic-petrologic
    consequences.' *J. Geophys. Res.* **91**, 8285-8294.
Hill, M.L. (1985) 'Geology of the Redcap Mountain area, Coast Plutonic
    Complex, British Columbia.' Unpublished Ph.D. Thesis, Princeton
    University, Princeton, NJ, 216 pp.
Hollister, L.S. (1977) 'The reaction forming cordierite from garnet,
    the Khtada Lake metamorphic complex, British Columbia.' *Can.
    Mineral.* **15**, 217-229.
Hollister, L.S. (1982) 'Metamorphic evidence for rapid (2 mm/yr)
    uplift of a portion of the Central Gneiss Complex, Coast Mountains,
    British Columbia.' *Canadian Mineralogist* **20**, 319-332.
Hollister, L.S. and Crawford, M.L. (1986) 'Melt-enhanced deformation:
    A major tectonic process.' *Geology* **14**, 558-561.
Hutchison, W.W. (1982) 'Geology of the Prince Rupert-Skeena map-area,
    British Columbia.' *Memoir Geological Survey of Canada* **394**, 1-116.
Kenah, C. and Hollister, L.S. (1983) 'Anatexis in the Central Gneiss
    Complex, British Columbia.' In *Migmatites, Melting and
    Metamorphism*, M.P. Atherton and C.D. Gribble, eds., Shiva Publ. Co.,
    142-162.
Lappin, A.R. and Hollister, L.S. (1980) 'Partial melting in the
    Central Gneiss Complex near Prince Rupert, British Columbia.'
    American Journal of Science **280**, 518-545.
McLellan, E.L. (1988) 'Migmatite structures in the Central Gneiss
    Complex, Boca de Quadra, Alaska.' *J. Meta. Geol.* **6**, 517-542.
Mehnert, K.R. (1975) 'The Ivrea zone — a model of the deep crust.'
    Neues Jahrb. Mineral. Abh., Part 2, **125**, 156-199.
Parrish, R.R. (1983) 'Cenozoic thermal evolution and tectonics of the
    Coast Mountains of British Columbia 1, Fission track dating,
    apparent uplift rates, and patterns of uplift.' *Tectonics* **2**,
    601-631.
Selverstone, J. and L.S. Hollister (1980) 'Cordierite-bearing
    granulites from the Coast Ranges, British Columbia: P-T conditions
    of metamorphism.' *Canadian Mineralogist* **18**, 119-129.
Sisson, V.B. (1985) 'Contact metamorphism and fluid evolution

associated with the intrusion of the Ponder Pluton, Coast Plutonic Complex, British Columbia, Canada.' Ph.D. Thesis, Princeton University, 345 pp.

Woodsworth, G.J. (1979) 'Geology of the Whitesail Lake map-area, British Columbia.' *Geological Survey of Canada Current Research*, Part A, Paper 79-1A, 71-75.

Woodsworth, G.J., Loveridge, W.D., Parrish, R.R. and Sullivan, R.W. (1983) 'Uranium-lead dates fron the Central Gneiss Complex and Ecstall pluton, Prince Rupert map area, British Columbia.' *Canadian Journal of Earth Science* 20, 1475-1483.

Wyllie, P.J. (1977) 'Crustal Anatexis: an experimental review.' *Tectonophysics* 43, 41-71.

A FIELD GUIDE TO THE KAPUSKASING UPLIFT, A CROSS SECTION THROUGH THE ARCHEAN SUPERIOR PROVINCE[1]

JOHN A. PERCIVAL
Lithosphere and Canadian Shield Division
Geological Survey of Canada,
588 Booth Street
Ottawa, Ontario K1A OE4

ABSTRACT. The Kapuskasing uplift provides a window on the deep crust of the Superior Province. Evidence in support of an exposed oblique crustal cross section includes: a systematic easterly increase in metamorphic grade from 3 kbar greenstones at Wawa to 8 kbar Kapuskasing granulites; structural and paleomagnetic data on gneisses and dykes indicating 10-15° of east-side-up block rotation; seismic refraction and reflection data tracing the high-velocity Kapuskasing zone from surface down dip at 15°W to about 20 km; and progressively slower cooling rates with paleodepth. The composite cross-section derived from the oblique exposure comprises an upper, 2-3 kbar greenstone-granite megalayer (2.89-2.67 Ga), an intermediate, 5-6 kbar tonalite gneiss megalayer (2.72-2.67 Ga) and a lower, 7-9 kbar, heterogeneous granulite gneiss megalayer (2.76-2.60 Ga). Regional cooling and uplift preceded thrust uplift along the Ivanhoe Lake fault 2.2-1.95 Ga ago. Steep normal faults transecting the uplift possibly reflect topographic collapse and are cut by 1.88 Ga carbonatites. Regional NW-SE shortening of about 70 km during the 2.2-1.88 Ga Kapuskasing event was accomodated above a mid-crustal decollement by 15 km of brittle uplift of granulites along a 15°-dipping fault and below it by 10-15 km of ductile thickening of the present lower crust over a 200-km-wide zone.

PART I: Geological Framework of the Kapuskasing Uplift

INTRODUCTION

The purpose of the trip is to examine the characteristics and interrelationships of Archean greenstone-granite and high-grade gneiss terranes of the Superior Province. A 300-km long west to east transect between Wawa and Timmins, Ontario will be used to illustrate regional-scale relationships.

Figures 1 and 2 show the major geological features of the central Superior Province; Figure 3 traces the trip route. On the first day we examine features of the Michipicoten belt, a dominantly metavolcanic

[1]Geological Survey of Canada Contribution No. 25788

M. H. Salisbury and D. M. Fountain (eds.), Exposed Cross-Sections of the Continental Crust, 227–283.

portion of the Wawa subprovince, intrusions of the Wawa gneiss terrane
and the Wawa-Kapuskasing boundary. On day two, features of the
Kapuskasing structural zone are visited, including the Shawmere
anorthosite complex and high-grade gneisses, before examining the
Ivanhoe Lake cataclastic zone separating rocks of the Kapuskasing zone
from those of the Abitibi Belt.

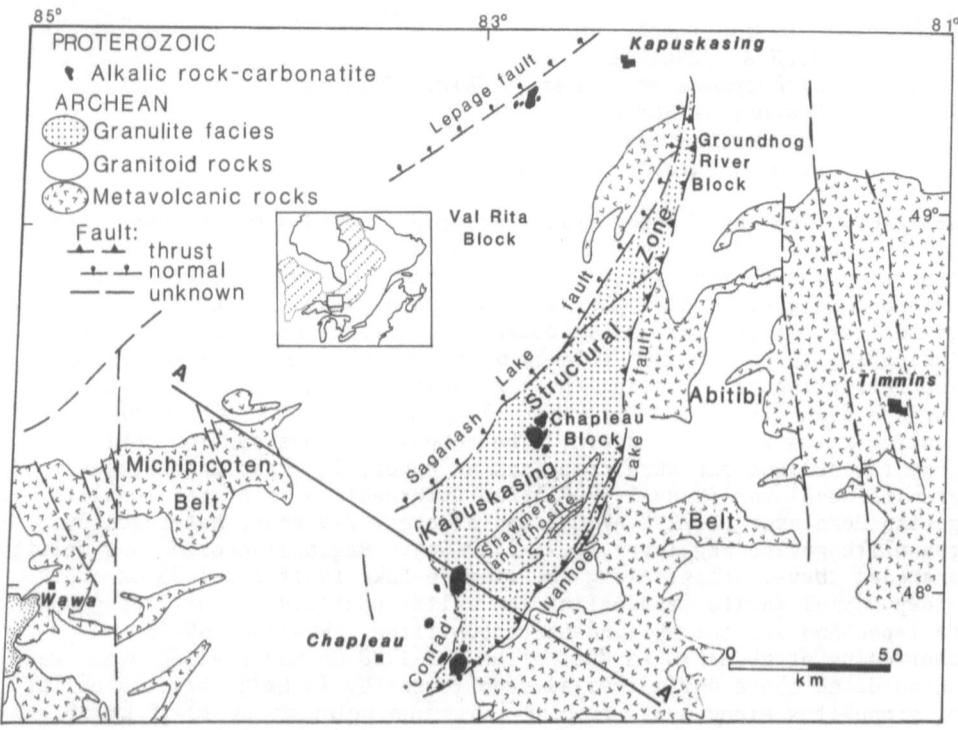

Figure 1. Lithotectonic map of central Superior Province showing
distribution of major features including the Abitibi and Wawa
subprovinces and Kapuskasing structural zone.

REGIONAL SETTING

The Superior Province is an Archean terrane composed of east-west
trending belts of alternate volcanic-rich and sediment-rich character,
termed subprovinces (Figs. 1,2). The continuity of the east-west belts
is interrupted by a northeast-trending zone of high-grade metamorphic
rocks, the Kapuskasing structural zone (Thurston et al., 1977). At its
southern end, the Kapuskasing structure is fault-bounded on the
southeast but the western contact is complex and gradational over
120 km to low-grade rocks of the Michipicoten belt near Lake Superior
(Percival and Card, 1983; 1985) (Figs. 1-3).
    The Kapuskasing "high", a prominent northeasterly gravity and
aeromagnetic anomaly, was interpreted by Wilson and Brisbin (1965) to

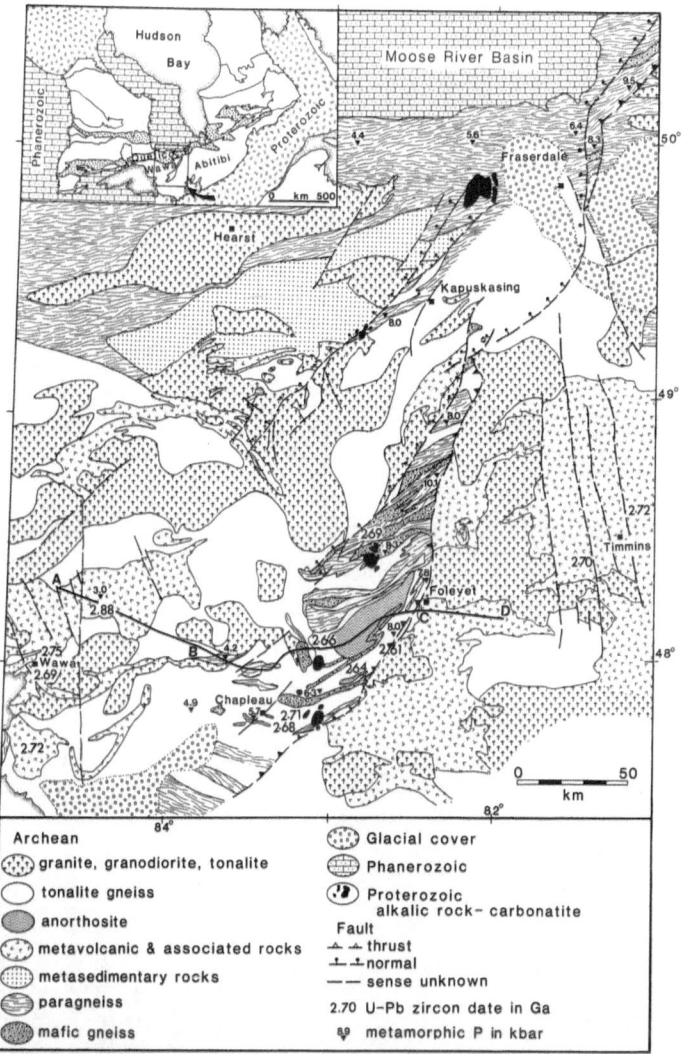

Figure 2. Generalized geological map of central Superior Province in the vicinity of the excursion. Section line A-B-C-D corresponds to the cross section of Fig. 19.

indicate pronounced upwarp of the Conrad discontinuity. Bennett et al. (1967) concluded that the Kapuskasing structure is a complex horst uplifted during the Proterozoic. The association of 1,100-1,000 Ma alkalic rock-carbonatite complexes led Burke and Dewey (1973) to suggest that the Kapuskasing structure is a failed arm of the Keweenawan rift structure. Watson (1980) postulated that the Kapuskasing zone was uplifted during late Archean or early Proterozoic sinistral transcurrent movement. The low-to-high-grade transition at the southern end of the structure has been interpreted as an oblique

230

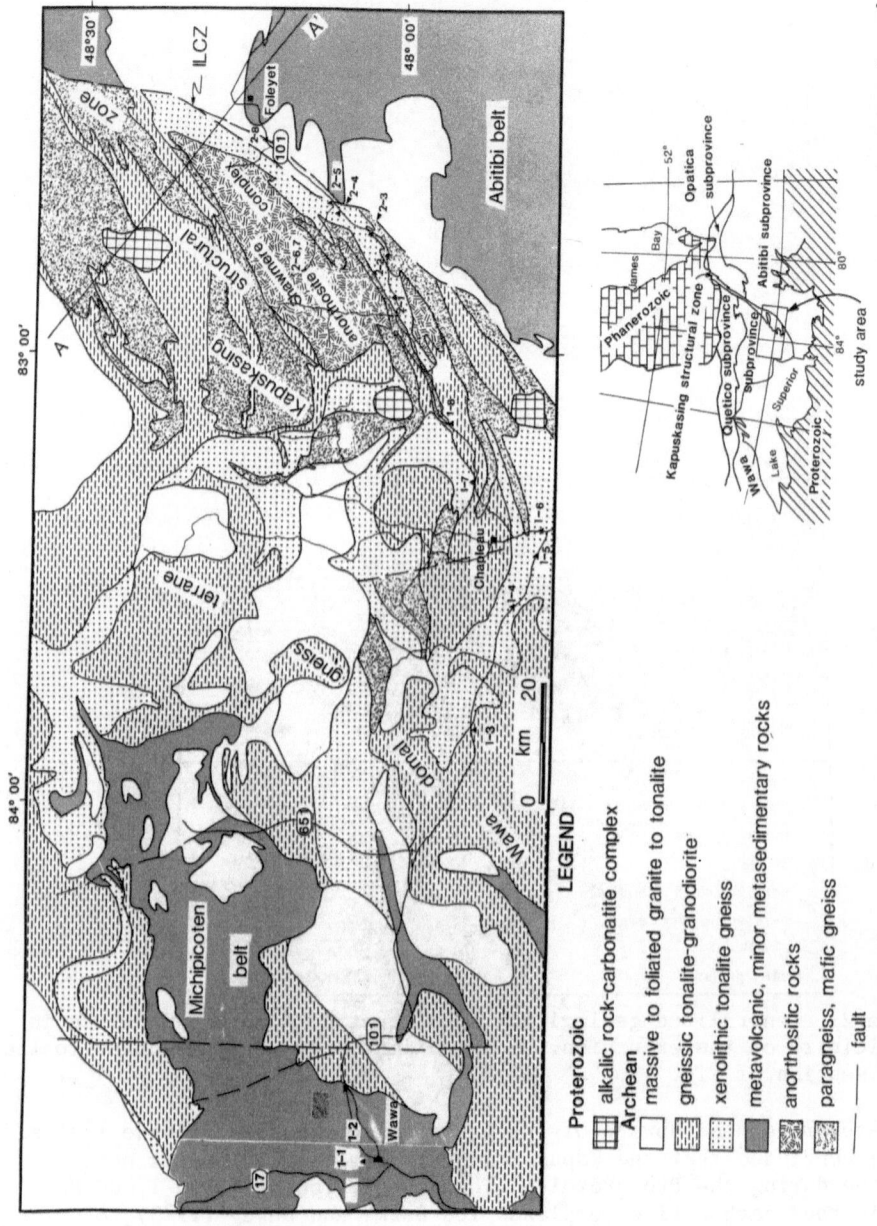

Figure 3. Geology of the Wawa-Abitibi area with location of field trip stops, Days 1 and 2.

LEGEND

Proterozoic

▦ alkalic rock-carbonatite complex

Archean

☐ massive to foliated granite to tonalite

gneissic tonalite-granodiorite

xenolithic tonalite gneiss

metavolcanic, minor metasedimentary rocks

anorthositic rocks

paragneiss, mafic gneiss

——— fault

cross-section of the crust uplifted along an east-verging thrust
(Percival and Card, 1983). Recent earthquakes in the region indicate
that the structure is still active (Forsyth et al., 1983).

GEOPHYSICAL CHARACTERISTICS OF SOUTH-CENTRAL SUPERIOR PROVINCE

A map showing apparent crustal thickness in the Lake Superior region,
based on seismic refraction studies, was presented by Halls (1982).
More recent data (Fig. 4; Boland and Ellis, 1986) show that the depth
to Moho increases dramatically to about 53 km beneath the Kapuskasing
structure from background values near 45 km in the Wawa area and about
35 km near Timmins. A step-like decrease in crustal thickness is
apparently associated with the eastern boundary of the Kapuskasing
zone. The crustal root beneath the high density (Percival, 1986),

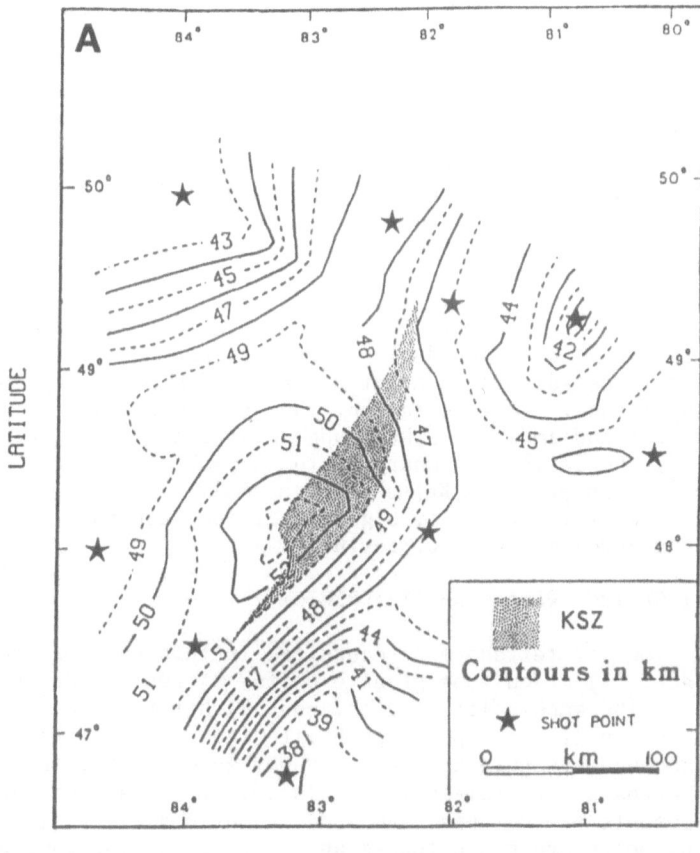

Figure 4. a) Depth to Moho in the Kapuskasing area (after Boland and
Ellis, 1989), based on the 1984 Lithoprobe refraction study (Northey
and West, 1986).

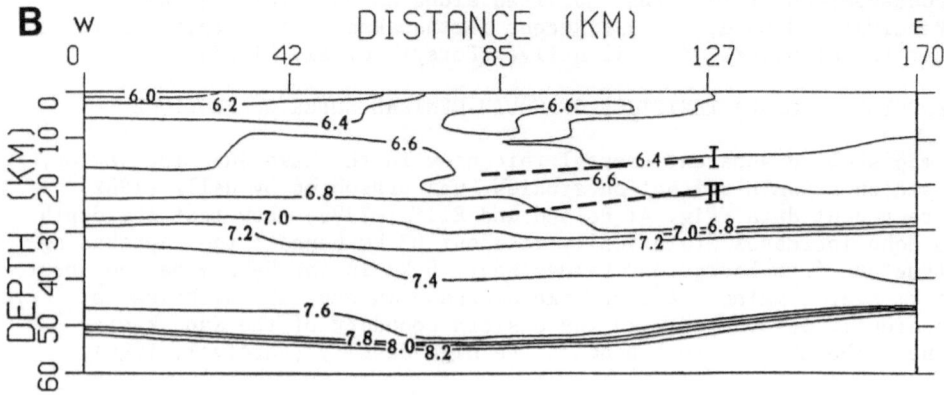

Figure 4. b) Refraction profile across the Kapuskasing uplift (after Boland et al., 1988).

high-velocity (Cook, 1985; Percival and Fountain, in press) Kapuskasing rocks may result from isostatic compensation.

The Bouguer gravity anomaly map for the Wawa-Timmins region is shown in Figure 5. In general, areas underlain by metavolcanic rocks have associated positive gravity anomalies and granitoid-gneissic rocks have negative anomalies. The Kapuskasing structural zone has an associated linear positive gravity anomaly extending from James Bay in the north to some 50 km southwest of Chapleau. In the Wawa-Chapleau-Foleyet area, the gradient is gradual on the west and abrupt on the east, suggesting a west-dipping contact between the Kapuskasing zone and Abitibi subprovince. In this region, the gravity profile (Fig. 15) shows a paired high-low anomaly. The trough of the low is coincident with the fault at the eastern boundary of the Kapuskasing zone.

To the north, the positive Kapuskasing anomaly broadens as it coalesces with an east-west gravity high associated with the Quetico-Opatica metasedimentary subprovince.

GENERAL GEOLOGY OF THE WAWA - ABITIBI REGION

Three distinct types of terrane are recognized in this part of the Superior Province: 1) greenstone-granite belts, 2) regions dominated by orthogneiss in the amphibolite facies, and 3) heterogeneous gneisses in the granulite to upper amphibolite facies. Differences in metamorphic grade and pressure between terranes suggest that the terrane types represent crustal-scale megalayers (Percival and Card, 1985), the components of the upper and middle continental crust. The Wawa and Abitibi belts are greenstone-granite terranes; the Wawa gneiss terrane is an example of terrane type 2 and the Kapuskasing zone represents type 3 terrane (Fig. 3).

Greenstone-Granite Terranes. The Michipicoten belt, part of the volcanic-rich Wawa subprovince, is composed mainly of metavolcanic rocks of ultramafic, mafic and felsic composition (Goodwin, 1962;

Sylvester et al., 1987), with intercalated greywacke, conglomerate, chert and iron formation, mainly siderite. Dome and basin structures (Goodwin, 1962) as well as downward-facing strata and overturned structures (Attoh, 1980) have been recognized. Metamorphic grade ranges from sub-greenschist to amphibolite facies (Fraser et al., 1978). Several suites of intrusive rocks include synvolcanic bodies ranging from peridotite to granodiorite, younger granodiorite batholiths, and still younger granite and syenite plutons (Card, 1982).

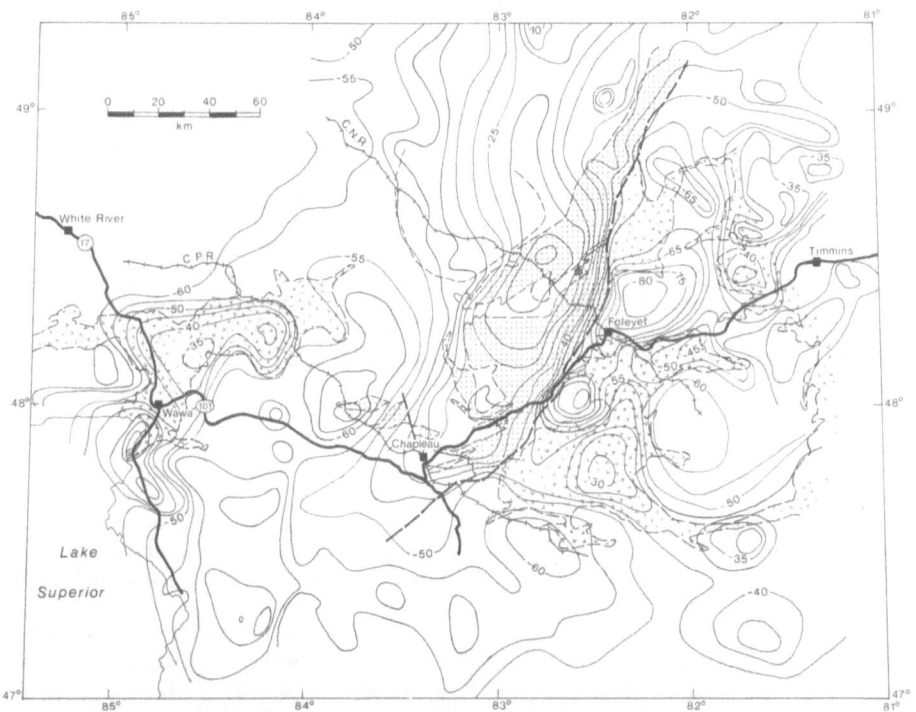

Figure 5. Bouguer gravity map superimposed on simplified geology. (Gravity values in mGal with 5 mGal contour interval, from Earth Physics Branch maps 44078, 44084, 48078 and 48084). Checks - greenstone belts; dots - Kapuskasing gneiss; unpatterned-undivided granitoid rocks.

The supracrustal rocks of the southern part of the belt were divided into three major cycles by Goodwin (1962). A lower cycle, consisting of mafic and felsic volcanics, is capped by Michipicoten-type iron formation, mainly siderite, but with lesser pyrite-, chert- and graphite-rich rocks. The associated Jubilee Stock, a high-level subvolcanic intrusion, was emplaced within a caldera structure (Sage, 1980). The middle cycle comprises mafic volcanics overlain by clastic metasediments and felsic tuffs and breccias. The clastic sediments, including the Dore conglomerate, wacke, siltstone, and crossbedded arkose, are the facies equivalents of the felsic pyroclastics and are formed mainly of detritus eroded from the felsic centres. The upper cycle comprises intermediate to felsic (andesite-dacite) tuffs and quartz-feldspar porphyry. Recent work by Sage (pers. comm., 1986) indicates only two cycles within the main volcanic pile, but an additional, lower cycle (Turek et al., 1988).

The Abitibi subprovince is dominated by a thick sequence of volcanic and sedimentary rocks of the Abitibi greenstone belt (Jensen, 1981; 1985). The supracrustal succession typically comprises sequences of ultramafic, mafic, and felsic volcanics. Intercalated turbiditic sedimentary rocks contain a high proportion of volcanic detritus. In the Abitibi belt, the uppermost group, the Timiskaming, is an unconformity-bounded sequence of alkalic volcanics and fluviatile sediments (Hyde, 1980) localized along major east-west fault zones.

Large areas of the Abitibi greenstone belt are metamorphosed to greenschist facies; subgreenschist, prehnite-pumpellyite facies rocks are common in the Timmins-Rouyn area and narrow aureoles of amphibolite facies rocks occur adjacent to plutonic bodies (Jolly, 1978).

The supracrustal rocks of the Abitibi subprovince display evidence of polyphase deformation in the form of major and minor structures of several ages and orientations. In the Abitibi greenstone belt, older northerly-trending folds are overprinted by east-west trending major and minor folds, forming major dome and basin structures (Pyke, 1982). The major isoclinal folds with east-west striking subvertical axial planes, steeply-plunging minor folds, subvertical axial plane foliation, and steeply plunging stretching lineation were probably formed under subhorizontal, generally north-south major compression. Toward the southern margin of Abitibi belt the major folds are overturned northward, and in the adjacent Pontiac subprovince, folds are recumbent. The Cadillac-Larder Lake fault zone, which constitutes the boundary between the Abitibi and Pontiac subprovinces, probably has both transcurrent and thrust components of movement.

Several suites of intrusive rocks in the Abitibi subprovince can be distinguished on the basis of composition, structural relationships, setting, and age (Card, 1982). The oldest suite includes synvolcanic sills, dykes and plutons ranging in composition from peridotite to granodiorite; the more felsic intrusions are typically quartz diorite and trondhjemite. Gneissic plutonic rocks of tonalite and granodiorite composition, commonly containing amphibolitic enclaves, occur in the northeastern and southwestern Abitibi subprovince. Massive felsic plutonic rocks intrude both the greenstones and the gneissic rocks in the form of simple and composite plutons and batholiths. They form several suites, including early granodiorites, younger granite batholiths, and still younger syenite-diorite plutons. Contacts between the plutons and the country rocks are commonly concordant and steeply dipping; dominant east-west structural trends are locally deflected about the intrusions.

A time framework for events in the Michipicoten and Abitibi belts can be constructed from U-Pb zircon dates. In the western Abitibi belt, volcanic rocks range in age from 2,802 to 2,703 Ma (Nunes and Pyke, 1980; Nunes and Jensen, 1980; Mortensen, 1987), with late alkaline volcanics (Timiskaming Group) at 2,685 Ma, and in the

Michipicoten belt, from 2,749 to 2,696 Ma, with synvolcanic plutons at 2,737, 2,744 (Turek et al., 1982) and 2,745 Ma (Sullivan et al., 1985). Rare volcanic and associated plutonic rocks of 2,890 Ma age are present locally (Turek et al., 1988). A number of late- to post-tectonic plutons from the Abitibi and Michipicoten belts have zircon dates within a few million years of 2,680 (Krogh et al., 1982; Frarey and Krogh, 1986). Thus the main Abitibi and Michipicoten supracrustal sequences and early intrusions developed between 2,750 and 2,700 Ma ago. The dates on volcanics and late plutons bracket the age of deformation and regional metamorphism at between 2,700 and 2,680 Ma ago. Major volcanic, plutonic, and tectonic events of relatively brief duration were essentially synchronous throughout the Abitibi and Wawa subprovinces, a region some 1,200 km long and 200 km wide. The lithologic and age similarities between the Abitibi and Wawa subprovinces strongly suggest original continuity, now interrupted by the Kapuskasing structural zone.

Diabase dyke swarms of late Archean and Proterozoic age are present throughout the region. The oldest dykes, the north-trending Matachewan-Hearst swarm of the Abitibi and Wawa subprovinces, have a U-Pb zircon age of 2,454±2 Ma (Heaman, 1988). Northwest-striking diabase dykes in Wawa subprovince are petrographically similar to and have been paleomagnetically correlated with the Matachewan swarm (Ernst, 1981; Ernst and Halls, 1984). Abitibi and Wawa subprovinces are thus inferred to have been tectonically stable cratons by this time. Northeast-striking tholeiitic dykes are about 2,150 Ma old (Gates and Hurley, 1973); northwest olivine diabase dykes (Sudbury swarm) are about 1,240 Ma old (Krogh et al., 1987); and east-northeast olivine diabase dykes (Abitibi swarm) are dated at 1,140 Ma by U-Pb on baddelyite (Krogh et al., 1987).

Wawa Gneiss Terrane. The Michipicoten belt is intruded to the southeast by tonalitic gneiss and plutons of the Wawa gneiss terrane (Figs. 2,3). The rocks in this region consist of at least four lithologic components (Fig. 6): (1) hornblende-plagioclase± clinopyroxene mafic and rare paragneiss xenoliths, ranging from centimetres to tens of metres in maximum dimension, making up 5 to 50% of individual outcrops, and enclosed in (2) the volumetrically most abundant phase, hornblende-biotite tonalitic gneiss which is cut by (3) concordant to discordant layers of foliated to gneissic biotite-hornblende granodiorite, which in turn are cut by (4) late discordant quartz monzonite pegmatite. Xenolith-rich tonalitic gneiss units alternate on a 5 to 10 km scale with xenolith-poor units and can be traced for distances of at least 50 km. Layering in mafic xenoliths is locally discordant to layering in enclosing gneiss. Small folds of layering in tonalitic gneiss are commonly truncated by layers of foliated granodiorite. Tonalite gneiss has a minimum U-Pb zircon age of 2,707 Ma, partly reset by the intrusion of granodiorite sheets at 2,677 Ma (Percival and Krogh, 1983). The layers of granodiorite on the cm to km scale in the gneiss terrane can be correlated by zircon geochronology with discordant plutons of 2,680 Ma age in the greenstone belts, suggesting that the plutons have deep roots in the gneissic terrane.

Figure 6. Tonalitic gneiss (Stop 1-3) showing components, present in different proportions, which make up the unit on the regional scale. Mafic gneiss inclusions (right side of photograph) are enclosed by thinly layered gneiss and concordant granodiorite sheets (foreground).

In the area between the Michipicoten belt and Kapuskasing zone (Fig. 3) the orientation of foliation, gneissosity and axial surfaces of small folds is transformed from dominantly upright at higher structural levels to subhorizontal, rolling geometry at depth (Moser, 1988). Previously interpreted as domal or antiformal culminations on a scale of 20 to 25 km, the flat structures now appear to be parts of lozenge structures (Moser, 1988), related to syn-magmatic extension, possibly a result of crustal-scale gravitational collapse. The Highbrush Lake and Racine Lake "domes" have cores of tonalite-granodiorite gneiss whereas the Chaplin Lake dome and Missinaibi Lake arch have granitic cores flanked by foliated to gneissic rocks. A planar fabric in the homogeneous granitic rocks, defined by lenticular quartz and biotite alignment, is generally concordant to gneissosity in mantling gneiss. The Robson Lake "dome", adjacent to the Kapuskasing structural zone, has a core of interlayered mafic gneiss, paragneiss and tonalitic gneiss in granulite facies, typical of the Kapuskasing lithological sequence.

A geobarometer for calc-alkaline igneous rocks is based on the Al content of hornblende (Hammarstrom and Zen, 1986):

$$P = -3.92 + 5.03 \, Al_{Total}$$

Application of the barometer to a suite of tonalites from the Wawa gneiss terrane suggests that the pressure of igneous crystallization increases from about 5 kbar in the central part of the terrane to over 6 kbar near the Kapuskasing zone (Fig. 7). These results are intermediate between independent pressure estimates for the Michipicoten belt of 2-3 kbar, based on sphalerite-pyrrhotite geobarometry (Studemeister, 1983) and for the Kapuskasing zone of 6-8 kbar based on garnet-pyroxene-plagioclase-quartz barometry (Percival, 1983). A value of 8 kbar near Wawa is derived from a biotite tonalite which contains hornblende only adjacent to contacts with amphibolite; the hornblende is probably xenocrystic and the derived pressure therefore meaningless.

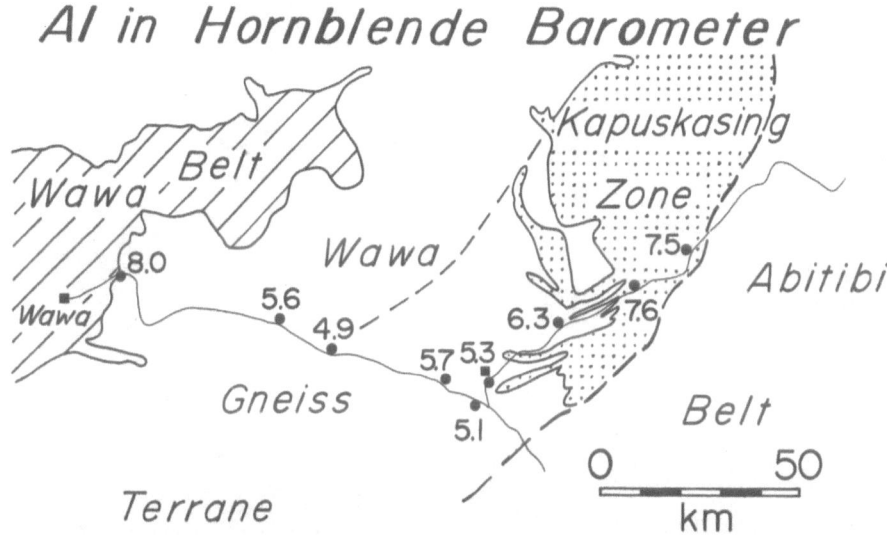

Figure 7. Map showing distribution of pressure estimated from the hornblende geobarometer on tonalitic rocks.

Metasedimentary rocks occur in two locations in the eastern Wawa subprovince. A discontinuous, antiformal to domal belt of paragneiss west of the Racine Lake "dome" may be continuous to the east with paragneiss of the Kapuskasing zone (Figs. 3 and 8). Stretched-pebble metaconglomerate occurs in association with quartz wacke and amphibolite in the vicinity of Borden Lake (Fig. 23). The polymictic (tonalite, granodiorite, meta-andesite, metasediments, amphibolite, vein quartz), clast-supported rock contains cobbles ranging from equant to constricted (1.5 m x 7 x 7 cm) with a prominent shallow northeast plunge. In cross-section the clasts vary from equidimensional to northwest-dipping ellipses.

Kapuskasing Structural Zone. The Kapuskasing structural zone comprises northeast-striking, northwest-dipping belts of paragneiss, mafic gneiss, gneissic and xenolithic tonalite, and rocks of the Shawmere anorthosite complex (Bennett et al., 1967; Thurston et al., 1977) (Figs. 3 and 8).

238

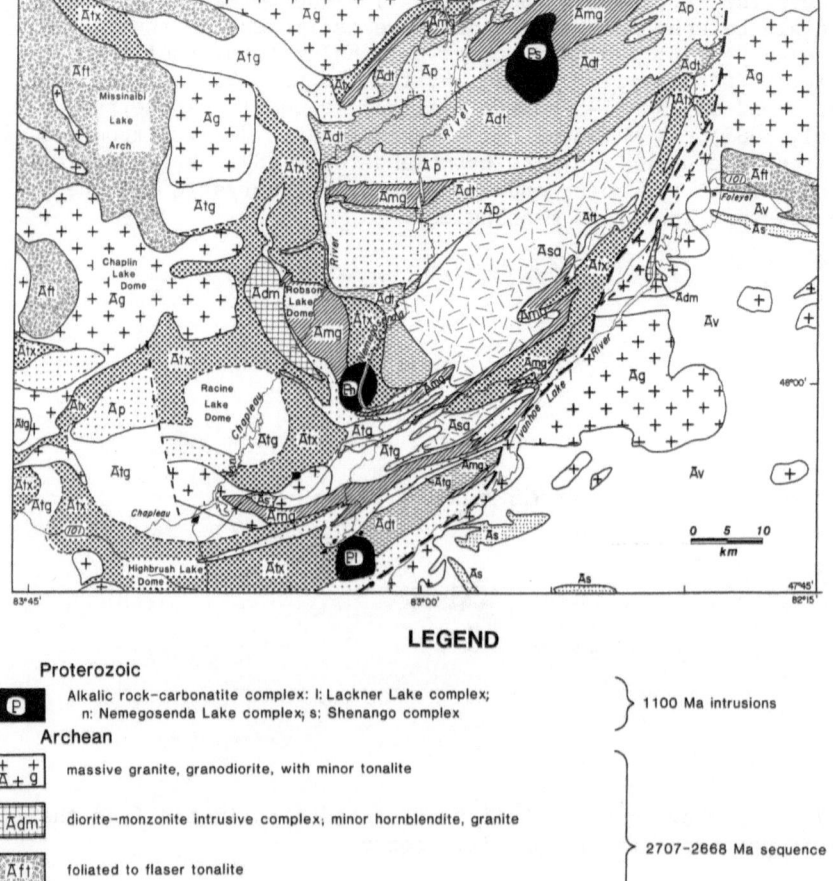

LEGEND

Proterozoic

 Alkalic rock-carbonatite complex: l: Lackner Lake complex;
n: Nemegosenda Lake complex; s: Shenango complex } 1100 Ma intrusions

Archean

A + g	massive granite, granodiorite, with minor tonalite
Adm	diorite-monzonite intrusive complex; minor hornblendite, granite
Aft	foliated to flaser tonalite
Atg;x	tonalite-granodiorite gneiss; xenolithic
Av	metavolcanic rocks, mainly metabasalt
As	metasedimentary rocks (includes metaconglomerate with tonalite cobbles with a U-Pb zircon date of 2664±12 Ma)
Adt	flaser diorite to mafic tonalite - includes minor gabbro, hornblendite, granodiorite
Asa	Shawmere anorthosite complex: metamorphosed gabbroic anorthosite, anorthosite, gabbro, minor tonalite
Amg	mafic gneiss: high Ca,Al basaltic composition, with tonalitic leucosome
Ap	paragneiss- quartz-rich composition, with up to 15% tonalitic leucosome

2707-2668 Ma sequence

2749-2696 Ma sequence

?

pre-2765 Ma sequence

fault; Ivanhoe Lake cataclastic zone

Figure 8. Geology of the Kapuskasing structural zone and vicinity.

Migmatitic paragneiss is compositionally layered with garnet, biotite, quartz-rich and rare graphitic varieties. Concordant tonalitic leucosome constitutes up to 20 per cent of many outcrops. Enclaves and layers of mafic gneiss in paragneiss occur on the 10 cm to 1 km scale. Migmatitic mafic gneiss is characterized by garnet-clinopyroxene-hornblende-plagioclase-quartz-ilmenite±orthopyroxene mineral assemblages and generally contains concordant tonalitic leucosome. Layering, on a 1 to 10 cm scale, is produced by variable proportions of minerals. Table 1 presents two sets of whole-rock analyses from adjacent anhydrous (garnet-clinopyroxene-plagioclase-quartz) and hornblende-bearing layers from mafic gneiss in two different locations. From the analyses it is unclear whether the layering is a preserved compositional heterogeneity or a product of metamorphic differentiation. The bulk composition corresponds to high calcium (10-15 wt% CaO), high alumina (13.4-17.2 wt% Al_2O_3) basalt (Table 1). Nickel and chromium abundances of mafic gneiss are in the 95-220 and 12-190 ppm ranges respectively and are not definitive in distinguishing between basaltic igneous and marly sedimentary parentage for the rock type.

In the area of Figure 8, four linear, northeast-striking bodies of flaser-textured to foliated diorite and mafic tonalite occur dominantly within paragneiss terranes. These medium- to coarse-grained, locally migmatitic rocks consist of hornblende, biotite and plagioclase, with up to 10 per cent quartz as well as orthopyroxene, clinopyroxene and rare garnet. Gabbro, hornblendite and rare pyroxenite occur locally as layers 10 cm to 2 m thick, generally within 2 km of paragneiss contacts.

Discrete belts of xenolithic and gneissic tonalite are present south of the main body of the Shawmere anorthosite complex and small bodies are present to the north. The southern belt is made up of coarse-grained garnet-hornblende-biotite-plagioclase-quartz tonalite containing enclaves of mafic gneiss, paragneiss, hornblendite and garnet-orthopyroxene-hornblende-biotite rocks. Southwest along this belt, garnet decreases in abundance and the composition is granodioritic. Inclusions in this area are amphibolite, hornblendite, and cummingtonite-hornblende-biotite rocks.

The Shawmere anorthosite complex (Thurston et al., 1977) consists of a main northern body, 15 x 50 km and a smaller mass, measuring 5 x 15 km. The bodies taper to the northeast and southwest and thus have concordant contacts. Gneissic textures prevail in the outer portions of the main body, whereas primary igneous minerals and textures are preserved in the interior (Simmons et al., 1980). The main body comprises four distinct lithological-textural units (Riccio, 1981; Fig. 9): (1) a border zone of migmatitic, foliated to gneissic garnetiferous amphibolite, (2) a banded zone consisting of 1 to 30 cm-thick layers of anorthosite, gabbro, garnet-rich, and ultramafic rock, (3) an anorthosite zone containing minor gabbro and (4) a megacrystic gabbroic anorthosite zone with plagioclase phenocrysts to 50 cm and minor anorthosite, anorthositic gabbro, gabbro and melagabbro. A 1 km wide body of foliated garnetiferous tonalite is present within the outcrop area of the anorthosite. Its genetic

Table 1: Whole rock chemical analyses of mafic gneiss from the Kapuskasing zone, with CIPW norms. Analyst: R. Charbonneau, GSC Lab. 1: granulite layer, P79-475 (Gt-Cpx-Pl-Qz, 5% Hb); 2: amphibolite layer, P-475 (Gt-Cpx-Pl-Qz, 25% Hb); 3: granulite layer, P79-371 (Gt-Cpx-Pl, tr Qz); 4: amphibolite layer, P79-371 (Hb 40%, Gt 15%, Cpx 15%, Pl 20%); 5: average of three mafic gneisses from the KSZ (79-84A, 123, 299); 6: high-alumina basalt (Ringwood, 1975).

	1	2	3	4	5	6
SiO_2	47.8	46.6	52.5	43.1	47.8	49.9
TiO_2	0.81	0.81	1.81	1.59	1.0	1.3
Al_2O_3	15.5	15.6	17.2	13.4	16.2	17.0
Fe_2O_3	1.3	2.2	2.2	5.7	3.4	1.5
FeO	9.1	9.4	8.5	12.8	8.5	7.6
MnO	0.27	0.19	0.32	0.3	0.32	0.2
MgO	4.53	5.29	3.64	9.25	5.41	8.2
CaO	15.4	14.2	11.2	10.0	13.50	11.4
Na_2O	2.0	2.4	2.8	1.6	2.3	2.8
K_2O	0.25	0.41	0.12	0.58	0.33	0.2
H_2O	0.5	1.1	0.3	1.6	0.8	
CO_2	2.3	2.0	0.4	0.1	0.6	
Ni	0.014	0.014	0.0095	0.0098	0.024	
Cr	0.019	0.018	0.018	0.014	0.015	
Total	100.0	100.4	100.6	100.4	100.2	100.1

CIPW Norm

	1	2	3	4	5	6
QZ	1.5		6.6			
OR	1.49	2.44	0.71	3.47	1.95	1.0
AB	17.02	20.46	23.63	13.72	19.8	23.5
AN	32.77	30.81	33.92	28.03	32.97	33.4
DI	11.42	11.52	6.99	10.20	13.77	18.9
HE	12.80	10.90	8.28	6.41	10.58	
EN	6.05	3.64	5.80	5.39	4.96	9.4
FS	7.83	3.95	7.89	3.89	4.43	
FO		3.01		9.26	1.54	9.3
FA		3.60		7.36	1.21	
MT	1.9	3.21	3.18	8.37	4.95	2.2
IL	1.55	1.55	2.24	3.06	1.90	2.5
AP	0.12	0.14	0.26	0.24	0.19	
CC	5.26	4.58	0.91	0.46	1.44	

relationship to the anorthosite complex is not clear although it
appears to be temporally related (Simmons et al., 1980). The southern
body consists dominantly of coarse gabbroic anorthosite.

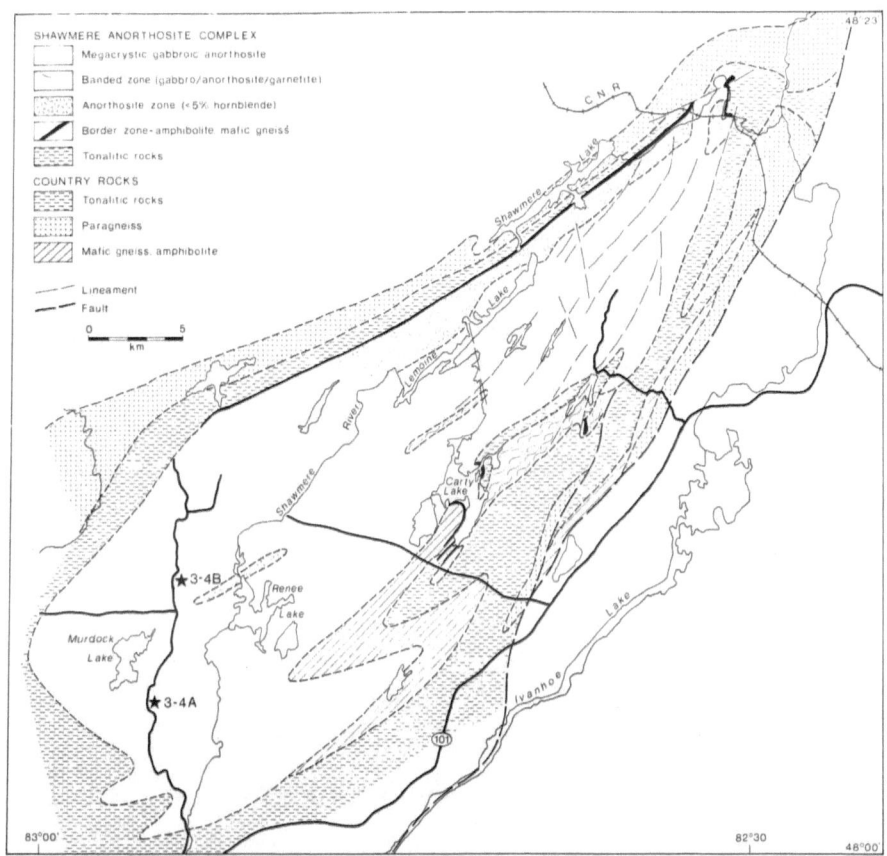

Figure 9. Geology of the Shawmere anorthosite complex (after Riccio,
1981 and Percival, 1981).

The orientation of gneissosity and lithological contacts make up
the prominent east-northeast structural grain of the Kapuskasing
structural zone. Gneissosity in all rock types is folded or warped
about gently-plunging (0-25°) northeast-trending axes. The folds vary
from isoclinal with consistent "Z" sense asymmetry when viewed toward
the east to northwest-facing monoclinal flexures. Axial surfaces are
rarely accompanied by a foliation defined by flattened quartz grains.
The trend of lineations and fold axes is northeast-southwest throughout
this part of the Kapuskasing zone, but plunge direction varies on a
regional scale from dominantly southeasterly in the south to
northeasterly in the north. Between these areas, lineations are within
10° of horizontal and abrupt changes in plunge direction occur on the

100 m scale. Both regional and local plunge reversals can be related to gently southeast-plunging warp axes.

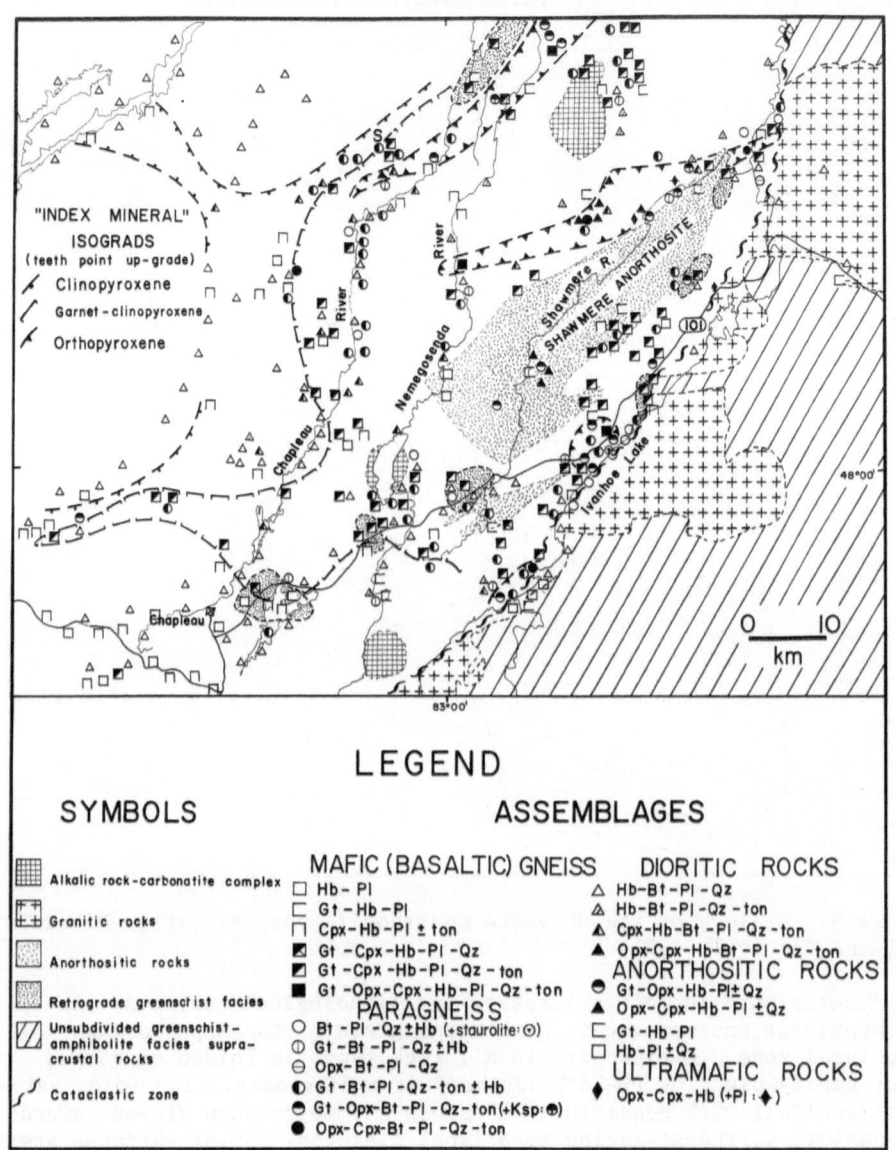

Figure 10. Metamorphic mineral assemblages and index mineral isograds for part of the Chapleau-Foleyet area. Gt-garnet; Opx-orthopyroxene; Cpx-clinopyroxene; Hb-hornlende; Bt-biotite; Pl-plagioclase; Ksp-feldspar; Qz-quartz; ton-tonalitic segregations. (after Percival, 1983).

Eight deformational phases have been recognized in the Kapuskasing zone in the Ivanhoe Lake area (Bursnall, 1989), including early, ductile (D_{1-4}) structures and D_{5-8} ductile-brittle, fault-related structures. The Archean structures are D_1 gneissic layering, folded by small-scale isoclines (D_2) and D_3 folds and shear zones; D_4 mylonite probably formed during decoupling related to initial uplift.

Two high-grade metamorphic zones can be distinguished in this part of the Kapuskasing structural zone. Assemblages characteristic of a lower-grade garnet-clinopyroxene-plagioclase zone are developed in mafic gneiss. Orthopyroxene, present in four areas in most rock types, is diagnostic of a higher-grade orthopyroxene zone (Fig. 10; Percival, 1983).

A continuous reaction resulting in decomposition of hornblende in mafic rocks to produce garnet and clinopyroxene may be written:

hornblende + plagioclase $\rightleftharpoons$
$$\text{garnet + clinopyroxene + quartz + } H_2O \qquad (1)$$

The coexistence over large areas of this divariant assemblage and tonalitic leucosome veinlets (Fig. 11) suggests that the reaction was anatectic and also produced a liquid over a range of P-T conditions (Fig. 12.):

hornblende + plagioclase $\rightleftharpoons$
$$\text{garnet + clinopyroxene + tonalite} \qquad (2)$$

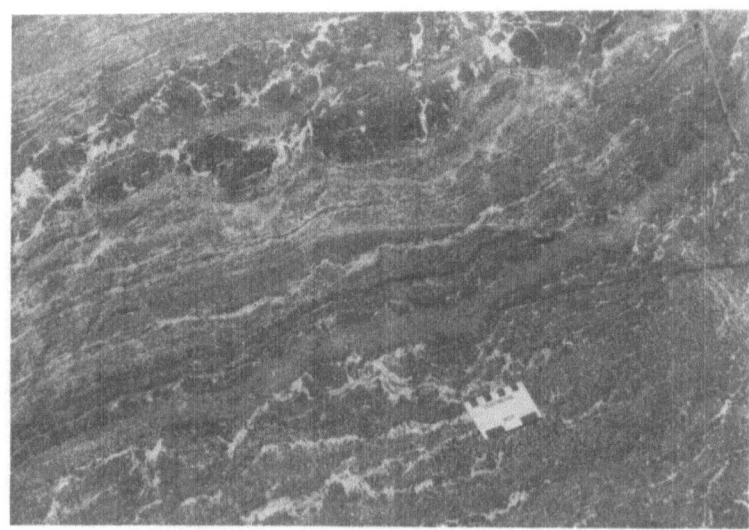

Figure 11. Typical mafic gneiss from the Kapuskasing structural zone. The darker grey layers contain abundant hornblende in contrast to lighter grey areas with garnet, clinopyroxene and plagioclase. White, concordant to amoeboid veins are quartz-plagioclase leucotonalite of inferred in situ anatectic origin.

A possible reaction leading to the production of orthopyroxene in mafic rocks is:

hornblende + garnet $\rightleftharpoons$
orthopyroxene + clinopyroxene + H_2O (3)

The evolved water would presumably have been taken up by anatectic liquids.

In paragneiss, a reaction producing orthopyroxene in the presence of anatectic melt is:

biotite + quartz + plagioclase $\rightleftharpoons$
orthopyroxene + granodioritic liquid   (4)

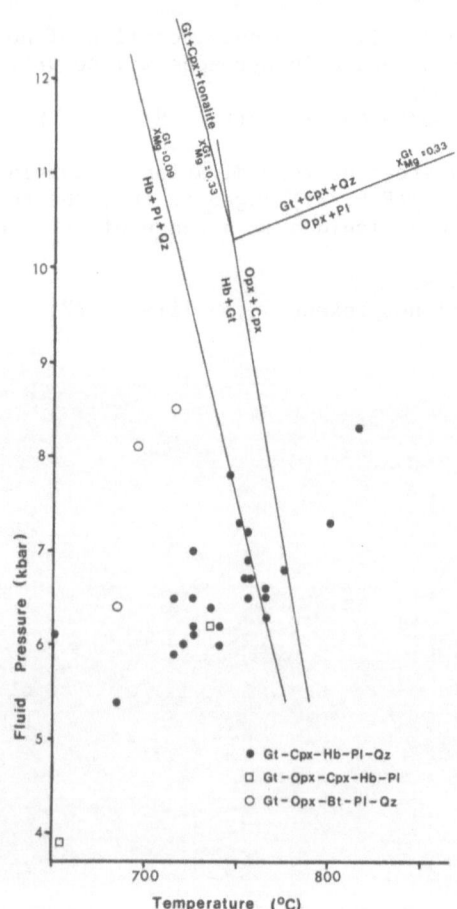

Figure 12. Summary of reactions applicable to mafic rocks and metamorphic pressure-temperature estimates. Temperatures are derived from the garnet-clinopyroxene thermometer (Ellis and Green, 1979) and pressures from garnet-pyroxene-plagioclase-quartz barometers (Newton and Perkins, 1982).

A P-T diagram summarizing continuous reactions in the mafic system and apparent metamorphic conditions based on various mineral geothermometers and geobarometers, is presented in Figure 12. Apparent pressures, based on Newton and Perkins' (1982) garnet-clinopyroxene-plagioclase quartz barometer, are plotted on a map in Figure 13 and have an average value of 6.3 kbar. This barometer underestimates pressure by at least 1-1.6 kbar (Newton and Perkins, 1982; Ghent et al., 1983) and possibly as much as 3 kbar (Moecher et al., 1988) and hence an average value in the 8-9 kbar range is more likely. Apparent temperatures, based on the Ellis and Green (1979) garnet-clinopyroxene thermometer (Fig. 12) are in the range 700-800℃. Metamorphic fluids were probably depleted in H_2O, based on water barometry (Percival, 1983) and the presence of carbonic fluid inclusions (Rudnick et al., 1984).

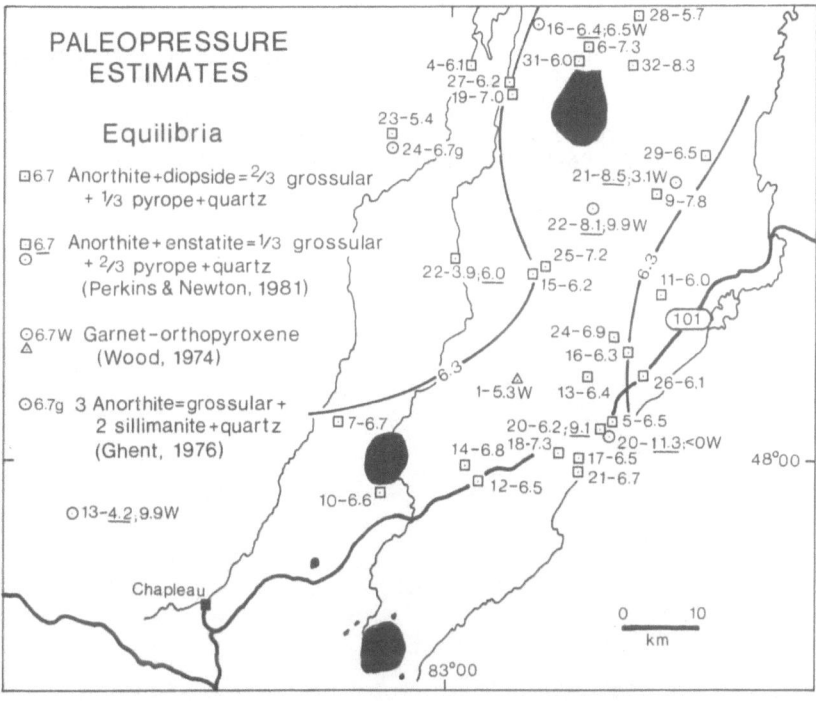

Figure 13. Paleopressure map of the Chapleau-Foleyet area. Symbols represent rock type (circles - paragneiss; squares - mafic gneiss; triangles - orthogneiss). Numbers to the right of the dash are pressure estimates (kbar) keyed to the equilibrium used to derive the value. The 6.3 kbar reference line is based on garnet-clinopyroxene-plagioclase-quartz equilibrium (after Percival, 1983).

The assemblage almandine garnet-clinopyroxene-plagioclase-quartz is diagnostic of the regional hypersthene zone according to Winkler (1979, p. 260, 267-268). de Waard (1965) and Green and Ringwood (1967)

suggested that this assemblage forms as an alternative to
orthopyroxene-plagioclase during high-pressure granulite-facies
metamorphism.   Turner (1981) attached a different significance to the
assemblage, regarding it as transitional from amphibolite to granulite
facies based on Binns' (1964) study.   In the present study area, the
location of the garnet-clinopyroxene-plagioclase zone between
hornblende-plagioclase±clinopyroxene rocks and orthopyroxene-bearing
rocks suggests that it characterizes the amphibolite-granulite facies
transition.   Although the assemblage is the same as that in the
Adirondacks (de Waard, 1965) and temperature conditions were similar
(cf. Bohlen and Essene, 1977), the path of metamorphism was different.
In the Grenville Province, the development of garnet-clinopyroxene
assemblages has been attributed to isobaric cooling of orthopyroxene-
plagioclase granulites (Martignole and Schrijver, 1971; Whitney, 1978)
whereas in the Kapuskasing zone, garnet and clinopyroxene formed during
prograde reactions.

Rounded zircons of probable metamorphic origin from Kapuskasing
mafic gneiss gave a concordant date of 2,650 Ma and from a leucosome
layer in paragneiss of 2,627 Ma (Percival and Krogh, 1983; Fig. 14).
Further work on metamorphic zircon has extended the range of
metamorphic dates to 2,696-2,584 Ma (Krogh et al., 1988).   A minimum

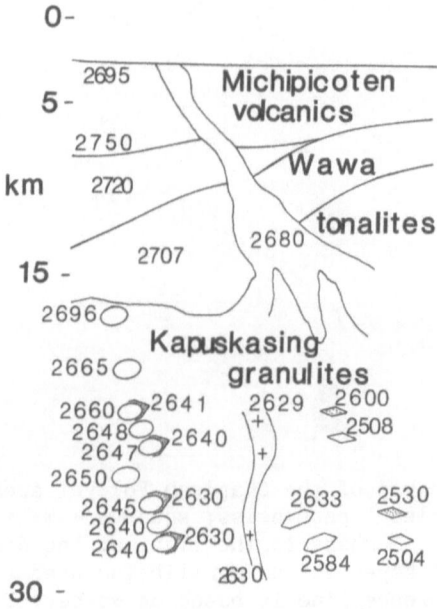

Figure 14.   Summary cartoon of U-Pb zircon and sphene dates from the
Kapuskasing uplift (after Percival and Krogh, 1983 and Krogh et al.,
1988).   Note younging with depth relationship in Kapuskasing zone.

age of emplacement for foliated tonalite from the Shawmere complex is provided by zircons (2,765 Ma) but the U-Pb system has been strongly affected by the high-grade metamorphism (Percival and Krogh, 1983). The rocks intruded by the tonalite are thus older than most of the volcanic rocks of the Abitibi and Michipicoten belts. However, Sm-Nd model ages for several units in the Kapuskasing zone range from 2.7-2.75 Ga, suggesting mantle extraction at approximately the same time as the high-level rocks of the greenstone belts (McNutt and Dickin, 1988).

Quartz-bearing gabbroic anorthosite contains zircon with two habits: red-stained resorbed grains with meta-igneous appearance, and equant, multifaceted, colourless grains of probable metamorphic origin. Both populations have a U-Pb age of 2,649 Ma (J.A. Percival and R.W. Sullivan, unpublished data), interpreted as the time of metamorphic cooling. The analysis of the coarsest resorbed grains plots slightly to the right of the 2,649 Ma discordia line, suggesting an older component.

At least two swarms of fresh mafic dykes transect metamorphic rocks of the Kapuskasing zone. East-northeast-striking, southeast-dipping Kapuskasing dykes are 1 to 10 m wide, sparsely plagioclase porphyritic, medium- to fine-grained, ophitic, green-grey gabbro. Biotite from country rock adjacent to a dyke gave a $^{40}Ar/^{39}Ar$ plateau of 2,043 Ma (Hanes et al., 1988), interpreted as the time of dyke emplacement. Northeast-trending olivine-bearing dykes may belong to the Abitibi swarm.

Several small alkalic rock-carbonatite complexes are associated with the Kapuskasing zone. The more northerly bodies have K-Ar dates of 1,655 to 1,720 Ma, whereas those in the south have dates of 1050 to 1100 Ma (Gittins et al., 1967). The Cargill complex in the north has conflicting U-Pb zircon dates of 1,888±3 (L.M. Heaman, pers. comm) and 1,907±20 Ma (Kwon, 1986). The Borden Township body has a Pb-Pb age of 1,872±13 Ma (Bell et al., 1987). Thin lamprophyre dykes and a rare diatreme breccia are associated with the complexes; biotite from a lamprophyre dyke in the Chapleau-Foleyet area gave a K-Ar date of 1,144±31 Ma (Stevens et al., 1982); a second lamprophyre contains perovskite dated by U-Pb at 1,144±4 Ma (L.M. Heaman, pers. comm., 1988). The Nemegosenda complex contains zircon of 1,107 Ma age (L.M. Heaman, pers. comm., 1988).

RELATIONSHIP OF KAPUSKASING STRUCTURAL ZONE TO ADJACENT SUBPROVINCES

The contact between the Kapuskasing structure and Abitibi subprovince is the <100 m wide Ivanhoe Lake fault zone (Bursnall 1989), that separates the two terranes of contrasting lithological, structural, and metamorphic character. The zone is defined in part by positive, linear north-northeast-trending aeromagnetic anomalies and coincides with the trough of a paired high (Kapuskasing) - low (Abitibi) gravity anomaly (Figs. 5 and 15).

The Ivanhoe Lake cataclastic zone is a 1-2 km wide zone of faulting and cataclasis, characterized by narrow veinlets of finely comminuted rock which form discontinuous, randomly-oriented pods and networks in gneisses of the Kapuskasing structural zone. Several types

248

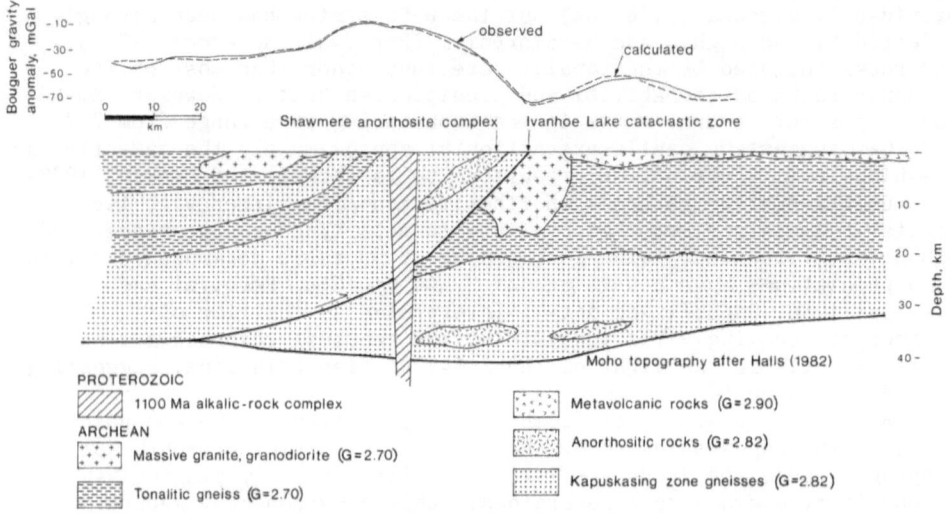

PROTEROZOIC

▨ 1100 Ma alkalic-rock complex

ARCHEAN

▦ Massive granite, granodiorite (G=2.70)

▦ Tonalitic gneiss (G=2.70)

▦ Metavolcanic rocks (G=2.90)

▦ Anorthositic rocks (G=2.82)

▦ Kapuskasing zone gneisses (G=2.82)

Figure 15. Generalized west-east cross-section from the Wawa domal gneiss terrane, through the Kapuskasing structural zone into the Abitibi subprovince, showing gross crustal structure. The gravity model based on the average rock densities: tonalitic gneiss and granite: 2.70; metavolcanics: 2.90; Kapuskasing structural zone and lower crust: 2.82 g/cm³.

of fault rocks can be distinguished. Foliated to massive, semi-opaque mylonite, cataclasite and blastomylonite, partly or totally recrystallized to fine grained epidote, chlorite, carbonate, and actinolite appears to be less common than cataclasite to pseudotachylite with aphanitic, almost opaque matrix and rounded, embayed monomineralic porphyroclasts. The different fault types form discrete or overlapping zones. Where distinct, each may give rise to seismic reflections (Fig. 16).

In a well exposed part of the Ivanhoe Lake fault zone, Bursnall (1989) recognized at least four sets of fault-related structures: D_5, partly overlapping D_4, involved production of small mylonite zones, as well as cataclasite and pseudotachylite. Open D_6 and D_7 folds are possibly related to reverse and normal fault movements respectively and D_8 is represented by post-lamprophyre faults.

The dip of the Ivanhoe Lake fault zone is not well constrained geologically. Although some fault-rock veinlets are parallel to gneissosity and therefore dip gently northwest, many others have random orientation. The juxtaposition of high-grade against low-grade rocks indicates reverse displacement across the fault zone. The associated paired gravity anomaly is characteristic of many well-documented overthrust terranes (Smithson et al., 1978; Fountain and Salisbury, 1981) and suggests that the Ivanhoe Lake fault zone is the surface expression of a northwest-dipping thrust fault (Fig. 15). A short (10 km) seismic reflection survey over the zone indicated a reflector

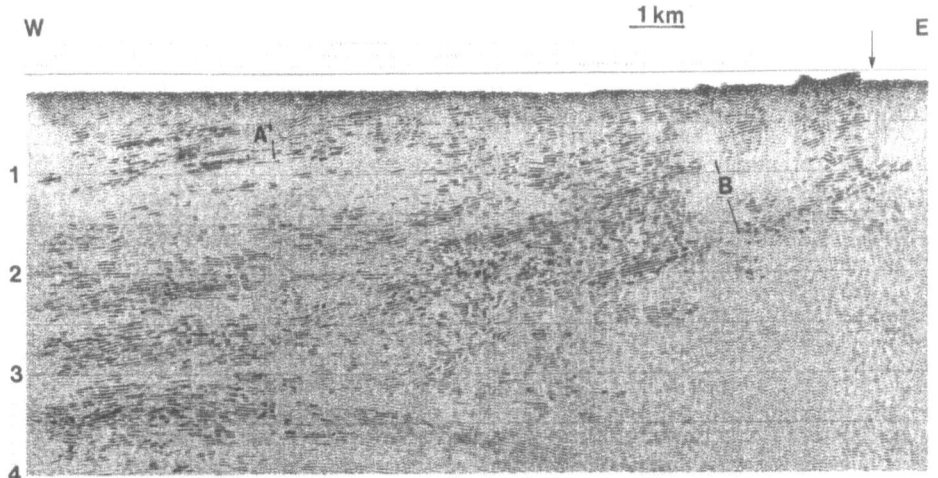

W 1 km E

Figure 16. Seismic reflection profile over the eastern Kapuskasing
zone. Reflections B are likely from the Ivanhoe Lake cataclastic zone;
reflections A correspond to lithological layering near the basal
contact of the Shawmere anorthosite complex. Arrow corresponds to
surface expression of a brittle fault of the Ivanhoe Lake cataclastic
zone.

in the appropriate position with a northwesterly dip of 38-40° (Cook,
1985). A subsequent high-resolution reflection profile (Fig. 16) shows
a zone of reflectors, rather than a discrete surface, dipping about
35°NW, corresponding to the Ivanhoe Lake zone.

The Wawa-Kapuskasing boundary varies in character over its length.
North of Bonar Lake, it is a fault, with distinct aeromagnetic
expression, which diverges westward into Wawa tonalites toward the
southwest. South of Bonar Lake, the boundary has gradational
lithogical, structural and metamorphic characteristics. Mafic gneiss
with minor paragneiss is typical of the Kapuskasing zone but also
occurs in the Robson Lake "dome" with characteristic structural style
of the Wawa subprovince. Garnet-clinopyroxene-hornblende-plagioclase
assemblages are common here, with rare orthopyroxene, suggesting that
the metamorphic grade is similar to that in the Kapuskasing structural
zone. The discontinuous paragneiss belt that extends for up to 30 km
into the Wawa subprovince may also be a part of the Kapuskasing
lithological sequence. Tonalitic gneiss can be traced eastward from
the Borden Lake area, where it has the complex structures
characteristic of the Wawa subprovince, into strongly foliated and
lineated gneiss typical of the Kapuskasing zone.

The change in structural style from subhorizontal, rolling
geometry in the eastern Wawa subprovince to linear ENE belts in the
Kapuskasing structural zone can be used to define a transitional
boundary zone between terranes with contrasting structural styles, but
no sharp line can be drawn on this basis. South of Chapleau, the
orientation of gneissic layering changes eastward from horizontal near

the Highbrush Lake "dome", to strong northeast-striking, northwest-dipping gneissosity. The transition to subhorizontal foliation may be a common feature of mid to lower crustal levels; the change in orientation to northwesterly dips may reflect rotation during uplift. A north-south-trending structural culmination coincides with the eastern "domes" of the Wawa subprovince. East of the culmination, lineations plunge easterly toward a structural depression into which southwest-trending lineations of the southern Kapuskasing zone also plunge. To the north, lineations plunging northeasterly off the northeastern flank of the Missinaibi Lake arch appear to be continuous with northeast-plunging, reclined folds in the northern Kapuskasing structural zone (Percival, 1981 a,b). This large scale structure may be explicable as an antiform overlying a ramp in the Ivanhoe Lake thrust. Cataclastic veinlets characterize the faulted contact between mafic gneiss and tonalitic gneiss southwest of Kapuskasing Lake. To the south, the gradational nature of lithological contacts as well as the structural and metamorphic continuity between tonalites and high-grade gneisses suggests that the contacts were established prior to metamorphism and doming, and that rock units of the Kapuskasing zone locally occur structurally below the Wawa tonalite-granodiorite gneiss. Based on the change in average rock density across this diffuse subhorizontal boundary, Percival (1986) suggested that it could represent an exposed mid-crustal (Conrad) discontinuity.

STRUCTURE OF THE KAPUSKASING CRUSTAL CROSS SECTION

The transition from the Michipicoten belt to the eastern boundary of the Kapuskasing zone can be interpreted as an oblique crustal cross section based on the following: 1) metamorphic grade increases eastward from low greenschist facies in the Michipicoten belt (Studemeister, 1983) through amphibolite facies in the Wawa domal gneiss terrane to upper amphibolite and granulite facies in the Kapuskasing zone; 2) the proportion of plutonic to supracrustal rocks increases eastward in the Wawa subprovince; 3) relatively old rocks (>2,765 Ma) are in the Kapuskasing zone at the inferred base of the section; 4) the gravity anomaly can be best modelled by using a west-dipping crustal slab; and 5) rocks with seismic velocities typical of the upper crust are not present in the Kapuskasing zone (Boland and Ellis, in press; Boland et al., 1988; Percival and Fountain, in press). The three major terrane types recognized in the Abitibi-Wawa region can be related to depth zones in the crust based on metamorphic evidence and consistent with known seismic velocity characteristics of the crust of the Superior Province. Thus the uppermost crust is made up of supracrustal rocks of the greenstone belts and discordant plutonic rocks. Beneath is a megalayer made up of variably deformed felsic to intermediate plutonic rocks, with large-scale "domal" geometry. With increasing depth within this layer, the attitude of gneissosity changes from sub-vertical, near greenstone contacts, to sub-horizontal, near the Kapuskasing zone. The lowermost exposed megalayer is represented by the Kapuskasing zone, made up of a heterogeneous lithological assemblage at high metamorphic grade. Moderate dips of lithological layering are interpreted as the dominant sub-horizontal lower crustal attitude rotated passively during uplift.

Construction of a generalized crustal cross-section (Fig. 17) requires several assumptions: 1) the dip of the crustal slab is constant; 2) pressure is a function of depth so that estimates of metamorphic pressure can be used to derive the thickness of the section; 3) the metamorphic assemblages are the product of a single metamorphic event; and 4) post-metamorphic vertical displacement on faults within the section is negligible. The highest-grade assemblage

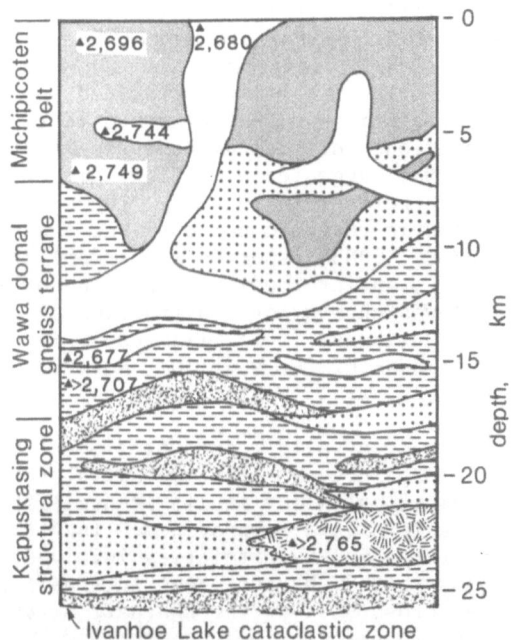

Figure 17. Restored vertical section through the Michipicoten belt, domal gneiss terrane and Kapuskasing zone. Numbers are zircon dates (±4 Ma) on igneous and meta-igneous rocks (after Percival and Card, 1983).

from the Wawa area is garnet-andalusite in metagreywacke (Ayres, 1969), indicating a maximum pressure of 3.3 kb and a depth of about 11 km (Carmichael, 1978). Similar pressures, in the 2-3 kb range, are based on sphalerite-pyrrhotite geobarometry on rocks from Gutcher Lake, 30 km northwest of Wawa (Studemeister, 1983). The range of pressures estimated from the Kapuskasing zone, based on Newton and Perkins' (1982) garnet-clinopyroxene-plagioclase-quartz barometer, is 5.4 to 8.4 kb (average of 6.3 kb, Percival, 1983) but the lower values may result from re-equilibration during cooling. These values correspond to depths of 18 to 28 km (average 21 km). The minimum erosion-level difference is therefore 7 km, but the difference is probably closer to

15 km. The minimum and maximum dip estimates over a constantly-dipping slab 120 km long are approximately 5° and 10°.

The dips of post-metamorphic dykes in the Kapuskasing zone and eastern Wawa subprovince may provide an independent estimate of the tilt of the slab in this area. Matachewan dykes dip NE at 75° to 85° and ENE Kapuskasing dykes dip SE at 70° to 85° based on measurements of dykes with vertical exposure in roadcuts. Post-metamorphic mafic dykes in the Shield generally have near-vertical orientations, as do Matachewan dykes in the Abitibi subprovince (Thurston et al., 1977; Milne, 1972). The consistent non-vertical dip may thus have resulted from large-scale crustal rotation. To restore the dykes of both swarms to vertical, a 14° counter-clockwise rotation about an axis trending 038° is necessary. Thus a 14° northwesterly dip is indicated in this eastern area. The difference in dip estimate provided by these two methods may be due to uncertainties in the data used in the calculations, faulty assumptions, or real differences in dip from east to west. Boland et al. (1988) suggested a westerly dip of ~15° based on interpretation of seismic refraction results. The overall dip must flatten to the northwest and is reversed northwest of the Michipicoten belt where Ernst (1981, p. 87; 1983) reported consistent 85° SW dips of Matachewan dykes. Therefore, an intermediate dip value of 10° perpendicular to the fault was chosen for construction of Figure 17. If dips flatten toward the northwest, this will result in over-estimation of the true thickness of the section. However, the seismic refraction results support an overall westerly dip on the order of 15° (Boland and Ellis, in press).

The generalized section is a valid representation provided that (1) a single regional metamorphic event affected all of these rocks, and (2) late vertical displacement along faults is negligible between the Kapuskasing zone and eastern Wawa subprovince. In view of the complex relationships described and uncertainties involved, these simplifications may be unwarranted; however, the information which can be derived from an exposed cross section through part of the crust is potentially valuable enough to permit some speculation.

The generalized crustal cross-section (Fig. 17), has at its base a sequence of upper amphibolite to granulite facies gneiss and anorthosite, the full thickness of which is unknown, and of which some 5 to 10 km is exposed in the Kapuskasing zone. Structurally above and separated by an analogue of the Conrad discontinuity (Percival, 1986) is an estimated 10 to 15 km thickness of tabular batholiths of gneissic and xenolithic tonalite. Massive granitic rocks occur as sheets and deep-rooted plugs at this structural level. In the upper 5-10 km, both granitic rocks and gneissic migmatitic haloes surround the low-grade Michipicoten belt. The interfaces between the adjacent, generally horizontal megalayers are undulating surfaces with several kilometres of relief, manifest as gneiss domes at intermediate structural levels and as intrusive bodies at higher levels.

In the western Superior Province, two seismic discontinuities at 16-19 and 21-22 km, define upper, middle and lower crust (Hall and Brisbin, 1982). Using the Kapuskasing model, the upper discontinuity corresponds to the boundary between a structurally higher granitoid gneissic layer and a subjacent heterogeneous high-grade gneiss complex, whereas the lower discontinuity, corresponding to the middle-lower crustal boundary, is probably a metamorphic isograd (orthopyroxene isograd?) within the heterogeneous gneiss.

Woods and Allard (1986) studied electrical conductivity in the Kapuskasing region with a large-scale magnetometer array. Although the lower crust is anomalously conductive in the area (Duncan et al., 1980), there is no conductivity anomaly associated with the Kapuskasing zone of mid- to lower-crustal origin. This was interpreted to indicate that the conductivity anomaly at depth is the result of in situ fluids which were lost during uplift of the Kapuskasing structure.

Similar models of mega-layered continental crust are based on seismic and gravity data (Smithson and Brown, 1977; Berry and Mair, 1980). Other inferred cross sections through the crust (Ivrea zone, Pikwitonei region, Musgrave, Fraser ranges; Fountain and Salisbury, 1981) have in common a downward increasing metamorphic grade and a thick, intermediate-depth, amphibolite-facies section of quartzofeldspathic gneiss, corresponding to the domal gneiss terrane of the Wawa subprovince. In the central Superior Province section, these gneisses intrude and assimilate both the overlying supracrustal succession and parts of the underlying complex. The entire section down to ~20 km was added to the crust in the interval between 2750 and 2680 Ma. The pre-existing crust may have, but need not have been as thick as present continental crust prior to the major thickening event. The high metamorphic grade in this older crust can be accounted for by burial, first by a volcanic pile and somewhat later by intrusion of tonalite sheets.

ARCHEAN EVOLUTION OF THE KAPUSKASING CRUSTAL STRUCTURE

Paragneiss and mafic gneiss of the Kapuskasing zone are considered to be part of a sedimentary-volcanic succession deposited prior to 2,765 Ma ago. The Shawmere anorthosite was emplaced into this succession, probably also prior to 2,765 Ma ago and probably as a stratiform body at depths of less than 20 km, as inferred from the presence of relict olivine (Thurston et al., 1977; Kushiro and Yoder, 1966). As suggested by Simmons et al. (1980), the intrusion may represent the differentiation product of tholeiitic basalt magmas which also erupted at surface.

Major eruption of volcanic rocks and deposition of sediments occurred between 2,749 and 2,696 Ma ago in the Michipicoten belt (Turek et al., 1982) and between 2,725 and 2,703 Ma ago in the western Abitibi belt (Nunes and Pyke, 1980). The lowermost volcanics are generally mafic and so have not been dated by the U-Pb zircon method.

Synvolcanic intrusions, including ultramafic, mafic, and trondhjemitic to granodioritic bodies, were intruded into the Michipicoten and Abitibi piles 2,750 to 2,700 Ma ago. Large volumes of tonalite intruded beneath and adjacent to the greenstone belts at this time. The minimum age of 2,707 Ma for Wawa tonalite (Percival and Krogh, 1983) is given by a nearly concordant point and is therefore probably close to the age of crystallization. The tonalites could be the subsurface expression of magmas that produced dacites in the upper parts of the volcanic piles. Tonalite intrusions, now gneissic, engulfed and detached fragments of the lower parts of the greenstone succession (now represented as mafic xenolith trains), possible older, tonalite basement enclaves and the western parts of the Kapuskasing zone which extend into the tonalite gneiss terrane. The tonalitic magmas may represent juvenile magmas derived from the mantle, or may be the products of partial melting of a heterogeneous lower crust similar to that exposed in the Kapuskasing zone. The tonalitic intrusions have imposed amphibolite-facies aureoles on metavolcanic host rocks; considering the volume of tonalite, the heat from these magmas was probably sufficient to account for most of the metamorphism of the volcanics. Tonalitic magmatism thus may have coincided with regional metamorphism and acted as the main agent of heat transfer into the upper crust. It may also have provided heat for high grade metamorphism in the Kapuskasing zone beneath (Figs. 14; 17; Wells, 1981; cf Wells, 1979).

The age of major deformation in the Abitibi and Wawa subprovinces is closely bracketed between 2,696 Ma, the approximate age of the youngest volcanics of the main pile, and 2,680 Ma, the approximate age of late- to post-tectonic plutons (Frarey and Krogh 1986). In supracrustal rocks at high crustal levels, this deformation produced upright to vertically-plunging structural features as well as thrusts and nappe-like structures (Poulsen et al., 1981; Gorman et al., 1978; Thurston and Breaks, 1978). At deeper structural levels, the deformation resulted in gneissosity and subsequent folds in plutonic rock and paragneiss, followed by later doming. Forceful emplacement of massive plutons also deflected structural trends in country rock into concordance with the margins of these bodies. Following intrusion of the massive plutons at 2,680 Ma, there was relative tectonic quiesence in Abitibi and Wawa subprovinces. There is evidence, however, of continued activity in the Kapuskasing zone.

High-grade metamorphic rocks of the Kapuskasing zone yield concordant U-Pb zircon dates of 2,696 to 2,616 Ma. U-Pb zircon dates are generally considered to record the age of crystallization of the zircons, which in this case are of metamorphic origin. This interpretation would imply that metamorphism in the Kapuskasing zone occurred 2,696 to 2,616 Ma ago, up to 60 Ma after tectonic stabilization of much of the rest of Superior province. A discrete burial and metamorphism event, restricted to the Kapuskasing zone, could explain the deformed, metamorphosed conglomerate cobbles from Borden Lake which have a zircon date of 2,664 Ma (Percival et al., 1981). However, an anomalously young zircon date of 2,552 Ma on a trondhjemitic cobble from the weakly metamorphosed >2,696 Ma Dore conglomerate near Wawa, reported by Turek et al. (1984), suggests that the date on the Borden Lake conglomerate cobble may not represent the age of crystallization of the source pluton. In addition, tectonic mechanisms which could lead to deep burial of the 500 km long x 50 km wide Kapuskasing "sliver" are unknown and seem to be unlikely after termination of the major tectonism in the Abitibi and Wawa subprovinces. It is more likely that a single protracted metamorphic event was responsible for producing the observed characteristics.

One must therefore examine the assumption that zircons are closed to lead loss immediately following crystallization, regardless of the cooling history. Slowly decreasing metamorphic temperatures from peak levels of ~800°C could result in lead diffusion out of zircon for several million years after crystallization, provided that there is some finite "blocking temperature" for zircon. A value of 700±50°C was estimated for zircon blocking by Mattinson (1978). This hypothesis however does not explain multiple zircon ages up to 60 Ma apart, from single samples (Fig. 14). Therefore, a long period of crystallization in response to changes in temperature and fluid circulation is inferred. The inverse age distribution, unrelated to major intrusive activity, suggests a heat source above the granulites, possibly overaccreted intrusions (Wawa gneisses), producing thermal consequences as modelled by Wells (1981).

The prominent east-northeast structural trends in the Kapuskasing zone, defined by the orientation of migmatitic and gneissic layering, are the result of relatively late tectonism. Structurally complex tonalitic gneiss units that can be traced from Wawa subprovince into the Kapuskasing zone have a strong ENE foliation and lineation in the Kapuskasing zone (Percival and Coe, 1981). Late Archean ductile extension appears to be the cause of the high-strain fabric (Moser, 1988); subsequent passive re-orientation by uplift on the Ivanhoe Lake fault could account for the present configuration.

UPLIFT OF THE KAPUSKASING STRUCTURE

The age of uplift of the Kapuskasing zone is not well constrained. Evidence of late Archean transcurrent movement was cited by Watson (1980) and Percival and Coe (1980), however its magnitude was probably small, judging by the minor apparent offset of the Abitibi-Opatica contact (Fig. 1). Geochronological evidence suggests a history of slow cooling following the metamorphic peak (Fig. 18). Minerals with blocking temperatures from above 700°C (zircon) to below 300°C (biotite Rb-Sr) show similar younging-with-depth patterns, suggesting cooling to below 300°C, prior to uplift (uplift of "hot" rocks would lead to rapid cooling and therefore an age "plateau"). Hence, uplift is inferred to post-date 1.95 Ga, the Rb-Sr age of the youngest biotite (Fig. 18).

Early Proterozoic carbonatites pin faults associated with the Kapuskasing uplift. The oldest such complex, the Cargill has U-Pb zircon ages of 1,907 and 1,888 Ma. The age of uplift therefore is bracketed between approximately 1,950 and 1,900 Ma.

The coincidence of Proterozoic events along the Kapuskasing structure with major orogenic activity elsewhere in the Shield suggests that the structure is an intracratonic basement uplift related to a distant compressional event, possibly an early Proterozoic collision in the Churchill Province to the northwest (Percival and McGrath, 1986).

The shortening may have been accomodated by both brittle and ductile processes (Percival and Green, 1988): the 15-20-km-thick upper crust of the Wawa-Kapuskasing plate moved eastward under brittle conditions about 70 km and upward 15-20 km over the Abitibi plate and

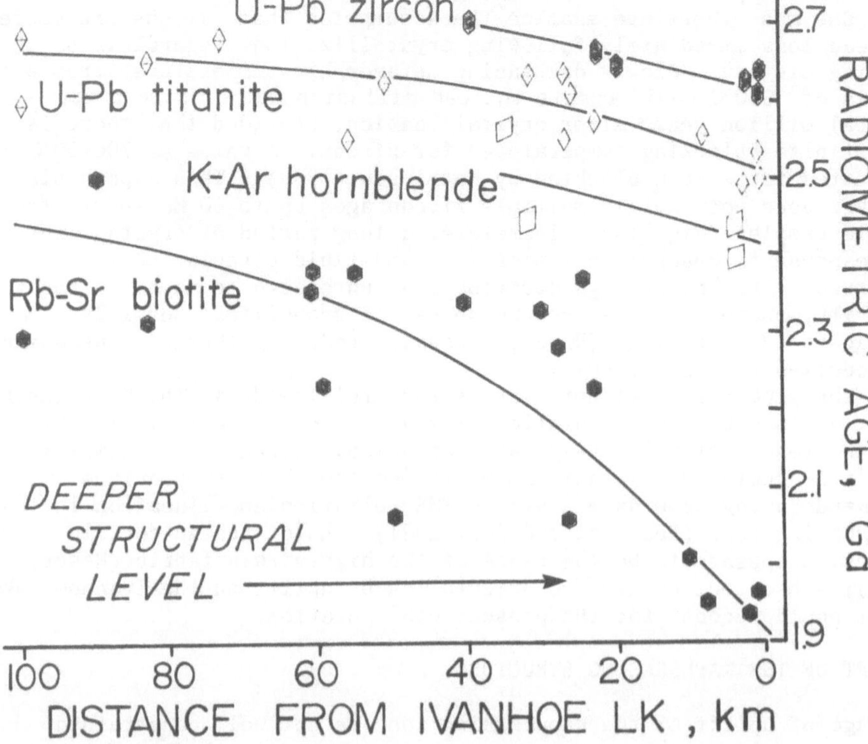

U-Pb zircon

U-Pb titanite

K-Ar hornblende

Rb-Sr biotite

RADIOMETRIC AGE, Ga

2.7
2.5
2.3
2.1
1.9

*DEEPER
STRUCTURAL
LEVEL* ⟶

100 80 60 40 20 0
DISTANCE FROM IVANHOE LK , km

Figure 18. Compilation of isotopic dates from the Kapuskasing uplift, plotted against distance from Ivanhoe Lake. Zircon and titanite (sphene) dates from T.E. Krogh; biottie Rb-Sr from Z.E. Peterman; K-Ar hornblende, biotite from Hunt and Roddick (1987).

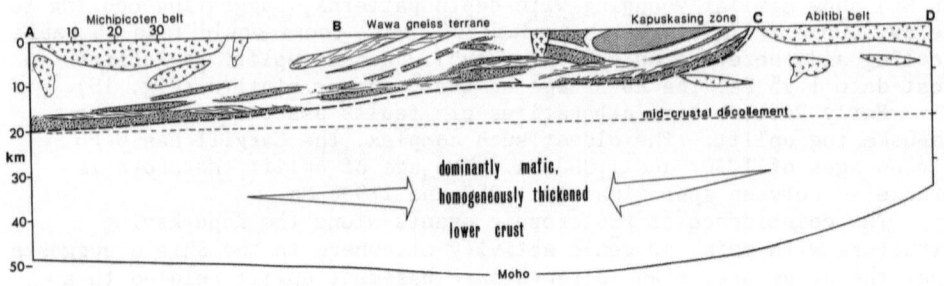

Figure 19. Cross section through the Kapuskasing uplift showing the structural position of granulite-facies rocks. The section is based on surface geology as well as subsurface information from siesmic refraction (Boland et al., 1988) and reflection (A.G. Green, pers. comm., 1988). See Fig. 2 for cross section location.

was removed by erosion or topographically subdued by normal faults. The same amount of shortening was accomodated by the crust below the 15-20 km decollement by flow into a crustal root (Fig. 19; Boland et al., 1988), either through homogeneous ductile thickening or along a discrete detachment. Thus the root not only isostatically compensates the upper crustal granulite load, but is a related effect of the same thickening process.

PART II: Road Log

SUMMARY

Various structural levels within the central Superior Province will be examined to demonstrate their characteristics and interrelationships. Starting with the lowest-grade rocks in the Michipicoten belt of the Wawa area, we will progress up-grade through an unbroken oblique crustal cross-section into tonalitic gneisses and granulites of the Kapuskasing zone.

DAY 1: GEOLOGY OF THE WAWA SUBPROVINCE, WAWA TO CHAPLEAU

The first day begins in well-preserved supracrustal rocks of the Michipicoten belt, passes through progressively deeper structural levels of the Wawa gneiss terrane and ends near the Wawa-Kapuskasing boundary (Fig. 3). Exposures near Wawa will demonstrate lithological, structural and metamorphic characteristics of the low-grade terrane. To the east, the transition to gneissic plutonic rocks and internal characteristics of the Wawa gneiss terrane will be examined.

Stop 1-1: Spherulitic rhyolite (100-200 m south of gate on McLeod Mine road)

This stop contains the best exposure of a spherulitic (hollow) flow banded felsic flow within the McLeod Mine area. The spherulitic unit is overlain by a flow breccia containing well developed fiamme and this unit is in turn overlain by a massive tuff with scattered lapilli-size clasts. This exposure lies in the lower felsic section of the 2,750 Ma volcanic cycle. Stratigraphic tops are north and the section is overturned, dipping south. Interpreted faulting along Wawa Lake and beach deposits at Wawa prevent an estimate of the stratigraphic height of this felsic section above the contact with the intermediate to mafic pillowed and massive metavolcanics.

Return to Hwy 101

00.0 km - Junction of Highway 101E and Broadway Ave, Wawa. Proceed east on Highway 101.

20.9 km - Stop 1-2: Mafic gneiss - tonalite contact zone (N. and S. sides of Hwy 101)

A large enclave of mafic gneiss is enclosed in and intruded by tonalitic gneiss in a migmatitic zone marginal to the Michipicoten greenstone belt. The hornblende-plagioclase gneiss is considered to represent deformed, metamorphosed Michipicoten volcanics and metagabbro. In this exposure it is cut by early tonalitic intrusions, late aplitic and pegmatitic dykes, and still later mafic and lamprophyric dykes. The gneiss displays subvertical foliation, mineral lineation, and tight steeply-plunging isoclinal minor folds.

45.1 km - Jct. Hwy. 651. Continue east.

68.5 km - Stop 1-3: Xenolithic tonalite gneiss at Budd Lake (N. and S. sides of Hwy 101)

This complex outcrop consists of several phases: 1) xenoliths of mafic gneiss, interpreted as rafts of the Michipicoten metavolcanic sequence, 2) gneissic tonalite with small, wispy mafic xenoliths, cut by 3) foliated to gneissic hornblende-epidote-biotite-sphene tonalite, all cut by 4) pink-white granitic pigmatite. Hornblende from foliated tonalite (3) contains 10.4 wt% Al_2O_3, indicating crystallization at approximately 5.6 kbar (Hammarstrom and Zen, 1985). Late faults may be the extension of the Saganash Lake fault, with about 10 km of normal displacement in the centre, diminishing to the north and south (Percival and McGrath, 1986; Leclair and Nagerl, 1988).

Continue east on 101

94.4 km - Stop 1-4: Tonalite gneiss and mafic dykes (Fig. 20) (N. and S. of Hwy 101)

Figure 20. Large (25 m) Hearst dyke cutting tonalitic gneiss (Stop 1-4). Note easterly dip to western dyke margin.

Tonalite gneiss is cut by northwest- and northeast-trending mafic dykes with chilled margins. The older northwest-striking Hearst dykes (Ernst and Halls, 1980, 1984) occur west of the Kapuskasing zone. The dykes have a similar trend and similar characteristics, including plagioclase phenocrysts and tholeiitic composition, to Matachewan dykes east of the Kapuskasing zone. Ernst and Halls (1980, 1984) also reported similar paleomagnetic poles for the two swarms. The Hearst dykes have a U-Pb zircon age of 2,454±2 Ma (Heaman, 1988). In a zone 50 km wide west of the Kapuskasing zone, the Hearst dykes average 4 m in width and have a consistent easterly dip of 80° (Ernst, 1983; Percival, 1981). The tonalitic gneiss is thinly layered and has sparse mafic xenoliths. Gneissosity appears to have chaotic orientation but is subhorizontal on average. There is evidence for at least two sets of structures: an older gneissosity is reoriented by younger subhorizontal foliation to give complex sigmoidal patterns.

Continue east

122.2 km - Stop 1-5: Golden Route Lozenges (N. and S. side of Hwy 101)

The two outcrops, A and B, which make up this stop exhibit a style of anastomosing ductile deformation found throughout the Wawa gneiss terrane and Chapleau Block of the Kapuskasing zone (Figs. 21, 22). Preliminary structural interpretation places it as the third and most dominant in a series of at least 3 regional deformation events which predate the Kapuskasing uplift (Moser, 1988). At this stop, tonalitic rocks are involved in this fabric whereas in other outcrops, some

Figure 21. Highly strained tonalitic gneiss showing turtleback form characteristic of parts of the Wawa gneiss terrane (Moser, 1989) (Stop 1-5A).

tonalitic phases crosscut similar structures. Temporally, therefore the deformation occurred sometime during a protracted period of tonalite intrusion which extended approximately from 2,720 to 2,680 Ma.

The geometry of the fabric is generally that of a three dimensional waveform, the amplitude and wavelength of which seem to vary with lithology, depth and, no doubt numerous other factors including strain rate, stress field orientation, temperature and partial melt content. Outcrops A and B, tonalitic and mafic gneisses respectively, demonstrate different responses to the same strain event.

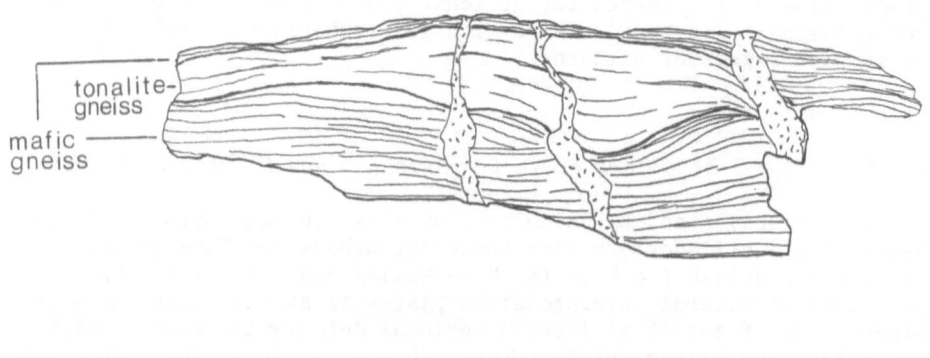

5 metres

Figure 22. Example of lozenge geometry in interlayered tonalitic and mafic gneiss (after Moser, 1988).

Outcrop A shows a highly strained and well-lineated tonalitic gneiss whose protolith was similar to the medium grained, xenolithic tonalite of Stop 1-4. One of the shallow domal culminations of the anastomosing fabric can be seen well in cross section at the east end of the largest outcrop. Low-angle discontinuities of layering on the flanks are characteristic of these structures.

Outcrop B shows fabrics in mafic gneiss anastomosing at shorter wavelengths and larger amplitudes than in the tonalitic rocks. As well, the median planes of lens-shaped bodies enclosed by the high strain surfaces ("lozenges") are extremely variable in orientation. This contrasts with the flatter fabric of outcrop A. Despite the difference in orientation, the gneissic layer thickness continues to increase away from lozenge crests.

This lozenge style of deformation is interpreted to indicate a normal sense of offset on the lozenge flanks and in turn these fabrics are postulated to represent a mid-crustal expression of extensional (bulk coaxial) deformation.

Continue east

131.6 km - Junction of Highway 101 and Highway 129. Follow 101E, 129N
    toward Chapleau

*132.9 km - Stop 1-6: Xenolithic tonalite with horizontal high-strain
    zones (W. side of Hwy 101-129)*

Xenolith alignment and gneissosity have sub-vertical orientations in
the central part of this road-cut. Layering becomes horizontal and
mafic xenoliths become thin mafic layers in 1/2 m-thick horizontal
shear zones near the top and bottom of the outcrop. The high-strain
zones are inferred to be lozenge boundaries at the margin of a lens of
less deformed rock.

139.5 km - Junction of Highway 101E and Highway 129 south of Chapleau.
    Proceed east on Highway 101.

152.5 km - Stop 1-7: Borden Lake conglomerate (S. side of Hwy 101)

This outcrop consists of stretched-pebble metaconglomerate with a
strong rodding lineation and weak, gently north-dipping foliation. The

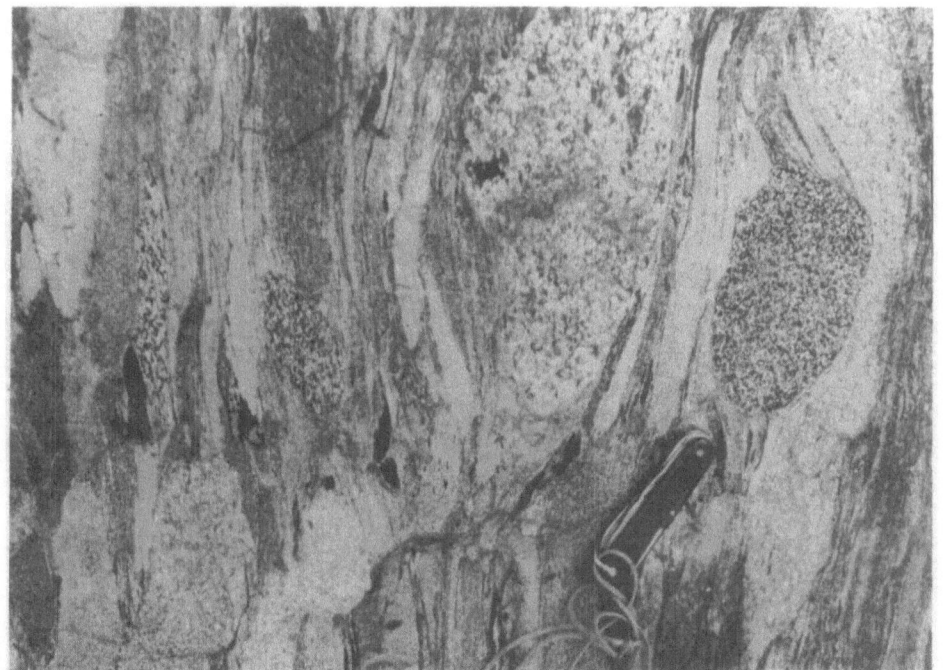

Figure 23. Stretched-pebble conglomerate from Borden Lake (Stop 1-7).
Note the equant shape of coarse plutonic clasts and elongate form of
fine-grained (metavolcanic?) cobbles.

rock is a clast-supported conglomerate containing ~10% matrix of garnet-hornblende-biotite-quartz. The cobbles, which range up to 1 m in length, are felsic metavolcanics, metasediments, granodiorite-tonalite, plagioclase-porphyritic meta-andesite and amphibolite, with rare hornblendite and vein quartz (Fig. 23). The metaconglomerate is spatially associated with amphibolite and paragneiss to the south on Borden Lake, and is cut by granite, however, the stratigraphic relations of the supracrustal rocks are unknown.

An aggegate of tonalitic cobbles extracted from the metaconglomerate yielded zircons dated at 2,664±12 Ma (Percival et al., 1981). The zircons have a corroded appearance and produced discordant data points and hence the interpretation of the data is open. Rather than recording the original crystallization age of the source pluton for the cobbles the zircons probably date a later deformation-metamorphic event. The source pluton for the cobbles may be similar to that which provided material for the Dore conglomerate of the Michipicoten belt near Wawa.

Continue east on 101

*165.4 km - Stop 1-8: Mafic gneiss xenoliths with amphibolitic margins
        (N. side of Hwy 101)*

This exposure demonstrates an important aspect of the boundary between the Kapuskasing zone and Wawa gneiss terrane. Aside from the large-scale structural contrast between the domal Wawa terrane and the

Figure 24. Exposure near the amphibolite-granulite boundary (Stop 1-8). Tonalite, similar to that of the Wawa gneiss terrane to the west, encloses mafic gneiss, similar to Kapuskasing rocks to the east. Mafic inclusions have amphibole-plagioclase assemblages with locally preserved garnet-clinopyroxene in xenolith cores (see Fig. 25).

linear belts in the Kapuskasing zone, intrusive relations are also
instructive. The outcrop consists of two main components: (1) coarse-
grained hornblende-biotite tonalite, the dominant rock type to the
west, and (2) medium-grained mafic gneiss consisting of garnet-
clinopyroxene-hornblende-plagioclase-quartz assemblages (Fig. 24).
Small xenoliths of mafic gneiss in tonalite have margins, up to several
cm thick, consisting of hornblende-plagioclase (Fig. 25). Dykes of
tonalite cutting mafic gneiss are bordered by mafic rock with
hornblende-plagioclase assemblages. The interpretation of age
relationships is that the high-grade metamorphism that produced the
garnet-clinopyroxene assemblages in mafic gneiss preceded the intrusion
of tonalite. Water in the tonalite magma was presumably released upon
crystallization and hydrated the adjacent less-hydrous mafic rock.
Tonalite at this outcrop was dated at 2,640±2 Ma (Krogh et al., 1988).
The high-grade metamorphism that produced garnet-clinopyroxene
assemblages, along with metamorphic zircon, is dated at 2,660 Ma (Krogh
et al., 1988), whereas zircon in hydration selvedges is 2,640 Ma.

Figure 25. Mafic inclusion in tonalite, Stop 1-8 (see Fig. 24).
Assemblages of clinopyroxene-plagioclase (light grey, under scale card)
are hydrated to hornblende-plagioclase (dark grey) adjacent to
intrusive tonalite (white, right side of photograph).

Return to Ivanhoe Lake

DAY 2: GEOLOGY OF THE KAPUSKASING STRUCTURAL ZONE

*0.0 km - Stop 2-1: Thinly-layered tonalitic gneiss and diatreme
    breccia (N. side of Hwy 101)*

Fine grained tonalitic gneiss at this exposure is strongly foliated and
layered on a 1-5 mm scale with garnet, hornblende and biotite-rich

layers (Fig. 26). Extremely attenuated intrafolial folds are present locally. Units characterized by extremely planar foliation such as this are relatively rare in the Kapuskasing zone. Although the orientation of foliation in this exposure is typical for the Kapuskasing zone, most Kapuskasing gneisses are medium- to coarse-grained and layered with distinctive leucocratic portions. In addition, the layering in the typical gneisses is warped about gently northeast or southwest-plunging axes. The fine grain size and thin planar layering in this outcrop suggest a relatively late, high-strain flattening or shearing event.

Figure 26. Fine-grained, thinly-layered tonalitic gneiss (Stop 2-1) showing highly attenuated mafic layers and lenses.

A thin diatreme dyke occurs in this same exposure (Fig. 27). It is part of a set of lamprophyre dykes of ~1,144 Ma age (L.M. Heaman, pers. comm., 1988) that occur in the Kapuskasing zone and are particularly common in the area between the Lackner and Nemegosenda Lake complexes. Both the matrix and fragments in the dyke are altered, but some fragments can be identified. These include tonalitic gneiss, spinel lherzolite and massive pink granite. As massive granite does not occur in the Kapuskasing zone, the granite fragments are relatively exotic. Their source was probably below the Kapuskasing zone, possibly in granite of the Abitibi belt, which according to gravity and seismic reflection data, lies vertically below at a depth of ~15 km.

Continue east on Hwy 101

15.1 km - Stop 2-2: Kapuskasing Gneiss (S. side of Hwy 101)

Layered mafic gneiss with concordant <u>in situ</u> tonalitic leucosome is the main rock type, and is crosscut by tonalitic and pegmatitic dykes.

Figure 27. Diatreme breccia margin to lamprophyric dyke, Stop 2-1. Inclusions, surrounded by white alteration rims, include local wall-rock fragments as well as exotic massive granites and mantle-derived spinel lherzolites.

Layering on the 5 to 10 cm scale is given by alternating hornblende-rich and garnet-pyroxene-rich layers (see analyses of similar layers in Table 1). Metre-scale blocks of mafic gneiss in breccia give parts of the outcrop a chaotic appearance (Fig. 28). These structurally complex panels are separated by m-scale high-strain zones with gently rolling, north-dipping, pronounced foliation (lozenge structures). Cataclasite

Figure 28. Mafic gneiss breccia (Stop 2-2) consisting of m-scale blocks separated by thin tonalitic veins or sharp discontinuities.

zones occur along some lithological contacts and probably relate to the Ivanhoe Lake fault, about 2 km to the southeast.

Continue east on Hwy 101

17.5 km - Kapuskasing gneiss (N. side of Hwy 101)

At this outcrop, mafic and ultramafic (garnet-hornblende-orthopyroxene-clinopyroxene) rocks are cut by tonalitic dykes with hydration selvedges. U-Pb dating of zircon suggests the high-grade metamorphism occurred at 2,640 Ma, whereas hydration and dyke intrusion were later at 2,630 Ma (Krogh et al., 1988).

22.8 km - Stop 2-3: Kapuskasing gneisses (N. and S. sides of Hwy 101)

There are several features of interest at this outcrop (Fig. 29):

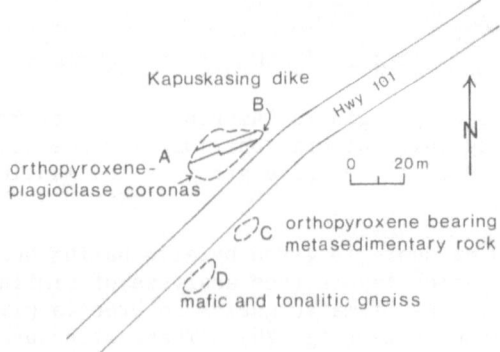

Figure 29.   Location of outcrops at Stop 2-3.

A. Mafic gneiss is present on the northwest side of the road. It is a coarse-grained rock consisting of three types of layers on the 5-100 mm scale (Fig. 30): i) relatively anhydrous mafic rock made up of garnet, clinopyroxene, plagioclase and quartz, with some hornblende (analogous to analyses 1 & 3, Table 1); ii) more hydrous layers containing less garnet and clinopyroxene and more hornblende (analogous to analyses 2 & 4, Table 1); and iii) tonalitic leucosome layers, both concordant to layering and transverse in the amphibole-rich mafic rocks. Note that the tonalite has no retrogressive effect on adjacent anhydrous mafic gneiss. The tonalitic leucosome veinlets are considered to be in situ anatectic melt segregations developed during prograde metamorphic reactions (see reaction 2). In the western end of the outcrop, submicroscopic symplectites of orthopyroxene-plagioclase identified by microprobe analyses, form barely-visible coronas around garnet, clinopyroxene and hornblende (Fig. 31). The rock contains three plagioclase compositions: An_{89} is present in coronas whereas worm-like intergrowths of An_{35} and An_{50} make up the matrix plagioclase. The mineral compositions yield estimates of 735°C using the Ellis and Green

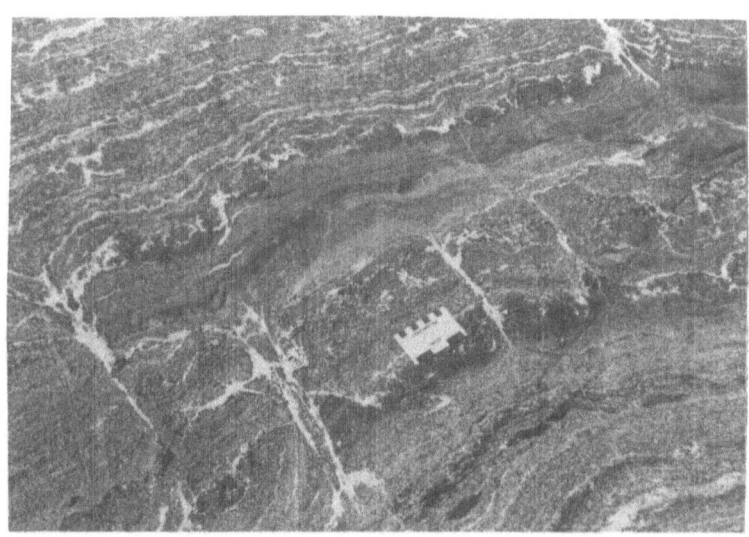

Figure 30. Mafic gneiss (Stop 2-3) consisting of cm-scale layers of different composition and mineralogy (shades of grey) and white tonalitic leucosome, both in concordant and axial surface orientations.

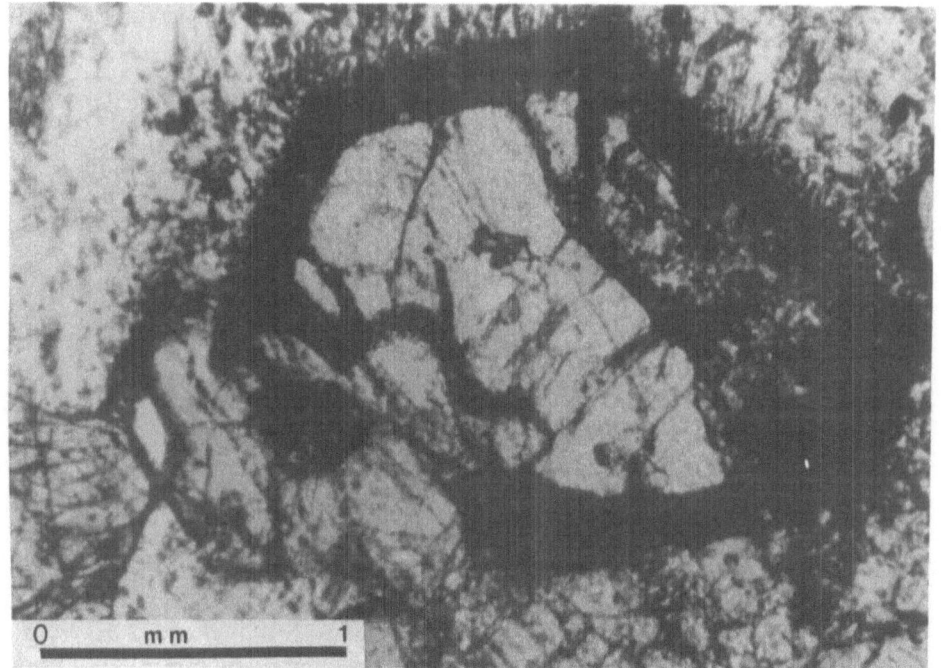

Figure 31. Photomicrograph of mafic gneiss showing garnet replaced by fine-grained orthopyroxene-plagioclase symplectite (Stop 2-3A). Matrix minerals include plagioclase, quartz, clinopyroxene and sphene.

(1979) garnet-clinopyroxene thermometer and 6.2 kbar using the garnet-clinopyroxene-plagioclase-quartz barometer (Newton and Perkins, 1982). At the same temperature the coronal minerals and matrix garnet yield 9.1 kbar with the garnet-orthopyroxene-plagioclase-quartz Newton and Perkins barometer.

B.   A Kapuskasing mafic dyke cuts the eastern end of the outcrop. The overall attitude of the dyke is 070/75 SE although the margin is offset by numerous small sinistral faults. The outer 2 cm of the margin is chilled. Sparse plagioclase phenocrysts are present in the dominantly medium grained ophitic olivine-bearing gabbro. Several dykes of this swarm have been dated by the whole-rock K-Ar method and yield "ages" between 2,367 and 3,649 Ma, indicating the presence of excess argon (Stevens et al., 1982). Hanes et al, (1986) estimated an age of 2,040-2,200 Ma based on $^{40}Ar/^{39}Ar$ analyses of a Kapuskasing dyke and its baked country rock.

C.   Homogeneous metasedimentary rock: South of the road is a flat outcrop of medium grained rock with the assemblage garnet-orthopyroxene-biotite-plagioclase-quartz. Plagioclase occurs as porphyroblasts to 2 cm and orthopyroxene is up to 5 mm. The rock has the same mineral assemblage as high-grade paragneiss in the Kapuskasing zone but lacks the migmatitic layering typical of paragneiss. Application of the garnet-orthopyroxene-plagioclase-quartz geobarometer yields values in excess of 11 kbar by both Newton and Perkins (1982) and Perkins and Chipera (1985) calibrations.

D.   Interlayered mafic and tonalitic gneiss: This outcrop demonstrates complex relations between mafic and tonalitic gneiss. Isoclinal folds

Figure 32.   Thin seams of ultramylonite cutting xenolithic tonalite gneiss (Stop 2-3D). Fault-related features are common in the 1-2-km-wide Ivanhoe Lake cataclastic zone.

of layering are truncated by tonalite pods and dykes, suggesting
multiple generations of tonalite. Thin mylonite seams (Fig. 32),
probably related to the Ivanhoe Lake fault, about 1 km to the east, are
subparallel to gneissic layers.

Continue east

30.0 km - Stop 2-4: Xenolithic tonalitic gneiss (S. side of Hwy 101)

This outcrop consists of medium- to coarse-grained tonalite made up of
garnet, hornblende, biotite, plagioclase and quartz. A variety of
xenoliths includes mafic gneiss (garnet-clinopyroxene-plagioclase-
quartz), amphibolite, biotite-rich schists and spinel pyroxenite.
Amphibole-rich rims characterize the high-grade inclusions.

Continue east

*32.2 km - Stop 2-5: Ivanhoe Lake cataclastic zone (Fig. 33) (S. side
    of Hwy 101)*

The outcrop is on the western, high-grade side of the cataclastic zone
and consists of migmatitic mafic gneiss with garnet-clinopyroxene-
hornblende-plagioclase-quartz assemblages. It is transected by
numerous small fault offsets and by one major cataclasite vein. In
thin section, this black aphanitic material is seen to consist mainly
of (recrystallized) fine actinolitic amphibole and of porphyroclasts of
hornblende. A $^{40}Ar/^{39}Ar$ whole-rock analysis of material from this vein
yielded an age plateau at 1,720 Ma. On the west side of the outcrop
are thin (3 cm) rusty-weathering lamprophyre dykes. Tonalitic rocks

Figure 33. Brittle offsets of layering in mafic gneiss at Stop 2-5.
Most of the outcrop has a dissected appearance.

cut by cataclasite and pseudotachylite are exposed north of the highway
on the "Fire Tower" road (Figs. 34, 35).

Figure 34. Pseudotachylite vein cutting tonalite, 2 km north of Stop
2-5.

Continue east

35.3 km - Turn north on logging road; follow main road (Road used for
high-resolution reflection survey (Fig. 16).

50.2 km - Stop 2-6: Shawmere gabbroic anorthosite (W. side of road)

The outcrop is mainly coarse-grained gabbroic anorthosite with
hornblende and rare garnet as mafic minerals. Ultramafic layers
consisting of ortho- and clinopyroxene with hornblende rims and sparse
plagioclase to 3 cm, occurs in layers and pods up to 4 m thick. Mafic
and ultramafic layers are locally folded into shallow NE-plunging
structures with a prominent lineation. One part of the outcrop is a
spectacular coronitic gabbroic anorthosite with football-sized
plagioclase megacrysts (Fig. 36). Coronas have orthopyroxene or
clinopyroxene cores; orthopyroxene has successive hornblende and pale
garnet rims; clinopyroxene has hornblende rims and rare orthopyroxene
cores.

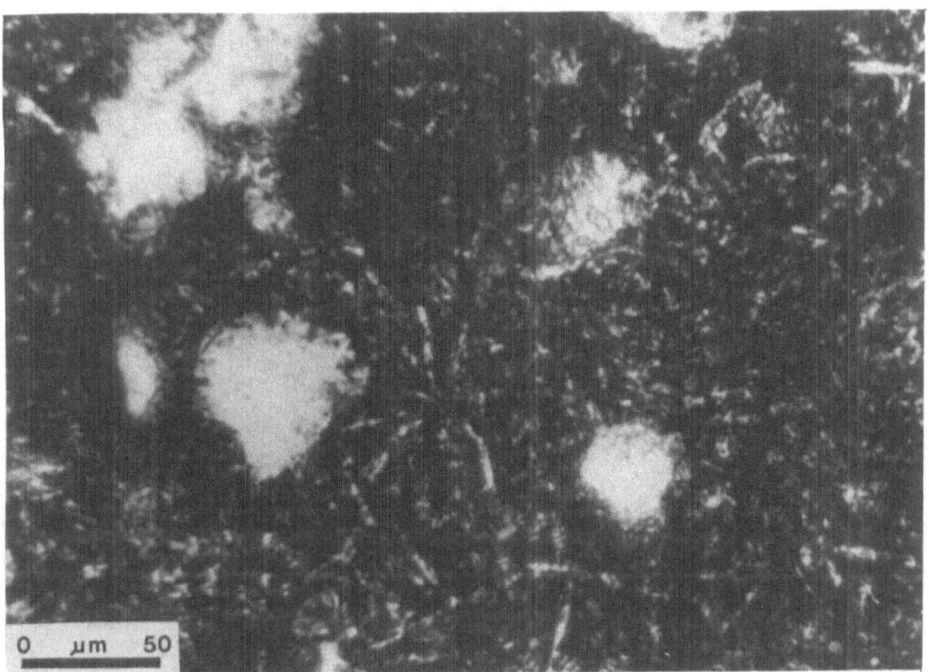

Figure 35. Photomicrograph of pseudotachylite in tonalite showing embayed porphyroclasts of plagioclase and quartz and plagioclase microlites suggestive of glass devitrification (same location as Fig. 34).

Figure 36. Plagioclase megacrysts (white), surrounded by garnet-hornblende coronas against primary pyroxenes, in the Shawmere anorthosite complex (Stop 2-6).

55.2 km - Return to small road to north

57.7 km - Stop 2-7: Shawmere layered sequence (N. side of road)

Northwest dipping, m-thick units of anorthosite, gabbroic anorthosite, gabbro, ferrogabbro and dunite are typical of the banded zone near the base of the Shawmere (Fig. 37). These rocks may represent cumulates and probably correspond to part of the seismically reflective sequence observed beneath homogeneous parts of the Shawmere. Alteration veins consisting of margarite-zoisite-calcite-quartz assemblages characterize parts of the outcrop. Equilibria indicate alteration conditions of 300-400°C, 3-4 kbar (Morrison and Valley, 1987).

Figure 37. Boudinaged peridotite layer (medium grey) in gabbroic anorthosite from the layered sequence of the Shawmere complex (Stop 2-7).

65.1 km - Return to Hwy 101, proceed east

76.7 km - Stop 2-8: Mylonitic rocks, Ivanhoe Lake Fault (?)

Two structures appear to characterize the Ivanhoe Lake fault in this area: a westerly brittle fault strand (cataclasite, pseudotachylite; Stop 2-5) and an easterly ductile branch. At this exposure, monzonite and pyroxenite are strongly foliated and lineated (SW plunge). This fault separates massive, high-level granites of the Abitibi subprovince to the east from foliated and gneissic amphibolite-facies rocks to the west that occur in a horse bounded by the splays of the Ivanhoe Lake zone). The structure may coincide with the basal reflector on the high resolution seismic reflection line (Fig. 16).

REFERENCES

Attoh, K., 1980 'Stratigraphic relations of the volcanic-sedimentary successions in the Wawa greenstone belt, Ontario' In Current Research, Part A, Geological Survey of Canada Paper 80-1A, 101-106.

Ayres, L.D., 1969 'Geology of Townships 31 and 30, Ranges 20 and 19' Ontario Department of Mines Geological Report 69, 100 p.

Bell, K., Blenkinsop, J., Kwon, S.T., Tilton, G.R. and Sage, R.P., 1987 'Age and rediogenic isotopic systematics of the Borden carbonatite complex, Ontario, Canada' Canadian Journal of Eath Sciences 24, 24-30.

Bennett, G., Brown, D.D., George, P.T. and Leahy, E.J. 1967 'Operation Kapuskasing' Ontario Department of Mines Miscellaneous Paper 10, 98 p.

Berry, M.J. and Mair, J.A., 1980 'Structure of the continental crust: a reconciliation of the seismic reflection and refraction studies' In The Continental Crust and its Mineral Deposits (Ed. D.W. Strangway) Geological Association of Canada Special Paper 20, 149-180.

Binns, R.A., 1964 'Zones of progressive regional metamorphism in the Willyama complex, Broken Hill District, New South Wales' Journal of the Geological Society of Australia 11, 283-330.

Bohlen, S.R. and Essene, E.J., 1977 'Feldspar and oxide thermometry of granulites in the Adirondack Highlands' Contributions to Mineralogy and Petrology 62, 153-169.

Bohlen, S.R., Wall, V.J. and Boettcher, A.L., 1983 'Experimental investigation and application of garnet granulite equilibria' Contributions to Minerology and Petrology 83, 52-61.

Boland, A.V. and Ellis, R.M., in press 'Velocity structure of the Kapuskasing uplift, northern Ontario, from seismic refraction studies' Journal of Geophysical Research

Boland, A.V., Ellis, R.M., Northey, D.J., West, G.F., Green, A.G., Forsyth, D.A., Mereu, R.F., Meyer, R.P., Morel-a-l'Huissier, P., Buchbinder, G.G.R., Asudeh, I. and Haddon, R.A.W., 1988 'Seismic delineation of upthrust Archean crust in Kapuskasing, northern Ontario, Canada' Nature 335, 711-713.

Burke, K. and Dewey, J.F., 1973 'Plume-generated triple junctions: Key indicators in applying plate tectonics to old rocks' Journal of Geology 81, 406-433.

Bursnall, J.T., 1989 "Structural sequence from the southern part of the Kapuskasing structural zone in the vicinity of Ivanhoe Lake, Ontario' In Current Research, Part C, Geological Survey of Canada Paper 89-1C, 405-411.

Card, K.D., 1982 'Progress report on regional geological synthesis, central Superior Province' In Current Research, Part A Geological Survey of Canada Paper 82-1A, 23-28.

Carmichael, D.M., 1978 'Metamorphic bathozones and bathograds: A measure of the depth of post-metamorphic uplift and erosion on the regional scale' American Journal of Science 278, 767-797.

Coates, M.E., 1968 'Stevens-Kagiano Lake area' Ontario Department of Mines, Geological Report 68, 22 p.

Cook, F.A., 1985 'Geometry of the Kapuskasing structure from a Lithoprobe pilot reflection survey' Geology 13, 368-371.

de Waard, D., 1965 'The occurrence of garnet in granulite-facies terrain of the Adirondack Highlands' Journal of Petrology 6, 165-191.

Duncan, P.M., Huang, A., Edwards, R.N., Bailey, R.C. and Garland, G.D., 1980 'The development and applications of a wide band electromagnetic sounding system using a pseudo-noise source' Geophysics 45, 1276-1296.

Ellis, D.J. and Green, D.H., 1979 'An experimental study of the effect of Ca upon garnet-clinopyroxene Fe-Mg exchange equilibria' Contributions to Mineralogy and Petrology 71, 13-22.

Ernst, R.E., 1981 'Correlation of Precambrian diabase dike swarms across the Kapuskasing structural zone, northern Ontario' Unpublished M.Sc. thesis University of Toronto, Toronto.

Ernst, R.E., 1983 'Structural characteristics of mafic dykes - a possible tool for mapping depth-of-exposure in the central Superior Province' (extended abstract). In Lunar and Planetary Institute Technical Report 83-03, 42-46.

Ernst, R.E. and Halls, H.C., 1984 'Paleomagnetism of the Hearst dike swarm and implications for the tectonic history of the Kapuskasing structural zone, northern Ontario' Canadian Journal of Earth Sciences 21, 1499-1506.

Ferry, J.M. and Spear, F.S., 1978 'Experimental calibration of the partitioning of Fe and Mg between biotite and garnet' Contributions to Minerology and Petrology 66, 113-117.

Forsyth, D.A., Morel, P., Hasegawa, H., Wetmiller, R., Adams, J., Goodacre, A., Nagy, D., Coles, R., Harris, J., and Basham, P., 1983 'Comparative study of the geophysical and geological information in the Timiskaming- Kapuskasing area' Atomic Energy of Canada Ltd. Technical Record 238.

Fountain, D.M. and Salisbury, M.H., 1981 'Exposed cross-sections through the continental crust; Implications for crustal structure, petrology, and evolution' Earth and Planetary Science Letters 56, 263-277.

Fountain, D.M. and Salisbury, M.H., 1986 'Seismic properties of the Superior Province crust based on seismic velocity measurements on rocks from the Michipicoten-Wawa-Kapuskasing terranes, Ontario' Geological Association of Canada Program with Abstracts 11, 69.

Frarey, M.J. and Krogh, T.E., 1986 'U-Pb zircon ages of late internal plutons of the Abitibi and eastern Wawa subprovinces, Ontario and Quebec' In Current Research Part A, Geological Survey of Canada Paper 86-1A, 43-48.

Fraser, J.A., Heywood, W.W. and Mazurski, M., 1978 'Metamorphic map of the Canadian Shield' Geological Survey of Canada, Map 1475A, scale 1;3 500 000.

Ganguly, J. and Saxena, S.K., 1984 'Mixing properties of aluminosilicate garnets: constraints from natural and experimental data, and applications to geothermo-barometry' American Minerologist 69, 88-97.

Gates, T.M. and Hurley, P.M., 1973 'Evaluation of Rb-Sr dating methods applied to the Matachewan, Abitibi, Mackenzie and Sudbury dike swarms in Canada' Canadian Journal of Earth Sciences 10, 900-919.

Ghent, E.D., 1976 'Plagioclase-garnet-Al_2SiO_5-quartz: a potential geobarometer-geothermometer' American Mineralogist 61, 710-714.

Ghent, E.D., Stout, M.Z., and Raeside, R.P., 1983 'Plagioclase-clinopyroxene-garnet-quartz equilibria and the geobarometry and geothermometry of garnet amphibolites from Mica Creek, British Columbia' Canadian Journal of Earth Sciences 20, 699- 706.

Gittins, J., MacIntyre, R.M. and York, D., 1967 'The ages of carbonatite complexes in eastern Canada' Canadian Journal of Earth Sciences 4, 651-655.

Goodwin, A.M., 1962 'Structure, stratigraphy and origin of iron formation, Michipicoten area, Algoma district, Ontario, Canada' Geological Society of America Bulletin 73, 561-586.

Gorman, B.E., Pearce, T.H. and Birkett, T.C., 1978 'On the structure of Archean greenstone belts' Precambrian Research 6, 23-41.

Green, D.H. and Ringwood, A.E., 1967 'An experimental investigation of the gabbro to eclogite transformation and its petrological applications' Geochimica et Cosmochimica Acta 31, 763-833.

Hall, D.H., and Brisbin, W. C., 1982 'Overview of regional geophysical studies in Manitoba and northwestern Ontario' Canadian Journal of Earth Sciences 19, 2049-2059.

Halls, H.C., 1982 'Crustal thickness in the Lake Superior region' In Geology and Tectonics of the Lake Superior Basin (Ed. R.J. Wold and W.J. Hinze) Geological Society of America Memoir, 156, 239-243.

Hammarstrom, J.M. and Zen, E-an, 1985 'An empirical equation for igneous calcic amphibole geobarometry' Geological Society of America Abstracts with Programs 602.

Hammarstrom, J.M. and Zen, E-an, 1986 'Aluminum in hornblende: an empirical igneous geobarometer' American Mineralogist 71, 1297-1313.

Hanes, J.A., Archibald, D.A., Queen, M. and Lee, J.K.W., 1988 '^{40}Ar/^{39}Ar geochronology of diabase dykes: implications for the tectonothermal evolution of the Kapuskasing uplift' In 1988 KSZ Lithoprobe workshop University of Toronto, 7-16.

Harley, S.L., 1984 'An experimental study of the partitioning of Fe and Mg between garnet and orthopyroxene' Contributions to Minerology and Petrology 86, 359-373.

Heaman, L.M., 1988 'A precise U-Pb zircon age for a Hearst dyke' Geological Association of Canada Program with Abstracts 13, A53.

Hunt, P.A. and Roddick, J.C., 1987 'A compilation of K-Ar ages' Geological Survey of Canada Paper 87-2, 143-210.

Hyde, R.S., 1980 'Sedimentary facies in the Archean Timiskaming Group and their implications, Abitibi greenstone belt, northeastern Ontario' Precambrian Research 12, 161-195.

Indares, A. and Martingole, J., 1985 'Biotite-garnet geothermometry in the granulite facies: the influence of Ti and Al in biotite' American Minerologist 70, 272-278.

Jensen, L.S., 1981 'A petrogenetic model for the Archaean Abitibi Belt in the Kirkland Lake area, Ontario' Unpublished Ph.D. thesis University of Saskatchewan.

Jensen, L.S., 1985 'Stratigraphy and petrogenesis of Archean metavolcanic sequences, southwestern Abitibi subprovince, Ontario' In Evolution of Archean Supracrustal Sequences (Ed L.D. Ayres, P.C. Thurston, K.D. Card and W. Weber) Geological Association of Canada Special Paper 20, 65-87.

Jolly, W.T., 1978 'Metamorphic history of the Archean Abitibi belt' In Metamorphism in the Canadian Shield (Ed J.A. Fraser and W.W. Heywood) Geological Survey of Canada Paper 78-10, 63-78.

Krogh, T.E., 1982 'Improved accuracy of U-Pb zircon ages by the creation of more concordant systems using an air abrasion technique' Geochimica et Cosmochimica Acta 46, 637-649.

Krogh, T.E., Corfu, F., Davis, D.W., Dunning, G.R., Heaman, L.M., Kamo, S.L. and Machado, N., 1987 'Precise U-Pb isotopic ages of diabase dykes and mafic to ultramafic rocks using trace amounts of baddeleyite and zircon' In Mafic Dyke Swarms (Ed H.C. Halls and W.F. Fahrig) Geological Association of Canada Special Paper 34, 147-152.

Krogh, T.E., Davis, D.W., Nunes, P.D., and Corfu, F., 1982 'Archean evolution from precise U-Pb isotopic dating' Geological Association of Canada Program with Abstracts 7, 61.

Krogh, T.E., Heaman, L.M. and Machado, N., 1988 'Detailed U-Pb chronology of successive stages of zircon growth at medium and deep levels using parts of single zircon and titanite grains' In 1988 KSZ Lithoprobe workshop University of Toronto, 243.

Kushiro, I. and Yoder, H.S. Jr., 1966 'Anorthite-forsterite and anorthite-enstatite reactions and their bearing on the basalt-eclogite transformation' Journal of Petrology 3, 337-362.

Kwon, S.T., 1986 'Lead-strontium-neodymium isotope study of the 100 to 2700 Ma old alkalic rock-carbonatite complexes in the Canadian Shield: Inferences on the geochemical and structural evolution of the mantle' unpublished Ph. D. thesis University of California at Santa Barbara, 226p.

Leclair, A. and Nagerl, P., 1988 'Geology of the Chapleau, Groundhog River and Val Rita blocks, Kapuskasing area, Ontario' In Current Research Part C Geological Survey of Canada Paper 88-1C, 83-91.

Martingole, J. and Schrijver, K., 1971 'Association of (hornblende) garnet-clinopyroxene (sub) facies of metamorphism and anorthosite masses' Canadian Journal of Earth Sciences 8, 698-704.

Mattinson, J.M., 1978 'Age, origin, and thermal histories of some plutonic rocks from the Salinian block of Californa' Contributions to Mineralogy and Petrology 67, 233-245.

McNutt, R.M. and Dickin, A.P., 1988 'Rb/Sr and Sm/Nd studies of four units in the vicinity of the Kapuskasing structural zone' In 1988-89 KSZ workshop II University of Toronto, 117-118.

Milne, V.G., 1972 'Geology of the Kukatush-Sewell area, District of Sudbury' Ontario Division of Mines, Geological Report 97, 116 p.

Moecher, D.P., Essene, E.J. and Anovitz, L.M., 1988 'Calculation and application of clinopyroxene-garnet-plagioclase-quartz barometers' Contributions to Mineralogy and Petrology 100, 92-106.

Morrison, J. and Valley, J.W., 1987 'Late fluids in the Kapuskasing structural zone' EOS 44, 1527-1528.

Mortensen, J.K., 1987 'U-Pb chronostratigraphy of the Abitibi greenstone belt' Geological Association of Canada Program with Abstracts 12, 75.

Moser, D., 1988 'Structure of the Wawa gneiss terrane near Chapleau, Ontario' In Current Research Part C Geological Survey of Canada Paper 88-1C, 93-99.

Moser, D., 1989 'Mid-crustal structures of the Wawa gneiss terrane near Chapleau, Ontario' In Current Research Part C, Geological Survey of Canada Paper 89-1C, 215-224.

Newton, R.C., and Haselton, G.T., 1981 'Thermodynamics of the garnet-plagioclase-Al$_2$SiO$_5$ geobarometer' In Thermodynamics of Minerals and Melts (Ed R.C. Newton, A. Navrotsky and B.J. Wood) Springer-Verlag, New York, 131-148.

Newton, R.C., and Perkins, D., 1982 'Thermodynamic calibration of geobarometers based on the assemblages garnet-plagioclase-orthopyroxene(clinopyroxene)-quartz' American Mineralogist 67, 203-222.

Northey, D.J. and West, G.F., 1986 'A crustal scale seismic refraction experiment over the Kapuskasing structural zone' Geological Association of Canada Program with Abstracts 11, 108.

Nunes, P.D., and Jensen, L.S., 1980 'Geochronology of the Abitibi metavolcanic belt, Kirkland Lake area - Progress Report' In Summary of Geochronology Studies, 1977-1979 (Ed E.G. Pye) Ontario Geological Survey Miscellaneous Paper 92, 40-45.

Nunes, P.D., and Pyke, D.R., 1980 'Geochronology of the Abitibi metavolcanic belt, Timmins-Matachewan area - Progress Report' In Summary of Geochronology Studies, 1977-1979. (Ed E.G. Pye) Ontario Geological Survey Miscellaneous Paper 92, 34-39.

Percival, J.A., 1981 'Geological evolution of part of the central Superior Province based on relationships among the Abitibi and Wawa subprovinces and the Kapuskasing structural zone' Unpublished Ph.D. thesis Queen's University, Kingston, Ontario 300 p.

Percival, J.A., 1981b 'Geology of the Kapuskasing Structural Zone in the Chapleau-Foleyet area' Geological Survey of Canada, Open File Map 763.

Percival, J.A., 1983 'High-grade metamorphism in the Chapleau-Foleyet area, Ontario' American Mineralogist 68, 667-686.

Percival, J.A., 1985a 'The Kapuskasing structure in the Kapuskasing-Fraserdale area, Ontario' In Current Research Part A Geological Survey of Canada Paper 85-1A, 1-5.

Percival, J.A., 1985b 'An estimate of the amount of displacement on the Quetico-Kapuskasing boundary by geothermobarometry' Geological Association of Canada Program with Abstracts 10, A48.

Percival, J.A., 1986 'A possible exposed Conrad discontinuity in the Kapuskasing uplift, Ontario' In Reflection Seismology: The Continental Crust (Ed M. Barazangi and L.D. Brown) American Geophysical Union Geodynamics Series 14, 135-141.

Percival, J.A., and Card, K.D., 1983 'Archean crust as revealed in the Kapuskasing uplift, Superior Province, Canada' Geology 11, 323-326.

Percival, J.A. and Card, K.D., 1985 'Structure and evolution of Archean crust in central Superior Province, Canada' In Evolution of Archean Supracrustal Sequences (Ed L.D. Ayres, P.C. Thurston, K.D. Card and W. Weber) Geological Association of Canada Special Paper 28, 179-192.

Percival, J.A., and Coe, K., 1980 'Geology of the Kapuskasing structural zone in the Chapleau-Foleyet area, Ontario' In Card, K.D., Percival, J. A., and Coe, K. Progress report on regional geological synthesis, Central Superior Province. In Current Research Part A Geological Survey of Canada Paper 80-1A, 61-68.

Percival, J.A. and Coe, K., 1981 'Parallel evolution of Archaean low- and high-grade terrane; a view based on relationships between the Abitibi, Wawa and Kapuskasing belts' Precambrian Research 14, 315-331.

Percival, J.A. and Fountain, D.M., in press 'Metamorphism and melting at an exposed example of the Conrad discontinuity' In Fluid Movements, Element Transport and the Composition of the Deep Crust (Ed. D. Bridgwater) NATO ASI Reidel.

Percival, J.A. and Green, A.G., 1988 'Towards a balanced crustal-scale cross section of the Kapuskasing uplift' In 1988-89 KSZ workshop II University of Toronto, 233-234.

Percival, J.A. and Krogh, T.E., 1983 'U-Pb zircon geochronology of the Kapuskasing structural zone and vicinity in the Chapleau-Foleyet area, Ontario' Canadian Journal of Earth Sciences 20, 830-843.

Percival, J.A. and McGrath, P.H., 1986 'Deep crustal structure and tectonic history of the northern Kapuskasing uplift of Ontario: an integrated petrological-geophysical study' Tectonics 5, 553-572.

Percival, J.A., Loveridge, W.D. and Sullivan, R.W., 1981 'U-Pb zircon ages of tonalitic metaconglomerate cobbles and quartz monzonite from the Kapuskasing structural zone in the Chapleau area, Ontario' In Rb-Sr and U-Pb Isotopic age Studies, Report 4. In Current Research Part C Geological Survey of Canada Paper 81-1C, 107-113.

Percival, J.A., Stern, R.A. and Digel, M.R., 1985 'Regional geological synthesis of western Superior Province, Ontario' In Current Research Part A Geological Survey of Canada Paper 85-1A, 385-397.

Perkins, D. and Chipera, S.J., 1985 'Garnet-orthopyroxene-plagioclase-quartz barometry: refinement and application to the English River subprovince and the Minnesota River Valley' Contributions to Minerology and Petrology 89, 69-80.

Pirie, J. and Mackasey, W.O., 1978 'Preliminary examination of regional metamorphism in parts of Quetico metasedimentary belt, Superior Province, Ontario' In Metamorphism in the Canadian Shield, Geological Survey of Canada Paper 78-10, 37-48.

Poulsen, K.H., Borradaile, G.J. and Kehlenbeck, M.M., 1981 'An inverted Archean succession at Rainy Lake, Ontario' Canadian Journal of Earth Sciences 17, 1358-1369.

Powell, R., 1978 'The thermodynamics of pyroxene geotherms' Philosophical Transactions of the Royal Society of London A288, 457-469.

Pyke, D.R., 1982 'Geology of the Timmins area, District of Cochrane' Ontario Geological Survey Geological Report 219, 141p.

Riccio, L., 1981 'Geology of the northeastern portion of the Shawmere anorthosite complex, District of Sudbury' Ontario Geological Survey, Open File Report 5338, 113 p.

Ringwood, A.E., 1975 Composition and Petrology of the Earth's Mantle McGraw-Hill, New York, 618 p.

Rudnick, R.L., Ashwal, L.D. and Henry, D.J., 1984 'Fluid inclusions in high-grade gneisses of the Kapuskasing structural zone, Ontario: metamorphic fluids and uplift/erosion path' Contributions to Minerology and Petrology 87, 399-406.

Sage, R.P., 1980 'Wawa area, District of Algoma' In Summary of Field Work, 1980 (Ed V.G. Milne, O.L. White, R.B. Barlow, J.A. Robertson and A.C. Colvine) Ontario Geological Survey Miscellaneous Paper 96, 47-50.

Sen, S.K. and Bhattacharya, A., 1985 'An orthopyroxene-garnet thermometer and its application to the Madras charnockites' Contribution to Minerology and Petrology 83, 64-71.

Simmons, E.C., Hanson, G.N., and Lumbers, S.B., 1980 'Geochemistry of the Shawmere anorthosite complex, Kapuskasing structural zone, Ontario' Precambrian Research 11, 43-71.

Smithson, S.B. and Brown, S.K., 1977 'A model for lower continental crust' Earth and Planetary Science Letters 35, 134-144.

Smithson, S.B., Brewer, J., Kaufman, S., Oliver, J., and Hurich, C., 1978 'Nature of the Wind River thrust, Wyoming, from COCORP deep-reflection data and from gravity data' Geology 6, 648-652.

Stevens, R.D., Delabio, R.N. and Lachance, G.R., 1982 'Age determinations and Geological Studies, K-Ar isotopic ages, Report 15' Geological Survey of Canada Paper 81-2, 56 p.

Studemeister, P.A., 1983 'The greenschist facies of an Archean assemblage near Wawa, Ontario' Canadian Journal of Earth Sciences 20, 1409-1420.

Sullivan, R.W., Sage, R.P. and Card, K.D., 1985 'U-Pb zircon age of the Jubilee stock in the Michipicoten greenstone belt near Wawa, Ontario' In Current Research Part B Geological Survey of Canada Paper 85-1B, 361-365.

Sylvester, P.J., Attoh, K. ad Schulz, K.J., 1987 'Tectonic setting of late Archean bimodal volcanism in the Michipicoten (Wawa) greenstone belt, Ontario' Canadian Journal of Earth Sciences 24, 1120-1134.

Thompson, A.B., 1976 'Mineral reactions in pelitic schists II. Calculation of some P-T-X phase relations' American Journal of Science 276, 425-454.

Thurston, P.C. and Breaks, F.W., 1978 'Metamorphic and tectonic evolution of the Uchi-English River subprovince' In Metamorphism in the Canadian Shield Geological Survey of Canada Paper 78-10, 49-62.

Thurston, P.C., Siragusa, G.M., and Sage, R.P., 1977 'Geology of the Chapleau area, Districts of Algoma, Sudbury and Cochrane' Ontario Division of Mines Geological Report 157, 293 p.

Turek, A., Smith, P.E., and Van Schmus, W.R., 1982 'Rb-Sr and U-Pb ages of volcanism and granite emplacement in the Michipicoten belt, Wawa, Ontario' Canadian Journal of Earth Sciences 19, 1608-1625.

Turek, A., Smith, P.E. and Van Schmus, W.R., 1984 'U-Pb zircon ages and the evolution of the Michipicoten plutonic-volcanic terrane of the Superior Province, Ontario' Canadian Journal of Earth Sciences 21, 457-464.

Turek, A., Van Schmus, W.R. and Sage, R.P., 1988 'Extended volcanism in the Michipicoten greenstone belt, Wawa, Ontario' Geological Association of Canada Program with Abstracts 13, p. A127.

Turner, F.J., 1981 Metamorphic Petrology; Mineralogical, Field and Tectonic Aspects Second edition. McGraw-Hill, New York, 524 p.

Wanless, R.K., 1969 'Isotopic age map of Canada' Geological Survey of Canada, Map 1256A.

Watson, J., 1980 'The origin and history of the Kapuskasing structural zone, Ontario, Canada' Canadian Journal of Earth Sciences 17, 866-876.

Wells, P.R.A., 1979 'Chemical and thermal evolution of Archean sialic crust, Southern West Greenland' Journal of Petrology 20, 187-226.

Wells, P.R.A., 1981 'Thermal models for the magmatic accretion and subsequent metamorphism of continental crust' Earth and Planetary Science Letters 46, 253-265.

Whitney, P.R., 1978 'The significance of garnet "isograds" in granulite facies rocks of the Adirondacks' In Metamorphism in the Canadian Shield Geological Survey of Canada, Paper 78-10, 357-366.

Wilson, H.D.B., and Brisbin, W.C., 1965 'Mid-North American ridge structure' (Abstract) Geological Society of America, Special Paper 87, 186-187.

Winkler, H.G.F., 1979 Petrogenesis of Metamorphic Rocks 5th edition. Springer-Verlag, Berlin, Heidelburg, New York.

Wood, B.J., 1974 'The solubility of alumina in orthopyroxene coexisting with garnet' Contributions to Mineralogy and Petrology 46, 1-15.

Woods, D.V. and Allard, M., 1986 'Reconnaissance electromagnetic
    induction study of the Kapuskasing structural zone: Implications
    for lower crustal conductivity' <u>Physics of the Earth and Planetary
    Interiors</u> **42**, 135-142.

MAJOR THRUST FAULTS AND THE VERTICAL ZONATION OF THE MIDDLE TO UPPER PROTEROZOIC CRUST IN CENTRAL AUSTRALIA

A.Y. Glikson
Division of Petrology and Geochemistry
Bureau of Mineral Resources, Geology and Geophysics
Canberra, A.C.T., Australia

C.G. Ballhaus
Department of Geology
University of Tasmania
Hobart, Tasmania

B.R. Goleby & R.D. Shaw
Division of Petrology and Geochemistry
Bureau of Mineral Resources, Geology and Geophysics
Canberra, A.C.T., Australia

ABSTRACT

The vertical crustal structure of, and correlations between, the Musgrave and Arunta blocks, central Australia, are considered in the light of tectonic, isotopic, thermobarometric and geophysical data. A crustal model is tentatively favoured including [1] granulite facies basal meta-supracrustals and orthogneiss intruded by mafic/ultramafic layered intrusions, namely the Giles Complex in the Musgrave Block and smaller-scale analogues in the Arunta Block; [2] amphibolite facies and retrogressed granulite facies supracrustals extensively invaded by granites. The major Bouguer gravity anomalies of central Australia, featuring amplitudes of up to 140 mgal, have been studied by seismic reflection surveys, indicating 'thick skinned' intra-sialic thrust tectonics which result in displacement of the MOHO at depths of 30-50 km and in the juxtaposition of infracrustal zone [1] rocks with mesocrustal zone [2] rocks. It is suggested that the granitic magmas formed by anatexis along the granulite-amphibolite metamorphic front. Rb-Sr ages show a wide scatter between 1.8-0.9 b.y. and may be interpreted in terms of different timing of thermal events in the Musgrave and Arunta blocks. However, in the lack of definitive U-Pb zircon data it is tentatively considered likely that the two blocks are broadly contemporaneous. Low to moderate initial Sr87/Sr86 ratios indicate short [<200 m.y.] crustal prehistories. The major thrusting events are considered to have culminated during the late Devonian-early Carboniferous. The deep level layered intrusions of the Giles Complex in the Musgrave Block and smaller scale analogues in the Arunta Block are considered to have triggered major anatectic events. These layered intrusions, emplaced at intermediate to deep crustal levels [above 6 kb], may signify the proximity of the MOHO in the upper Proterozoic.

M. H. Salisbury and D. M. Fountain (eds.), Exposed Cross-Sections of the Continental Crust, 285–304.
© 1990 *Kluwer Academic Publishers.*

1. INTRODUCTION

This paper reviews aspects of the geology of the Musgrave and
Arunta blocks in central Australia with reference to the temporal
and spatial relationships between their major components, their
crustal structure and alternative models of tectonic evolution.
The structure of the middle to late Proterozoic Arunta and
Musgrave blocks, which are exposed respectively north and south
of the late Proterozoic to early Carboniferous Amadeus Basin
[Wells et al., 1970], is dominated by thrust and thrust-nappe
tectonics. Major east-west striking thrust faults include the
Napperby thrust [Wells and Moss, 1983; Stewart et al., 1980],
Redbank-Mt Zeil thrust [Glikson, 1987a, 1987b], Harry Creek
deformed zone [Shaw et al., 1984], thrust-nappes of the Harts
Range [Ding and James, 1985], Mt Sonder-Mt Razorback thrust,
Blatherskite nappe, Arltunga nappe complex [Forman, 1966;
Stewart, 1967; Shaw et al., 1984], and the Woodroffe thrust and
related faults [Major, 1973; Collerson et al., 1972] [Figs
1,2,3,7]. Thrusting is commonly directed toward late
Proterozoic-Palaeozoic basins, namely the Amadeus Basin, Ngalia
Basin, Officer Basin and Georgina Basin. Major deformations
responsible for this structural pattern culminated in the early
Carboniferous [Alice Springs Orogeny] and the Cambrian [Petermann
Ranges Orogeny] [Wells et al., 1970; Wells and Moss, 1983].
Upper Proterozoic faulting and Tertiary fault reactivation are
believed to have been important, but are less well defined.
Multiple thrusting has resulted in the juxtaposition of thrust
slices of contrasted lithology, structural style, metamorphic
grade and isotopic ages, derived from different crustal levels.

2. TECTONIC/LITHOSTRATIGRAPHIC DIVISIONS

Musgrave Block: Collerson et al. [1972] delineated major faults
[Woodroffe thrust, Davenport shear] separating amphibolite
facies, granulite facies and amphibolite-granulite facies blocks
in the Musgrave Range, and defined different deformation patterns
and phases in each of these domains [Fig. 1]. Mafic igneous
rocks of the Giles Complex are associated mainly with the central
granulite block and form much of the hanging wall of the
Woodroffe thrust. The northern and southern amphibolite facies
domains are intruded by extensive granitic gneiss and by late
tectonic granite and adamellite. Thomson [1975] classified the
older metamorphic rocks of the Musgrave Block in terms of [1] the
Musgrave-Mann metamorphics, which comprise hornblende granulite
and orthopyroxene granulite [charnockite] facies metasediments
[quartzofeldspathic to mafic granulites and gneisses, garnet-
bearing quartzite, marbles, trace iron formations] intruded by
granitic gneisses [Wataru gneiss, Olia gneiss]; [2] younger
granites [Kulgera and Pottoyu granites]; [3] the layered
mafic/ultramafic intrusions of the Giles Complex [Fig. 3a]; [4]
little metamorphosed felsic volcanics and sediments [Bentley
Group], associated with cauldron subsidences and high level
granites; [5] unconformable arenites [Townsend and Dean
Quartzites].

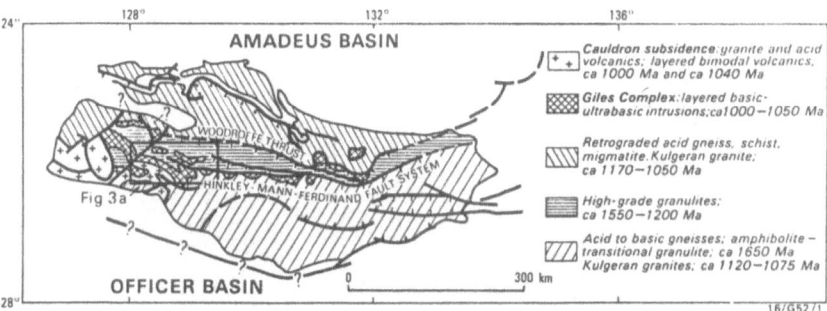

Fig. 1 - Geological sketch map of the Musgrave Block, showing the northern amphibolite facies zone, central granulite zone with Giles Complex intrusions and southern mixed amphibolite-granulite zone.

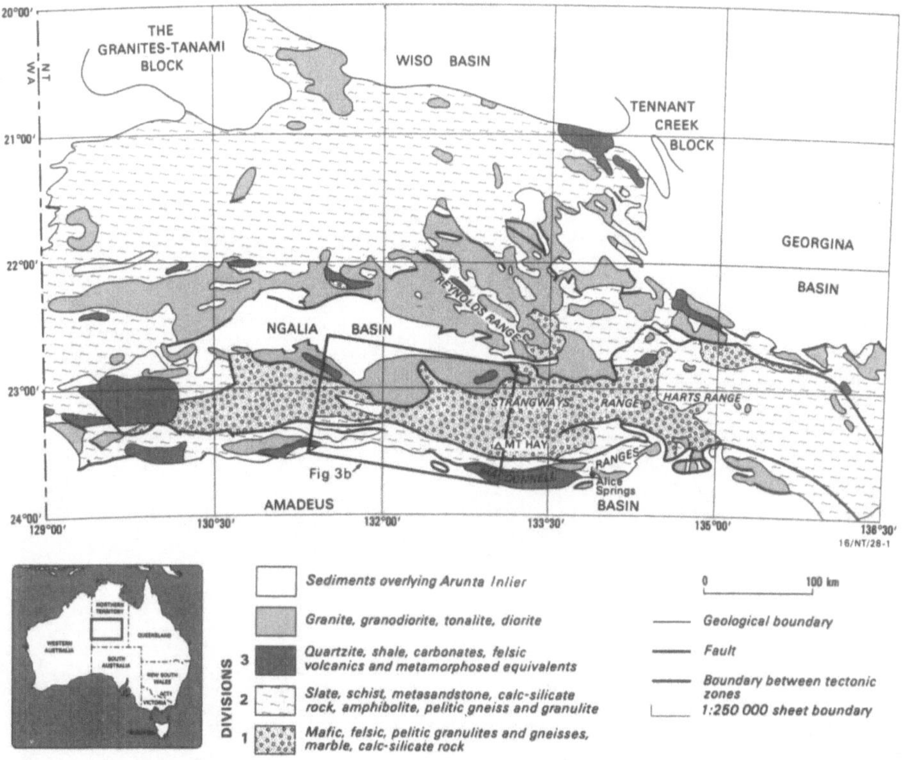

Fig. 2 - Geological sketch map of the Arunta Block [from Shaw et al., 1984], showing the distribution of the three principal lithological/metamorphic divisions [see text].

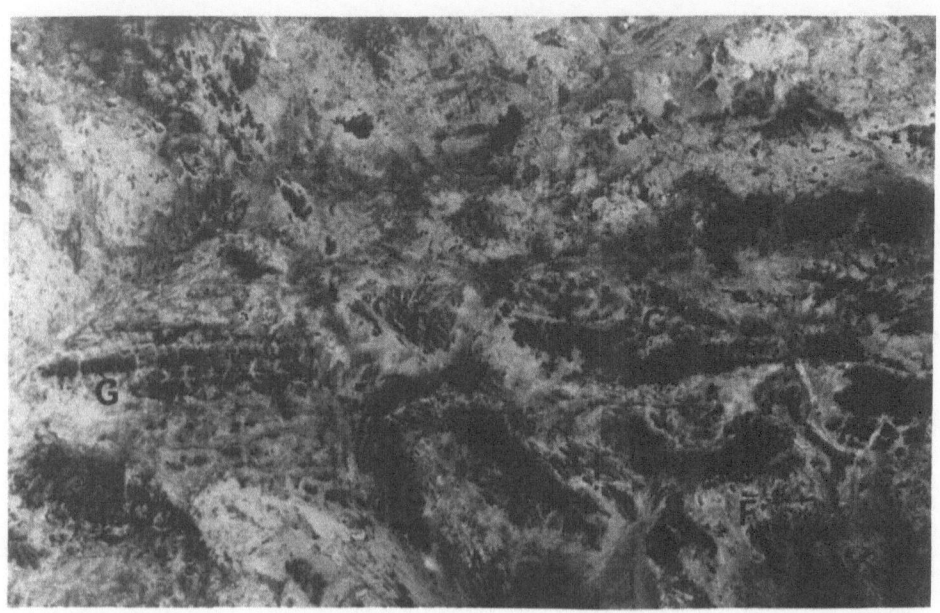

Fig. 3a - LANDSAT imagery of the western part of the Musgrave
Block [Tomkinson Range], showing the Giles Complex [G] and
felsic granulite/gneiss units [F][for location refer to frame in
Fig. 1];

Fig. 3b - LANDSAT imagery of the southwestern part of the Arunta
Block, showing the Redbank-Mt Zeil thrust zone [RTZ], granulite
massifs to the north [G] and granite/amphibolite facies
paragneiss terrain to the south [P][for location refer to frame
in Fig. 2].

Arunta Block: Shaw et al. [1984] and Stewart et al. [1984] divided the Arunta Block into three latitudinal tectonic provinces [northern, central and southern] marked by different lithological, metamorphic and structural histories and bounded by major deformed zones [Fig. 2]. Stratigraphic correlations between these domains are uncertain. Nevertheless, three major lithostratigraphic divisions are recognized, as follows [in probable ascending stratigraphic order]: [1] felsic and mafic granulites and calcareous and aluminous metasediments of the central province; [2] aluminous and siliceous metasediments of the northern and southern provinces, with minor occurence in the central province; [3] silicic metasediments [Fig. 2]. The northern, central and southern provinces are dominated respectively by greenschist to amphibolite [with associated granulite], amphibolite to granulite, and amphibolite facies assemblages. Granitoid intrusions are widespread in the northern and southern provinces but scarce in the central province. The more extensive batholiths preferentially intrude amphibolite facies terrains. In some instances the amphibolite facies rocks are clearly retrogressed from granulite-facies rocks and contain relics of the latter. For example, in the Reynolds Range area the extensive Napperby orthogneiss intrudes retrogressed amphibolite facies metasediments which are overthrust by granulites along the Anmatjira fault [Stewart et al., 1980]. An example of a late tectonic granite intruding prograde amphibolite facies paragneisses is the Teapot Granite located south of the Redbank-Mount Zeil thrust in the southwestern Arunta Block [Fig 3b] [Glikson, 1987a, 1987b].

The above examples portray a similar structural pattern in the Arunta and the Musgrave blocks, namely, granulite facies blocks thrust over terrains dominated by amphibolite facies metasediments which are intruded by extensive granites. In a number of places the structurally deepest segments of the hanging wall granulites include layered mafic-ultramafic intrusions, i.e. the Giles Complex [Nesbitt et al., 1970; Daniels, 1974] and the Mt Hay layered gabbro-anorthosite complex [Glikson, 1984; Glikson, 1987a, 1987b]. The extension of the thrust faults which separate the granulite facies-dominated blocks from the amphibolite facies dominated-blocks in depth is considered below.

3. ISOTOPIC AGE RELATIONSHIPS

Musgrave Block: The oldest Rb-Sr isochron ages are for granulite [1614+/-168 m.y.] in the central part of the block [Webb, 1978] [Fig. 4]. Gray and Compston [1978] and Gray [1978] conducted a detailed Rb-Sr study of felsic metasedimentary granulites at Mt Aloysius and granulites associated with the Giles Complex in the western part of the Musgrave Block. The isochrons outline two broad thermal episodes: [1] about 1.55 b.y., interpreted as the age of the supracrustal precursors and [2] about 1.2 b.y., interpreted as a metamorphic age. The age of the Kalka layered intrusion, part of the Giles Complex, is defined between 1.2 b.y. [in view of proposed intrusive relations between the mafic/ultramafic intrusion and the felsic granulites] and 1.1 b.y. [biotite age of the mafic intrusion]. However, it is possible that the layered complex is significantly older if an analogy with the Fraser Complex in the Albany-Fraser belt is valid. Fletcher and Myers [pers. comm., 1988] have measured a Sm-Nd mineral isochron age of 1291+/-21 m.y. and Rb-Sr biotite-whole rock pair age of 1295+/-45 m.y. on least-deformed gabbro from the Fraser Complex. In the Giles Complex a wide range of initial Sr87/Sr86 ratios, including high values [0.7049-0.7088], but near-constant initial Nd143/Nd144 ratios typical of continental basalts, indicate extensive contamination of the basic magma by felsic granulites [Gray et al., 1981].

Webb [1985] summarized the isotopic evidence for the Musgrave Block in terms of the following episodes:
1.7 - 1.6 b.y.: interpreted as a period of high grade metamorphism, with low initial Sr87/Sr86 ratios [about 0.706], suggesting short crustal prehistory of precursor materials.
about 1.12 b.y.: extensive formation of granites with low initial Sr87/Sr86 ratios [0.705-0.707] and less commonly higher values.
1.0 - 0.53 b.y.: K-Ar ages of biotite and hornblende from granitic gneiss, representing cooling and thermal resetting, the latter possibly related to movements along the Woodroffe thrust.

The differences between the above schemes may be in part regional and in part due to secondary thermal overprinting [see below].

Arunta Block: The oldest well defined age reported from the Arunta Block is a Sm-Nd isochron age of 2015+/-120 measured on bimodal granulites from the Strangways Range [Windrim and McCulloch, 1986]. Black and McCulloch [1984] reported similar Sm-Nd model ages, which possibly represent a mantle differentiation event. Rb-Sr isochron ages concentrate about 1.8-1.7 b.y., with low to moderate initial Sr87/Sr86 ratios [0.706-0.716] [Black et al., 1983], suggesting limited crustal residence times for precursor materials. The available data allow a broad grouping of the following thermal episodes:

about 2.0 b.y.: defined by Sm-Nd isochrons for mafic and felsic granulites of the Strangways Range [Windrim and McCulloch, 1986].
1.8 - 1.7 b.y. [Strangways Event]: defined by Rb-Sr isochrons of granulites and granites of the central province, possibly representing a metamorphic peak.
about 1.7 b.y. [Aileron Event]: defined by Rb-Sr and K-Ar ages of metamorphics, leucogranites and pegmatites accompanied by isotopic resetting in amphibolite to granulite facies metamorphics.
about 1.45 b.y. [Anmatjira Event]: Rb-Sr and K-Ar ages in the northern province and Rb-Sr ages in the central province, from sites associated with deformation zones.
about 1.2 - 0.9 b.y. [Ormiston Event]: Rb-Sr ages of migmatites and alkaline bodies in the southern province.
about 0.9 b.y.: intrusion of mafic dykes followed by deposition of platform cover arenites [Black et al., 1983].

292

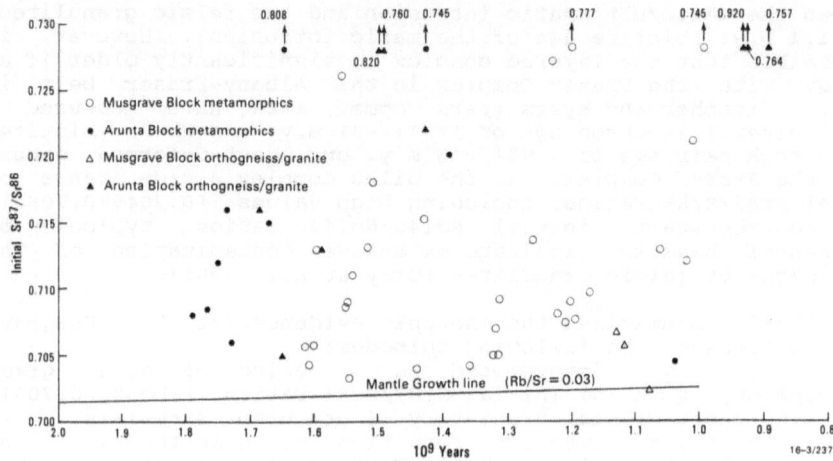

Fig. 4 - Rb-Sr isochron ages plotted against initial Sr87/Sr86 ratios for Musgrave and Arunta rocks [data from Webb, 1985; Gray, 1978; Gray and Compston, 1978; Black et al., 1983]

Late Ordovician and late Devonian to early Carboniferous: major deformations, involving thrusting, affecting Arunta basement and the Amadeus and Ngalia basins.

The wide scatter of Rb-Sr isotopic ages [Fig. 4], coupled with the uncertain geological meaning of the Rb-Sr and K-Ar systematics, renders uncertain any conclusions regarding the contemporaneity or otherwise of major thermal episodes in the Musgrave and Arunta blocks. These questions must remain open pending U-Pb zircon age measurements.

4. THERMOBAROMETRY

Musgrave Block: Goode and Moore [1975] presented evidence for high pressure crystallization of the Kalka, Ewarara and Gosse Pile intrusions of the Giles Complex, including subsolidus reactions between olivine [ol] and plagioclase [plg], orthopyroxene [opx] and plg and opx and spinel. Other evidence includes spinel exsolution in opx, clinopyroxene [cpx] and plg, high Al and Cr contents in both opx and cpx, high Al^{VI} in cpx, dominance of opx as an early crystallization phase, high distribution coefficients of Mg-Fe for coexisting pyroxene pairs and thin chill margins. The pressures were estimated at about 10-12 kb. A detailed study of mineral parageneses in the layered basic-ultrabasic Wingellina Hills intrusion [Ballhaus and Glikson, in press] indicates lower pressures of about 6+/-1.4 kb [C. Ballhaus, pers. comm., 1989]. The parageneses include [1] opx-cpx-pleonaste symplectites along olivine-calcic plagioclase contacts; [2] ol-[opx-cpx]-chromiferous spinel symplectites along olivine-sodic plagioclase contacts; [3] cpx-[plg]-spinel symplectites surrounding cumulus chromite, and [4] groundmass neoblasts olivine, orthopyroxene, clinopyroxene, plagioclase and spinel. The near-isobaric cooling path defined by these reactions suggests the intrusion of the Giles Complex at intermediate to infra-crustal levels.

Arunta Block: Thermobarometry of pyroxene granulites and garnet-two pyroxene granulites from the Mt Hay-Mt Chapple terrain along the Redbank-Mt Zeil thrust zone suggests PT values in the range of 600-900° and 6.5-9.0 kb [Glikson, 1987b]. Detailed thermobarometric studies of granulites in the central and northern provinces of the Arunta Block [Warren, 1983; Warren and Stewart, 1989; Warren and Hensen, 1989] indicate that the lower amphibolite to granulite facies rocks formed by prograde high-temperature [<900°] variable pressure [<9 kb] metamorphism associated with extensive migmatization under water-undersaturated conditions. This is indicated by sapphirine-bearing rocks, where prograde reactions produced sapphirine-orthoclase-orthopyroxene assemblages [estimated PT 8+/-1 kb, 850-920°C], hydration produced sapphirine-phlogopite assemblages and isobaric cooling resulted in reaction of sapphirine and cordierite to form orthopyroxene and sillimanite [Warren, 1983]. Retrograde metamorphism has taken place along hydrated deformed zones under isobaric cooling conditions. Peak metamorphic conditions are correlated with the approximately 1.8 b.y. old thermal event whereas the low-T high-P retrogression may be as late as the early Carboniferous Alice Springs Orogeny.

5. GRAVITY ANOMALIES AND SEISMIC REFLECTION PROFILES

The steep Bouguer gravity gradients associated with major tectonic lineaments in central Australia, i.e. the Redbank-Mt Zeil, Anmatjira, Woodroffe and Fraser faults [Anfiloff and Shaw, 1973; Mathur, 1976; Wellman, 1978], may signify sharp changes in crustal density and deep seated mantle anomalies. Forman and Shaw [1973] and Mathur [1976] interpreted the Papunya gravity high [+50 mgal] in terms of upward dislocation of the MOHO by thrusting or mantle doming. Anfiloff and Shaw [1973] modelled the gravity anomalies in terms of intracrustal density contrasts between dense granulite-dominated blocks and granite and/or metasediment-dominated blocks. The different models are portrayed in Fig. 5. Wellman [1978] considered this problem in terms of crustal density and thickness variations and the role of compensating masses. Shaw et al. [1984] attributed the large gravity highs to concentrations of basic igneous rocks formed in an early Proterozoic extensional tectonic environment.

The gravity anomalies [<140 mgal amplitude] indicate a non-isostatic regime, despite the lack of evidence for major deformation since the Carboniferous. In an attempt to understand the cause of the gravity anomalies, approximately 500 km of deep seismic reflection data were recorded by the Bureau of Mineral Resources, Geology and Geophysics [Goleby et al., 1987, 1989] [Fig. 6]. The outstanding feature of deep crustal profiles through the Arunta Block is a series of moderately northward dipping faults which penetrate the crust to depths of at least 30 km and in the case of the 35-40° dipping Redbank-Mount Zeil thrust zone possibly as deep as 50 km - where a wedge of lithosphere appears to over-ride an underthrusted crustal wedge with near-flat reflections [Fig. 7]. This supports models which suggest 'thick skinned' tectonics, as contrasted to 'thin skinned' tectonics, since no mid-crustal ramp structures are shown by the seismic data. The Bouguer anomaly pattern over the southwestern Arunta Block shows that the Papunya gravity 'high' is offset northward from the Redbank thrust zone and the overthrust granulites by over 50km, suggesting a deep seated mantle origin of this anomaly [Goleby et al., 1989]. The combination of seismic, gravity and surface structural data gives rise to a concept of the Amadeus Basin as a regional ramp fault structure, involving upthrusting of the mantle along major boundary faults [Fig. 7] [Shaw et al. in prep., 1989].

Lambeck [1983] invoked a long-term compressional stress field in central Australia in order to explain the origin and repeated reactivation of thrust faults between uplifted basement domes and subsiding basins, leading to variations in crustal thickness between these domains. In this model the crust is relatively thin beneath uplifted basement terrains, as indicated by Mathur's [1976] gravity model, and thickens below the Amadeus Basin including up to 14 km of late Proterozoic-Palaeozoic sediments. This suggestion is supported by teleseismic travel-time anomalies of up to 1.5s measured along two lines of instruments, which suggest a possible MOHO uplift north of the Redbank-Mount Zeil

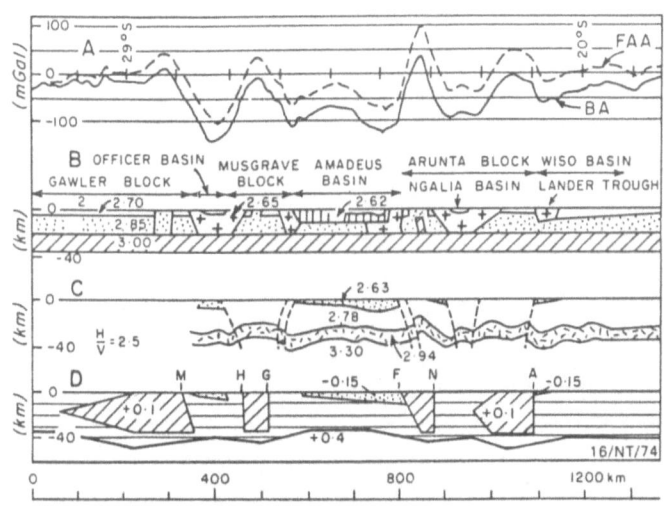

Fig. 5 - Crustal models along a N-S profile approximating longitude 133° in central Australia:

A - free air anomalies [FAA] and Bouguer anomalies [BA];
B - a crustal model by Anfiloff and Shaw [1973];
C - a crustal model by Mathur [1976];
D - a crustal model by Wellman [1978], giving density differences from a standard column.

296

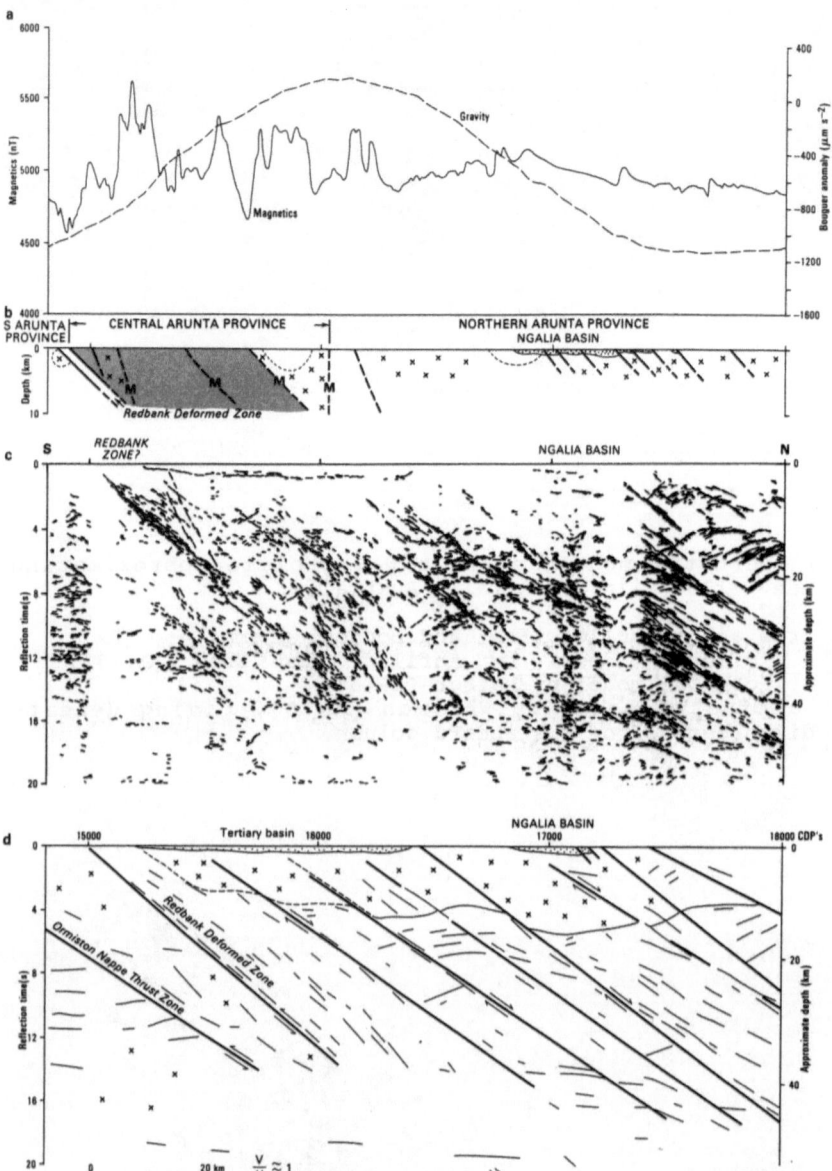

Fig. 6 - N-S geophysical profiles across the western part of
the Arunta Block: [a] Bouguer gravity and aeromagnetic
profiles; [b] interpreted geological cross-section [M - fault
line interpreted from magnetics]; [c] line drawing of deep
seismic reflection data from BMR's transect. the lines
represent continuous reflction segments of at least 10 traces
width; [d] interpreted crustal model.

thrust zone, in agreement with the Bouguer anomalies [Lambeck et al., 1988]. The development of thrusting in central Australia was confined to a major wedge-like block by major NE and NW striking faults, namely the Lake Mackay, Ennugan, Walabanba and Woolanga faults, delineated by geophysical anomalies and structural evidence [Anfiloff and Shaw, 1973; Shaw et al., 1984].

6. MODELS OF CRUSTAL STRUCTURE AND EVOLUTION

Any account of the evolution of the middle to late Proterozoic crust in central Australia must explain the following:

[1] Both the Musgrave and the Arunta blocks consist of [A] granulite-facies slices associated with relatively small bodies of orthogneiss, and [B] amphibolite facies metamorphics extensively intruded by orthogneiss and late tectonic granites. A-type domains are generally thrust over B-type domains.

[2] The granulites of both the Musgrave and Arunta blocks include mainly supracrustal assemblages, dominated by meta-arenites and meta-pelites with a probably important volcanic component in the Strangways Range and Mt Chappell Range. Granulites of both blocks include layered mafic-ultramafic intrusions, namely the Giles Complex and the Mt Hay intrusion.

[3] Confident definition of thermal events must await application of isotopic U-Pb zircon measurements; however, existing Rb-Sr data allow a tentative comparison between the following episodes in the two blocks: [A] Major thermal events accompanied by granitic activity @ 1.8-1.6 b.y. in the Arunta Block and @ 1.7-1.5 b.y. in the Musgrave Block; [B] Major metamorphic events accompanied by granitic activity @ 1.0-0.9 b.y in the Arunta Block and @ 1.2-1.0 b.y in the Musgrave Block. A scatter of Rb-Sr ages between these clusters counsels caution regarding their geological meaning.

[4] The generally low to moderate initial Sr87/Sr86 ratios shown by the Rb-Sr isochrons [Fig. 4] place limits on the crustal prehistory and/or the Rb/Sr ratios of crustal precursors of both the meta-supracrustals and the granites.

It is thus possible to tentatively regard the Musgrave and Arunta blocks as parts of a broadly contemporaneous and possibly originally contiguous Proterozoic terrain. The extent to which these blocks consist of accreted terrains associated with plate tectonic processes is unclear, requiring the identification of diagnostic volcanic and sedimentary assemblages. The different degrees of manifestation of thermal events may reflect lateral variations in tectonic/thermal activity or may be an artifact of biased sampling and of insufficient isotopic data, particularly in the Musgrave Block.

Two different interpretations can be advanced regarding the original spatial relationships between the granulite facies blocks and the granite-invaded amphibolite facies blocks [Glikson, 1987a, 1987b]:

I. The granulite facies zones and the amphibolite facies zones were originally laterally separated from each other, forming distinct zones characterized by different histories and bounded by faults [? precursors of the thrust faults].

II. The amphibolite facies zones originally overlaid the granulite facies zones and the extensive granites emplaced in the former are the products of anatexis associated with prograde metamorphism along the granulite-amphibolite facies transition zone or metamorphic front.

Shaw et al. [1984] proposed an evolutionary model for the Arunta Block which includes elements of both vertical zonation and lateral zonation. In their interpretation Division 1 supracrustals, including an important volcanic component, formed within a linear extension zone or rift and were subsequently unconformably overlain by the Division 2 and 3 arenites. Interthrusting of the three sequences due to repeated development of compressional regimes, accompanied by anatexis and granite formation, resulted in the present structural configuration. Several extension-compression cycles correlated with thermal peaks indicated by Rb-Sr ages were envisaged by these authors.

An interpretation of the evidence in terms of original vertical zonation could involve the following evolutionary stages: [1] deposition of supracrustals, including volcanics, throughout the central Australian region between 2.2 - 1.6 b.y.; [2] peak metamorphism about 1.8-1.6 b.y. at crustal depths of about 20-30 km, with decreasing metamorphic grade upwards and extensive anatexis along the granulite-amphibolite transition zone; [3] intrusion of major layered basic/ultrabasic complexes about 1.1 b.y. ago, providing heat for extensive anatexis presumably associated with an introduction of volatiles, granite formation and age resetting of the granulites; [4] interthrusting of the granulite and amphibolite facies zones, mainly during the Palaeozoic, resulting in their lateral juxtaposition.

The role of the major layered basic/ultrabasic intrusions of the Giles Complex is currently under study. The gabbros mostly occur as fault slices and were extensively recrystallized and migmatized under granulite facies conditions, especially along their margins, as for example in the western part of the Hinckley Range, Western Australia. Primary contacts with the felsic granulites are mostly obliterated, although intrusive contacts are reported from localities in South Australia [Nesbitt et al., 1971]. Whether these contacts formed by primary magmatic intrusion or by loci of syn-metamorphic anatexis is unclear and the possibility remains that the layered intrusions are older that the 1.1 b.y. old thermal event. Similar relations occur in the Fraser Complex, a probable analogue of the Giles Complex within the Albany-Fraser belt. Fletcher and Myers [Pers.comm., 1988] describe the Fraser Complex as an interthrusted assemblage which consists of

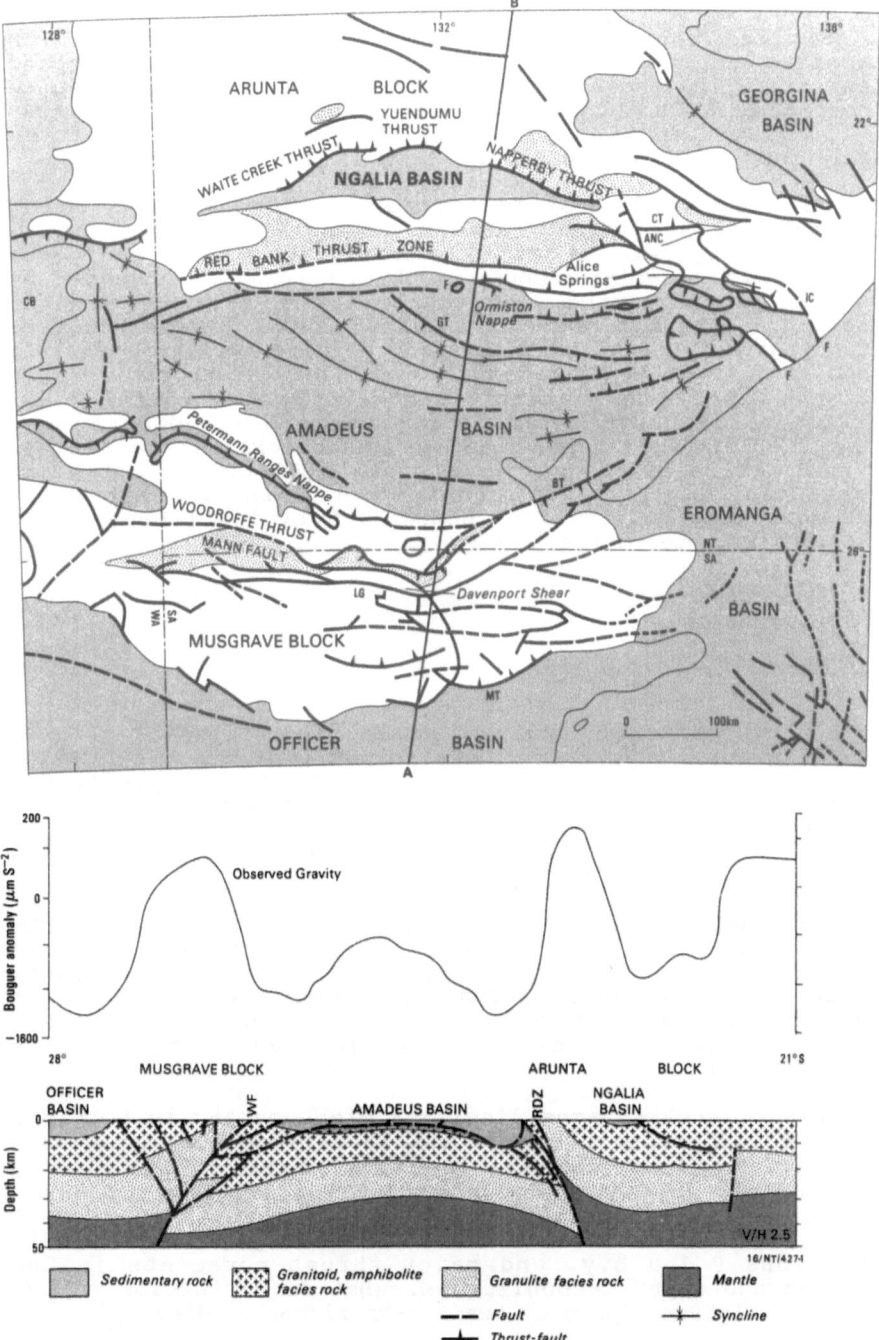

Fig. 7 - A structural sketch map of central Australia and a N-S transect showing the Bouguer gravity profile and an interpreted crustal cross-section [Shaw et al., 1989, in prep., after Wells and Moss, 1983].

subvertical tectonic slices of gabbro interleaved with quartzite and gneiss which were deformed and metamorphosed under granulite facies conditions and subsequently thrust to higher crustal levels within a short time [<25 m.y.] at @ 1295 m.y., as deduced from mineral Sm-Nd isochron data. It is likely that the Fraser and Giles complexes are of similar age.

The nature and age of the basement beneath the thick supracrustal sequences remains unknown. The dominantly quartzo-feldspathic to pelitic composition of the sediments point to an intra-sialic environment, a conclusion supported by the lack of ophiolites. It is surprising perhaps that the floor of a system formed within about $200X10^6$ years prior to metamorphism is nowhere identified, despite the strong uplift associated with the thrusting. A possible interpretation is that the Giles Complex signifies the proximity of the MOHO. Indicators of plate tectonic processes have nowhere been identified in the Musgrave and Arunta blocks, which renders uniformitarian hypotheses questionable [Glikson, 1981] and opens the way for alternative models in terms of intracontinental extension-compression cycles [Shaw et al. 1984] and rift tectonics [Etheridge et al., 1987].

7. CONCLUSIONS

[1] Both the Musgrave and the Arunta blocks consist of deep crustal granulite/gneiss blocks thrust over granite-invaded upper crustal amphibolite facies blocks along major thrust faults representing a 'thick-skinned' tectonic pattern which has involved upthrusting of the lithosphere within depth ranges of 30-50 km in the case of the Redbank-Mount Zeil thrust zone.

[2] The granulite-facies lower crustal zones are dominated by supracrustal rocks of predominantly quartzofeldspathic composition with a possibly important volcanic component. The supracrustals were intruded by major layered complexes under pressures of over 6 kb. Major anatectic events produced granites along the granulite-amphibolite boundary zones.

[3] Rb/Sr isochron ages display a wide scatter in both the Musgrave and Arunta blocks. Mantle differentiation events are indicated by Sm-Nd isochron ages @ 2.0 b.y., supracrustal deposition by Rb-Sr data pre-1.8 b.y., major metamorphism and anatexis @ 1.8-1.7 b.y., repeated thermal and magmatic events between 1.8-0.9 b.y., major basic intrusions @ 1.3 b.y. and major thrust movements in the Cambrian and upper Carboniferous. Better definition of the temporal relationships must await U-Pb zircon studies.

[4] The Arunta and Musgrave supracrustals, as well as the younger upper Proterozoic to Palaeozoic basins, have formed within ensialic environments deformed and thrusted in conjunction with the development of intracontinental compressional stress regimes, with the consequent uplift and exposure of deep cross sections of the middle to upper Proterozoic sialic crust.

[5] The Giles Complex has been intruded at intermediate to infra-crustal levels under pressures above 6 kb and may signify the proximity of the MOHO in the upper Proterozoic. This possibility needs testing by further field studies and U-Pb zircon analysis.

Acknowledgements

A.Y. Glikson wishes to thank D.M. Fountain and M.H. Salisbury for their kind invitation to present this review at the NATO Advanced Study Institute course on Cross Sections through the Continental Crust, at Killarney, Ontario, September, 1988. We thank H.L. Harrington, A.J. Stewart and two anonymous referees for their comments and criticism. Published with the permission of the Director, Bureau of Mineral Resources, Geology and Geophysics.

References

ANFILOFF, W. & SHAW, R.D., 1973 - The gravity effects of three large uplifted granulite blocks in separate Australian shield areas. Symp. Earth's Gravity Field and Secular Variations in Position, Proc., pp.273-289.

BALLHAUS, C. & GLIKSON, A.Y., 1989 - Multiple intrusion and magma mixing in the Wingellina Hills layered intrusion, Giles Complex, central Australia. J. Petrol. [in press].

BLACK, L.P., SHAW, R.D. & STEWART, A.J., 1983 - Rb-Sr geochronology of Proterozoic events in the Arunta inlier, central Australia. BMR J. Aust. Geol. Geophys., 8:129-138.

BLACK, L.P. & McCULLOCH, M.T., 1984 - Sm-Nd ages of the Arunta, Tennant Creek and Georgetown inliers of northern Australia. Aust. J. Earth Sci., 31:49-60.

COLLERSON, K.D., OLIVER, R.L. & RUTLAND, R.W.R., 1972 - An example of structural and metamorphic relationships in the Musgrave orogenic belt, central Australia. J. geol. Soc. Aust., 18:379-393.

DANIELS, J.L., 1974 - The Geology of the Blackstone Region, Western Australia. Geol. Surv. W. Aust. Bull. 123.

DING, P. & JAMES, P.R., 1985 - Structural evolution of the Harts Range area and its implication for the development of the Arunta Block, central Australia. Precambrian Res., 27: 251-276.

ETHERIDGE, M.A., RUTLAND, R.W.R. & WYBORN, L.A.I., 1987 - Orogenesis and tectonic processes in the early to middle Proterozoic of northern Australia. in: Kroner, A. [ed.] Proterozoic Lithospheric Evolution. Am. Geophys. Union Geodynamic Ser., 17:131-147.

FORMAN, D.J., 1966 - The geology of the southwestern margin of the Amadeus Basin, central Australia. Aust. Bur. Miner. Resour. Geol Geophys. Rep. 87.

FORMAN, D.J. & SHAW, R.D., 1973 - Deformation of crust and mantle in central Australia. Aust. Bur. Miner. Resour. Geol. Geophys. Bull. 144.

GLIKSON, A.Y., 1981 - Geochemical, isotopic and palaeomagnetic tests of early sial-sima patterns: the Precambrian crustal enigma revisited. Geol. Soc. Am. Mem. 161: 95-118.

GLIKSON, A.Y., 1984 - Granulite-gneiss terrains of the southwestern Arunta Block, central Australia: Glen Helen, Narwietooma and Anburla 1:100 000 sheet areas. Aust. Bur. Miner. Resour. Geol. Geophys. Rec. 1984/22.

GLIKSON, A.Y., 1987 - Regional structure and evolution of the Redbank-Mt Zeil thrust zone: a major lineament in the Arunta inlier, central Australia. BMR J. Aust. Geol. Geophys. 10:89-107.

GLIKSON, A.Y., 1987 - An upthrusted early Proterozoic basic granulite-anorthosite suite and anatectic gneisses, southwest Arunta Block, central Australia: evidence on the nature of the lower crust. Trans. geol. Soc. S. Africa, 89: 263-283.

GOLEBY, B.R., SHAW, R., WRIGHT, C. & KENNETT, B., 1987 - Thick skinned tectonics in central Australia: deep seismic results from a multidisciplinary study. Aust. Bur. Miner. Resour. Res. Newsletter, 7: 8-9.

GOLEBY, B.R., SHAW, R.D, WRIGHT, C., KENNET, B.L.N. & LAMBECK, K., 1989 - Geophysical evidence for 'thick-skinned' crustal deformation in central Australia. Nature, 337:325-330.

GOODE, A.D.T. & MOORE, A.C., 1975 - High pressure crystallization of the Ewarara, Kalka and Gosse Pile intrusions, Giles Complex, central Australia. Contr. Miner. Petrol., 51: 77-97.

GRAY, C.M., 1978 - Geochronology of granulite facies gneisses in the western Musgrave Block, central Australia. J. geol. Soc. Aust., 25:403-414.

GRAY, C.M., CLIFF, R.A. & GOODE, A.D.T., 1981 - Neodymium-strontium isotopic evidence for extreme contamination in a layered basic intrusion. Earth Planet. Sci. Lett. 56:189-198.

GRAY, C.M. & COMPSTON, W., 1978 - A Rb-Sr chronology of the metamorphism and prehistory of central Australia granulites. Geochim. Cosmochim. Acta, 42:1735-1748.

LAMBECK, K., 1983 - Structure and evolution of the intracratonic basins of central Australia. R. Astron. Soc. Geophys. J., 74:843-886.

LAMBECK, K., BURGESS, G. & SHAW, R.D., 1988 - Teleseismic travel time anomalies and deep crustal structure in central Australia. Geophys. J., 94: 105-124.

MAJOR, R.B., 1973 - Woodroffe, South Australia, 1:250 000 Geol. Sheet Expl. Notes, Geol. Surv. S. Aust.

MATHUR, P., 1976 - Relationships of Bouguer anomalies to crustal structure in southwestern and central Australia. BMR J. Aust. Geol. Geophys., 1:177-186.

NESBITT, R.W., GOODE, A.D.T., MOORE, A.C.& HOPWOOD, T.P., 1970 - The Giles Complex, central Australia: a stratified sequence of mafic and ultramafic intrusions. spec. Publ. geol. Soc. S. Afr., 1:547-564.

304

SHAW, R.D., STEWART, A.J. & BLACK, L.P., 1984 - The Arunta inlier: A complex ensialic mobile belt in central Australia. Part 2: tectonic history. Aust. J. Earth Sci., 31: 457-484.

SHAW, R.D, ETHERIDGE, M.A. & LAMBERT, K., 1989 - The late Proterozoic to early Palaeozoic intracratonic Amadeus Basin, central Australia: a key to tectonic forces in plate interiors. Tectonics [submitted].

STEWART, A.J., 1967 - An interpretation of the structure of the Blatherskite nappe, Alice Springs, Northern Territory. J. geol. Soc. Aust., 14:175-184.

STEWART, A.J., OFFE, L.A., GLIKSON, A.Y., WARREN, R.G. & BLACK, L.P., 1980 - Geology of the northern Arunta Block, Northern Territory. Aust. Bur. Miner. Resour. Geol. Geophys. Rec. 1980/83.

THOMSON, B.P., 1977 - The Musgrave Block. in: C.L. Knight [ed.], Economic Geology of Australia and Papua New Guinea. Aust. Instit. Min. Met. Mono. no. 5:451-460.

WARREN, R.G., 1983 - Metamorphic and tectonic evolution of granulites, Arunta Block, central Australia. Nature, 305:300-303.

WARREN, R.G.& STEWART, A.J., 1989 - Isobaric cooling of Proterozoic high-temperature metamorphics in the northern Arunta Block, central Australia: implications for tectonic evolution. Precamb. Res. 40/41:175-198.

WARREN, R.G. & HENSEN, B.J., 1989 - The PT evolution of\ the Proterozoic Arunta Block, central Australia., and implications for tectonic evolution. spec. Publ. geol. Soc. London [in press].

WEBB, A.W., 1985 - Geochronology of the Musgrave Block. Min. Resour. Rev. S. Aust. Dept. Mines & Energy, 155:23-37.

WELLMAN, P. 1978 - Gravity evidence for abrupt changes in mean crustal density at the junction of Australian crustal blocks. BMR J. Aust. Geol. Geophys., 3:153-162.

WELLS, A.T., FORMAN, D.J., RANFORD, L.C. & COOK, P.J., 1970 - Geology of the Amadeus Basin, central Australia. Aust. Bur. Miner. Resour. Bull. 100.

WELLS A.T. & MOSS, J.F., 1983 - The Ngalia Basin, Northern Territory: stratigraphy and structure. Aust. Bur. Min. Resour. Geol. Geophys. Bull. 212.

WINDRIM, D.P. & McCULLOCH, M.T., 1986 - Nd and Sr isotopic systematics of central Australian granulites: chronology of crustal development and constraints on the evolution of the lower crust. Contr. Mineral. Petrol., 94:289-303.

THE LATE ARCHEAN HIGH-GRADE TERRAIN OF SOUTH INDIA AND THE DEEP STRUCTURE OF THE DHARWAR CRATON

R.C. Newton
Department of the Geophysical Sciences
University of Chicago
5734 South Ellis Avenue
Chicago, IL 60637, USA

ABSTRACT. The Dharwar Craton of southern India is an Archean continental fragment showing a continuous exposed crustal section from low-grade gneisses and greenstone basins in the north to granulites in the south with more than eight kilobars paleopressure, corresponding to nearly thirty kilometers depth. Low-potassium acid to intermediate "gray gneiss" was massively accreted 3.4 to 3.0 Ga ago and was capped by tholeiitic to komatiitic flows and their sedimentary detritus 3.0 to 2.6 Ga ago. Juvenile crust formation recommenced 2.6 Ga ago, probably in the form of granodioritic Andean margin and island arc plutonics. Termination of this quasi-modern magmatism by suturing of the island arc(s) coincided with large-scale wrenching of the South India continent, which created channelways for the rise of metasomatizing solutions from the mantle. These CO_2-H_2O fluids destabilized biotite and amphibole at deep levels in the thickened southern cratonal region, extending the development of the pyroxene granulite facies, which may have existed in the deep crust from earlier metamorphic episodes. Liberated H_2O and alkalis were implanted in the middle crust in granitic liquid and CO_2-bearing fluids caused widespread carbonate metasomatism at higher crustal levels, with emplacement of the Kolar gold ores. The Craton as a whole has been nearly inert since latest Archean, though the tectonized southern region was affected by Proterozoic remobilization. This plate tectonic interpretation suggests that the exposed granulites of the southern cratonal region should not be supposed to represent the normal deep crust under the low-grade gneiss-greenstone terrain, but rather may be the middle levels of a crust thickened in continental collision. The inertness, hence great longevity, of the northern cratonal interior results from low radioactivity at deep levels and massive, unsutured foundations. Exposed Archean granulite facies terrains may represent deformed margins of ancient continental nucleii.

Low-Grade and High-Grade Cratons

Extensive terrains of Archean rocks are exposed in the large continental shield areas and in smaller areas of strong uplift and deep dissection. Studies of the petrology and structures of these cratonal areas have led to a first-order classification into low-grade and high-grade terrains. The low-grade terrains ("gneiss-greenstone-granite terrains") are dominantly composed of felsic to intermediate low-K gneisses of probable plutonic origin, with associated supracrustal rocks. A minor component of the supracrustals is metamorphosed and attenuated older relics intercalated within the gneisses. The most abundant supracrustals form a capping, probably largely unconformable, as "greenstone" troughs or

M. H. Salisbury and D. M. Fountain (eds.), Exposed Cross-Sections of the Continental Crust, 305–326.
© 1990 *Kluwer Academic Publishers.*

basins, upon the gneisses. They consist of voluminous mafic to ultramafic volcanic rocks and dominantly graywacke metasediments, with secondary argillites, ferruginous quartzites and acid volcanics. A third cratonal component is K-rich granites which commonly cross-cut gneisses and greenstones and are therefore somewhat younger (Anhaeusser et al., 1969). The low-grade gneiss-greenstone terrains may be bordered by highly deformed and highly metamorphosed tracts, either from a continuous progression of metamorphic grade, as in the northern margin of the Kaapvaal Craton, South Africa (Van Reenen et al., 1988) or separated by a structural discontinuity. The high-grade belts commonly feature an association of shelf-type (quartzite-carbonate-argillite) metasediments (Shackleton, 1976) with conformable amphibolites, but lithologies of the gneiss-greenstone association may be dominant. Large gabbro-anorthosite plutonic complexes are associated with the metasediments in southern Africa, southern India and southwest Greenland.

The metamorphism of the high-grade portions of the cratons commonly reached the granulite facies, with temperatures in excess of H_2O-saturated solidi. The metamorphism must therefore have been H_2O-deficient, either because of pre-metamorphic desiccation or removal of H_2O in anatectic liquids or in CO_2-rich metamorphic vapors (Winkler, 1976). A conspicuous feature of some very high-grade granulites is extreme depletion in the large-ion lithophile elements (LILE) relative to common crustal rocks of similar lithologies (Heier, 1973). This depletion must have resulted from upward removal in either anatectic liquids or metamorphic vapors.

An outstanding property of the cratons is their inertness, attested to by long-term survival. The gneiss-greenstone terrains have remained near-sea-level or slightly positive areas throughout most of their histories, as indicated by dominantly low metamorphic grade, closed isotopic systems and, commonly, thin, undeformed platformal cover, with preservation in some areas of ancient surficial deposits, such as the 2.4 Ga Gowganda tillite of Ontario (Mustard and Donaldson, 1987). Continental rejuvenation or rifting largely avoided these old continental nuclei. As an illustration of cratonal stability, opening of the southern Atlantic proceeded along suture zones of Proterozoic or early Paleozoic ages, preserving intact older cratonal segments (Sykes, 1978).

The resistance to change of the cratons, particularly of the Archean low-grade terrains, suggests that their infracontinental roots are different from, and more durable than, those of younger continental terrains. Indeed, the distinct differences in infrastructure may extend below the crust well into the lithosphere, as indicated by the significantly lower mantle-derived heat flow in cratonal areas (Ballard and Pollack, 1988), and by unique seismic transmission properties of subcratonal mantle (Silver and Chan, 1988).

The relationship of the low- and high-grade terrains is a major problem. The high-grade terrains may represent suture zones of continental collision (Hoffman, 1988) or zones of intracontinental disruption (Kröner, 1984). Many were reactivated at various times: they commonly show mineralogic evidence of polymetamorphism, disturbed isotope systematics, and reactivation of major faults, which may change character from transcurrent to reverse to normal motions in successive tectonic events. An important question of the relationship is whether the high-grade terrains pass beneath the low-grade cratons as the normal cratonal roots, as suggested by Kröner (1980) for the North Finland and southern African craton-mobile belt associations and by Percival and Card (1983) for the relationship of the granulite facies Kapuskasing terrain of central Ontario and the lower-grade gneiss-greenstone terrain to the west. The low radioactivity of some granulite facies rocks from cratonal high-grade terrains is suitable for the lower cratonal crust (Heier, 1973), yet the great stability of the low-grade portions of cratons contrasts with the proneness of high-grade terrains to reactivation, suggesting that they are fundamentally different entities.

Prominent metasomatized shear zones, some of great length, commonly traverse cratonal areas. An example is the early Proterozoic Nordre Stromfjord zone of SW Greenland

(Sorensen, 1983), which varies in width from 5 to 30 km and has a 120-km left-lateral offset. Composite gneiss studies inside and outside of the zone indicate substantial granitization associated with the shearing, much of it at granulite grade. Similar shear zones are the Laxfordian zones of NW Scotland (Beach, 1976) and the Athabasca Mobile Zone of Western Canada (Burwash and Krupicka, 1970). Late Archean shear zones through the greenstone belts host most of the major gold deposits in quartz-carbonate veins. An important example is the Kolar area of the Dharwar Craton (Fig. 1), where gold-bearing veins have the characteristic north-south strike of the Craton as a whole (Hamilton and Hodgson, 1986). Major shear zones and their metasomatism can usually be characterized as late-tectonic in terms of the Archean continent-forming events or post-tectonic and retrogressive, constituting a major rejuvenation of some high-grade terrains.

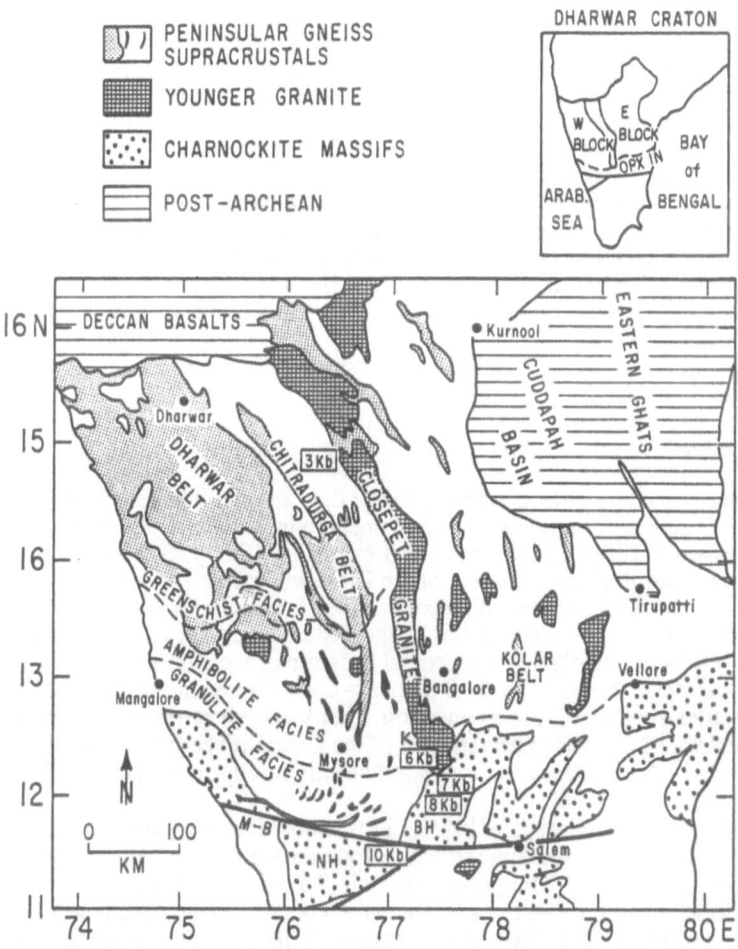

Figure 1. Geology of the Dharwar Craton, from Karunakaran (1974) and Drury and Holt (1980). Metamorphic facies boundaries from Pichamuthu (1965). Paleopressures (rectangles) from Harris and Jayaram (1982), Hansen et al. (1984) and Raith et al. (1988).

The Dharwar Craton

LITHOLOGY AND METAMORPHISM

The Dharwar Craton (Fig. 1) is an Archean terrain which displays the classical features of an early continent. The major components are low-K tonalitic to trondhjemitic gneiss (Peninsular Gneiss) with minor older supracrustal enclaves (Sargur Group) and extensive volcanic-sedimentary basins (Dharwar Supergroup) which, in large part, unconformably overlie the gneiss (Chadwick et al., 1981). These "greenstone" basins are much more abundant in the western craton. The youngest rocks are K-rich granites and charnockitic gneisses. The largest granite exposure is the Closepet Granite, which extends nearly 500 km north-south across the Craton. Numerous other relatively young granite bodies occur east of the Closepet belt, and a few to the west. In the southern portion of the Craton the gneisses and their supracrustal associations develop orthopyroxene in an apparently continuous progression to the granulite facies. Granulites are thus properly part of the craton and among the youngest components, although there may be an older generation of granulites in the southern high-grade terrain (Pichamuthu and Srinivasan, 1984). Extensive summaries of the petrology of the Dharwar Craton are contained in Naqvi and Rogers (1983) and Newton and Anderson (1986).

The southern high-grade portion of the Dharwar Craton is the culmination of an apparently continuous progressive metamorphic series from low greenschist to granulite facies. It is not certain whether all of the metamorphic rocks are of late Archean age; some may be older Archean (Radhakrishna and Naqvi, 1986). Facies boundaries as depicted by Pichamuthu (1965) and Subramaniam (1967) are shown in Fig. 1. Pichamuthu suggested that the facies series represents a depth zone arrangement, with deeper crustal levels exposed to the south. Mineralogic paleopressures shown in Fig. 1 are in accord with this interpretation.

STRUCTURES

A prominent generally N-S structural grain of the Craton is apparent in the outcrop pattern. The entire Craton was subjected to N-S elongation in the late Archean (Mukhopadhyay, 1986), perhaps in shearing along near-vertical planes (Drury and Holt, 1980). This deformation controlled emplacement of the Closepet Granite and gave rise to the elongate and hook-like patterns of some of the volcanic-sedimentary basins. The more equant forms of the northernmost greenstone basins suggests that the degree of deformation decreases from south to north, paralleling the general decrease of metamorphic grade of the Dharwar supracrustal rocks. Shearing on N-S planes is concentrated in thin zones of high strain which broaden southward (Drury and Holt, 1980). These zones may coincide with steep through-crustal faults in the northern part of the Craton revealed by deep seismic sounding (Kaila et al., 1979). Some of these major dislocations appear to offset the Mohorovicic discontinuity.

According to Mukhopadhyay (1986), deformation of the Dharwar rocks took place in two major phases: first, isoclinal upright-to-overturned folding with variable axial orientation, and second, folding upon nearly N-S axes with rotation of earlier-formed axes into this direction and pronounced extension, giving rise to the general grain of the Craton. According to Naha et al. (1986), all components of the Craton were folded together in these deformations, including gneisses, older and younger supracrustals, and granites, implying a fundamental structural unity of the Craton. Drury and Holt (1980) cited LANDSAT image and ground observations as evidence for large recumbent folds in the southern part of the Craton. They advocated large-scale north-south wrench faults, mostly sinistral, as a cause of the elongation of the Dharwar basins and the second generation folds, and Drury et al. (1984) emphasized this shearing as a fundamental event in cratonal history, but suggested dominantly dextral displacements. Other workers (Mukhopadhyay, 1986; Naha et al., 1986) have reserved judgement on these suggested large-scale features in the lack of sufficiently detailed field work.

CHRONOLOGY

The Closepet Granite approximately separates a West Block from an East Block (Swami Nath et al., 1976), distinguishable petrologically by relative abundances of granites and greenstones. The West Block appears to be the more stable ensialic portion, with no basement gneiss younger than about 3.0 Ga and some as old as 3.4 Ga (Rogers, 1986), and has been given the name "Karnataka Nucleus" (Radhakrishna and Naqvi, 1986). The inert character of the West Block of the craton may in part result from upward metasomatic transport of U from the deep crust to shallow levels at about 3.0 Ga ago (Callahan and Rogers, 1987).

The Dharwar succession of ultramafic to mafic lavas and graywacke-dominated sediments commenced as a cap on the craton about 3.0 Ga ago, as dated by whole-rock Sm-Nd methods on komatiites from the Kudremukh Belt (Drury et al., 1983). Some of the earliest Dharwar units may have been intruded by the last of the trondhjemite plutons (Rogers, 1986). As the Dharwar basins evolved, sediments and acid volcanics became more abundant. Rhyolites from the Chitradurga Belt have been dated at 2.57 Ga by whole-rock U-Pb methods (Taylor et al., 1984) and have approximately the same age as a number of small granite bodies in the central West Block, such as the Arsikere Granite (Rogers, 1988). Stabilization of the West Block as an early cratonal sedimentary platform was well advanced by 2.6 Ga ago (Srinivasan and Ojakangas, 1986).

Granodioritic gneisses of late Archean age occur in the Kolar area of the East Block (Krogstad et al., 1988; Krogstad et al., 1989). The association of gneisses of age 2.65-2.55 Ga on the west side of the Kolar Schist Belt (Fig. 2) with Nd and Sr isotopes indicative of involvement of older cratonal material of at least 3.2 Ga age suggests Andean margin magmatism. The Kolar Belt tholeiitic and komatiitic schists have primitive mantle signatures with mantle derivation ages of 2.75-2.65 Ga. It is significant that tholeiites from the eastern and western portions of the schist belt appear to be derived from different mantle sources, based on differences in Pb isotopes and rare-earth element patterns. Associated metasediments could have marginal basin characteristics. The eastern side of the Kolar Belt is underlain by granodioritic gneisses of 2.55-2.53 Ga ages with no discernible isotopic record of prior crustal residence. An Aleutian arc setting could be a modern analog for these rocks. The gneisses and schists are co-deformed in much the same way as the craton as a whole (Hanson et al., 1988). Early flat isoclinal folds are refolded into tight upright folds of a N-S axial strike. Hanson et al. (1988) state that left-lateral shearing on N-S planes is evident, although Hamilton and Hodgson (1986) observed that the shearing in the Kolar gold mines is invariably right-lateral. Shearing and emplacement of the quartz-carbonate gold ores along N-S structures is thus younger than 2.55 Ga and is probably a characteristic part of the late Archean events in the Craton.

The isotopically unique character of the several elements of the Kolar Belt strongly suggest that several crustal terrains with different mantle provenances were juxtaposed by plate tectonic processes in the latest Archean (Krogstad et al., 1989), much in the manner of Phanerozoic terrain accretion in western North America (Saleeby, 1983). The near-coincidence of isotopic ages of many tectonothermal events in the southern portion of the Dharwar Craton with those of the continental accretion events in the Kolar strongly suggests plate tectonics as the major causal agency of the high-grade terrain.

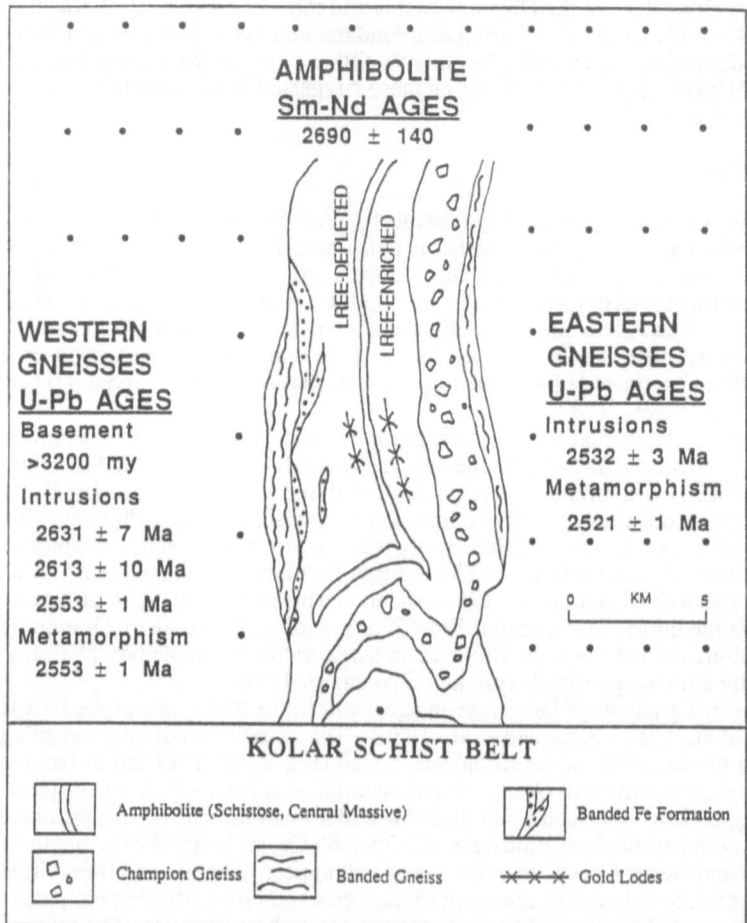

Figure 2. Geology and geochronology of the Kolar area, from Hanson et al. (1988) and Krogstad et al. (1988). The amphibolitic schist belt contains some metasedimentary units, the Champion Gneiss and banded Fe formations. Volcanic rocks of the eastern and western portions of the schist belt, although of the same age, may have had different mantle sources, based on light rare earth element (LREE) abundances (see text).

The Closepet Granite was intruded at 2.50-2.53 Ga in the Kabbal, Karnataka area (area K in Fig. 1), based both on zircon (Buhl et al., 1983) and allanite (Grew and Manton, 1984) U-Pb ages. Incipient and massive charnockites close to the orthopyroxene isograd in the Krishnagiri area directly south of the Kolar area (Fig. 1) give Rb-Sr and Sm-Nd whole rock ages of 2.47 Ga (Peucat et al., 1987), as do associated amphibolite facies gneisses, with very short prior crustal residence time at 2.47 Ga ago. Isotopic ages of 2.56 Ga ago of the Madras granulites (Bernard-Griffiths et al., 1987), 2.62 Ga from the Koorg charnockite massif of western Karnataka (Spooner and Fairbairn, 1970) and 2.50 Ga from the Nilgiri Hills massif of northern Tamil Nadu (Raith et al., 1988) demonstrate the widespread nature of the high-grade event. After this finale the consolidated Craton remained nearly inert.

SHEARING AND METASOMATISM

Apart from the consideration of whether shearing motions account for the second generation of N-S folds, there is much evidence for lateral-motion faulting in various places. The author observed consistent right-lateral shears on N-S-striking planes at Kabbal. Emplacement of incipient charnockite (orthopyroxene-bearing felsic granulite), undoubtedly by a fluid alteration process, was principally guided by these shears, and is very closely associated with the Closepet Granite (Friend, 1981). K-feldspar-epidote-calcite veins and albitite veins of the southern Closepet Granite (Radhakrishna, 1958), and extensive carbonate-chlorite shear veins closely associated with the 2.6 Ga Chitradurga Granite just west of the Closepet (Chadwick et al., 1981) should probably be classed with the latest-Archean generation of metasomatic veins. Hanson et al. (1988) assigned an age of about 2.4 Ga to the Kolar veins. Shearing and lower-grade metasomatism thus seem to represent the waning stages of the high-grade event. There is the suggestion of a late and weakened front of thermal activity penetrating northward into the Craton from the southern high-grade area at the close of the Archean.

Retrogression along large E-W shear systems, associated with the emplacement of younger granites, affected the southern portion of the Craton and most of the southern end of India in Post-Archean times (Drury et al., 1984). The Moyar-Bhavani line (M-B in Fig. 1) may form an effective southern margin of the Craton; it is not certain if the Nilgiri Hills charnockite massif (NH in Fig. 1) is an integral part of the Craton.

The High-Grade Terrain-- Granite-Charnockite Relations

A significant feature of the Dharwar Craton is that the southern granulite facies terrain is structurally contiguous with the low-grade part of the Craton. This is in contrast to some other Precambrian terrains, such as the North Finland area (Hörmann et al., 1980), where the high-grade portion is separated from lower-grade areas by a structural dislocation, and thus harder to relate to them. That the granulite province of southern Karnataka and northern Tamil Nadu is a contiguous part of the lower-grade Craton is evidenced in several ways:

1) Lithologic components of the low-grade northern Craton are represented in the southern granulite areas. These include tonalitic-trondhjemitic gneiss, older supracrustal enclaves, and, possibly, metasediments representing the lower Dharwar sequence (Pichamuthu and Srinivasan, 1984).

2) LANDSAT images unmistakably show unbroken linear trends extending from amphibolite facies Peninsular Gneiss to uniformly charnockitic areas. This is most evident in the Sivasamudrum and Satnur-Halaguru areas on LANDSAT Image #E077-001 (10 Jan, 1977) in areas mapped, respectively, by Mahabaleswar and Naganna (1981) and Devaraju and Sadashivaiah (1969).

3) No abrupt lithologic or structural breaks were found separating amphibolite facies and granulite facies areas by Devaraju and Sadashivaiah (1969) or by Condie et al. (1982) in the Krishnagiri area.

4) Geothermometry and geobarometry profiles of Raase et al. (1986) and Hansen et al. (1984) show no breaks across the facies transition, but only smooth increases of temperature and pressure.

312

5) The regional gravity map of Subrahmanyam (1978) shows no anomalies at the cratonal high-grade margin.

6) Naha et al. (1986) found that structural styles of the high-grade and low-grade parts of the craton are identical and concluded that they were deformed together. Since the high-grade metamorphism is largely syn- to late-kinematic (Naha et al., 1986), it is an overprint on the Craton.

Field relations near the zone of transition from amphibolite facies to granulite facies are especially significant. Numerous studies (Friend, 1983; Janardhan et al., 1982; Stähle et al., 1987; Hansen et al., 1987) have shown that incipient orthopyroxene appears in the migmatitic gneisses replacing amphibole in shear zones, and all studies concluded that fluids promoting granulite facies assemblages moved along the shears. Outcrops of Closepet Granite in the transition zone show the same type of deformation: early and late shear zones in migmatitic gneiss and in discrete granite bodies show that emplacement of granite and charnockite were essentially coeval, in accord with the interpretation of Friend (1981).

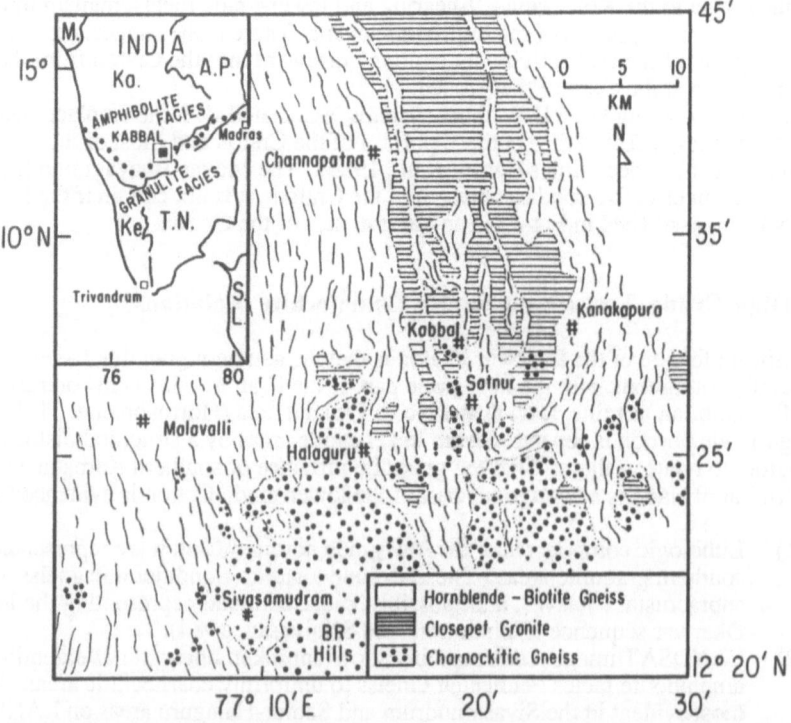

Figure 3. Outcrop distribution of Peninsular Gneiss, Closepet Granite and charnockite at the orthopyroxene isograd in the Kabbal-Satnur-Sivasamudram area, Karnataka. The southern terminus of the Closepet Granite coincides with the incoming of the high-grade terrain. Geology from Devaraju and Sadashivaiah (1969), Mahabaleswar and Naganna (1981), Suryanarayana (1960), and Mahabaleswar et al. (1986). Abbreviations: Ka. = Karnataka, A.P. = Andhra Pradesh, T.N. = Tamil Nadu, S.L. = Sri Lanka, BR = Biligirirangan, M. = Maharashtra, Ke. = Kerala.

The close association of charnockite and granite at Kabbal contrasts with the almost complete mutual avoidance of these lithologies, regionally. Figure 3 shows that the Closepet outcrop does not continue southward into the charnockitic terrain, beyond the latitude where Peninsular Gneiss itself becomes charnockitic. This relationship, combined with the geochronology, suggests that charnockite and granite migmatitic gneiss are complementary products of a process which occurred about 2.53 Ga ago; their present geographic distributions, together with the paleopressure gradient from north to south across the region, further suggest that the relationship was depth-controlled, with granite emplacement in the middle crust and upper crust and charnockite formation in the lower crust. Kabbal would mark an arrested interface between granite and charnockite. Such a relationship could be explained by the classic hypothesis that granulite-facies terrains are the residues of deep-crustal anatexis and removal upward of granitic liquid (Fyfe, 1973; Nesbitt, 1980; Pride and Muecke, 1980). However, for reasons outlined below, it is concluded that this hypothesis does not account for the Closepet-charnockite relations.

At least four hypotheses could be erected for the origin of the Closepet Granite. These are:

1) Juvenile mantle-derived magmas. The high potassium content of much of the granite makes it an unlikely candidate for direct mantle derivation (Wyllie et al., 1976), and the elongate outcrop and shear-zone relationships are not reminiscent of more modern Andean margins or island arc associations.

2) In-situ anatectic magmas. Although this hypothesis was favored by Friend (1984), a large volume of highly potassic liquid is unlikely to have been generated from the regional low-K gneisses without input of additional potassium (Pichamuthu and Srinivasan, 1984).

3) Melts removed from an underlying granulite terrain. There are several major problems with this hypothesis. First is the "room problem": the problem of accommodation of the large volumes of synkinematic granite displacing Peninsular Gneiss is exacerbated by the isoclinal foliation of both gneiss and granulites, arguing against a tensional regime. Second is the metasomatic aspect of much of the granite, with degree of alteration of Peninsular Gneiss varying from development of sporadic pink K-feldspar megacrysts to complete remobilization (Friend, 1981). Third, major element chemistry of many of the granulite facies gneisses of the high-grade margin are not those of the restites of partial melting, but are themselves plausible liquid compositions (Weaver and Tarney, 1983). Finally, the older ideas that Rb and U depletion in high-grade terrains like the Biligirirangan Hills (BH in Fig. 1) could be accounted for by removal of anatectic melts is losing favor (Lamb et al., 1986; Harrison et al., 1986).

4) Metasomatic melting. Mid-crust melting as a result of infiltration of H_2O and K has been suggested repeatedly to account for compositional patterns in migmatites (Olsen, 1985). This hypothesis seems much preferable to the others in terms of the ameliorated room problem, the observed metasomatism and the shear zone association of the Closepet. The relationship between granulites and granites implied by this hypothesis is that fluids moving upward along shear zones transported K and H_2O, liberated in deeper-level granulite facies breakdown of biotite and amphibole, to shallower levels, emplanting these components metasomatically into solid rocks and into a granitic liquid. The relations seen at Kabbal are those of the dividing interface between K-undersaturation and K-oversaturation of the fluids. It is probable that charnockites formed in deep levels of the crust are depleted in U and Rb by this process. The incipient charnockites of the granulite facies boundary in the

middle crust are not greatly depleted in the large-ion lithophile elements, however (Stähle et al., 1987).

The origin of the postulated fluids may be the deep crust or the upper mantle. The great length of the Closepet terrain and the shear zone association suggest that the fluids are of subcrustal origin, as has been repeatedly suggested for other large metasomatic shear zones in Precambrian terrains, such as the Nordre Stromfjord, Laxfordian, and Athabasca zones.

Interpretation of Dharwar Craton Tectonics

BOUNDARY CONDITIONS

The interpretations which follow hinge on the acceptance of several key postulates which seem to be in the most general accord with the available observations:

1) Coherence of the high-grade terrain. The N-S deformation and metamorphism are overprinted onto pre-existing craton in a narrow time interval of the latest Archean.

2) Prevalence of transcurrent motion offsets in the Craton. These may be either right-lateral or left-lateral. They are concentrated in zones associated with enhanced petrogenetic activity, including the Closepet Granite, some charnockitic hill ranges like the Biligirirangan and Shevaroy Hills, and zones of lower-grade metasomatic vein activity, as near Chitradurga and in the Kolar Belt.

3) Four distinct episodes of magmatic activity without significant overlap: tonalitic to trondjhemitic plutons at 3.4 to 3.0 Ga ago, ultramafic to mafic and lesser felsic lava flows at 3.0-2.6 Ga ago, calc-alkaline plutons in the Kolar area at 2.65-2.55 Ga ago, and K-rich Closepet-type granites at 2.53 Ga ago or somewhat later. All of these magma types require different tectonic settings.

4) Copious action of fluids in the crust during the latest Archean events.

PLATE TECTONIC ANALOGY

The many lines of evidence for horizontal tectonism virtually require some form of plate motions. It is apparent that some of the accretionary and orogenic products in the Dharwar Craton were different from those ascribed to modern plate tectonics. The massive low-K gneisses, ultramafic lavas and evidence for extreme metasomatism of the crust are nearly unrepresented in post-Archean times. Pervasive deformation of a continental interior, pictured here in the large-scale shearing motions of the low-grade Dharwar Craton, is not actually in keeping with the notion of rigid plates. Nevertheless, more modern analogies can be sought in attempting to explain features such as ensialic volcanic basins, calc-alkaline arc magmatism, and regional metamorphism, all of which features have been treated by concepts of modern plate tectonics.

E-W VERSUS N-S CONVERGENCE

The principle N-S alignment of fold axes in the Dharwar Craton has suggested to Radhakrishna and Naqvi (1986) and Krogstad et al. (1989) that convergence in the Craton as a whole and in the Kolar area, respectively, was east to west. This evidence is supported by some lineations in the Chitradurga area which have vertical components suggestive of steep thrusts (Chadwick et al., 1981). Hanson et al. (1988) and Krogstad et al. (1989) recognized transcurrent faults, but Radhakrishna and Naqvi (1986) did not emphasize this possibility. In contrast, Drury (1983), Drury et al. (1984) and

Gopalakrishna et al. (1986) suggested dominantly N-S convergence, largely in order to explain the position of the high-grade terrain as a collisional zone consistently in the southern portion of the Craton.

The principal difficulties with E-W convergence are those of explaining transcurrent motions at high angles to the direction of convergence, and alignment of metamorphic isograds at low angles to convergence. The latter problem is more fundamental, since more modern orogens commonly display isograds generally parallel to overthrusts and nappe axes, as in the Blue Ridge of the southern Appalachians (Carpenter, 1970). On the other hand, large transcurrent motions generally parallel to the inferred direction of convergence have been suggested for the Pan-African collision in the western Arabian Shield (Stoeser and Camp, 1985: their Fig. 6c). A third possibility is convergence oblique to the north-south shear motions. The present author's preference for north-south convergence in the Dharwar Craton, following Drury (1983), is based largely on the north-south metamorphic gradient in the Craton. The north-south elongation of the Craton is tentatively attributed to large-scale transcurrent faulting, in the form of intracontinental transform faults, as discussed below. These transforms coincide with the zones of high strain of Drury and Holt (1980), and may reconcile N-S elongation with N-S convergence in the same way that segments of ocean floor move independently along parallel transform faults in the direction of spreading.

OBDUCTION VERSUS SUBDUCTION

The presence of apparently juvenile late Archean continental terrains in the east Kolar, Krishnagiri and Nilgiri Hills areas suggests that they were formed on oceanic crust (i.e. were island arcs) which were sutured to the Craton at about 2.5 Ga ago. As a possible analogy, successive coalescence of island arcs was considered to have built up the Archean Superior Province of southern Canada by Langford and Morin (1976). Obduction (overthrusting) is the accretion mechanism most commonly advocated in Phanerozoic tectonics (e.g. Templeman-Kluit, 1979; Pudsey et al., 1985). However, Molnar and Gray (1979) would consider an immature island arc, with roots of less than about 15 km subductable, especially in fast convergence. The high-grade metamorphism of the granulites south of Krishnagiri could be explained either way: very thick nappe-stacking of obducted slices or wholesale subduction of the newly-formed terrains. The heat for metamorphism could have been supplied by a number of mechanisms, including late-stage magmatism, radioactive self-heating of a thickened crust (England and Bickle, 1984) or advection of fluids (Harris et al., 1982), perhaps from the deep subducted slab (Hoisch, 1987). The north-south gradients in paleotemperature and paleopressure, together with the apparent decrease in intensity of deformation northward, are easier to explain by top-loading from the south, with subsequent erosion and rebound, greatest for the thickened crust near the inferred continental margin. The north-south gradients thus seem more consistent with the obduction hypothesis.

The present north-south elongation of the inferred suture zone, the Kolar Belt, is suggested to be a consequence of the transform motions which succeeded the north-south convergence. The original strike of the collapsed Kolar basin is inferred to have been east-west.

TRANSITION FROM THRUSTING TO TRANSCURRENT FAULTING

Thickening of the southern margin of the Craton must have been accomplished by thrusting. However, little evidence of this has been brought forth from detailed field studies, and the paucity of demonstrable thrust traces would have to be explained by transcurrent motions as a principle cause of the second generation folds, the deformation sequence was overthrusting and nappe formation followed by large-scale strike-slip faulting. Seemingly similar transitions with time from overthrusting to strike-slip motions are not uncommon in Precambrian metamorphic terrains. Other examples are the Savonranta District of E. Finland (Halden and Bowes, 1984), the Buksefjorden area of S.W. Greenland (Chadwick and Nutman, 1979), the W. Uusimaa area of S. Finland (Schreurs and Westra, 1986), the northern Yilgarn Block of W. Australia (Platt, 1980), the Bamble region of S. Norway (Starmer, 1985), and the southern margin of the Limpopo Belt, southern Africa (McCourt and Vearncombe, 1987).

Various possible tectonic mechanisms for this kind of deformation sequence are conceivable. A change in the direction of sea-floor spreading may be a consequence of jamming a subduction zone by island arc collision. This mechanism could account for the change from westward underflow of North America in the Mesozoic to transform motion in the Cenozoic (Ernst, 1984). Oblique subduction is another possibility, in which a general northwest convergence direction in southern India could account for both thickening of the southern margin and N-S transcurrent motion. A third possibility is lateral extrusion between rigid colliding blocks (the "toothpaste effect"). A fourth possibility is rotation of earlier-formed strain features into the direction of convergence, as pictured by McLelland and Isachsen (1986) for the Adirondack Mountains high-grade terrain of New York. This idea seems quite consonant with the synthesis of Mukhopadhyay (1986) in which rotation of first-generation isoclinal structures was accomplished by second-generation deformation.

The concept of intracontinental transform faulting adds yet another possibility. Very large strike-slip motions in the North American Craton were explained with this concept by Hoffman (1988, p. 548) and in the Western Arabian Shield by Stern (1985, p. 497). In the latter example, the strike-slip motion of the Najd fault system, perhaps totalling 1200 km, is parallel to the inferred convergence direction of island arc accretion which preceded the transform faulting. Platt (1980) mapped rotated overthrust traces coincident with large strike-slip faults in the Yilgarn Block and suggested that the latter were intracontinental transform faults reflecting motions in the underlying mantle (p. 148). This idea is appealing for the Dharwar Craton in that it allows for a consistent N-S convergence direction, and both left-lateral and right-lateral transcurrent faults. The large length-scale of the transcurrent faults is consonant with a large depth-scale (through-crustal and perhaps through-lithosphere); such faults could provide conduits for fluids of deep-seated origin.

In the present tentative plate-tectonic reconstruction, the zones of high finite strain of Drury and Holt (1980) could be large continental transforms similar to the Nordre Stromfjord shear zone. This linear belt and those of the Bamble region and the Great Slave shear zone of northern Canada (Hanmer, 1988) are about 25 km wide in their highest-grade portions. The Nordre Stromfjord and Great Slave belts taper considerably in their lower-grade extensions, which may be ascribed to a more brittle response to the same forces (Hanmer, 1988). It is postulated here that the high-grade zones showing strike-slip offsets, such as the Closepet-Kabbal zone (Gopalakrishna et al., 1986) and its probable southward continuation in the charnockitic Biligirirangan Hills terrain, are such linear shear belts. This is identical to the picture presented by Drury and Holt (1980), in which thin zones of high strain become distributed over broader areas at deeper crustal levels (hence further south). Although some of the Closepet Granite is well-foliated, the smaller amount of deformation relative to adjacent banded gneisses suggests a late-kinematic emplacement.

ORIGIN OF METASOMATIC FLUIDS

The Kabbal-type transgressive charnockites, the K-rich Closepet Granite, and the lower-grade metasomatic shear veins all point to active fluids in the zones of high strain. This further enhances the analogy with the Nordre Stromfjord zone (Hickman and Glassley, 1984), where large quantities of fluid caused pronounced major and trace element and isotopic changes at granulite grade. The great length scales of the Closepet Granite and the high strain zones suggest commensurate depth scales, i.e. mantle depths. Numerous authors have shown that metasomatism at granulite grade must involve fluids rich in CO_2 (e.g. Hansen et al., 1984). It is not likely that a sufficient crustal source of CO_2 exists, which suggests mantle sources. Large mantle sources of CO_2 are implicated in alkaline magmatism (Rock, 1976), especially in LILE-enriched provinces (Menzies et al., 1985), and in low- to medium-grade metasomatized fault zones such as the South Tien Shan carbonated zone (Baratov et al., 1984), the Tatar Deep Fault of Siberia (Lapin et al., 1987) and the Umberatana Zone of southern Australia (Lottermoser, 1988). All cited authors have implicated enriched mantle or asthenospheric sources of fluids. The nearly invariable association of lamprophyres with Archean gold deposits, and the mantle-like carbon-isotope ratios of these deposits, suggest a mantle source of Au-depositing fluids (Groves et al., 1988; Rock and Groves, 1988; Wyman and Kerrich, 1988).

The enriched-mantle-asthenosphere source must have been replenished in volatiles in areas like the Dharwar Craton where previous episodes of basaltic and komatiitic volcanism would have depleted it. The most probable replenishing agency was subduction of surface volatiles in the form of hydrothermally altered oceanic crust and marine carbonate. Koster van Groos (1988) estimated that the rate of CO_2 recycling in the late Archean was several times that of today, which has been estimated to be of the order of 10^{14} gm/yr. This is a much larger source for metasomatic fluids than any conceivable crustal source in the form of limestone bodies, which are generally a small component of high-grade terrains.

POSSIBLE SEQUENCE OF CRATON-BUILDING EVENTS

Figure 4 suggests a sequence of events consistent with N-S plate convergence over a protracted period of the Archean. Northward underflow of oceanic lithosphere for most of this period was culminated by island arc formation and collision. It is acknowledged that consistent northward subduction over a period of nearly one billion years is unlikely simplicity in what was undoubtedly a complex mantle convection regime. However, the northward convergence postulated for the late Archean orogeny ties in with the scheme suggested by Drury (1983) for the volcanics of the Dharwar basins. In the period 3.4 to 3.0 Ga ago, low-angle subduction with remelting of hydrated ocean crust (amphibolite) generated large quantities of low-K granitoids, the Peninsular Gneiss. These materials were initially low in radioactive components and were further depleted in U at deep levels of the West Block by upward passage of fluids associated with the late stages of the magmatism at 3.0 Ga ago. This low radioactivity is considered to be a primary reason for the long-term stability of the West Block of the Dharwar Craton. The East Block of the Craton, which lacks large volcanic-sedimentary basins and which was rejuvenated to a greater degree by the terminal Archean events, may not have been as extensively processed by the ~3.0 Ga events and may have evolved somewhat independently of the West Block.

Steepening of the subduction zones, perhaps as a consequence of the newly-accreted continental mass, resulted in generation of tholeiites at a deep level, and komatiites at a still deeper level. Upward mantle motions associated with the magmatism could have put the Craton under tension, allowing for fast rise and extrusion of the mafic-ultramafic magmas, and the formation of intracratonal basins. Erosional debris of the volcanic rocks contributed to the basin fill. Drury (1983) envisioned the tectonic setting of the Dharwar basins as analogous to modern marginal basins.

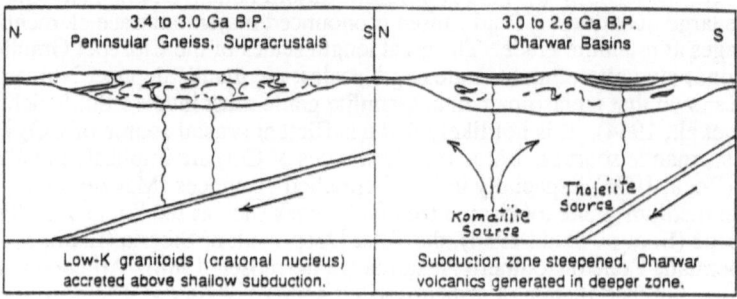

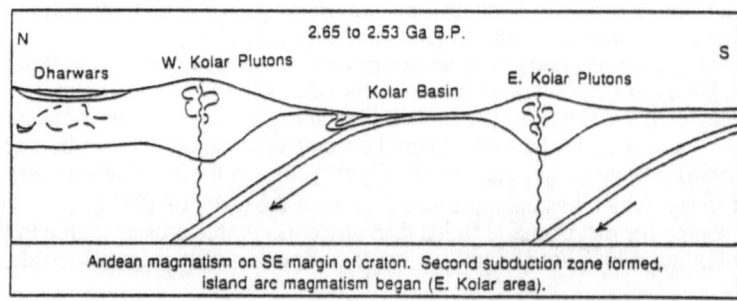

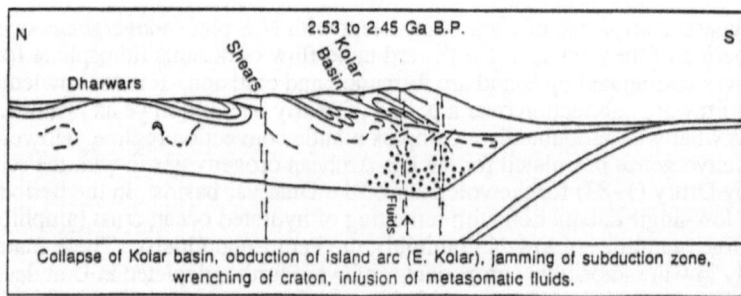

Figure 4. Interpretation of the evolution of the Dharwar Craton in terms of consistent northward subduction in the late Archean.

Andean margin magmatism, followed by island arc magmatism, ensued in the late Archean in the Kolar area, as suggested in Fig. 4. The Kolar Schist Belt is interpreted by Krogstad et al. (1989) as an intervening ocean basin. Figure 4 suggests tandem subduction zones such as are pictured for the Newfoundland Appalachians by Colman-Saad (1980). By this means the East Kolar arc could have approached the West Kolar Andean margin and superseded it. Obduction of the island arc close to 2.5 Ga ago could have terminated the magmatism.

Thrust slices of the obducted arc thickened the continental margin and initiated deep-seated metamorphism. It is possible that the transition from overthrusting to transcurrent motions resulted from increased resistance to deformation of the continental margin after extreme thickening. Figure 4 suggests that fluids of deep provenance, either from a subducted slab, from enriched lower lithosphere, or from the asthenosphere, were tapped by the shearing motions. These fluids were instrumental in charnockitic metamorphism of the lower crust, metasomatic melting to form the Closepet Granite in the middle crust, and emplacement of metasomatic veins, including the Kolar gold ores, in this middle to upper crust.

Implications for the Deep Structure of the Craton

Figure 5 is an interpretation of the structure of the Craton, including a SE to NW vertical section, consistent with the foregoing tectonics discussion. The basic framework of the Craton is thought to be the massive gneisses with their supracrustal enclaves. Figure 5 suggests that the lower portions of the crust are somewhat more mafic than gneisses of the exposed low-grade terrain, from original fractionation of hornblende and pyroxenes. The

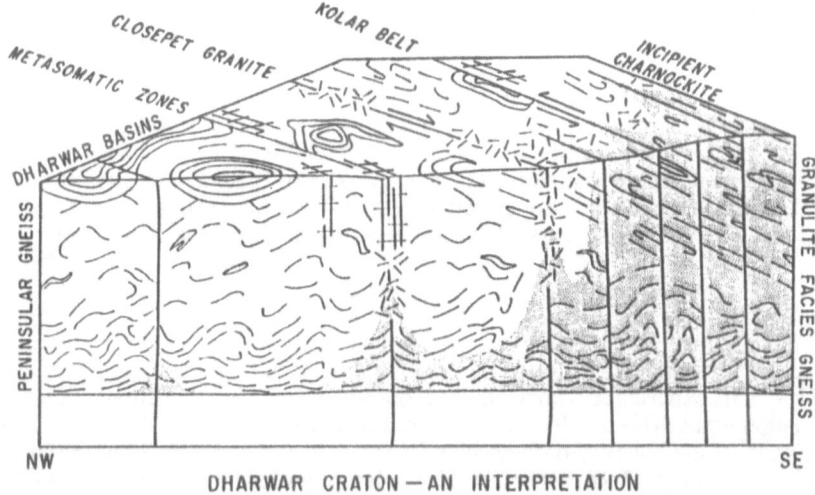

Figure 5. Interpretation of the deep structure of the Dharwar Craton. Right-lateral shears in the southern Closepet Granite region based on the author's observations. Left-lateral shears in the Kolar region from Hanson et al. (1988). The schematic diagram suggests that the highly deformed and metasomatized charnockitic gneisses result from a disturbance at the southern margin of the Craton, possibly island arc obduction, and that the deep structure of the northern cratonal interior may not resemble the charnockitic terrain.

overthrusted SE terrain includes metamorphosed original cratonal elements as well as obducted juvenile terrain, and the crust is somewhat thicker there, as shown by isostatic

320

compensation of the charnockitic hill ranges (Subrahmanyam, 1978). Deformation of cratonal gneiss and Dharwar basins decreases northward away from the high-grade terrain. Shearing on near-vertical N-S planes transposes the foliation associated with the earlier thrusting. Refolding of the Dharwars, increasing in intensity towards the south, results in dome-and-basin patterns and attenuated outcrops. The density of shear surfaces is suggested to be greater towards the south. The shears control the location of metasomatic charnockite, which pervades the Craton at the southern margin, and of granite and metasomatic veins at successively higher crustal levels.

Figure 5 suggests that it is unlikely that the intensely-deformed high-grade terrain exposed in the southern portion of the Craton passes beneath it as its normal roots, but suggests rather that the high-grade terrain is a marginal alteration of the Craton produced by continental collision. There is no compelling reason to believe that the cratonal roots are normally charnockitic, though they may contain pyroxene cumulates of their magmatic stage. Widespread upward motion of U at 3.0 Ga ago could have contributed to the stability of the cratonal roots, but does not necessarily signify a granulite-facies event of the magnitude of the event 2.5 Ga ago.

The highly tectonized southern portion of the Dharwar Craton was susceptible to subsequent rejuvenation during the Proterozoic. The interior cratonal root may have been, from its inception, a more massive and inert entity which has not been disturbed by comparable tectonism. Such early Archean sialic nucleii may not have been emulated in younger continental accretion. It may be also true that greater fluid activity during crustal processes in the Archean was a factor which partially accounts for the stabilities of the ancient cratons, in promoting upward movement of radioactivity and H_2O from deeper levels. Such fluid streaming might have occurred during massive crustal accretion in volatile exsolution from gneiss precursors, as pictured for the Nuk Gneiss of Greenland by Wells (1979), or during subsequent mantle degassing below the continents. If the CO_2 recycling rate was much greater in the Archean than at present (Koster van Groos, 1988), an additional continental "ripening" factor was present which has not operated on the same scale in the more recent past. It may be that episodes of fluid activity associated with early crustal accretion are to be distinguished in their effects from episodes of fluid activity associated with orogenic events which subsequently modified the cratonal margins.

Acknowledgements

The author's research is supported by a National Science Foundation grant, #EAR-8707156. His experience with southern India was sponsored by National Science Foundation grant #INT-8219140. Much benefit was derived from discussions with participants of the NATO Conference on Exposed Deep Crustal Sections, Killarney, Ontario, September 17-27, 1988. David Rowley provided helpful information on modern tectonics and must not be blamed if it is applied wrongly here. Helpful criticisms of two anonymous reviewers are acknowledged. In particular, the 19 critical comments of one reviewer created substantial revision of this paper.

References

Anhaeusser, C.R., Mason, R., Viljoen, M.J. and Viljoen, R.P., 1969, 'A reappraisal of some aspects of Precambrian shield geology.' *Geol. Soc. Amer. Bull.* **80**, 2175-2200.

Ballard, S. and Pollack, H.N., 1988, 'Modern and ancient geotherms beneath southern Africa.' *Earth, Plan. Sci. Lett.* **88**, 132-142.

Baratov, R.B., Gnutenko, N.A. and Kuzemko, V.N., 1984, 'Regional carbonization connected with the epi-hercynian tectonogenesis in the southern Tien Shan.' *Dokl. Akad. Nauk S.S.S.R.* **274**, 124-126.

Beach, A., 1976, 'The interrelations of fluid transport, deformation, geochemistry and heat flow in early Proterozoic shear zones in the Lewisian complex.' *Phil. Trans. Roy. Soc. Lond.* **A-280**, 569-604.

Bernard-Griffiths, J., Jahn, B.-M. and Sen, S.K., 1987, 'Sm-Nd isotopes and REE geochemistry of Madras granulites, India: an introductory statement.' *Precamb. Res.* **37**, 343-355.

Buhl, D., Grauert, B. and Raith, M., 1983, 'U-Pb zircon dating of Archean rocks from the South Indian Craton: results from the amphibolite to granulite facies transition zone at Kabbal Quarry, southern Karnataka.' *Fort. Mineral.* **61**, 43-45.

Burwash, R.A. and Krupicka, J., 1970, 'Cratonic reactivation in the Precambrian basement of western Canada. Part II. Metasomatism and isostacy.' *Canad. J. Earth Sci.* **7**, 1275-1294.

Callahan, E.J. and Rogers, J.J.W., 1987, 'Thorium and uranium contents of gneisses and trondhjemites in the Western Dharwar Craton, India.' *Canad. J. Earth Sci.* **24**, 934-940.

Carpenter, R.A., 1970, 'Metamorphic history of the Blue Ridge Province of Tennessee and North Carolina.' *Geol. Soc. Amer. Bull.* **81**, 749-762.

Chadwick, B. and Nutman, A.P., 1979, 'Archaean structural evolution in the northwest of the Buksefjorden Region, southern West Greenland.' *Precamb. Res.* **9**, 199-226.

Chadwick, B., Ramakrishnan, M. and Viswanatha, M.N., 1981, 'The stratigraphy and structure of the Chitradurga Region: an illustration of cover-basement interaction in the late Archaean evolution of the Karnataka Craton, Southern India.' *Precamb. Res.* **16**, 31-54.

Colman-Saad, S.P., 1980, 'Geology of south-central Newfoundland and evolution of the eastern margin of Iapetus.' *Amer. J. Sci.* **280**, 991-1017.

Condie, K.C., Allen, P. and Narayana, B.L., 1982, 'Geochemistry of the Archean low- to high-grade transition zone, southern India.' *Contr. Min. Pet.* **81**, 157-167.

Devaraju, T.C. and Sadashivaiah, M.S., 1969, 'The charnockites of Satnur-Halaguru area, Mysore State.' *Ind. Mineral.* **10**, 67-88.

Drury, S.A., 1983, 'The petrogenesis and setting of Archaean metavolcanics from Karnataka State, South India.' *Geochim. Cosmochim. Acta* **47**, 317-330.

Drury, S.A., Harris, N.B.W., Holt, R.W., Reeves-Smith, G.J. and Wightman, R.T., 1984, 'Precambrian tectonics and crustal evolution in South India.' *J. Geol.* **92**, 3-20.

Drury, S.A. and Holt, R.W., 1980, 'The tectonic framework of the South Indian Craton: a reconnaissance involving landsat imagery.' *Tectonophys.* **65**, T1-T15.

Drury, S.A., Holt, R.W., Van Calsteren, P.C. and Beckinsale, R.D., 1983, 'Sm-Nd and Rb-Sr ages for Archean rocks from western Karnataka, South India.' *J. Geol. Soc. India* **24**, 254-259.

England, P.C. and Bickle, M.J., 1984, 'Continental thermal and tectonic regimes during the Archaean.' *J. Geol.* **92**, 353-368.

Ernst, W.G., 1984, 'Californian blueschists, subduction, and the significance of tectonostratigraphic terranes.' *Geology* **12**, 436-440.

Friend, C.R.L., 1981, 'The timing of charnockite and granite formation in relation to influx of CO_2 at Kabbaldurga, Karnataka, South India.' *Nature* **294**, 550-552.

Friend, C.R.L., 1983, 'The link between charnockite formation and granite production: evidence from Kabbaldurga, Karnataka, Southern India.' In Atherton, M.P. and Gribble, C.D., Eds., *Migmatites, Melting and Metamorphism.* Nantwich (U.K.): Shiva Publ. Co., pp. 264-326.

Friend, C.R.L., 1984, 'The origins of the Closepet granites and the implications for the crustal evolution of Southern Karnataka.' *J. Geol. Soc. India* **25**, 73-84.

Fyfe, W.S., 1973, 'The generation of batholiths.' *Tectonophys.* **17**, 273-283.

Gopalakrishna, D., Hansen, E.C., Janardhan, A.S. and Newton, R.C., 1986, 'The southern high-grade margin of the Dharwar Craton.' *J. Geol.* **94**, 247-260.

Grew, E.S. and Manton, W.I., 1984, 'Age of allanite from Kabbaldurga Quarry, Karnataka.' *J. Geol. Soc. India* **25**, 193-195.

Groves, D.I., Golding, S.D., Rock, N.M.S., Barley, M.E. and McNaughton, N.J., 1988, 'Archaean carbon reservoirs and their relevance to the fluid source for gold deposits.' *Nature* **331**, 254-257.

Halden, N.M. and Bowes, D.R., 1984, 'Metamorphic development of cordierite-bearing layered schist and mica schist in the vicinity of Savonranta, eastern Finland.' *Geol. Soc. Finland Bull.* **56**, 3-23.

Hamilton, J.V. and Hodgson, C.J., 1986, 'Mineralization and structure of the Kolar Gold Field, India.' In MacDonald, A.J., Ed., *Gold '86.* Willowdale, Ontario: Konsult Int. Inc., pp. 270-283.

Hanmer, S., 1988, 'Great Slave Lake Shear Zone, Canadian Shield: reconstructed vertical profile of a crustal-scale fault zone.' *Tectonophys.* **149**, 245-264.

Hansen, E.C., Janardhan, A.S., Newton, R.C., Prame, W.K.B.N. and Ravindra Kumar, G.R., 1987, 'Arrested charnockite formation in southern India and Sri Lanka.' *Contr. Min. Pet.* **96**, 225-244.

Hansen, E.C., Newton, R.C. and Janardhan, A.S., 1984, 'Fluid inclusions in rocks from the amphibolite-facies gneiss to charnockite progression in southern Karnataka, India: Direct evidence concerning the fluids of granulite metamorphism.' *J. Meta. Geol.* **2**, 249-264.

Hanson, G.N., Krogstad, E.J. and Rajamani, V., 1988, 'Tectonic setting of the Kolar Schist Belt, Karnataka, India.' *J. Geol. Soc. India* **31**, 40-42.

Harris, N.B.W., Holt, R.W. and Drury, S.A., 1982, 'Geobarometry, geothermometry, and late Archean geotherms from the granulite facies terrain of South India.' *J. Geol.* **90**, 509-528.

Harris, N.B.W. and Jayaram, S., 1982, 'Metamorphism of cordierite gneisses from the Bangalore region of the Indian Archean.' *Lithos* **15**, 89-98.

Harrison, T.M., Watson, E.B. and Rapp, R.P., 1986, 'Does anatexis deplete the lower crust in heat-producing elements?: Implications from experimental studies.' *EOS* **67**, 386.

Heier, K.S., 1973, 'Geochemistry of granulite facies rocks and problems of their origin.' *Phil. Trans. R. Soc. Lond.* A-**273**, 429-442.

Hickman, M.H. and Glassley, W.E., 1984, 'The role of metamorphic fluid transport in the Rb-Sr isotopic resetting of shear zones: evidence from Nordre Strømfjord, West Greenland.' *Contr. Min. Pet.* **87**, 265-281.

Hoffman, P.F., 1988, 'United plates of America, the birth of a craton.' *Ann. Rev. Earth Planet. Sci.* **16**, 543-603.

Hoisch, T.D., 1987, 'Heat transport by fluids during channelized flow and thermal consequences for regional metamorphism.' *Geol. Soc. Amer. Abstr. w/Prog.* **19**, 704.

Hörmann, P.K., Raith, M., Raase, P., Ackermand, D. and Seifert, F., 1980, 'The granulite complex of Finnish Lappland: petrology and metamorphic conditions in the Ivalojoki-Inarijärvi area.' *Geol. Surv. Finland Bull.* **308**, 1-95.

Janardhan, A.S., Newton, R.C. and Hansen, E.C., 1982, 'The transformation of amphibolite facies gneiss to charnockite in southern Karnataka and northern Tamil Nadu, India.' *Contr. Min. Pet.* **79**, 130-149.

Kaila, K.L., Roy Chowdury, K., Reddy, P.R., Krishna, V.G., Narain, H., Subbotin, S.I., Sollogub, V.B., Chekunov, A.V., Kharetchko, G.E., Lazarenko, M.A. and Ilchenko T.V., 1979, 'Crustal structure along Kavali-Udipi profile in the Indian peninsular shield from deep seismic sounding.' *J. Geol. Soc. India* **20**, 307-333.

Karunakaran, C., 1974, 'Geology and mineral resources of the states of India. Pt. VI: Tamil Nadu and Pondicherry.' *Geol. Surv. India Misc. Publ.* **30**, 1-28.

Koster van Groos, A.F., 1988, 'Weathering, the carbon cycle, and differentiation of the continental crust and mantle.' *J. Geophys. Res.* **93**, 8952-8958.

Krogstad, E.J., Hanson, G.N. and Rajamani, V., 1988, 'U-Pb ages and gneisses near the Kolar Schist Belt: Evidence for the juxtaposition of discrete Archean terranes.' *J. Geol. Soc. India* **31**, 60-62.

Krogstad, E.J., Balakrishnan, S., Mukhopadhyay, D.K., Rajamani, V. and Hanson, G.N., 1989, 'Plate tectonics 2.5 billion years ago: Evidence at Kolar, South India.' *Science* **243**, 1337-1340.

Kröner, A., 1980, 'New aspects of craton-mobile belt relationships in the Archaean and Early Proterozoic: examples from southern Africa and Finland.' In Closs, H., Gehlen, K.v., Illies, H., Kuntz, E., Neumann, J. and Seibold, E., Eds., *Mobile Earth*. Harald Boldt Verlag, pp. 225-234.

Kröner, A., 1984, 'Changes in plate tectonic styles and crustal growth during the Precambrian.' *Soc. géol. France Bull.* **26**, 297-319.

Lamb, R.C., Smalley, P.C. and Field, D., 1986, 'P-T conditions for the Arendal granulites, southern Norway: implications for the roles of P, T and CO_2 in deep crustal LILE-depletion.' *J. Meta. Geol.* **4**, 143-160.

Langford, F.F. and Morin, J.A., 1976, 'The development of the Superior Province of northwestern Ontario by merging island arcs.' *Amer. J. Sci.* **276**, 1023-1034.

Lapin, A.V., Ploshko, V.V. and Malyshev, A.A., 1987, 'Carbonatites of the Tatar deep-seated fault zone, Siberia.' *Int. Geol. Rev.* **29**, 551-567.

Lottermoser, B.G., 1988, 'A carbonatite diatreme from Umberatana, South Australia.' *J. Geol. Soc. London* **145**, 505-514.

Mahabaleswar, B. and Naganna, C., 1981, 'Geothermometry of Karnataka charnockites.' *Bull. Minéral*. **104**, 848-855.

Mahabaleswar, B., Vasant Kumar, I.R. and Friend, C.R.L., 1986, 'Geochemistry of the Archaean gneiss complex and associated rocks of the Kanakapura area, Karnataka, South India.' *J. Geol. Soc. India* **27**, 282-297.

McCourt, S. and Vearncombe, J.R., 1987, 'Shear zones bounding the central zone of the Limpopo Mobile Belt, southern Africa.' *J. Struct. Geol.* **9**, 127-137.

McLelland, J.M. and Isachsen, Y.W., 1986, 'Synthesis of geology of the Adirondack Mountains, New York, and their tectonic setting within the southwestern Grenville Province.' *Geol. Assoc. Canada Spec. Pap.* **31**, 75-94.

Menzies, M., Kempton, P. and Dungan, M., 1985, 'Interaction of continental lithosphere and asthenospheric melts below the Geronimo Volcanic Field, Arizona, USA.' *J. Petrol.* **26**, 663-693.

Molnar, P. and Gray, D., 1979, 'Subduction of continental lithosphere: Some constraints and uncertainties.' *Geology* **7**, 58-62.

Mukhopadhyay, D., 1986, 'Structural pattern in the Dharwar Craton.' *J. Geol.* **94**, 167-186.

Mustard, P.S. and Donaldson, J.A., 1987, 'Early Proterozoic ice-proximal glaciomarine deposition: The lower Gowganda Formation at Cobalt, Ontario, Canada.' *Geol. Soc. Amer. Bull.* **98**, 373-387.

Naha, K., Srinivasan, R. and Naqvi, S.M., 1986, 'Structural unity in the early Precambrian Dharwar Tectonic Province, Peninsular India.' *Quart. J. Geol. Min. Met. Soc. India* **58**, 219-243.

Naqvi, S.M. and Rogers, J.J.W., Eds.,1983, 'The Precambrian of South India.' *Geol. Soc. India Mem.* **4**, 1-575.

Nesbitt, H.W., 1980, 'Genesis of the New Quebec and Adirondack granulites: evidence for their production by partial melting.' *Contr. Min. Pet.* **72**, 303-310.

Newton, R.C. and Anderson, A.T., Eds., 1986, 'The Dharwar Craton of South India: An Archean protocontinent.' *J. Geol.* **94**, 127-299.

Olsen, S.N., 1985, 'Mass balance in migmatites.' In Ashworth, J.R., Ed., *Migmatites.* Glasgow and London: Blackie, pp. 145-179.

Percival, J.A. and Card, K.D., 1983, 'Archean crust as revealed in the Kapuskasing uplift, Superior province, Canada.' *Geology* **11**, 323-326.

Peucat, J.J., Vidal, P.H., Bernard-Griffiths, J. and Condie, K.C., 1987, 'Sr, Nd, and Pb systems across the amphibolite to granulite facies transition in southern India.' *Terra Cognita* **7**, 333.

Pichamuthu, C.S., 1965, 'Regional metamorphism and charnockitization in Mysore State, India.' *Ind. Mineral.* **6**, 119-126.

Pichamuthu, C.S. and Srinivasan, R., 1984, 'The Dharwar Craton.' *Ind. Nat. Sci. Acad., Perspective Report Ser.* **7**, 3-34.

Platt, J.P., 1980, 'Archaean greenstone belts: a structural test of tectonic hypotheses.' *Tectonophys.* **65**, 127-150.

Pride, C. and Muecke, G.K., 1980, 'Rare earth element geochemistry of the Scourian Complex, N.W. Scotland - Evidence for the granite-granulite link.' *Contr. Min. Pet.* **73**, 403-412.

Pudsey, C.J., Coward, M.P., Luff, I.W., Shackleton, R.M., Windley, B.F. and Jan, M.Q., 1985, 'Collision zone between the Kohistan arc and the Asian plate in NW Pakistan.' *Trans. Roy Soc. Edinburgh, Earth Sci.* **76**, 493-479.

Raase, P., Raith, M., Ackermand, D. and Lal, R.K., 1986, 'Progressive metamorphism of mafic rocks from greenschist to granulite facies in the Dharwar Craton of South India.' *J. Geol.* **94**, 261-282.

Radhakrishna, B.P., 1958, 'On the nature of certain brick-red zones in Closepet granite.' *Mysore Geol. Dept. Records* **48**, 61-73.

Radhakrishna, B.P. and Naqvi, S.M., 1986, 'Precambrian continental crust of India and its evolution.' *J. Geol.* **94**, 145-166.

Raith, M., Hengst, C., Nagel, B., Bhattacharya, A. and Srikantappa, C., 1988, 'Metamorphic conditions in the Nilgiri granulite terrane and the adjacent Moyar and Bhavani shear zones: a reevaluation.' *J. Geol. Soc. India* **31**, 112-113.

Rock, N.M.S., 1976, 'The role of CO_2 in alkali rock genesis.' *Geol. Mag.* **113**, 97-192.

Rock, N.M.S. and Groves, D.I., 1988, 'Can lamprophyres resolve the controversy over mesothermal gold deposits?' *Geology* **16**, 538-541.

Rogers, J.J.W., 1986, 'The Dharwar Craton and the assembly of Peninsular India.' *J. Geol.* **94**, 129-144.

Rogers, J.J.W., 1988, 'The Arsikere Granite of southern India: magmatism and metamorphism in a previously depleted crust.' *Chem. Geol.* **67**, 155-163.

Saleeby, J.B., 1983, 'Accretionary tectonics of the North American cordillera.' *Ann. Rev. Earth Planet. Sci.* **15**, 45-73.

Schreurs, J. and Westra, L., 1986, 'The thermotectonic evolution of a Proterozoic, low pressure, granulite dome, West Uusimaa, SW Finland.' *Contr. Min. Pet.* **93**, 236-250.

Shackleton, R.M., 1976, 'Shallow and deep-level exposures of Archaean crust in India and Africa.' In Windley, B.F., Ed., *The Early History of the Earth.* Wiley-Interscience, pp. 317-321.

Silver, P.G. and Chan, W.W., 1988, 'Implications for continental structure from seismic anisotropy.' *Nature* **335**, 34-39.

Sorensen, K., 1983, 'Growth and dynamics of the Nordre Stromfjord shear zone.' *J. Geophys. Res.* **88**, 3419-3438.

Spooner, C.M. and Fairbairn, H.W., 1970, 'Strontium 87/strontium 86 initial ratios in pyroxene granulite terranes.' *J. Geophys. Res.* **75**, 6706-6713.

Srinivasan, R. and Ojakangas, R.W., 1986, 'Sedimentology of quartz-pebble conglomerates and quartzites of the Archean Bababudan Group, Dharwar Craton, South India: Evidence for early crustal stability.' *J. Geol.* **94**, 199-214.

Stähle, H.J., Raith, M., Hoernes, S. and Delfs, A., 1987, 'Element mobility during incipient granulite formation at Kabbaldurga, southern India.' *J. Petrol.* **28**, 803-834.

Starmer, I.C., 1985, 'The evolution of the South Norwegian Proterozoic as revealed by the major and mega-tectonics of the Kongsberg and Bamble sectors.' In Tobi, A.C. and Touret, J.L.R., Eds., *The Deep Proterozoic Crust in the North Atlantic Provinces.* Dordrecht: Reidel, pp. 259-290.

Stern, R.J., 1985, 'The Najd Fault System, Saudi Arabia and Egypt: a Late Precambrian rift-related transform system.' *Tectonics* **4**, 497-511.

Stoeser, D.B. and Camp, V.E., 1985, 'Pan-African microplate accretion of the Arabian Shield.' *Geol. Soc. Amer. Bull.* **96**, 817-826.

Subrahmanyam, C., 1978, 'On the relation of gravity anomalies to geotectonics of the Precambrian terrains of the South Indian Shield.' *J. Geol. Soc. India* **19**, 251-263.

Subramaniam, A.P., 1967, 'Charnockites and granulites of southern India: a review.' *Dan. Geol. Foren.* **17**, 473-493.

Suryanarayana, K.V., 1960, 'The Closepet granite and associated rocks.' *Ind. Mineral.* **1**, 86-100.

Swami Nath, J., Ramakrishnan, M. and Viswanatha, M.N., 1976, 'Dharwar stratigraphic model and Karnataka craton evolution.' *Rec. Geol. Surv. India* **107**, 149-175.

Sykes, L.R., 1978, 'Intraplate seismicity, reactivation of preexisting zones of weakness, alkaline magmatism, and other tectonism postdating continental fragmentation.' *Rev. Geoph., Space Sci.* **16**, 621-688.

Taylor, P.N., Moorbath, S., Chadwick, B., Ramakrishnan, M. and Viswanatha, M.N., 1984, 'Petrography, chemistry, and isotopic ages of Peninsular gneiss, Dharwar acid volcanic rocks, and the Chitradurga granite with special reference to the late Archean evolution of the Karnataka craton, southern India.' *Precamb. Res.* **23**, 349-375.

Templeman-Kluit, D.J., 1979, 'Transported cataclasite, ophiolite and granodiorite in Yukon: evidence of arc-continent collision.' *Can. Geol. Surv. Pap.* **79-14**, 16-27.

Van Reenen, D.D., Roering, C., Smit, C.A., Barton, J.M. and Van Schalwyk, J.F, 1988, 'The high-grade margin of the northern Kaapvaal Craton, South Africa.' *J. Geol.* **96**, 549-560.

Weaver, B.L. and Tarney, J., 1983, 'Elemental depletion in Archaean granulite-facies rocks.' In Atherton, M.P. and Gribble, C.D., Eds., *Migmatites, Melting and Metamorphism*. Nantwich (UK): Shiva Publ., Ltd., pp. 250-263.

Wells, P.R.A., 1979, 'Chemical and thermal evolution of Archaean sialic crust, southern West Greenland.' *J. Petrol.* **20**, 187-226.

Winkler, H.G.F., 1976, *Petrogenesis of Metamorphic Rocks*. Heidelberg: Springer-Verlag, 4th Ed., 334 pp.

Wyllie, P.J., Huang, W.-L., Stern, C.R. and Maaløe, S., 1976, 'Granitic magmas: possible and impossible sources, water contents, and crystallization sequences.' *Canad. J. Earth Sci.* **13**, 1007-1019.

Wyman, D. and Kerrich, R., 1988, 'Alkaline magmatism, major structures, and gold deposits: Implications for greenstone belt gold metallogeny.' *Econ. Geol.* **83**, 454-461.

AN OBLIQUE CROSS SECTION OF ARCHEAN CONTINENTAL CRUST AT THE NORTHWESTERN MARGIN OF SUPERIOR PROVINCE, MANITOBA, CANADA

WERNER WEBER
Manitoba Geological Survey
535-330 Graham Avenue
Winnipeg, Manitoba
R3C 4E3
Canada

and

KLAUS MEZGER
Department of Earth and Space Sciences
State University of New York
Stony Brook, New York 11794
U.S.A.

ABSTRACT.   An oblique cross section of ca. 20 km of middle to lower Archean continental crust is exposed over a distance of ca. 100 km at the northwest margin of Superior Province in Manitoba.   U-Pb mineral ages from garnet and zircon indicate 5 periods of metamorphism in the lower crust (Pikwitonei granulite domain), at ca. 3000 Ma, 2744 - 2738 Ma, 2700 - 2687 Ma, 2660 - 2637 Ma and 2629 - 2591 Ma. These events have been tentatively correlated with distinct depositional/intrusive events in the adjacent upper crust (Cross Lake and Oxford-Knee-Gods Lake greenstone belts).

Pressure and temperature estimates based on experimentally calibrated geobarometers and geothermometers, textural information and zoning of garnets suggest peak conditions during the last interval of pervasive regional metamorphism at ca. 2640 Ma, from southeast to northwest: 575°C/3 kb at Atik Lake, 750°C/7 kb at Cauchon Lake, 830°C/7.5 - 8.0 kb at Natawahunan Lake, and 9 kb close to the Thompson belt. P-T-t paths derived from these data indicate that the major source for heating during metamorphism was the intrusion and underplating of hot magmas. In combination with tectonic compression these magmas lead to crustal thickening and cratonization. Subsequent late Archean to Early Proterozoic tectonic thinning in a continental margin setting, prior to rifting and onset of the Trans-Hudson orogeny, is the most likely cause for exposing the lower crust along the NW margin of the Superior Province.

1. Introduction

Seismic profiles indicate that continental crust shows a pronounced layering.   This layering of stable continental crust may be caused by changes in composition and/or metamorphic grade as a function of depth.   In order to evaluate the layering and to describe the processes that lead to the formation of a stable layered continental crust, it is essential to study possible changes in physical properties, composition as well as metamorphism within an intact crustal cross section.   Such cross sections through the

M. H. Salisbury and D. M. Fountain (eds.), Exposed Cross-Sections of the Continental Crust, 327–341.
© 1990 *Kluwer Academic Publishers.*

continental crust are rare but in the last decade have been recognized in several young and ancient orogenic belts (e.g. Fountain and Salisbury, 1981; this volume).

In this paper we will discuss the geologic history of an *oblique* cross section through ca. 20 km of middle to lower continental crust, exposed in north central Manitoba, Canada. The cross section is located along the northwestern margin of the Archean Superior Province (Fig. 1) where rocks show an increase in metamorphic grade from east to northwest from greenschist facies to upper amphibolite facies in the Gods Lake domain and to granulite facies in the Pikwitonei domain. Concomitant with the increase in metamorphic grade the northwest margin of the Superior Province shows a decrease in the ratio of supracrustal to granitoid rocks on a regional scale.

2. Geology of northwestern Superior Province

The Gods Lake domain (Fig. 1) comprises ca. 80% granitoid rocks and ca. 20% greenstone belts. The granitoids are mainly gneissic or foliated tonalites to granodiorites and minor porphyritic to equigranular granites. The supracrustal rocks form east trending, relatively thin (max. 50 km wide), continuous (max. 150 km long) and discontinuous belts and occur as inclusions of various sizes in granitoid rocks. Inclusions are particularly abundant in marginal zones of granitoid batholiths that intruded supracrustal rocks.

The larger greenstone belts of the Gods Lake domain comprise two major lithostratigraphic subdivisions (Table 1), an older (2830 - 2700 Ma) komatiitic-tholeiitic to calc-alkaline, subaqueous, volcanic-sedimentary succession and a younger (2710 - 2690 Ma) unconformably overlaying, subaerial, sedimentary and high-K volcanic succession. In the Hayes River drainage basin, these successions are subdivided into the Hayes River and Oxford Lake groups, respectively, while in the Cross Lake region these successions are defined as the Pipestone Lake and Cross Lake groups, respectively. U-Pb ages of supracrustal and magmatic rocks of these two greenstone belt areas are listed in Table 1. The 3.0 Ga platform succession (Thurston et al., 1987) has not been identified yet, but is suspected to occur based on the presence of 3.0 Ga old granitoid crust.

The Pikwitonei granulite domain comprises ca. 90% inclusion-rich enderbitic orthogneiss and enderbite, and ca. 10% discontinuous belts and trails of supracrustal rocks that are considered high grade equivalents of greenstones exposed in the Atik Lake, Cross Lake and Gods Lake greenstone belts. These belts in the Pikwitonei domain include mafic and minor ultramafic rocks, locally with preserved pillow structures, banded oxide and silicate facies iron formation, anorthosite and layered gabbro complexes, and siliceous, aluminous and calcareous supracrustal rocks. The latter two probably represent, at least in part, hydrothermally altered mafic volcanic rocks, in analogy to alterations described from Atik Lake in the Gods Lake domain by Weber (1974) and Bernier and MacLean (1989).

Toward the northwest the Pikwitonei domain grades into the Churchill-Superior boundary zone. This zone consists primarily of Pikwitonei-type crust that was overprinted and reworked during the Hudsonian orogeny (Weber and Scoates, 1978) as a result of collision between the Archean Superior Province craton and the juvenile Proterozoic magmatic arc of the Reindeer zone along the eastern margin of the Trans-Hudson orogen (Weber, in press).

Field mapping indicated that the gradient of regional metamorphism in the northwestern Superior Province is the result of a late Archean prograde metamorphic overprint (Weber, 1980, Weber and Scoates, 1978; Hubregtse, 1980). This interpretation was based on the observation that in the transition zone from upper amphibolite to

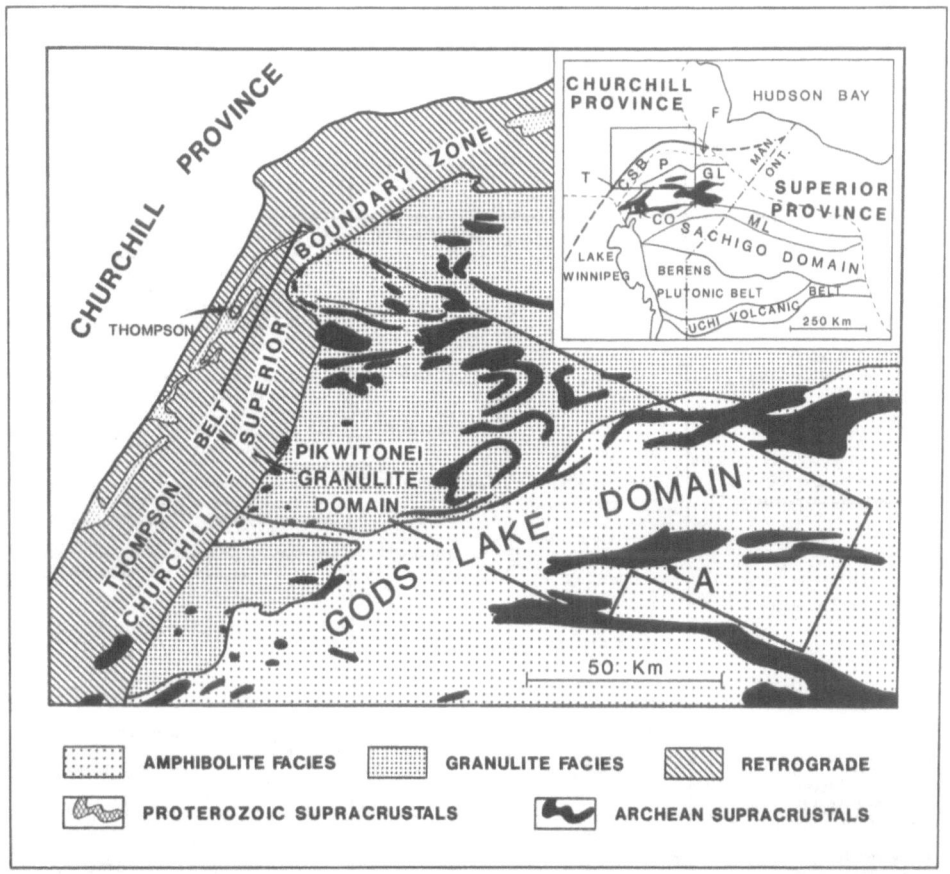

Fig. 1: Generalized geological map of the NW Superior Province (after Manitoba Mineral Resources, 1979). Areas in black represent Archean supracrustal units that are mixed with granitoid rocks in the Pikwitonei domain. The cross-hatched areas represent Proterozoic supracrustals (Ospagwan Group). Outlined area is enlarged in Fig. 2. **A** Atik Lake (formerly Utik Lake), **CSB** Churchill-Superior boundary zone, **C** Cross Lake greenstone belt, **F** Fox River belt, **GL** Gods Lake domain, **ML** Molson Lake domain, **O** Oxford-Knee-Gods Lake greenstone belt, **P** Pikwitonei domain, **T** Thompson belt.

TABLE 1: GEOLOGICAL EVENTS IN THE UPPER TO LOWER CRUST, NORTHWEST MARGIN OF SUPERIOR PROVINCE, MANITOBA

UPPER CRUST			LOWER TO MIDDLE CRUST	
Gods Lake domain (GL) Churchill-Superior boundary zone (CSB)	Cross Lake greenstone belt (C) Oxford-Knee-Gods Lake greenstone belt (O) Thompson belt (T) Fox River belt (F)		Pikwitonei domain (P) Churchill-Superior boundary zone (CSB) Atik Lake greenstone belt (A), northern Gods Lake domain (GL)	
Deposition, intrusion, extrusion	Age, Ma	Age, Ma	Metamorphic-magmatic event	
		z,m,t 1820-1720[8,9]	M5:	Collision (underthrusting) of Superior Province with Churchill Province during Trans-Hudson orogeny along CSB, overprinting of granulites in CSB. Maximum upper amphibolite facies (T):700°C/6-7 kb). Intrusion of granites and pegmatites.
Fox River sill, volcanics (F) * Molson dyke swarm (GL, P, CSB) Ospagwan group (T)	z 1883[1] i 1883[1] i ≥1883[2]	i 1883[2,8]		Uplift of 30 km of granulite (to surface) since 2600 Ma (CSB)
		z 2605-2591[10]	M4:	Amphibolite facies (P), hydrous granites, pegmatites; start of uplift
peraluminous alkaligranite (C), rare-element enriched pegmatites (C)	z 2653-2634[3,4]	g,z 2660-2637[10-13]	M3:	Amphibolite facies (A)600°C/3 kb; granulite facies (P):730°C/8 kb, partial melting.
Magill Lake granite (O)	z 2692[5]	g,z 2700-2687[10-12]	M2:	Upper amphibolite facies (P): 700°C intrusion of hot felsic magma, generation of partial melts
Cross Lake, Oxford Lake groups: sediments (O) sediments (C) shoshonitic volcanics (O)	i >2692[6] d ≤ 2709[3] z 2706[5]			
UNCONFORMITY				
tonalites (C)	z 2719-2747[3]	g 2744-2738[10]	M1:	Upper amphibolite to granulite facies (P): 800°C (locally).
tonalitic gneiss (O) calc-alkaline volcanics (C) Pipestone Lake group (C): anorthosites tholeiites Hayes River group (O): calc-alkaline volcanics tholeiites tonalitic gneiss (C) tonalitic gneiss (O)	z 2730[5] z 2732-2729[3,4] z 2758-2763[3] i >2763[3] z 2830[5] i >2830[5,6] z 2830[4] z 2883[5]			
granitoids (C,T)	z ~3000[3,7]	g 2984[10]	M0:	Amphibolite facies (P).

1 Heaman et al., 1986b; 2 Macek, 1987; 3 Corkey et al., 1988; 4 Davies, pers. comm., 1989; 5 Davis, 1987; 6 Hubregtse, 1985; 7 Machado et al., 1988; 8 Weber, in press; 9 Bleeker, in press; 10 Mezger et al., 1989; 11 Heaman et al.; 1986a; 12 Krogh et al., 1986.
Ages are U-Pb ages for zircon z, garnet g, monazite m, titanite t. Analytical errors of U-Pb ages are less than 3 Ma except reference 7.
i indirect age based on field relationship with a dated rock unit
d detrital single zircon age.
* Abbreviations in parentheses designate the domain or belt in which the lithology or event occurred.
Key to abbreviations in heading of Table 1 and Figure 1.

granulite facies late leuco-tonalitic mobilizates are observed cutting earlier planar structures in supracrustal and granitoid rocks. These mobilizates contain amphibole and clinopyroxene in the Gods Lake domain and orthopyroxene in the Pikwitonei domain. The boundary of the Pikwitonei domain with the Gods Lake domain was defined based on the appearance of orthopyroxene in these mobilizates. The orthopyroxene isograd roughly parallels the greenstone belt, but in several localities crosscuts the supracrustal belts at an oblique angle (e.g., Cauchon Lake; Fig. 1). The increasing metamorphic grade is the result of prograde metamorphism. The Pikwitonei-Gods Lake area is therefore a rare example of an Archean terrane where supracrustal belts can be traced almost continuously from greenschist into granulite facies. However, younger (late Archean or Proterozoic) shear zones and mylonite zones near the orthopyroxene isograd, such as the mylonite zone shown in Fig. 2 may be responsible for locally steepening the metamorphic gradient (see below).

Following the last Archean metamorphic episode and prior to the collision of the Superior Province with the Churchill Province, mafic to ultramafic dykes of the Molson dyke swarm intruded into rocks of the Superior Province (Scoates and Macek, 1978). These dykes range in width from a few centimeters to several tens of meters. They strike in a north-northeast direction and do not follow any metamorphic fabric. In the Churchill-Superior boundary zone these mafic to ultramafic dykes were overprinted by early Proterozoic, Hudsonian metamorphism.

These earlier interpretations of the metamorphism and the relative timing of events were based on geological mapping and have since been confirmed by geochronological studies as well as regional studies of the metamorphic pressures and temperatures (Paktunc and Baer, 1986; Mezger et al., 1989a, 1989b, 1990; Heaman et al., 1986a, b) and are discussed in more detail in this paper.

3. Tectonic history

The reconstruction of the tectonic history of the northwestern margin of the Superior Province is based mainly on regional mapping by Hubregtse and Weber (see Hubregtse, 1978, 1980, 1985; Weber, 1974, 1976, 1977, 1984; Weber and Scoates, 1978; and Corkery et al., 1988) supplemented by more recent geothermobarometry data by Mezger et al. (1990) and U-Pb geochronology data by Heaman et al. (1986a), Mezger et al. (1989a, 1989b), Davis (1987), and Corkery et al. (1988).

Granitoid rocks formed during several distinct periods at ca. 3000 Ma, 2883 - 2830 Ma, 2760 - 2719 Ma and 2692 - 2634 Ma. The earliest evidence for geologic activity along the NW margin of the Superior Province comes from scattered U-Pb zircon and garnet ages of ca. 3.0 Ga. These ages from granitoid rocks at Cross Lake (Corkery et al., 1988) Cauchon Lake (Mezger et al., 1989a) and the Thompson belt (Machado et al., 1988) indicate the presence of old (pre-greenstone?) sialic crust, the extent of which is unknown at present. This period was probably dominated by the formation of granitoids, and it was associated with metamorphism (M0) as documented by garnet xenocrysts in a peraluminous granite from Cauchon Lake and possibly deformation (D0). The extent and duration of this early episode are still poorly constrained.

U-Pb zircon ages indicate that most granitoids probably intruded between 2760 - 2719 Ma and are contemporaneous with mafic to felsic volcanism. Layered glomeroporphyritic gabbro to anorthosite and layered mafic-ultramafic complexes are cogenetic with the mafic volcanism. Metamorphism (M1) varying from greenschist to amphibolite facies, associated with the granitoid intrusions was identified in the Oxford Lake area (Hubregtse, 1985); kyanite rather than sillimanite formed during the highest

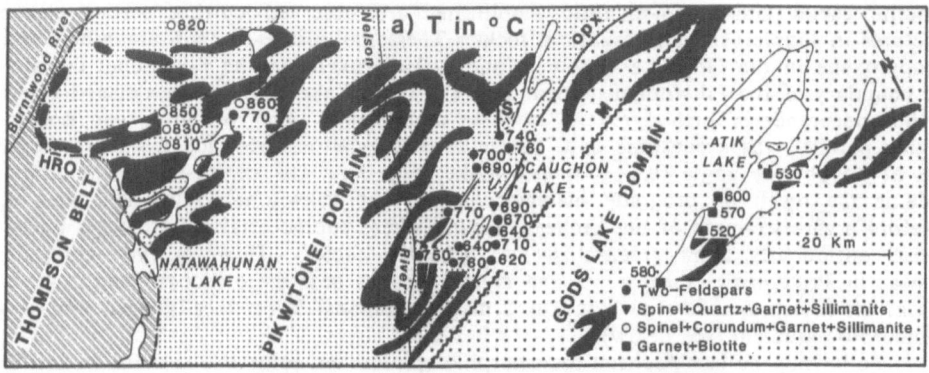

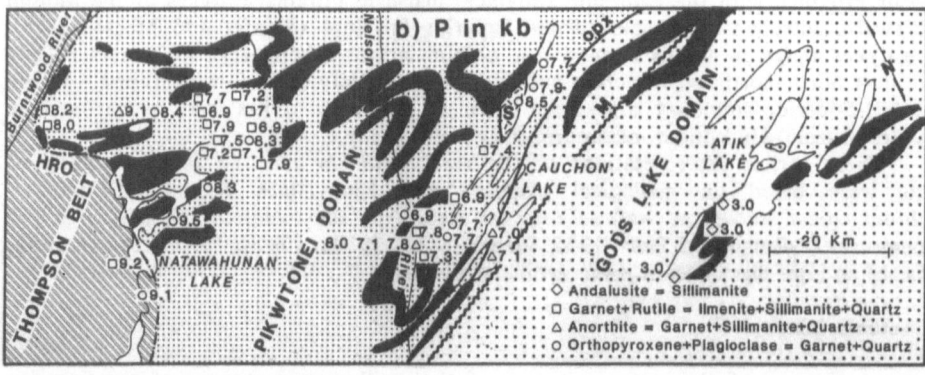

Fig. 2: Maps showing an enlarged area of Fig. 1, from Atik Lake in the southeast to the Burntwood River in the northwest. Temperatures during late-Archean metamorphism (Fig. 2a) increased from ca. 575°C at Atik Lake to ca. 830°C at Natawahunan Lake and pressures (Fig. 2b) increased from ca. 3 kb at Atik Lake to ca. 9 kb close to the Thompson belt of the Churchill-Superior boundary Zone where Archean granulites were overprinted and retrogressed during early Proterozoic Trans-Hudsonian orogeny. These increasing metamorphic conditions indicate that the area shown represents an oblique cross section through ca. 20 km of Archean crust. **HRO** limit of Hudsonian retrograde overprint; **opx** orthopyroxene isograd; **M** mylonite zone.

grade conditions indicating relatively high P-T ratio. At Cauchon Lake in the southern Pikwitonei domain (Fig. 1, 2), amphibolite facies conditions with temperatures >520°C prevailed at ca. 2740 Ma, and at Natawahunan Lake granulite facies conditions of 700 - 800°C are indicated by the reaction

staurolite + quartz = garnet + sillimanite + H₂O

at 2744 ± 2 Ma and the reaction:

staurolite = garnet + sillimanite + spinel + H₂O

at 2738 ± 5 Ma (Mezger et al., 1989a; Fig. 3). However, mineral assemblages, folds and planar structures related to this event are difficult to identify because they were largely overprinted and re-oriented by the subsequent tectonic episode.

The last major period of granitoid intrusions led to the formation of rare-element enriched pegmatites in the Cross Lake area and occurred from 2702 - 2650 Ma during and after high-K volcanism (Table 1). The synchronous dynamometamorphic episode (D2, M2; ca. 2700 - 2687 Ma) probably produced a large part of the metamorphic mineral assemblages in the Gods Lake domain and southern Pikwitonei domain. At Oxford Lake in the central Gods Lake domain, conditions ranged from greenschist to amphibolite facies, 635°C/4 kb (Hubregtse, 1985). At the same time titanite grew in amphibolites at Atik Lake indicating amphibolite facies conditions. At Cauchon Lake higher grade conditions (T>700°C) were reached at 2700 - 2687 Ma (Mezger et al., 1990; Fig. 3). A U-Pb zircon age of 2695 ± 5 Ma from an orthopyroxene-bearing leucosome (Heaman et al., 1986a) and ages of 2700 - 2687 Ma from metamorphic garnets (Mezger et al., 1989a) indicate that high grade conditions (above stability of staurolite + quartz) prevailed in the Cauchon Lake area at that time. The intrusion of schollen-enderbites is associated with the formation of voluminous orthopyroxene-bearing pegmatoids. Metasedimentary xenoliths in the schollen-enderbite contain inverted and exsolved pigeonites and ternary feldspars that indicate that the host magma had a temperature in excess of 1100°C and therefore could have been a major heat source that promoted high grade metamorphism (Mezger et al., 1990).

A third prograde dynamometamorphic event (D3, M3) is recognized in the Pikwitonei domain during 2660 - 2637 Ma (Heaman et al., 1986a; Krogh et al., 1986; Mezger et al., 1989). It produced characteristic minor and major folds with southeast trending axial traces (Hubregtse, 1980). Mobilizates composed of orthopyroxene + plagioclase ± quartz and containing typically high uranium zircons (U >2000 ppm) were emplaced into S3 axial planes and S2 layering structures. At Cauchon Lake this event led to the formation of extensive migmatites in metapelitic and mafic gneisses as well as in the surrounding enderbitic gneisses. In the amphibolite-granulite transition zone the rocks generally retained M2 amphibolite facies mineral assemblages but are crosscut by orthopyroxene-bearing S3 structures. This clearly documents that M3 is the highest grade event. Conditions in the amphibolite-granulite facies transition zone at Cauchon Lake were estimated to have been 700 - 750°C/7 kb at ca. 2640 Ma, and in the Natawahunan Lake area in the central Pikwitonei domain, conditions reached ca. 830°C at 7.5 - 8 kb, and 9 kb close to the Thompson belt (Mezger et al., 1990).

Although not readily recognizable because mineral assemblages were probably essentially formed during M2, this third event also led to peak conditions in the northern Gods Lake domain, e.g., at Atik Lake, where maximum conditions were ca. 575°C/3 kb and new titanites grew at 2658 ± 1 Ma (Mezger et al., 1990; Bernier and MacLean, 1989).

In the Cauchon Lake area peak conditions were followed by near isobaric cooling (Fig. 3). As indicated by U-Pb zircon and garnet ages, crosscutting pegmatites intruded from 2629 - 2598 Ma and hydrous garnet- and sillimanite-bearing granites of minimum melt composition around 2600 Ma. The associated partial hydration of country rocks reset the two-feldspar temperatures to ca. 650°C and retrogressed orthopyroxene to amphibole (Mezger et al., 1990). It is likely that during this time, southwest trending S3A augengneiss shear zones developed in the Cauchon Lake area and throughout the Pikwitonei domain (and Gods Lake domain). These shear zones may represent extensional faults leading to crustal thinning and resulting isostatic uplift (see below).

U-Pb rutile ages of 2430 Ma (Cauchon Lake) and 2363 - 2290 Ma (Natawahunan Lake) reflect conditions of ca. 430 - 380°C at that time (Mezger et al., 1989b). This indicates that following the last high grade metamorphism the terrane cooled at a rate of ca. 1.5°C/Ma and the uplift rate was slow, in the range of 25 - 70 m/Ma (Fig 3.). At 1883 Ma the ultramafic to mafic Molson Dyke swarm intruded (Heaman et al., 1986b).

Mylonite zones, trending southwest and east-southeast (Fig. 2), indicate ductile deformation at low grade conditions during late Archean or early Proterozoic times; some may represent foreland shear zones related to the Trans-Hudson collision along the Churchill-Superior boundary. The Hudsonian collisional event led to extensive structural overprint and retrogression of granulite grade gneisses to greenschist and amphibolite facies rocks adjacent to the Pikwitonei granulite domain, and thus strongly altered what might have been the deepest crustal rocks exposed in the Churchill-Superior boundary zone (Fig. 1).

4. P-T-t paths and implications for tectonic history

Geochronologic information, combined with pressure and temperature estimates, textural observations and zoning in garnets were used to construct quantitative P-T-t paths for the Cauchon Lake and the Natawahunan Lake areas (Fig. 3). The paths for both areas are similar, with the major difference that granulite grade conditions were reached at Natawahunan Lake already at 2740 Ma (Mezger et al., 1990). Based on reaction textures in sapphirine bearing gneisses described by Arima and Barnett (1984), Hensen (1987) deduced an anti-clockwise P-T-t path also for the granulite facies metamorphism in the Sipiwesk Lake area in the southern Pikwitonei domain. As indicated by Fig. 3, the temperatures and pressures shown in Figs. 2a and 2b correspond to the last thermal peak around 2640 Ma and therefore represent conditions for only a short time interval within the long metamorphic history of the northwestern Superior Province.

As shown in Table 1, certain periods of magmatism and metamorphism may have been only of local importance. The clustering of ages from metamorphic zircon and garnet indicates that mineral reactions did not occur continuously but were punctuated, lasting only for relatively short intervals, during the >150 Ma metamorphic history of the terrane (Mezger et al., 1989a). Such periods of mineral growth were the result of changing physical conditions, particularly increase in temperature possibly related to magmatism. Periods of mineral growth may have been separated by relative quiescence, during which the terrane cooled and no new mineral growth took place. This probably was the case between ca. 2720 Ma and 2700 Ma based on a major unconformity in the Oxford and Cross Lake greenstone belts. In this area tonalites intruded between 2747 and 2719 Ma, and were brought to the surface and eroded between 2709 and 2692 Ma (Table 1; cf. Corkery et al., 1988).

The anticlockwise P-T-t paths (Fig. 3) determined by Mezger et al. (1990) and Hensen (1987) for the northwestern Superior Province indicates heating of the crust prior or synchronous with thickening during late Archean. This suggests the addition of

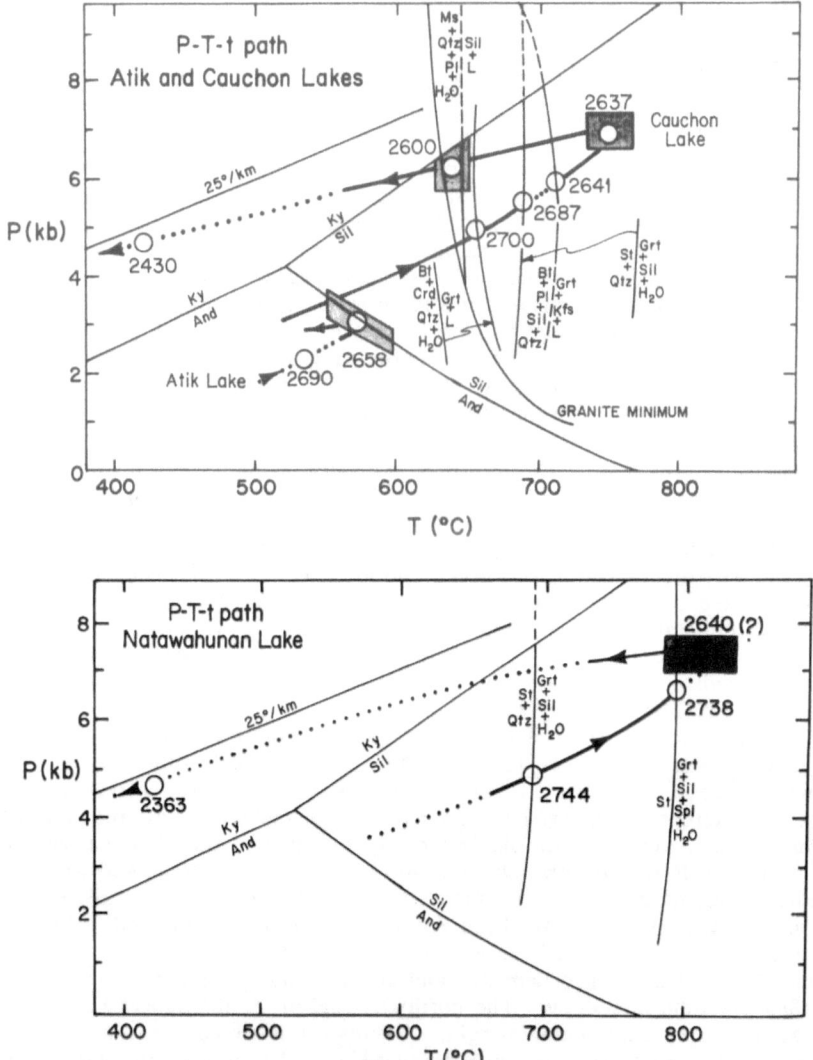

Fig. 3: P-T-t paths for Atik, Cauchon and Natawahunan Lake areas (Mezger et al., 1990). For this diagram the information from reaction textures, mineral zoning, geothermometry and geobarometry was combined with geochronologic information from rutile (low temperature points) and garnet (all others). The path may not be continuous but may have excursions towards lower temperatures between the periods of mineral growth. *(circles indicate ages, numbers represent ages in Ma, boxes represent P-T estimates, solid lines refer to parts of the path where pressure, temperature and age constraints on the metamorphic history are available, the dotted lines are tentative).*

hot magmas to the lower crust as the most likely heat source, rather than heating caused by burial as a result of tectonic stacking. The most likely magmatic heat sources are schollen-enderbite and opdalite that have reached temperatures in excess of 1100°C at ca. 2690 Ma (Mezger et al., 1990). Such rocks and their metamorphic derivatives, deformed during D3/M3 to inclusion rich enderbitic gneisses, comprise the main lithology of the Pikwitonei domain (>50%).

It is to be expected that hot, possibly mantle derived magmas, part of which crystallized as enderbites in lower crustal levels, ascended into higher crustal levels and possibly to the surface. High-K, mafic to felsic volcanics occurring in several greenstone belts and dated at 2706 Ma in the Oxford Lake greenstone belt (Table 1), may represent the extrusive portions of these hot magmas. It is conceivable that the process which resulted in hot magmas intruding lower crustal levels is a continuation of the magma generating process that yielded the 2760 - 2719 Ma old high level granitoids. Such a long lived process would lead to increased crustal thickness and eventually to cratonization. The slow cooling rate of ca. 1.5°C/Ma for the Pikwitonei domain is consistent with the model of Wells (1980) for the formation of granulites in a terrane where the thermal budget is controlled largely by magmatism.

Regionally consistent stress indicators suggest a compressional tectonic regime for the Pikwitonei domain during this D3/M3 episode. However, this compression combined with the magmatic thickening did not lead to the formation of a very thick continental crust as indicated by the inferred erosion (<70 m/Ma) and cooling rates (ca. 1.5°C/Ma) following this last high grade metamorphic episode (Fig. 3). This also indicates that the terrane must have been near isostatic equilibrium after ca. 2640 Ma. Based on U-Pb ages of Fe-Ti minerals from the Burntwood River area, the western part of the terrane was ca. 300 - 400°C at ca. 2175 Ma, therefore uplift must have occurred long after high grade metamorphism and may have been unrelated to the tectonic processes that caused the metamorphism (Mezger et al., 1990).

Field evidence indicates that the Pikwitonei granulites must have been exposed close to their present level before 1883 Ma. This is based on the observations that in the Thompson belt (Fig. 1, Table 1) Pikwitonei-type gneisses form a true stratigraphic basement to Proterozoic Ospwagan Group supracrustal rocks (Bleeker and Macek, 1988, Weber, in press). Since 1883 Ma old Molson dykes not only intruded the basement but also the overlying deformed Ospwagan Group (Macek, 1987, Bleeker and Macek, 1988), the basement must have been at the surface before 1883 Ma. This implies that crust of at least 30 km thickness must have been removed between 2640 Ma and 1883 Ma which is equivalent to a time-integrated erosion rate of 40 m/Ma.

Based on lithology, geochemistry and geochronology a potential scenario for the 2640 - 1883 Ma time interval is: The northwest margin of the Superior protocontinent evolved as a passive continental margin. During this process, extensional tectonics associated with low angle normal faults produced crustal thinning and isostatic uplift of lower crust (cf. Wernicke, 1985). Subsequently, the thinned crust rifted initiating the Trans-Hudson orogeny. Thinned crust along the stretched continental margin subsided and marginal basins formed (i.e. Ospwagan Group). The thinned continental margin was underplated by mafic igneous material of oceanic origin. This igneous material intruded and extruded along the continental margin forming the Molson dyke swarm, Fox River sill and associated volcanics.

Reconnaissance refraction seismic data indicate constant crustal thickness of ca. 30 km for the entire northwestern Superior Province (Mereu and Hunter, 1969) suggesting isostatic equilibrium. The margin of the northwestern Superior Province was probably thinner during initiation of the Trans-Hudson orogeny, but may have been telescoped and thickened during the Hudsonian collision. Additional uplift of the granulite terrane possibly resulted from this collision.

In contrast to the anticlockwise P-T-t path in the Superior Province margin (Fig. 3) the path in the Thompson belt, in the Churchill-Superior boundary zone, has a clockwise sense (Bleeker, in press) confirming conclusions based on geological and geophysical data (cf., Weber, 1987 and in press) that tectonic loading, i.e. overriding of the Proterozoic Reindeer zone onto the Archean Superior craton margin is the cause for the metamorphic overprint in the Churchill-Superior boundary zone. The thrusting took place during a transpressional, final phase of the Trans-Hudson orogeny under maximal upper amphibolite facies conditions (Bleeker, in press). U-Pb ages of zircons, monazites and titanites from a variety of rock units bracket this final event between 1822 and 1720 Ma (Machado et al., 1988; Weber, in press).

5. Conclusions

The NW margin of the Superior province exposes at least 20 km of continental crust from ca. 3kb at Atik Lake to ca. 9kb at the Burntwood River. This Archean continental crust had a complex igneous and metamorphic history lasting from >3.0 Ga to 1720 Ma. U-Pb mineral ages from rocks between Atik Lake and Natawahunan Lake indicate 5 distinct periods of pre-Hudsonian (older than 1820 Ma) metamorphism that can be correlated with magmatism in the higher crustal levels (Table 1).

Integration of the P-T-t data and derived interpretations with geological data from greenstone belts in the NW Superior Province indicates the following succession of geological events in the lower and upper crust (Table 1).

M0, (amphibolite facies) dated at ca. 3.0 Ga, is the oldest event and is only locally detectable. This early event is considered to be related to formation of granitoids which preceded the dated supracrustal rocks of the northwestern Superior Province.

M1, (amphibolite to granulite facies in Pikwitonei domain) dated at 2744 - 2738 Ma, is only locally preserved. This metamorphism followed deposition of tholeiitic to calc-alkaline volcanic rocks which make up the bulk of the greenstone belts of northwestern Superior Province. This metamorphism is related to voluminous granitoid magmas that intruded lower to upper crustal levels.

M2, (upper amphibolite facies in Pikwitonei domain) dated at 2700 - 2687 Ma, was associated with the intrusion of hot granitic magmas in the lower crust and led to formation of partial melts. High-K volcanics (2706 Ma) and granitoids (2692 Ma) in the upper crust probably were derived from these magmas and partial melts, respectively.

M3, dated at 2660 - 2637 Ma, reached amphibolite facies in the NW Gods Lake domain and granulite facies with widespread partial melting in the Pikwitonei domain. Rising partial melts cystallized as peraluminous granitoids and rare-element-enriched pegmatites in higher crustal levels.

M4, (amphibolite facies in Pikwitonei domain) dated at 2605 - 2591 Ma, registered the start to lower grade conditions, which was most likely related to uplift. Hydrous granitoids and pegmatites formed in the lower crustal Pikwitonei domain.

Between 2600 Ma and 1883 Ma the northwestern Superior probably went through a period of tectonic thinning which eventually led to rifting near its present margin and initiation of the Trans-Hudson orogeny. During this period granulites were brought to the surface in the Churchill-Superior boundary zone and formed the basement for Proterozoic supracrustal rocks in the Thompson belt and Fox River belt. Subsequently, collision of the Superior craton with a collage of Proterozoic arcs of the Churchill Province, during the transpressional, final stage of Trans-Hudson orogeny (1822 - 1720 Ma), led to overriding of the Proterozoic Reindeer zone of the Churchill Province onto the Superior craton along the Churchill-Superior boundary zone, and reworking and overprinting of Archean granulites under maximal upper amphibolite facies conditions.

6. Acknowledgments

This paper is based on results of field work done between 1974 and 1980 by W. Weber and J.J.M.W. Hubregtse with the Geological Survey of the Manitoba Department of Energy and Mines, combined with results from a Ph.D. study between 1984 - 1988 by K. Mezger with the Department of Earth and Space Sciences, State University of New York at Stony Brook (funded by NSF grants EAR 84-16250 and EAR 86-15714 to S.R. Bohlen and EAR 86-07973 to G.N. Hanson) Additional results are from U-Pb isotope studies by the Royal Ontario Museum (funded by the Geological Survey of Canada under the 1984 - 1989 Canada-Manitoba Mineral Development Agreement).

7. References

Arima, M. and Barnett, R.L.
    1984:   Sapphirine bearing granulites from the Sipiwesk Lake area of the late Archean Pikwitonei granulite terrain, Manitoba, Canada; *Contributions to Mineralogy and Petrology*, **88**, p. 102-112.

Baragar, W.R.A. and Scoates, R.F.J.
    1981:   The circum-Superior belt: a Proterozoic plate margin?: in *Precambrian Plate Tectonics;* Kroener, A., ed.; Elsevier, Amsterdam, p. 297-330.

Bernier, L.R. and MacLean, W.M.
    1989:   Gold and sulphide-bearing chert: Geochemical evolution of volcanogenic alteration pipes, Utik Lake, Manitoba; *Canadian Journal of Earth Sciences*, 26, in press.

Bleeker, W.
    in press: New structural and metamorphic constraints on Early Proterozoic oblique collision along the Thompson Nickel belt, northern Manitoba, Canada; in *Early Proterozoic Trans-Hudson orogen*, Lewry, J.F. and Stauffer, M.R., eds., *Geological Association of Canada, Special Paper*.

Bleeker, W. and Macek, J.J.
    1988:   Thompson nickel belt project - Pipe pit mine; in *Manitoba Energy and Mines, Report of Field Activities* **1988**, p. 111-115.

Corkery, M.T., Lenton, P.G., Breedveld, M. and Davis, D.W.
    1988:   Cross Lake geological investigations; in *Manitoba Energy and Mines, Report of Field Activities* **1988**, p. 106-110.

Davis, D.M.
    1987:   In: Oxford House, 1:250 000; *Manitoba Energy and Mines, Bedrock Geology Compilation Map Series*, NTS 53L.

Fountain, D.M. and Salisbury, M.
    1981:   Exposed cross sections through the continental crust: implications for crustal structure, petrology, and evolution; *Earth and Planetary Scicience Letters*, **56**, p. 263-77.

Heaman, L., Machado, N., Krogh, T. and Weber, W.
    1986a: Preliminary U-Pb zircon results from the Pikwitonei granulite domain, Manitoba; in *Geological Association of Canada-Mineralogical Association of Canada, 1986, Programme with Abstracts*, **11**, p. 79.

Heaman, L.M., Machado, N., Krogh, T.E. and Weber, W.
    1986b: Precise U-Pb zircon ages for the Molson dyke swarm and the Fox River sill: Implications for early Proterozoic crustal evolution in NE Manitoba, Canada; *Contributions Mineralogy and Petrology*, **94**, p. 82-89.

Hensen, B.
    1987:   P-T grids for silica-undersaturated granulites in the system MAS (n+4) and FMAS (n+3) -- tools for the determination of P-T paths of metamorphism; *J. Metam. Geol.*, **5**, p. 255-271.

Hubregtse, J.J.M.W.
    1978:   Sipiwesk Lake-Landing Lake-Wintering Lake area; in *Manitoba Mineral Resources Division Report of Field Activities* **1978**, p. 54-62.

    1980:   The Archean Pikwitonei granulite domain and its position at the margin of the northwestern Superior Province, central Manitoba; *Manitoba Energy and Mines, Geological Paper* **GP80-3**, 16 pp.

    1985:   Geology of the Oxford Lake-Carrot River area; M*anitoba Energy and Mines, Geological Report* **GR83-1A**, 73 pp.

Krogh, T., Heaman, L., Machado, N., Davis, D. and Weber, W.
    1986:   U-Pb geochronology program: Pikwitonei-Thompson-Cross Lake area; in *Manitoba Energy and Mines, Report of Field Activities* **1986**, p. 178-180.

Macek, J.J.
    1987:   Geological mapping at Pipe Mine; in *Manitoba Energy and Mines, Report of Field Activities* **1987**, p. 136-137.

Machado, N., Heaman, L. and Weber, W.
    1988:   U-Pb geochronology program: Thompson belt and northwest Superior Province; in *Manitoba Energy and Mines, Report of Field Activities* **1988**, p. 124.

Manitoba Mineral Resources Division
    1979:   Geological Map of Manitoba, scale 1:1 000 000; *Map* **79-2**.

Mereu, R.T. and Hunter, J.A.
    1969:   Crustal and upper mantle structure under the Canadian Shield from Project Early Rise data; *Bulletin Seismological Society of America*, **59**, p. 147-165.

340

Mezger, K., Hanson, G.N. and Bohlen, S.R.
    1989a: U-Pb systematics of garnet: Dating the growth of garnet in the late Archean Pikwitonei granulite domains at Cauchon and Natawahunan Lakes, Manitoba, Canada; *Contributions to Mineralogy and Petrology*, 101, p. 136-148.

    1989b: U-Pb ages of metamorphic rutiles: Application to the cooling history of high grade terranes; *Earth and Planetary Sciences Letters*, in press.

Mezger, K., Bohlen, S.R. and Hanson, G.N.
    1990:   Metamorphic history of the Archean Pikwitonei granulite domain and the Cross Lake subprovince, Superior Province, Manitoba, Canada; *Journal of Petrology*, in press.

Paktunc, A.D. and Baer, A.J.
    1986:   Geothermobarometry of the northwestern margin of the Superior Province: implications for its tectonic evolution. *J. Geol.*, 94, p. 381-94.

Scoates, R.F.J. and Macek, J.J.
    1978:   Molson dyke swarm; *Manitoba Mineral Resources Division, Geological Paper* 78-1, 53 pp.

Thurston, P.C., Cortis, A.L. and Chivers, K.M.
    1987:   A reconnaissance re-evaluation of a number of northwestern greenstone belts: evidence for an early Archean sialic crust; in *Summary of Field Work and Other Activities 1987, Ontario Geological Survey Miscellaneous Paper* 137, p. 4-24.

Weber, W.
    1974:   Utik Lake-Bear Lake project; in *Manitoba Mineral Resources Division, Summary of Geological Fieldwork 1974, Geological Paper* 2/74, p. 27-32.

    1976:   Cauchon, Partridge Crop and Apussigamasi Lakes area; in *Manitoba Mineral Resources Division, Report of Field Activities 1976*, p. 54-57.

    1977:   Odei-Burntwood Rivers Region; Cauchon Lake Region; in *Manitoba Mineral Resources Division Report of Field Activities 1977*, p. 58-61.

    1980:   The Pikwitonei granulite domain: edge of the Superior Province craton; *EOS, Transactions American Geophysical Union*, 61(17), p. 386-397 (Abstract).

    1984:   The Pikwitonei granulite domain: a lower crustal level along the Churchill-Superior boundary in central Manitoba; in *Workshop on "A cross section of Archean crust", Lunar and Planetary Institute, Technical Report* 83-03, p. 95-97.

    1987:   The Churchill-Superior boundary zone: constraints for the tectonic evolution of the southeastern margin of the Trans-Hudson orogen; in *Geological Association of Canada - Mineralogical Association of Canada, 1987, Programme with Abstracts*, 12, p. 100

in press: The Churchill-Superior boundary zone: southeast margin of Trans-Hudson Orogen, a review; in *Early Proterozoic Trans-Hudson orogen*, Lewry, J.F. and Stauffer, M.R., eds., in *Geological Association of Canada, Special Paper*.

Weber, W. and Scoates, R.F.J.
1978: Archean and Proterozoic metamorphism in the northwestern Superior Province and along the Churchill-Superior boundary, Manitoba; in *Metamorphism in the Canadian Shield, Geological Survey of Canada, Paper* **78-10**, p. 5-16.

Wells, P.R.A.
1980: Thermal models for the magmatic accretion and subsequent metamorphism of continental crust; *Earth and Planetary Scicience Letters*, **46**, p. 253-65.

Wernicke, B.
1985: Uniform-sense normal simple shear of the continental lithosphere; *Canadian Journal of Earth Sciences*, **22**, p. 108-135.

TWO TRANSECTS ACROSS THE GRENVILLE FRONT, KILLARNEY AND TYSON LAKE
AREAS, ONTARIO

A. Davidson
Lithosphere and Canadian Shield Division
Geological Survey of Canada
588 Booth Street
Ottawa, Ontario K1A 0E4
Canada

ABSTRACT. The Grenville Front marks the northwest limit of the Grenville Province. It is a tectonite front of major significance, separating older structural provinces of the Canadian Shield from the
1.3 - 1.0 Ga Grenville orogen to the southeast. An historical perspective of the whole front is given, followed by a more detailed appraisal
of relationships across the front in Ontario. Of the various relationships found along this 200 km segment of the front, two have been
chosen for illustration on field excursions: 1, a section along the
north shore of Georgian Bay, where a 1740 Ga granite and rhyolite complex lies between the Grenville orogen and folded Huronian sedimentary
rocks of the Southern Province and is involved in Grenvillian tectonism; 2, an oblique section, 10 to 20 km inland from Georgian Bay, which
demonstrates abrupt changes in structural style and metamorphic grade
on entering the Grenville Province. In both sections, severe modification within the Grenville orogen has been superimposed on rocks that
were already deformed and metamorphosed, leading to extremely complex
geological relationships. The only previously unaffected rock unit is
diabase of the 1.24 Ga Sudbury dyke swarm, the effects of the Grenvillian orogeny on which is examined in some detail. The evidence
presented on the field excursions supports the contention that the
northwest margin of the Grenville orogen is a strongly uplifted, compressed zone in which progressively deeper crustal levels are exposed
at the present surface.

PART I: THE GRENVILLE FRONT

Introduction

Accepted by most as the northwest limit of the Grenville Structural
Province of the Precambrian Shield in Canada, the Grenville Front
(Fig. 1) has long been recognized, at a regional scale, as the surface
expression of a major crustal discontinuity that truncates structures
characteristic of the other, older Shield provinces to the northwest.

M. H. Salisbury and D. M. Fountain (eds.), Exposed Cross-Sections of the Continental Crust, 343–400.
© 1990 Kluwer Academic Publishers.

The front is the locus of faulting and mylonitization, of metamorphic change and, in places, of truncation of major rock units. It is well-expressed geophysically, particularly by its aeromagnetic signature but also for much of its length by a broad Bouguer gravity 'low'. It is also expressed by resetting of some radioisotope systems in rocks and minerals, notably K-Ar.

Regardless how it is defined, the Grenville Front is situated at or close to the northwest margin of the Grenville Orogen that is generally recognized to have been the site of major orogeny between 1.3 and 1.0 Ga ago. As an orogenic front, however, it has certain peculiarities: (1) the adjacent foreland to the northwest does not retain, if it ever had, a cover of supracrustal rocks that can be related in terms of time or paleoenvironment to the Grenville Orogen, with one possible exception - the Seal Lake Group in central Labrador (10 in Fig. 1); (2) it does not mark the appearance of plutonic rocks of Grenvillian age which, in fact, are for the most part restricted to the southeast half of the Grenville Province and are noticeably absent near the front; (3) it separates the older Shield provinces from a broad tract of gneisses of which some are recognized as reworked equivalents of rocks northwest of it and others, even though pre-Grenvillian in age, have no obvious counterparts.

This field excursion examines several aspects of the Grenville Front at two localities in Ontario where mylonitic rocks are prominently displayed. Among these aspects are the nature and correlation of rocks on either side of the front, the type and style of deformation encountered as the front is approached and how these are influenced by the rock types involved, the kinematic indication in the rocks of displacement sense, and the change in metamorphic grade across the front. Discussion will no doubt embrace topics such as how best to define the front, timing of tectonism along the front with respect to the Grenvillian Orogeny, and mechanisms of cataclasis and mylonitization. The two areas visited (see Fig. 2) are (1) the Killarney area, where mid-Proterozoic volcanic and granitoid plutonic rocks lie between deformed Huronian strata of the Southern Province and gneisses of the Grenville Province, and (2) the Tyson Lake area, where metasedimentary schist and gneiss in the Grenville Province, questionably correlated with the Huron Supergroup, are interlayered with metagranitoid rocks, both being cut by deformed and metamorphosed diabase dykes that are confidently correlated with the Sudbury diabase dyke swarm in the Southern Province.

Historical Perspective

Although the Grenville Front is recognized as a major tectonic feature of the Canadian Shield, its nature and precise location have been subjects of some dispute, mostly due to the fact that various positions have been taken by different authors concerning what parameters should be used to define it (see Gower et al., 1980). Historically, it was first noted near Georgian Bay as a marked change in structural trend and rock type by Collins (1925, Fig. 8), who referred to it as simply 'the (eastern) limit of Huronian strata'. The northwest boundary of the Grenville Province as we know it today was stated to be the locus

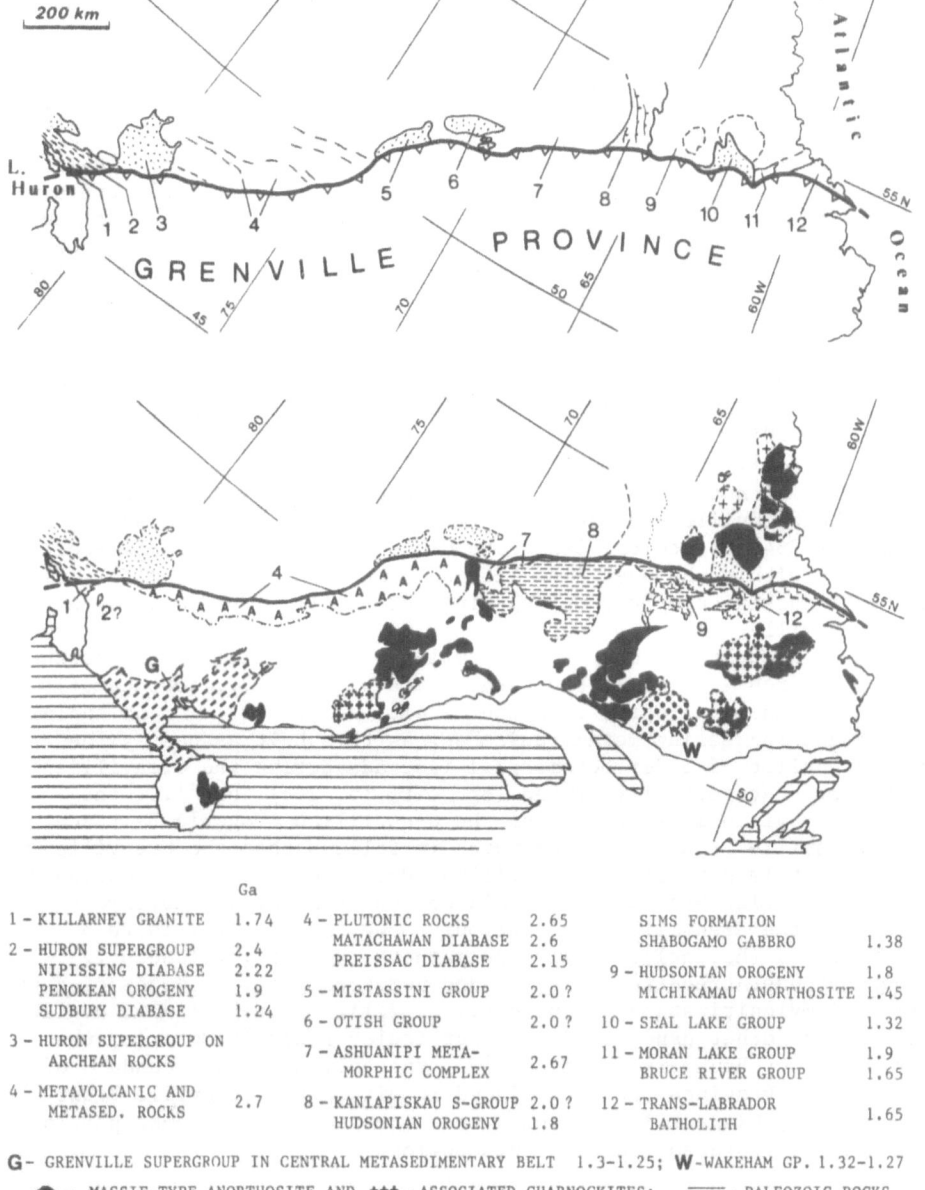

		Ga			
1 – KILLARNEY GRANITE	1.74		4 – PLUTONIC ROCKS	2.65	SIMS FORMATION
			MATACHAWAN DIABASE	2.6	SHABOGAMO GABBRO 1.38
2 – HURON SUPERGROUP	2.4		PREISSAC DIABASE	2.15	
NIPISSING DIABASE	2.22				9 – HUDSONIAN OROGENY 1.8
PENOKEAN OROGENY	1.9		5 – MISTASSINI GROUP	2.0 ?	MICHIKAMAU ANORTHOSITE 1.45
SUDBURY DIABASE	1.24		6 – OTISH GROUP	2.0 ?	10 – SEAL LAKE GROUP 1.32
3 – HURON SUPERGROUP ON			7 – ASHUANIPI META-	2.67	11 – MORAN LAKE GROUP 1.9
ARCHEAN ROCKS			MORPHIC COMPLEX		BRUCE RIVER GROUP 1.65
4 – METAVOLCANIC AND	2.7		8 – KANIAPISKAU S-GROUP	2.0 ?	12 – TRANS-LABRADOR 1.65
METASED. ROCKS			HUDSONIAN OROGENY	1.8	BATHOLITH

G– GRENVILLE SUPERGROUP IN CENTRAL METASEDIMENTARY BELT 1.3–1.25; **W**–WAKEHAM GP. 1.32–1.27
◆ – MASSIF-TYPE ANORTHOSITE AND +++ – ASSOCIATED CHARNOCKITES; ═══ – PALEOZOIC ROCKS

Figure 1. The Grenville Province and correlation of rocks across the
Grenville Front. Approximate ages are current best estimates.

of a zone of faulting by Gill (1948), who also noted that '... a rapid change from low grade metamorphics to high grade gneisses occurs along this fairly straight zone, ...', which J. Tuzo Wilson (1949, Fig. 7) labelled the 'Huron-Mistassini fault zone'. The term 'Grenville Front' crept unobtrusively into the literature after it was used on a tectonic map of Canada (Derry, 1950), on which its nature was not defined. Different opinions were subsequently voiced on whether it represents a fault or fault zone (McLaughlin, 1954), a metamorphic boundary (Hewitt, 1957) or even, in part, '... a phenomenon of the draughtsman's table ...' (Engel, 1956, p. 81). Recognizing that mylonite, shear zones and faults mark parts but apparently not all of the Grenville Front, Wynne-Edwards (1972, p. 318) defined it on geochronological grounds as '... the place where the K/Ar dates assume the 1000 m.y. value associated with the Grenvillian Orogeny.' Stockwell (1982), although using isotope geochronology to help define it, realized the need for a structural basis for definition of a structural province; however, he preferred to locate parts of the front at the biotite isograd spatially associated with northeast-trending faults, and also inferred part of it to be an intrusive contact (Killarney area).

Whether or not the Grenville Front is taken as the northwest boundary of the Grenville Province also varies from author to author. Stockwell (1964) placed the Seal Lake Group of the Naskaupi fold-belt in Labrador within the Grenville Province, and later (1982) also included the Mistassini and Otish groups in Quebec, all of these being north of the Grenville Front. Wynne-Edwards (1972, p. 270) defined the Grenville Foreland Belt, a zone 20 to 60 miles wide parallel to and northwest of the front, where supracrustal rocks display '... northeast-trending, southeast-dipping cleavage, faults and folds ...'; he explained southeast-increasing metamorphism across this belt as partly the result of uplift of older metamorphic isograds during the Grenvillian Orogeny, and categorically stated (op. cit., p. 266) that '... the Grenville Province is that part of the Grenvillian Orogenic Belt south of the Grenville Front...' (from which it follows that the Grenville Foreland Belt is part of the Grenvillian Orogenic Belt but is excluded from the Grenville Province). He placed the front at the northwest boundary of the Grenville Front Tectonic Zone, a zone '... 10 to 50 miles wide ... (having) ... very low magnetic relief ... (and) ... strong northeast-trending foliation and numerous parallel zones of cataclasis and mylonitization.' (op. cit., p. 271). Lumbers (1975, p. 122; 1978), on the other hand, defined a particular fault in Ontario, the Grenville Front Boundary Fault, as '... the only mappable boundary of the Grenville Province ...'; he placed it within the Grenville Front Tectonic Zone, his version of which thus incorporates at least part of Wynne-Edwards' Foreland Belt in this region.

Grenville Front Facts and Interpretations

Where a major fault or mylonite zone marks the Grenville Front, it is known in most places that such structures are steep, although invariably inclined southeast toward the Grenville Province. It is generally accepted that the Grenville side of the front has been uplifted with respect to the northwest side. This interpretation is supported by, among other evidence, the fact that metamorphic grade is uniformly high on the Grenville side whereas it is variable on the other, and of an age related to the Shield orogen within which it is found. Metamorphism may be virtually nonexistent a few kilometres northwest of the front in supracrustal rocks that cover the older Shield orogens, e.g., the little-disturbed Huron Supergroup (Douglas, 1980) (<2.5>2.2 Ga) near Lake Timiskaming, the Mistassini and Otish groups (~2.15 Ga) northeast of Chibougamau and the Seal Lake Group (~1.3 Ga) of central Labrador (3, 5, 6 and 7 in Fig. 1A). In some places, too, it has been shown that Grenvillian magnetic overprinting does not extend more than 2 to 3 km beyond the front (Temagami area, Hyodo et al., 1986). In such places it is hard to justify the 20 to 60 mile width accorded by Wynne-Edwards (1972) to the Grenville Foreland Belt. The cover rocks show increasingly intense folding toward the front, culminating in tight folds overturned to the north or northwest. These rocks, however, are not recognized southeast of the front in the adjacent Grenville Front Tectonic Zone. Being veneers only a very few kilometres thick on the older Shield provinces, their non-appearance on the Grenville side is readily explained by uplift. The attitude of their folds next to the front and the southeasterly inclination of the front structures imply, if not long distance overthrusting, then at least substantial reverse fault displacement.

On the other hand, rocks of the older Shield provinces that are deeply rooted in the crust, such as plutonic rocks and metamorphic complexes, the steeply folded Archean greenstone and metagreywacke formations of the Superior Province and the Proterozoic supracrustal rocks of the Labrador Trough, can be traced across the front into the Grenville Province (4, 7 and 8 in Fig. 1A and 1B). For example, easily recognized iron formation of the Labrador Trough is known to extend at least 100 km south of the front (Rivers and Chown, 1986). Among plutonic rocks, Superior Province granitoids may be recognized by their retention of pre-Grenvillian ages determined by analysis of the least readily disturbed radioisotope systems. Middle Proterozoic anorthosite and related plutonic rocks occur on both sides of the northeastern part of the front, and although no individual massif is transected by the front, it is reasonable to assign them to the same plutonic province (Fig. 1B). Massifs of this suite within the Grenville Province, with the exclusion of those of truly Grenvillian age (~1.1 Ga), are deformed, at least at their margins, and original intrusive relationships with older rocks are rarely observed. Plutonic rocks of similar ages north of the front are not disrupted in this way (Emslie, 1978; Duchesne, 1984).

The nature, significance and duration of tectonic activity along the Grenville Front have been variously interpreted. Gill (1948, p. 103) suggested the possibility that '... movements on low angle thrust faults carried the Grenville rocks to their present position ...', but admitted that evidence for such thrusts along this boundary had yet to be demonstrated. In fact, outlying klippen of Grenville rocks are not recognized northwest of the front. Lumbers (1971) described the Grenville Front Tectonic Zone in Ontario as one in which Grenville rocks

'... were compressed and thrust against the relatively rigid Southern
and Superior Provinces ...', although he later presented a model that
attempted to account for the observed highly strained rocks by a mechanism involving relatively minor vertical displacement and no overthrusting at the front (Lumbers, 1975, Fig. 8). In a series of cross
sections, Wynne-Edwards (1972, Fig. 4) illustrated the front as a moderately southeast-dipping thrust fault and implied that the now-exposed
gneisses above this thrust are reworked Archean rocks that are basement
to Proterozoic supracrustal rocks to the northwest.

When plate tectonic theory was first introduced, the front was
considered to be a potential candidate for the site of a suture between
two continental masses (e.g., Dietz, 1966; Krogh and Davis, 1971; Gibb
et al., 1980). There are, however, no features preserved at the present surface that would support this interpretation, although a paired
gravity anomaly along the northeastern part of the front has been taken
as possible evidence for an ancient suture nearby (Thomas and Tanner,
1975). Mitigating against the whole Grenville Province being an accreted continental mass is the fact, already stated, that some rock
units of older Shield provinces to the northwest can be recognized to
continue for considerable distances southeast of the front. Suture
hunters thus prefer to seek more favourable ground in the core of the
exposed Grenvillian Orogen (Rondot, 1978; Anderson and Burke, 1983), or
even southeast of it where it has been suggested that a Grenvillian
suture may have been reworked in the Appalachian Orogen (Dewey and
Burke, 1973; Baer, 1976; see also Thomas and Gibb, 1985). The Grenville Front in these later models is thus relegated to the status of an
orogenic limit spatially removed from the locus of continent-continent
collision of which the Grenvillian Orogeny was the consequence.

Tectonic activity at the Grenville Front has certainly occurred
within the time span usually assigned to the Grenvillian Orogeny, or
Orogenic Cycle (Moore and Thompson, 1980): the Seal Lake Group in
Labrador (ca. 1.32 Ga; Baragar, 1981) and the northwest-trending Sudbury diabase dykes (ca. 1.24 Ga; Krogh et al., 1987) are strongly
deformed at the front. Some authors have considered that the front was
a long-lived tectonic entity whose history extends back to the early
Proterozoic (Lumbers, 1975) or even the late Archean (Stockwell, 1982),
thus implying activity, albeit perhaps intermittent, for as long as
1 to 1.5 Ga. If this is so, one might expect there to be some evidence of this activity recorded in the rocks immediately northwest of
the front, evidence such as facies changes in sedimentary formations or
particular distribution and perhaps type of igneous rocks. Neither
supracrustal nor plutonic upper Archean rocks of the Superior Province
offer such evidence - the east-west greenstone belts, the intervening
higher grade metawacke (Pontiac Group) terranes and the plutonic rocks
that invade them show the same characteristics far from the front as
close to it. In Ontario, Lumbers (1978; Sims et al., 1981) interpreted
certain gneisses in the Grenville Front Tectonic Zone to be deep water
facies of lower Huronian sediments, specifically the Mississagi Formation. He suggested (1978, p. 356) '... that a relatively deep trough
or linear depression developed along what is now the northwestern margin of the Grenville Province ...' which '... subsequently influenced

the location and development of the Grenville Front Tectonic Zone ...'.
However, most other Huronian formations, although recognized to thicken
southward (Young, 1983), do not show facies changes in the direction of
the Grenville Province. Moreover, the gneisses in question may equally
well be interpreted as reworked metawackes of the Archean Pontiac Group
which is found adjacent to the front beneath Horunian formations in Ontario and can be traced across the front in Quebec to the northeast
(Davidson, 1986a, Fig. 2). Farther northeast along the front, neither
the Mistassini and Otish groups nor the Kaniapiskau Supergroup, all of
which are likely correlated (Chown and Caty, 1973), show any evidence
of relationship to a developing Grenvillian orogen. Facies changes and
orientation in the Labrador Trough are related to the ensuing Hudsonian
Orogeny (southwestward thrusting toward the Archean basement) whose
effects are truncated at the Grenville Front (north- to northwestward
thrusting; Rivers, 1983).

Interpretation that the Grenville Front has been a long-lived
tectonic entity inmplies that the Grenville Province has been in existence for a long time and is not therefore solely the product of the
Grenvillian Orogeny. This raises the question: to what orogeny or
orogenies can the supposed pre-Grenvillian tectonic activity along the
front and within the Grenville Province be ascribed? Stockwell (1964)
outlined different parts of the Grenville Province that he identified
as having been affected by earlier orogenies (Kenoran, 2.6-2.5 Ga;
Hudsonian, ca. 1.8 Ga; Elsonian, ca. 1.45 Ga) whose type areas lie in
other parts of the Shield. He subsequently added the Killarnean (ca.
1.55 Ga), whose type area he placed within the southwestern Grenville
Province (Stockwell, 1982). To these may be added the newly described
Labradorian Orogeny (ca. 1.65 Ga), related to which is the Trans-
Labrador batholith which is transected by the Grenville Front in eastern Labrador (Thomas et al., 1985) (Fig. 1, unit 12). The wide range
in age of these orogenic events and the uneven distribution of their
vestiges within the Grenville Province is hard to reconcile with a
particular tectonic zone that had been active at those times. It is
therefore preferred to attempt to interpret the Grenville Front as
having a direct and unique relationship with the Grenvillian Orogeny
(1.3 - 1.0 Ga), and to explain evidence of earlier orogeny within the
Grenville Province in terms of former histories that did not necessarily have anything to do with the subsequent development of the Grenvillian orogen.

The Grenville Front in Ontario

The northwest edge of the Grenville Province in Ontario is defined by
the Grenville Front Boundary Fault (Lumbers, 1975; Card and Lumbers,
1977). From southwest to northeast the immediately adjacent rocks
(Fig. 2) are 1), from Killarney to southeast of Sudbury, the Killarney
complex, Bell Lake, Chief Lake and allied plutons (ca. 1.75 - 1.45 Ga)
with rafts and screens of Huronian formations; 2), from Sudbury to
northwest of River Valley, Huronian formations (mainly Mississagi),
Nipissing diabase and, near Riber Valley, leucogabbro and anorthosite;
3), along the segment south of Temagami, Archean metagreywacke and

granitoid rocks. All of these rock units are highly deformed in the immediate vicinity of the Boundary Fault, and are overprinted by meta- morphism that increases abruptly toward it, generally attaining lower amphibolite facies at the front. This frontal metamorphism is restric- ted to a zone 5 km wide or less in rocks not previously metamorphosed (the Huronian sediments and Nipissing diabase of the Cobalt Embayment - 3 in Fig. 1), but it is less easy to recognize its limits in formerly metamorphosed rocks. Grenville front-related metamorphism can be re- cognized by the association and orientation of its minerals with south- east-dipping, front-parallel foliation and with dip-parallel lineation, but recognition may be difficult where earlier foliations are reorient- ed toward parallelism with the Grenvillian trend.

The Foreland Zone close to the front contains numerous faults and narrow mylonite, ultramylonite and breccia zones, indicative of mainly brittle deformation. In some places the rocks have undergone an earlier stage of ductile deformation before brittle disruption. The Boundary Fault itself is marked in many places by a pronounced mylonite zone several tens of metres thick, adjacent to which the rock units of the Foreland Zone are extremely attenuated. In places, for example south- east of Sudbury and also adjacent to the Bell Lake granite (Fig. 2), mylonite zones fork and rejoin, enclosing lenses of less deformed rock, and in such places it is not easy to define an individual boundary fault.

Grenville Province rocks next to the Boundary Fault have undergone penetrative ductile deformation and recrystallization at amphibolite facies and usually do not closely resemble those in the adjacent Fore- land Zone. However, metasedimentary gneisses in the Grenville Province east and northeast of Killarney bear enough resemblance to what might be interpreted as highly deformed and metamorphosed Huronian formations that attempts have been made to recognize Huronian stratigraphy among them (Quirke and Collins, 1930). Nevertheless, recent work has failed to establish this correlation beyond doubt (Frarey, 1985). In the same area, on the other hand, granitoid orthogneisses are quite similar to the plutonic rocks next to the Boundary Fault and have similar pre- Grenvillian U-Pb zircon ages (ca. 1.7 Ga; Krogh et al., 1971). Farther northeast, the uncertainty of the correlation of gneissic rocks in the Grenville Front Tectonic Zone with Huronian sediments or Archean meta- greywackes has been mentioned already. Near River Valley, however, a body of gabbro and anorthosite has been mapped on both sides of the Boundary Fault (Lumbers, 1973), and still farther northeast, granitoid gneisses in the Tectonic Zone south of Temagami retain Archean U-Pb zircon ages (Krogh and Davis, 1968) and are justifiably equated with tonalite and granodiorite of the Ingall Lake batholith in the Superior Province to the north.

The Grenville Front Boundary Fault and allied structures are everywhere moderately to steeply inclined toward the southeast quad- rant. Foliation and gneissosity within the adjacent Tectonic Zone are similarly oriented, with local perturbations, and tend to become more shallow southeastward. Associated in most places with this planar element is a well developed, southeast-plunging mineral alignment, clearly recognizable as a stretching lineation, especially near the

Boundary Fault. Mesoscopic and microscopic structures in the mylonitic rocks associated with the front commonly indicate upward displacement of the southeast (Grenville) side, as will be shown during this excursion. A similar sense of displacement has been documented in several major shear zones in the interior of the Grenville Province to the southeast (Davidson, 1984; Hanmer and Ciesielski, 1984; Hanmer et al., 1985).

A recent seismic reflection survey in Georgian Bay along a line crossing the southwestern extrapolation of the Grenville Front 85 km south of exposure on the north shore, and covered beneath the water by flat-lying lower Paleozoic sediments, has identified a remarkable array of moderately southeast-dipping reflectors within the Grenville Front Tectonic Zone (Green et al., 1987, 1988) (Fig. 3). The westernmost reflector coincides closely with the position of the front identified on potential field maps of Georgian Bay and Lake Huron. These reflectors penetrate deep in the crust, perhaps to Moho, and shallow from 35°

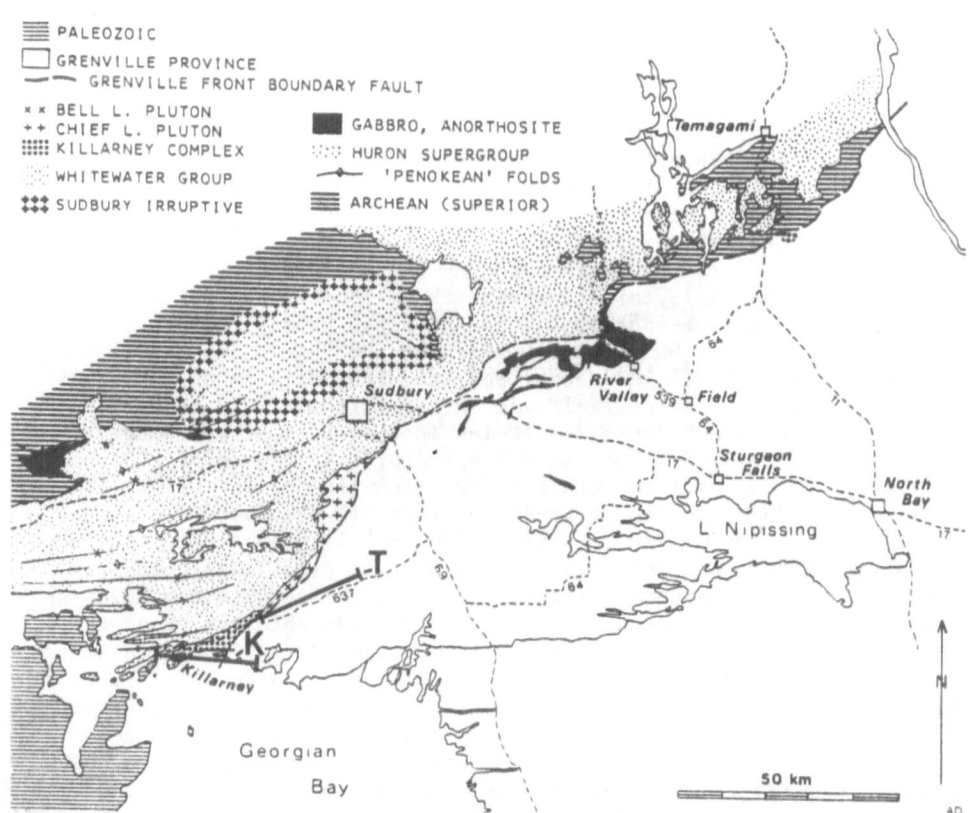

Figure 2. Geology marginal to the Grenville Front in Ontario. K and T identify excursion sections in the Killarney and Tyson Lake areas respectively.

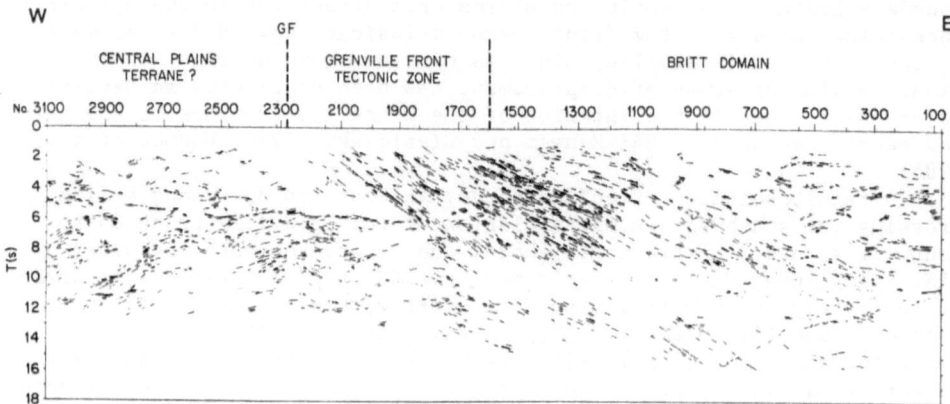

Figure 3. Seismic reflections across the Grenville Front in Georgian Bay recorded on GLIMPCE line J, 80 km south of Killarney (after Green et al., 1988).

to 25° toward the southeast. Follow-up examination of the nearest exposures of the Grenville Front Tectonic Zone, the previously poorly understood north shore of Georgian Bay (Davidson and Bethune, 1988), revealed various geologic units sliced and bounded by mylonite and high strain zones, all dipping moderately southeast, although at some-what higher angles (30 - 60°) than the seismic reflectors. It is not known whether the reflectors represent the mylonites themselves or are due to juxtaposition across such zones of rocks having different re-flective properties. On the north shore, a pronounced dip-parallel stretching lineation is well developed, and kinematic sense, where it could be identified in the field, is one of northwestward thrust dis-placement. Geothermobarometric studies in this part of the Grenville Province (Anovitz, 1987) suggest paleopressures in the range 800 to 1000 Mpa (8 to 10 kbar) to within a few kilometres of the Grenville Front. These data and the observations that, in the Grenville Front Tectonic Zone, some metasedimentary rocks contain hypersthene within 3 km of the front, and deformed Sudbury diabase dykes have coronitic assemblages of hypersthene, clinopyroxene and garnet within 8 km of the front (Bethune and Davidson, 1988), imply considerable uplift on the Grenville side, particularly when it is considered that Huronian sedi-mentary rocks in the Cobalt Embayment (3 in Fig. 1) have only anchizone assemblages within 4 km of the front (Davidson, 1986c). The Grenville Front Tectonic Zone can therefore be interpreted as exposing a con-tracted section of deep continental crust, thanks to deep-rooted thrust uplift at the northwest margin of the Grenville Orogen. What is seen, however, is not an intact deep crustal cross section but one of short-ened crust in which tilted panels and slices from lower levels are now stacked side by side.

PART II: THE KILLARNEY VOLCANO-PLUTONIC COMPLEX AND ITS RELATIONSHIP
TO THE GRENVILLE OROGEN

Geologic Setting

The region between Killarney and Sudbury (Fig. 4) contains that part of
the Grenville Front adjacent to the Southern Province where the Huron
Supergroup formations and Nipissing diabase sills are folded about east-
trending anticlines and synclines with gently plunging axes. In the
vicinity of the Grenville Front, Huronian formations from the Mississagi
(upper Hough Lake Group) to the Bar River (uppermost Cobalt Group) and
the Nipissing diabase are abruptly truncated by granite plutons (ca.
1.74 and 1.47 Ga) whose northwestern contacts are intrusive in some
places but faulted in others. Mylonite zones lie within and/or along
the southeast sides of the various, distinctive plutonic units. Metasedimentary rocks, highly deformed but still recognizable as belonging
to specific Huronian formations, occur locally at the southeast sides
of these plutons, as well as forming rafts and inclusions within them.
To the southeast, mylonitic, protomylonitic and gneissic granitoid rocks
are common and are probably deformed and metamorphosed age-equivalents
of the granites near the front. Also present are narrow, northeast-
trending belts of metasedimentary schist and gneiss and minor amphibolite and mafic gneiss whose relationships to the Huron Supergroup and
Nipissing diabase are unceratin (Frarey, 1985). Southeast-trending,
vertical olivine diabase dykes of the Sudbury swarm (ca. 1.24 Ga) cut
the Southern Province rocks and the granites near the front but are
deformed and metamorphosed in the Grenville Province. East-trending
diabase dykes of the Grenville swarm, probably early Paleozoic in age,
cut cleanly across the Grenville Front.

   The southwesternmost granite pluton adjacent to the Grenville
Front (Killarney granite *sensu lato*) has been interpreted recently as a
high-level volcanic-plutonic association, correlative in age and type
with similar rocks in mid-continental North America (Van Schmus and
Bickford, 1981; Bickford et al., 1986; Davidson, 1986b; van Breemen and
Davidson, 1988). Referred to as the Killarney complex, it was deformed
and metamorphosed before the Grenvillian Orogeny, raising doubt about
the interpretation that this complex, and by inference the other 'Grenville Front granites', are somehow related to the development of the
Grenville Orogen. Before elaborating the new data, it is worth summarizing briefly the previous interpretation of Grenville Front relationships in the Killarney area.

History of Geologic Interpretation

The granite around Killarney was recognized to intrude Huronian sedimentary strata to the northwest by Bell (1898). During mapping in the
1920s, Collins (1925) and Quirke and Collins (1930) identified most of
the Huronian formations that are recognized today. They attempted to
deal with the fact that this thick succession (see Table I in Part III) is
truncated by the belt of granitoid rocks between Killarney and Sudbury,
does not appear to continue on the other side of these plutons. They

recorded that the granitoid rocks southeast of the contact become gneissic and enclose metasedimentary relics that might be interpreted as highly altered equivalents of the Huronian rocks. Wholesale granitization was called upon to explain this 'disappearance of the Huronian', and the intrusive relationship along the northwest edge of the granite and gneiss terrane was accounted for by local melting of feldspathized quartzite.

Once the Grenville Province had been defined (Gill, 1948), Collin's (1925) 'limit of Huronian strata' became part of the Grenville Front and the belt of granite plutons along the front in this region was considered to be a manifestation of Grenvillian Orogeny, the loci of the plutons being controlled by a major fault zone. This contention was initially taken to be supported by K-Ar age determinations on micas that gave Grenvillian ages, interpreted as ages of intrusion. Subsequent dating by Rb-Sr and U-Pb methods, however, have proven that these granites are 500 to 750 Ma older than the ca. 1000 Ma originally assigned to them (Krogh et al., 1971; Wanless and Loveridge, 1972), the K-Ar ages representing resetting during Grenvillian thermal overprinting. The older ages were the basis for maintaining that the fault zone, interpreted to have controlled granite emplacement, was in existence long before the Grenvillian Orogeny took place.

The Grenville Front has been differently placed on recent geologic maps of the Killarney region. Frarey and Cannon (1969; Frarey, 1985) assigned the 'Grenville Front granites' to the Grenville Province, placing the front along their mutual northwest contact. This was justified on the basis that parts of this contact are sites of mylonitization and brecciation associated with major faults. On the other hand, Lumbers (1975; Card and Lumbers, 1977) traced the Grenville Front Boundary Fault along the southeast side of or within this belt of granites. Lumbers (1975, p. 9) stated that similar granitoid rocks southeast of the Boundary Fault are '... invariably metamorphosed by high rank, Late Precambrian regional metamorphism that has no spatial or temporal association with these plutons ..', but also that the 'Killarney batholithic complex' is related '... spatially and possibly temporally to the Grenville Front Tectonic Zone ...', thereby also implying a pre-Grenvillian history for the front.

In order to try to resolve which, if either, of these two locations of the Grenville Front is the most acceptable, detailed work was undertaken recently in the Killarney area (Davidson, 1986b; Clifford, 1986; van Breemen and Davidson, 1988). Three features of the geology west of Lumbers' Boundary Fault were given particular attention: (1) the northwest contact of the Killarney granite, known to be faulted in some places but intrusive in others, (2) included masses mapped as gneiss and agmatite within the Killarney granite (sensu stricto), itself asserted to have a gneissic character in places (Frarey and Cannon, 1969; Frarey, 1985), and (3) relics of quartzite within the Killarney complex, reported by Quirke and Collins (1930) as having escaped conversion by feldspathization to granite. The findings of this reexamination are incorporated in the following description of the Killarney complex.

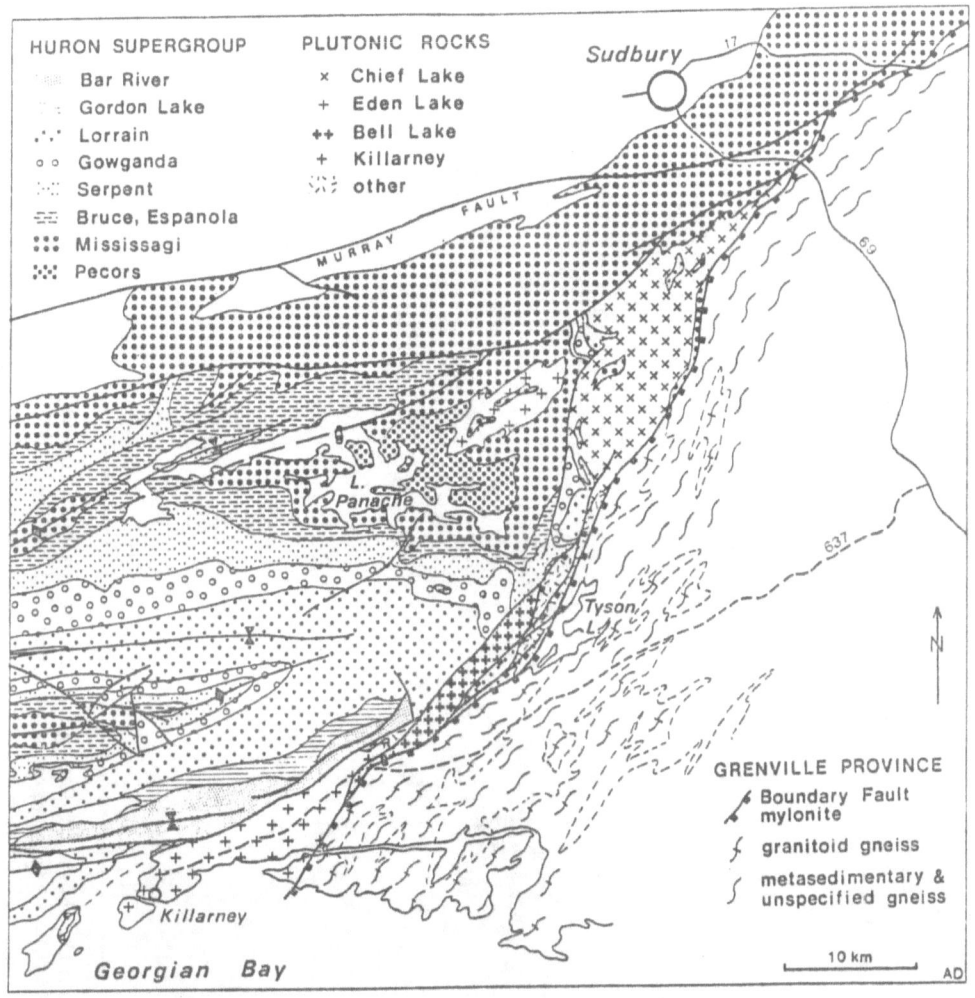

Figure 4. Geology of the Grenville Front region between Sudbury and
Killarney (after Card and Lumbers, 1977, and Frarey, 1985). Nipissing
diabase intrusions in the Huronian succession are omitted.

Geology of the Killarney Area

Immediately northwest of the Killarney granite, bedrock comprises the
upper three formations of the Huronian Cobalt Group (Lorrain, Gordon
Lake and Bar River; see Table I). On the islands and peninsulas of
Killarney Bay (Fig. 5), the Killarney granite (s.s.) clearly dykes and
includes the Gordon Lake and Bar River formations. Northeast from
Killarney Bay the contact is rarely exposed, being concealed in a
narrow valley occupied in part by a chain of lakes; in a few places,

356

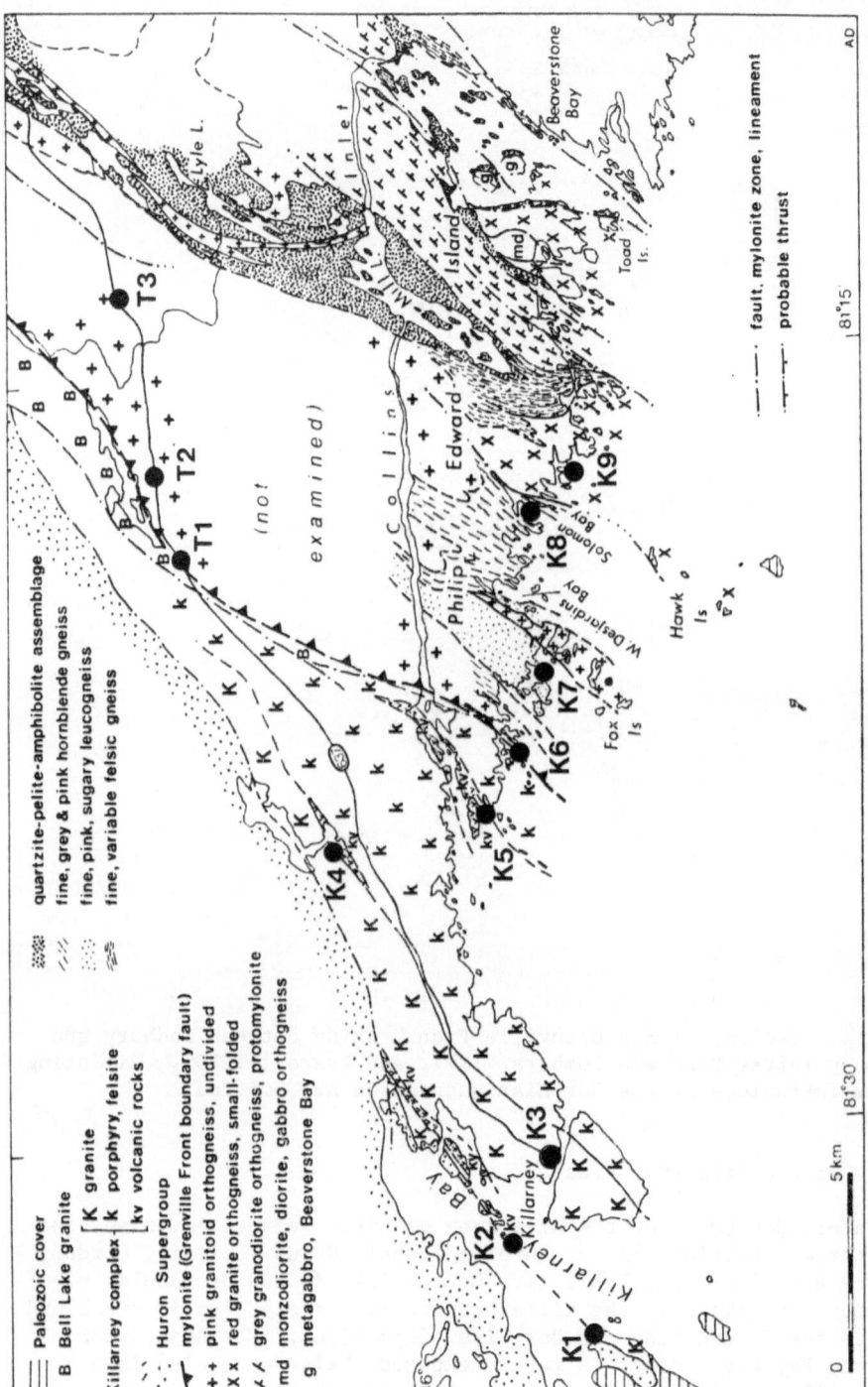

Figure 5. Geology along the north shore of Georgian Bay between Killarney Bay and Beaverstone Bay (after Davidson and Bethune, 1988). Field excursion stops K1 to K9 and T1 to T3 are shown.

however, granite dykes are present in the Bar River and Lorrain quartzites northwest of this lineament. Olivine diabase dykes of the Sudbury
swarm cut the northwest contact of the Killarney granite.

Volcanic rocks, first reported by Card (1976), occur in the Killarney Bay area as inclusions and screens within the Killarney granite
(Stops K1 and K2). To the east, screens and inclusions recorded as
gneiss by Frarey and Cannon (1969; Frarey, 1985) are found to be without doubt also volcanic rocks (Stops K4 and K5). In some places these
rocks are very little deformed and, although hornfelsed, preserve
primary features such as quartz-filled amygdules, quartz and feldspar
phenocrysts, and equant, angular fragments of rhyolitic rocks (Fig. 6A)
forming both homo- and heterolithic breccias. In other places, as at
George Lake (Stop K4) and along the coast of Georgian Bay (Stop K5),
screens of similar rocks are foliated and have flattened clasts (Fig.
6B). Card (1976) considered that the volcanic rocks in Killarney Bay
might represent an as yet undefined Huronian formation lying stratigraphically above the Bar River Formation. Nowhere, however, are the
volcanic rocks in direct contact with Huronian rocks, the two being invariably separated by granite.

Near its northwest contact and particularly west of Killarney
village, the Killarney granite (s.s.) is massive, uniform, rose-pink,
medium grained and leucocratic (Stop K3). It is locally subporphyritic, containing subhedral perthitic K-feldspar phenocrysts in a seri-
ate quartz and two-feldspar matrix, along with small amounts of biotite,
very rare partly altered hornblende, accessory magnetite, apatite,
zircon and fluorite. Minor secondary muscovite is common. In most
places the granite is virtually free from inclusions, save for scarce,
isolated and generally rounded xenoliths of pink felsite. Sharp-walled,
narrow felsite or porphyritic felsite dykes occur locally. Southeast
of a line extending from Killarney northwest through George Lake, the

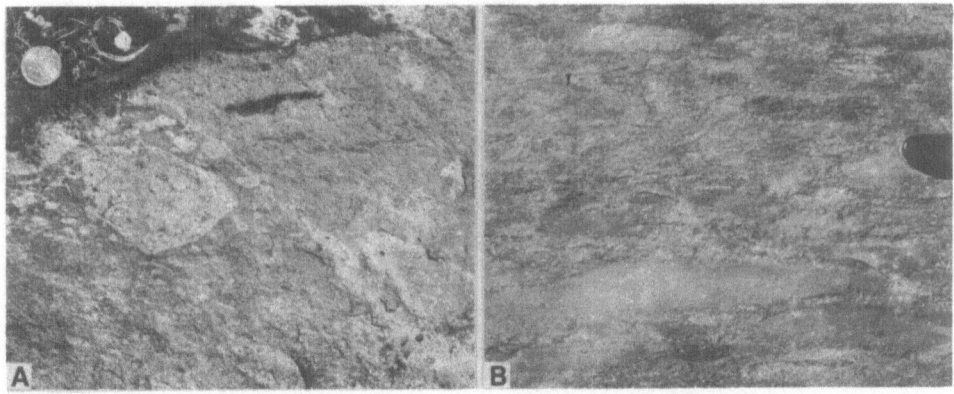

Figure 6. A, pyroclastic rock with angular rhyolite fragments; head of
Killarney Bay (unit kv in Fig. 5). B, deformed pyroclastic rock with
flattened rhyolite fragments; George Lake (Stop K4).

lenticular polycrystalline aggregates. The first appearance of this deformation can be seen in the granite at Killarney village (Stop K3), where discontinuous hair-line shears occur sporadically. Almost coincident with the appearance of this foliation is a contact with pink, porphyritic, fine-grained felsite, referred to as the 'Lighthouse porphyry' by Frarey (1985), which in fact constitutes the bulk of the Killarney granite (s.l.). The nature of this contact is debatable; where observed, it appears to be gradational within a few metres, but in one place at least it is known to be sharp. The porphyritic felsite southeast of the granite is not known to be cut by granite dykes; neither is it cut by the narrow felsite dykes that intrude the adjacent granite, and it could be construed that these dykes and the Lighthouse porphyry are related which, if proven, would mean that the porphyry is younger than the granite, a contention originally put forward by Jones (1930). Porphyritic felsite in the Killarney Bay area, however, is cut by granite dykes emanating from the main mass (Davidson, 1986c).

The Lighthouse porpyritic felsite, although foliated, is mainly uniform at any one place. In different parts of its extent, however, it varies in colour, in mafic content and in the proportion of small phenocrysts of K-feldspar, plagioclase and quartz. This unit hosts screens of flattened, heterolithic volcanic breccia near the west end of Collins Inlet (Fig. 5, Stop K5). In a zone extending from the eastern tip of George Island for 5 km northeast along the shore, porphyritic felsite grades into what can only be described as a clastic phase of the same rock, composed of subangular, elongate fragment (1 to 10 cm) of porphyritic felsite crowded in a similar but 'bleached' matrix that contains more muscovite than the clasts. In places this autoclastic breccia is laced with a network of sealed fractures, picked out by what look like diffusion bands up to 2 cm wide parallel to the fractures. This feature develops first into a marked bleaching of the wall rock on either side of quartz-muscovite seams, with quartz in the centres, and then, progressively, into a quartz-rich rock with fine muscovite, this rock having the appearance of quartzite. Quirke and Collins (1930) described this progression in the reverse order, maintaining that the quartz rock represents relics of Lorrain quartzite that had survived feldspathization to make granite, of which process the porphyritic felsite represents the intermediate stage. They interpreted the quartz phenocrysts in the felsite as relict sedimentary quartz grains, and the feldspar phenocrysts as porphyroblasts. However, in thin sections of the least deformed, unaltered felsite, the larger quartz grains commonly show the embayed, subhedral habit typical of volcanic or hypabyssal quartz phenocrysts (Fig. 7A), and some rocks are seen to have a granophyric matrix (Fig. 7B). These features attest to an igneous origin, and the association of these rock with undoubted volcanic rocks suggests that the Killarney complex is a high-level volcanic-plutonic association; it is this reasonable to interpret the quartz rock east of Killarney as the product of silicification.

The Killarney granite (s.s.) has a Rb-Sr whole rock age of ca. 1625 Ma (Wanless and Loveridge, 1972). More recently, U-Pb zircon ages have been obtained for the Killarney granite (1741±1.4 Ma) and the Lighthouse porphyry (1732+7-6 Ma) (van Breemen and Davidson, 1988).

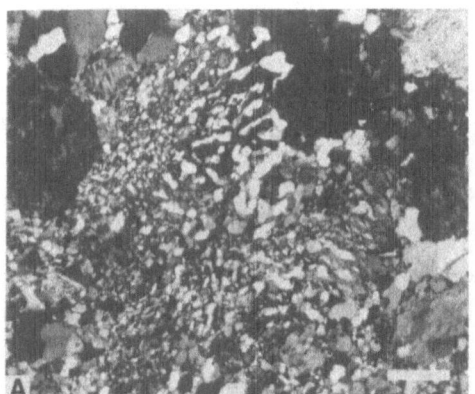 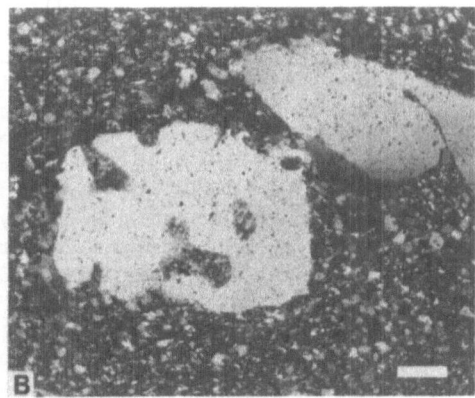

Figure 7. Photomicrographs of textures in Lighthouse porphyritic felsite, crossed polarizers; scale bar is .1 mm. A, embayed quartz pheno-
cryst; B, granophyric groundmass.

The U-Pb zircon ages for the Killarney complex accord well with ages
obtained for older granites and rhyolites in the subsurface of mid-
continental North America (Van Schmus and Bickford, 1981), and the
younger Rb-Sr age is similar to Rb-Sr ages from that same region which
Van Schmus (1980; personal communication, 1984) attributes to an un-
specified regional resetting event.

The Grenville Front Boundary Fault in this area is shown on the
1:253 440 scale Sudbury - Cobalt geological map (Card and Lumbers, 1977)
as passing along the southeast side of the granite plutons. Its southwestern extension has been traced from the south boundary of this map
to the coast of Georgian Bay (Figs. 4 and 5), where it is well exposed
on the south shore of Philip Edward Island (Stop K6b). Mylonitization
associated with the Grenville Front Boundary Fault adjacent to the
Killarney complex has produced a foliation that dips moderately to
steeply (35 - 75°) southeast. All the Grenville Province rocks to the
southeast carry this foliation, associated with which is a well developed, dip-parallel stretching lineation. Orientation of biotite flakes
and quartz lenticles is the common expression of this foliation; elongation of the quartz lenticles and other mineral aggregates gives the
lineation. Highly deformed pegmatite, probably rotated from former
orientations, forms tectonic layering parallel to the foliation in
parts of the granitoid gneiss lenses. All the precursors in the more
intense high strain zones are reduced to mylonite and ultramylonite in
which small feldspar porphyroclasts are all that remain of the original
minerals. Preferred orientation of deformation structures in the mylonitic rocks, such as S-C fabric, shear band foliation and rotated
feldspar porphyroclasts, indicate a sense of shear in accord with reverse displacement parallel to the regional stretching lineation. In
the Boundary Fault zone itself, pseudotachylite occurs as veinlets that
cut across and locally brecciate the mylonites.

Foliations within the Killarney complex strike northeast like
those in the rocks southeast of the Boundary Fault, and it would seem
from a cursory glance at map patterns that the two are related. However, an analysis of foliation and lineation attitudes on either side
of the Boundary Fault clearly shows a difference in structure (Fig. 8).
Whereas foliation southeast of the Boundary Fault invariably dips
southeast, that in the Killarney complex dips to either side of vertical. Lineations in the two domains are more obviously at variance:
lineations at and east of the Boundary Fault are dip-parallel and plunge
to the southeast; those in the Killarney complex plunge moderately to
gently between east and northeast. Surface trajectories suggest that
the Killarney foliation turns toward parallelism with Grenvillian
structure as the front is approached; the same is true for east-trending folds in the Huronian rocks farther northeast. Age relationships
between the two structural domains are given by the following evidence:
foliation and lineation in the Killarney complex are cut by straight-
walled, non-deformed pegmatite dykes (Stop K6a, Fig. 14 A and B) that
have been dated at ca. 1400 Ma (van Breemen and Davidson, 1988). These
in turn are cut by southeast-trending olivine diabase dykes of the ca.
1240 Ma Sudbury swarm (Stop K6b). Both types of dyke are deformed in
the mylonite zone that marks the Grenville Front Boundary Fault at the
coast of Georgian Bay (Stop K6b) where, however, it can be observed
that deformation in the pegmatite dykes preceded diabase intrusion and
was ductile, whereas later deformation that affected both diabase and
pegmatite was brittle. In this region, therefore, there are (at least)
four periods of deformation, manifested by (1) folding of the Huron
Supergroup, generally attributed to the Penokean Orogeny (ca. 1.9 Ga)
and predating intrusion of the Killarney granite, (2) foliation in the
Killarney complex adjacent to the Grenville Province that is post-
Penokean but pre-Grenvillian (and pre-ca. 1.4 Ga pegmatite), (3) intense mylonitization (ductile) at the margin of the Grenville Orogen
(post-pegmatite but pre-diabase), and (4) brittle deformation at the
Grenville Front Boundary Fault (post-diabase). The last phase of deformation becomes ductile within a few kilometres southeastward in the
Grenville Province. The nature and extent of the second and third
of these deformations is not well understood. However, it is known
that the volcanic-granite terrane which extends in the subsurface across most of the mid-continent south of the Penokean fold-belt is composed of two major age groupings: an earlier, deformed granite-rhyo-
lite association (1.8 - 1.6 Ga) and, mainly farther to the south, non-
deformed, younger granite plutons (1.5 - 1.35 Ga) (Bickford and Van
Schmus, 1981). It may be that the Killarney complex is part of the
former association, the granitoid gneisses to the east being equivalent
rocks reworked during the Grenvillian Orogeny. It is worth noting that
granitoid orthogneiss bodies farther east and southeast in the Grenville Province have U-Pb zircon ages commensurate with the younger mid-
continent grouping (van Breemen et al., 1986).

In light of the above, it seems more logical to place the margin
of the Grenville Province in the Killarney area at the mylonite zone
that passes southeast of the Killarney complex rather than along its
northwest contact. The Killarney granite contact diverges from the

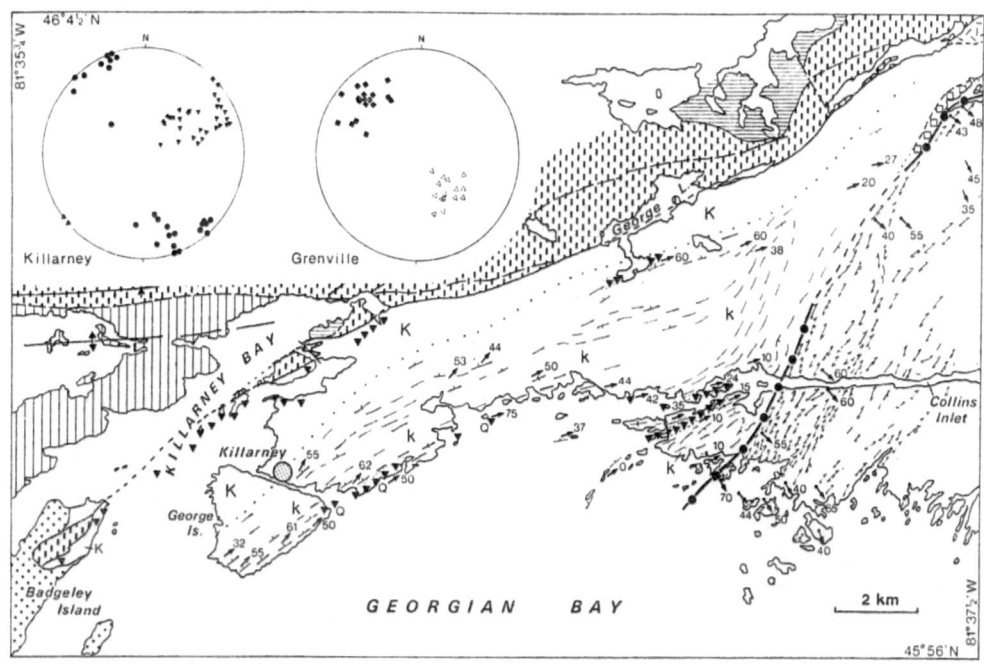

Figure 8. Structural geology of the Killarney complex and adjacent Grenville Front Tectonic Zone. Stereoplots are lower hemisphere projections. Solid symbols = poles to foliations, triangles = lineations.

region in which deformation of Grenvillian orientation is extant. Its southwest continuation is buried beneath Paleozoic cover, but aeromagnetic patterns suggest the presence of plutonic rocks south of the Huronian terrane and west of the extension of the Grenville Front beneath Lake Huron; boreholes on Manitoulin Island, southwest of Killarney, intersect 'Killarnean' granite. If the Grenville Front is placed at the northwest edge of the Killarney granite, it would not be the boundary of the intense effects of orogeny of Grenvillian age. Although technically constituting a structural discontinuity, as pointed out by Frarey (1985), and thus satisfying Stockwell's (1964) criteria for defining a structural province boundary, the boundary between the folded Huronian rocks of the Southern Province and the Killarney granite does not have the same style or age as the mylonite zone that defines Lumbers' Grenville Front Boundary Fault, with its associated northwesterly reverse displacement. When the extent and nature of middle Proterozoic orogeny in mid-continental North America has been more fully evaluated, with its own 'front' against older rocks to the north, the Killarney complex and granites of similar age adjacent to the Grenville Front will likely be found to represent the northeasternmost extent of this major province. Stockwell (1982) foresaw this development in his definition of the 'Killarnean Orogeny'; the fact that he placed the type area within the Grenville Province merely reflects its former pre-Grenvillian extent to the east. This recognition helps to explain the 'disappearance of the Huronian'. The Killarney–Grenville relationship has analogies elsewhere along the Grenville Front, for example in Labrador where granites ascribed to the ca. 1.65 Ga Labradorian Orogeny (Thomas et al., 1985) occur within the Grenville Province and have counterparts among the granites and volcanic rocks of the Makkovik Province to the north (12 in Fig. 1).

East of the Grenville Front Boundary Fault on the Georgian Bay coast, superb, clean exposures allow close field examination of the complex gneissic rocks that underlie the northwest part of the Grenville Orogen. Along the south shore of Philip Edward Island (Fig. 5), fine-grained, mainly light pink and grey streaky gneisses are interleaved with lenticular masses of deformed granitoid rocks. None of the gneisses is recognizably derived from sedimentary rock of characteristic lithology (quartzite, pelite, calc-silicate) that might be expected to have survived severe deformation and metamorphism, whereas farther north, even as close to the outer coast as Mill Lake and Beaverstone Bay (Fig. 5), such metasedimentary rocks are well preserved (and are discussed in Part III of this guide). However, at two places so far discovered in the otherwise nondescript gneisses (Davidson and Bethune, 1988), one of which is Stop K7, deformed coarse clastic structure is preserved and may represent volcaniclastic rocks like those found in the Killarney complex. If this is true, the enclosing gneisses may be deformed equivalents of felsic volcanic and hypabyssal rocks, and the whole assemblage, including the granitoid masses, may have been derived from volcano-plutonic complexes. Allowing that part of the Killarney complex had been severely deformed long before the Grenvillian Orogeny, it is little wonder that protoliths of such rocks are obscure southeast of the Grenville Front. A marked Bouguer gravity low is located astride the Grenville Front in this region, indicating that low density rocks are present to considerable depth on both sides of the front (McGrath et al., 1988); these data are in accord with the interpretation that the felsic gneisses and granitoids southeast of the front were derived from rhyolitic volcanic rocks and high level granitoid intrusions.

Stop Descriptions

The field excursion in the Killarney area will allow examination of some characteristic features of the Killarney complex, mylonitization at the Grenville Front Boundary Fault, and possible equivalents of the Killarney complex in the Grenville Province, as follows (see Fig. 5 for stop locations):

K1 - Inclusion-rich border phase of the Killarney granite at its northwest contact with the Gordon Lake Formation;

K2 - Hornfelsed volcanic rocks in a large screen within the Killarney granite;

K3 - Massive, homogeneous Killarney granite with scattered xenoliths, felsite dykes and minor brittle deformation;

K4 - Deformed volcanic rocks at the southwest contact of the Killarney granite;

K5 – Deformed volcaniclastic rocks and fine-grained hypabyssal granite in the interior part of the Killarney complex;

K6 – a: Non-deformed pegmatite cutting foliation in hypabyssal granite;
　　 b: Development of mylonite and ultramylonite at the Grenville Front Boundary Fault;
　　 c: Penetrative ductile deformation southeast of the Grenville Front Boundary Fault;

K7 – Fine-grained felsic gneisses in the Grenville Province, possibly derived from Killarney complex volcanic rocks;

K8 – Style of folding in fine-grained felsic gneiss with thin amphibolite layers;

K9 – Deformed granitoid rocks in the Grenville Province.

Note: Stops K1 and K2, and K5 to K9, can only be visited by boat.

Stop K1. Contact between hornfelsed siltsone of the Gordon Lake Formation and the Killarney granite; Maxwell Point, Badgeley Island, 4.5 km west-southwest of Killarney village.

The Gordon Lake Formation is a thin-bedded siltstone, exposed here on the steep, southeast-facing limb of a large, upright, east-plunging anticline whose axial trace lies 4 km across the water to the north. It overlies quartzite of the Lorrain Formation and is overlain by similar quartzite of the Bar River Formation which forms the large hill to the south where it is quarried for silica. Numerous small-scale sedimentation features (graded bedded, ripples, shallow channel scours, mud cracks and loading structures like sand balls and flames) clearly indicate younging to the southeast. The aluminous nature of the siltstones has given rise to porphyroblasts of cordierite, now altered to dull greenish chloritic material, and andalusite (small pink prisms); these minerals are attributed to contact metamorphism adjacent to the Killarney granite.

The northwest contact of the Killarney granite is discordant, cutting at a shallow angle across the contact between the Gordon Lake and Bar River formations. In northeast Killarney Bay an apophyse of granite penetrates between these two formations. At this stop the border phase of the granite is crowded with xenoliths of several types (Fig. 9, A and B). Interestingly, very few of them appear to have been derived from the adjacent Gordon Lake siltstones, and quartzite xenoliths derived from the nearby Bar River Formation are absent. The granite itself is pink, massive, leucocratic, medium to fine grained and essentially equigranular. It includes equant masses of an earlier, grey, porphyritic granite phase that itself contains xenoliths of other rock types (Fig. 9, B). It also includes individual xenoliths of these other types, namely dark, uniform blocks rich in biotite and amphibole, dark to medium grey, thinly layered rocks (tuffs?), light grey or violet-pink felsite with small feldspar phenocrysts, and grey, streaky felsic rocks that may have been pumice tuffs or ignimbrites.

The absence of wall-rock xenoliths is puzzling, but may be explained as follows. The association of the Killarney granite with predominantly felsic volcanic rocks and hypabyssal porphyritic felsite, all of similar age and composition, allows this assemblage to be interpreted as a high level volcano-plutonic complex (Davidson, 1986b).

Intrusive upwelling of granite magma into its own volcanic or hypabyssal carapace may have been accompanied by down-going magma flow along the outer contact, the magma carrying with it xenoliths of volcanic rocks and of earlier porphyritic granite that had already incorporated volcanic inclusions.

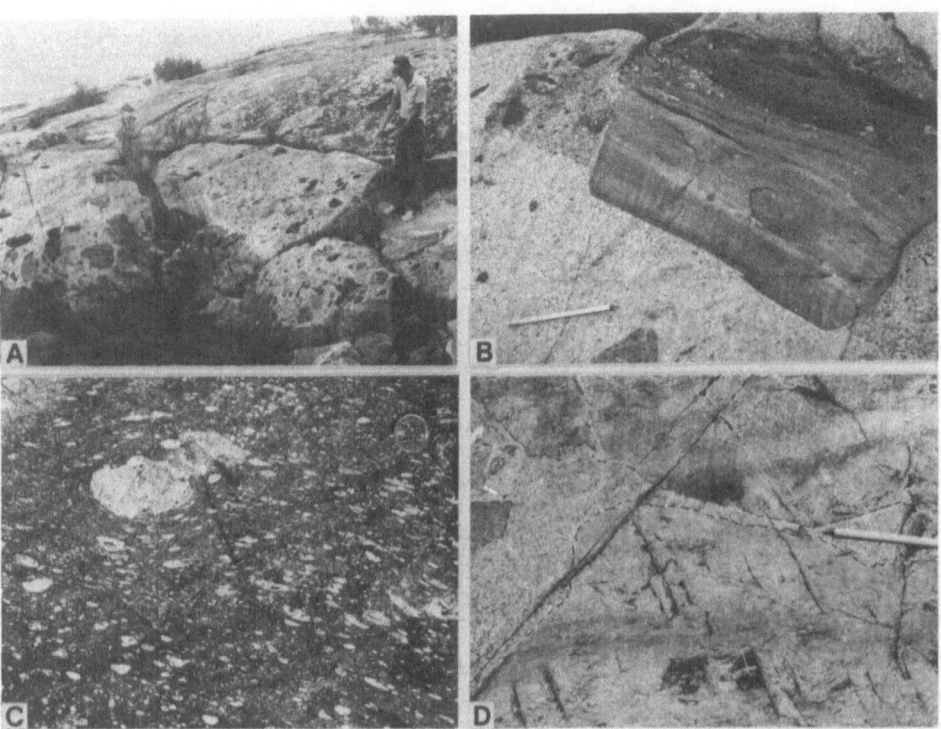

Figure 9. A, xenolithic border phase of the Killarney granite at Maxwell Point, Badgeley Island; B, xenoliths within xenolith: large xenolith of layered hornfels protruding from xenolithic grey porphyry inclusion in Killarney granite; pencil is 15 cm long; C, porphyritic andesite at Stop K2; coin is 19 mm in diameter; D, granite dykelet penetrates contact between porphyritic andesite (above) and dacitic lapillae tuff (below pencil); Stop K2.

Stop K2. Volcanic rocks of the Killarney complex; islands in Killarney Bay, 2 km northwest of Killarney village.

Volcanic rocks are exposed among these islands and extend northeastward in a large screen within the Killarney granite. They are everywhere separated from the Huronian country rocks, in this area the Bar River quartzite, by a narrow interval of granite. The clean outcrops here illustrate several lithologies and their relationships. The southern part of this small island, 250 m southwest of the lighthouse

on Partridge Island, is underlain by dark grey hornfelsed andesite with
small plagioclase phenocrysts (Fig. 9, C). On the east side this unit
is in contact with dacitic lapillae tuff and both are cut by dykelets
of Killarney granite (Fig. 9, D). The northern part of the island is
composed of felsic volcanic or hypabyssal rocks, also invaded by granite which at one place includes an angular fragment of an intermediate
volcaniclastic rock.

Stop K3. Killarney granite outcrops in Killarney village.
     The Killarney granite may be examined at a number of places in the
village of Killarney where it is exposed in low, clean whale-backs,
some of which have been blasted during street construction. The outcrop behind the Sportsmans' Inn is a good example. Pink, massive,
leucocratic granite here contains approximately equal parts of sodic
plagioclase and K-feldspar, 25% quartz and 2% biotite. Some parts of
the Killarney granite are sub-porphyritic (Fig. 10, A). Outcrops by
the shore just west of the Sportsmans' Inn are cut by narrow dykes of
pink felsite. The surface of the blasted outcrop on the street to the
north shows the first evidence of deformation in the form of dark hair-
line shears and brittle fractures. Outcrops at the shore of Killarney
Bay to the west do not show this deformation, but those farther east,
for example at Killarney Mountain Lodge, show a closer spacing of these
shears, which also cut the felsite dykes.
     The first blasted outcrop on the road to the airport, a few blocks
north of the waterfront, exposes weakly foliated granite with a faint
mineral lineation plunging steeply northeast; rare felsite xenoliths
are elongate in the foliation plane (Fig. 10, B). Before reaching the
airport, low outcrops near a fork to the right of the road expose
uniform pink feldspar porphyry ('Lighthouse porphyry' of Frarey, 1985)
the fork leads to a lighthouse at the east entrance to Killarney
Channel, where the autoclastic porhyritic felsite and the effects of
silicification described in the foregoing text can be examined.

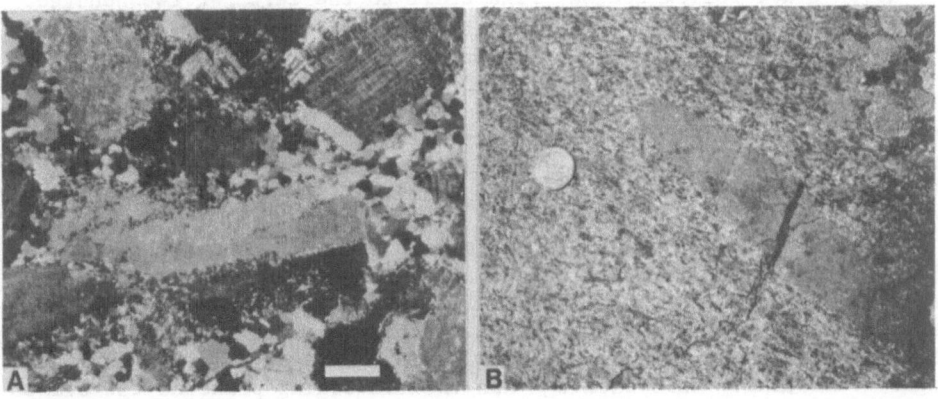

Killarney granite, airport road.

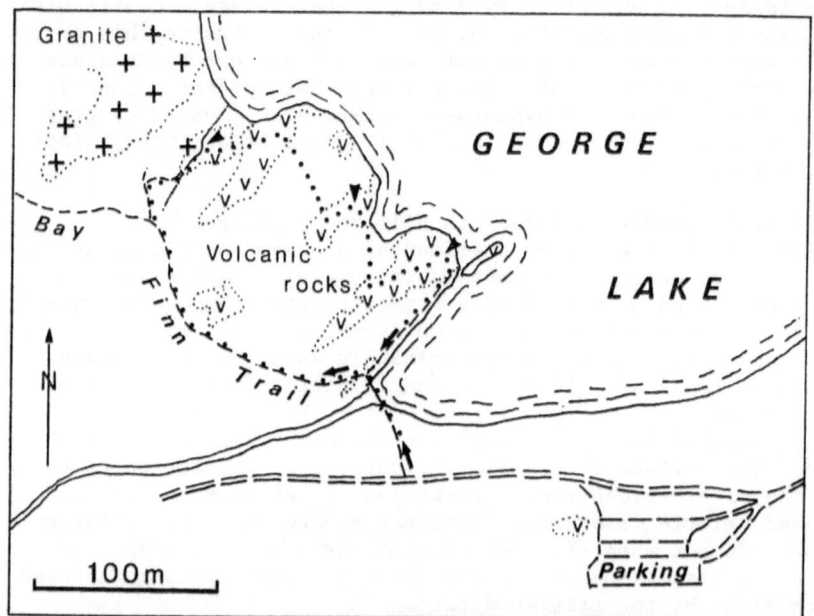

Figure 11. Outcrop map, area of Stop K4.

Stop K4. Deformed felsic volcanic rocks on the southeast side of the
Killarney granite. Outcrops are north of the outlet of George Lake,
Killarney Provincial Park, 11 km by road northeast of Killarney.
    Walk from the parking area (Fig. 11) to the west end of the south
bay of George Lake and cross the weir that controls the lake level. On
the north side, follow the Bay Finn trail to the left for two hundred
metres, then take a secondary trail that forks to the right, and keep
to the right until coming out in the open at the top of the hill. Exposed here is the southeast contact of the Killarney granite with light
grey felsite (metamorphosed compacted ash-flow tuff?). Note the bulb-
ous projections of granite into the felsite. Narrow dykes of pink,
sparsely porphyritic felsite intrude both granite and grey felsite.
The hills of white rock to the north are underlain by quartzite of the
Bar River Formation.
    Go down the south side of the outcrop close to the lake and cross
the wooded, covered interval to the next rock ridge. The first outcrop
is felsic, heterolithic coarse clastic rock with flattened clasts of
different kinds of porphyritic rhyolite (Fig. 6, B). At the lake shore
thin-bedded tuffs display a well developed cleavage at a very shallow
angle to bedding. This cleavage and the flattening plane of the clasts
dip northwest at about 55°, toward the granite contact.
    Walk southward again through the woods and climb the next rock
ridge, angling to the left. The rock on this spur is white-weathering,
light grey quartz-feldspar porphyry that locally shows a clastic structure of uncertain origin (primary brecciation or the result of deform-

ation?). These rocks have a well developed cleavage whose surfaces shine with fine muscovite. The cleavage contains a prominent, streaky, northeast-plunging lineation. A notable feature is the presence of small quartz grains that are neither flattened nor elongated. In a rock that appears so deformed, it would seem unusual for quartz grains to have remained so equant unless the matrix was softer than quartz during deformation. This may be due to the low temperature of deformation, but it is also possible that this rock was once a quartz vitrophyre with a glassy or finely recrystallized glass groundmass.

Follow the trail west along the lakeshore back to the weir and return to the parking area, where nearby outcrops are light grey-pink porphyritic felsite, typical of much of the Lighthouse porphyry, and thought to represent hypabyssal intrusive rocks.

Stop K5. Folded, flattened felsic volcaniclastic rocks of the Killarney complex. Island in bay on southwest shore of west end of Philip Edward Island; 9 km east of Killarney and 1.75 km northwest of the Grenville Front Boundary Fault.

This island exposes clean outcrops of a variety of felsic volcaniclastic rocks that are highly flattened and display pre-Grenvillian deformation typical of the southern part of the Killarney complex. The rocks comprise homolithic fragmental felsite with rare, large clasts of coarser porphyry (Fig. 12, A), finely bedded tuffs, and heterolithic fragmental rocks with both felsic and mafic clasts now drawn out into thin plates (Fig. 12, B). The attitude of primary layering can be seen locally to dip moderately to the north. Layering is cut by a steep cleavage that in places is itself crenulated. Clasts are not only flattened but are also strongly elongate, defining a shallow east-northeast-plunging extension lineation. The steep cleavage and shallow lineation are related to major folds in the Killarney complex; the rocks here lie on the south limb of a tight synform (syncline?) whose axial trace lies within Collins Inlet, 750 m to the north. These volcaniclastic rocks flank a domal mass of uniform, subporphyritic, fine

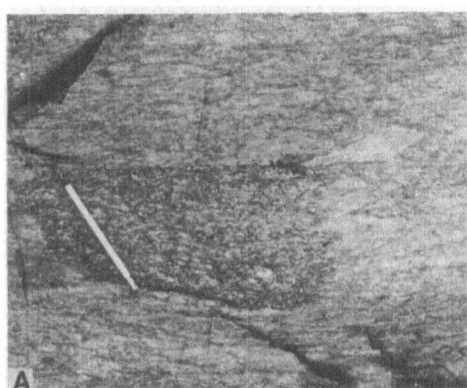

Figure 12. Stop 5. A, coarse grained clast in felsite breccia; B, highly flattened clasts in heterolithic fragmental unit.

grained granite, exposed on the south side of this island. This hypabyssal granite contains small plagioclase and K-feldspar phenocrysts but none of quartz; it has a somewhat more mafic matrix than the typically leucocratic Lighthouse porphyry.

Stop K6. Grenville Front Boundary Fault and adjacent rocks; islands at the southwest shore of Philip Edward Island, 11 km east of Killarney.

The Grenville Front Boundary Fault crosses the Georgian Bay coast near the west end of Philip Edward Island, where clean outcrop surfaces at the coast and on the many small islands and rock reefs display relationships to advantage. The geology around this stop is given in Figure 13. West of the Boundary Fault the Killarney complex is repre-

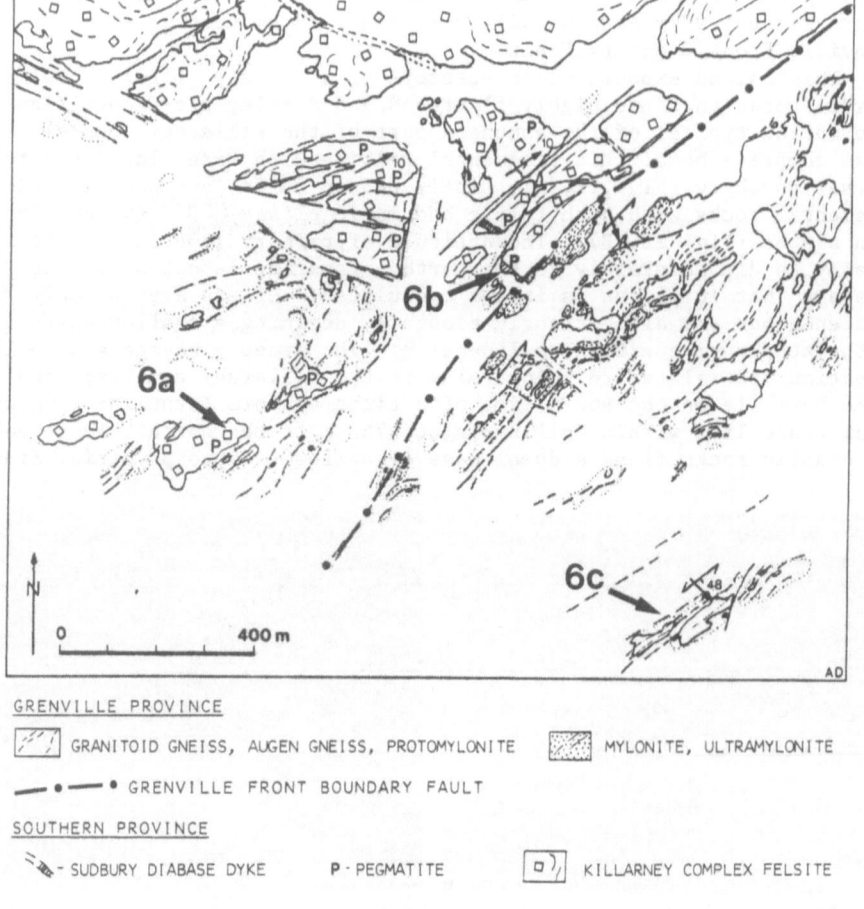

GRENVILLE PROVINCE

▨ GRANITOID GNEISS, AUGEN GNEISS, PROTOMYLONITE   ▨ MYLONITE, ULTRAMYLONITE

—•—• GRENVILLE FRONT BOUNDARY FAULT

SOUTHERN PROVINCE

↘ - SUDBURY DIABASE DYKE   P - PEGMATITE   ▢ KILLARNEY COMPLEX FELSITE

Figure 13. Sketch map of the area around Stop K6, south shore, Philip Edward Island, illustrating relationships at the Grenville Front.

sented by the same hypabyssal granite seen at Stop K5. Its easterly-striking foliation has variable dip; its lineation plunges gently east-northeast and may be coaxial to folds that, however, still require to be defined by numerous measurement of the foliation, because marker units are not present to help outline the folds. Pegmatite dykes cut the foliation and lineation in the fine grained granite and are not deformed as close as 100 m to the Boundary Fault. A northwest-trending vertical diabase dyke of the Sudbury swarm is partly exposed along the southwest shore of Philip Edward Island and cuts cleanly across contacts between pegmatite and granite. The granite is metamorphosed to greenschist facies (albite – epidote – chlorite – biotite – ?actinolite – titanite). The diabase is altered, but retains some clear original plagioclase and cores of augite rimmed by secondary amphibole, biotite and chlorite; its olivine is entirely altered to chloritic material heavily dusted with fine magnetite.

The Boundary Fault itself is an intense mylonite zone several tens of metres wide that dips steeply southeast and truncates contacts between the rock types observed in the Killarney complex; it thus formed later than intrusion of the Sudbury diabase dyke. Immediately north-west of the mylonite zone, brittle deformation effects are rampant at all scales. In the zone itself, felsite, pegmatite and diabase are all reduced to flinty ultramylonite.

Southeast of the Boundary Fault, other mylonite zones of similar orientation separate large panels of variably deformed quartzofelds-pathic rocks whose moderately southeast-dipping foliation carries a pronounced dip-parallel stretching lineation, typical of Grenvillian orogenic overprint. Deformation style is ductile, and the rocks contain metamorphic mineral assemblages indicative of at least middle amphibolite facies.

At this stop, three separate locations are used to illustrate the relationships described above.

Stop K6a. The outcrop on this island, just west of the Boundary Fault, clearly shows truncation of foliation in the fine grained granite by straight-walled, non-deformed pegmatite dykes (Fig. 14, A and B), and subsequent deformation in the form of narrow mylonitic to cataclastic shears. On the islets just to the southeast both pegmatite and granite are brittly deformed, the pegmatite being broken into small segments. Pegmatite from the larger island 500 m to the north-northeast has been dated (U-Pb, highly discordant zircons) at 1400 ± 50 Ma (van Breemen and Davidson, 1988), confirming that the earlier deformation in the Killarney complex is pre-Grenvillian.

Stop K6b. At the southwest corner of this island, just west of the Boundary Fault mylonite zone, a brittly deformed Sudbury diabase dyke truncates hypabyssal granite and pegmatite, both of which are deformed; note the ductile deformation of quartz in the pegmatite next to the exposed contact of the diabase dyke. To the east along the shore, narrow ultramylonite zones trending northeast and dipping steeply southeast offset the diabase dyke, one with sinistral and the other with dextral sense. In this area, deformed amphibolite dykes, known also in places in the Killarney complex to the west, give rise to black ultramylonite bands within dark pink, flinty ultramylonite derived from the hypabyssal granite. A few metres farther east all of the rocks, including pegmatite, are converted to flinty mylonite and ultramylonite with feldspar porphyroclasts of various size (Fig. 14, C). These rocks show a variety of folds, common in this type of mylonite zone and indicating extreme plasticity during mylonitization; these folds are considered to have formed during the mylonitization process, not to be the result of a distinctly later phase of deformation. X-Z oriented thin sections of these rocks display rotated, bent and mechanically broken feldspars, and extremely thin quartz lamellae with oblique subgrain orientation (Fig. 14, D), both of which imply northwest-directed thrust displacement.

These outcrops speak for themselves; if you must hammer the mylonitic rocks, protect your eyes and have regard for unwary bystanders. Mylonites tend to fly apart when struck, releasing high-speed, sharp-edged missiles.

Stop K6c. These small islands, less than 800 m from the mylonite zone that marks the Boundary Fault, display penetrative ductile deformation in thoroughly recrystallized, fine grained, pink and grey quartzofeldspathic rocks. The rocks carry a pronounced dip-parallel stretching lineation on 45° southeast-dipping surfaces (Fig. 14, E). A major component of flattening is illustrated by extraordinarily well developed ptygmatic folding of narrow pegmatite seams (Fig. 14, F). Note that both 'S' and 'Z' fold forms are present along individual seams where viewed on surfaces normal to the lineation. In thin section, these rocks show a clean, relatively equigranular assemblage of strained quartz, oligoclase, microcline, biotite and hornblende, with rare garnet, in contrast to the lower grade assemblages just northwest of the Boundary Fault, and indicating a marked jump in metamorphic grade across the front.

Stop K7. Fine grained quartzofeldspathic gneiss, possibly correlative with Killarney complex volcanic rocks; small island east of Solomon Island, 1.8 km southeast of the Boundary Fault.

Transposed quartzofeldspathic gneiss and deformed pegmatite are exposed on this island. The fine grained gneiss is fairly typical of gneisses between deformed granitoid masses that underlie the south shore of Philip Edward Island between the Boundary Fault and Beaverstone Bay, 12 km east of here. Southeast-dipping foliation and layering carry the ubiquitous dip-parallel lineation characteristic of the Grenville Front Tectonic Zone, though it is not as strongly developed here as elsewhere. At a few places on the west side of the island a faint, stretched clastic character is visible in the gneiss; at one place, apparently in a low-strain, intrafolial fold hinge, the clasts are more equant in down-plunge view and are of varied composition (Fig. 15, A). May this be the deformed equivalent of the flattened heterolithic volcaniclastic rocks seen at Stop K5?

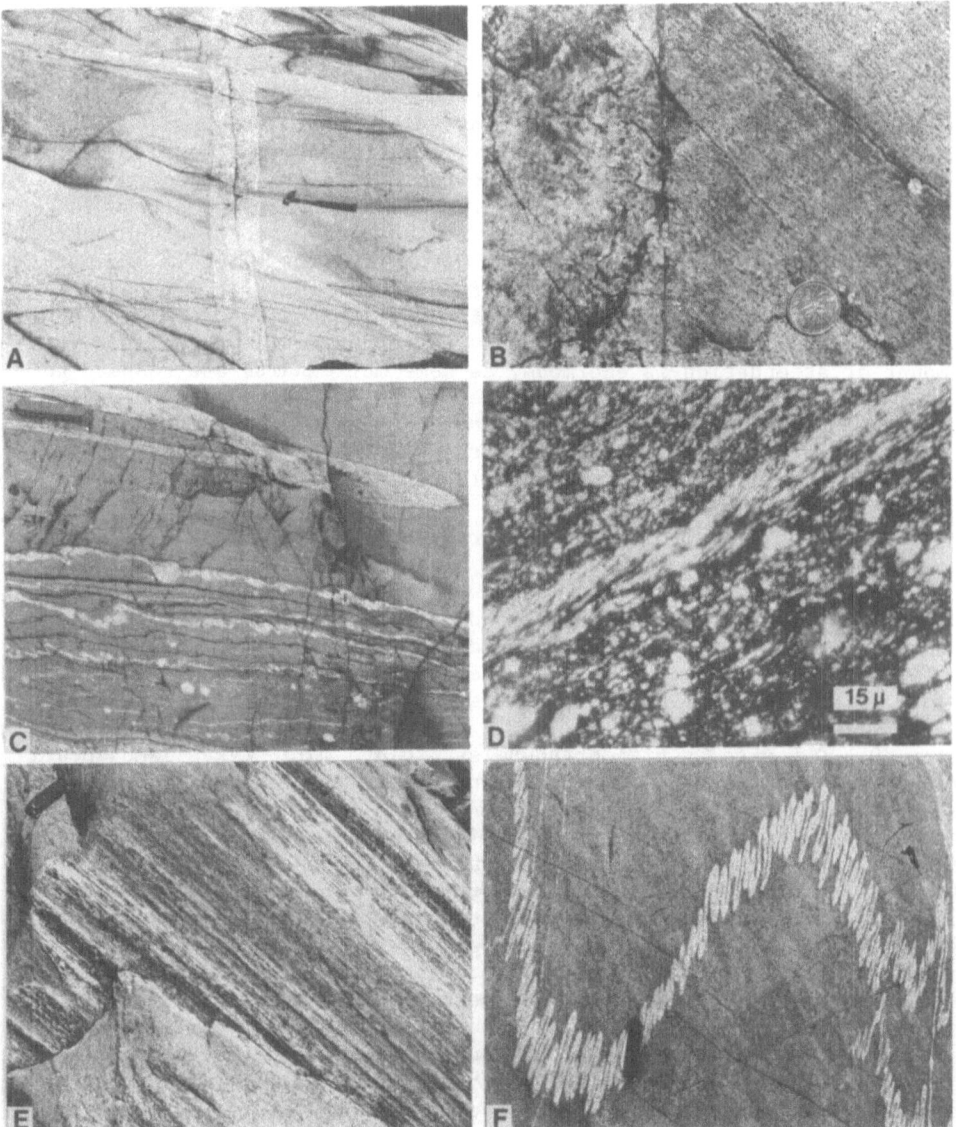

Figure 14. Stop K6a: A, straight-walled pegmatite dyke cutting hypabyssal granite of the Killarney complex; B, foliation in granite is
cut by pegmatite. Stop K6b: C, mylonite derived from hypabyssal granite and pegmatite at the Grenville Front; D, photomicrograph, crossed
polarizers, oblique foliation of subgrains in quartz lens in mylonite.
Stop K6c: E, intense southeast-plunging stretching lineation on the
surface of pegmatite seam in fine grained quartzofeldspathic gneiss;
F, extreme development of ptygmatic folds in pegmatite seam.

Stop K8. Multiple deformation in quartzofeldspathic gneiss with thin
amphibolite layers; Campbell Island, west side of Solomon Bay, 5 km
southeast of the Boundary Fault.

This island is underlain by pink to violet-grey, uniformly fine
grained quartzofeldspathic gneiss containing minor hornblende, biotite
and magnetite. Layering is not everywhere evident, but locally is
developed well enough to display mesoscopic folds of two generations.
Vaguely-bounded layered units trend roughly southeastward across the
island and show early folds that are refolded about southeast-plunging
axes. Alignment of small hornblende porphyroblasts (Fig. 15, B) parallels later fold axial planes and crosses the hinges of first phase
folds. These structures are accentuated by the presence of thin amphibolite seams and layers that probably represent mafic dykes; if so, the
amount of attenuation is enormous, as some mafic units are reduced to
long seams of millimetre thickness before being pinched out altogether.
Note the 'fuzzy' projections along the contacts of some amphibolite
layers (Fig. 15, C); are these features tiny folds whose curved traces
reflect distortion of the stress field near the interfaces during
folding?

These rocks may represent Killarney-age felsic volcanic rocks, but
no primary features remain to give a clue to their protoliths. The
layering, whatever its origin, was undoubtedly folded about easterly
trending axial surfaces at an unknown time before superposition of
phase two folds that have 'Grenvillian' orientation. Perhaps the early
folds are a manifestation of the same deformation(s) that affected the
southern exposed part of the Killarney complex northwest of the front.

Stop K9. Deformed granite orthogneiss, folded mafic dykes, and shear
zones; southeast part of Silver Island, vicinity of Big Rock Bay, 7 km
southeast of the Boundary Fault.

Bedrock in this area is granitoid orthogneiss, part of a body that
is in sharp contact with the fine grained quartzofeldspathic gneiss
examined at Stop K8, 800 m to the west-northwest. In its least deformed state it exhibits an east- to southeast-trending foliation with
small intrafolial leucosomes. In most places this early foliation is
small-folded and a new leucosome has developed in the axial planes of
the younger folds (Fig. 15, D). Superimposed on this structure are
shear zones in which the earlier structures are 'tightened' into parallelism with the shear plane, developing protomylonitic to mylonitic
fabric (Fig. 15, F). Note that the younger foliation is at a shallow
angle to the shear zones (east side of the southeastern peninsula of
the island).

Mafic dykes cut the earlier foliation and leucosomes, are folded
by the second phase deformation (Fig. 15, E) and are distorted again in
the late shear zones (Fig. 15, F), which have 'Grenvillian' trend.
Some dyke segments, a few metres wide, have preserved chilled margins,
exhibit relict diabasic texture in their cores, and contain scattered
plagioclase phenocrysts near their contacts. Most dykes, however, are
highly attenuated, are converted to fine grained amphibolite, and outline high-amplitude reclined isoclines, in some of which the limbs are
represented by thin biotite-hornblende seams that weather out.

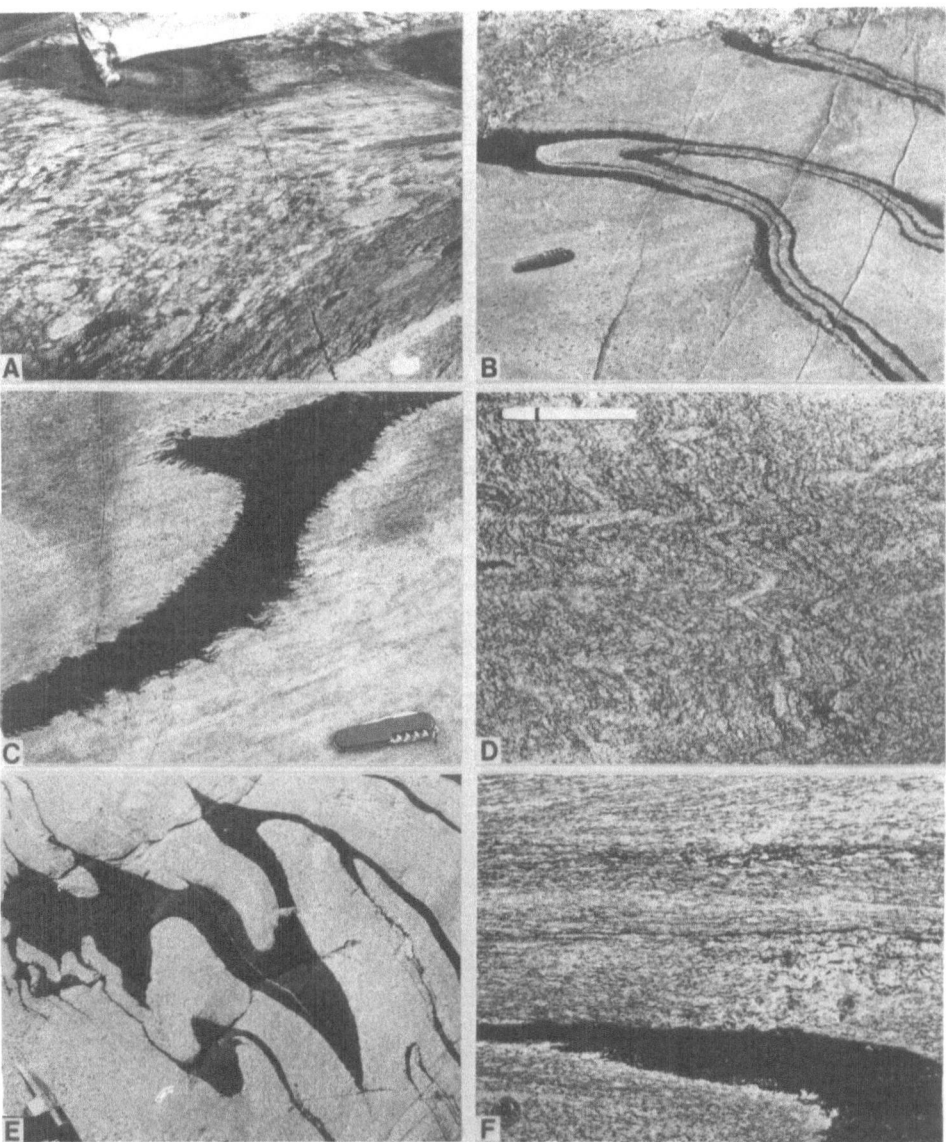

Figure 15. Stop K7: A, heterolithic clastic rock preserved in low strain lens within highly attenuated quartzofeldspathic gneiss and pegmatite. Stop K8: B, fold in layered gneiss with amphibole porphyroblasts aligned in the axial plane; C, deformed metamafic dyke in quartzofeldspathic gneiss. Stop K9: D, folded foliation in granitoid orthogneiss with leucosomes developed parallel to the axial planes of the second phase folds; E, folded metamafic dykes in orthogneiss; F, margin of shear zone showing development of 'Grenvillian' protomylonite from migmatitic orthogneiss and metamafic dyke.

These structures result from ductile deformation that could only have taken place relatively deep in the crust. The granite orthogneiss has not been dated. Chemical analysis of well preserved, diabase-textured mafic rock from a dyke remnant on a nearby island suggests that it is chemically dissimilar to Sudbury diabase; it is olivine-normative, but has more Al, Mg and Ca, and less Ti, Fe, K, P and Zr. Similar narrow dykes, everywhere folded, are common along the coast of Georgian Bay to the east and extend into Britt domain southeast of the Grenville Front Tectonic Zone. On the basis of their style of deform-ation and more complete recrystallization, they are judged to be older than Sudbury diabase dykes, which are preserved as coronitic metadia-base dyke remnants in the same region; exposed contact relationships between the two, however, have not so far been found.

PART III: GEOLOGY OF THE TYSON LAKE AREA - TECTONIC CONSTRAINTS GIVEN
BY THE SUDBURY DYKES

Introduction

A broad section across the Grenville Front in the Tyson Lake area
(Fig. 16) passes, from northwest to southeast, through (1), upper
Huronian strata with Nipissing diabase sills, folded about east-west
axes, (2), granites of the Bell Lake and Annie Lake plutons, flanked
to the southeast by intensely deformed Huronian formations, and (3),
gneisses of sedimentary and plutonic origin. Present in all three
zones are olivine diabase dykes of the Sudbury swarm. These vertical
dykes, up to 100 m wide, trend southeast and pass uninterrupted through
the first two zones, except for minor dislocation at faults. In the
third zone, i.e. within the Grenville Province, they are represented by
folded dykes and dyke segments which show progressive deformation and
metamorphism toward the southeast (Frarey, 1985; Bethune and Davidson,
1988). The Grenville Front Boundary Fault is therefore placed between
zones 2 and 3 where the major break occurs in the metamorphic and
structural state of the dykes. This break is the locus of intense
mylonitization and also marks the southeast limit of metasedimentary
rocks that can be correlated to specific Huronian formations with some
confidence.

The Huron Supergroup Adjacent to the Grenville Front

The Huron Supergroup, comprising four groups with an aggregate thickness of the order of 15 km, is represented in the Southern Province
west of Tyson Lake by the uppermost part of the Hough Lake Group
(Mississagi Formation), the Quirke Lake Group (Bruce, Espanola and
Serpent formations) and the Cobalt Group (Gowganda, Lorrain, Gordon
Lake and Bar River formations). Characteristic lithologies and estimated thicknesses, summarized from Frarey (1985), are given in Table I;
the total estimated thickness for this region close to the front is
between 9 and 12 km. This enormously thick succession is folded about
gently east-plunging axes, and although sheared in places and everywhere metamorphosed to at least middle greenschist facies, retains
ample evidence of its various sedimentary deposition environments
through preserved primary compositions and sedimentary structures.
Individual formations are for the most part distinctive and usually
have abruptly gradational, conformable contacts; an exception is the
contact between the Serpent and Gowganda formations, which is knife-
sharp and may represent a disconformity in this area. (Northeast of
Sudbury the Gowganda Formation successively overlaps older Huronian
formations until it lies directly on Archean basement.) In any event,
where some confusion may arise as to whether, for example, a particular
quartzite belongs to the Serpent, Lorrain or Bar River Formation, its
assignation is left in no doubt once its characteristic neighbouring
formations are identified.

All Huronian formations are intruded by thick sills and some small
dykes of Nipissing diabase, a tholeiitic metagabbro with metamorphic

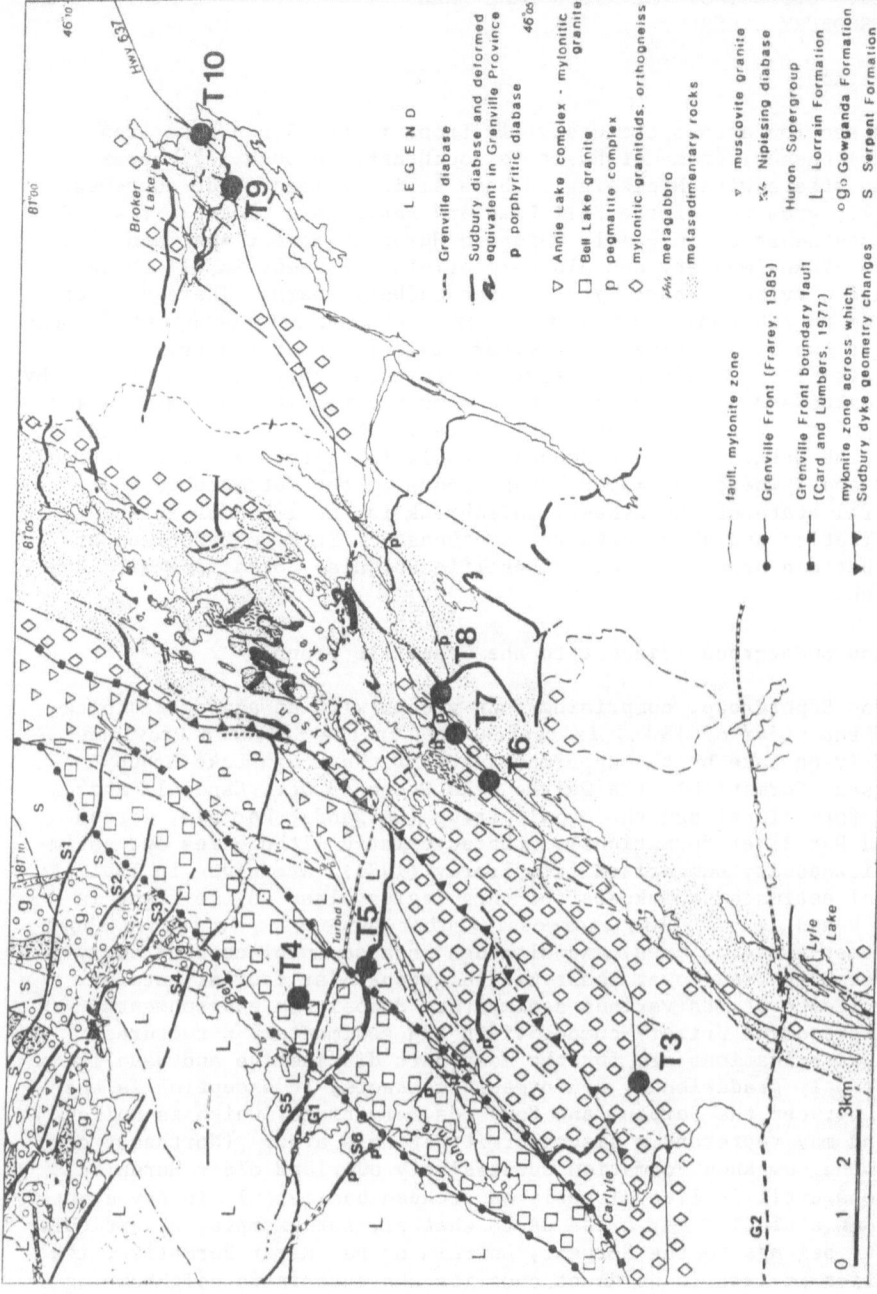

Figure 16. Geology of the Tyson Lake area (after Bethune and Davidson, 1988). Locations of field excursion stops T3 to T10 are shown.

TABLE I. Huron Supergroup formations in the Southern Province west of Tyson Lake (after Frarey, 1985).

Cobalt Group

Bar River Formation	Orthoquartzite, hematitic quartzite and siltstone	>910 – 1220 m (top unknown)
Gordon Lake Formation	Argillite, siltstone, quartzite	1060 – 1280 m
Lorrain Formation	Orthoquartzite, feldspathic quartzite, arkose	2140 – 1740 m
Gowganda Formation	Matrix-supported conglomerate, laminated and massive argillite, siltstone, feldspathic quartzite	1220 – 2100 m

Quirke Lake Group

Serpent Formation	Feldspathic quartzite, arkose, minor conglomerate, siltstone	1060 – 2100 m
Espanola Formation	Thin dolomite member at top, calcareous siltstone and shale, basal limestone member (< 45 m)	~430 m
Bruce Formation	Matrix-supported conglomerate	175 – 600 m

Hough Lake Group

Mississagi Formation	Cross-bedded feldspathic sandstone, minor argillite	1500 – 2600 m

(Minimum thickness 8.8 km)

minerals commensurate with metamorphic assemblages in the Huronian country rocks. A sill of this suite in the Cobalt Group north of Sudbury has been dated at 2.22 Ga (Corfu and Andrews, 1986). Volcanic rocks near the base of the Huronian succession (Elliot Lake Group) are ca. 2.45 Ga old (Krogh et al., 1984). These ages bracket the age of the Huron Supergroup.

Bell Lake and Annie Lake Granites and their Country Rocks

These two northeast-elongate plutons cut sharply across the east-trending formations of the Quirke Lake and Cobalt groups. Although their common northwest contact is the locus of faulting, their intrusive relationship with the Huronian is established by granite dykes in adjacent sedimentary rocks, sedimentary xenoliths in the plutons, and a contact aureole in suitable sediments (e.g. cordierite- and garnet-bearing hornfels developed in Gowganda argillite). Both granites are characterized by coarse megacrysts of K-feldspar, rendering these rocks markedly inequigranular and thereby distinct from the Killarney granite. Both have a greater content of mafic minerals (biotite and minor hornblende) and contain more plagioclase than the Killarney granite, their compositions locally grading to granodiorite. The Annie Lake granite contains several phases, for the most part less mafic and a little finer grained than the Bell Lake granite, which it intrudes.

The Bell Lake granite locally shows a weak alignment of K-feldspar megacrysts and of mildly elongate xenoliths, in places trending southeast. This foliation is truncated by protomylonitic foliation trending northeast and dipping southeast, parallel to typical Grenvillian trends. Both foliations, however, are cut by pegmatite and aplite dykes and by Sudbury diabase dykes. Krogh et al. (1971) obtained a U-Pb zircon age of 1525 Ma for the Bell Lake granite. More recently, a precise U-Pb zircon age of 1471 ± 3 Ma was obtained for the same granite (van Breemen and Davidson, 1988). The discrepancy between these ages may be explained if the older age was determined using zircons with inherited older cores. The Annie Lake granite, also deformed, has not been dated. The contact between the Bell Lake and Killarney granites is not exposed. At its southwest end the southeast side of the Bell Lake granite lies adjacent to the Grenville Front Boundary Fault, where it is reduced to an augen protomylonite. In this area its western contact with the Killarney complex may be a splay from the Boundary Fault.

In the area immediately west of Tyson Lake, feldspathic, micaceous and pure quartzites are confidently correlated with the Lorrain Formation on the northwest side of the Bell Lake granite (Frarey, 1985). Thin slivers of biotite-rich schist and amphibolite adjacent to the quartzite are likely correlative with the Gowganda Formation and Nipissing diabase respectively. Farther north along the same zone and lying east of the Annie Lake granite, fine grained quartzite, calcareous schist and minor marble, and metaconglomerate with scattered, grey granite pebbles are probably correlative with the Serpent, Espanola and Bruce formations of the Quirke Lake Group. These metasedimentary rocks are distributed among fault-bounded panels that dip southeast but which internally show mesoscopic folding of bedding within east-trending enveloping surfaces. A strong, moderately to steeply southeast-plunging stretching or rodding lineation is well developed, and interference patterns attest to more than one phase of deformation.

Southwest of Tyson Lake these metasediments are pinched out between the Bell Lake granite and pink, foliated biotite granite (Johnnie Lake granite) that lies to the southeast. The latter has given a U-Pb zircon age of ca. 1.73 Ga (Krogh et al., 1971), making it age-correlative with the Killarney complex. The two granites are separated by an intense mylonite zone at Carlyle Lake (Stop T1) which can be traced to the Boundary Fault on Philip Edward Island (Stop K6b). Northeast of Carlyle Lake the Boundary Fault steps diagonally across the Johnnie Lake granite to the southwest part of Tyson Lake (southeastern toothed line in Fig. 16).

Metasedimentary Schist and Gneiss and Deformed Plutonic Rocks of the Grenville Province

The zone southeast of the Boundary Fault (Fig. 16) is underlain by metasedimentary rocks and deformed granitoids in lenticular strips

oriented northeast. Pervasive southeast-dipping foliation and pronounced dip-parallel lineation are present almost everywhere. Relationships
between granites and metasedimentary rocks are usually obscured by mylonitization along contacts, but stretched metasedimentary inclusions
within the granitoids confirm the formerly intrusive nature of the meta-
plutonic rocks. Pegmatite is ubiquitous, usually as lenses and fila-
ments with K-feldspar augen within the foliation plane and more rarely
as less deformed veins and dykes at an angle to foliation.

Metasedimentary rock types in the vicinity of Tyson Lake include
micaceous (muscovite) quartzite, brownish grey feldspathic quartzite
with biotite, garnet-biotite gneiss and schist, in places with sillimanite and rare relict kyanite, amphibolite, and pink, fine grained
feldspathic gneiss, locally associated with thin layers of calc-silicate gneiss. Individual types rarely amount to more than a few tens of
metres in thickness across the strike of foliation and layering. In
some outcrops between discrete mylonite zones, however, relict bedding
is discernible and trends easterly within enveloping surfaces defined
by the hinges of mesoscopic reclined folds whose axial planes and axes
are parallel to the regional foliation and lineation respectively. As
strain increases toward the mylonite zones that bound such lenses, bedding is progressively transposed to disrupted layers in the foliation
plane.

These metasedimentary rocks extend southward toward Georgian Bay
but are not exposed along the outer coast of Philip Edward Island.
They appear to terminate, in Mill Lake and Beaverstone Bay (see Fig. 5),
in the hinges of large reclined folds with steep south-plunging axes,
overthrust from the southeast by metagranitoid rocks (Davidson and
Bethune, 1988). Until the structure of these metasedimentary belts is
more fully known, it will not be possible to suggest direct correlation
with Huronian formations. In addition, complicating factors must be
considered, such as the possible presence of formations lower in the
Huronian succession than are present in the adjacent part of the Southern Province, and changes in facies and thicknesses if the metasediments represent distal Huronian equivalents now telescoped toward the
northwest.

Fate of the Sudbury Dykes in the Grenville Province

The southeast-trending olivine diabase dykes of the 1240 Ma Sudbury
swarm play an important rôle in deciphering the tectonic history of
this area. These dykes pass unaffected, except for minor brittle deformation and minimal offset at discrete faults, through the Bell Lake
and Annie Lake granites and through the deformed but probable Huronian-
correlative metasedimentary rocks immediately to the southeast. As
they near the prominent mylonite zone located in the western part of
Tyson Lake, they swing northeastward and cannot be individually traced
farther east; it appears that they are cut off by faults marked in the
country rocks by mylonite and ultramylonite, and at a few places in
Tyson Lake the diabase itself is strongly mylonitized. Northwest of
this discontinuity the olivine diabase shows no sign of strain or meta-

morphism (Fig. 17, A), except at the brittle faults already mentioned
where low grade minerals such as chlorite, albite, epidote and actinolite may be present.

Southeast of this line in a zone up to 3 km wide, olivine diabase
occurs as discontinuous dyke segments in different orientations, many
of which are curved and some of which have the form of isolated, horse-
shoe-shaped folds. Thin sections show internal deformation in the form
of bent plagioclase laths, and the beginning of metamorphism: secondary biotite, titanian pargasitic amphibole and garnet have formed next
to Fe-Ti oxide grains, and coronas of hypersthene rimmed by pargasite –
spinel symplectite occur around olivine in contact with plagioclase
(Fig. 17, B). For these reasons, the Grenville Front Boundary Fault is
placed along the mylonite zone in western Tyson Lake across which this
marked change occurs, as it represents the locus of greatest uplift of
Grenvillian (post-diabase) age.

Farther southeast again, olivine metadiabase dykes can be traced
continuously for several kilometres, having trends that are slightly
more easterly than the equivalent dykes northwest of the front. Many
dykes follow sinuous courses, and some have short tracts with northeast
orientation. The state of internal deformation increases progressively
southeastward, eventually developing to a fully penetrative stretching
lineation that plunges southeast in concert with that in the country
rocks. Narrow mylonite zones are present within these dykes and are
apparently restricted to them. Metamorphism increases southeastward,
with widening of coronas and a change from pargasite – spinel symplectite to clinopyroxene and garnet outside hypersthene that has partly or
wholly replaced olivine (Fig. 17, C). In the internal mylonites these
secondary minerals form very fine grained, streaky aggregates of polygonal grains enclosing augen of twisted plagioclase and primary clinopyroxene (Fig. 17, D). In the lineated metadiabase farthest from the
front, spinel-clouded plagioclase laths are increasingly replaced by
fine, polygonal plagioclase aggregates, releasing spinel to recrystallize as interstitial grains or to contribute to further garnet growth.

Despite these severe modifications, the metadiabase is still recognizable as dykes with chilled margins that cut across country rock foliation and layering. In detail, some exposed dyke contacts show small-
scale folding with the host gneisses, coaxial with the regional lineation, but many do not. Similarly, local deflections have been noted in
a few places where the dykes are particularly curved, but orientation
of country rock structure does not seem to be modified at dyke contacts
in most places. To explain these observations, it seems necessary that
the dykes did not originally have straight courses but also that their
courses have been subsequently modified (folded) by shortening toward
the front with marked uplift of the Grenville side.

Because no single dyke can be traced continuously across the
Southern Province margin into the Grenville Front Tectonic Zone, chemical analyses and geochronological studies have been undertaken to
establish whether or not the diabase dykes on either side of the front
are the same. Sudbury diabase has a characteristic chemistry (Palmer
et al., 1977; Condie et al., 1987), being of alkali olivine basalt
affinity; it is enriched in Fe, K, P, Ba and Zr, and depleted in Mg,

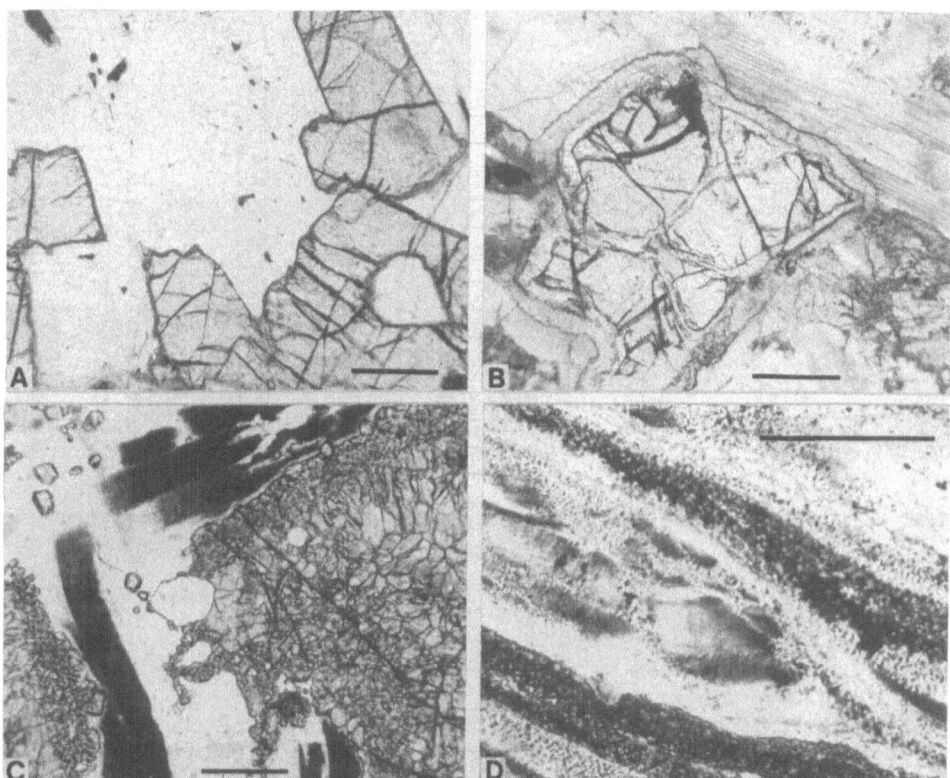

Figure 17. Photomicrographs (plane light) illustrating progressive metamorphism of Sudbury diabase dykes. A, unaltered olivine diabase, Southern Province; B, narrow corona of hypersthene and pargasite - spinel symplectite around olivine in diabase, Tyson Lake; C, complete replacement of olivine by hypersthene, clinopyroxene and garnet, with intense spinel clouding in the cores of plagioclase laths, east of Broker Lake; D, recrystallized corona minerals are dispersed along the foliation plane and surround plagioclase augen in mylonite from an internal shear zone in metadiabase, Stop T9. Scale bar = 0.1 mm.

Ca and Sr with respect to the tholeiitic composition of the diabase of most other swarms in the Canadian Shield. Table II allows comparison of analyses of diabase and metadiabase from excursion stops T5, T8, T9 and T10 with average Sudbury diabase from northwest of the front, and also with tholeiitic diabase from a dyke of the ca. 0.5 Ga Grenville swarm in the excursion area. The analyses show that the metamorphosed diabase in the Grenville Province cannot be distinguished chemically from fresh Sudbury diabase; the high Ba:Sr ratio is particularly diag-nostic. Coronitic olivine metadiabase masses in the neighbouring part of northwest Britt domain to the southeast, entirely isolated within ductile gneisses, also have Sudbury diabase chemistry. If these are

TABLE II.   Analyses of Diabase and Metadiabase

Weight %	1	2	3	4	5	6
SiO_2	47.5	47.7	47.6	48.2	48.0	49.4
TiO_2	2.92	2.66	3.51	2.63	2.94	1.35
Al_2O_3	14.7	16.2	13.3	14.5	15.7	14.2
Fe_2O_3	3.1	2.4	2.8	2.9	3.5	2.9
FeO	12.3	11.5	13.9	11.9	11.2	8.8
MnO	0.21	0.20	0.24	0.21	0.20	0.19
MgO	5.48	4.96	4.60	5.72	5.03	7.37
CaO	7.60	8.00	7.77	7.74	8.38	11.65
Na_2O	3.25	3.5	3.3	3.2	3.4	1.9
K_2O	1.47	1.25	1.61	1.19	1.11	0.30
P_2O_5	0.61	0.64	0.83	0.57	0.50	0.11
S	0.09	0.07	0.11	0.11	0.08	0.02
CO_2	0.1	0.0	0.2	0.1	0.1	0.1
H_2O	0.7	0.4	0.6	0.4	0.4	1.1
Total	100.03	99.48	100.37	99.37	100.54	99.39
ppm						
Ba	695	761	820	593	653	99
Sr	315	401	251	325	344	173
Zr	221	188	298	192	175	58

1. Average of 14 analyses, Sudbury diabase northwest of the front.
2. Stop T5. 3. Stop T8b. 4. Stop T9. 5. Stop T10.
6. Grenville diabase, dyke between stops T4 and T5.

Analyses done at the Geological Survey of Canada.

indeed highly modified relics of the Sudbury swarm, then the Grenville
Front Tectonic Zone and at least the northwest part of the adjacent
interior gneiss terrane were part of the North American craton before
ca. 1240 Ma ago.
    The relatively high content of Zr in Sudbury diabase (Table II)
is reflected by the presence of primary igneous baddeleyite (ZrO_2) in
most samples in large enough quantity to be extracted for U-Pb isotopic
dating. A preliminary U-Pb baddeleyite age of 1235 ± 15 Ma has been
determined at the geochronological laboratories of the Geological Survey
of Canada, using baddeleyite fractions from four diabase and metadiabase
occurrences, respectively 19 km and 2.5 km (Stop T5) northwest of the
Boundary Fault, and 2.5 km (Stop T8b) and 5 km southeast of it. This age
is an upper concordia intercept (MSWD 0.87) and agrees well with the
baddeleyite age of 1238 ± 4 Ma reported by Krogh et al. (1987) for Sudbury diabase near Espanola, 45 km northwest of the front. In addition,
baddeleyite in the two samples collected southeast of the front is
rimmed by secondary zircon in small radiating crystals having the same
form as the corona minerals around olivine and oxide grains in these
rocks. A metamorphic U-Pb zircon age is not yet available for the

sample from Stop T8b, but the other sample, collected east of Sudbury, has a $^{207}Pb/^{206}Pb$ age of ca. 1005 Ma, indicating that metamorphism was significantly later that intrusion. The implications of the U-Pb baddeleyite age are (1), the deformed metadiabase dykes in the Grenville Province are the same as the Sudbury dykes in the Southern and Superior provinces to the northwest, and (2), major deformation and metamorphism preceded diabase intrusion at ca. 1240 Ma, but presumably post-dated the formation of pegmatite dykes at ca. 1400 Ma (see description of the Killarney area, and Stop K6a). As mentioned above, this pre-diabase, post-pegmatite deformation has the characteristic Grenvillian attitude, i.e., southeast-dipping foliation and dip-parallel stretching lineation. It is interesting to note that ductile deformation in the interior of the Grenville Province at the margins of major lithotectonic domains (ca. 1169 Ma for the Parry Sound shear zone, van Breemen et al., 1986; ca. 1080 Ma for the northwest boundary of the Central Metasedimentary Belt, van Breemen and Hanmer, 1986) appears to be younging progressively southeastward. It is also important to note that Sudbury diabase dyke intrusion is coeval with plutonism and the younger of two periods of volcanism represented in the Grenville Supergroup (Lumbers et al., 1988; Davis and Bartlett, 1988). Sudbury dyke intrusion is older than primary crystallization of olivine gabbro, now coronitic metagabbro, in the Algonquin region (U-Pb baddeleyite age ca. 1170 Ma; Davidson and van Breemen, 1988), itself almost coeval with major deformation at Parry Sound and with granite-monzonite plutonism in the Frontenac Axis (Marcantonio et al., 1988). Evidently the Grenville orogen, like Rome, was not built in a day ...

Stop Descriptions

The field excursion in the Tyson Lake area follows highway 637 northeast from Killarney, starting at the Grenville Front Boundary Fault and proceeding diagonally into the Grenville Province. A side trip along the Bell Lake road allows examination of the Bell Lake granite, deformed Lorrain quartzite and non-deformed Sudbury diabase northwest of the Boundary Fault. Emphasis is placed on the nature of the metasedimentary rocks, their increasing metamorphic grade southeastward, and the state of deformation of the Sudbury metadiabase dykes within the Grenville Province. The stops are as follows (Figs. 5 and 16):

T1 - Grenville Front Boundary Fault;
T2 - Kinematic indicators in mylonitic granite and pegmatite;
T3 - Granitoid orthogneiss with kinematic indicators;
T4 - Bell Lake granite;
T5 - Deformed Lorrain quartzite at lower amphibolite grade and non-deformed Sudbury diabase;
T6 - Metasedimentary gneiss and schist, middle amphibolite grade;
T7 - Hypersthene-bearing granitoid;
T8 - a, Sudbury metadiabase dykes with plagioclase phenocrysts, cutting layering in metasedimentary gneiss;
     b, metasedimentary gneisses of various protoliths at upper amphibolite to granulite grade;

T9 - Sinuous course of Sudbury metadiabase dyke within metasedimentary
      gneiss, and its internal deformation;
T10 - Strongly lineated Sudbury metadiabase.
Note: at the time of preparing this field guide, reconstruction of
highway 637 along the stretch containing stop T1 to T3 was about to
begin. Most of Stop T1 will not be adversely affected, but the small-
scale features in the roadside outcrops of stop T2 and T3 may well be
blown into the bush or buried in the new roadbed, and it may be necessary to seek these features in newly blasted roadcuts. In addition,
Stop T4 on the Bell Lake granite lies within an area recently staked
for building stone quarrying and is slated for further clearing and
stripping late in 1988; this may or may not prove to be an advantage.

Stop T1. Grenville Front Boundary Fault between the Bell Lake and
Johnnie Lake granites; east boundary of Killarney Provincial Park,
19.5 km from Killarney.

The outcrop at the crest of the low hill is dark grey, porphyroclastic mylonite and augen schist derived from the Bell Lake granite.
The outcrop width of this unit is only a few hundred metres here, but
it expands northeastward to a maximum of 2 km, where the granite is not
severely deformed and locally not deformed at all. It can be traced as
a narrow train of augen mylonite for several kilometres along the
Boundary Fault to the southwest. Foliation at the roadside outcrop
dips 65° southeast and contains a dip-parallel lineation.

From the east end of this outcrop, walk southeast through the
bush, crossing a valley. The low cliff on the other side exposes at
its base flinty ultramylonite with tiny feldspar porphyroclasts. This
grades up structural section to mylonite and then to protomylonitic
granite whose precursor is not the same as the Bell Lake granite.
Mylonitic foliation dips moderately southeast and carries a prominent
dip-parallel stretching lineation. Note that the lineations at this
stop have the typical 'Grenvillian' orientation, markedly different to
the northeast to easterly plunging lineations in the Killarney complex.
X-Z oriented thin sections of the mylonitic rocks here show highly
attenuated quartz plates with blocky subgrains, the majority of which
remain at extinction between crossed polarizers, indicating a strongly
preferred C-axis orientation parallel to Y (Fig. 18, A). C' foliation
along which biotite flakes are concentrated, and rotated feldspar augen
both indicate reverse sense of displacement. The ultramylonitic rocks
along the base of this outcrop ridge show evidence of later brittle
deformation, and it is likely that the valley below is the site of a
major late fault, probably related to the last stage of uplift along
the Grenville Front.

Follow the base of the outcrop ridge northeast back to the road
and go to the top of the hill to the east. Blasted outcrops of protomylonitic and foliated 'Johnnie Lake granite' show weak evidence of the
same kinematic sense as the mylonites that lie structurally below.

The gravel track that meets the highway near this outcrop leads
to Carlyle Lake, where clean exposures of mylonitic Bell Lake granite
showing good examples of strain gradients can be examined.

<u>Stop T2.</u> S-C fabric and rotated feldspar augen in mylonitic granite and pegmatite; outcrop on north side of highway 637, 21.8 km from Killarney.

The same granitoid unit as at Stop T1 here contains mylonitized pegmatite parallel to the northeast-trending foliation. The pegmatite contains good examples of rounded and rotated K-feldspar augen with tails ('winged' porphyroclasts) with delta form (Passchier and Simpson, 1986), illustrated in Figure 18, B. (Please do not attempt to hack out samples from the in-place X-Z sections; suitable sample material can be found among the loose rubble.) At the top left of the vertical rock face by the cottage access road, which is an X-Z section, note the development of S-C fabric that confirms the rotation sense of the feldspar augen - anticlockwise in this section, i.e., top side up the lineation to the northwest. On the top surface at the other end of the outcrop, an earlier foliation in the granitoid rock strikes approximately southeast and is folded about northeast-trending shears.

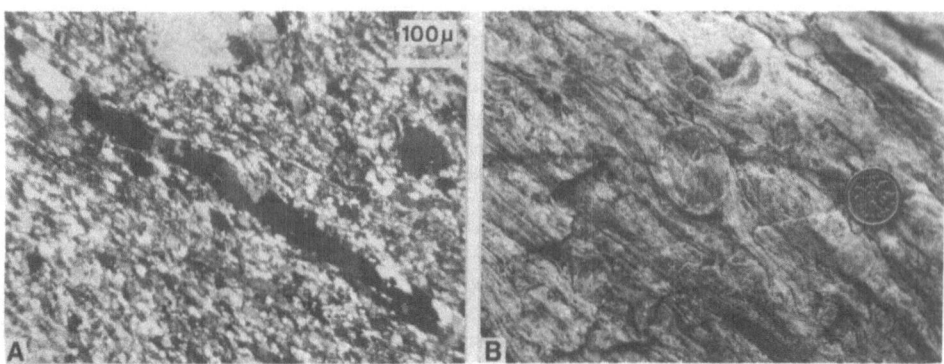

Figure 18. A, photomicrograph, crossed polarizers, of oriented quartz fabric in mylonite at Stop T1; B, rotated K-feldspar augen, Stop T2.

<u>Stop T3.</u> Granitoid gneiss; rock cut on the north side of highway 637, 26.7 km from Killarney and 2.2 km southeast of the Boundary Fault.

This stop illustrates penetrative deformation in granitoid rocks, here in the form of well foliated orthogneiss. Rocks of different grain size are interlayered parallel to the through-going foliation, which here dips 50° southeast and again carries the 'Grenville' lineation. Note the lenticular form of quartz in these rocks. Augen granitoid at the west end of the outcrop has S-C fabric and shear bands. Pegmatites are inclined more steeply than foliation, but are themselves deformed; this orientation suggests rotation toward the foliation plane and, along with the fabric elements, implies reverse displacement.

Continue east along the highway for 5.0 km and turn left on the Bell Lake road. 7.8 km along this road take the right fork down hill across a linear valley (underlain by a diabase dyke of the Grenville swarm; Table II, no. 6). At 8.6 km the granite quarry is on the right.

<u>Stop T4</u>. Bell Lake granite, 3.5 km northwest of the Boundary Fault.
    The Bell Lake granite forms a pluton at least 18 km long, elongate
northeast, and only 2 km wide. The stripped outcrop at this stop
(Fig. 19) exemplifies many aspects of this 1470 Ma old granite: its
inequigranular nature, with K-feldspar megacrysts (phenocrysts) in
blocky rectangular crystals up to 2 cm long; its uniform character as a
matrix for numerous xenoliths ranging from a few centimetres to more
than a metre across; alignment of both megacrysts and xenoliths in
linear fashion, here plunging easterly at 45° (best seen at 'd' in
Fig. 19). Most of the xenoliths are darker and more biotite-rich than
the granite, and a few carry 'dents de cheval' of K-feldspar. There
are rare xenoliths of pink, leucocratic granite and of quartzite ('e'
in Fig. 19); large rafts of quartzite are included in the granite
nearby, however. The granite is cut by a few, straight, narrow aplite
and pegmatite dykes ('a').

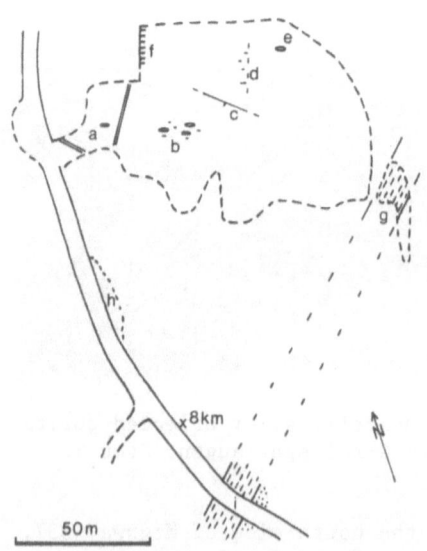

Figure 19. Sketch map of area cleared for granite quarrying at Stop T4.
Marked locations are: a, aplite and pegmatite veins; b, elongate xenoliths; c, late fault surface with epidote and slickensides; d, K-feldspar megacryst orientation; e, quartzite xenolith; f, quarried face;
g, mylonite zone; h, dyking relationship between two granite phases;
i, mylonite zone and part of large quartzite raft.

    At the southeast edge of the cleared area ('g'), and not in exposed
contact with typical Bell Lake granite to the northwest, is a zone of
augen mylonite striking northeast and dipping very steeply <u>northwest</u>,
i.e., not a typical Grenvillian orientation. It contains an angular

block of coarser protomylonite and is in contact with Bell Lake granite to the southeast. The same mylonite is exposed in road cuts ('i') just south of the quarry site, and there it can be seen to have developed in a finer, more leucocratic granite that intrudes the typical megacrystic Bell Lake granite; at this place its southeast contact is with glassy quartzite, part of an included raft. Between this outcrop and the entrance to the cleared area ('h') the finer granite can be seen to have cut xenoliths in the Bell Lake granite, clearly establishing the age relationship. The mylonite at 'i' contains kinematic indication of northwest side down, normal in sense with respect to the attitude of the foliation.

Return southward along the Bell Lake road and stop at the crest of the hill 1.4 km from Stop T4.

Stop T5. Non-metamorphosed Sudbury diabase dyke and metamorphosed Lorrain quartzite; 2.3 km northwest of the Boundary Fault.

The road cut exposes grey, plagioclase-rich olivine diabase typi-cal of the coarser parts of thick Sudbury dykes. This east-southeast-trending dyke crosses foliation in both the Bell Lake granite to the northwest and in quartzite to the southeast, and exhibits chilled margins. It is offset with sinistral sense along the faulted contact of the Bell Lake granite (Fig. 20). The diabase shows minor internal

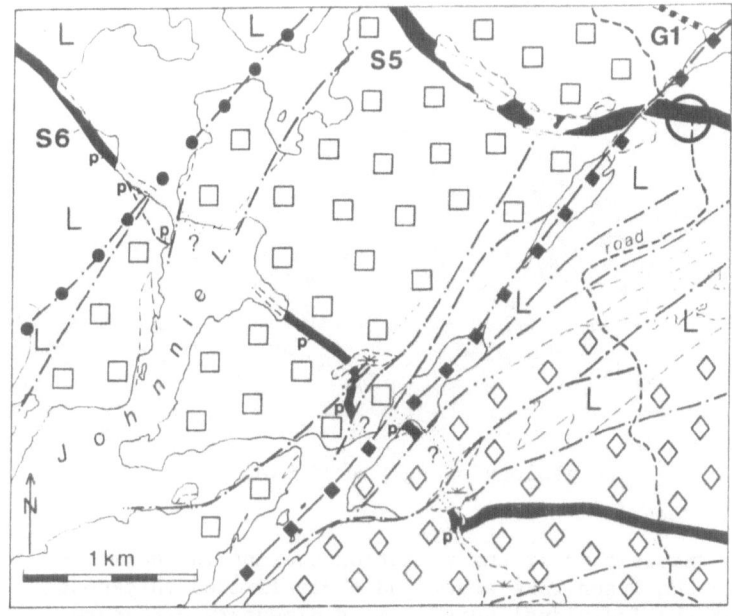

Figure 20. Geology southwest of Stop T5 (circled at top right). The legend is the same as in Figure 16.

cataclasis in places but is not metamorphosed. An analysis of this
diabase is given in Table II, column 2; its plagioclase-rich nature is
reflected in higher than normal Al, Ca and Na contents and lower Mg and
Fe than average Sudbury diabase (compare with Table II, column 1). It
is useful to collect a small sample of diabase from this outcrop in
order to compare it with metamorphosed equivalent examples at following
stops.

Walk down the hill to the south. Micaceous quartzite is exposed
at the bottom of the hill; its foliation strikes northeast and dips
southeast. Outcrops of the same unit on the hill to the south contain
large boudins of amphibolite, perhaps derived from Nipissing diabase.
At the crest of the hill a more feldspathic quartz schist with biotite
and muscovite is exposed, and is cut by a granite dyke. In all these
metasedimentary rocks muscovite is stable with quartz and sillimanite
is not present; metamorphic grade is probably lower amphibolite facies.
The quartz-rich rocks are correlated with the Lorrain Formation, which
is the country rock on the opposite (northwest) side of the Bell Lake
granite.

Continue back to highway 637 and turn left. The highway crosses a
narrow lens of pink foliated granite and then passes into metasedimentary rocks. Stop at roadside outcrops 2.1 km along the highway from the
intersection with the Bell Lake road.

Stop T6. Metasedimentary schists – are these Huronian rocks? 33.8 km
from Killarney, and 2.0 km southeast of the Boundary Fault.

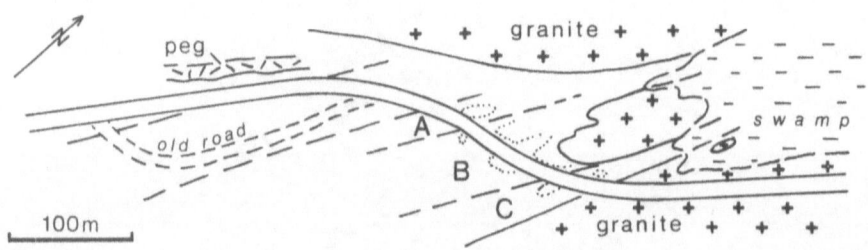

Figure 21. Sketch map of the geology along the highway at Stop T6.

This stop illustrates some of the variation present in the metasedimentary rocks in the Tyson Lake area, and their limited thickness
across the strike of the individual units. No sedimentary structures
are preserved. Unit A (Fig. 21) is composed of garnet – biotite schist
interlayered with quartz – feldspar – 2 mica schist; both locally contain
sillimanite. Unit B is predominantly quartz schist (quartzite) containing minor feldspar and mica; it is at most 65 m thick. Unit C is a

calc-silicate rock (quartz – plagioclase – hornblende – diopside), and is
in contact (faulted?) with foliated pink granitoid to the southeast.
On the north side of the highway a lobe of granite crosses the contact
between quartzite and calc-silicate rock. The occurrence of sillimanite in the presence of stable quartz and muscovite (unit A) suggests
middle amphibolite facies.

In the Huronian succession the Espanola Formation is the only
dominantly calcareous unit (see Table I). If the calc-silicate rocks
(unit C) at this stop represent the Espanola Formation and the other
units present are in correct stratigraphic sequence, the quartzite
(unit B) should be Serpent Formation, the garnet – biotite schist (unit
A) should be Gowganda Formation, and the succession would be facing
(younging) northwest. The next unit a few tens of metres northwest,
however, is rusty-weathering biotite – quartz – feldspar schist and impure quartzite of the type that in Tyson Lake Quirke and Collins (1930)
equated with Mississagi Formation. Calc-silicate rocks occur again a
few tens of metres northwest. Not only do these units seem unreason-
ably thin to represent the Huronian formations mentioned, but they are
in the wrong stratigraphic order. If they do represent highly attenuated Huronian rocks, they must be in the form of stacked, out-of-order
slices.

Continue east along the highway for 1.2 km and stop by the gravel
road on the north side.

Stop T7. Hypersthene-bearing granitoid; 35.0 km from Killarney and
2.1 km southeast of the Boundary Fault.
The lens of pink foliated granite that lies southeast of the
schists examined at the last stop grades into a brownish green phase
at this locality. It is a biotite – hornblende quartz monzonite with
small amounts of hypersthene, partly altered (weathered?) to brownish
secondary material and minor carbonate. Magnetite, apatite, allanite
and zircon are plentiful accessory minerals. Secondary biotite defines
a weak foliation and is associated with minor amounts of metamorphic
garnet. Quartz grains are somewhat elongate polygonal aggregates, but
the rock as a whole seems to be very little deformed and must lie in
a low strain zone within the granitoid lens. The hypersthene occurs as
large subhedral grains and has only weak pleochroism in thin section.
It appears to be a primary igneous mineral, and the rock can thus be
classified as a charnockite of igneous origin. The implications of
this occurrence so close to the Grenville Front are open for discussion.

Continue 0.7 km along the highway and park just past the bridge
over the Mahzenazing River.

Stop T8. Relationship between metasedimentary gneiss and Sudbury
metadiabase dykes; 35.7 km from Killarney, and 2.5 to 3.0 km southeast
of the Boundary Fault.

Metasedimentary rocks on the southeast side of the charnockite/granite
lens of Stop T7 are gneisses with metamorphic mineral assemblages indicative of upper amphibolite to granulite facies. At the large road
cut here they are nondescript, dull pink-grey quartzofeldspathic

gneisses of indefinite origin, but garnet- and sillimanite-bearing
varieties are present at the west side of the outcrop and in the bed of
the Mahzenazing River above the bridge.

This stop has two purposes: (i), to illustrate the crosscutting
relationship between Sudbury metadiabase dykes and gneissic layering
within the Grenville Province (Stop T8a), and (ii), to examine metamorphic mineral assemblages in metasedimentary gneisses that indicate
attainment of granulite facies conditions (Stop T8b).

Stop T8a. Follow the flagged trail into the bush just west of the
rock cut on the north side of the highway. Steep outcrops on either
side of a narrow draw display exposed contacts of vertical dykes of
metadiabase with large phenocrysts of black plagioclase. On the right
(southeast) side of the draw, note the small-scale corrugations at the
contacts, the steep alignment of the phenocrysts and the occurrence of
elongate spongy garnet in the metadiabase close to its contacts.

Climb up the steep slope to the left of the dykes. At the top of
the slope is the edge of an abandoned road-metal quarry, on the south
wall of which the same dykes can be seen. Make your way to the head of
the quarry and descend carefully to the rock piles on the quarry floor
near the south wall - be wary of loose blocks and do not attempt to
climb the rock faces. Fresh exposure of the dykes and their cross-
cutting contacts (Fig. 22, A) can be examined at the quarry face, but
please collect your samples from fallen blocks; several large pieces
have dyke contacts in them.

Similar coarsely porphyritic diabase is known from several other
localities. Crowded phenocrysts are confined to the margins of some
larger dykes, but narrow offshoots may either be porphyritic throughout or may contain isolated patches rich in phenocrysts. To the west,
marginal phenocrysts occur in a larger dyke of which the dykes viewed
here appear to be offshoots. Roughly in line northwest of the Boundary
Fault, a large dyke with marginal phenocrysts has been traced northwestward across the Johnnie Lake and Bell Lake granites (marked 'P' in
Fig. 16), where its phenocrysts are clouded with fine epidote and ser-
icite, and for several kilometres into the Huronian sediments of the
Southern Province, where clear phenocrysts occur. Phenocrysts south of
the Boundary Fault are invariably black on account of numerous submicroscopic inclusions, probably spinel. In thin sections of less
deformed diabase they are seen to be compositionally homogeneous with
thin, sharply zoned outer rims surrounding rounded (resorbed) cores
which are somewhat more anorthitic than the matrix plagioclase laths.
At this locality the phenocrysts include small grains of olivine that
are surrounded by double coronas of columnar hypersthene and radiating
pargasite - spinel symplectite, similar to that shown in Figure 17, B.

Despite the non-deformed appearance of these dykes, thin sections
show a remarkable degree of internal deformation (see Frarey, 1985,
Fig. 35) in the form of bent plagioclase laths and augen of primary
minerals in a mesh of fine shear surfaces along which very fine grained
metamorphic minerals (biotite, amphibole, garnet) are distributed. The
rock is not a mylonite per se, but internal cataclastic structure
mimics that of porphyroclastic mylonite (Fig. 22, B). The plagioclase

phenocrysts have curved cleavage faces and in thin sections are seen to
have curved twin planes and splays of deformation twin lamellae adjacent to brittle microshears and dislocations; there is no visible development of subgrains.

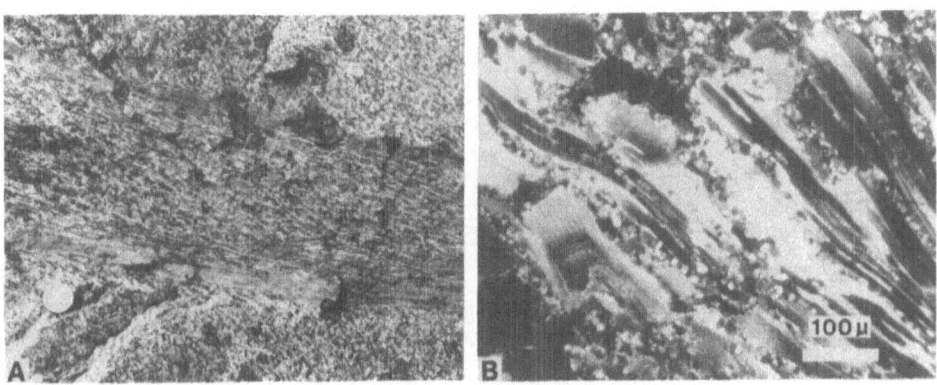

Figure 22. Stop T8a. A, preserved chilled margin of metadiabase dyke
cutting layering in quartzofeldspathic gneiss; B, photomicrograph,
plane light, of sheared structure of metadiabase in matrix surrounding
plagioclase phenocrysts.

Leave by the quarry entrance and walk along the highway across the
embankment to the roadcuts on the hill to the east.

Stop T8b. The variety of metasedimentary gneisses is best appreciated
on the top surface of the rock cuts on the left side of the highway.
Note the migmatitic nature of some of the units, the presence of narrow
deformed mafic dykes (amphibolite) and crosscutting pegmatites, and
look for layers of calc-silicate rock with diopside, pale grossularite
and wollastonite; thin layers of diopside marble are also present.
Halfway up the hill, a thick, non-porphyritic metadiabase dyke crosses
the highway at an angle to the strike of the gneisses, but its contacts
are not exposed. This dyke (Table II, column 3) is somewhat richer in
Fe and Ti than average Sudbury diabase (Table II, column 1), due to a
higher Fe-Ti oxide content. It also has a higher content of Zr; baddeleyite rimmed by fine polycrystalline zircon has been extracted from
this rock for U-Pb isotopic dating, now in progress.

Metasedimentary gneisses east of the dyke are similar to those
already examined at this stop. Muscovite is absent and sillimanite
occurs with K-feldspar in garnet – biotite gneiss. Thin sections show
that strongly pleochroic hypersthene occurs with garnet, biotite, plagioclase and quartz, with or without hornblende, in some of the darker
gneisses (Fig. 23), and that the calc-silicate rocks contain quartz,
scapolite, diopside, grossularite, wollastonite and titanite. The
second rock cut on the south side of the highway (end of the left hand
curve) displays early, small-scale folds in garnetiferous gneiss that

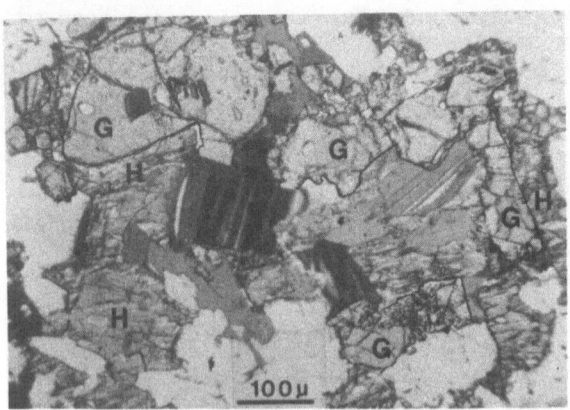

Figure 23. Photomicrograph, plane light, showing hypersthene (H) associated with garnet (G, outlined) and biotite in metasedimentary gneiss at Stop 8b.

are cut through cleanly by a narrow, metamorphosed mafic dyke which itself is internally foliated.

Continue along the highway for 11.3 km; stop by the roadside where a grassy swamp is visible to the left. Figure 24 shows the locations of stops T9 and T10, whose purpose is to examine the state of deformation of Sudbury metadiabase dykes in the Broker and Attlee lakes area, some 8 km southeast of the Boundary Fault.

Stop T9. Sinuous course of Sudbury metadiabase dyke and its internal deformation; 47.0 km from Killarney and 8 km southeast of the Boundary Fault.

Small outcrops of metadiabase are present on both sides of the highway, and a low, rounded outcrop of metadiabase, likely the same dyke, is exposed on the other side of the grassy flat to the north. Walk to this outcrop, keeping to the left side of the swamp, where the country rock outcrops show garnet-bearing leucosomes in metasedimentary gneiss that includes both quartz-rich and sillimanite-bearing varieties. Cross a small beaver dam to reach the metadiabase outcrops, and note that layering in the gneisses strikes at right angles to the dyke. The metadiabase outcrop has some clean surfaces which can be used to see the metamorphic minerals with a hand lens. The pale, centimetre-size patches near the north contact are spongy garnet aggregates; these can be seen to be elongate and plunging moderately southeast on appropriate surfaces.

Return to the highway along the other side of the swamp; low outcrops of fine grained grey gneiss have narrow calc-silicate lenses. Cross the highway and follow the flagged trail into the bush to the south. The trail follows the dyke, which curves toward the left and comes out on the highway again. Look for narrow mylonite zones within the dyke in the rock cuts at the highway, and note that these do not penetrate the country rocks (Fig. 25, A). Note also that the metadia-

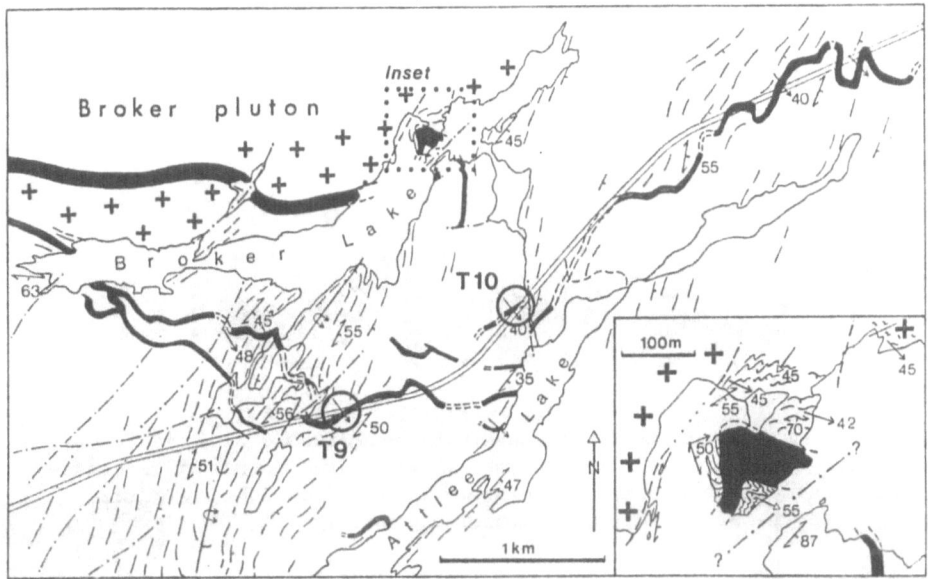

Figure 24. Distribution of Sudbury metadiabase dykes in the Broker and Attlee lakes area (stops T9 and T10), illustrating sinuous courses of the dykes. (From Bethune, 1989.)

base shows a subtle but through-going lineation in most places. The same, pale, spongy garnet aggregates are prominent near the contact on the south side of the highway. The metadiabase is cut locally by small aplitic dykes. Chemical analysis of this dyke (Table II, column 4) confirms its Sudbury affinity. In thin section, plagioclase laths are bent and twisted and their former oscillatory compositional zoning is outlined by spinel dust that is preferentially concentrated in the originally more calcic zones. Spinel clouding in plagioclase in coronitic olivine metagabbro is the result of egress of Ca to form the grossular component of garnet and to make clinopyroxene; its place is taken by Fe, Mg and Na that migrate inward to make spinel and a more sodic plagioclase (Grant, 1988):

$$3CaAl_2Si_2O_8 - 3Ca + 2Na + 2(Mg,Fe) = 2NaAlSi_3O_8 + 2(Mg,Fe)Al_2O_4.$$

In this rock, olivine has been almost wholly replaced by hypersthene surrounded by coronas of clinopyroxene and garnet (see Fig. 17, C).

The dyke contact is exposed on the south side of the road where it is seen to dip steeply southeast; it cuts gneissic layering which in this vicinity dips more shallowly to the southeast than at previous stops. The trend to shallower dips continues eastward.

Continue along the highway for another 1.5 km and park near the private driveway on the right just before crossing a small stream.

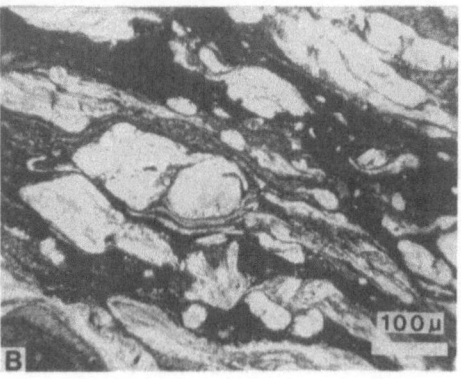

Figure 25. A, narrow mylonitic shear in metadiabase at Stop T9; B,
photomicrograph, crossed polarizers, showing fine mosaic of recrystallized plagioclase within and between spinel-clouded plagioclase
laths; note the spinel clouding mimicking zoning.

Stop T10. Strongly lineated Sudbury metadiabase; 48.5 km from Killarney
and 9 km southeast of the Boundary Fault.
     In contrast to the last stops, metadiabase here is strongly lin-
eated throughout. This is a mineral stretching lineation and it is
accentuated by the presence of markedly elongate spongy garnet aggregates. The down-plunge view seen on the top surface of the rock cut
shows that this rock is an L-tectonite - there is no obvious planar
fabric when the rock is viewed directly down the lineation. Deformed
metadiabase farther east from this stop have a similar linear fabric.
In thin section this rock shows recrystallization of plagioclase to a
fine mosaic in bands within and between the original laths (Fig. 25, B).
In the highly strained parts of this outcrop, recrystallization has
gone to completion, the end product being a very fine grained mafic
granulite as seen adjacent to narrow zone of ultramylonite.
     This lineated metadiabase is a far cry from the fresh Sudbury
metadiabase examined at Stop T5, but it has the same chemistry (Table
II, column 5). A complete gradation in metamorphic mineral assemblages
and fabric development has been documented in the intervening terrane,
and it is apparent that there was little or no chemical change in the
dykes. This is probably due in large part to the dry nature of the
metamorphism, the dykes having intruded already metamorphosed rocks
from which water was not available; hence the Sudbury diabase dykes
were not 'amphibolitized'.

Geological Survey of Canada Contribution 25888

REFERENCES

Anderson, S.L., and Burke, K., 1983, 'A Wilson Cycle approach to some
    Proterozoic problems in eastern North America'; in Proterozoic
    Geology: Selected Papers from an International Proterozoic
    Symposium, L.G. Medaris, Jr., C.W. Byers, D.M. Mickelson and
    W.C. Shanks, eds., Geological Society of America, Memoir 161,
    p. 75-93.
Anovitz, L.M., 1987, Pressure-temperature-time Constraints on the
    Metamorphism of the Grenville Province, Ontario; Ph.D. thesis,
    University of Michigan, Ann Arbor, Michigan, 479 p.
Baer, A.J., 1976, 'The Grenville Province in Helikian times: a pos-
    sible model of evolution'; Philosophical Transactions of the Royal
    Society of London, series A, 280, p. 499-515.
Baragar, W.R.A., 1981, Tectonic and Regional Relationships of the Seal
    Lake and Bruce River Magmatic Provinces; Geological Survey of
    Canada, Bulletin 314, 72 p.
Bell, R., 1898, 'Report on the geology of the French River sheet, Ont-
    ario'; Geological Survey of Canada, Annual Report, 9, Part 1,
    29 p.
Bethune, K.M., 1989, 'Diabase dykes and the Grenville Front southwest
    of Sudbury, Ontario: further constraints on the nature and timing
    of deformation and metamorphism'; in Current Research, Part C,
    Geological Survey of Canada, Paper 89-1C, p. 19-28.
Bethune, K.M., and Davidson, A., 1988, 'Diabase dykes and the Grenville
    Front southwest of Sudbury, Ontario'; in Current Research, Part C,
    Geological Survey of Canada, Paper 88-1C, p. 151-159.
Bickford, M.E., Van Schmus, W.R., and Zeitz, I., 1986, 'Proterozoic
    history of the midcontinent region of North America'; Geology, 14,
    p. 492-496.
Card, K.D., 1976, Geology of the McGregor Bay - Bay of Islands Area,
    Districts of Sudbury and Manitoulin; Ontario Division of Mines,
    Geoscience Report 138, 63 p.
Card, K.D., and Lumbers, S.B., 1977, 'Sudbury - Cobalt'; Ontario Geol-
    ogical Survey, Geological Compilation Series, Map 2361, scale
    1 : 253 440.
Chown, E.H., and Caty, J.L., 1973, 'Stratigraphy, petrography and
    paleocurrent analysis of the Aphebian clastic formations of the
    Mistassini - Otish basin'; in Huronian Stratigraphy and Sediment-
    ation, G.M. Young, ed., Geological Association of Canada, Special
    Paper 12, p. 49-71.
Clifford, P.M., 1986, 'Petrological and structural evolution of the
    rocks in the vicinity of Killarney, Ontario: an interim report';
    in Current Research, Part B, Geological Survey of Canada, Paper
    86-1B, p. 147-155.
Collins, W.H., 1925, North Shore of Lake Huron; Geological Survey of
    Canada, Memoir 143, 160 p.
Condie, K.C., Bobrow, D.J., and Card, K.D., 1987, 'Geochemistry of Pre-
    cambrian mafic dykes from the southern Superior Province of the
    Canadian Shield'; in Mafic Dyke Swarms, H.C. Halls and W.F. Fahrig,
    eds., Geological Association of Canada, Special Paper 34, p. 95-108.

Corfu, F., and Andrews, A.J., 1986, 'A U-Pb age for mineralized Nipissing diabase, Gowganda, Ontario'; Canadian Journal of Earth Sciences, 23, p. 107-109.

Davidson, A., 1984, 'Identification of ductile shear zones in the southwestern Grenville Province of the Canadian Shield'; in Precambrian Tectonics Illustrated, A. Kröner and R. Greiling, eds., E. Schweizerbart'sche Verlagsbuchhandlung, Stuttgart, West Germany, p. 263-279.

Davidson, A., 1986a, 'New interpretations in the southwestern Grenville Province'; in The Grenville Province, J.M. Moore, A. Davidson and A.J. Baer, eds., Geological Association of Canada, Special Paper 31, p. 61-74.

Davidson, A., 1986b, 'Grenville Front relationships near Killarney, Ontario'; in The Grenville Province, J.M. Moore, A. Davidson and A.J. Baer, eds., Geological Association of Canada, Special Paper 31, p. 107-117.

Davidson, A., 1986c, A New Look at the Grenville Front in Ontario; Geological Association of Canada, Field Trip 15, guidebook, 31 p.

Davidson, A., and Bethune, K.M., 1988, 'Geology of the north shore of Georgian Bay, Grenville Province of Ontario'; in Current Research, Part C, Geological Survey of Canada, Paper 88-1C, p. 135-144.

Davidson, A., and van Breemen, O., 1988, 'Baddeleyite - zircon relationships in coronitic metagabbro, Grenville Province, Ontario'; Contributions to Mineralogy and Petrology, 100, p. 291-299.

Davis, D.W., and Bartlett, J.R., 1988, 'Geochronology of the Belmont Lake Metavolcanic Complex and implications for crustal development in the Central Metasedimentary Belt, Grenville Province, Ontario'; Canadian Journal of Earth Sciences, 25, p. 1751-1759.

Derry, D.R., 1950, 'A tectonic map of Canada'; Proceedings of the Geological Association of Canada, 3, p. 39-53.

Dewey, J.F., and Burke, K.C.A., 1973, 'Tibetan, Variscan, and Precambrian basement reactivation: products of continental collision'; Journal of Geology, 81, p. 683-692.

Dietz, R.S., 1966, 'Passive continents, spreading sea floors, and collapsing continental rises'; American Journal of Science, 264, p. 177-193.

Douglas, R.J.W., 1980, Proposals for Time Clasification and Correlation of Precambrian Rocks and Events in Canada and Adjacent Areas of the Canadian Shield. Part 2: A Provisional Standard for Correlating Precambrian Rocks; Geological Survey of Canada, Paper 80-24, 19 p.

Duchesne, J.-C., 1984, 'Massif anorthosites: another partisan review'; in Feldspars and Feldspathoids, W.L. Brown, ed., D. Reidel Publishing Company, Dordrecht, The Netherlands, p. 411-433.

Emslie, R.F., 1978, 'Anorthosite massifs, rapakivi granites, and late Proterozoic rifting of North America'; Precambrian Research, 7, p. 61-98.

Engel, A.E., 1956, 'Apropos the Grenville'; in The Grenville Problem, J.E. Thomson, ed., Royal Society of Canada, Special Publication 1, p. 74-96.

Frarey, M.J., 1985, Proterozoic Geology of the Lake Panache - Collins
    Inlet Area, Ontario; Geological Survey of Canada, Paper 83-22, 61p.
Frarey, M.J., and Cannon, R.T., 1969, Notes to Accompany a Map of the
    Proterozoic Rocks of Lake Panache - Collins Inlet Map-areas, Ont-
    ario (41I/3, H/14); Geological Survey of Canada, Paper 68-63, 5 p.
Gibb, R.A., Thomas, M.D., and Mukhopadhyay, M., 1980, 'Proterozoic
    sutures in Canada'; Geoscience Canada, 7, p. 149-154.
Gill, J.E., 1948, 'Mountain building in the Canadian Shield'; 18th
    International Geological Congress, Part XIII, p. 97-104.
Gower, C.F., Ryan, A.B., Bailey, D.G., and Thomas, A., 1980, 'The pos-
    ition of the Grenville Front in eastern and central Labrador';
    Canadian Journal of Earth Sciences, 21, p. 678-693.
Grant, S.M., 1988, 'Diffusion models for corona formation in metagabbros
    from the western Grenville Province, Canada'; Contributions to
    Mineralogy and Petrology, 98, p. 49-63.
Green, A.G., Milkereit, B., Davidson, A., Spencer, C., Hutchinson, D.R.,
    Cannon, W.R., Lee, M.W., Agena, W.F., Behrendt, J.C., and Hinze,
    W.J., 1988, 'Crustal structure of the Grenville Front and adjacent
    terranes'; Geology, 16, p. 788-792.
Green, A., Milkereit, B., Davidson, A., Spencer, C., Morel, P., Teskey,
    D., Cannon, W., Hutchinson, D., Behrendt, J., Lee, M., and Agena,
    W., 1987, 'Crustal structure of the Grenville Front' (abstract);
    Eos, 68, p. 1356.
Hanmer, S.K., and Ciesielski, A., 1984, 'Structural reconnaissance of
    the northwest boundary of the Central Metasedimentary Belt, Gren-
    ville Province'; in Current Research, Part B, Geological Survey of
    Canada, Paper 84-1B, p. 128-132.
Hanmer, S., Thivierge, R.H., and Henderson, J.R., 1985, 'Anatomy of a
    ductile thrust zone: part of the northwest boundary of the Central
    Metasedimentary Belt, Grenville Province, Ontario'; in Current
    Research, Part B, Geological Survey of Canada, Paper 85-1B, p. 1-5.
Hewitt, D.F., 1957, 'The Grenville Province'; in The Proterozoic in
    Canada, J.E. Gill, ed., Royal Society of Canada, Special Publi-
    cation 2, p. 132-140.
Hyodo, H., Dunlop, D.J., and McWilliams, M.O., 1986, 'Timing and extent
    of Grenvillian overprinting near Temagami, Ontario'; in The Gren-
    ville Province, J.M. Moore, A. Davidson and A.J. Baer, eds.,
    Geological Association of Canada, Special Paper 31, p. 119-126.
Jones, W.A., 1930, 'The petrography of the rocks in the vicinity of
    Killarney, Ontario'; University of Toronto Studies, Geological
    Series, no. 29, p. 39-60.
Krogh, T.E., and Davis, G.L., 1968, 'Geochronology of the Grenville
    Province'; Carnegie Institution of Washington, Yearbook, 67,
    p. 224-230.
Krogh, T.E., and Davis, G.L., 1971, 'The Grenville Front interpreted
    as an ancient plate boundary'; Carnegie Institution of Washington,
    Yearbook, 70, p. 239-240.

Krogh, T.E., Corfu, F., Davis, D.W., Dunning, G.R., Heaman, L.M., Kamo, S.L., Machado, N., Greenough, J.D., and Nakamura, E., 1987, 'Precise U-Pb isotopic ages of diabase dykes and mafic to ultramafic rocks using trace amounts of baddeleyite and zircon'; in Diabase Dyke Swarms, H.C. Halls and W.F. Fahrig, eds., Geological Association of Canada, Special Paper 34, p. 147-152.

Krogh, T.E., Davis, D.W., and Corfu, F., 1984, 'Precise U-Pb zircon and baddeleyite ages for the Sudbury area'; in The Geology and Ore Deposits of the Sudbury Structure, E.G. Pye, A.J. Naldrett and P.E. Giblin, eds., Ontario Geological Survey, Special Volume 1, p. 431-446.

Krogh. T.E., Davis, G.L., and Frarey, M.J., 1971, 'Isotopic ages along the Grenville Front in the Bell Lake area, southwest of Sudbury, Ontario'; Carnegie Institution of Washington, Yearbook, 69, p. 337-339.

Lumbers, S.B., 1971, 'Some aspects of the northwestern margin of the Grenville Province between Sudbury and Lake Timiskaming, Ontario'; Geological Association of Canada, Annual Meeting, Sudbury 1971, Abstracts of Papers, p. 36-37.

Lumbers, S.B., 1973, 'River Valley area, Districts of Nipissing and Sudbury'; Ontario Division of Mines, Preliminary Map P.844, Geological Series.

Lumbers, S.B., 1975, Geology of the Burwash Area, Districts of Nipissing, Parry Sound, and Sudbury; Ontario Division of Mines, Geological Report 116, 158 p.

Lumbers, S.B., 1978, 'Geology of the Grenville Front Tectonic Zone in Ontario'; in Toronto '78, Field Trips Guidebook, A.L. Currie and W.O. Mackasey, eds., Geological Association of Canada, p. 347-371.

Lumbers, S.B., Heaman, L.M., and Vertolli, V.M., 1988, 'Middle Proterozoic magmatic history of the CMB, Grenville Province, Ontario'; Geological Association of Canada, Program with Abstracts, 13, p. A75.

Marcantonio, F., Dickin, A.P., Heaman, L.M., and McNutt, R.H., 1988, 'The age and origin of granites from the Westport - Gananoque area, Grenville Province, Ontario'; Geological Association of Canada, Program with Abstracts, 13, p. A78.

McGrath, P.H., Halliday, D.W., and Felix, B., 1988, 'An extension of the Killarney complex into the Grenville Province based on a preliminary interpretation of a new gravity survey, Georgian Bay, Ontario'; in Current Research, Part C, Geological Survey of Canada, Paper 88-1C, p. 145-149.

McLaughlin, D.B., 1954, 'Suggested extension of the Grenville orogenic belt and the Grenville front'; Science, 120, p. 287-289.

Moore, J.M., Jr., and Thompson, P.H., 1980, 'The Flinton Group: a late Precambrian metasedimentary succession in the Grenville Province of eastern Ontario'; Canadian Journal of Earth Sciences, 17, p. 1685-1707.

Palmer, H.C., Merz, B.A., and Hayatsu, H., 1977, 'The Sudbury dikes of the Grenville Front region: paleomagnetism, petrochemistry, and K-Ar age studies'; Canadian Journal of Earth Sciences, 14, p. 1867-1887.

Passchier, C.W., and Simpson, C., 1986, 'Porphyroclast systems as kinematic indicatots'; Journal of Structural Geology, 8, p. 831-843.

Quirke, T.T., and Collins, W.H., 1930, The Disappearance of the Huronian; Geological Survey of Canada, Memoir 160, 129 p.

Rivers, T., 1983, 'The northern margin of the Grenville Province in western Labrador - anatomy of an ancient orogenic front'; Precambrian Research, 22, p. 41-73.

Rivers, T., and Chown, E.H., 1986, 'The Grenville orogen in eastern Quebec and central Labrador - definition, identification and tectonometamorphic relationships of autochthonous, parautochthonous and allochthonous terranes'; in The Grenville Province, J.M. Moore, A. Davidson and A.J. Baer, eds., Geological Association of Canada, Special Paper 31, p. 31-50.

Rondot, J., 1978, 'Stratigraphie et métamorphismes de la région du Saint-Maurice'; in Metamorphism in the Canadian Shield, J.A. Fraser and W.W. Heywood, eds., Geological Survey of Canada, Paper 78-10, p. 329-352.

Sims, P.K., Card, K.D., and Lumbers, S.B., 1981, 'Evolution of early Proterozoic basins in the Great Lakes region'; in Proterozoic Basins of Canada, F.H.A. Campbell, ed., Geological Survey of Canada, Paper 81-10, p. 379-397.

Stockwell, C.H., 1964, 'Fourth report on structural provinces, orogenies, and time-classification of rocks of the Canadian Precambrian Shield'; Geological Survey of Canada, Paper 64-17, part II, p. 1-21.

Stockwell, C.H., 1982, Proposals for Time Classification and Correlation of Precambrian Rocks and Events in Canada and Adjacent Areas of the Canadian Shield. Part 1: A Time Clasification of Precambrian Rocks and Events; Geological Survey of Canada, Paper 81-19, 135 p.

Thomas, A., Nunn, G.A.G., and Wardle, R.J., 1985, 'A 1650 Ma orogenic belt within the Grenville Province of northeastern Canada'; in The Deep Proterozoic Crust in the North Atlantic Provinces, A.C. Tobi and J.L.R. Touret, eds., NATO ASI Series, series C, 158, D. Reidel Publishing Company, Dordrecht, The Netherland, p. 151-161.

Thomas, M.D., and Gibb, R.A., 1985, 'Proterozoic plate subduction and collision: processes for reactivation of Archean crust in the Churchill Province'; in Evolution of Archean Supracrustal Sequences, L.D. Ayres, P.C. Phurston, K.D. Card and W. Weber, eds., Geological Association of Canada, Special Paper 28, p. 263-279.

Thomas, M.D., and Tanner, J.G., 1975, 'Cryptic suture in the eastern Grenville Province'; Nature, 256, p. 392-394.

van Breemen, O., and Davidson, A., 1988, 'Northeast extension of Proterozoic terranes of mid-continental North America'; Geological Society of America Bulletin, 100, p. 630-638.

van Breemen, O., and Hanmer, S., 1986, 'Zircon morphology and U-Pb geochronology in active shear zones: studies on syntectonic intrusions along the northwest boundary of the Central Metasedimentary Belt, Grenville Province, Ontario'; in Current Research, Part B, Geological Survey of Canada, Paper 86-1B, p. 775-784.

van Breemen, O., Davidson, A., Loveridge, W.D., and Sullivan, R.W, 1986, 'U-Pb geochronology of Grenville tectonites, granulites and igneous precursors, Parry Sound, Ontario'; in The Grenville Province, J.M. Moore, A. Davidson and A.J. Baer, eds., Geological Association of Canada, Special Paper 31, p. 191-207.

Van Schmus, W.R., 1980, 'Chronology of igneous rocks associated with the Penokean orogeny in Wisconsin'; in Selected Studies of Archean Gneisses and Lower Proterozoic Rocks, Southern Canadian Shield, G.B. Morey and G.N. Hanson, eds., Geological Society of America, Special Paper 182, p. 159-168.

Van Schmus, W.R., and Bickford, M.E., 1981, 'Proterozoic chronology and evolution of the midcontinent region, North America'; in Precambrian Plate Tectonics, A. Kröner, ed., Elsevier, Amsterdam, The Netherlands, p. 261-296.

Wanless, R.K., and Loveridge, W.D., 1972, 'Rubidium-strontium isochron age studies, report 1'; Geological Survey of Canada, Paper 72-23, p. 45-47.

Wilson, J.T., 1949, 'Some major structures of the Canadian Shield'; Transactions of the Canadian Institute of Mining and Metallurgy, 52, p. 231-242.

Wynne-Edwards, H.R., 1972, 'The Grenville Province'; in Variations in Tectonic Styles in Canada, R.A. Price and R.J.W. Douglas, eds., Geological Association of Canada, Special Paper 11, p. 263-334.

Young, G.M., 1983, 'Tectono-sedimentary history of early Proterozoic rocks of the northern Great Lakes region'; in Early Proterozoic Geology of the Great Lakes Region, L.G. Medaris, Jr., ed., Geological Society of America, Memoir 160, p. 15-32.

BASALTIC COMPOSITION XENOLITHS AND THE FORMATION,
MODIFICATION AND PRESERVATION OF LOWER CRUST

R. W. Kay[1, 2] and S. Mahlburg Kay[1]
[1]INSTOC
Snee Hall
Cornell University
Ithaca, NY 14853
[2]Department of Geological Sciences
Snee Hall
Cornell University
Ithaca, NY 14853

ABSTRACT. Basaltic intrusives add to the inventory of lower continental
crust and modify existing lower crust in regions of tectonic divergence
and convergence. Therefore, it is not surprising that basaltic
composition xenolithic fragments of lower crust are common in volcanic
rocks of rift zones and magmatic arcs. Due to "flake tectonics"
(Oxburgh, 1972) (in which the crust splits at midcrustal levels under
compression), the deepest levels of the crust are infrequently exposed
and these exposures yield scanty information on lower crustal
composition. Evidence based on xenoliths is direct, but is also
circumstantial as the xenolith populations are non-representative. The
xenoliths, and magmas derived from the lower crust, may be more
instructive in indicating mechanisms for the evolution of the lower
crust, than in directly revealing lower crustal composition. In
particular, preservation of the basaltic lower crust of convergent
margins (as sampled in active island arcs) is problematic, due to
possible wholesale lower crustal delamination, aided by segregation and
upward migration of less dense melts leaving more dense residues, at
convergent zones with structurally thickened crust. Both xenoliths and
lower crustally-derived silicic magmas are used as evidence to support
formation of the dense garnet-bearing granulite and eclogite required
for lower crustal delamination, a process that leaves andesitic
composition upper crust. The removal of more mafic lower crust, leaving
less mafic upper crust is a possible mechanism for creating an andesitic
crust without deriving large amounts of andesite directly from the
mantle. No analogous delamination mechanism exists for basaltic
underplates in rifted continental margins or intraplate localities, and
these underplates may have a higher preservation probability than the
convergent margin lower crustal basalts.

1.   INTRODUCTION

Volcanic rocks frequently entrain xenoliths of crustal and mantle
origin. In a recent survey of the lower crustal xenolith suites that is
not repeated here, Griffin and O'Reilly (1987a) are impressed by the
predominance of "... mafic granulite with subordinate amounts of felsic
metaigneous rocks and rare metasediments." They point to intrusion of
mafic magmas (basalts) as a "major factor in the evolution of the lower
crust" in both continental rift zones and orogenic belts, a viewpoint

401

M. H. Salisbury and D. M. Fountain (eds.), Exposed Cross-Sections of the Continental Crust, 401–420.
© 1990 *Kluwer Academic Publishers.*

with which we concur (e.g. Kay and Kay, 1981). Information on tectonic and lithologic context, as well as actual depth of origin, have been lost during transportation of the xenoliths, and some of their textural, mineralogical, and geochemical features have been destroyed. While recognizing the great diversity of compositional and mineralogical types we focus on the basaltic composition xenoliths for two reasons in addition to their ubiquitous occurrence: they represent additions of new crust from the mantle, and the basaltic intrusives (of which the xenoltihs are samples) are carriers of thermal energy that exercise primary control over the rheology of the lower crust through both heating and partial melting of preexisting lower crust.

Xenolith studies concerned with the physics of xenolith entrainment and chemical interactions between the xenoliths and its host are not considered here. We focus instead on the pre-transportation conditions and history of the xenoliths. The context for this class of topics is external to the host magma and to the xenoliths themselves, and is a tectonic context in a broad sense. We ask what the basaltic xenoliths tell us about processes operating in the lower crust; lower crustal history and lithology; and relationships of upper crust, lower crust, and mantle. These, as with many fundamental questions, are often not as well constrained as we might hope. Considered separately are the two principal regimes of lower crustal formation and modification: rift zones and convergent margins. Finally, two persistently recurring topics relevant to the lower crust are considered: lower crustal preservation and volatile sources. The question of lower crustal composition is not directly answered by xenolith studies--in particular by averaging xenolith analyses--but is usefully considered by tectonic region, and in terms of processes like lower crustal delamination.

2. Rift Zones

2.1 Tectonic Background

The origin of the basalt that is commonly erupted in rift zones is easily explained by decompression melting of upwelling mantle peridotite (e.g. White et al. 1987). Indeed, it is difficult to account for the high geothermal gradients in rifts, and for the associated crustally-derived silicic melts without basaltic intrusion (Sass and Lachenbruch, 1979; Huppert and Sparks, 1988). Models that increase mantle heat flux to attain temperatures sufficient for the melting and segregation of silicic melts in the crust require upwelling of hot asthenospheric mantle. Generation of basalt by decompression melting of mantle asthenosphere, rapid segregation of the melt, and its intrusion into overlying mantle and crustal lithosphere invariably accompanies asthenospheric upwelling.

Lower crustal temperatures in rift zones often exceed the melting temperature of preexisting crust (both salic and mafic) that has hydrous minerals. For example, muscovite-bearing pelitic rocks partially melt at temperatures near 700°C, and biotite and amphibole-bearing rocks at temperatures near 800° and 900° respectively (Clemens and Vielzeuf, 1987, Whitney, 1988). Recrystallization accompanying penetrative deformation in partially molten rocks results in formation of mineral

assemblages appropriate to the ambient P-T conditions. Mineral
isochrons will be reset partially or completely to the time of rifting.
The Pyrenees crustal section described by Wickham (1987) and Wickham and
Oxburgh (1987) provides a good example of an overall crustal response to
basaltic underplating accompanying rifting, where the crust has abundant
hydrated pelitic rocks. More generally, Clemens and Vielzeuf (1987) and
Huppert and Sparks (1988) provide discussions of phase petrological and
fluid dynamical constraints on melting and melt segregation. From these
studies it is apparent that the thermal history and the abundance and
distribution of volatiles are critical to the amount of crustal melting
and to crustal rheology in rift zones.

2.2 Xenolith Studies

Xenolith suites in the late Tertiary volcanic rocks from eastern
Australia and from the Rhein Graben (Germany) and Rio Grande (U.S.)
rifts are among the best-studied examples representing rift
environments. Using these suites, some of the propositions outlined in
the "Tectonic Background" section have been tested.

   Xenolith studies can help to define the thermal history of rifts,
knowledge of which is essential for the development of structural
models, and for determining the role of basaltic underplating. Ideally,
the deep thermal structure of rifting zones is defined by the P-T
conditions of a series of crustal and mantle xenoliths from different
depths, provided that the mineral equilibria used to determine P and T
were attained at the time of rifting. To determine ambient conditions
from xenoliths recovered from volcanic rocks of active rifts, isotopic
homogenization of Sr and Nd, indicating zero age is essential within the
chemically homogeneous regions of coexisting minerals used to determine
P and T. This follows from the expectation that the diffusion rates of
major element cations (Mg, Fe, Ca, Al) in minerals used to determine P
and T are comparable to those of Sr and Nd (Sneeringer et al., 1984).
Different results have been found in studies of Sr and Nd isotope
systems in xenoliths from the Rio Grande Rift, Eastern Australia, the
Rhein Graben and the Colorado Plateau.

   At Kilbourne Hole, Rio Grande Rift, Reid (1987) determined a two-
mineral internal isochron age of about 30 ma (within error, almost
coincident with eruption time) for a metapelitic granulite xenolith that
plots on a 1.6 ba whole rock isochron. The conjecture that mineral
equilibria reflect a very recent geotherm (e.g. Padovani and Carter,
1977) is thus justified. Furthermore, the very high temperatures at
shallow depths require a significant thickness of basaltic underplating
(Sass and Lachenbruch, 1979). Whether any of the basaltic composition
granulite xenoliths at Kilbourne Hole represent this basalt has yet to
be determined. In contrast to the Rio Grande case, Stosch and Lugmair
(1984) report that coexisting granulite facies minerals in lower crustal
xenoliths brought to the surface by Pleistocene alkali basalts in the
Eifel region of the Rhein Graben give a mineral isochron of 172 ma.
Clearly, fossil metamorphic conditions are recorded in these xenoliths.
Similar results are reported by Arculus et al (1988) who found isotopic
disequilibrium between minerals in eclogite and amphibolite xenoliths
included in a 25 ma latite from Chino Valley, Arizona, and in granulite

xenoliths from a 170 ma nephelinite from Delegate, Australia. At Chino
Valley and Delegate, isotopic differences between minerals are too large
to have developed in the time between eruption of the host lava and
present. Xenolith minerals that have not achieved diffusion-controlled
homogenizations of Sr and Nd at the time of eruption are not expected to
comprise equilibrium assemblages at that time and the xenoliths can not
represent basalts underplated immediately preceding eruption of the host
lava. A series of such xenoliths will not yield information on
geothermal gradients. Arculus also reports that minerals from an
eclogite from Delegate were isotopically identical at the time of
eruption, a finding that he attributes to higher T (corresponding to
greater depth) of the eclogite, compared to the granulites.

Evidence of the thermal history of deep crustal rocks is erased if
they are thoroughly recrystallized immediately preceding their
entrainment as xenoliths. However, minerals in some granulite xenoliths
are not completely recrystallized, (e.g. Harte et al., 1981, Kay and
Kay, 1983) and zonation of mineral assemblages, and of mineral
compositions can be used to infer P-T paths, in much the same way as is
done for exposed granulites (e.g. Bohlen, 1987). A key observation for
some Queensland basaltic granulites (e.g.Kay and Kay, 1983) is that
isobaric cooling from garnet-free to garnet-bearing mineral assemblages
occurred, consistent with slow relaxation of a thermal pulse, such as
that caused by lithospheric thinning and crustal underplating by
basaltic magmas (Bohlen, 1987). The thermal pulse is not solely due to
localized, and therefore short-lived, heating in the crust, or to
entrainment of the xenoliths in their basaltic host magma, both of which
have easily recognizable mineralogical effects (e.g. melting of garnet,
formation of spinel symplectites).

The amount of basaltic composition lower crust created by intrusion
and underplating of basalt during a period of rifting remains difficult
to define (e.g. McKenzie and Bickle, 1988). In several eastern
Australia localities, two-pyroxene granulites of basaltic composition
are the most frequent feldspathic lithology, and to reflect this fact,
crustal cross sections of Griffin and O'Reilly (1987 a,b) show
lithosphere from 20 to 55 km dominated by mafic granulites (presumably,
many are rift-related). As to the total amount of rifting (beta, the
ratio of final to inital surface area) is small, abnormally hot mantle
(compared with mantle under mid oceanic ridges) is required (e.g.
McKenzie and Bickle, 1988). However, as summarized above, dating of
mafic granulite xenoliths from Delegate shows that intrusion age of
basalts destined to recrystallize into some of the granulites far
preceded rifting, at which time a second batch of basaltic intrusions
was formed. At some xenolith localities, chemical groupings of mafic
xenoliths (representing different periods of basaltic volcanism)
indicate the same thing (e.g. Kay and Kay, 1983). Thermal models of
Sass and Lachenbruch (1979) indicate that basaltic intrusion into the
upper crust over a several million year period explains both the present
heat flow and the rapid decrease in heat flow following cessation of
volcanism in different regions in eastern Australia at different times.
In eastern Australia, the xenoliths which can mineralogically constrain
the P-T models are contained in low-volume alkaline intrusions that
probably formed at greater depth and are associated with smaller

lithospheric thermal anomalies than the larger volume of tholeiitic
basalts that erupted subsequently at many localities. Given this complex
and protracted igneous history, the meaning of a xenolith-defined
geotherm constructed in the absence of chronological control is
questionable. The history of deep crustal intrusion is correspondingly
uncertain.

Crystal-liquid fractionation processes are certain to occur in any
slowly-cooled deep-crustal basaltic intrusive. Contamination of the
magma by wall-rock interaction is likewise probable. Yet Arculus and
Johnson (1981), Stosch et al. (1986), Stosch and Lugmair (1984), and
Arculus et al. (1988) point out that the geochemistry of many rift-
related basaltic composition xenoliths is not explained by igneous
processes alone (melting, assimilation, crystal fractionation). A
subsolidus fluid appears to be responsible for the introduction of Ba,
Sr and alkali metals to many of the mafic (and ultramafic) xenoliths in
the Eifel locality described by Stosch and Lugmair (1984). Amphibole in
the mafic granulites is textually secondary and appears to have formed
by introduction of isotopically distinct metasomatic fluids (melts or
aqueous fluids), probably immediately preceding, and broadly related to
the recent rift-related basaltic volcanism that entrained the xenoliths.

Given the polymetamorphic history of many mafic and ultramafic
xenoliths, it is perhaps not surprising that trace element and major
element fractionation trends in many xenoliths do not appear to have
been controlled by the mineral phases present in the xenolith when it
erupted. For instance, Arculus et al. (1988) show that the large rare
earth fractionations characteristic of garnet are absent in whole rock
rare earth abundance patterns of eclogite xenoliths from Australia and
the Colorado Plateau. The major element and trace element fractionation
between samples appears to have occurred before the final metamorphic
stage, at a time when the rocks had a garnet-free (e.g. gabbroic)
mineralogy.

2.3 Summary and Commentary

The xenolith (or "inclusion")fragments brought up by basaltic extrusives
in continental rift zones provide direct and indirect evidence for the
importance of basaltic intrusives in the lower crust and lithospheric
mantle. The direct evidence is the abundant basic granulites:
clinopyroxene-plagioclase rocks with or without orthopyroxene, garnet,
spinel, biotite, and various minor phases. Often these granulites fall
in more than one compositional group, for instance a "cognate" group
compositionally related to the host lava, and related lavas, and an
"accidental" group representing pre-existing crust. Some trace elements
appear to have been added to (especially) the "accidental" group; this
subsolidus metasomatic event is commonly reflected in textually
secondary hydrous phases: biotite and amphibole. Often the minerals in
the basaltic composition granulites are not in either textural or
isotopic equilibrium at the time of their removal from the lower crust
or upper mantle. This lack of equilibrium is also true of non-basaltic
xenoliths, although the non-metasomatic minerals of some xenoliths of
both ultramafic and mafic composition are at isotopic equilibrium --

perhaps due to their higher temperature (see Harte et al. 1981). It should be noted that, for many non-metasomatic minerals, time resolution is not very precise for the Rb-Sr and Nd-Sm systems and chronology of pre-metasomatic events is difficult to resolve.

Ideally, pressure and temperature calculated from a series of xenoliths that have attained equilibrium at various depths at the same time can be used to define geotherms. More practically, detailed studies sufficient to resolve thermal relaxation times consistent with basaltic intrusion over times of even tens of millions of years have not been done. The strong depth dependence of thermal relaxation of deep-seated thermal perturbation such as that caused by asthenosphere upwelling or basaltic underplating that accompanies the asthenospheric upwelling (e.g. Furlong and Fountain, 1986) creates an additional unresolved problem.

3. CONVERGENT MARGINS

3.1 Tectonic Background

At convergent margins, and in particular under the volcanoes of the associated magmatic arcs, the tectonics of the mantle, especially as related to the origin of arc basalt, is a controversial topic. It is sufficient for our purpose to observe that olivine tholeiite basalts and liquidus olivines, clinopyroxenes and spinels derived from them are present in many volcanoes of convergent plate margins. These olivine tholeiites are most straightforwardly interpreted as the products of partial melting of mantle peridotite in the "mantle wedge", the mantle located above the subducting oceanic lithospheric plate. The peridotite of the mantle wedge is an open-system reservoir-- it receives mass from the underlying plate and discharges mass to the overlying lithospheric mantle and crust. Magmas other than olivine tholeiite are also mantle-derived contributors to crustal mass: high-Al basalts, Mg-andesites, boninites, and picrites. Among these other magmas only Mg-andesites and boninites are difficult to relate to olivine tholeiite by standard petrogenetic processes (e.g. assimilation, crystal fractionation) operating at lower crustal or uppermost mantle pressures. Both Mg-andesites and boninite-series magmas are rare products of active volcanoes; therefore unless petrologic sampling has been grossly unrepresentative, a present-day perspective leads us to conclude that a basaltic magma ranging from olivine-rich (picrite) to olivine-poor (hi-Al basalt) is the predominant magma type to enter the crust at the magmatic arcs of convergent margins. Much as at continental rift zones, which at crustal levels are structurally akin to some arcs (e.g. Stern, 1987), it is likely that the melting temperatures of some preexisting crustal rock types are exceeded at lower crustal depth. Melting, assimilation and fractional crystallization are competing and interacting processes that account for the composition of volcanic products, which are often more siliceous than basalt. The volcanic products themselves are very poor indicators of the bulk composition of mass added to the arc crust (which is basalt: e.g. Hyndman and Foster, 1988), but are very good indicators, by mass balance, of the bulk composition of mass added to the lower crust (which is pyroxenitic to gabbroic,complementary to andesitic mass added to the upper crust). The bulk composition of lower crust differs at oceanic and continental margins, but this is not due to differences in the new magmatic additions to the crust, which are basaltic in both cases, but to differences in the preexisting crust and perhaps to differences in structural redistribution of crustal lithologies ("subcretion" of oceanic sediment for example). It has become increasingly clear that the localization of deformation in space and time is strongly dependent on crustal chemical heterogeneities, especially of H_2O, which in extreme cases fluxes crustal zones rendering them very ductile (e.g. Hollister and Crawford, 1986).

None of the structural and magmatic processes thought to operate in the crust is observable in the field at active convergent margins-- the level of exposure rarely exceed several kilometers in active or recently active arcs. The deepest crustal levels of an arc, as well as its shallow-level volcanic products, appear to be exposed in tectonically exhumed sections such as those in the Ivrea-Strona Ceneri-Lugano zone, Italian Alps; (e.g. Pin and Sills, 1986; Cumming et al, 1987; Voshage et al, 1987; Stille and Buletti, 1987) and in Kohistan, Pakistan (e.g. Hamilton, 1988, Khan et al., 1990). In the Ivrea zone, complexity is revealed, with the exposed lower crust resulting from two magmatic periods, the first associated with an oceanic setting at about 600 my, and the second a convergent margin at about 300 my. This degree of complexity may be normal in deeper levels of arcs.

It is noteworthy that associated with the Ivrea zone, arc volcanic activity of ~300 ma contains very few magmatic rocks that are not hybrids of mantle-derived basalt and fusible quartz-feldspar-bearing sediment. In some other arcs much less crustal interaction has occurred, or perhaps is less apparent because the overall section is thought to be less siliceous, and offers less compositional contrast to the mantle-derived magmas, as in oceanic arcs like the Aleutians (e.g. Kay and Kay, 1985a).

3.2 Xenolith Studies

Xenoliths relevant to identification of lower crustal lithologies and processes at convergent plate margins occur in volcanic rocks from the magmatic arcs themselves, and in later tectonically unrelated magmas that cut earlier magmatic arcs. These two types of xenoliths yield complementary information, for the lower crusts they represent have had quite different histories.

As an aside, up to only several years ago it was "common knowledge" that mantle and lower crustal xenoliths are absent in arc magmas and that liquidus minerals in arc magmas crystallized in the upper crust. It was thought that "rare" xenocrysts of Mg-rich olivine in some andesites were accidental fragments from mantle depth. It now appears that the widespread xenolithic and xenocrystic fragments in both andesite and basaltic arc magmas originate in the lower crust and upper mantle and contain otherwise unobtainable information that might support or refute hypotheses advanced in the above "Tectonic Background" section.

First, mass balance considerations combined with the results of experimental investigations indicate that mafic to ultramafic crystal cumulates representing the liquidus phases of olivine tholeiite at lower crustal pressure might be expected among xenoliths in calc-alkaline andesites and basaltic rocks. The xenoliths from Itinomegeta, Japan (Sakuyama and Koyaguchi, 1984; Aoki and Fujimaki, 1982) and Adak and Kanaga, Aleutian Islands (Conrad and Kay, 1984, Swanson et al., 1987) include examples of these deep-level cumulates (see Table 1), largely corroborating the deductions made on the basis of mass-balance and experimental phase relationships.

Table 1   MINERALOGY OF MAFIC AND ULTRAMAFIC ROCKS OF THE LOWER CRUST AND UPPER MANTLE OF CONVERGENT MARGINS

Ol	Cpx	Opx	Hbl	Plag	M-C-R	Gar	Qtz	Ep
Xenoliths - Adak Is, Aleutians [2] Xenoliths (All rocks, except last entry, are not deformed)								
			+	+	M			
	+		+	+	M			
	+	+		+	M		+	
+	+		+	+	M			
+	+		+		C			
+	+		(+)		C			
Kanaga Is., Aleutians, Xenoliths [3] (all rocks are deformed)								
	+	+		+	M			
+	+			+	M			
+	+				C			
Big Creek, Sierra Nevada, U.S. [4] (All rocks are deformed)								
	+			+		+		
	+		+	+		+	+	
	+	+				+		
+	+	+						
Crustal Terrane - J. Jal Complex, Kohistan [5] (All rocks are deformed)								
+		+	+		R	+	+	
+			+		R	+	+	+
+		+	+		R	+	+	+
			+		R	+		
+		+			R	+		
+	+	+			R	+		
Ultramafic, Unrelated to Above (All rocks are deformed)								
	+	+			M			
+	+	+			C			
+		+			C			

1. Minerals Olivine (Ol), Clinopyroxene, (Cpx), Orthopyroxene (Opx), Hornblende (Hbl), Plagioclase (Plag), Magnetite (M), Chromite (C), Rutile (R), Garnet (Gar), Quartz (Qtz), Epidote (Ep)
2. Conrad and Kay (1984), DeBari et al., (1987); 3. Pope (1983), Johnston (1983); 4. Dodge et al., (1988); 5. Jan and Howie (1981).

We note particularly the presence on Adak of ultramafic xenoliths with both tectonized (DeBari et al. 1987) and cumulate (Conrad and Kay, 1984) textures, that represent the early fractionation stages of olivine tholeiite magma. The Sr and Nd isotopic ratios in xenoliths and arc magmas are the same (Kay et al. 1986) as they must be if the xenoliths are related to the magmas. These ultramafic fractionates are thought to accumulate at the crust-mantle boundary, and to form a renewable crust-mantle boundary (see Kay and Kay, 1985a). The residual magma, a high-Al basalt (Crawford et al, 1987) is a common extrusive lava in the arc.

The fractionation at lower crustal depths of plagioclase-bearing liquidus phases amounting to half of the mass of the high-Al basalt yields another common extrusive magma: calc-alkaline andesite. Cumulate xenoliths corroborating this process are also represented in the xenolith population (Conrad and Kay, 1984).

Xenoliths that represent deeper crust and upper mantle under extinct convergent margin terranes appear to be common in some kimberlite suites (e.g. Rogers and Hawkesworth, 1982) and in some suites from alkali basalt (e.g. Kay and Kay, 1983). For example, the xenoliths from Miocene andesite (contaminated alkali basalt) that intrudes the Sierra Nevada batholith (Dodge et al, 1988; Domenik et al, 1983) include samples of lower crust from directly under this extinct magmatic arc. Comparison of the Sierran xenoliths with those from Adak, Aleutian arc is instructive. As shown on Table 1, some of the mineralogical features of the calc-alkaline gabbroic xenoliths are similar: the presence of the hydrous phase hornblende for instance. However, the dense phase garnet, absent in the Adak xenoliths, is widespread in the Sierra mafic granulites (which include eclogites), and is also found in some peridotites. Therefore, the mafic lithologies in the lower crust and upper mantle of the Sierras are denser than those in compositionally equivalent Aleutian lower crust and upper mantle. The density contrast is mainly due to the denser mineral assemblage, not to thermal contraction corresponding to lower in situ temperature. Although no information is available, it seems likely that the P-T conditions recorded by Sierran xenolith minerals refer to the Mesozoic-coincident with the arc-related Sierran plutonism- not the Miocene.

Finally, much as in the case of rift xenoliths, the presence of secondary ("metasomatic") textures and geochemical features are present in some volcanic arc xenoliths (Arculus and Johnson, 1981; Kay and Kay, 1989.

3.3 Summary and Discussion

As basaltic additions to lower crust at convergent margins crystallize, they leave cumulate residues, while concurrently assimilating soluble fractions of wall-rock. The residual magmas may erupt at various stages of magmatic evolution. If the preexisting crust is relatively hydrous and especially if it contains micaceous quartzo-feldspathic rocks, the heat released by crystallizing basaltic magmas is sufficient for partial melting of the crust. The hotter the crust, the more that will melt, up to a mass roughly equal to the intruded mass (the case for crust at its solidus before intrusion). Crustal melt, wholly or partly segregated then can rise as water-undersaturated silicic intrusive or extrusive

magmas (see Huppert and Sparks, 1988). Residual lower crust,
complementary to the segregated silicic magma, will be a common
constituent of the lower crust in regions with abundant crustally-
derived silicic magma. Intrusives of basaltic composition will also be
abundant in the lower crust.

When the lower crust is relatively dry and lacking in quartz plus
feldspar, basaltic intrusion may result only in a heating episode that
is recorded by crustal rocks as a distinctive isobaric cooling path for
metamorphic minerals. Bohlen (1987) recognizes this P-T history in many
exposed granulites, and it has also been observed in mafic granulite
xenoliths that represent oceanic crustal basement and early arc crust of
the Aleutian arc (Johnston 1983). Metasomatic additions to the lower
crustal rocks, sometimes with accompanying high-T hydration may
accompany the cooling of the rocks. The chronologies of these P-T-time
paths in xenoliths have not been investigated.

4. CRUSTAL COMPOSITION, EVOLUTION AND PRESERVATION

4.1 The Problem of Crustal Composition

The composition of continental crust appears to be "andesitic"
(Poldervaart, 1955) yet, as outlined in the preceding sections of this
paper, present-day additions appear to be "basaltic". Resolution of
this paradox is central to any understanding of crust and mantle
evolution. Two alternatives will be examined: perhaps mantle-derived
additions to the crust in the early stages of earth history were
andesitic, not basaltic; perhaps, the"andesitic" upper crust of
compositionally stratified young "basaltic" crust has a greater
preservation probability than the mafic-ultramafic lower crust.

4.2 Andesites from the mantle

Experimental petrology has defined conditions, including the presence of
water, for the generation of andesites from mafic mantle sources (e.g.
Tatsumi, 1982). To be inventoried as new continental crustal material,
the andesite's mafic source, it must be emphasized, needs to be in the
mantle, not in the crust. Melting of andesite out of a basaltic layer
at the base of the crust, for example, is merely redistribution of
crustal mass, not andesite addition to the crust. The introduction of
hydrated basaltic rocks into the mantle by megathrusts is commonplace:
it occurs at subduction zones. In order for oceanic crust to descend
into the mantle, the basalt must have transformed to its dense
equivalent, eclogite. Broadly speaking, partial melting of eclogite at
depths of about 100 km yields andesite (under the volcanic front of
magmatic arcs). Why, then, is eclogite melting an unpopular mechanism
for production of orogenic andesites? First, the compositions of
partial melts of eclogite do not resemble orogenic andesite compositions
very closely (e.g. Stern and Wyllie, 1978). Second, the trace element
contents (especially, REE, alkali metals) of common orogenic andesites
do not resemble those of partial melts of eclogites with oceanic crustal
composition (Gill, 1974). Note that high Ni, and Cr contents of some
andesites make an origin by melting of a basaltic composition source

unlikely as this process would result in a low Ni, low Cr melt (see Kay, 1978, for example). Third, orogenic andesites are often spatially and temporally associated with basalts, and often can be related by crustal fractionation and melting processes to the basalts. Even where the andesites are not accompanied by basalts, the andesites often contain mafic liquidus minerals from basalts (e.g. Kay and Kay, 1985b), in which case the andesites are thought to originate by mixing of mafic and felsic magmas. For instance, the mantle xenolith-bearing calcalkaline andesites from Itinomegata, interpreted by Aoki and Fujimaki (1982) as originating in the mantle, have been reinterpreted as mixtures of basalt and dacite (Sakuyama and Koyaguchi, 1984). Finally, while partial melting of a garnet-bearing mafic source in the mantle has been invoked to explain (for example) Archean trondhjemite-tonalite-granodiorite series rocks with high La/Yb ratios, it is not clear how to determine whether this melting occurs in the mantle (e.g. Martin, 1986) or in the lower crust (e.g. Jahn et al., 1981).

Under hydrous conditions at pressures corresponding to shallow mantle depths, peridotite melting yields a distinctive andesite (Tatsumi, 1982) with high MgO (5-10% at 58-60% SiO_2) and Mg number (consistent with equilibrium with high Mg number olivine in the mantle-- e.g. Nicholls, 1974). The experimental melt compositions are similar to a broad class of andesites called magnesian andesites (Kay, 1978), sanukitoids (Stern, et al., 1989), high-Mg andesites (Tatsumi and Ishizaka, 1982), high MgO andesites (Stern et al., 1984), bajaites (Saunders et al., 1987), and boninites (Meijer, 1980). This class of andesites is geochemically and isotopically extremely diverse: La/Yb ratios for the types of the preceding sentence range from 1.0 to about 40. The only chemical characteristics these andesites seem to share is their high Mg, Ni, and Cr and high Mg/Fe ratios. These characteristics appear not to be the result of mixing (of mafic basalt and dacite or rhyolite, for example). Having grouped these volcanic rocks, it remains to note that they all occur at active convergent margins where subducted crust is young and hot, or in extensional regions, again involving hot mantle, that are spatially coincident with recently active convergent margins. The coincidence with hot mantle that has a plausible hydration mechanism (transfer of water from the subducted plate) with magnesian andesite is strong and we believe, significant.

Are these Mg-andesites a common mantle-derived magma at present? Does fractionation of these Mg-andesites yield the common calc-alkaline andesite of volcanic arcs? The answer to both questions is probably no. Based on an inventory of recent lavas whose tectonic affinities can be reliably accessed, the Mg-andesites are not frequent among arc volcanic rocks, including submarine lavas that represent the early stages of island arcs (e.g. Bloomer et al., 1989). It may be concluded that the Mg-andesites are not volumetrically important as lavas. Fractionation of mafic phases (olivine, orthopyroxene, clinopyroxene, and amphibole) from the Mg-andesites at mid-to lower-crustal pressure could yield calc-alkaline andesite, but generally a boninite fractionation series, distinct from the calc-alkaline series, develops (Meijer, 1980). For the Aleutian case in which basalts associated with calc-alkaline andesites share common Sr, Nd and Pb isotopic compositions, and in which observed liquidus phases match those necessary for crystal fractionation of basalt to andesite, it seems perverse to call on an unobserved primary Mg-andesite magma as a parent to the andesites. Among Aleutian magmas the only observed Mg-andesites (Kay, 1978) have uniquely high e_{Nd} values, and uniquely low radiogenic to non-radiogenic Pb and Sr isotope ratios. Orthopyroxene-bearing cumulates of the boninite fractionation series are missing from Aleutian xenolith populations.

It remains to note that if somewhat extreme (at present) tectonic conditions are required for genesis of andesites from mantle, these conditions may have been more common in the early stages of earth history. The most voluminous Mg-andesites at present, probably those of the boninite series, have trace element contents that clearly eliminate them from consideration as the parental magmas of, for instance, Archean tonalites and monzodiorities (broadly, andesites) cited by Martin (1986) and Stern et al. (1989). On the other hand, magma types like the Mg-andesites from the Sentouchi volcanic belt (Tatsumi and Ishizaka, 1982), Baja California (Saunders et al., 1987) and the Aleutians (Kay, 1978) are plausible parental magmas to the Archean tonalites and monzodiorites. But melting of lower crustal mafic rocks that are compositionally complementary to the voluminous Archean calc-alkaline, intrusives that frequently precede the less voluminous "post tectonic" Mg-andesite related magma types seems an equally plausible, and to us a more likely mechanism than melting of the mantle. In the "andesites from the mantle" hypothesis for the formation of the continental crust, the burden of proof rests with its proponents, who so far have only demonstrated its plausibility.

4.3 ANDESITES FROM THE LOWER CRUST - AND DELAMINATION OF THEIR RESIDUES

If direct melting of mantle to yield andesite remains an "revelatory - if true" hypothesis, melting of basaltic lower crust, and fractionation of basalt at crustal levels to yield andesite are in many cases "commonplace- and true" hypotheses. But, in reference to a process that results in the net addition of andesite to the crust, compositional differentiation of basalt within the crust requires that the basaltic crystalline residues or fractionates are either removed to the mantle or, by phase transition, become dense enough to be identified as mantle on seismic refraction surveys. That is, the processes of andesite differentiation, if they operate in the crust, require a complementary process of recycling of mafic crust to mantle. For mantle derivation of andesite no such process is necessary: the residues are already in the mantle. The transfer of mass from crust-to-mantle is envisioned as one of two possibilities: segregation of ultramafic fractionates, (or formation of ultramafic residues), which are already dense enough to be inventoried as mantle at liquidus temperatures, or metamorphic transformation of relatively less dense basic fractionates or residues (e.g. gabbro) to dense equivalents (e.g. eclogite) that become mantle. The second process requires high pressure (thick crust) and may be accompanied by melting with melt segregation enhanced by water, because both the percentage of melt and its fluidity are greater with more water. Xenoliths and crustally derived magmas provide evidence for both types of recycling. Some olivine-pyroxene (ultramafic) xenoliths in basaltic rocks are commonly interpreted as "cognate": comprised of accumulated liquidus minerals of a magma (commonly basalt: e.g. Conrad and Kay, 1984) similar to the one that transported them to the surface. Basaltic phase relationships at lower crustal pressure show that olivine and pyroxenes are the initial liquidus phases of olivine tholeiite (e.g. Gust and Perfit, 1987). As the pressure of crystallization of a basaltic melt decreases, and as its water content increases, the basalt yields a greater mass proportion of ultramafic fractionates. If the fractionation occurs at the crust-mantle boundary (e.g. Herzberg et al 1983) and if the fractionated crystals sink, new mantle will form, and the fractionated magma will be less mafic. Crustal components in the magma (either through assimilation or subduction-related recycling) may be transfered to the mantle in the fractionated crystals. Generally, as soon as plagioclase becomes a liquidus phase the mean density of fractionated minerals will be less than that of the mantle: the fractionated minerals will form new lower crust. From simple mass balance considerations, it is impossible to form andesitic crust from crystallization of only ultramafic minerals from olivine tholeiite. High-Al basalt, a common magma type in arc volcanoes, is likely to be the most fractionated product of olivine tholeiite involving separation of only mafic minerals. About 50% of additional crystallization - involving both plagioclase and mafic phases - is required to evolve andesite from hi-Al basalt. These fractionates are of crustal density, (see Kay and Kay, 1985a).

Coexiting garnet and pyroxene are not liquidus phases of hi-Al basalt until pressures exceeding those at the base of the thickest crust of island arcs (Green, 1982): this pressure constraint accounts for the absence of garnet-bearing lithologies in the xenolith populations of oceanic island arcs (of Table 1). A survey of eclogitic xenoliths (Arculus et al. 1988) indicates that many have basaltic composition and their compositional diversity appears to have been established when garnet was not present. In other words they are recrystallized gabbros that used to have crustal density, but recrystallized metamorphically to mantle density. Kay and Kay, (1986, 1988) have outlined the case for this transformation to have occurred at a continental collision, where hydrous basaltic lower crust occurs at the base of abnormally thick crust in a relatively cool environment. The proposition is that the "basaltic" lower crust is dense enough to sink into the underlying mantle, and is therefore no longer present at times greater than several million years following collision. Its compositional complement remains: less dense "andesitic" mid to upper crust. Note that mantle lithosphere may also delaminate, but that the main lithospheric density maximum may be located (depending on how much recrystallization occurs) within the basaltic lower crust. This process is thought to operate at present in the Alps (e.g. Laubscher, 1988, see Figure 1), where the basaltic lower crust is removed into the mantle. In Alpine analogues that have been tectonically extinct for more than only a few million years recently erupted xenolith suites provide no test of the delamination hypothesis.

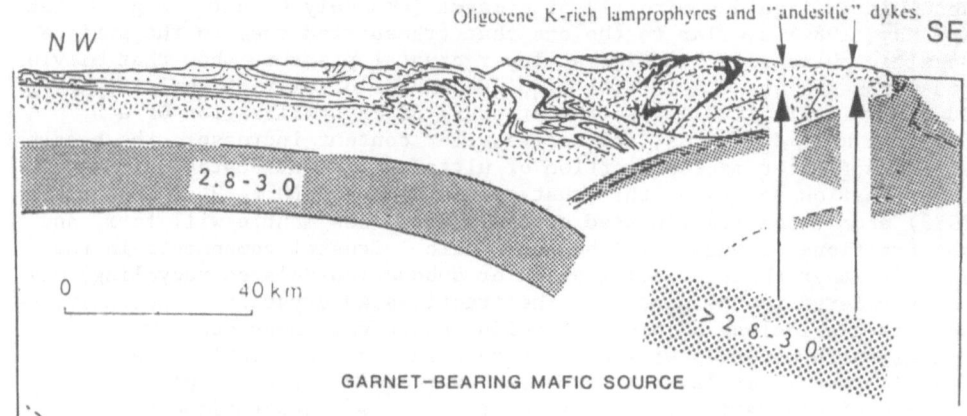

Figure 1.   The deep structure of the Alps (after Laubscher, 1988).
Attention is drawn to the density between 2.8 and 3.0, which apparantly
increases to mantle values in the deeper parts of the underthrust plate.
The dikes indicated refer to underthrusting not at present, but earlier
in the collisional history (Venturelli et al., 1984).

Magmas generated in the lower crust and upper mantle at the time of
or immediately preceding delamination provide unique information on the
delamination process. First, the common post-collisional basalts and
slight extension may be common consequences of mantle upwelling to ~35-
40 km that accompanies delamination, as outlined by England and Houseman
(1989). Note that Bird's (1979) and Houseman et al's (1981)
lithospheric delamination models involve only the mantle part of the
lithosphere, whereas here, we propose that both lower crust and mantle
delaminate. Second, many intermediate composition ("andesitic") post
collisional magmas, including rocks such as the Alpine "appinites", and
various lamprophyres have trace element abundances that indicate
equilibration with a garnet-bearing, plagioclase-free residue (either
olivine-bearing or olivine-free). The density of these restites can
easily exceed that of the underlying peridotitic mantle -- as required
for delamination.

A second site for delamination may be at the base of tectonically
collapsed and thickened crust of some continental convergent margins.
The Andean margin is perhaps best constrained, with large differences in
crustal thickness occurring along the margin at present. Structural and
petrologic modelling (e.g. Isacks, 1988; Kay et al., 1987) calls for
relatively rapid changes in crustal thickness in the Tertiary. In this
context, the most direct evidence for the presence of dense mafic
residues (dominated by garnet, and clinopyroxene or amphibole) in
regions of thickened crust like the Puna-Altiplano and the northernmost
part of the Southern Volcanic Zone is the trace element patterns of
melts that are derived from the thickened lower crust. The most
diagnostic trace element signature of such melts is their low heavy rare
earth contents (Kay et al., 1987).

4.4   Final Observations

Lower crustal xenoliths provide abundant evidence for intrusion of
basaltic composition magmas into lower continental crust in a diversity
of tectonic regimes. Probably, the greatest accumulations occur at
convergent plate margins. However, the basaltic lower crust that
originates by crustal underplating at rift zones and intraplate
localities may have a higher preservation probability than that created
at convergent margins, for which a removal mechanism exists --
delamination at the base of structurally "overthickened" zones. As the
removal mechanism leaves andesitic upper crust as the net composition of
crustal addition, lower crustal delamination offers an alternative to
the extraction of andesitic composition melt from the mantle to explain
the andesitic composition of the continental crust. As advocated by
Nelson (1990), and O'Nions and Oxburgh (1989), however, non-convergent
zone basaltic underplates, by virtue of their high preservation
probability, may be more important in accounting for the lowermost
crustal regions in fully stabilized (cratonic) crusts.

REFERENCES

Aoki, K.I., and Fujimaki, H., 1982. Petrology and geochemistry of calc-alkaline andesite of presumed upper mantle origin from Itinome-gata, Japan. Amer. Mineral., 67: 1-13.

Arculus, R.J., and Johnson, R.W., 1981. Island-arc magma sources: a geochemical assessment of the roles of slab-derived components and crustal contamination. Geochem. J., 15: 109-133.

Arculus, R.J., Ferguson, J., Chappell, W., Smith, D., McCulloch, M.T., Jackson, I., Hensel, H.D., Taylor S.R., Knutson, J., and Gust, D.A., 1988. Eclogites and granulites in the lower continental crust: Examples from eastern Australia and southwestern U.S.A., in Eclogites and Related Rocks, D.C. Smith (ed.), Elsevier, N.Y.

Bird, P., 1979. Continental delamination and the Colorado Plateau. Jour. Geophys. Res. 84: 7561-7571.

Bloomer, S., Stern, R., Fisk, E., and Geschwind, C., 1989. Shoshonitic volcanism in the northern Mariana arc, I, Mineralogic and major and trace element characteristics, Jour. Geophys. Res., 94: 4469-4496.

Bohlen, S.R., 1987. Pressure-temperature-time paths and tectonic models for the evolution of granulites. Jour. Geol., 95: 617-632.

Clemens, J.D., and Vielzeuf, D., 1987. Constraints on melting and magma production in the crust. Earth and Plant. Sci. Let. 86: 287-306.

Conrad, W.K., and Kay, R.W., 1984. Ultramafic and mafic inclusions from Adak Island: crystallization history, and implications for the nature of primary magmas and crustal evolution in the Aleutian Arc. Jour. Petrol., 25: 88-125.

Crawford, A.J., Fallon, T.J., and Eggins, S., 1987. The origin of island arc high alumina basalts. Contrib. Mineral. Petrol., 97: 417-430.

Cumming, G.L., Koppel, V., and Ferrario, A., 1987. A lead isotope study of the northeastern Ivrea Zone and the adjoining Ceneri zone (N-Italy): evidence for a contaminated subcontinental mantle. Contrib. Mineral Petrol., 97: 19-30.

DeBari, S., Kay, S.M., and Kay, R.W., 1987. Ultramafic xenoliths from Adagdak Volcano, Adak, Aleutian Islands, Alaska:Deformed igneous cumulates from the Moho of an island arc. Jour. Geology, 95: 329-341.

Dodge, F.C.W., Lockwood, J.P., and Calk, L.C., 1988. Fragments of the mantle and crust from beneath the Sierra Nevada batholith: xenoliths in a volcanic pipe near Big Creek, California. Geol. Soc. Amer. Bull., 100: 938-947.

Domenick, M.A., Kistler, R.W., Dodge, F.C.W., and Tatsumoto, M., 1983. Nd and Sr isotopic study of crustal and mantle inclusions from the Sierra Nevada and implications for batholith petrogenesis. Geol. Soc. Amer. Bull., 94: 713-719.

England, P., and Houseman, G., 1989. Extension during continental convergence, with application to the Tibetan Plateau. Jour. Gephys. Res., 94: 17,561-17,579.

Furlong, K.P., and Fountain, D.M., 1986. Continental crust underplating: Thermal considerations and seismic-petrologic consequences. Jour. Geophys. Res., 91: 8285-8294.

Gill, J., 1974. Role of underthrust oceanic crust in the genesis of a

Fijian calc-alkaline suite. Contrib. Mineral. Petrol., 43: 29-45.

Green, T.H., 1982, Anatexis of mafic crust and high pressure crystallization of andesite in Andesites: Orogenic Andesites and Related Rocks, R.S. Thorpe (ed.), 465-487, Wiley and Sons, Chichester, UK.

Griffin, W.L., and O'Reilly, S.Y., 1987a. The composition of the lower crust and the nature of the continental Moho-xenolith evidence, in Mantle Xenoliths, P.H. Nixon, (ed.), 413-431.

Griffin, W.L., and O'Reilly, S.Y, 1987b. Is the continental Moho the crust-mantle boundary?, Geology, 15: 241-244.

Gust, D.A., and Perfit, M.R., 1987. Phase relations of a high-Mg basalt from the Aleutian island arc: implications for primary island arc basalts and high-Al basalts. Contrib. Mineral. and Petrol., 97: 7-18.

Hamilton, W.B., 1988. Plate tectonics and island arcs. Geol. Soc. Amer. Bull., 100: 1503-1527.

Harte, B., Jackson, P.M., and Macintyre, R.M., 1981. Age of mineral equilibria and granulite facies modules from kimberlites. Nature, 291: 147-148.

Herzberg, C.T., Fyfe, W.S., and Carr, M.J., 1983. Density constraints of the formation of the continental Moho and crust. Contrib. Mineral. Petrol., 84: 1-5.

Hollister, L.S., and Crawford, M.L., 1986. Melt-enhanced deformation: A major tectonic process. Geology, 14: 558-561.

Houseman, G., McKenzie, D., and Molnar, P., 1981. Convective instability of a thickened boundary layer and its relevance for the thermal evolution of continental convergent belts. Jour. Geophys. Res., 86: 6115-6132.

Huppert, H.E., and Sparks, R.S.J., 1988. The generation of granitic magmas by intrusion of basalt into continental crust. Journ. Petrol., 29: 599-624.

Hyndman, D.W., and Foster, D.A., 1988. The role of tonalites and mafic dikes in the generation of the Idaho batholith. Jour. Geol., 96: 31-46.

Isacks, B.L., 1988. Uplift of the central Andean plateau and bending of the Bolivian orocline. Jour. Geophys. Res., 93: 3211-3231.

Jahn, B.M., Glikson, A.Y., Peucat, J.J., and Hickman, A.H., 1981. REE geochemistry and isotopic data of Archean silicic volcanics and granitoids from the Pilbara Block, Western Australia: implications for the early crustal evolution, Geochim. et Cosmochim. Acta., 45: 1633-1652.

Jan, M.Q., and Howie, R.A., 1981. The mineralogy and geochemistry of the metamorphosed basic and ultrabasic rocks of the Jijal Complex, Kohistan, NW Pakistan. Jour. Petrol., 22: 85-126

Johnston, L., 1983. The petrology of gabbroic xenoliths from Kanaga Island, the central Aleutians, Alaska. Senior Thesis, Cornell University, Ithaca, NY, 34 p.

Kay, R.W., 1978. Aleutian magnesian andesites: melts from subducted Pacific ocean crust. Jour. of Vol. and Geotherm. Res., 4: 117-132.

Kay, R.W., and Kay, S. Mahlburg, 1989, Recycled continental crustal components in Aleutian arc magmas: Implications for crustal growth and mantle heterogeneity. 145-162 in Crust/Mantle

Recycling at Convergence Zones, S.R. Hart and L. Gulen (eds). NATO, ASI Series, Kluwer Academic Pub.

Kay, R.W., and Kay, S. Mahlburg, 1988. Crustal recycling and the Aleutian Arc: Geochim. et Cosmochim. Acta, 52: 1351-1359.

Kay, R.W., and Kay, S. Mahlburg, 1986. Petrology and geochemistry of the lower continental crust: An overview: Geological Society of London Special Publication, 24: 147-159.

Kay, R.W., and Kay, S. Mahlburg, 1981. The nature of the lower continental crust: inferences from geophysics, surface geology, and crustal xenoliths. Rev. of Geophys. and Space Phys., 19: 271-297.

Kay, R.W., Rubenstone, J.L., and Kay, S. Mahlburg, 1986. Aleutian terranes from Nd isotopes. Nature, 322: 605-609.

Kay, S. Mahlburg, and Kay, R.W., 1983. Thermal history of the deep crust inferred from granulite xenoliths, Queensland, Australia, Amer. Jour. Sci., 283A: 486-513.

Kay, S. Mahlburg, and Kay, R.W., 1985a. Role of crystal cumulates and the oceanic crust in the formation of the lower crust of the Aleutian arc. Geology, 13: 461-464.

Kay, S. Mahlburg, and Kay, R.W., 1985b. Aleutian tholeiitic and calc-alkaline magma series I: The mafic phenocrysts, Contrib. Mineral. Petrol., 90: 276-290.

Kay, S. Mahlburg, Maksaev, V., Mpodozis, C., Moscoso, R., and Nasi, C., 1987. Probing the evolving Andean lithosphere: Mid-late Tertiary magmatism in (Chile 29°-30° 30'S) over the modern zone of subhorizontal subduction. Jour. Geophys. Res., 92: 6173-6189.

Khan, M., Jan, M., Windley, B., and Tarney, J., 1990. The Chilas mafic igneous complex: the root of the Kohistan island arc in the Himalayas of N. Pakistan. Geol. Soc. Amer. Bull., (in press).

Laubscher, H., 1988. Material balance in Alpine orogeny. Geol. Soc. Amer. Bull., 100: 1313-1328.

Martin, H., 1986. Effect of steeper Archean geothermal gradient on geochemistry of subduction-zone magmas. Geology, 14: 753-756.

McKenzie, D., and Bickle, M.J., 1988. The volume and composition of melt generated by extension of the lithosphere. Jour. Petrol., 29: 625-679.

Meijer, A., 1980. Primitive arc volcanism and a boninite series: examples from western Pacific island arcs. SEATAR Vol. Am. Geophys. Union, 631-657.

Nelson, K.D., 1990. A unified view of craton evolution motivated by recent deep seismic reflection and refraction results. Geophys. Jour. Roy. Astr. Soc. (in press).

Nicholls, I.A., 1974. Liquids in equilibrium with peridotitic mineral assemblages at high water pressures. Contr. Mineral. and Petrol., 45: 289-316.

O'Nions, R.K., and Oxburgh, E.R., 1988. Helium, volctile fluxes and the development of continental crust. Earth. Plant. Sci. Lett., 90: 331-347.

Oxburgh, E.R., 1972. Flake tectonics and continental collision. Nature, 239: 202-204.

Padovani, E.R., and Carter, J.L., 1977. Aspects of the deep crustal evolution beneath south central New Mexico, in The Earth's Crust: Its Nature and Physical Properties, Geophys. Monogr. Ser. 20: 19-55, J.G. Heacock (ed.), Amer. Geophys. Union.

Pin, C. and Sills, J.D., 1986. Petrogenesis of layered gabbros and ultramafic rocks from Val Sesia, the Ivrea zone, NW Italy: trace element and isotope geochemistry. in The Nature of the Lower Crust. J.B. Dawson (ed.). Geol. Soc. Lond. Sp. Publ. 25: 231-249.

Pope, R., 1983, The petrology of ultramafic and mafic xenoliths from Kanga Island, the central Aleutians. M.S. thesis, Cornell University, Ithaca, NY, 121 p.

Poldervaart, A., 1955. Chemistry of the earth's crust. Spec. Paper Geol. Soc. Amer., 62: 119-144.

Reid, M.R., 1987. Chemical stratification in the crust: isotopic trace element and major element constraints from crustally contaminated lavas and lower crustal xenoliths, Ph.D. thesis Mass. Inst. Tech. 297 p.

Rogers, N.W., Hawkesworth, C.J., 1982. Proterozoic age and cumulate origin for granulite xenoliths. Nature, 299: 409-413.

Sass, J.H., and Lachenbruch, A.H., 1979. Thermal regime of the Australian continental crust, in The Earth: its Origin, Structure, and Evolution, 301-351 M. McElhinney (ed.), Academic Press, London.

Sakuyama, M., and Koyaguchi, T., 1984. Magma mixing in mantle xenolith-bearing calc-alkalic ejecta, Ichinomegata volcano, northwestern Japan. Jour. of Volcanol. and Geotherm. Res., 22: 199-224.

Saunders, A.D., Rogers, G., Marriner, G.F., Terrell, D.J., and Verma, S.P., 1987. Geochemistry of Cenozoic volcanic rocks, Baja California, Mexico: Implications for the petrogenesis of post-subduction magmas. Jour. of Volcanol. and Geotherm. Res., 32: 223-245.

Sneeringer, M., Hart, S.R., and Shimizu, N., 1984. Strontium and samarium diffusion in diopside. Geochim. Cosmochim. Acta, 48: 1589-1608.

Stern, C.R., Futa, K., and Muehlenbachs, K., 1984. Isotope and trace element data for orogenic andesites from the Austral Andes. in Andean Magmatism: Chemical and Isotopic Constraints: 31-46, R. Harmon and B. Barreiro (eds.). Shiva, Cheshire, U.K.

Stern, C.R., and Wyllie, P.J., 1978. Phase compositions through crystallization intervals in basalt-andesite-H_2O at 30 kbar with implications for subduction zone magmas. Amer. Mineral., 63: 641-663.

Stern, R., Hanson, G., and Shirey, S., 1989. Petrogenesis of mantle-derived, LILE-enriched Archean monzodiorites and trachyandesites (sanukitoids) in the southwestern Superior Province. Can. J. Earth Sci., 26: 1688-1712.

Stern, T., 1987. Asymmetric back-arc spreading, heat flux and structure associated with the Central Volcanic Region of New Zealand. Earth Planet. Sci. Lett., 85: 265-276.

Stille, P. and Buletti, M., 1987. Nd-Sr isotopic characteristics of the Lugano volcanic rocks and constraints on the continental crust formation in the South Alpine domain (N-Italy-Switzerland). Contrib. Mineral. Petrol., 96: 140-150.

Stosch, H.G., and Lugmair, G.W., 1984. Evolution of the lower continental crust: granulite facies xenoliths from the Eifel, West Germany. Nature, 311: 368-370.

420

Stosch, H.G., Lugmair, G.W., and Seck, H.A., 1986. Geochemistry of granulite-facies lower crust xenoliths: implications for the geological history of the lower continental crust below the Eifel, West Germany. Geol. Soc. Lond. Spec. Publ., 24: 309-317.

Swanson, S., Kay, S. Mahlburg, Brearley, M., and Scarfe, C., 1987. Arc and back-arc xenoliths in Kurile-Kamchatka and western Alaska, in Mantle Xenoliths, P.H. Nixon (ed.), Wiley 303-318.

Tatsumi, Y., 1982. Origin of high-magnesian andesites in the Setouchi volcanic belt, southwest Japan, II. Melting phase relations at high pressures. Earth and Plant. Sci. Let. 60: 305-317.

Tatsumi, Y., and Ishizaka, K., 1982. Origin of high-magnesian andesites in the Setouchi volcanic belt, southwest Japan, I. Petrographical and chemical characteristics. Earth and Plant. Sci. Let., 60: 293-304.

Venturelli, G., Thorpe, R.S., Dal Piaz, G.V., Del Moro, A, and Potts, P.J., 1984. Petrogenesis of calc-alkaline, shoshonitic and associated ultrapotassic Oligocene volcanic rocks from the Northwestern Alps, Italy. Contrib. Mineral. Petrol., 86: 209-220.

Voshage, H, Hunziker, J.C., Hofmann, A.W., and Zingg, A., 1987. A Nd and Sr isotopic study of the Ivrea zone, Southern Alps, N-Italy. Contrib. Mineral. Petrol., 97: 31-42.

White, R.S., Spence, G.D., Fowler, S.R., McKenzie, D.P., Westbrook, G.K., and Bowen, A.N., 1987. Magmatism at rifted continental margins. Nature, 330: 439-444.

Whitney, J.A., 1988. The origin of granite: the role and source of water in the evolution of granitic magmas. Geol. Soc. Amer. Bull., 100: 1886-1897.

Wickham, S.M., 1987. The segregation and emplacement of granitic magmas. Jour. Geol. Soc., 144: 281-297.

Wickham, S.M., and Oxburgh, E.R., 1987. Low-pressure regional metamorphism in the Pyrenees and its implications for the thermal evolution of rifted continental crust. Phil. Trans. Roy. Soc. Lond., A321: 219-243.

AVERAGE COMPOSITION OF LOWER AND INTERMEDIATE CONTINENTAL CRUST, KAPUSKASING STRUCTURAL ZONE, ONTARIO.

M.G.TRUSCOTT and D.M.SHAW
Department of Geology
McMaster University
Hamilton, Ontario L8S 4M1
Canada

ABSTRACT. Estimates of the average compositions of the Archean lower and intermediate continental crust (LCC and ICC) in the region of the Kapuskasing Structural Zone (KSZ) were made using the proportions and average compositions of lithologic units. The division between LCC and ICC has been taken as the garnet-clinopyroxene isograd. Patterns of incompatible elements are similar for most of the lithologic units, with mild to strong depletion of Rb, U, and Th compared with upper crust estimates. Generally there is more marked depletion in the mafic gneisses than in other units. Heat production is low, 0.5 μWm^{-3} for the LCC, in close agreement with other estimates. Average REE patterns are similar for ICC and LCC, but with HREEs markedly enriched in the LCC due to the predominance of mafic gneisses in the LCC and of HREE-depleted tonalitic and granitic rocks in the ICC. Average LCC and ICC closely resemble Lewisian granulite and amphibolite models for many elements, however, the KSZ lower crust model is rather mafic compared with most granulite models. The intermediate composition of the LCC (SiO_2=56 wt.%) shows close agreement with Taylor and McLennan's (1985) LCC model for major elements but with very different concentrations of trace elements, and has greater similarities to lower crust models constructed from mafic xenoliths in volcanic rocks and kimberlites. Seismic information from the deep crust below the KSZ suggests that this region is very mafic. This mafic crust likely consists of two major components: underplated mantle-derived material (possible source of tonalites) and mafic residual materials resulting from anatexis of deep crust which produced felsic melts intruded into the KSZ and WDG.

1. Introduction

The Kapuskasing Structural Zone (KSZ) and Wawa Domal Gneiss Terrane (WDG) are located in the central Superior Province of the Canadian Shield (Figure 1) and are the granulite and amphibolite-grade parts respectively of a 120 km wide transition zone from low-grade rocks of the Michipicoten Greenstone Belt on the west to high-grade rocks terminated by the Ivanhoe Lake Cataclastic Zone (ILCZ) on the east. This transition zone is considered by Percival (1983) and Percival and Card (1983) and others to represent an oblique section through about 20 km of Archean crust, with exposure down into the upper part of the lower continental crust.

The lower continental crust (LCC) is generally interpreted to be a heterogeneous region with masses of igneous and metamorphic rocks of average intermediate composition (Smithson and Brown, 1977). Fountain and Salisbury (1981), comparing five exposed crustal sections, concluded that the lower crust consists of granulite facies rocks of mafic to intermediate composition. However, the predominant rock types in those sections differ considerably from each other. It is evident that an understanding of the proportions and

M. H. Salisbury and D. M. Fountain (eds.), Exposed Cross-Sections of the Continental Crust, 421–436.
© 1990 *Kluwer Academic Publishers.*

compositions of the dominant rock types is necessary for the development of a geochemical model for average lower continental crust and for the evaluation of the assumption that granulite terranes can be used to represent the lower crust.

Investigations of various aspects of the geochemistry of the KSZ have been reported in an expanding literature for the region. For example, Simmons et al. (1980) described the Shawmere Anorthosite. Ashwal et al. (1984, 1987) and Truscott and Shaw (1986) calculated heat production. Truscott and Shaw (1984) discussed the boron geochemistry of the KSZ rocks and Shaw et al. (1988) described variations in boron and lithium in rocks and mineral separates in the KSZ and WDG. Rudnick et al. (1984) determined that aqueous, saline, and carbonic fluids were involved in metamorphism in the KSZ region. Truscott and Shaw (1986,1987) calculated average compositions for the KSZ crust. Truscott (1987, 1988) described characteristic patterns of REE and incompatible elements in the major lithologic units of the KSZ region and in melt segregations and granitic pegmatite veins. Analyses of a few samples from some major units were presented by Rudnick and Taylor (1986) and Taylor et al.(1986).

The present study provides new geochemical data for the major lithologic units of the KSZ, examines the chemical fractionation of some incompatible elements, and, using the relative abundances of the lithologic units, provides a geochemical model for intermediate and lower crust, enabling comparison with other granulite terranes and known lower crust sections.

2. Geologic Setting

The rocks in the KSZ, shown in Figure 2, consist of ENE-trending belts predominantly of paragneiss, mafic gneiss, gneissic and xenolithic tonalite, anorthositic rocks of the Shawmere Anorthosite Complex, and monzonites and granitic plutonic rocks. Samples for this study were taken from the southern part of the KSZ and the adjacent WDG in the Chapleau Block, from the map region west of Chapleau to east of Foleyet.

Percival and Krogh (1983) determined the age of regional metamorphism to be 2696-2680 Ma, using U-Pb zircon geochronology. Young alkalic-carbonatite complexes occur in the region (ages from 1092 to 1872 Ma, from Gittins et al., 1967, Bell and Blenkinsop, 1980, and Bell et al., 1987), but these young plutons were intruded after exhumation of the crustal section and therefore were not sampled.

Estimates of geothermometry and geobarometry by Percival (1983) give temperature and pressure ranges in the KSZ and adjacent WDG from 600 to 825 °C and 3.1 to 9.9 kbar with most values between 650 and 775 °C and 6.0 to 7.3 kbar. Percival distinguished three metamorphic zones in the region; his garnet-clinopyroxene-plagioclase isograd is used in this study to separate the higher-grade granulite rocks of the lower crust portion (LCC) from the amphibolite-grade rocks of the intermediate crust section (ICC). This isograd is shown on the map in Figure 2, along with sample locations (taken from Shaw et al., 1988).

3. Analytical Methods

Major oxides and many trace elements were analyzed by X-ray Fluorescence Spectrometry. Thorium and rare earth elements (except Gd) were analyzed by Instrumental Neutron Activation Analysis (INAA). The U analyses were done by delayed INAA by Nuclear Activation Services Ltd. Gadolinium and B were analyzed by Prompt Gamma Neutron Activation Analysis (PGNAA) using the techniques described in Higgins et al. (1984).

Figure 1. Map of the Superior Province of Canada (after Thurston *et al.*, 1977).

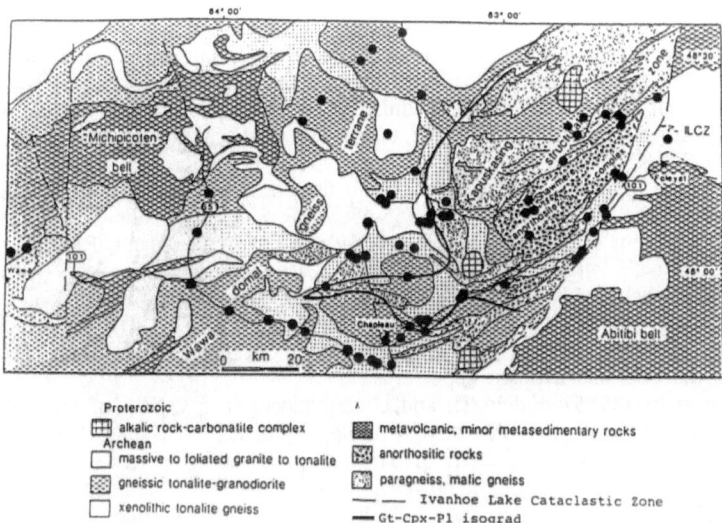

Figure 2. Map of the Kapuskasing Structural Zone and adjacent Wawa Domal Gneiss terrane with the garnet-clinopyroxene-plagioclase isograd dividing granulite-grade rocks on the east up to the Ivanhoe Lake Cataclastic Zone from the amphibolite-grade rocks on the west (after Percival and Card, 1983 and Percival, 1983). Sample locations are shown as black dots (from Shaw *et al.*, 1988).

Lithium was determined by Atomic Absorption Spectroscopy. Complete analyses of individual samples and further analytical information and precision and accuracy measurements will be given in a subsequent paper (in preparation).

4. Geochemical Results

Table I shows the average compositions calculated for the major lithologic units from the most representative samples in each group. The weighting factors used in calculating averages for the KSZ region (LCC) and the WDG region (ICC) are given on Table I below the unit averages, and were obtained by determining the relative areal proportions of the lithologic units on the Chapleau-Foleyet geologic map of Percival (1981). These averages are approximate because of uncertainties in positions of rock contacts due to obliterating effects of glaciation and lake coverage and the impossibility of estimating the abundance of mafic xenoliths in xenolithic tonalite gneiss. Kapuskasing and Hearst diabase dikes have not been included in the calculation because of their trivial volumetric importance and the difficulty of estimating their abundance accurately. However, including 2% diabase has only minor effects in the second decimal point in the average LCC abundances.

4.1. INCOMPATIBLE ELEMENTS

Many granulite terranes have been shown to display depletion of the large ion lithophile elements (LILE), particularly K, Rb, U, and Th, with increases in ratios of K/Rb and Th/U compared with upper crust averages.

Figure 3a shows a plot of K/Rb for rocks from the study area, with the weighted average compositions for ICC and LCC indicated. Figure 3b shows the field for granulites (from Rudnick *et al.*, 1985), with the approximate average trend superimposed. As observed by Rudnick *et al.* (1985), the majority of granulite analyses fall above the Main Trend line for igneous rocks (from Shaw, 1968), implying general relative depletion of Rb. From the few available analyses of KSZ rocks, Rudnick *et al.* (1985) concluded that the K/Rb ratio was not high for the KSZ. The average trends for world granulites and KSZ granulites diverge, with the KSZ trend indicating generally lower K values.

As shown in Figure 3a, the KSZ rocks do not show marked depletion of K and Rb. The majority of the KSZ intermediate (55-65 wt.% SiO_2) and silicic (>65 wt.% SiO_2) samples in the present study fall just above Shaw's Main Trend (near K/Rb 230). However, the basic rocks (<55 wt% SiO_2) plot between K/Rb 260 and 800, thus displaying the increase in K/Rb below 1 wt.% K noted for Archean granulites by Rudnick *et al.* (1985). The weighted average compositions for the LCC and the ICC have K/Rb ratios of 280 and 260 respectively. Thus, overall the LCC in this region shows only minor Rb depletion compared to the ICC and average upper crust.

Rudnick *et al.* (1985) studied Th and U abundances in a variety of granulite terranes and concluded that most granulites display U depletion, and that some have depletion of Th as well. In Figure 4, the pattern of Th and U in the KSZ rocks is similar to the granulites of Rudnick *et al.* (1985), however many of the KSZ samples with U below 1 ppm show relatively low Th: the Th/U ratio is lowest at low U values and Th appears to be considerably depleted in many rocks. The average Th/U reported by Rudnick *et al.* (1985) for most granulites is >4, and the ratios for the ICC and LCC calculated for the KSZ are 8.5 and 4.1 respectively, concomitant with the decrease in Th from 7.9 to 2.54 ppm and in U from 0.93 to 0.62 ppm from intermediate to lower crust. For comparison, upper crust averages for Th and U have been calculated by Taylor and McLennan (1985)

UNIT	A1	A2	A3	A4	A7g	A7bdh	A9ab	A10b	A10c	A1	A2	A4	A7bdh	A7g	A8	A9ab	A9de	A10b	A10c		LCC	LCC,AN	ICC
SIO2	61.058	49.237	46.242	52.787	65.927	67.250	50.342	70.705	73.092	65.150	49.270	52.787	69.079	66.400	67.485	50.342	75.210	70.705	73.092		57.783	55.789	68.912
TIO2	0.790	0.942	9.072	1.007	0.613	0.552	0.760	0.302	0.214	0.760	0.831	1.007	0.332	0.448	0.345	0.760	0.220	0.302	0.214		0.836	0.691	0.367
AL2O3	16.904	14.997	23.274	15.175	16.163	15.875	11.565	16.715	15.764	15.590	15.183	15.175	16.240	17.620	16.777	11.565	14.590	16.715	15.764		15.769	17.132	16.275
FE2O3	8.412	13.647	6.310	9.975	5.007	5.133	10.640	2.320	1.926	6.900	12.914	9.975	3.434	4.162	3.735	10.640	1.890	2.320	1.926		9.095	8.616	3.547
MNO	0.094	0.270	0.098	0.147	0.053	0.073	0.175	0.025	0.004	0.090	0.216	0.147	0.042	0.030	0.043	0.175	0.020	0.025	0.004		0.138	0.131	0.035
MGO	3.300	6.656	8.760	6.307	2.477	1.778	8.995	0.785	0.528	3.580	7.679	6.307	1.227	1.245	1.632	8.995	0.340	0.785	0.528		4.312	3.24	1.366
CAO	4.444	11.468	13.236	9.410	4.463	4.045	12.045	3.048	2.08	2.820	10.829	9.410	3.593	3.830	3.867	12.045	3.226	3.048	2.080		7.15	8.204	3.404
NA2O	2.644	2.003	1.496	2.727	3.236	3.475	2.035	3.667	3.166	2.690	2.326	2.727	4.180	3.635	3.422	2.035	3.050	3.667	3.166		2.707	2.497	3.604
K2O	2.100	0.671	0.180	1.662	1.780	1.702	2.485	2.265	3.096	2.260	0.625	1.662	1.684	2.922	2.440	2.485	4.350	2.265	3.096		1.633	1.379	2.325
P2O5	0.252	0.104	0.046	0.807	0.283	0.113	0.958	0.150	0.122	0.160	0.133	0.807	0.163	0.108	0.245	0.958	0.100	0.150	0.122		0.376	0.315	0.161
B	1.050	4.225	4.196	3.580	1.137	1.375	10.650	2.350	2.634	3.950	2.583	3.580	2.176	1.088	4.370	10.650	1.500	2.550	2.634		2.667	2.92	2.378
LI	16.840	13.440	4.464	14.083	20.967	18.417	21.525	40.525	16.04	16.100	17.486	14.083	26.744	35.750	54.000	21.525	20.400	40.523	16.040		16.057	14.042	75.215
LA	71.180	5.556	0.300	44.200	33.300	17.757	57.625	17.410	26.92	49.000	6.214	44.200	15.367	30.525	33.847	57.625	27.500	17.410	26.920		26.43	21.83	24.533
CE	47.000	14.160	1.598	98.200	72.967	34.715	165.825	36.325	43.42	86.400	14.593	98.200	27.053	57.650	64.325	165.825	51.800	36.325	43.420		58.54	48.45	44.149
ND	22.420	8.380	1.790	61.633	31.367	17.328	67.900	14.775	28.42	28.900	8.309	61.633	11.152	29.250	34.325	67.900	18.000	14.775	20.420		32.2	36.0	21.305
SM	3.956	2.304	0.375	10.885	4.177	2.976	16.447	2.445	1.952	4.500	2.310	10.885	2.085	3.185	4.755	16.447	3.050	2.445	1.952		5.68	4.74	2.747
EU	1.260	0.911	0.251	2.780	1.037	0.953	3.482	0.735	0.516	1.070	0.761	2.780	0.687	0.825	0.905	3.482	1.020	0.735	0.516		1.6	1.36	0.754
GD	1.090	2.999	0.373	8.567	2.927	1.945	11.500	1.647	1.08	3.400	2.687	8.567	1.603	2.135	3.367	11.500	2.200	1.647	1.080		4.59	1.83	1.92
TB	0.080			1.130			0.347	1.145	0.267	0.119	0.310	0.314				0.510		0.310	0.267	0.119			
DY	2.302	2.954	0.512	5.286	2.337	2.083	6.478	1.752	0.982	2.900	3.169	5.286	1.220	1.170	2.237	6.478	0.500	1.752	0.982		3.31	2.81	1.411
YB	1.356	1.759	0.343	1.522	0.993	1.025	1.237	0.632	0.363	1.730	1.912	1.522	0.454	0.520	0.638	1.237	0.550	0.632	0.363		1.39	1.21	0.558
LU	0.248	0.360	0.079	0.260	0.125	0.193	0.258	0.147	0.056	0.290	0.377	0.260	0.103	0.090	0.128	0.258	0.150	0.147	0.056		0.75	0.22	0.106
TH	3.664	0.939	0.375	2.720	4.557	3.812	6.643	4.438	14.62	12.700	0.610	2.720	3.369	6.753	10.768	6.643	6.790	4.438	14.620		2.991	2.519	7.912
U	0.712	0.544	0.070	0.817	0.533	1.007	2.292	1.455	1.54	0.700	0.255	0.817	0.537	0.715	1.050	2.292	0.970	1.455	1.540		0.745	0.622	0.934
RB	52.200	17.957	4.620	40.333	72.000	53.283	95.500	83.850	80.28	85.000	11.000	40.333	61.256	84.667	82.000	95.500	154.000	83.850	80.280		48.755	40.894	73.655
GRADE	G	G	G	G	G	G	G	G	G	A	A	A	A	A	A	A	A	A	A				
UNIT	A1	A2	A3	A4	A7g	A7bdh	A9ab	A10b	A10c	A1	A2	A4	A7bdh	A7g	A8	A9ab	A9de	A10b	A10c		LCC	LCC,AN	ICC
PROP.LCC	0.2691	0.1842		0.3014	0.1221	0.1136	0.0036	0.0036	0.0024														
LCC,AN	0.2214	0.1516	0.1772	0.248	0.1004	0.0935	0.003	0.003	0.002														
PROP.ICC										0.0492	0.0118	0.0164	0.3707	0.1812	0.031	0.0091	0.0073	0.0401	0.2832				

Table I. Average Compositions of Map Units and Lower and Intermediate Continental Crust, Kapuskasing Structural Zone and Wawa Domal Gneiss Terrane.

GRADE : G - granulite;   A - amphibolite.
UNIT : map units taken from Percival, 1981. A1 - paragneiss; A2 - mafic gneiss; A3 - anorthositic rocks; A4 - dioritic plutonic rocks; A7g - xenolithic tonalite gneiss; A7b - tonalitic gneiss; A8 - quartz monzonite orthogneiss; A9 - Floranna Lake intrusive complex: A9ab - mafic units, A9ed - felsic units; A10b - quartz monzonite to granodiorite intrusive rocks; A10c - granodiorite to granite intrusive rocks.
PROP,LCC : proportions of lithologic units in granulite-grade area, excluding anorthosite, based on areal proportions on map (Chapleau-Foleyet, from Percival, 1981).
PROP,ALC : proportions of lithologic units in granulite-grade region, including anorthosite..
PROP,ICC : proportions of lithologic units in amphibolite-grade region.
LCC : weighted average for lower continental crust excluding anorthosite.
LCC,AN : weighted average for lower continental crust including 17% anorthosite.
ICC : weighted average for intermediate continental crust.

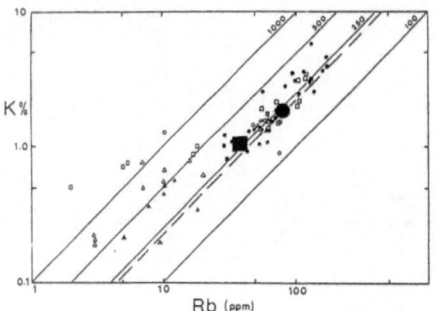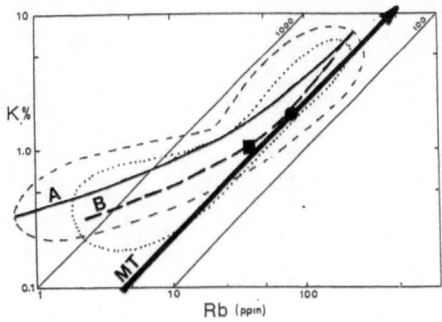

Figure 3a. Plot of K% v.s. Rb (ppm) for Archean granulites from the KSZ area: basic rocks (<55wt.% SiO_2) △ ; intermediate rocks (55-65 wt.% SiO_2) ☐ ; silicic rocks (>65 wt.% SiO_2) O . Closed symbols: amphibolites. Open symbols: granulites. ● average for intermediate continental crust.
■ average for lower continental crust. Dashed line: Main trend for igneous rocks from Shaw, 1968.
Figure 3b. Dashed line encircles field for Archean granulites from Rudnick *et al.* (1985); Dotted line encircles field for KSZ rocks; Solid line A: approximate average trend for Archean granulites; Broken line B: approximate average trend for KSZ rocks; MT: Main trend for igneous rocks from Shaw (1968).

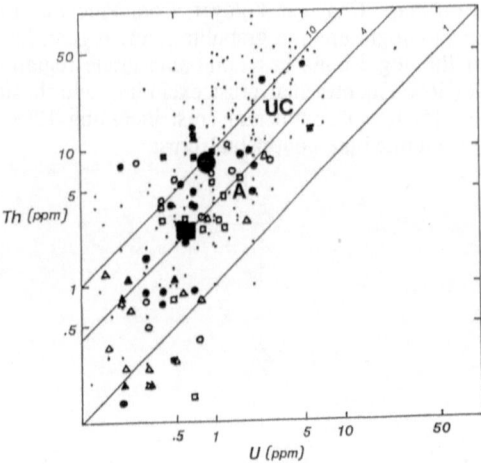

Figure 4. Plot of Th (ppm) v.s. U (ppm). Small dots: Archean granulites (from Rudnick *et al.*, 1986); Rocks from the study area have symbols as in Figure 3. UC: average upper crust, and A: average Archean upper crust from Taylor and McLennan (1985).

as 10.7 and 2.7 ppm for average upper crust and 5.7 and 1.5 ppm for average Archean crust. These values for Archean upper crust are lower than the ICC here, and could reflect a high component of mafic volcanic rocks in the upper crust estimate.

Heat production for the KSZ and WDG was calculated by Ashwal *et al.* (1984, 1987) as 0.44 μWm^{-3} and 1.37 μWm^{-3} respectively, from passive gamma-ray spectrometry analyses. Using INAA methods and densities of 2.75 gm/cm^3 for ICC and of 2.87 gm/cm^3 for LCC, Truscott and Shaw (1986) determined an average heat production for the KSZ-WDG region of 0.45 μWm^{-3}. The new averages calculated for the LCC and ICC parts of the KSZ region and based on additional samples are 0.52 and 0.92 μWm^{-3} respectively, reflecting the decrease in abundance of heat-producing radiogenic elements from intermediate to lower crust. Although the values for heat production for LCC are close in both determinations, the Ashwal *et al.* (1987) value for the WDG is higher by almost 30% due to their inclusion of two granite samples with considerably higher levels of Th and U than determined in any granites or pegmatites in the present study.

4.2. RARE-EARTH ELEMENTS (REE)

Patterns of average REEs are shown in Figure 5 for the lithologic units predominant in the KSZ. Chondrite-normalizing values were taken from Evensen *et al.* (1978), recalculated to be volatile-free as recommended by Taylor and McLennan (1985).

The paragneisses (Unit A1), consisting of micaceous quartzo-feldspathic rocks with garnet in places, have fractionated LREE-enriched patterns, that become nearly horizontal in the heavy rare earth elements (HREE), similar to sedimentary rocks from Isua and Barberton (McLennan *et al.*, 1983, 1984). La_N and Yb_N vary from about 34 to 134 and from 5 to 10 respectively, and La_N/Yn_N varies from about 4 to 19 . The average paragneiss, used in calculations of overall KSZ composition, has La_N 58, Yn_N 5.5, and La_N/Yn_N 10.6, and a very small positive Eu anomaly. Three Kapuskasing paragneisses analyzed by Taylor *et al.* (1986) show similar patterns, and they considered that two of their samples with small positive Eu anomalies have had a small fraction of partial melt removed.

Mafic gneisses (Unit A2), consisting of hornblende, clinopyroxene, garnet, plagioclase, and quartz, and with orthopyroxene in places, have nearly horizontal REE patterns, similar in slope and size to patterns from mafic gneisses reported from many other granulite terranes, e.g. the Lewisian (Weaver and Tarney, 1981), the low REE basic granulites from Madras (Weaver, 1980), and mafic granulites from the Adirondack Mountains, Pikwitonei and other regions (our labortory; unpublished data). The mafic gneisses vary from La_N 4 to 37 and Yn_N 4 to 9, with La_N/Yn_N ratios of 0.9 to 8. The average mafic gneiss for the KSZ has La_N 14.6, Yn_N 7.1, and La_N/Yn_N 2 and closely resembles Archean Abitibi tholeitic and calc-alkaline volcanic rocks (Smith, 1980).

In contrast, mafic granulites elsewhere may show rather fractionated patterns and high total REEs, particularly the Qianxi Group, Hebei Province, China which, according to Jahn and Zhang (1984, p.224) "may suggest an origin of their protoliths by partial melting of LREE-enriched mantle sources", an origin not applicable to produce the flat patterns in the KSZ rocks.

The geochemistry of the Shawmere Anorthosite Complex Unit A3) was discussed in detail by Simmons *et al.* (1980). The average REE pattern for this unit is nearly flat, with low total REE content, and a positive Eu anomaly, similar to many Archean anorthosite complexes (Taylor and McLennan, 1985).

The dioritic plutonic rocks (Unit A4) and quartz monzonite orthogneiss (Unit A8) have the highest total REE and show fractionated patterns similar to the paragneisses with HREE depletion, like the Archean augen gneisses from eastern Finland (Martin *et al.*, 1983).

428

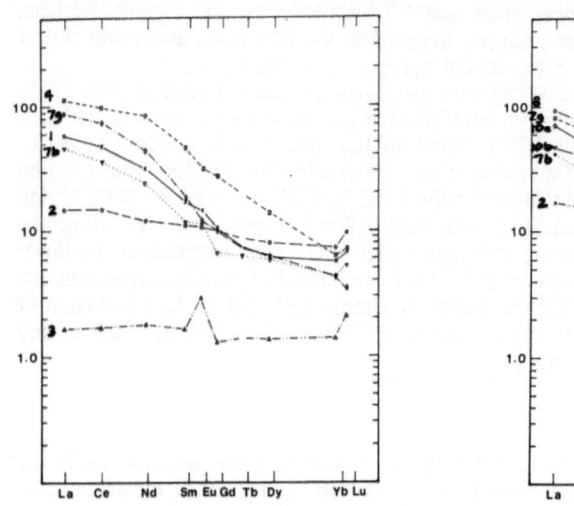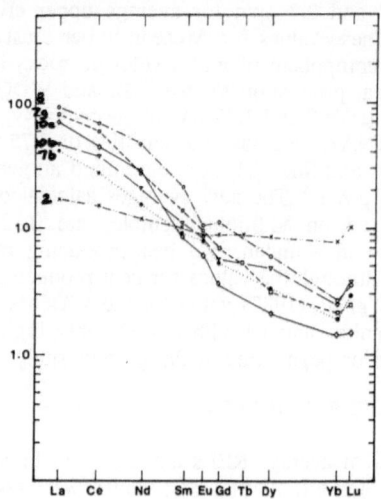

Figure 5. REE patterns of average compositions for the major lithologic units in the map area. a. Granulite-grade region (KSZ). b. Amphibolite-grade region (WDG).

A1 - paragneiss
A2 - mafic gneiss
A3 - anorthosite
A4 - diorite plutonic rocks
A7b- tonalite gneiss
A7g- xenolithic tonalite gneiss
A8 - quartz monzonite orthogneiss
A10- granite, quartz monzonite intrusive rocks

The average composition of quartz monzonite orthogneiss also has a small negative Eu anomaly. These patterns are similar to the tonalite-granodiorite orthogneisses (Unit A7) and granite-quartz monzonite intrusive rocks (Unit A10) which are the dominant rock types in the WDG. Three samples classified in the tonalite-trondhjemite group were analyzed by Rudnick and Taylor (1986) and show this same fractionated pattern with small positive anomalies. Small positive Eu anomalies are shown by some of the tonalite-granodiorite rocks in the KSZ, and the average composition for A7b has a small positive Eu anomaly. Although they plot almost exclusively in the tonalite-trondhjemite field of Barker (1979) and show fractionated REE patterns, few Unit A7 samples have positive Eu anomalies as accentuated as the typical Archean tonalite-trondhjemite rocks described by Rudnick and Taylor (1986), or the strongly fractionated patterns and high positive Eu anomalies shown in rocks with similar SiO_2 contents from southwest Finland (Arth *et al.*, 1978). Rudnick and Taylor (1986) consider that melting of a garnet granulite or eclogite source best fits the tonalite REE patterns in the KSZ.

5. Comparison with Other Geochemical Models for the Lower and Intermediate Continental Crust

Table II compares the weighted KSZ and WDG averages with other geochemical models. Figure 6a shows the REE patterns for three LCC models and for the KSZ model, which has been calculated with and without anorthosite. Although anorthosite contributes 17% to the KSZ model as determined from the exposed oblique cut, it is unclear whether anorthosite comprises such a large proportion of the crust at depths below the surface.

The inclusion of 17% anorthosite in the KSZ model results in a decrease in SiO_2 from 58% to 56% and minor adjustments in contents of other major element oxides, and in a small decrease in total REE content with no change in slope. The REE pattern for KSZ+AN compares closely with the Lewisian granulite model, although other trace elements vary considerably, and the Lewisian granulites are more felsic (SiO_2=61%), as is the average for world granulites (Shaw *et al.*, 1986). Taylor and McLennan's (1985) model (in which LCC is the residue from removing an upper crust composition from a calculated "andesite" bulk crust) and Rudnick and Taylor's (1987) average from the McBride crustal xenoliths compare closely with each other for many elements, but differ considerably from the KSZ. The McBride model is rather mafic compared to some other xenolith averages (e.g. Bournac), however eastern Australia xenoliths would seem to indicate an extremely mafic lower crust there (Griffin and O'Reilly, 1986).

The difference between the KSZ type of model, based on a granulite-grade terrane, and the Taylor and McLennan and McBride models can be reconciled by a variation in the proportion of mafic gneiss. Only the upper part of the lower crust is exposed in the KSZ and it may be quite realistic to expect more mafic gneiss at depth. Seismic refraction models for the KSZ (Boland *et al.*, 1988 and Boland and Ellis, 1988) indicate that there may be a considerable thickness of high velocity material in the deep crust below the delamination zone of the Kapuskasing uplift determined from seismic reflection data by Percival and Green (1988a, 1988b). This deep crust can best be modelled using dense materials such as very garnet-rich pyroxene granulites and ultramafic rocks, with seismic velocities of 7.2 to 7.6 km/sec. Fountain and Salisbury (1986) report laboratory measurements of velocities of 7.0 to 7.4 and 7.2 km/sec for KSZ mafic gneiss and anorthosite respectively at 6 kbars pressure. However, more garnet and pyroxene-rich mafic granulite material could provide the required velocities above 7.2 km/sec.

Comparisons of ICC estimates are difficult because of the paucity of estimates. The

	A	B	C	D	E	F	G	H	I	J	K
SiO2	54.4	61.2	56.3	61.5	57.78	55.79	50.29	57.4	66.7	68.2	68.91
TiO2	1	0.5	1.1	0.8	0.83	0.69	1.35	1.37	0.34	0.38	0.37
Al2O3	16.1	15.6	17.1	14.9	15.77	17.13	16.64	16	16	16.1	16.28
Fe2O3			8.8	8	9.09	8.62		2.05	3.6	4	3.55
FeO	10.6	5.3					12.13	7.3			
MnO	0.216	0.08	0.108	0.12	0.14	0.13	0.22	0.15	0.04	0.07	0.04
MgO	6.3	3.4	5	4.1	4.51	5.24	7.98	5.65	1.4	1.8	1.37
CaO	8.5	5.6	5.5	5.2	7.15	8.2	9.01	7.1	3.2	3.4	3.4
Na2O	2.8	4.4	2.1	3.1	2.71	2.5	1.6	1.95	4.9	3.8	3.6
K2O	0.34	1	1.42	1.6	1.63	1.38	0.45	0.93	2.1	2.3	2.33
P2O5		0.18	0.16	0.13	0.38	0.32			0.14		0.16
L.O.I.			2.2	0.5							
B	8.3			9	2.67	2.92					2.38
Li	11		7		16.06	14.04					25.2
Th	1.06	0.42	5.8	7.7	2.99	2.54	0.54		8.4		7.91
U	0.28	0.05	0.57	1.1	0.75	0.62	0.21		0.45		0.93
Rb	5.3	11	27	41	48.8	40.89	11.8		74		73.7
La	11	22	26		26.43	21.83	12.3		36		24.5
Ce	23	44	57		58.54	48.45	28		69		44.1
Nd	12.7	18.5			32.3	26.8	16		30		21.3
Sm	3.17	3.3	5.8		5.68	4.74	4.1		4.4		2.75
Eu	1.17	1.18	1.5		1.6	1.36	1.36		1.09		0.75
Gd	3.13				4.59	3.85	4.31				1.92
Tb	0.59	0.43	0.78				0.79		0.41		
Dy	3.6				3.31	2.81	5.05				1.41
Yb	2.2	1.2	2.98		1.39	1.21	3.19		0.76		0.56
Lu	0.29		0.48		0.25	0.22					0.11
H.P.	0.38	0.13	0.72	1.03	0.58	0.48	0.14		0.9		1.02

Table II: Estimates of lower and intermediate crust composition. H.P.: heat production in μWm^{-3}. A. Lower crust, Taylor and McLennan, 1985; B. Lewisian granulites, Sheraton, 1970, Weaver and Tarney, 1981; C. Bournac xenoliths, Dupuy *et al.*, 1979; D. Lower crust, granulites, Shaw *et al.*, 1986; E,F. Lower crust, Kapuskasing Structural Zone, from this paper (E - excluding anorthosite, F - including anorthosite); G. McBride xenoliths, Rudnick and Taylor, 1987. H. Lower crust, Ivrea Zone, Mehnert, 1975. I,J. Lewisian amphibolites, Weaver and Tarney, 1981. K. Intermediate crust, Wawa Domal Gneiss Terrane, this paper.

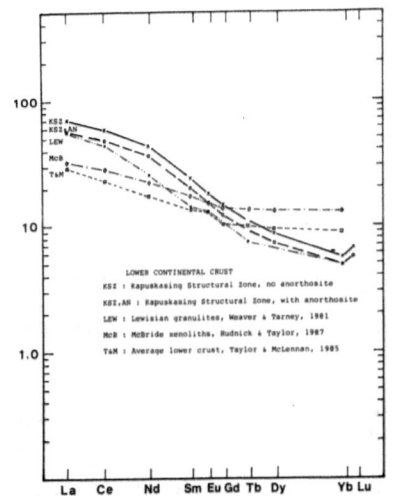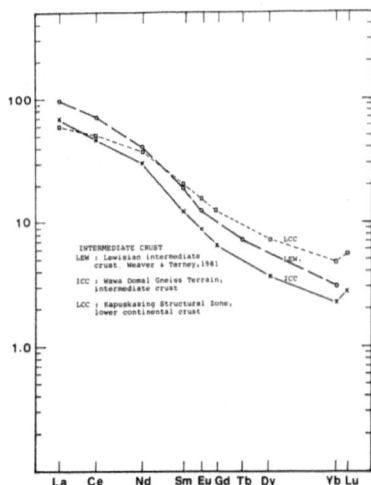

Figure 6. REE patterns of estimates for average crustal composition: a. Lower continental crust. b. Intermediate continental crust.

agreement between major oxides and some trace elements (e.g. Th, Rb) is close for the Lewisian ICC models and the WDG model in Table II. The agreement is also close for LREEs with differences in relative abundances of the HREEs, as shown in Figure 6b. The Wawa ICC shows similar LREE abundances to the Kapuskasing LCC composition, but with much lower HREEs; this difference is caused by the predominance of the intermediate and silicic gneisses in the WDG region, which have the most fractionated patterns.

The differences between the LCC and ICC models shown in Table II illustrate the problems of determining a unique composition for these regions of the crust. Granulite terranes provide some information about lower crust compositions, but were equilibrated mostly at shallow depths in the lower crust (pressures usually about 7 to 9 kbar and rarely down to 13 kbar, corresponding to depths of 20 to 27 km and occasionally deeper). They generally appear to be rather felsic representatives of deep crust, compared with the overwhelming abundance in xenolith suites of very mafic material. The major problem in determining the average composition of the lower crust from granulites is the considerable variation in compositions of these terranes. For example, the anomalously low heat production value for the Lewisian granulites shown in Table II result from an extreme depletion of Th and U which is not shown consistently in other granulite regions or in models based on xenoliths.

The Kapuskasing region, as a granulite terrane, provides a very mafic model for the lower crust, similar to that from some lower crustal xenoliths and also the very mafic Ivrea Zone section. However, the Kapuskasing model may be applicable to only the uppermost part of the lower crust, likely above 25 km. The lowermost crust in the KSZ may consist of two major components: (1) a thick mafic underplated zone (as suggested by Holland and Lambert, 1975; Glikson, 1986; Rudnick and Taylor, 1986) would provide material for the abundant tonalites and trondhjemitic rocks in the KSZ and WDG; (2) mafic residual material would result from anatexis in the deep crust which produced the voluminous felsic melts intruded into the KSZ and WDG. If the unexposed lowermost crust in the KSZ is represented by mafic gneiss and garnet-pyroxene granulite as suggested from seismic velocities, or by residual mafic material, the average LCC for this region could be considerably more mafic than the surface granulites suggest, and calculations of composition using mafic gneisses as the deep crust give an overall average for the LCC of 52 wt.% SiO_2 or less compared to the present 56wt%.

6. Acknowledgements

This project was supported by grants from the Canadian Lithoprobe Project and the Natural Science and Engineering Research Council to D.M. Shaw. Irradiations for INAA and PGNAA analyses were done at the McMaster Nuclear Reactor. File word processing was done by J.M. Richardson. Contribution no. 105: Canadian Lithoprobe Project and contribution no. 162: McMaster Isotopic, Nuclear, and Geochemical Studies Group Research Papers

7. References

Arth, J.G., Barker, F., Peterman, Z.E., and Friedman, I, 1978. Geochemistry of the gabbro-diorite-tonalite-trondhjemite suite of southwest Finland and its implications for the origin of tonalitic and trondhjemitic magmas. Jour. Petrol. 19: 289-316.

Ashwal, L.D., Morgan, P., Kelley, S.A., and Percival, J.A., 1984. An Archean crustal radioactivity profile: the Kapuskasing Structural Zone, Ontario. Geol. Soc. Am. Abstr. with Prog. 16, No.6: 433.

Ashwal, L.D., Morgan, P., Kelley, S.A., and Percival, J.A., 1987. Heat production in an Archean crustal profile and implications for heat flow and mobilization of heat-producing elements. Earth Planet. Sci. Letters 85: 439-450.

Barker, F., 1979. Trondhjemites: definition, environment, and hypotheses of origin. In Trondhjemites, Dacites, and Related Rocks (ed. F. Barker). Elsevier, Amsterdam : 1-12.

Bell, K. and Blenkinsop, J., 1980. Ages and initial 87Sr/86Sr ratios from alkalic complexes in Ontario. Ont. Geol. Survey Misc. Paper 93: 16-23.

Bell, K., Blenkinsop, J., Kwon, J.T., Kwon, S.T., Tilton, G.R., and Sage, R.P., 1987. Age and radiogenic isotopic systematics of the Borden carbonatite complex, Ontario, Canada. Can. Jour. Earth Sci. 24: 24-30.

Boland, A.V. and Ellis, R.M., 1988. Velocity structure of the Kapuskasing Uplift, Northern Ontario, from seismic refraction studies. J. Geophys. Res. (submitted).

Boland, A.V., Ellis, R.M., Northey, D.J., West, G.F., Green, A.G., Forsyth, D.A., Mereu, R.F., Meyer, R.P., Moel-a-l'Huissier, P., Buchbinder, G.G.R., Asudeh, I. and Haddon, R.A.W., 1988. Seismic delineation of upthrust Archaean crust in Kapuskasing, Northern Ontario. Nature 335: 711-713.

Dupuy, C., Leyreloup, A., and Vernieres, J., 1979. The lower continental crust of the Massif Central (Bournac, France) - with special references to REE, U, and Th composition, evolution, heat-flow production. Phys. Chem. Earth 11: 401-415.

Evensen, M.N., Hamilton, P.J. and O'Nions, R.K., 1978. Rare-earth abundances in chondritic meteorites. Geochim. et Cosmochim. Acta 42, No. 8: 1199-1212.

Fountain, D.M. and Salisbury, M.H., 1981. Exposed cross sections through the continental crust: implications for crustal structure, petrology, and evolution. Earth Planet. Sci. Lett. 56: 263-277.

Fountain, D.M. and Salisbury, M.H., 1986. Seismic properties of the Superior Province based on seismic velocity measurements on rocks from the Michipicoten-Wawa-Kapuskasing terranes, Ontario. GAC-MAC Prog. with Abstr. 11: 69.

Gittins, J, MacIntyre, R.M. and York, D., 1967. The ages of carbonatite complexes in eastern Canada. Can. Jour. Earth Sci. 4: 651-655.

Glikson, A.Y., 1986. An upthrusted early Proterozoic basic granulite-anorthosite suite and anatectic gneisses, south-west Arunta Block, central Australia: evidence on the nature of the lower crust. Trans. Geol. Soc. S. Afr. 89: 263-283.

Griffin, W.L. and O'Reilly, S.Y., 1986. The lower crust in eastern Australia: xenolith evidence. In The Nature of the Lower Continental Crust (eds. Dawson, J.B. et al.), Geol.

434

Soc. London Sp. Pub. 24: 363-374.

Higgins, M.D.,Truscott, M.G., Shaw, D.M., Bergeron, M., Buffet, G.H., Copley, J.R.D., and Prestwich, W.V., 1984. Prompt-gamma neutron activation analysis at McMaster Nuclear Reactor. In Proc. Int. Symp. on Use of Low and Medium Flux Research Reactors (eds. O.K. Harling, L. Clark Jr., and P. von der Hardt). Suppl. to Atomkern energie Kerntechnik, Vol. 44: 690-697. K. Thiemig, Graphische Kunstanst. u. Buchdruckerei AG, Munchen, F.D.R.

Holland, J.G., and Lambert, R. St.J., 1975. The chemistry and origin of the Lewisian gneisses of the Scottish Mainland: The Scourie and Inver assemblages and sub-crustal accretion. Precambrian Res. 2: 161-188.

Jahn, B.M. and Zhang, Z.Q., 1984. Archean granulite gneisses from eastern Hebei Province, China: rare earth geochemistry and tectonic implications. Contrib. Mineral. Petrol. 85: 224-243.

Martin, H., Chauvel, C., and Jahn, B.M., 1983. Major and trace element geochemistry and crustal evolution of Archaean granodioritic rocks from eastern Finland. Precambrian Res. 21: 159-180.

McLennan, S.M., Taylor, S.R., and Eriksson, K.A., 1983. Geochemical evolution of the Archean shales from South Africa. I. The Swaziland and Pongola Supergroups. Precambrian Res. 22: 93.

McLennan, S.M., Taylor, S.R., and McGregor, V.R., 1984. Geochemistry of Archean metasedimentary rocks from West Greenland. Geochim. et Cosmochim. Acta 48: 1-13.

Mehnert, K.R., 1975. The Ivrea Zone - a model of the deep crust. Neues Yahrb. Mineral. Abh. 125, Part 2: 156-199.

Percival, J.A., 1981. Preliminary map, Geology of the Kapuskasing structural zone in the Chapleau-Foleyet area, Ontario. Geol. Survey of Can. , Open File Map 763.

Percival, J.A., 1983. High-grade metamorphism in the Chapleau-Foleyet area, Ontario. Am. Mineral. 68: 667-686.

Percival, J.A. and Card, K.D., 1983. Archean crust as revealed in the Kapuskasing uplift, Superior Province, Canada. Geology 11: 323-326.

Percival, J.A. and Green, A.G., 1988(a). Lithoprobe seismic profiles of exposed lower crust in the Kapuskasing uplift, Superior Province, Canada. Geol. Soc. Am. Abstr. with Prog. 20. No. 7: A234-A235.

Percival, J.A. and Green, A.G., 1988(b). Towards a balanced crustal-scale cross section of the Kapuskasing uplift. Project Lithoprobe Kapuskasing Structural Zone Workshop II, University of Toronto, Nov. 1988: 233-234.

Percival, J.A. and Krogh, T.E., 1983. U-Pb zircon geochronology of the Kapuskasing structural zone and vicinity in the Chapleau-Foleyet area, Ontario. Can. Jour. Earth Sci. 20: 830-843.

Rudnick, R.L., Ashwal, L.D., and Henry, D.J., 1984. Fluid inclusions in high-grade gneisses of the Kapuskasing structural zone, Ontario: metamorphic fluids and uplift/erosion path. Conrib. Mineral. Petrol. 87: 399-406.

Rudnick, R.L., McDonough, W.F., McCulloch, M.T., and Taylor, S.R., 1986. Lower crustal xenoliths from Queensland, Australia: evidence for deep crustal assimilation and fractionation of continental basalts. Geochim. et Cosmochim. Acta 50: 1099-1115.

Rudnick, R.L., McLennan, S.M., and Taylor, S.R., 1985. Large ion lithophile elements in rocks from high-pressure granulite facies terrains. Geochim. et Cosmochim. Acta 49: 1645-1655.

Rudnick, R.L. and Taylor, S.R., 1986. Geochemical constraints on the origin of Archaean tonalitic-trondhjemitic rocks and implications for lower crustal composition. In The Nature of the Lower Continental Crust (eds. Dawson, J.B. *et al.*), Geol. Soc. London Sp. Pub. 24: 179-191.

Rudnick, R.L. and Taylor, S.R., 1987. The composition and petrogenesis of the lower crust: a xenolith study. Jour. Geophys. Res. 92: 13981-14005.

Shaw, D.M., 1968. A review of K-Rb fractionation trends by covariance analysis. Geochim. et Cosmochim. Acta 32: 573-601.

Shaw, D.M., Cramer, J.J., Higgins, M.D., and Truscott, M.G., 1986. Composition of the Canadian Precambrian Shield and the continental crust of the earth. In The Nature of the Lower Continental Crust (eds. Dawson, J.B. *et al.*), Geol. Soc. London Sp. Pub. No. 24: 275-282.

Shaw, D.M., Truscott, M.G., Gray, E.M., and Middleton, T.M., 1988. Boron and lithium in rocks and minerals from the Wawa-Kapuskasing region, Ontario. Can. Jour. Earth Sci. 25: 1485-1502.

Sheraton, J.W., 1970. The origin of the Lewisian gneisses of northwest Scotland, with particular reference to the Drumbeg area, Sutherland. Earth Planet. Sci. Lett. 8: 301-310.

Simmons, E.C., Hanson, G.N., and Lumbers, S.B., 1980. Geochemistry of the Shawmere Anorthosite Complex, Kapuskasing Structural Zone, Ontario. Precambrian Res. 11: 43-71.

Smith, I.E.M., 1980. Geochemical evolution in the Blake River Group, Abitibi Greenstone Belt, Superior Province. Can. Jour. Earth Sci. 17: 1292-1299.

Smithson, S.B. and Brown, S.K., 1977. A model for lower continental crust. Earth Planet. Sci. Lett. 35: 134-144.

Taylor, S.R. and McLennan, S.M., 1985. The Continental Crust: its Composition and Evolution. Blackwell Scientific Publications, Oxford, 312 p.

Taylor, S.R., Rudnick, R.L., McLannan, S.M., and Ericksson, K.A.,1986. Rare earth element patterns in Archean high-grade metasediments and their tectonic significance. Geochim. et

Cosmochim. Acta 50: 2267-2279.

Thurston, P.C., Siragusa, G.M., and Sage, R.P., 1977. Geology of the Chapleau area, Districts of Algoma, Sudbury and Cochrane. Ont. Div. of Mines Geol. Rept. 157: 293 p.

Truscott, M.G., 1987. REE and other trace elements in the lower continental crust, Kapuskasing Structural Zone. GAC-MAC Prog. with Abstr. 12: 97.

Truscott, M.G., 1988. Preliminary studies of geochemical fractionation studies in partial melts and pegmatites, Kapuskasing Structural Zone. Project Lithoprobe Kapuskasing Structural Zone Workshop II, University of Toronto, Nov., 1988: 99-100.

Truscott, M.G. and Shaw, D.M., 1984. Boron abundance and localization in granulites. Geol. Soc. Amer. Abstr. with Prog. 16: 677-678.

Truscott, M.G. and Shaw, D.M., 1986. Preliminary estimate of Lower Continental Crust composition in the Kapuskasing Structural Zone. Geol. Assoc. Can.-Mineral. Assoc. Can. Ann. Mtg. Prog. with Abstr. 11: 138.

Truscott, M.G. and Shaw, D.M., 1987. Composition of lower and intermediate crust, Kapuskasing Structural Zone, Ontario. Geol. Soc. Am. Abstr. with Prog. 19, No.7: 871-872.

Truscott, M.G., Shaw, D.M., and Cramer, J.J., 1986. Boron abundance and localization in granulites and the lower continental crust. Bull. Geol. Soc. Finland 58, Part 1: 169-177.

Weaver, B.L., 1980. Rare-earth element geochemistry of Madras granulites. Contrib. Mineral. Petrol. 71: 271-279.

Weaver, B.L. and Tarney, J., 1981. Lewisian gneiss geochem-istry and Archaean crustal development models. Earth Planet. Sci. Lett. 55: 171-180.

FLUID–ROCK INTERACTIONS IN THE IVREA ZONE AND THE ORIGIN OF HIGH LOWER CRUSTAL CONDUCTIVITIES

A.J. BAKER
Grant Institute of Geology,
West Mains Road,
Edinburgh,
EH9 3JW,
U.K.

ABSTRACT. The preservation of premetamorphic stable isotope heterogeneities in Ivrea Zone metasedimentary lithologies implies large amounts of pervasive fluid flow did not occur. Infiltration of externally derived fluid, in particular of mantle carbon bearing fluid, must have been limited. The persistence of large stable isotopic gradients at lithological boundaries throughout metamorphism in the Ivrea Zone suggests that isotope exchange by diffusion was not effective over distances greater than a few tens of centimetres. A large fluid–filled porosity was not present for long periods of time. The quantities of fluid present during retrograde metamorphism in the Ivrea Zone are insufficient to explain the high lower crustal conductivities observed in many lower crustal regions. Petrological and stable isotopic data, from the Ivrea Zone and many other lower crustal terrains, suggest mechanisms of lower crustal conductivity other than conduction through a pore fluid.

Introduction

The quantity, composition and distribution of fluid present during the prograde and retrograde evolution of the lower crust is obviously important to our understanding of the kinetics of various geochemical and deformational processes and interpretation of geophysical data for the lower crust. Various patterns of fluid flow have been suggested. Flow may be of a single pass nature with no downward flow of fluid and no circulation (Wood and Walther, 1986). Alternatively extensive fluid recirculation may occur (Etheridge et al., 1983). As surface fluids at hydrostatic pressure are unlikely to circulate downwards into the zone of lithostatically pressured fluids, it was suggested that the lithostatically pressured fluids were isolated by an impermeable zone in the midcrust (Etheridge et al., 1983). Fluid flow may be strongly channelised or pervasive on a grain boundary scale. If fluid flow is of a single pass nature or fluids are strongly channelised, stable isotopic heterogeneities will be preserved during metamorphism. If fluid flow is pervasive or extensive convection and recirculation occurs, it is expected that stable isotope homogenisation will occur.

The extent to which mantle derived CO_2 contributes to the lower crustal fluid budget is uncertain. The flushing of some granulite terrains with mantle derived

M. H. Salisbury and D. M. Fountain (eds.), Exposed Cross-Sections of the Continental Crust, 437–452.
© 1990 *Kluwer Academic Publishers.*

CO_2 has been suggested on the basis of field and fluid inclusion studies (see review by Newton, 1986). Some stable isotopic evidence is consistent with, but does not require such a hypothesis (Jackson et al., 1988; Baker and Fallick, 1988). However, evidence from carbon isotopes (Vry et al., 1988; Baker, 1988) and $C/^3He$ ratios (O'nions and Oxburgh, 1988) indicates that large mantle to crust fluid fluxes are not a general feature of lower crustal metamorphism. This might not be the case along active margins where release of fluids from the subducted slab could give rise to significant fluid fluxes.

The present lower crust is thought to have equilibrated in the amphibolite and granulite facies according to evidence from presently exposed lower crustal sections such as the Ivrea Zone (Fountain and Salisbury, 1981) and xenoliths (e.g. Griffin and O'Reilly, 1986). Much of the present lower crust will now have cooled to temperatures below those of peak metamorphism. Inferences of the quantities of fluid present during this lower crustal history, subsequent to peak metamorphism, can be based either on petrological observation of uplifted lower crustal terrains or on geophysical data. Petrological studies usually imply fluid absence while geophysical studies have often inferred the presence of significant fluid filled porosities.

The preservation of peak metamorphic mineral assemblages in many inferred lower crustal terrains has been invoked as evidence that hydrous fluids were absent during their retrograde evolution (see Kay and Kay, 1981; Yardley, 1986). However, the preservation of peak metamorphic mineral assemblages could be consistent with the presence of CO_2 and NaCl enriched fluids during retrograde evolution. Post metamorphic CO_2 rich fluid inclusions are known from several high grade metamorphic areas (Lamb et al., 1987; Morrison and Valley, 1988). Fluid absence has also been inferred from the slow progress of retrograde reactions (e.g. Rubie, 1986).

Zones of high high lower crustal conductivity have been recognised in many geophysical studies. These zones may correspond either to the whole of the lower crust (e.g. Gough, 1986) or to a thin highly conducting zone at mid crustal levels (Jones, 1987). A number of explanations have been proposed for these high conductivities including the presence of free carbon, hydrated minerals, iron oxides, sulphides and fluids (see Haak and Hutton, 1986). It seems to have been generally concluded that free aqueous brines are the commonest cause of enhanced conductivity (Connerney et al., 1980; Shankland and Ander, 1983). Application of Archies Law has lead to estimates of lower crustal fluid filled porosity of the order of 1 % (e.g. Shankland and Ander, 1983; Hall, 1986). These porosities would also be capable of explaining fine scale variations in lower crustal seismic velocity and the high reflectivity of the lower crust (Gough, 1986, Matthews and Cheadle, 1986).

This paper reviews recent stable isotopic data from the Ivrea Zone (Baker, 1988; Baker, submitted), a sample of extended (Brodie and Rutter, 1987) lower crust from near the crust–mantle transition. Data from this zone and other metamorphic terrains is used to discuss the long term distribution of fluid in the mid to lower crust.

The Ivrea Zone

GEOLOGICAL SITUATION

The general geology of the Ivrea Zone has been reviewed by Zingg (1983). The

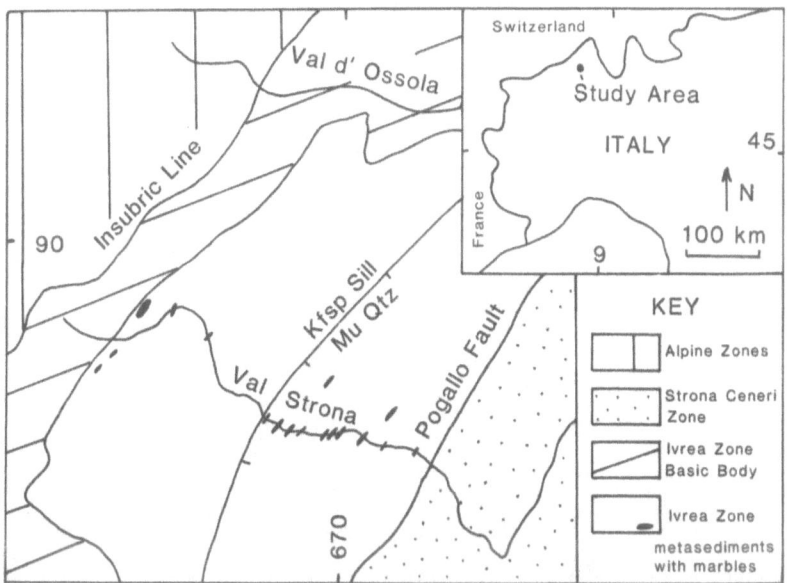

Figure 1. Geological sketch map of the study area.

zone consists of a series of metabasites and metasediments intruded by a large basic body which may have been partially responsible for the amphibolite to granulite facies metamorphism. Metasediments include pelites which are commonly graphite bearing and a very varied sequence of marbles. The zone has suffered a long lower crustal history and has probably remained near the mantle –crust boundary for a long period. A gravity anomaly suggests the presence of mantle material close beneath the present surface (see Zingg, 1983) which is represented by some slices of mantle peridotite within the zone. Pelites range from muscovite + quartz to K feldspar + sillimanite to locally orthopyroxene bearing while metabasites contain clinopyroxene or orthopyroxene + clinopyroxene at the highest grades. Marbles were infiltrated by considerable quantities of water during prograde and early retrograde metamorphism. Stable isotope systematics have been examined from a prograde sequence across the Ivrea Zone (the Val Strona – see figure 1) and implications for mantle–crust fluid transfer and patterns of fluid flow considered (Baker, 1988).

STABLE ISOTOPE DATA

Pelites. Oxygen isotope data for pelites is shown in figure 2. The oxygen isotope composition of pelite quartz shows no systematic variation across the zone. The range of compositions is for the most part similar to likely premetamorphic isotope compositions ($\delta^{18}O \sim + 15$ permil). The lack of oxygen isotope homogenisation precludes extensive fluid circulation within the zone. High $\delta^{18}O$ values for quartz in pelites near marble bodies is indicative of some marble–pelite oxygen isotope exchange. Infiltration of substantial quantities of mantle fluid are ruled out by lack

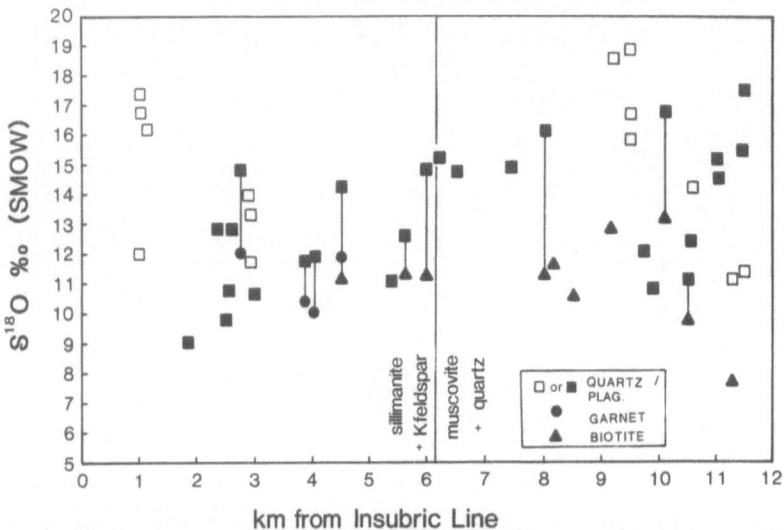

Figure 2 Pelite oxygen isotope data from the Val Strona versus distance from the Insubric Line which roughly corresponds to decreasing metamorphic grade. Open symbols are for samples from near marble bodies.

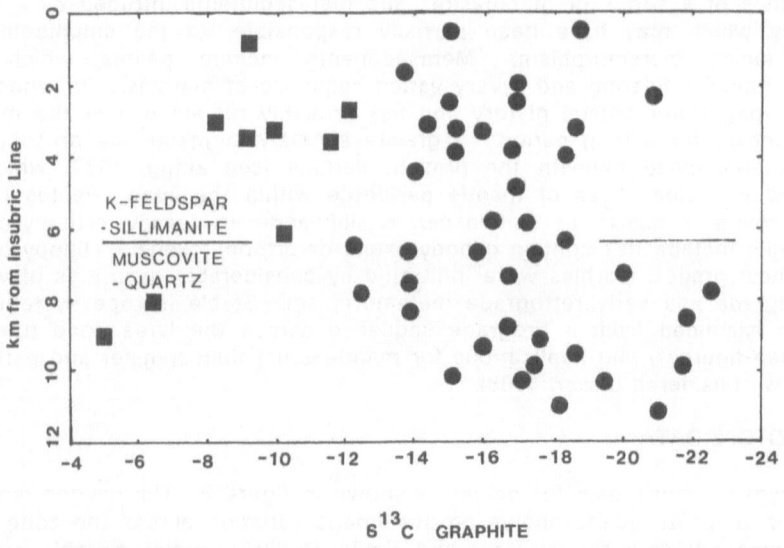

Figure 3 The carbon isotopic composition of graphite in pelites versus metamorphic grade in the Val Strona (after Baker, 1988). Squares are samples within one metre of a marble body.

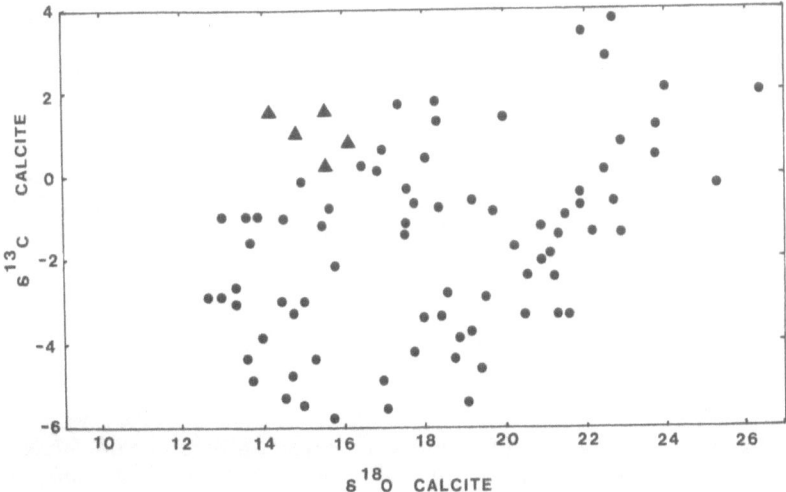

Figure 4 $\delta^{18}O$ versus $\delta^{13}C$ for calcites from marbles in the Val Strona. Triangles are for calcite poor metabasites (after Baker, 1988).

of isotopic shifts towards $\delta^{18}O \sim + 6$ permil (Baker, submitted).

Graphite Carbon isotope data for graphite in pelites is shown in figure 3 (after Baker, 1988). There is no systematic variation of graphite $\delta^{13}C$ with metamorphic grade; at all grades $\delta^{13}C$ varies from $- 10$ to $- 25$ per mil. The heavier graphite in the vicinity of marbles is an indication of some marble-pelite carbon isotope exchange or the deposition of graphite from marble derived fluids. The graphite is, in general, not equilibrated with marble carbon with $\delta^{13}C \sim 0$ per mil. The general lack of carbon isotope equilibration, both heterogeneity in graphite isotope composition and lack of marble-graphite equilibration, is an argument against pervasive fluid flow through the zone. Graphite is not shifted towards the average mantle carbon isotope composition of $- 7$ per mil. Significant quantities of mantle derived carbon bearing fluid cannot therefore have infiltrated the zone. This test is particularly sensitive as pelites contain such a small proportion of graphite (Baker, 1988).

Marbles The carbon and oxygen isotope composition of calcites from marbles is shown in figure 4. Calcites show correlated shifts from $\delta^{18}O \sim 25$ permil, $\delta^{13}C \sim 2$ permil to lighter values as the result of the infiltration of metasediment derived fluids. Some marbles retain close to premetamorphic isotope compositions while others are extensively shifted. These observations suggest a large component of channelised fluid flow. If all fluid flow was pervasive, substantial isotopic homogenisation should occur.

Stable Isotope Gradients There are stable isotope gradients across the margins of Ivrea Zone marbles in both $\delta^{13}C$ and $\delta^{18}O$ of up to 5 permil per metre over distances of 10 cm to 1 m (Baker, submitted). The preservation of these gradients places a limit on the amount of time for which a given fluid filled porosity might have been present. It is normally considered that oxygen isotope exchange over distances of more than a few centimetres occurs as the result of isotopic exchange between minerals and a flowing fluid (e.g. Rumble and Spear, 1983). Oxygen transport through a static pore fluid is not generally an important transport mechanism over large distances, but the rate of such transport may be used to constrain the maximum time for which a given amount of fluid was present. Some crude estimates have been performed following Nagy and Parmentier (1982) of the extent of diffusion controlled equilibration through a pore fluid for various water filled porosities. Bulk rock diffusivity may be calculated from

$$D = \sigma_f \alpha_f \phi D_f / \sigma_r \alpha_r$$

where σ is the density and α the number of oxygen atoms per unit mass for fluid and rock, ϕ is the porosity and D_f the diffusivity of the fluid. At about 6 kb and 500°C, $\sigma_f \alpha_f / \sigma_r \alpha_r \sim 0.5$ and $D_f \sim 10^{-4}$ cm^{2}s^{-1} (Walther and Wood, 1984). For a 1 % porosity $D = 5 \times 10^{-7}$ cm^{2}s^{-1} and for a 0.01% porosity $D = 5 \times 10^{-9}$ cm^{2}s^{-1}. For a 1 % porosity the half width of the diffusive exchange profile will be 2.5 m in 10^3 years and 250 m in 10^7 years and for a 0.01 % porosity, 25 cm in 10^3 years and 25 m in 10^7 years.

Other processes may limit the rate of oxygen isotope exchange between a mineral and an adjacent fluid such as rates of chemical reaction or intracrystalline diffusion. Where the fluid is in disequilibrium with the adjacent rock, reactions will occur and oxygen isotope exchange will be quite rapid (Cole and Ohmoto, 1986). This is likely to be the case across marble–pelite margins. The slowest possible step limiting fluid-rock oxygen isotope exchange is likely to be intracrystalline diffusion. A 1mm grain with $D \sim 10^{-17}$ cm^{2}s^{-1} will equilibrate with a surrounding fluid in about 10^7 years. For this situation and a 1 % porosity, diffusion through the pore fluid will only be the slowest step in oxygen isotope exchange at distances over about 250 m. For a 0.01 % porosity diffusion through the fluid will only be the slowest step for distances greater than 25 m. Since oxygen isotope gradients are preserved on a scale of metres or less, porosities of the order of 0.01 to 1% cannot have persisted for timescales of the order of a few million years across marble margins. These constraints are very much an upper limit as much isotope exchange is likely to occur as the result of fluid flow and rates of fluid–rock isotope exchange are probably faster than predicted by rates of oxygen isotope intracrystalline diffusion. Over a long time very extensive isotopic homogenisation would be expected. If fluid porosities of 1 % were present for the order of 50 Ma, isotope equilibration on a scale of kilometres would occur.

Summary. Preservation of small scale oxygen and carbon isotope heterogeneities in the Ivrea Zone precludes the long term presence of significant fluid porosities in and around marbles. This observation applies both to the prograde and retrograde evolution of the zone. However, it may be that marbles are atypical lithologies being fairly closed to metamorphic fluids. This possibility is supported by new data on the wetting characteristics of H_2O–CO_2 fluids in calcite which suggest that such fluids do not generally form an interconnected network when in textural equilibrium (Hay and Evans, 1988). However, it is not clear what implications this data has for fluid-rock interactions during prograde metamorphism of marbles when hydrofracture and reaction enhanced permeability are likely to be important processes. It is expected that the progress of devolatilisation reactions in marbles will produce a fracture porosity at least for short periods of time. Particularly heavy carbon and oxygen in pelites close to the margins of marble bodies (figures 2 and 3) suggests that extensive diffusive homogenisation has not occurred within pelites and that significant isotope gradients exist in pelites on a scale of metres. This implies that pervasive and large porosities did not exist for a long time in pelites.

Isotope gradients persist on a small scale in many metamorphic rocks (see below), suggesting that, in general, large fluid porosities are not developed even during prograde metamorphism in most terrains. If highly permeable zones, isolated from the surface exist for long periods of time, it is possible that convective circulation would occur (Etheridge et al., 1983). This would promote very extensive isotope homogenisation, much greater than that predicted by the diffusive exchange model.

The stable isotope data imply channelisation of fluids and are not consistent with extensive fluid input from external sources. Extensive input of mantle derived fluids is precluded despite proximity to the mantle. The heterogeneity is inconsistent with Etheridge et al.'s (1983) model for widespread pervasive fluid circulation. The observations may be explained by channelised flow of fluid during prograde metamorphism by single pass flow. Only small quantities of fluid were retained within the zone during its retrograde history. Even during the prograde history large quantities of fluid ($\sim$ 0.1 %) cannot have been pervasively present in and around marbles. As discussed above the preservation of peak metamorphic mineral assemblages in the Ivrea Zone also rules out the presence of hydrous fluids during retrograde metamorphism. However, this observation is not inconsistent with the presence of CO_2 or NaCl enriched fluids during the retrograde history.

Other Terrains

Some other high grade terrains show evidence for heterogeneous stable isotope systematics. For instance Valley and O'Neil (1984) report granulite facies marbles from the Adirondacks with very variable isotopic compositions. Similar heterogeneous stable isotope compositions are reported from Lofoten amphibolite and granulite facies marbles (Baker and Fallick, 1988) and Rogaland granulite facies marbles (Sauter, 1983). Deeper high grade rocks from the Pyrenees also show evidence for stable isotope heterogeneity (Wickham and Taylor, 1987). Many terrains from lower metamorphic grades show a lack of isotopic homogenisation (see Valley, 1986). A number of studies have focussed particularly on stable isotope gradients. In the Adirondacks gradients up to 5 per mil m^{-1} are found in $\delta^{18}O$ and $\delta^{13}C$ at marble-pelite and anorthosite-skarn contacts (Valley and O'Neil, 1982; 1984). Similar gradients are found at the margins of many amphibolite facies marbles (Sheppard and Schwarcz, 1970; Rye et al., 1976). Large oxygen isotopic gradients are also reported from the margins of other lithologies from a range of metamorphic grades. Rumble and Spear (1983) record the absence of oxygen isotope equilibration over length scales of less than 1 cm in rocks which had not undergone extensive devolatilisation, while Anderson (1967) observed scales of oxygen isotopic equilibrium in a variety of rocks of the order of 1cm to 1m. Fluid porosities present during prograde metamorphism in many metamorphic terrains were evidently insufficient to produce stable isotopic homogenisation over short distances by diffusive processes.

Isotopic homogenisation is a feature of some terrains, for instance in low pressure high temperature rocks from the Pyrenees (Wickham and Taylor, 1985) and in granulite facies rocks from the Canadian Shield (Shieh and Schwarcz, 1974). This homogenisation is generally attributed to the action of infiltrating fluids during prograde metamorphism rather than to isotopic exchange due to the long term presence of a fluid (Wickham and Taylor, 1985; see Rumble and Spear, 1983). If such isotopic homogenisation was a function of the long term presence of a pore fluid during the retrograde evolution, these terrains should show evidence for extensive retrograde hydration and reaction (see below). Since isotopically homogeneous terrains have not undergone extensive retrograde reaction, substantial quantities of pore fluid are unlikely to have been present during the retrograde evolution. Even if extensive hydration were evident this implies no more than transient fluid presence. Fluid might often be supplied from crystallising partial melts or from underthrusting lower grade rocks.

Rates of Retrograde Reaction

Rates of metamorphic reactions are a strong function of the quantities of hydrous fluid present. Recently Rubie (1986) has attempted to quantify the kinetics of retrograde reactions as applied to the long term fluid history of the lower crust. He observed that the low rate of transformation of jadeite + quartz to albite in some alpine eclogite facies granites only permitted the presence of a free fluid for a very short space of time. He used additional evidence from fine grained quartz mylonites which had not undergone grain growth to further argue for fluid absence.

Walther and Wood (1984) have reviewed data on the rates of metamorphic reaction. Progress of retrograde reactions could be limited by rates of nucleation, surface reaction or grain boundary film diffusion (see Rubie, 1986; Walther and Wood, 1984). Surface reaction is geologically rapid (Rubie, 1986; Walther and Wood, 1984). In the presence of a fluid, even a monomolecular fluid film, diffusion is so rapid that reactions limited by such a process, proceed to completion on times of the order of 10^5 years or less (Rubie, 1986; Walther and Wood, 1984). Nucleation of some phases is very rapid (Rubie, 1986) although in general rates of nucleation are not well quantified under geological conditions. However, the common preservation of reaction assemblages, with reactions not having gone to completion, in metamorphic rocks is evidence for fluid absence which does not depend on nucleation rate data. Examples of such incomplete reaction are abundant in mid to deep crustal rocks e.g. partially amphibolitised eclogites, corona textures in granulites and isograd assemblages in many pelitic rocks.

The kinetic data suggests that any hydrous fluid present should rapidly react. Experimental evidence indicates that the reaction H_2O + andalusite + K feldspar = muscovite + quartz goes rapidly to completion at temperatures of the order of 500°C (Schramke et al., 1987). If a multicomponent fluid is present, for instance one that contains CO_2 and NaCl, retrograde reactions should in general enrich the fluid in these non-hydrous components. This fluid could remain in equilibrium with a cooled lower crustal rock. However, these fluids should enhance rates of retrograde reaction for reactions not involving fluid, e.g. aluminosilicate polymorph transformations and the reaction jadeite + quartz = albite, as discussed by Rubie (1986). The chemical kinetic evidence does not appear consistent with the presence of even small quantities of fluid for timescales of millions of years. Even a pervasive monomolecular grain boundary hydrous film will have a significant effect on reaction rates (Rubie, 1986).

Fluid Distribution

The isotopic data on the Ivrea Zone suggest that fluid porosities of the order of one percent were not present in and around marbles for timescales of more than the order of a million years. This evidence confirms evidence based on retrograde reaction kinetics, that fluids were not pervasively present during the retrograde evolution. Isotopic and petrological data from other terrains suggests that this conclusion applies to many high grade metamorphic rocks. The observations are consistent with the presence of fluid only in spaced fractures. However, if fluid filled fractures do exist and are interconnected over a significant distance, the fractures are likely to propagate resulting in fluid loss. If the vertical gradient of the least principal stress differs significantly from the unit weight of the fluid the fracture will not be stable for more than a given vertical distance, perhaps of the order of tens to hundreds of metres (Secor and Pollard, 1975). Fluid in fractures is thus unlikely to be retained at depth for long periods of time unless the fractures are of small vertical extent. Another possible process is that over long periods of time fluid in fractures will infiltrate surrounding rocks by solution-reprecipitation mechanisms. Some data indicates that such processes might occur at quite high rates (Stevenson, 1983; Watson and Brenan, 1987). If this is so, surrounding rocks will not remain fluid absent.

Pelites produce of the order of 15 % of their own volume of fluid during metamorphism. This increase in volume will result in hydrofracture and loss of fluid (Walther and Orville, 1982). Along some retrograde P-T paths fluid will contract so that there is no further driving force for fluid loss (see Hall, 1986). It is consequently possible to envisage the retention of very small amounts of fluid in deep crustal rocks which would probably be enriched in non hydrous components during retrograde reaction. These might later be released during expansion consequent upon uplift, perhaps contributing towards the formation of the microveinlets observed in some high grade rocks (Morrison and Valley, 1988). However, quantities of such fluid must be small in order to preserve stable isotope gradients and to inhibit retrograde reaction. Small quantities of CO_2 rich fluid are likely to be isolated in pores and not interconnected if the wetting properties of fluid control the fluid distribution (Watson and Brenan, 1987). Such isolated fluids would provide an insignificant contribution to measured conductivities.

Lower Crustal Conductivities

ESTIMATION OF CONDUCTIVITIES

Available experimental work on the conductivities of rock at high pressure and temperature indicates that dry rocks are insufficiently conducting to explain high lower crustal conductivity (Olhoeft, 1981; Kariya and Shankland, 1983). However, the presence of a pore fluid could explain this discrepancy (e.g. Shankland and Ander, 1983). Attempts to estimate lower crustal porosities have been made from observed conductivities and Archies Law :

$$\gamma_r/\gamma_f = \phi^n$$

where γ_f is the conductivity of fluid, γ_r of the fluid-saturated rock and ϕ is the porosity. n is 2 where the porosity is randomly distributed. Where the porosity lies in spaced fractures parallel to the direction of conductivity measurement n = 1 (Hyndman and Drury, 1976). For fractures in natural rocks where fractures may not be interconnected perfectly and fracture orientation may vary, n is probably greater than one. For equilibrium grain boundary fluid distributions in many metamorphic rocks $1 < n < 2$ (Hyndman, 1988).

For a brine with a resistivity of $0.02\Omega m$ (one molar NaCl), a porosity of 1 % is required to achieve a bulk resistivity of $100\Omega m$ if n = 2. If the porosity is a fracture porosity and n = 1 then a porosity of 0.02 % would be sufficient. An estimate may be made for the conductivity of rocks which are undergoing active devolatilisation. Regional metamorphic fluid fluxes are of the order of 9 x $10^{-10} gs^{-1} cm^{-2}$ of cross sectional area (see Walther and Orville, 1982). Required crack lengths per unit area to accommodate this fluid flux may be estimated from the equation

$$q = d^3 lP/12\mu$$

where q is the fluid flux, d the crack width, l the length of crack per unit area, P the viscous pressure gradient and μ the dynamic viscosity (see Walther and Orville, 1982). For a crack width of 2 x 10^{-5} cm, 1cm of crack is required per cm^2 of cross sectional area to accommodate this fluid flow. A one molar NaCl brine with a resistivity of about 0.02 Ωm gives a bulk resistivity of about 1000 Ωm if n = 1. For a crack width of $0.02\mu m$, 1000cm of fracture are required per cm^2 which gives a bulk resistivity of $10\Omega m$. This is equivalent to grain boundary fluid flow for a small grain size and is inconsistent with Ivrea Zone stable isotope systematics. Stable isotope evidence from the Ivrea Zone indicates that fluid is strongly channelised. Walther and Wood (1984) suggest that fracture spacings are typically of the order of 1m, equivalent to 10^{-2} cm of fracture per cm^2 of cross sectional area. This implies a crack width of 2 x 10^{-4} cm and gives a bulk resistivity of 10,000 Ωm. These calculations suggest that, even in actively devolatilising rocks, it may not be possible to account for observed lower crustal conductivities solely by conduction through a pore fluid.

DISCUSSION

A fluid-filled porosity of 1.5 % is required, if pervasively distributed, to explain the low resistivities if conduction through the pore fluid is the only mechanism operative. Such a high fluid filled porosity is inconsistent with the preservation of stable isotope heterogeneities. It would have to exist for hundreds of millions of years if it is to be sampled by geophysical studies. Fluids in fractures will not be retained for such long periods of time if the fractures are of large vertical extent. If they are not extensive and interconnected they are unlikely to make a significant contribution to observed conductivities. It appears that even in actively devolatilising regions, quantities of fluid may not be sufficient to explain observed conductivities.

Since fluids are apparently not present in sufficient quantities to account for high lower crustal conductivities solely by pore fluid conduction it seems that other mechanisms must be important. These mechanisms could involve the presence of fluid in isolated pores or as a hydrated grain boundary phase, but not in the quantities predicted by Archies Law. Experimental data does suggest the operation of conduction mechanisms other than conduction through a pore fluid in metabasites (Lee et al., 1983), in basalts (Drury and Hyndman, 1979) and in serpentinites (Stesky and Brace, 1973). At room temperature and 4kb metabasite conductivities were one to two orders of magnitude higher than predicted by Archies Law. At 300°C the increase in conductivity relative to room temperature conductivity was more than one order of magnitude greater than predicted by Archies Law (Lee et al., 1983). Behaviour at temperatures in excess of this has not been measured. It is clear that the additional conduction path makes a very substantial contribution to the bulk conductivity and that at temperatures of the order of 500°C and pressures of 6 to 7 kb metabasites might have quite high conductivities. Additional conduction paths might be produced by surface effects particularly in fine grained or altered rocks (Lee et al., 1983), bulk mineral conduction (e.g. Brace, 1971), the presence of magnetite (Stesky and Brace, 1973) graphite or hydrogen ion mobility (Elphick and Graham, 1988). Some ferric iron bearing amphiboles also show high conductivities (Parkhomenko, 1982) although these are not typical of those in lower crustal basic material. Further experimental data is currently needed for probable lower crustal rock types at appropriate pressures and temperatures in the presence of a trace of fluid.

MODELS OF THE LOWER CRUST

Any model for the lower crust must account for the observed resistivity variations within the crust. The resistive upper crust with $\rho > 100$ Ωm commonly overlies a conductive zone at 15–20 km depth with $\rho \sim 10 - 100$ Ωm. The vertical extent of some of these conductive zones is often poorly constrained while the product of vertical extent and conductivity is relatively well known (Jones, 1987). Hence it is uncertain to what extent the whole of the lower crust is conductive or whether the conductive zone is a relatively thin mid crustal layer. The upper limit of the conductive lower crust is believed to be also the upper limit of the reflective lower crust and may coincide with an increase in seismic velocities (Gough, 1986; Jones, 1987). This upper limit is believed to lie at temperatures in excess of 300 to 400°C (Klemperer, 1987) or between 400 and 500°C (Jones, 1987). It is interesting to note that in at least one instance the lower crust is conductive, but not reflective (Jones et al., 1988).

It is observed in a number of areas that the lower crust is either reflective or conductive at depth, but apparently similar rocks exposed at the surface do not show these properties (Wood and Allard, 1986; Klemperer et al., 1987). This supports the origin of these anomalies from some transient lower crustal feature, generally considered to be fluids (Klemperer, 1987). However, high grade rocks exposed at the surface may not be representative of the lower crust. Few exposed granulite terrains are dominantly basic although xenolithic evidence (e.g. Griffin and O'Reilly, 1986) and seismic evidence (Hall, 1986) suggests that the lower crust is often dominantly basic. Metamorphic pressures in granulite facies rocks are rarely high enough to be consistent with equilibration in unstretched lower crust.

A number of hypotheses have related the high conductivities to the presence of fluid. For instance, Gough (1986) suggested that fluid was present throughout the whole crust, but was at hydrostatic pressure in closed cracks in the upper crust and was at lithostatic pressure in interconnected cracks in the lower crust. However, it is likely that hydrostatically pressured fluid can only be maintained in rocks near the surface (see Wood and Walther, 1986). Rock strengths are insufficient to maintain fluids at hydrostatic pressure below a few kilometres depth. Except at very low temperature it is likely that fluid distribution will be controlled by equilibrium wetting properties of fluid throughout the whole crust (see Watson and Brenan, 1987). Hydrous fluids, if present, should be interconnected in quartz bearing rocks throughout most of the crust.

Jones (1987) suggested that there was an impermeable layer at midcrustal depths of about 15 to 20 km, which trapped fluids in a thin highly conducting mid crustal zone. Unretrogressed dry granulite facies rocks could be located beneath this zone and dry upper crust above. Such a hypothesis requires the retention of fluids at these depths for long periods of time. If the conductive zone is maintained at constant depth during uplift and erosion of the crust it would be expected that a large proportion of the uplifted crust should at some stage experience these fluid present conditions.

Hyndman (1988) suggested that fluids are trapped in rocks between about 400 and 700°C, while fluid is absent from granulite facies rocks. A pervasive distribution of fluid was postulated, but with a low permeability which, it was argued, could maintain high lower crustal porosities for long periods. This model is inconsistent with the observed isotopic heterogeneity in rocks from most greenschist to amphibolite facies terrains. Few such terrains show evidence for extensive isotopic and chemical equilibration and retrograde reaction.

There appear to be considerable problems with hypotheses which advocate pervasive fluids. Alternative suggestions in which fluids are localised in fractures or zones of deformation appear more attractive. However, in the long term it is hard to explain how fluids are retained within these zones (see above and Hall, 1986). Shear zones continuously recharged with fluid might be conductive, but this explanation could only apply in areas where prograde metamorphism is occurring at deeper crustal levels. Shear zones have also been suggested as a possible cause of lower crustal reflectivity. However, seismic evidence suggests that deep crustal reflectivity is not the result of shear zones or fluids in at least some areas (McCarthy and Thompson, 1988).

An alternative hypothesis is that high lower crustal conductivities are related to some compositional contrast. The lower crust is believed to be have a large basic component on the evidence of some exposed examples (Fountain and Sailsbury, 1981), lower crustal xenoliths (e.g. Griffin et al., 1986) and seismic velocities (Hall, 1986; Fountain et al., 1987). Lithological variations from basic to metasedimentary rocks on an appropriate scale may explain the high reflectivity of the lower crust (Fountain et al., 1987; McCarthy and Thompson, 1988). Experimental data suggests that the conductivities of some metabasites depart considerably from Archies Law and it may be that metabasites in the presence of very small quantities of fluid are capable of explaining high lower crustal conductivities. Other potential contributing mechanisms might be conduction through serpentinised ultramafics or sulphide bearing layered intrusions. The contribution of metabasites to high lower crustal metabasites has been suggested by Lee et al. (1983) who observed that the same departures from Archies Law were not evident in many quartz bearing rocks. The presence of a large number of metabasites in the lower crust is consistent with P and S wave seismic data (Hall, 1986; McCarthy and Thompson, 1988). In at least some areas increases in seismic velocity coincide with high conductivity anomalies (Jones, 1987) or the beginning of the reflective lower crust (see discussion in McCarthy and Thompson, 1988). A transition to a basic lower crust could explain the near coincidence of the upper limit of the reflective and conductive lower crusts. Futher possibilities which might be important in at least some cases include the presence of sulphides or oxide minerals, graphite or serpentinites (e.g. DeBeer et al., 1982).

The inference that high conductivities are not generally the result of conduction through pore fluids does not apply to all conductivity anomalies. Near the surface fluids may circulate downwards and large fluid filled porosities might be expected. High conductivities in geothermal fields presumably do result from the presence of fluids. Nearer the surface rocks may be strong enough to retain fluids in highly permeable zones and very near the surface will support hydrostatically pressured fluids. Shear wave splitting in the upper 10 km or so of the earths crust does suggest the almost ubiquitous presence of fluid filled cracks in this region (Crampin and Atkinson, 1985; cf. Gough, 1986). It is noted that much of this upper crust is not anomalously conductive despite apparent fluid presence and that the seismic anisotropy indicative of the presence of vertical fluid filled cracks appears to be restricted to the upper 10 km of the crust (Kaneshima et al., 1988).

Conclusions

The preservation of large stable isotope gradients and isotopic heterogeneities in the Ivrea Zone and other metamorphic terrains is not consistent with the long term pervasive presence of quantities of fluid sufficient to explain high lower crustal conductivities. This supports previous inferences based on retrograde reaction kinetics. Fluids in spaced fractures or in highly permeable zones provide more likely locations for lower crustal fluids, but it is hard to explain how fluids are retained within these zones for long periods. Evidence from the Ivrea Zone and elsewhere indicates that in general a substantial mantle to crust fluid flux does not continuously recharge lower crustal fluids. Conduction mechanisms other than conduction solely through a pore fluid are required, such as those in some experimental studies of basic rocks.

ACKNOWLEDGEMENTS
The author was in receipt of a Royal Society of Edinburgh Research Fellowship during the course of this work. M. Mareschal and an anonymous reviewer are thanked for helpful criticisms.

REFERENCES

Anderson, A.T. (1967) 'The dimensions of oxygen isotope equilibrium attainment during prograde metamorphism.' *J. Geol.* **75**, 323–332.
Baker, A.J. (1988) 'Stable isotopic evidence for limited fluid infiltration of deep crustal rocks from the Ivrea Zone, Italy.' *Geology* **16**, 492–495.

Baker, A.J. 'Stable isotopic evidence for fluid rock interactions during high grade metamorphism in the Ivrea Zone, Italy.' Submitted to *Journal of Petrology.*

Baker, A.J. & Fallick, A.E. (1988) 'Evidence for CO_2 infiltration in granulite facies marbles from Lofoten-Vesteralen, Norway.' *Earth Planet. Sci. Lett.* **91**, 132-140.

Brace, W.F. (1971) 'Resistivity of saturated crustal rocks to 40 km based on laboratory measurements.' In Heacock, J.G. (ed.) *Structure and physical properties of the earths crust. Geophys. Monogr. Ser.* **14**, 243-255, AGU, Washington, D.C.

Brodie, K.H. & Rutter, E.H. (1987) 'Deep crustal extensional faulting in the Ivrea Zone of norhtern Italy.' *Tectonophysics* **140**, 193-212.

Cole, D.R. & Ohmoto, H. (1986) 'Kinetics of isotopic exchange at elevated temperatures and pressures.' In Valley, J.W., Taylor, H.P. & O Neil, J.R. (eds.) *Stable isotopes in high temperature geological processes* pp 41-87, Min. Soc. Am.

Connerney, J.E.P., Nekut, A. & Kuckes, A.F. (1980) 'Deep crustal electrical conductivity in the Adirondacks.' *J. Geophys. Res.* **85**, 2603-2614.

Crampin, S. & Atkinson, B.K. (1983) 'Microcracks in the earths crust.' *First Break* **3**, 16-20.

DeBeer, J.H., Van Zijl, J.S.V. & Gough, D.I. (1982) 'The southern cape conductive belt (South Africa), its composition, origin and tectonic significance.' *Tectonophysics* **83**, 205-225.

Drury, M.J. & Hyndman, R.D.(1979) 'The electrical resistivity of oceanic basalts.' *J. Geophys. Res.* **84**, 4537-4545.

Elphick, S. & Graham, C.M. (1988) 'The effect of hydrogen on oxygen diffusion in quartz, evidence for fast proton transients.' *Nature* **335**, 243-245.

Etheridge, M.A., Wall, V.J. & Vernon, R.H. (1983) 'The role of the fluid phase during regional metamorphism and deformation.' *J. metamorphic Geol.* **1**, 205-226.

Fountain, D.M. & Salisbury, M.H. (1981) 'Exposed cross sections through the lower continental crust : implications for crustal structure, petrology and evolution.' *Earth Planet. Sci. Lett.* **56**, 263-277.

Fountain, D., McDonough, P. & Gorham, J. (1987) 'Seismic reflection models for the continental crust based on metamorphic terrains.' *Geophys. J. Roy. astr. Soc.* **89**, 61-66.

Gough, D.I. (1986) 'Seismic reflectors, conductivity, water and stress in the continental crust.' *Nature* **323**, 143-144.

Griffin, W.L. & O'Reilly, S.Y. (1986) 'The lower crust in eastern Australia : xenolithic evidence.' In Dawson J.B., Carswell, D.A., Hall, J. & Wedepohl, K.H. (eds.) *The nature of the lower continental crust. Spec. Publ. geol. Soc. London* **24**, 363-374.

Haak, V. & Hutton, R. (1986) 'Electrical resistivity in the continental lower crust.' in Dawson, J.B., Carswell, D.A., Hall, J. & Wedepohl, K.H. (eds.) *The nature of the continental crust. Spec. Publ. geol. Soc. London* **24**, 35-49.

Hall, J. (1986)' The physical properties of layered rocks in deep continental crust.' In Dawson, J.B., Carswell, D.A., Hall, J. & Wedepohl, K.H. (eds.) *The nature of the lower continental crust. Spec. Publ. geol. Soc. London* **24**, 51-62.

Hay, R.S. & Evans, B. (1988) 'Intergranular distribution of pore fluid and the nature of high angle grain boundaries in limestone and marble.' *J. Geophys. Res.* **93**, 8959-8974.

Hyndman, R.D. (1988) 'Dipping seismic reflectors, electrically conducting zones and trapped water in the crust over a subducted plate.' *J. Geophys. Res.* **93**, 13, 391-13405.

Hyndman, R.D. & Drury, M.J. (1976) 'The physical properties of oceanic basement rocks from deep drilling on the mid Atlantic ridge.' *J. Geophys. Res.* **81**, 4042-4052.

Jackson, D.H., Mattey, D.P. & Harris, N.B.W. (1988) 'Carbon isotope compositions of fluid inclusions in charnockites from southern India.' *Nature* **333**, 167-171.

Jones, A.G. (1987) 'MT and reflection : an essential combination.' *Geophys. J. R. astr. Soc.* **89**, 7-18.

Jones, A.G., Kurtz, R.D., Oldenburg, D.W., Boerner, D.E. & Ellis, R. (1988) 'Magnetotelluric observations along the lithoprobe southeastern Canadian Cordilleran transect.' *Geophys. Res. Lett.* **15**, 677-680.

Kaneshima, S., Ando, M. & Kimura, M. (1988) 'Evidence from shear wave splitting for the restriction of seismic anisotropy to the upper crust.' *Nature* **335**, 627-629.

Kariya, K.A. & Shankland, J. (1983) 'Electrical conductivity of dry lower crustal rocks.' *Geophysics* **48**, 52-61.

Kay, R.M. & Kay, S.M. (1981) 'The nature of the lower continental crust : inferences from geophysics, surface geology and crustal xenoliths.' *Rev. Geophys. Space Phys.* **19**, 271-297.

Klemperer, S.L. (1987) 'A relation between crustal heat flow and the seismic reflectivity of the lower crust.' *J. Geophys.* **61**, 1-11.

Klemperer, S.L. & BIRPS group (1987) 'Reflectivity of the crystalline crust : hypotheses and tests.' *Geophys. J. Roy. astr. Soc.* **89**, 217-222.

Lamb, W.M., Valley, J.W. & Brown, P.E. (1987) 'Post metamorphic CO_2 rich fluid inclusions in granulites.' *Contrib. Mineral. Petrol.* **96**, 485-495.

Lee, C.D., Vine, F.J. & Ross, R.G. (1983) 'Electrical conductivity models for continental crust based on laboratory measurements on high grade metamorphic rocks.' *Geophys. J. Roy. astr. Soc.* **72**, 353-372.

Matthews, D. & Cheadle, M. (1986) 'Deep reflections from the Caledonides and Variscides west of Britain and comparison with the Himalayas.' In *Reflection seismology, a global perspective Geodynamics Series* **15**, 5-19, AGU, Washington D.C. .

McCarthy, J. & Thompson, G.A. (1988) 'Seismic imaging of extended crust with emphasis on the western United States.' *Bull. geol. Soc. Am.* **100**, 1361-1374.

Morrison, J. & Valley, J.W. (1988) 'Post granulite facies fluid infiltration in the Adirondack Mountains.' *Geology* **16**, 513-516.

Nagy, K.L. & Parmentier, E.M. (1982) 'Oxygen isotopic exchange at an igneous intrusive contact.' *Earth Planet. Sci. Lett.* **59**, 1-10.

Newton, R.C. (1986) 'Fluids and granulite facies metamorphism'. In Walther, J.V. & Wood, B.J. (eds.) *Fluid-rock interaction during metamorphism,* pp. 36-59, Springer-Verlag, New York.

Olhoeft, G.R. (1981) 'Electrical properties of granite with implications for the lower crust.' *J. Geophys. Res.* **86**, 931-936.

O'nions, R.K. & Oxburgh, E.R. (1988) 'Helium, volatile fluxes and the development of the continental crust' *Earth Planet. Sci. Lett.* **90**, 331-347.

Parkhomenko, E.I. (1982) 'Electrical resistivity of minerals and rocks at high temperature and pressure.' *Rev. Geophys. Space Phys.* **20**, 193-218.

Rubie, D.C. (1986) 'The catalysis of mineral reactions by water and restrictions on the presence of aqueous fluid during metamorphism.' *Min. Mag.* **50**, 339-415.

Rumble, D. & Spear, F.S. (1983) 'Oxygen isotope equilibration and permeability enhancement during regional metamorphism.' *J. geol. Soc. London* **140**, 619-628.

Rye, R.O., Schuiling, R.D., Rye, D.M. & Jansen, J.B.H. (1976) 'Carbon, oxygen and hydrogen isotope studies of the regional metamorphic complex at Naxos, Greece.' *Geochim. Cosmochim. Acta* **40**, 1031-1049.

Sauter, P.C.C. (1983) ' Metamorphism of siliceous dolomites in the high grade Precambrian of Rogaland, SW Norway.' *Geologica Ultraiectina* **32**, pp 143.

Schramke, J.A., Kerrick, D.M. & Lasaga, A.C. (1987) 'The reaction muscovite + quartz

= andalusite + K feldspar + water. Part 1 Growth kinetics and mechanisms.' *Am. J. Sci.* **287**, 547–559.

Schwarcz, H.P., Clayton. R.N. & Mayeda, T. (1970) 'Oxygen isotopic studies of calcareous and pelitic metamorphic rocks, New England.' *Bull. geol. Soc. Am.* **81**, 2299–2316.

Secor, D.T. & Pollard, D.D. (1975) 'On the stability of open hydraulic fractures in the earths crust.' *Geophys. Res. Lett.* **2**, 510–513.

Shankland, T.J. & Ander, M.E. (1983) 'Electrical conductivity, temperature and fluids in the lower continental crust.' *J. Geophys. Res.* **88**, 9475–9484.

Sheppard, S.M.F. & Schwarcz, H.P. (1970) 'Fractionation of carbon and oxygen isotopes between coexisting metamorphic calcite and dolomite.' *Contrib. Mineral. Petrol.* **26**, 161–198.

Shieh, Y.N. & Schwarcz, H.P. (1974) 'Oxygen isotope studies of granite and migmatite, Grenville Province of Ontario, Canada.' *Geochim. Cosmochim. Acta* **38**, 21–45.

Stesky, R.M. & Brace, W.F. (1973) 'Electrical conductivity of serpentinised rocks to 6kb.' *J. Geophys. Res.* **78**, 7614–7621.

Stevenson, D.J. (1983) 'On the role of surface tension in the migration of melt and fluids.' *Geophys. Res. Lett.* **13**, 1149–1152.

Valley, J.W. & O'Neil, J.R. (1982) 'Oxygen isotope evidence for shallow emplacement of Adirondack anorthosite.' *Nature* **300**, 497–500.

Valley, J.W. & O'Neil, J.R. (1984) 'Fluid heterogeneity during granulite facies metamorphism in the Adirondacks : stable isotope evidence.' *Contrib. Mineral. Petrol.* **85**, 158–173.

Valley, J.W. (1986) 'Stable isotope geochemistry of metamorphic rocks.' In Valley, J., Taylor, H.P. & O'Neil, J.R. (eds.) *Stable isotopes in high temperature geochemical processes.* pp 445–490, Min. Soc. Am.

Vry, J., Brown, P.E., Valley, J.W. & Morrison, J. (1988) 'Constraints on granulite genesis from carbon isotope compositions of cordierite and graphite.' *Nature* **332**, 66–68.

Walther, J.V. & Orville, P.M. (1982) 'Volatile production and transport in regional metamorphism.' *Contrib. Mineral. Petrol.* **79**, 252–257.

Walther, J.V. & Wood, B.J. (1984) 'Rate and mechanism in prograde metamorphism.' *Contrib. Mineral. Petrol.* **88**, 246–259.

Watson, E.B. & Brenan, J.M. (1987) 'Fluids in the lithosphere, 1 Experimentally determined wetting characterisitics of CO_2–H_2O fluids and their implications for fluid transport, host rock physical properties and fluid inclusion formation.' *Earth Planet. Sci. Lett.* **85**, 497–515.

Wickham, S.M. & Taylor, H.P. (1985) 'Stable isotopic evidence for large scale seawater infiltration in a regional metamorphic terrain : The Trois Seigneurs Massif, Pyrenees, France.' *Contrib. Mineral. Petrol.* **91**, 122–137.

Wood, B.J. & Walther, J.V. (1986) 'Fluid flow and its implication for fluid–rock ratio.' In Walther, J.V. & Wood, B.J. (eds.) *Fluid rock interactions during metamorphism.* pp. 60–88, Springer Verlag, New York.

Wood, D.V. & Allard, M. (1986) 'Reconnaisance electromagnetic induction study of the Kapuskasing structural zone.' *Phys. Earth Planet. Inter.* **42**, 135–142.

Yardley, B.W.D. (1986) 'Is there water in the deep continental crust ?' *Nature* **323**, 111.

Zingg, A. (1983) 'The Ivrea and Strona-Ceneri Zones (Southern Alps, Ticino and N. Italy) – a review.' *Schweiz. miner. petrog. Mitt.* **63**, 361–392.

ELECTRICAL CONDUCTIVITY: THE STORY OF AN ELUSIVE PARAMETER, AND OF HOW IT POSSIBLY RELATES TO THE KAPUSKASING UPLIFT (LITHOPROBE, CANADA)

Marianne MARESCHAL
IREM, Ecole Polytechnique
C.P. 6079, Succ. "A"
Montréal, Canada H3C 3A7

ABSTRACT. Possible sources of crustal conductivity are reviewed, with special emphasis on shield environments. Typical lower crustal conductivities are interpreted in terms of brines and/or graphite films precipitated at grain-boundaries under cooling after peak metamorphism. The nature of the conductive films may strongly depend on the age of the crust: grain-boundary graphite may be present in most lower crusts, while brines would be found in the more active and younger regions. Electromagnetic investigations of the Kapuskasing Uplift confirm that exposed lower crustal rocks are very resistive. Water does not seem to play any significant role in the present lower crust of the uplift, where no correlation between deep seismic reflectors and conductors can apparently be found. Interpretation is only preliminary, however, it also suggests that brines may exist deep in the upper crust (~ 5km). Other shallow conductors could be related to lithologic boundaries, faults or detachment zones.

1. Introduction

Electrical conductivity, σ, is one of the few physical parameters of the deep earth that can be estimated from surface measurements. Optimists tend to consider it a very good marker because its range of variation is extremely wide (resistivity, ρ, the reciprocal of conductivity can vary from approximately 10^{-1} to 10^{10} Ωm or more, while, for example, seismic velocities vary only from 0 to approximately 15 km/s). Pessimists, on the other hand, tend to point out that the processes by which rocks become conductive at depth are poorly known and that little consensus exists on the interpretation of deep crustal anomalies.

The first section of this paper reviews possible sources of crustal conductivity, with special emphasis on shield environments. Near surface (< 1 km) conductors (wet sediments, graphite, various mineral deposits, etc.) are ignored in this review; their effects are usually well known, particularly because of their potential impact on mineral exploration research.

The second section presents briefly the field measurements that were used to estimate the electrical conductivity of the Kapuskasing uplift and summarizes preliminary results. The picture can only be fragmentary since much work is still in process.

M. H. Salisbury and D. M. Fountain (eds.), Exposed Cross-Sections of the Continental Crust, 453–468.
© 1990 *Kluwer Academic Publishers.*

2. Possible Sources of Crustal Conductivity in Shield environments

The conductivity of rocks is usually extremely weak, so the electrical features of the crust are commonly described in terms of resistivities (measured in Ωm) rather than conductivities (measured in S/m).

One of the major problems to plague geo-electrical studies is the discrepancy of several orders of magnitude between laboratory measurements and field data. No part of the crust has ever displayed resistivities as high as individual dry rocks do in laboratory experiments. Wet rocks, on the other hand, usually undergo too many chemical reactions, when brought abruptly to high temperatures and pressures, to allow reliable measurements of electrical conductivity (e.g., Duba, 1976). Thus it has become common practice to rely on the conductivities measured in the field by various electromagnetic methods to define the normal electrical signatures of various tectonic environments (e.g., Haak and Hutton, 1986). Shields belong to the most resistive end of the spectrum: their upper crust is often characterized by resistivities on the order of $10^5 \Omega$m, while the resistivity of their lower crusts never decreases below 100 Ωm, a value which is approximately an order of magnitude larger than the resistivities of many younger lower crusts.

Electrical conductivity in most minerals, with the exception of metals, is a thermally activated process: the number and mobility of charge carriers increase with temperature. Obviously, this mobility is higher in molten rocks, and the effective conductivity is then representative of the volume ratio of melt to the solid state matrix, as well as of the degree of melt interconnection ("Archie's law" or more complex relations; e.g. Shankland and Waff, 1977). Zones of partial melt may reach resistivities as low as 1 Ωm (e.g., Haak, 1980) but are not expected in old, stable tectonic areas.

In shields, tectonic activity and active metamorphism are long complete, surface heat flow is low, and zones of "moderate" conductivity (resistivities on the order of a few 100's to 1000 Ωm) cannot be explained by the effect of temperature alone. For example, most dry silicate minerals would have resistivities on the order of 10^6 Ωm or more at 500°C, an upper limit to temperatures assumed to prevail at Moho depths in the Canadian Shield (e.g., Rao and Jessop, 1975). Pressure, which initially lowers resistivity by approximately an order of magnitude in bringing conductive grains into better contact, has no further effect except that of causing the collapse of pores in wet rocks (e.g., Brace and Orange, 1968); thus high pressures tend to be detrimental to conductivity. On the other hand, mineralogy alone cannot play a primary role since exposed lower crustal rocks are usually very resistive (e.g., Woods and Allard, 1986).

It is often argued that the absence of volatiles (lost by evaporation when rocks are brought to surface) explains the discrepancy between the extremely high resistivities measured in laboratory experiments and the rather low resistivities observed in the field. However, if fluids provide the electrical signature, essentially only substances that are both highly polar as gases and liquids and are also electrically conductive as liquids must be considered. Carbon dioxide and nitrogen would have little effect while water, (particularly if saline) and sulfur compounds would strongly affect the electrical properties of rocks (e.g., Olhoeft, 1981).

The problem is to explain why the upper crust, which, even in shields, is known to include numerous brines in fractures and cracks (e.g., Frape and Fritz (1987) for the Canadian Shield; Bayuk et al. (1984) for the Russian Shield) is much more resistive than the lower crust where pore space would be reduced by pressure. Gough (1986) suggests that the behavior can be explained by a model in which the entire crust contains saline water: this water would reside in separate cavities with little or no electrical contact in the compressively stressed rocks of the upper crust, but would form an interconnecting film on crystal surfaces in the lower crust. His argument is based on the fact that a "typical" upper crust is electrically resistive and seismically transparent, contains nearly all intracontinental earthquake hypocentres, and responds to stress elastically, with brittle fractures. A "typical" lower crust, on the other hand, is electrically conductive and seismically reflective, includes no earthquake hypocentres and shows ductile responses to stress.

The idea that fluid-filled fractures or pores can account for seismic reflections (e.g., Crampin and Atkinson, 1985; Mathews, 1986) has given rise to several discussions on the porosity (and, to a lesser extent, the salinity) required to explain observed coincident electrical and seismic reflectors in specific tectonic environments. For instance, the case for a continental crust overlying a subducting plate is quite convincing (e.g., Hyndman, 1988; Green et al., 1987). However, Hyndman and Shearer (1989) take the argument one step further: they suggest that a porosity of a few percent at grain boundaries is sufficient to explain many zones of coincident enhanced electrical conductivity and seismic reflectivity in lower Phanerozoic crusts. This would be particularly true if the brines were salty.

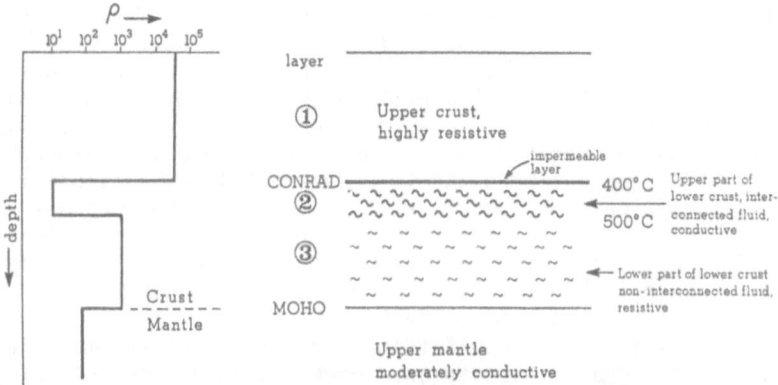

Fig.1: A mechanism possibly responsible for enhanced electrical conductivity in the upper part of the lower crust; brines are trapped underneath an impermeable layer (Jones, 1987).

Because of problems inherent in most electromagnetic methods, it is often difficult to estimate the depth of the conductive zones accurately. In the younger crusts, it seems that the top of the layer would tend to be observed at P-T conditions in which temperatures would range from 350° to 400°C. These conditions could represent an impermeable boundary determined by hydration reactions at the upper greenschist facies and/or by precipitation of silica (e.g, Etheridge et al. 1983; Jones, 1987; Fig.1). They could also be associated with the minimum temperature for ductile behaviour and equilibrium grain-boundary pore configurations. Hyndman and Klemperer (1989) claim further that the discrepancy between seismic refraction velocities measured in the field (particularly in Phanerozoic regions) and estimates of a primarily mafic lower crustal composition can be reconciled by porosities of a few percent. However, some of these discrepancies may only be apparent: they would simply result from the wide range of velocities commonly observed in mafic rocks in laboratory experiments (e.g., Fountain, 1989).

Of course, there is no reason to believe that a single mechanism is responsible for all deep electrical features. Electrical conductivity is a snapshot of the present crust: to preserve water over long periods of time without replenishment would require an extremely low permeability (which is difficult to reconcile with the porosity) or some widespread layering of the lower crust in a sequence of alternating, very thin, permeable and impermeable zones (e.g., Hyndman and Shaeffer, 1989). Pervasive saline water can certainly explain coincident lower crustal conductivities and reflectivities; but distinct causes for the reflectivity (such as lithologic layer in gneisses, mylonites, etc.; e.g., Green et al., 1989) and connected solid conductors for the electrical features (graphite, sepentinized rocks; e.g., Stesky and Brace, 1973) could achieve the same result. With large multidisciplinary programs such as LITHOPROBE in Canada, it is becoming progressively apparent that coincident reflectivity and conductivity is not an absolute rule (e.g., Green et al., 1989). In crystalline terranes, it may never be the case. In fact, both Gough (1986) and Hyndman and Shearer (1989) note that water-related mechanisms are unlikely to prevail in shields where the lower crust was probably subject to high temperatures without subsequent additions of water (few retrograde reactions, isotope heterogeneity, etc.).

Mineral deposits located along preferred conduits at what must originally have been intermediate to deep crustal levels indicate that brittle deformation can occur at great depths under conditions of rapid changes in pore fluid pressure or strain rate (e.g., Colvine et al., 1988). However, these conditions should be quite localized and, in shields, can only be relicts of a more active past. Thus, one consensus has been reached at last: that a major component of lower crustal conductivity must result from films of conductive material in pore spaces at grain boundaries. However, while the role of grain boundaries is usually acknowledged, the nature of the conductors is still debated. One of the problems is that any alternative conductor to brines must also be able to explain the discrepancy between laboratory and field data and, in particular, why rocks are usually quite resistive where they show a high degree of metamorphism upon exposure or near exposure to the surface. Frost and co-workers present a simple argument: solid films (~1000 Å thick) formed at lower crustal conditions should break under decompression during uplift and thus lose the connection needed for conductivity. This physical destruction would probably be non-reversible: once the continuity of the film is broken, it would not heal, even if the rock were submitted to high pressures (Frost et al., 1989).

The obvious material for solid films is graphite because it is characterized by a very high electrical conductivity (e.g., Duba and Shankland, 1982; Duba et al, 1988); fluid inclusions in granulites are certainly indicative of an original environment rich in CO_2 (e.g., Touret, 1971). The proposal is quite attractive, but can only explain the observed extensive zones of enhanced lower crustal conductivity, if graphite precipitation is a widespread mechanism in rocks that crystallize at high temperature, rather than being limited to rather highly reducing environments during peak metamorphism (e.g., Mathez and Delaney, 1981). Supporting evidence is finally emerging: films of graphite, too fine to be detected by traditionnal techniques, have recently been uncovered at the grain boundaries of three rocks from the Laramie Anorthosite Complex, using scanning Auger spectrometry (Frost et al., 1989). In two of the rocks, graphite was stable during igneous crystallization, but it was not in the other. The rock had originally formed at oxygen fugacities above those of the graphite saturation surface. In this case, interoxide equilibria would have driven the oxygen fugacity to graphite saturation during cooling and graphite would then have precipitated with accompanying release of oxygen. If this interpretation is valid, graphite precipitation is likely to have operated in many lower crustal rocks and, indeed, almost all exsolved and reconstituted oxide compositions from granulites plot on the graphite saturation surface. Secondary graphitization, involving precipitation of both graphite and magnesite, may also have played a role in producing carbon films.

Frost and co-workers show that grain-boundary graphite films with thicknesses of approximately 1000 Å can produce electrical conductivities typical of those observed in the lower crust: lower crustal rocks need only have an average grain size of 1 cm or less (Fig. 2). Although the same process could produce enhanced conductivities in the upper crust, the authors suggest that it would tend to be more effective at greater depths. The first cause would be graphite stabilization which is greatly enhanced by pressure. Thus, a mafic rock crystallizing in a shallow environment may not reach graphite saturation, even on cooling, whereas the same lithology cooling at greater depth would be likely to develop grain-boundary graphite. The second cause concerns the breakage of films during decompression associated with uplift, as already mentionned. Without physical connection, no electrical response can be produced. In fact, this explains why many rocks, even though they include graphite in their primary mineral assemblages, are resistive; the mineral assemblage itself prevents electrical contact between conductive grains.

458

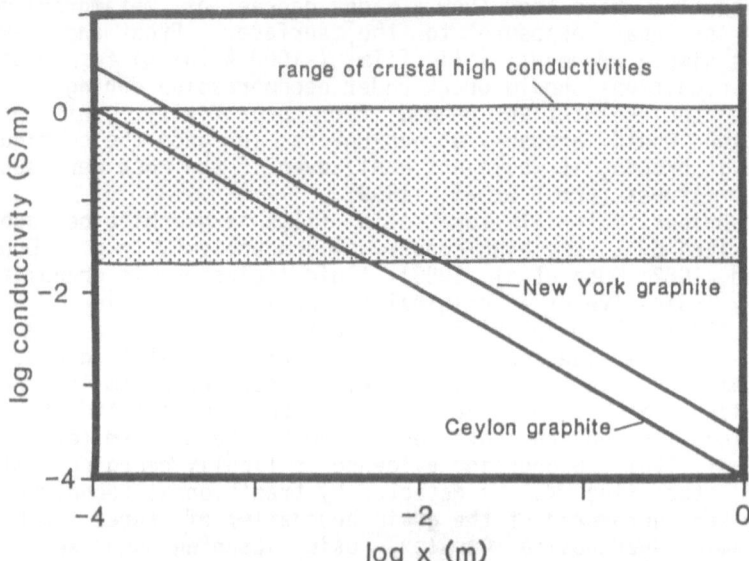

Fig.2: Calculated conductivity for rocks that contain a 1000 Å thick graphite film on the grain boundaries. Rocks having an average grain size of 1 cm or less will have conductivities equivalent to those seen in zones of high crustal conductivity (Frost et al., 1989).

At this point, it is interesting to note that many intermediate to deep conductive zones seem to be best represented by a highly resistive crust which would occasionally become more conductive in one distinct direction (Schmucker, 1988). This type of crustal anisotropy appears to be reasonably well modelled by that of a series of conductive, vertical dikes (in the Ωm range), due to high contents of graphite or aligned fluid-filled microcracks, lying parallel to each other in an otherwise resistive environment. The favored orientation would be regulated by the ambient horizontal stress field; the alignment of fluid-filled fractures possibly tending to be representative of more recent stress fields than that of solid conductors. Indeed, one would expect water to diffuse along grain boundaries or to open new brittle cracks in response to changes in stress conditions. Solid conductors, on the other hand, may be more apt at retaining their original overall distribution.

Summary. Lower crustal conductors may not be restricted to brines and graphite alone, but the important role that these features may play is beginning to be fully recognized. Interlocking fractures filled with brines, or solid conductors such as those found in the upper crust, are not necessarily needed to explain intermediate to lower crustal conductivities. Grain boundary films should also play a primary role at these levels. The nature of the conductive films may strongly depend on the age of the region. Interconnected solid conductors (predominantly graphite) may be present at the grain boundaries of most lower crustal rocks. In more active and younger regions, the presence of brines would likely improve the overall electrical conductivity, possibly by an order of magnitude or more.

3. Electromagnetic investigations of the Kapuskasing Uplift: Limitations of the field methods and preliminary results.

Electromagnetic investigations in the Kapuskasing uplift have been undertaken at various scales: first, a large scale geomagnetic depth-sounding (GDS) reconnaissance survey; second, some dense magnetotelluric (MT) profiling along selected seismic reflection lines; and finally, high-resolution transient electromagnetic (TEM) surveys in key target areas.

GDS investigations are based on the simultaneous recording of magnetic field variations over large arrays. In the Kapuskasing area, 30 recording instruments were used, with an average spacing of 30 to 40 km. The method is very good for mapping lateral variations in the concentration of electrical currents and it can usually delineate good conductors quite well. However, the exact depth of currents is difficult to estimate. Another problem occasionally arises from the fact that the electrical conductivity of a geological feature may not be directly related to the intensity of the currents that it carries; some additional currents, unrelated to the local electromagnetic field, may use the conductive feature as a channel between removed conductive areas.

MT interpretations are aimed at quantitative estimates of the variations of electrical conductivity with depth. The principle is simple: variations in the full electromagnetic response (magnetic plus electric) with source frequency, from high frequencies (shallow penetration) to low frequencies (deep penetration), should produce an accurate electrical profile from the surface down to the upper mantle. Because the method relies on natural ionospheric and magnetospheric variations, its frequency spectrum and its depth of penetration are very large. The method is reasonably good in layered environments, but the interpretation becomes increasingly difficult as lateral effects accumulate. Even in the simplest situation, it may be impossible to differentiate between the response of a thick, moderately conductive layer and a thin, highly conductive one. MT is more quantitative than GDS because it is based on simultaneous measurements of magnetic and electric fields (rather than on magnetic variations alone). However, since electric fields are highly affected by superficial heterogeneities, MT is also less robust than GDS. Its responses may be deformed by differences in soil composition, etc. in unpredictable ways, and may thus lead to erroneous estimates of depth and thickness ("static shift"). Of course, the combination of MT with other geo-investigations (in particular, with seismics, the only other deep geophysical method; e.g., Jones, 1987) may help to constrain the electrical parameters, but the most satisfying combination to avoid static shift is probably that of MT with a "controlled-source" investigation.

460

Controlled-source electromagnetic methods, as their name implies, use man-made transmitters to generate their source field. Because the source is precisely known, these methods can give reliable estimates of depth in reasonably layered environments. However, because it is man-made, its spectrum of frequencies is narrow and limited to the higher components. Thus, their minimum and maximum depths of penetration are strongly limited by practical factors, in addition to the usual detrimental effects of highly conductive layers which dissipate the electromagnetic energy in creating currents. For example, it is only because the region was very resistive that the UTEM system was able to image electrical structures down to approximately 15 km in the Kapuskasing uplift. As in many other controlled-source techniques, UTEM also has the advantage of relying on magnetic field measurements alone; the electrical sections produced are thus quasi-unaltered by static shift. Note finally that neither GDS, MT, nor UTEM responds to resistive zones; thus, these electromagnetic methods are only useful in defining conductive targets.

The Kapuskasing uplift was recently selected as one of the major LITHOPROBE targets in Canada, and has been subject to numerous investigations in the past few years. Many geophysical interpretations are still in process, implying that the results reported here can only be fragmentary. The structure is a 500-km long, NE-striking belt of high-grade metamorphic rocks that transect the Canadian Shield

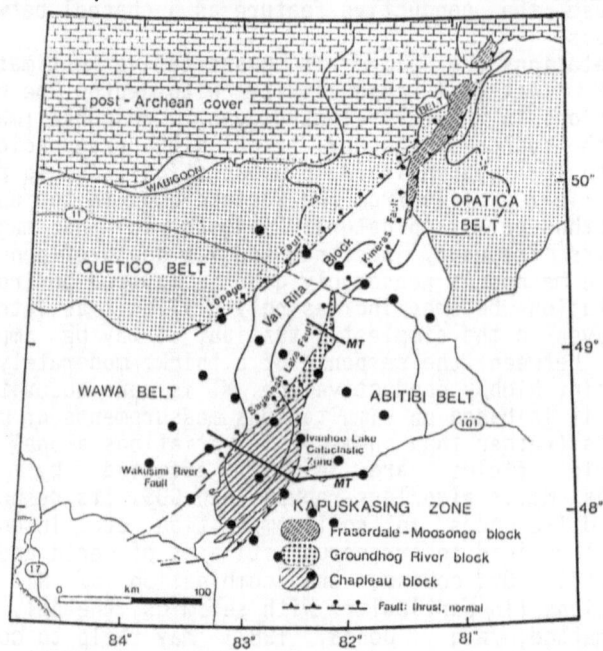

Fig.3: Regional tectonic elements of the Kapuskasing Uplift. The regional gravity high (-25 mGal) is outlined. Dots define the GDS array, MT lines are shown(adapted from Leclair and Nagerl, 1988).

approximately from Lake Superior to James Bay (Fig.3). The Ivanhoe Lake Fault (ILF), a continuous feature interpreted to be a thrust of Proterozoic age, bounds the Kapuskasing zone against lower grade rocks to the east (Abitibi). Faults to the west are discontinuous and have variable magnitudes of normal displacement (e.g., Percival, 1989).

The uplift is usually divided into three blocks. From south to north, these include the Chapleau block (CB), the Groundhog River block (GRB) and the Fraserdale-Moosonee block (FMB) (Fig 3). In the Chapleau block, in which offsets along discrete normal faults are minor, a continuous west-to-east transition is recognized, extending from greenschist, through amphibolite, to granulite facies. Changes in metamorphic pressure across this 120-km wide transition led to its interpretation as an exposed crustal section through a greenstone belt down to approximately 30 km (e.g., Percival and Card, 1983). The resulting model of crustal structure (Fig. 4, after Percival, 1989) is supported by seismic refraction (Boland et al., 1988) and seismic reflection (Green et al., 1989).

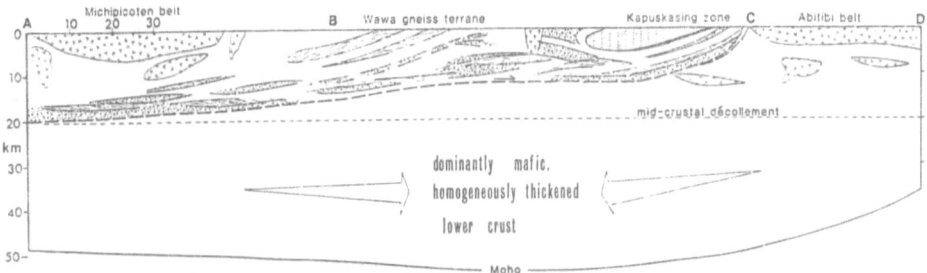

Fig.4: Approximately east (D) - west (A) cross-section through the Chapleau block of the Kapuskasing Uplift and adjacent terranes. The thrust fault and associated cataclastic zone (ILCZ) are located in C (Percival, 1989).

The Groundhog River block, on the other hand, is characterized by extremely limited exposures (e.g., Leclair and Nagerl, 1988) and was originally defined on the basis of its very strong aeromagnetic signature alone. It is less than 30-km wide, and is bound to the west by the brittle Saganash Lake fault (SLF). Its uplift appears to have been of lesser amplitude than that of the Chapleau block, but the resulting structure may be much more disrupted, because of the complex interplay of several normal faults (Leclair and Poirier, 1989). The geological model relies heavily on geophysical information, particularly on the results of very recent gravity and seismic reflection surveys (Nkwate and Salisbury, 1988; Milkereit et al., 1989).

The Fraserdale-Moosonee block is still less known. It has been subject to very few investigations, and is ignored for the rest of this paper.

The first electromagnetic investigation of the Kapuskasing uplift was a GDS reconnaissance survey which included both the Chapleau and the Groundhog River blocks (Woods and Allard, 1986; the array location is shown on Fig.3). At the low frequencies used, current concentrations are averaged over very large crustal volumes, and, thus, only major features could be recognized. However, since this array included a much wider area than that of the two Kapuskasing blocks, a lateral variation should have been observed if a wide zone (20 km or more) of conductive lower crustal material ($\rho \sim 100$ Ωm) had been thrusted toward surface along a very conductive fault (Fig.5).

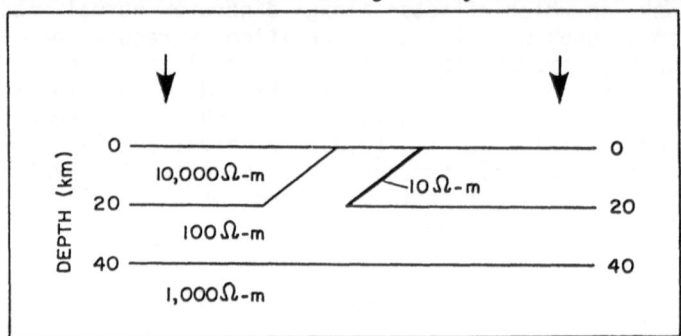

Fig.5: The schematic uplift model tested by the GDS experiment. The lateral extent of the array is indicated by arrows (Woods and Allard, 1986).

This response was not observed. In fact, there was no overall anomalous response associated with the Kapuskasing uplift; only a slight enhancement of conductivity, possibly corresponding to the location of the Ivanhoe Lake fault, was noted. Thus, Woods and Allard (1986) concluded that enhanced lower crustal conductivity must be related to intrinsic conditions at depth rather than to mineralogical composition. Even though it is now beleived that the deepest rocks exposed in the Chapleau block were uplifted from a source much further west than is schematically represented in Fig. 5 (see Fig. 4), these rocks are still thought to have originated at crustal depths of ~ 30 km (Percival, 1989), confirming the conclusions of Woods and Allard.

Rocks exposed today in the Chapleau block are primarily magmatic and were added in the interval between 2765 and 2680 Ma (Percival, 1989). They may not be related to present sources of lower crustal conductivity in the Kapuskasing uplift. However, some relation may exist if the conductivity observed today is partly due to graphite films precipitated at grain-boundaries during cooling right after the peaks of the metamorphic events. This alternative would become particularly attractive if rocks exposed in the Chapleau block from the deepest levels were shown to carry the broken relics of Archean graphite films.

More recently, dense MT profiling has also been undertaken along approximately E-W trending lines in the Chapleau and Groundhog River blocks (Kurtz et al., 1988; Mareschal et al., 1988; see locations on Fig. 3). While only preliminary results may be stated this stage, it seems fairly certain that the two blocks display similar variations of bulk conductivity with depth, and that little deviation from subhorizontality can be observed. Both blocks show a highly resistive upper crust (~ 10^5 Ωm), overlying progressively less resistive deeper levels. An approximately 4000 Ωm intermediate layer is followed by a more conductive zone in the ~100's Ωm range at lower crustal or upper mantle levels (more exact depth estimates are difficult to achieve at this stage). This more conductive zone appears to be anisotropic or of limited lateral extent beneath the Chapleau block. The lack of significant GDS response may support anisotropy, however, more work is needed before a final interpretation is proposed (e.g., Kurtz et al., 1988; 1989b).

So far, no significant relation seems to exist between lower crustal reflections and electrical conductivity in the Kapuskasing uplift (Green et al., 1989; Milkereit et al., 1989), an observation quite consistent with little fluid influence in the lower crusts of shields. Rather, the prominent reflectivity (exampled in the reflection zones of Fig. 7, but found at all levels in the Kapuskasing crust, e.g., Milkereit et al., 1989) are thought to originate from gently dipping zones of high strain (e.g., mylonites) and layered gneissic rocks (Green et al., 1989). The dominantly mafic composition suggested by Percival (1989) for the lower crust (Fig. 4) is supported by the high velocities observed in the seismic refraction survey (Boland et al., 1988). Thus, conditions may have been ideal for the precipitation of carbon films at grain boundaries during original cooling (Frost et al., 1989), possibly explaining lower crustal conductivities in the Kapuskasing uplift.

High resolution UTEM surveys in the Chapleau block (Kurtz et al., 1989a; Bailey et al., 1989) confirm the general electrical subhorizontality of the upper crust (Fig. 6), but the electrical image shows more features than the single resistive layer defined by MT.

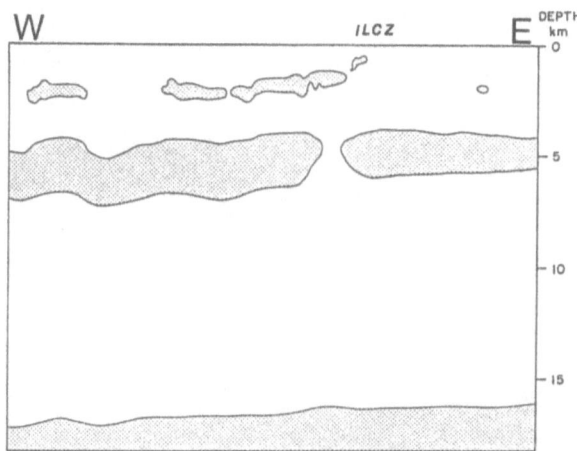

Fig.6: Electrical image of the Chapleau block as defined by UTEM. White background is extremelly resistive (10^6 Ωm), grey zones are slightly less resistive. Width is 25 km (Bailey et al., 1989).

464

For example, two relatively extensive zones of lower than average resistivity are observed in the very upper crust (Bailey et al., 1989). The first feature is a set of narrow disconnected conductors appearing at a depth of about 2 km. There is a small one to the west and an extensive central one apparently extending to the surface near the thrust fault (in the Ivanhoe Lake cataclastic zone; ILCZ). However, processing of UTEM data, like that of conventional seismic reflections, assumes shallow dips. Thus, high dip features may not be reliably rendered. The second feature is a narrow quasi continuous band at about 5 km with a gap approximately in the vicinity of the ILCZ. The conductivity enhancement at ~ 15 km corresponds to the intermediate MT layer.

Neither shallow feature is very conductive, in fact, both are simply less resistive than the very resistive background (this is why they are not observed with MT). Bailey and co-authors show that if this slight variation in resistivity is due to the presence of brines (which are commonly observed in the top two kilometers of the Canadian Shield; e.g., Frape and Fritz, 1987), the connected porosity should be on the order of .5%. They also contend that since no obvious correlation exists between the electrical features and the lithology as it is apparently defined by the seismic reflections, the porosity

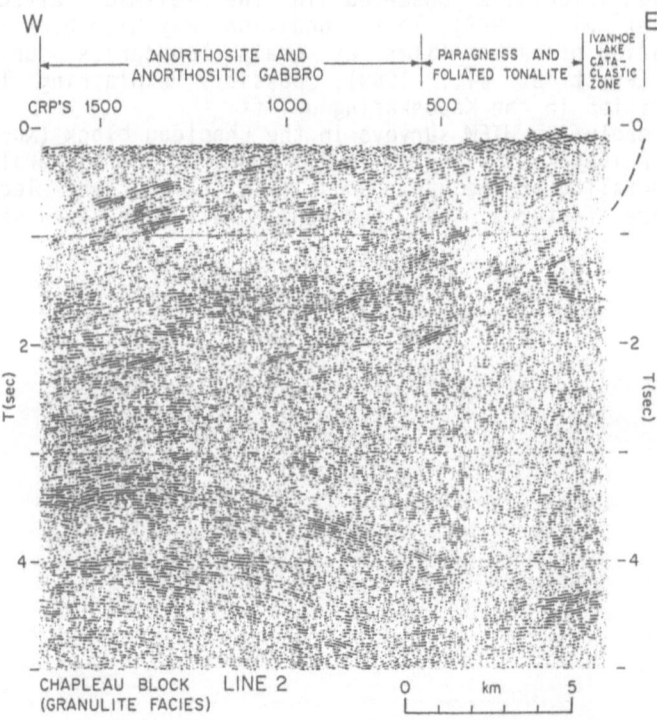

Fig.7: High-resolution seismic reflections along the UTEM survey line (west of ILCZ) in the Chapleau block. Possible thrust zone is extrapolated to surface (dashed line) (Green et al., 1989).

should postdate thrusting (the seismic section shown in Fig.7 and the left section of the UTEM image (left of ILCZ) shown in Fig.6 were recorded along the same line; notice that the scales are different). They suggest that any water with access to dipping conduits such as might be associated with the thrust fault would escape, possibly leading to the observed disruptions at the fault location in the two conductive zones. However, they also point out that the UTEM data may be explained equally well by the presence of shallow subhorizontal lithologic boundaries, faults or detachment zones. They argue that this alternative could reasonnably apply to the 2 km conductor (which may bear some relation to the shallow dipping reflectors observed at 1 sec or less), but that it seems less plausible for the extensive and more horizontaly continuous ~ 5 km conductor.

Shallow MT responses and induction vectors recorded in the Groundhog River block are indicative of important lateral effects, possibly related to the existence of iron formations in the Saganash Lake Belt, a metasedimentary belt part of the Val Rita Block (Fig.3; Leclair and Poirier, 1989; Mareschal et al., 1988). The data also seem to respond to shallow sediments infiltrated with water in the near-surface expression of several faults thought to prevail in the region (Leclair and Poirier, 1989). A high resolution UTEM survey, planned for the automn of 1989, will help to resolve some of these shallow features. In particular, it will attempt to verify possible correlations between shallow dipping reflectors (Milkereit et al., 1989) and upper crustal conductors (Mareschal et al., 1988).

Summary: Preliminary results of the electromagnetic investigations recently undertaken in the Kapuskasing uplift confirm the high resistivities of exposed lower crustal rocks. The present lower crust is moderatly conductive (reaching resistivities possibly as low as 100 Ωm). However, the lack of correlation between seismic reflectors and conductors at these levels suggests that fluids should only play a very minor role in explaining the enhancement of conductivity. The present upper crust, on the other hand, is extremely resistive. Two slightly less resistive features have recently been discovered at 2 and 5 km by a high-resolution UTEM survey in the Chapleau block. While the lower feature is thought to respond to brines, the upper one could be explained equally well by the presence of shallow subhorizontal lithologic boundaries such as faults or detachment zones. Shallow electrical responses are more complex in the Groundhog River block where they may be particularly affected by the presence of iron formations in the nearby Saganash Lake belt and/or by wet sediments accumulated in the upper expression of faults.

ACKNOWLEDGMENTS

I would like to thank R.C. Bailey, B.R. Frost, A. Green, R.D. Kurtz and J.A. Percival for their unconditional help. I am also grateful to Alex Brown for kindly editing the last version of this paper. This is Lithoprobe contribution # 104.

REFERENCES

BAILEY, R.C., CRAVEN, J.A., MACNAE, J.C., and POLZER, B.D.,1989. Imaging of deep fluids in Archean crust, Nature, in press.
BAYUK, E.I., BELIKOV, B.P., VERNIK, L.I., VOLAROVITCH, M.P., KUZNETSOV, Y.I., KUSMENKOVA, G.E., and PAVLOVA, N.N.,1984. In "The Superdeep Well of the Kola Peninsula", Y. Koziovsky et al., Eds., 332-337, Springer-Verlag.
BOLAND, A.V., ELLIS, R.M., NORTHEY, D.J., WEST, G.F., GREEN, A.G., FORSYTH, D.A., MEREU, R.F., MEYER, R.P., MOREL-A-L'HUISSIER, P., BUCHBINDER, G.R., ASUDEH, R.A., and HADDON, W., 1988. Seismic delineation of upthrust Archean crust in Kapuskasing, Northern Ontario, Nature, 335, 711-713.
BRACE, W.F., and ORANGE, A.S., 1968. Further studies of the effects of pressure on electrical resistivity of rocks, J.G.R., 73, 5407-5420.
COLVINE, A.C., FYON, J.A., HEATHER, K.B., MARMONT, S., SMITH, P.M., and TROOP, D.G., 1988. Archean lode gold deposits in Ontario. Part I. Depositional model., OGS Miscellaneous paper 139.
CRAMPIN, S., and ATKINSON, B.K.,1985. Microcracks in the Earth's crust, Leading Edge, First Break,3, 16-20.
DUBA, A., 1976. Are laboratory electrical conductivity data relevant to the earth?, Acta Geodaet., Geophys. et Montanist. Acad. Sci. Hung., 11, 485-495.
DUBA, A., and SHANKLAND, T.J., 1982. Free carbon and electrical conductivity in the earth's mantle, Geophys. Res. Lett., 9, 1271-1274.
DUBA, A., HUENGES, E., NOVER, G., WILL, G. and JODICKE, H., 1988. Impedance of black shale from Munsterland 1 borehole: an anomalously good conductor?, Geophys. J., 94, 413-419.
ETHERIDGE, M.A., WALL, V.J., and VERNON, R.H., 1983. The role of the fluid phase during regional metamorphism and deformation, J. Metamorph. Geol., 1, 205-226.
FOUNTAIN, D.M., 1989. Seismic properties of the lower continental crust - A review., CGU meeting, Montreal, May 1989.
FRAPE, S.K. and FRITZ, P., 1987. Geochemical trends for groundwaters from the Canadian Shield, In " Saline Water and Gases in Crystalline Rocks", Geol. Ass. of Canada, Special Paper 33, 19-38.
FROST, B.R., FYFE, W.S., TAZAKI, K., and CHAN, T., 1989. Grain-boundary graphite in rocks from the Laramie Anorthosite Complex: implications for lower crustal conductivity, Nature, in press.
GOUGH, D.I., 1986. Seismic reflectors, conductivity, water and stress in the continental crust, Nature, 323, 143-144.
GREEN, A.G., MILKEREIT, B., MAYRAND, L., SPENCER, C., KURTZ, R., and CLOWES, R.M., 1987. Lithoprobe seismic reflection profiling across Vancouver Island: results from reprocessing, Geophys. J. R. Astr. Soc., 89, 85-90.
GREEN, A.G., MILKEREIT, B., PERCIVAL, J., DAVIDSON, A., PARRISH,R., COOK, F., GEISS, W., CANNON, W., HUTCHINSON, D., WEST, G., and CLOWES, R., 1989. Origin of deep crustal reflections: results from

seismic profiling across high-grade metamorphic terranes in Canada, Tectonophysics, in press.

HAAK, V., 1980. Relations between electrical conductivity and petrological parameters of the crust and upper mantle, Geophys. Surveys, 4, 57-69.

HAAK, V., and HUTTON, R., 1986.Electrical resistivity in continental lower crust, in "The Nature of the Lower Continental Crust", J.B. Dawson et al., Eds., Geol. Soc. Lond. Spec. Pub. 24, 35-49.

HYNDMAN, R.D., 1988. Dipping seismic reflectors, electrically conductive zones, and trapped water in the crust over a subducting plate, JGR, 93, 13,391-13,405.

HYNDMAN, R.D., and KLEMPERER, S.L., 1989. Lower-crustal porosity from electrical measurements and inferences about composition from seismic velocities, Geophys. Res. Lett., 16., 255-258.

HYNDMAN, R.D., and SHEARER, P.M., 1989. Water in the lower continental crust: modelling magnetotelluric and seismic reflection results, Geophys. J., in press.

JONES, A.G., 1987. MT and reflection: an essential combination, Geophys. J. R. Astr. Soc., 89, 7-18.

KURTZ, R.D., NIBLETT, E.R., CRAVEN, J.A., STEVENS, R.A., and MACNAE, J.C., 1988. Electromagnetic studies over the Kapuskasing Structural Zone, in "Lithoprobe Workshop KSZ'88", G. West, Ed., 169-176, Toronto.

KURTZ, R.D., MACNAE, J.C., and WEST, G.F., 1989a. A controlled-source, time-domain electromagnetic survey over an upthrust section of Archean crust in the Kapuskasing Structural Zone ,Geophys. J., in press.

KURTZ, R.D., JONES, A.D., and GREEN, A.G., 1989b. Long period MT, GDS, and horizontal spatial gradient (HSG) data from the Kapuskasing window and comparison with seismic reflection and refraction results, IAGA Symposium I9, Exeter, July 1989.

LECLAIR, A.D., and NAGERL, P.,1988. Geology of the Chapleau, Groundhog River and Val Rita Blocks, Kapuskasing area, Ontario, in " Current Research, part C", Geological Survey of Canada, Paper 88-1C, 83-91.

LECLAIR, A.D., and POIRIER, G.G., 1989. The Kapuskasing uplift in the Kapuskasing area, Ontario, in "Current Research, Part C", Geological Survey of Canada, Paper 89-1C, 225-234.

MARESCHAl, M., CHAKRIDI, R., and CHOUTEAU, M., 1988. A magnetotelluric survey across the Groundhog River Block: progress report on the pseudo 1-D interpretation, in "Lithoprobe Workshop II KSZ'88-9", G. West, Ed., 77-72, Toronto.

MATTHEWS, D.H.,1986. Seismic reflections from the lower crust around Britain, in "The Nature of the Lower Crust", A. Dawson et al., Eds., Geol. Soc. Lond. Spec. Pub.,24, 11-22.

MATHEZ, E.A., and DELANEY, J.R., 1981. The nature and distribution of carbon in submarine basalts and peridotite nodules, Earth Planet. Sc. Let., 56, 217-232.

MILKEREIT, B., GREEN, A., COOK, F., and WEST, G., 1989. Lithoprobe seismic reflection profiles across the Kapuskasing Structure, Geological Survey of Canada, Open File Report, in press - June 1989.

NKWATE, E.A. and SALISBURY, M.H., 1988. New gravity mapping of the Kapuskasing structure in the Val Rita and Groundhog blocks: preliminary results, in "Lithoprobe Workshop II KSZ 88'9", G. West, Ed., 55-64, Univ. of Toronto.

OLHOEFT, G.R., 1981. Electrical properties of granite with implications for the lower crust, J.G.R., 86, 931-936.

PERCIVAL, A.J., 1989. Archean tectonic setting of granulite terranes of the Superior Province, Canada: aview from the bottom, in NATO-ASI volume on "Granulites and crustal differentiation, D. Vielzeuf, Ed., Kluwer Acad. Publ., in press.

PERCIVAL, J.A., and CARD, K.D., 1983. Archean crust as revealed in the Kapuskasing uplift, Superior Province, Canada., Geology, 11, 323-326.

RAO, R., and JESSOP, A.M., 1975. A comparison of the thermal characteristics of shields, Can. J. Earth. Sc., 12, 347-360.

SCHMUCKER, U.,1988. Regionally stable telluric directions, presented at the 9th Workshop on Electromagnetic Induction in the Earth and Moon, Sochi, URSS.

SHANKLAND, T.J., and WAFF, H.S., 1977. Partial melting and electrical conductivity anomalies in the upper mantle, J.G.R., 82, 5409-5417.

STESKY, R.M., and BRACE, W.F., 1973. Electrical conductivity of serpentinized rocks to 6 Kbars, JGR., 78, 7614-7621.

TOURET, J., 1971. Le facies granulite, Norvège Méridionale: II Les inclusions fluides, Lithos, 4, 423-436.

WOODS, D.V., and ALLARD, M., 1986. Reconnaissance electromagnetic induction study of the Kapuskasing Structural Zone: implications for lower crustal conductivity, Phys. Earth. Planet. Int., 42, 135-142.

DEFORMATION SEQUENCE IN THE SOUTHEASTERN KAPUSKASING STRUCTURAL ZONE, IVANHOE LAKE, ONTARIO, CANADA.

J.T. Bursnall
Geophysics Laboratory
Department of Physics
University of Toronto
Toronto
Ontario, M5S 1A1,
Canada

ABSTRACT. The southeastern part of the Kapuskasing Structural Zone has a complex deformational history that can be subdivided into eight phases. These are grouped into an Archean set, accompanied by high-pressure granulite metamorphism and migmatization representing deep-seated conditions within the Archean crust, and a second set that developed during and following the progressive uplift and emplacement of the high-grade rocks during the early to mid-Proterozoic. Ductile conditions and high strains occurred in the Archean phases whereas a progressive ductile to brittle transition is exbibited by the uplift-related structures.

1. Introduction

New exposures within a recently developed logging area between Foleyet and Chapleau, Ontario, provide improved coverage across the high-grade deep crustal Archean gneisses of the Kapuskasing Structural Zone (KSZ: Thurston et al., 1977; Percival, 1981a, b) and their faulted boundary with lower-grade rocks of the Abitibi Subprovince (Fig. 1, 2). This fault zone (the Ivanhoe Lake Fault Zone) encompasses a number of major discrete faults and subsidiary narrow brittle to brittle-ductile shear zones and occurs within a belt of intense fracturing known as the Ivanhoe Lake Cataclastic Zone (ILCZ; Percival, 1981a), which developed during Proterozoic uplift and SE-directed thrust emplacement of the 25-30 km deep Kapuskasing high-grade gneisses (Percival and Card, 1983, 1985; Cook, 1985; Geis et al., 1988). This report describes the complex sequential structural evolution of these multiply deformed Archean rocks, as presently understood, and the structure of their contact zone with the Abitibi belt to the southeast. Preliminary work within the area indicates at least eight distinct stages that may be regionally significant in the structural evolution of these sequences, and also that some of the youngest ILCZ structures are most likely related to post-emplacement faulting, which in detail has significantly modified and displaced the earlier structures. The present surface

M. H. Salisbury and D. M. Fountain (eds.), Exposed Cross-Sections of the Continental Crust, 469–484.

geology of the Ivanhoe Lake boundary zone may not, therefore, provide a true representation of the nature of the emplacement shear system that presumably exists at depth beneath the Kapuskasing plate and which may have been identified as strong reflectors on a LITHOPROBE high resolution seismic reflection survey (Geis et al., 1988; Percival, this volume).

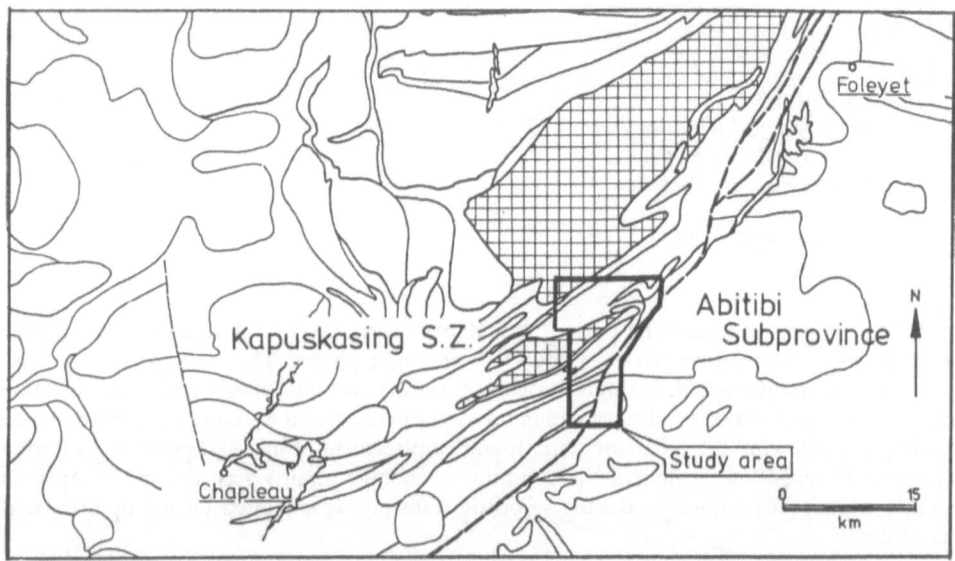

Figure 1: Location of study area. Ivanhoe Lake fault zone indicated by heavy dashed line. Shawmere anorthosite complex ornamented. Geological contacts taken from Percival and Card (1985)

2. Lithology

2.1. KAPUSKASING STRUCTURAL ZONE

The KSZ within the mapped area encompasses a compositionally wide range of rock-types disposed in discontinuous east-west belts that are seemingly slightly deflected towards the northeast as the Ivanhoe Lake fault zone is approached (Fig. 1). The lithological units of the present study, and used in the accompanying map (Fig. 2), are essentially those of Percival (1981a) and represent of the dominant rock types in that zone. Most lithologies are medium- to coarse-grained gneisses that typically have a lenticular to moderately persistent compositional layering which varies in thickness from a few millimetres to 50 cm or more. This regional planar fabric is in part made up of rotated and transposed leucosome veinlets in addition to fine lenses of aplitic leucosome segregations developed parallel to compositional layering. Foliation fish of the type described by Hanmer (1986) and low angle discontinuities are locally common, particularly within east-west trending late-

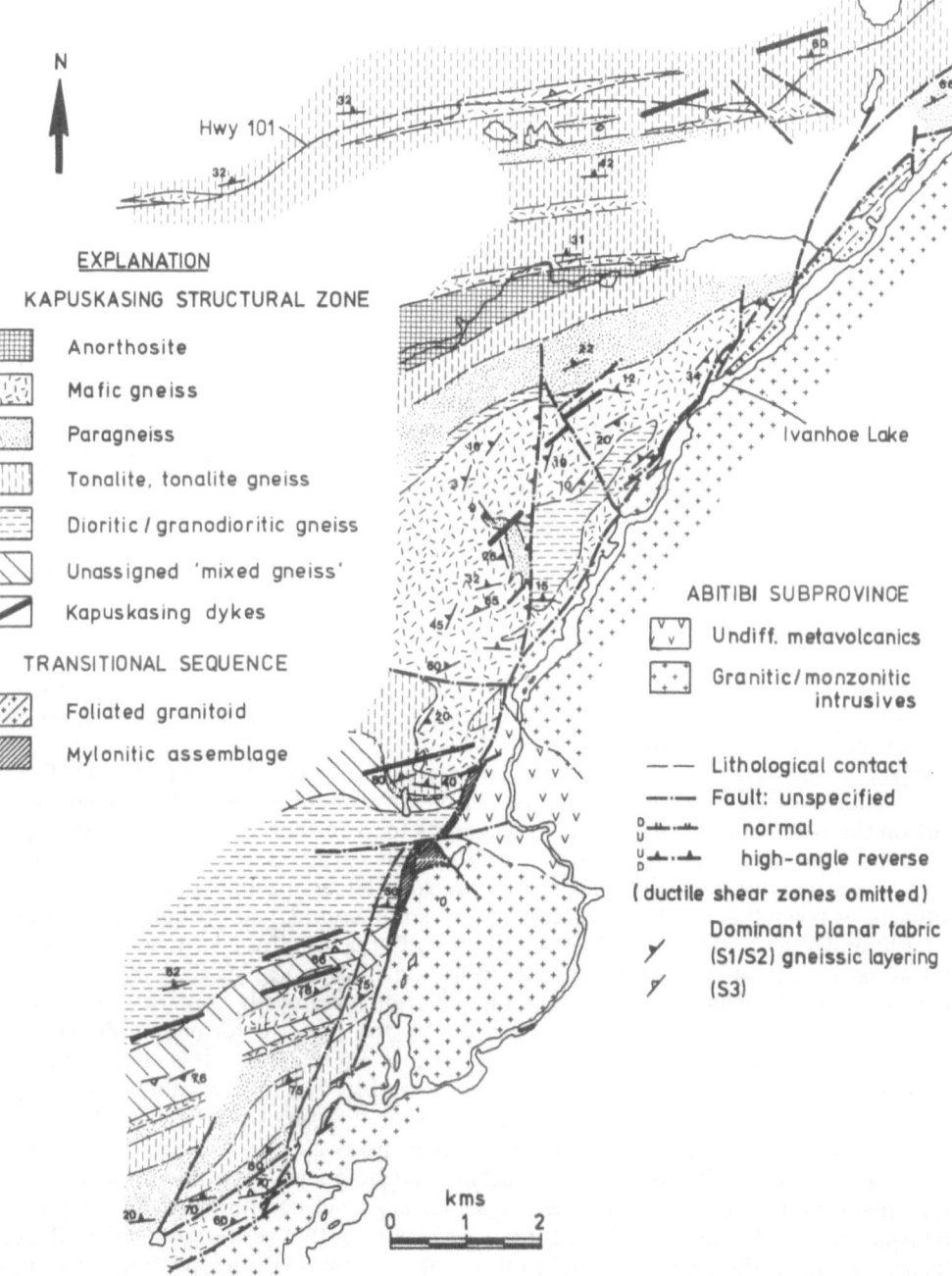

Figure 2: Geological map of the study area. Unornamented areas are unmapped.

metamorphic high strain zones that occur at intervals throughout the section; these may be recognized by slightly finer grain size than the contiguous gneisses, apparently much thinned and regular layering, and the presence of relict feldspar 'porphyroclasts' and mineral aggregate augen. The lenticular character of the gross compositional layering is also enhanced through boudinage and necking where ductility contrast is high (for example, within mixed mafic gneiss with tonalite). Granoblastic textures dominate so that pronounced preferred dimensional grain orientations are relatively rare: exceptions occur within late-metamorphic high strain zones (late-D$_3$ age in the structural sequence, see below) and as locally pronounced syn-metamorphic mineral lineations.

Tonalitic gneiss, granodioritic to dioritic gneiss, paragneiss (including polymict conglomerate), 'mixed gneisses', and a variety of mafic gneisses are all present (Fig. 2). Common occurrence of clinopyroxene within garnetiferous mafic gneiss, ortho-pyroxene within paragneiss, two-pyroxene mafic gneisses within the southern and easternmost (Percival, 1981a, 1983) part of the map area, indicate that uppermost amphibolite to granulite facies conditions accompanied the early tectonic evolution of the area. U-Pb zircon ages for high-grade metamorphism and coeval widespread migmatization are within the range ca. 2700 to ca. 2600 Ma (Percival and Krogh, 1983; Percival, this volume). Retrograde assemblages are mainly present in areas affected by the ILCZ, where chlorite and epidote are major constituents of the microshear zones.

Well-foliated <u>anorthosite</u> containing fragments of layered anorthositic gabbro is exposed along Hellyer Creek, 2 km south of Highway 101. The highly strained envelope rocks converge eastwards at this locality indicating the termination of a 15 x 5 km body that is considered part of the Shawmere anorthosite complex (Fig. 1: Thurston et al., 1977; Percival, 1981a, b): whether this smaller body represents a tectonically isolated segment or was originally independent of the main complex 8 km to the north is not known.

<u>Mafic gneiss</u> occurs in three distinct easterly to northeasterly trending belts. The most extensive occurs in the central part of the area about 3 km south of Highway 101 on the Aubé main haulage road and extends for a further 6 km to the south (Fig. 2). Here they structurally underlie a narrow belt of metasediments, varieties of dioritic gneiss, and homogeneous to weakly foliated quartz-feldspar gneiss. The superficial resemblance of this assemblage to some supracrustal lithologies within Abitibi and Wawa greenstone belts emphasises the possibility of a volcanic origin for at least some of of the mafic rocks within the KSZ, and that they may in part represent relict enclaves of supracrustal material tectonically and magmatically incorporated into the lower Archean crust during the earlier deformation phases.

The <u>metasedimentary rocks</u> contain polymict conglomerate lenses within inter-layered pelite and psammite (Fig. 3a), which resemble, and are probably equivalent to, those at Borden Lake (Percival,1981a, b; Moser, 1989) 35 km to the west. Clast compositions are broadly representative of local lithologies and similar to those in metaconglomerate at Borden Lake: quartzite, dioritic gneiss, quartz-feldspar gneiss, rare mafic and amphibolitic gneiss are set in a garnet-amphibole-biotite-quartz-feldspar matrix. Some felsic clasts contain quartz veins that seemingly predate their incorporation in the conglomerate; no unambiguous pre-existing foliation exists in any of the clasts. Other likely metasedimentary rocks, based on mineralogical

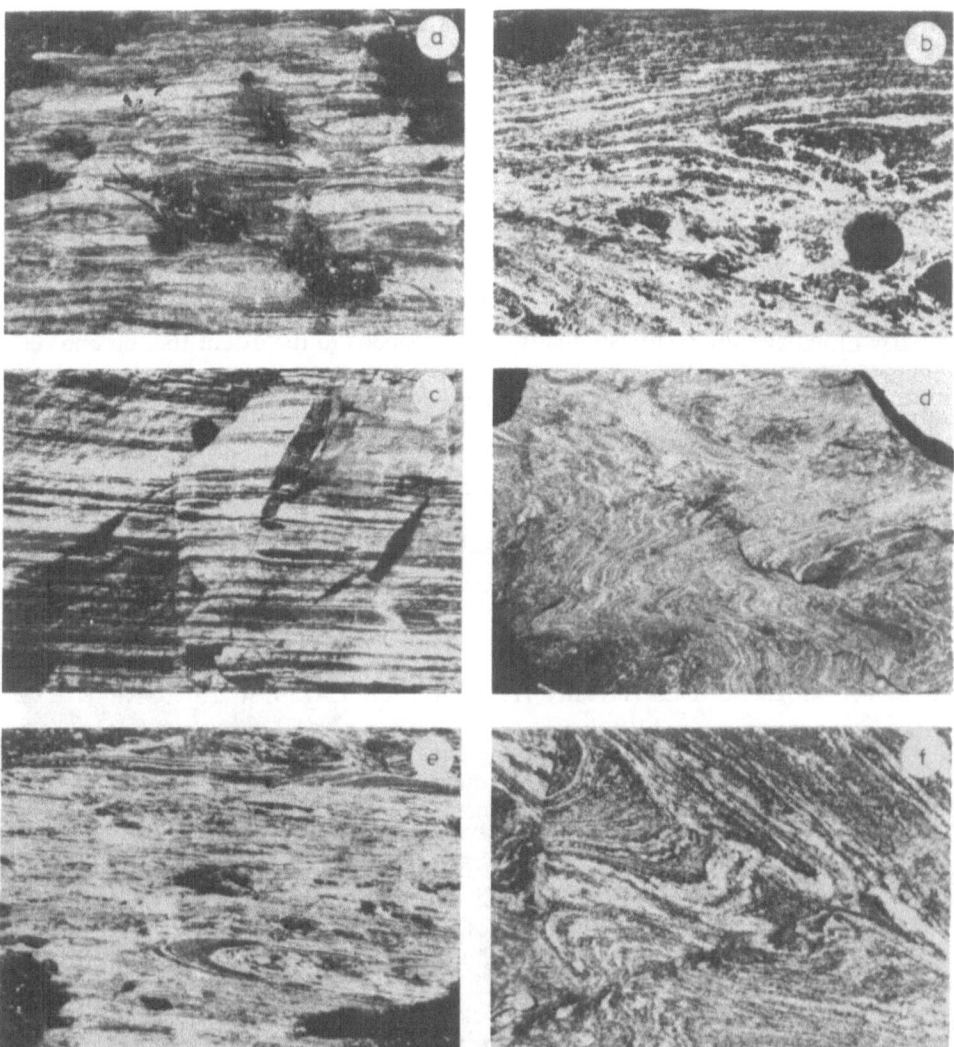

Figure 3: (a) Polymict conglomerate, central mafic gneiss belt. Large clast at top left is about 30 cm long. (b) Tonalite gneiss with mafic inclusions. Protomylonitic shear trends from bottom right to central left: exhibits internal structure suggesting dextral displacement. Lens-cap ~5 cm. (c) Strongly deformed inclusions in tonalitic gneiss. Field of view ~1.5 m. (d) D_3 folds in granodioritic gneiss. Note mafic boudin train to right. Field of view ~2 m. (e) Reclined intrafolial D_2 fold of mafic gneiss. Lens-cap bottom centre. (f) D_3 folds and fold interference patterns in mafic gneiss. D_3 axial surface traces oriented top left to bottom right. Hand-lens bottom left.

composition, occur in a number of additional belts and are included under 'paragneiss' in Figure 2.

Tonalite occurs as extensive areas, localized homogeneous patches or as thin aplitic to pegmatitic veins and clots that permeate the mafic gneisses and other lithologies and is almost ubiquitous. Tonalite generation occurred prior to the main, second phase, deformation (D_2) and continued at least to the early part of the third phase (D_3). Large tracts of tonalitic gneiss, in places crowded with mafic inclusions, are particularly common in the northern part of the area, straddling Highway 101, and through dramatically thinned inclusions qualitatively demonstrates the very high cumulative strains that obtained during the syn-metamorphic deformation phases (Fig. 3b, c).

'Mixed gneisses' are compositionally heterogeneous to the extent that no one rock type predominates. Typically, however, they are composed of interlayered and approximately equal volumes of mafic gneiss, tonalite and paragneiss that cannot be separated at map scale. Characteristically they grade into adjacent compositionally dominant units and may locally be a product of unit boundary transposition.

Rare pre- to syn-metamorphic mafic sheet intrusives, now severely disrupted and recognizable as trains of sub-angular boudins, have been observed within all KSZ units except the dioritic gneisses and weakly foliated tonalite in the southern part of the area (Fig. 2, 3b, d). Tonalitic to granitic pegmatites are common and were emplaced at various times throughout and closely following high-grade metamorphism.

2.2. ROCK ASSEMBLAGES WITHIN THE IVANHOE LAKE FAULT ZONE

A number of rock types and rock assemblages are peculiar to the area within and close to the Ivanhoe Lake fault zone; they are referred to as the 'Transitional Sequence' in Figure 2. Because of their location they are all typically disrupted by the pervasive fracturing and small-scale ductile shears of the ILCZ that exist in this zone. In a number of places they are intermediate in structural/textural and compositional character between the KSZ and the Abitibi belt and their status is therefore somewhat ambiguous: two of these are described below.

An assemblage of highly strained mylonitic mafic and felsic rocks that in part seem to be tectonically reconstituted KSZ mafic gneiss and in part Abitibi mafic metavolcanics are contained in a fault block within the Ivanhoe Lake fault. At this locality a 50-75 m thick zone of fine-grained and finely laminated mafic and interlayered quartzofeldspathic material locally contain and are deflected around relict small lozenges of coarse-grained mafic gneiss. These fine-grained rocks are strongly mylonitic and it is apparent that the mafic gneiss inclusions were degraded at their boundaries: it therefore seems that the mafic material may be wholly derived by the transformation of Kapuskasing mafic gneisses. In close proximity across unexposed ground and on strike with them are outcrops of regularly layered and somewhat flaggy medium-grained amphibolite, identical to Abitibi mafic rocks 1 km to the southeast, and suggesting that there may have been some tectonic interleaving between the two assemblages.

Reconnaissance mapping along the southern section of Ivanhoe Lake indicates that there may not be an obvious compositional break between KSZ gneisses and the

Abitibi subprovince granitoid plutons as is present elsewhere in the map area. In this section, where the Ivanhoe Lake Fault Zone obliquely traverses the broad central tract of KSZ mafic gneiss a number of seemingly anomalous and equivocal features exist. As this contact is approached from the northwest thin (a few centimetres to about five metres) concordant medium-grained felsic tonalitic to quartz monzonitic sheets increase in abundance and the mafic rocks locally are medium to fine grained (but seemingly maintain a mineralogical composition similar to their coarse-grained equivalents). Contiguous (but nowhere with observed contact) coarse-grained and gneissic 'quartz monzonite' contains abundant mafic schlieren and is superficially similar in composition to structurally isotropic quartz monzonite southeast of Ivanhoe Lake, which contains rare foliated mafic xenoliths. At one locality fine-grained mafic 'gneiss' with clinopyroxene porphyroblasts is separated from weakly foliated quartz monzonite by a screen of pegmatite. A metre-sized xenolith of similar mafic material contains a pre-inclusion tight fold and has been refolded during mild deformation of the host.

The precise location of the KSZ-Abitibi subprovince boundary in this area is therefore unclear. In addition to the possibility of tectonic interleaving of the two assemblages within the fault zone, the overall 'transitional' nature of these rocks could result from: 1) grain reduction and mild retrogression of KSZ mafic and associated gneisses or, 2) the involvement of deeper level Abitibi rocks in the fault zone

The simplest explanation for these relationships seems to be that the mafic rocks represent moderately high-grade Abitibi metavolcanics, intruded by granitic material, and subsequently transported as a fault-bounded slice to shallower depth during emplacement of the KSZ: the foliation within the granitoid host could have been produced at this time, or it may be a relict of an earlier event. Two subparallel faults defining the Ivanhoe Lake fault zone to the northeast that may coincide with two strong reflectors on a LITHOPROBE high resolution seismic reflection line (Geis et al., 1988; Percival, this volume) enclose an amphibolite grade package of foliated granitoids and other rocks (Percival, 1981b). Since the magnetic signature of these faults may be traced southwestward into the study area using shadow-enhanced aero-magnetic compilation maps (G. West, personal communication 1988), and enclose the area described above, a fault slice origin of deeper Abitibi crust is further supported. The increase in felsic material within the KSZ with structural depth at this locality and elsewhere is not yet fully understood.

2.3. ABITIBI SUBPROVINCE

Layered amphibolite and medium- to fine-grained felsic rocks within the Abitibi metavolcanic belt outcrop along the Ivanhoe River in the extreme south of the map area (Fig. 2). These are thinly interlayered in places and are intruded by granite/quartz monzonite. The mafic lithologies vary texturally and compositionally from fine- to medium-grained true amphibolite to plagioclase rich varieties and are strongly foliated: garnet is rarely present in schistose amphibolite close to the boundary with the KSZ and clinopyroxene has been identified in presumed Abitibi amphibolite 10 km to the northeast (Percival, 1986). Elsewhere the Abitibi contact rocks are granitic and include quartz monzonite of the Biggs Lake Pluton.

2.4. DYKES

A variety of post-metamorphic mafic dykes and granitic pegmatites occur within the field area. The most common are lamprophyric but rare sparsely plagioclase porphyritic types of the Kapuskasing suite (Card et al., 1981) are present throughout the map area (Fig. 2): trends are variable but fall in the range N40E-N88E and thicknesses are typically 10 to 15 metres. One northwest-trending mafic dyke was observed within quartz monzonite south of the Ivanhoe Lake fault zone. Only the lamprophyres were observed to cut structures related to the ILCZ (a similar lamprophyre 15km to the northeast gave a ca.1140ma 40Ar/39Ar age, M.Queen, personal communication).

Relict pre-metamorphic dykes may be represented by boudinaged and folded mafic sheets that are subparallel to or cross-cut gneissic layering at a low angle (e.g., Fig. 3d).

3. Structural History

Structures within this area may be considered as two temporally and geometrically distinct groups that were developed under distinctly different mechanical conditions. Those that pre-date or developed during or soon after the pervasive late Archean high-grade metamorphism and associated migmatization characteristically result from highly ductile behaviour of the rock mass and demonstrate the effects of complex multiple folding and locally very high strain levels which resulted in structural complexity of the mid- to lower part of the Archean crust in this region (Fig. 3).

The second group of structures post-dates high-grade regional metamorphism and includes those produced during and following the Kapuskasing uplift. These structures commonly exhibit progressive ductile to brittle transition in any one outcrop - from narrow (< 30 cm) protomylonite-mylonitic zones, typically at a low angle to gneissic layering of the host rock, to variably oriented cataclasites and and faults on all scales, in addition to minor folding: rare cross-cutting pseudotachylite veinlets also occur. Cumulative displacement along the emplacement surfaces is not known but must be at least 25-30 km, based on assumed slab dip from LITHOPROBE seismic reflection experiments (Cook, 1985; Geis et al., 1988; Green 1988a, b) and the required pressure difference from the contrasting metamorphic grade across the Ivanhoe Lake fault system (Percival, 1983). These structures reach their greatest expression within a few hundred metres of the Ivanhoe Lake fault zone but are present up to a distance of several kilometres from the fault trace. Their age is uncertain but the most recent modifications of this zone affect ca. 1140 Ma lamprophyric dykes (although it is possible that a suite of ca. 1800 Ma lamprophyres also occurs in this area (J.Hanes, personal communication, 1988)). Mylonite from the ILCZ produced a 40Ar/39Ar date of 1.72 Ga (Percival, 1981a) and a model based on isobaric cooling followed by rapid uplift suggests a ca. 1.95 Ga age for the lower bracket to the main uplift event (Parrish, 1987; Percival, this volume). Kapuskasing dykes, which are affected by uplift structures (Percival, 1981a) and locally invaded by pseudotachylite elsewhere (A. Leclair, personal communication)

TABLE I: Summary of Deformation Sequence:

Phase 1 (D_1)   Gneissic layering developed following some anatexis: minor folds of uncertain origin.

Phase 2 (D_2)   Development of small-scale folds, now tight to isoclinal and of variable trend (Fig. 3e, 5b); Locally very high strains and strong planar fabric & mineral lineation; Continued migmatization and pegmatite generation.

Phase 3 (D_3)   Early: tight gently inclined folds about moderately northward-dipping axial surfaces (Fig. 3d, f, 5c); associated weak fabric and mineral and intersection lineation.
Late: ductile shear zones with local mild retrogression; locally strong grain dimensional planar fabric (Fig. 4a).

Phase 4   Early decoupling stage of KSZ and initiation of ILCZ; presumed relative age of development of mylonitic and reconstituted mafic gneiss sequence within the Ivanhoe Lake Fault Zone (Fig. 4b, c)

Phase 5   Development of the ILCZ. Widespread production of small-scale mylonitic shear zones and planar fabrics succeeded by cataclasis and invasion of cataclasite veins and rare development of pseudotachylite (Fig. 4d).

Phase 6-7*   NW-trending gentle to open large-scale upright to moderately inclined horizontal folds in part related to east-directed high angle reverse faults.

Phase 6-7*   NE-trending, upright to moderately inclined, gently plunging folds adjacent to Ivanhoe lake. Normal faulting (relative down-throw to NW) of unknown extent along KSZ-Abitibi boundary possibly related.

Phase 8   Post-lamprophyre faulting.

*Note: Evidence for relative age of some structures assigned to Phase 6 and Phase 7 is equivocal.

were likely intruded at ca. 2040 Ma based on $^{40}Ar/^{39}Ar$ studies by Hanes et al. (1988). The younger age limit is seemingly constrained by U-Pb zircon ages of 1907 and 1888 Ma for the Cargill carbonatite complex that cuts faults related to the uplift to the north of the study area (Percival, this volume). Details of the deformation sequence so far determined for the KSZ are given below and summarized in Table I.

3.1. THE EARLY DEFORMATION PHASES (ARCHEAN)

The essentially syn-metamorphic Archean events, here designated D_1-D_3, collectively imparted a regionally pervasive strong planar compositional anisotropy, gneissic layering, and resulted in the dominant east-west regional trend of lithological units, dominant fabrics and low dips (Fig. 2, 5a, b)).   The earliest structures recognised in the map area are rare fold-like structures of uncertain origin and a stromatic gneissic layering within mafic inclusions in tonalitic gneiss; both are assigned to the D_1 phase. Other compositional layering within mafic gneiss may be a relict of primary compositional variation.   These structures are succeeded by two systematic fold sets. The earliest (D_2) are small-scale, north- to northest-trending, tight to isoclinal folds of gneissic layering and have low to moderate plunge; they are typically reclined (Fig. 3e).   Attendant axial surface fabrics are uncommon but many are true intrafolial folds, indicating that, locally at least, the prominent gneissic layering in outcrop is predominantly a secondary (D_2) composite structure, the earlier, folded, gneissosity being assigned to the D_1 phase.

Generally high ductile strains occurred during the second deformation phase and were accompanied or closely followed by peak metamorphism. Seemingly multiple invasion of granitoid veins, anatectic leucosome, and pegmatite that pre-date third-phase structures suggests that the structures ascribed to to D_2 could be further subdivided. Boudinage of early mafic dykes (Fig. 3d), and a locally strong mineral elongation lineation occurred during this period.

The third deformation phase (D_3) represents the latest of the high-grade deformations and may be subdivided into two stages. Early-D_3 folds are tight with gentle to moderately inclined north- to northwest-dipping axial surfaces and moderate to low west to northeast plunge (Fig. 3d, 5): a subparallel mineral lineation and, more commonly, a rodding lineation is present. They are seen to refold second phase or earlier folds (Fig. 3f) whose former variable trend is indicated by a range of interference patterns over short outcrop distance; a poorly developed axial planar fabric is developed in places. These folds locally possess curved hinge-lines that are thought to have developed at least in part during a period of pervasive ductile shearing that is tentatively placed in the structural sequence as late-D_3 in age (Fig. 4a): there is no evidence to suggest that a major interval separates the two sets of structures. These zones are best displayed along Highway 101 close to the Aubé logging road entrance. Some appear to have developed through the continued rotation and thinning of gently NNW-dipping long limbs of the D_3 folds, without having disrupted the continuity of the folded fabric. Elsewhere, however, zone boundaries are extremely abrupt indicating very high strain gradients (Fig. 4a). A strong grain dimensional planar fabric within the zones is a common feature and is enhanced by subparallel quartz lenticles. Syn- to post-kinematic garnet growth occurred within mafic gneiss and the zones are locally cut by pegmatite. The zones therefore most likely developed during the later stages of regional metamorphism.

Approximate cumulative local strain magnitude for the Archean phases are given by axial ratios of prolate clasts in metaconglomerate that exceed 12:1, and high axial ratios for deformed mafic inclusions in tonalitic gneiss that exceed 20:1 in sections close to the principal finite extension direction. It is likely that even higher cumulative strains occur elsewhere within discrete high strain zones (both D_2 and

Figure 4: (a) Late-D$_3$ shear zone in mafic gneiss. Field of view ~1m. (b) Mylonitic mafic rocks from fault block within Ivanhoe Lake fault zone. North to right; prominent foliation is cut by light-coloured microshears (c) Early ductile shear in dioritic gneiss within ILCZ and 1 km from fault zone. (d) Pseudotachylite veins (dark grey) in granodioritic gneiss. Field of view ~1.5 m.

D$_3$ in age): such ductile shear zones may be concentrated within tonalitic host that has invaded mafic and other compositions of gneiss. Extensional ductile shears (strain bands) of late-D$_3$ age are also developed locally but do not seem to be so extensive or of such large-scale as those being investigated by Moser (1988, 1989) southwest of Chapleau.

3.2. THE LATER DEFORMATION PHASES (PROTEROZOIC)

Unequivocally post-metamorphic events (deformation phases 4 to 8) are likely lower to mid-Proterozoic and younger. The earlier of these (phases 4-5) evolved during the progressive uplift of the KSZ and exhibit distinct stages that may well represent a continuum of changing mechanical behavior, from ductile to brittle, in response to uplift; they have been subsequently modified by the later stages of movement within the Ivanhoe Lake Fault Zone. These effects are therefore somewhat restricted to the vicinity of the fault zone within the ILCZ and are concentrated within a moderately narrow (500 m) zone adjacent to the KSZ-Abitibi

480

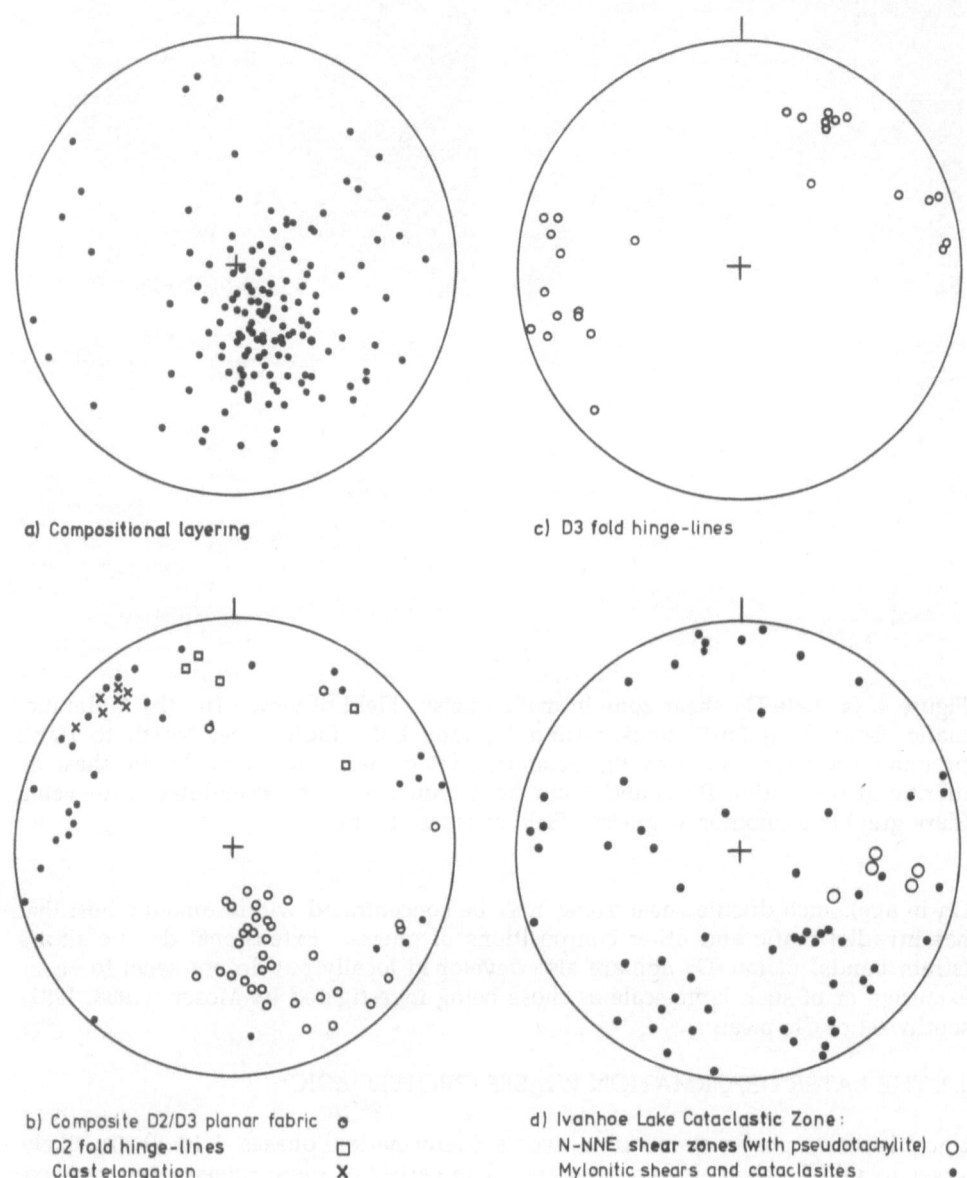

a) Compositional layering

c) D3 fold hinge-lines

b) Composite D2/D3 planar fabric ⊙
   D2 fold hinge-lines   □
   Clast elongation   ✗

d) Ivanhoe Lake Cataclastic Zone:
   N-NNE shear zones (with pseudotachylite) ○
   Mylonitic shears and cataclasites   ●

Figure 5: Equal-area orientation diagrams: (a) Compositional layering, whole are; (b) D$_2$ structural elements.; (c) D$_3$ fold hinge-lines; (d) Structures within the ILCZ.

contact. Here they consist of a close-spaced, ramifying network of fine to 1-2 cm wide ductile shears that in places develop a strong mylonitic fabric: these characteristically are curved and pinch and swell along their trace. Fine-scale shears marked by fine dark chloritic smears along compositional layer boundaries and parallel to other strong planar fabrics is a characteristic feature. These may thicken into black protomylonite and cataclasite zones (Fig. 3b) that may reach 20 cm in width; both sinistral and dextral displacement in the horizontal plane has been observed on mylonitic shears. These are typically cut by green epidotic cataclasite veinlets and both structures are displaced by microfaults. No regional systematic orientation for ILCZ structures has yet been determined but most dip steeply to the north and northwest.

The contact between Kapuskasing gneisses and Abitibi rocks (quartz monzonite, felsic and mafic metavolcanics) has been mapped to a few metres at a number of localities - but nowhere are contacts fully exposed. In the southernmost part of the map area, attitudes of layering and prominent fabric (gneissosity and cleavage) within the KSZ are moderately steep to subvertical with strikes ranging from highly discordant (east-west) to subparallel to the trace of the Ivanhoe Lake Fault Zone (Fig. 2); to the northeast, however dips are moderate to shallow and only slight discordance prevails. A steeply plunging mineral aggregate lineation exists in Abitibi granitoids (quartz monzonite?) within a few metres of the contact in the extreme south of the area: no unambiguous kinematic indicators were observed. It is apparent that the contact within the southern part of the section and subparallel faults 1 km to the west within the KSZ have been displaced dextrally along a number of northeast-trending faults (Fig. 2): the age of these is unknown but may equate with late shear zones that affect possible ca. 1140 Ma lamprophyre dykes within the Ivanhoe Lake Fault Zone elsewhere in the map area (see previous section on dykes).

Progressive ductile to brittle transition that may be exhibited by the sequential development of structures at any one locality may be viewed as response to the progressive uplift towards the southeast and unroofing of the KSZ plate. This would imply that most, if not all, structures falling in each of the recognized phases in a single outcrop are necessarily diachronous with depth, and also probably along the exposed length of the fault zone. It should be emphasized, however, that not all localities exhibit this ductile to brittle sequence and in many places field relations are equivocal. Episodic variations in uplift rate would explain the local transition from brittle to ductile shears and could explain some of the complexities present. The existence of pseudotachylite (Fig. 4d) at numerous localities would suggest rapid displacement at relatively shallow depths (Sibson, 1975, 1977).

Rocks representing the fundamental transport zone along which mid-crustal decoupling occurred have yet to be confirmed, but may be represented by the highly strained mylonitic mafic and felsic rocks within shear-bounded areas in the Ivanhoe Lake fault zone. A weak amphibole lineation is present within these and somewhat equivocal kinematic indicators suggest a southeast transport direction. The mylonitic lamination is cut by fine wispy microshears (Fig. 4b) and up to 3 cm-thick cataclasite veins that in places carry small lithic clasts. These in turn are cut by moderately north-dipping reverse faults and associated south-verging close folds. Later, northeast to east-west, probable strike-slip shear zones further dissect the sequence and deform coarse pegmatite and the lamprophyric dykes, representing the youngest observed structures within the zone.

It is impossible at the present mapping detail to determine the precise position in the structural sequence for some of the larger-scale structures that have been tentatively assigned to deformation phases 6 and 7. Two fold sets result in low amplitude dome and basin interference patterns on the regionally low northerly dips of the lithological units in the central part of the area and close to the Ivanhoe Lake Fault Zone (Fig. 2). One of these is responsible for slight deviations in regional strike along northwesterly-trending axes as the Ivanhoe Lake fault zone is approached. Minor subhorizontal northwest-trending folds with steep east-dipping axial surfaces and subparallel weak mica fabric within the metasediments of the central part of the map area may be of the same generation. They occur close to a steep, west-dipping reverse fault that contains narrow zones of foliated ultracataclasite invaded by pseudotachylite. The relationship of this fault to the pervasive cataclasite and thin mylonitic small-scale shear zones within this part of the ILCZ is not known.

A second set of northeast-trending, upright and gently plunging folds occurs close to Ivanhoe Lake: they fold cataclasites and mylonitic seams of the ILCZ and in part may be related to inferred normal faulting within the Ivanhoe Lake fault zone (Fig. 2).

4. Conclusion

The granulitic Kapuskasing Structural Zone has a long and complex deformation history, particularly close to its faulted boundary with the lower grade supracrustal/plutonic rocks of the Abitibi subprovince where repeated movement has resulted in considerable modification of the high grade Archean structures. The sequential structural development of the zone is still incompletely understood, particularly with respect to details of the precise timing of the uplift related events and the potential subdivision of the second phase Archean deformations. Regional structural correlations remain to be made. It is not entirely clear that the basal transport rocks of the Kapuskasing plate are represented by the mafic mylonitic sequence within the Ivanhoe Lake Fault Zone and the parentage of these and other 'transitional' rocks within the fault zone has yet to be demonstrated.

5. Acknowledgments

Thanks are due to J. A. Percival for his interest in suggesting and continued support during this project and for manuscript review, and to J.A. Hanes and D.Moser for additional discussions in the field. Funding for the 1988 season was through Hanes' Phase 2 Kapuskasing LITHOPROBE NSERC Grant at Queen's University, the receipt of which is gratefully acknowledged. D.Moser, J.A. Percival, and an anonymous reviewer are thanked for their corrections and suggestions for improvement to an earlier version of the manuscript.

6. References

Card, K.D., Percival J.A., LaFleur, J. and Hogarth, D.D., 1981: 'Progress Report on Regional Geological Synthesis, Central Superior Province'; *Geological Survey of Canada, Paper 81-1A*, p. 77-93.

Cook, F.A., 1985: 'Geometry of the Kapuskasing structure from a Lithoprobe pilot reflection survey'; *Geology*, **13**, p. 368-371.

Geis, W.T., Green, A.G., and Cook, F.A., 1988: 'Preliminary results from the high resolution seismic reflection profile: Chapleau block, Kapuskasing Structural Zone'; in *1988 KSZ Lithoprobe Workshop*, University of Toronto, p. 155-165.

Hanes, J.A., Archibald, D.A., Queen, M., and Lee, J.K.W., 1988: '40Ar/39Ar geochronology od diabase dykes: implications for the tectonothermal evolution of the Kapuskasing uplift'; in*1988 KSZ Lithoprobe Workshop*, Geophysical Laboratory, University of Toronto, p. 7-15.

Hanmer, S., 1986: 'Asymmetrical pull-aparts and foliation fish as kinematic indicators'; *Journal of Structural Geology*, v. 8, p. 111-122.

Moser, D., 1988: 'Structure of the Wawa gneiss terrane near Chapleau, Ontario'; in *Current Research, Part C, Geological Survey of Canada, Paper 88-1C*, p. 93-99.

Moser, D., 1989: 'Mid-crustal structures of the Wawa gneiss terrane near Chapleau, Ontario'; in *Current Research, Part C, Geological Survey of Canada, Paper 89-1C*, p. 215-224.

Parrish, R., 1987: 'Applications of numerical modeling of heat transfer to P-T-t path determination of exhumed mountain belts'; *Geological Association of Canada Program with Abstracts* **12**, p. 78.

Percival, J.A., 1981a: *Geological evolution of part of the central Superior Province base on relationships among the Abitibi and Wawa subprovinces and the Kapuskasing structural zone*; Ph.D. thesis, Queen's University, Kingston, Ontario, 300 p.

Percival, J.A., 1981b: 'Geology of the Kapuskasing structural zone in the Chapleau-Foleyet area'; *Geological Survey of Canada, Open File Map 763.*

Percival, J.A., 1983: 'High-grade metamorphism in the Chapleau-Foleyet area, Ontario'; *American Mineralogist*, **68**, p. 667-686.

Percival, J.A., 1986: 'The Kapuskasing uplift: Archean greenstones and granulites'; *Geological Association of Canada, Field Trip Guide 16.*

Percival, J.A., 1989: 'The Kapuskasing uplift: cross section through the Archean Superior Province'; this volume.

Percival, J.A. and Card, K.D., 1983: 'Archean crust as revealed in the Kapuskasing uplift, Superior Province, Canada'; *Geology*, **11**, p. 323-326.

Percival, J.A. and Caed, K.D., 1985: 'Structure and evolution of Archean crust in central Superior Province, Canada'; in *Evolution of Archean Supracrustal Sequences* Ed. L.D.Ayres, P.C.Thurston, K.D.Card, and W.Weber, Geological Association of Canada Special Paper 28, p. 179-192.

Percival, J.A. and Krogh, T.E., 1983: 'U-Pb zircon geochronology of the Kapuskasing structural zone and vicinity in the Chapleau-Foleyet area, Ontario'; *Canadian Journal of Earth Sciences*, **20**, p.830-843.

Sibson, R.H., 1975: 'Generation of pseudotachylite by ancient seismic faulting'; Geophysical Journal of the Royal Astronomical Society, **43**, p. 775-794.

Sibson, R.H., 1977: 'Fault rocks and fault mechanisms'; *Journal of the Geological Society of London*, **133**, p. 191-213.

Thurston, P.C., Siragusa, G.M., and Sage, R.P., 1977: 'Geology of the Chapleau area, Districts of Algoma, Sudbury, and Cochrane'; *Ontario Division of Mines, Geoscience Report 157*, 293 p.

West, G., Green, A., Cook, E., Milkereit, B., Hurley, P., and Geis, W., 1988: 'Kapuskasing Seismic Reflection Survey Programme'; in *1988-89 KSZ Lithoprobe Workshop*, Geophysical Laboratory, University of Toronto,

THE EXHUMATION OF CROSS SECTIONS OF THE CONTINENTAL CRUST: STRUCTURE, KINEMATICS AND RHEOLOGY

M. R. HANDY
Geologisches Institut, Universität Bern
Baltzerstrasse 1, 3012 Bern, Switzerland

KEYWORDS. continental crust, uplift, P-T paths, shear zone, lithospheric rheology

ABSTRACT. The exhumation of coherent cross sections of the continental crust is a selective, usually multistage tectonic process that involves special kinematic and geothermal evolutions. Three general types of exhumation history are distinguished: (1) uplift at low to moderate geothermal gradients in a convergent intracratonic setting; (2) extensional uplift at high geothermal gradients followed by emplacement near an active plate margin; (3) uplift at high-temperature conditions in the core of a deeply eroded intraplate or plate-margin orogen. The degree of structural coherence among different crustal levels of a crustal section is generally greater in sections with type 1 and 2 histories than in sections with a type 3 history. Thermal disequilibrium during shearing led to asymmetrical strain and fabric distribution across large shear zones that accommodated uplift of the crustal sections. The subduction, underplating and thickening of continental crust beneath exposed crustal cross sections produced anomolous lithospheric structures that have isostatically balanced dense lower crustal rocks at or near the surface over long periods of time. Relict crustal features that predate the exhumation are best preserved in crustal sections where cooling induced a pronounced partitioning of stress and strain already during the early stages of uplift.

1. Introduction

Exposed cross sections of the continental crust are large terrains in which most levels of the crust have been brought to the surface as a coherent package. Rocks exposed at the surface vary systematically from unmetamorphosed to weakly metamorphosed upper crustal assemblages to highly metamorphosed, dense lower crustal units. The interpretation of these terrains as crustal cross sections is based on the observation that cross-sectional trends in their lithologic, metamorphic, structural and geophysical characteristics are consistent with petrologic models of the crust and with the imaged geophysical properties of unexposed continental crust (Fountain and Salisbury, 1981). It is the general continuity among the different crustal levels exposed at the surface that distinguishes these terrains from other suites containing lower crustal rocks and makes exposed crustal cross sections attractive for studying evolutionary crustal processes in the field. In order to determine whether such crustal cross sections are really representative samples of

M. H. Salisbury and D. M. Fountain (eds.), Exposed Cross-Sections of the Continental Crust, 485–507.
© 1990 *Kluwer Academic Publishers.*

unexposed continental crust, we must determine how entire sections of the crust are brought to the surface and assess the effect that uplift has on their original internal structure, metamorphism and composition.

Many popular models of crustal exhumation pertain either to (1) medium- to high-grade rocks in the cores of eroded orogens, or to (2) the special case of high-pressure metamorphic rocks within subduction melanges (see brief reviews in Fountain and Salisbury, 1981; Platt, 1986). However, recent progress in reconstructing the history of some crustal cross sections (reviews in this volume) reveals that cross sections of the continental crust are exposed in a much wider range of tectonic settings than are considered in these tectonic models. Certainly, applying uplift models for high-pressure metamorphic terrains to explain the exhumation of coherent crustal cross sections is unsatisfactory because the tectonothermal history of lower crustal granulites in crustal cross sections (next section) differs considerably from that of blueschist facies terrains (Ernst, 1988). These exhumation mechanisms result in extensive imbrication of rocks with contrasting metamorphic histories, rather than the preservation of structural and metamorphic continuity among different crustal levels.

This paper reviews the uplift history of several exposed crustal cross sections and relates the kinematics of their exhumation to the evolution of thermobarometric conditions and deformational style in different crustal levels. Experimental and theoretical rock-mechanical concepts are used to constrain the long-term, high-strain lithospheric rheology during uplift. Consideration of how stress and strain partition during uplift is crucial to understanding why lower crustal rocks survive the exhumation process and sections of continental crust maintain their structural integrity.

2. The Exhumation History of Some Exposed Crustal Cross Sections

Tables I and II summarize the uplift characteristics of cross sections where the tectonometamorphic evolution is reasonably well constrained. There are three main types of exhumation history: (1) cold intacratonic uplift along thrusts rooted in the lower crust; (2) hot extensional uplift of the section to depths of 10 to 15 km, followed by cold emplacement to shallow depths near a convergent plate boundary; (3) repeated uplift under high-temperature conditions in the core of an intracratonic or a plate boundary orogen. It is important to note that not all of the cross sections in Tables I and II are equally well exposed or well studied. Also, many more crustal cross sections probably exist in other areas than are currently reported in the literature or even recognized in nature. Nevertheless, some generalizations can be made about their evolution and exhumation.

The exposed crustal cross sections in Table I experienced type 1 or 2 uplift histories and have a high degree of structural integrity. Regardless of age and tectonic setting, these crustal cross sections all show (1) a granulite facies lower crust consisting of sill-like mafic intrusive rock and felsic rock of supracrustal origin, (2) a middle crust containing amphibolite to greenschist facies metasediment and intermediate to felsic intrusive rock, and sometimes (3) an unmetamorphosed to weakly metamorphosed upper crust. Generally, the ratio of mafic to felsic rock increases with paleodepth in the sections. Complex polyphase deformation associated with regional metamorphism clearly predates shearing and retrograde metamorphism during uplift (references, Table I). Therefore, the overall structure of the crustal cross sections in Table I was established at high geothermal gradients well before exhumation began. The structurally deepest parts of the cross sections yield peak geobarometric estimates of 8 to 12 kilobars and contain only volumetrically minor, laterally discontinuous outcrops of ultrabasic rock. This indicates that lower crustal rock presently exposed at the surface lay no deeper than 30 to 45 kilometers at the time of regional metamorphism. Either the original crustal thickness did not exceed this amount and

Table I Uplift Characteristics of Some Relatively Coherent Crustal Cross Sections

Crustal Section	Plate Tectonic Setting	Major Fault(s) active during Uplift	Fault Rocks & Metamorphic Conditions	Age of Uplift
Kapuskasing (6, 7, 8)	Intracratonic: Superior Province (Ontario, Canada)	Ivanhoe Lake fault zone: thrust at base of section	retrograde greenschist f. mylonite & cataclasite	2690 - 2450 Ma.? 2200 - 1900 Ma.
Ivrea (4, 9, 11)	Plate Margin: European/Apulian plates (southern Alps, N. Italy, Europe)	Pogallo shear zone in the basal intermediate crust and other extensional mylonitic zones in the lower crust	retrograde amphibolite f. to greenschist f. mylonite in footwall grading to cataclasite in hangingwall	270 - 160 Ma. transtension & extension
		Fault at base of section truncated by Insubric line	retrograde mylonite & cataclasite?	75 - 60 Ma. Eo-Alpine stage
		Insubric line: strike-slip & thrust fault at base of section	cataclasite; greenschist f. mylonite at northern rim of section	27 - 17 Ma. Insubric phase
Fiordland (1, 5)	Plate Margin: Pacific/Tasman plates (Southern Alps, South Island, New Zealand)	Doubtful Sound shear zone: extensional shear zone within the lower crust	retrograde amphibolite f. & greenschist f. mylonites in footwall grading to cataclasite in hangingwall	100 - 80 Ma.
		Alpine fault: strike-slip & thrust fault near base of section	cataclasite	ca. 40 - 15 Ma.
Pikwitonei (2, 3, 10)	Plate Margin: Superior/Churchill blocks (Manitoba, Canada)	extensional shear zones in the lower crust	amphibolite f. mylonite?	2400 - 2300 Ma.
		Thompson mobile belt: strike-slip & thrust fault at base of section	retrograde amphibolite f. to greenschist f. mylonite & cataclasite	1900 - 1700 Ma.

References: (1) Gibson et al., 1988; (2) Green et al., 1985; (3) Hubregtse, 1980; (4) Hurford, 1986; (5) Oliver & Coggon, 1979; (6) Percival & Card, 1985; (7) Percival & McGrath, 1986; (8) Percival et al., 1989; (9) Schmid et al., 1987; (10) Weber & Scoates, 1978; (11) Zingg et al., in press

TABLE II Uplift Characteristics of Some Crustal Sections in the Cores of Deeply Eroded Orogens

Crustal Section	Plate Tectonic Setting	Major Fault(s) active during Uplift	Fault Rocks & Metamorphic Conditions	Age of Uplift
Arunta (3, 6, 7)	Intracratonic: Arunta & Amadeus blocks (Central Australia)	Redbank fault: thrust at base of section & shear zones within section bounding basement nappes	amphibolite facies mylonites? retrograde amphibolite f. mylonite in hangingwall grading to greenschist f. mylonite & cataclasite in footwall	>1450-1100 Ma. 400 - 300 Ma. Alice Springs orogeny
Grenville Front (1, 2, 8)	Intracratonic: Grenville Province (Ontario, Canada)	Grenville Front tectonic zone: thrust at base of section & shear zones within section	granulite and amphibolite f. mylonite in hangingwall grading to greenschist f. mylonite & cataclasite in footwall	1400 - 1100 Ga. Grenville orogeny
Kasila (9, 10)	Plate Margin: W. African craton/Guayana shield (Sierra Leone, Africa)	Todi shear zone: thrust at base of section & shear zones within section thrusts & folds (reactivation of Todi s.z.)	granulite and amphibolite f. mylonite in hangingwall grading to greenschist f. mylonite & cataclasite in footwall cataclasite	2700 - 2200 Ma. ca. 500 Ma.
Central Gneiss Complex (4, 5)	Plate Margin: Alexander/Stikine terrains, (Coast Mountains, British Columbia, Canada)	low-angle thrusts: shear zones within & at base of crustal imbricates	granulite & amphibolite f. mylonite & synkinematic melt in hangingwall grading to greenschist f. mylonite & cataclasite in footwall	Rapid uplift: 98 to 90 Ma. 60 to 48 Ma.

References: (1) Davidson, 1988; (2) Green et al., 1988; (3) Goleby et al., (1989); (4) Hollister, 1982; (5) Hollister & Crawford, 1986; (6) Obee & White, 1985; (7) Teyssier, 1985; (8) Van Breemen & Davidson, 1988; (9) Williams, 1988, (10) Williams & Culver, 1988

detachment occurred at the base of the crust (unlikely in every case) or the crust was thicker and a major detachment horizon was located just below this range of depths within the crust.

Significant structural differences between crustal cross sections are related to their exhumation history. Cross sections that experienced a type 3 uplift history (Table II) were internally sheared and imbricated under high-grade metamorphic conditions during uplift and so are neither as complete nor as continuous as the cross sections with type 1 or 2 histories. Large thrust faults that bound all exposed crustal cross sections (Fountain and Salisbury, 1981) separate high-grade rock at the base of the cross section from lower grade rock outside of the cross section. In intracratonic uplifts, these lower grade rocks are lithological and metamorphic equivalents to the middle and upper crustal levels of the exposed cross section (e.g., Abitibi and Michipicotan belts separated by the Kapuskasing uplift; Percival and Card, 1985), whereas adjacent to crustal cross sections uplifted at or near plate boundaries these lower grade rocks sometimes have a tectonometamorphic history that contrasts with that of the exposed crustal section (e.g., the pre-Alpine metamorphosed Strona-Ceneri zone and the Alpine-metamorphosed Sesia zone separated by the Ivrea zone; Zingg et al., in press). In either case, the structural and metamorphic asymmetry across the boundary fault often coincides with paired gravity anomalies (Fig. 3 in Fountain and Salisbury, 1981). Gravity "highs" over the lower crustal base of the uplifted cross sections compliment gravity "lows" above the thickened crust. The highest positive Bouguer anomalies are measured above cross sections that underwent type 2 uplift (e.g., Ivrea geophysical body, Berckhemer, 1968; Fiordland section, Oliver and Coggan, 1979). Geophysical modelling of these large gravimetric and seismic anomalies suggests that some of the cross sections are underlain by large slices of mantle material at relatively shallow depths (ca. 10 to 20 km). In contrast, the cross section that experienced relatively cold intracratonic uplift (i.e., the Kapuskasing section) is underlain by thickened lower crust (Boland et al., 1988).

3. Kinematic and Rheologic Evolution during Uplift

3.1. COLD INTRACRATONIC UPLIFT

To date, the Kapuskasing section is the only documented example of cold intracratonic uplift that resulted in the exposure of lower crust at the surface. The deep crustal structure beneath the Kapuskasing section depicted in Figure 1a is based on the contoured velocity profiles of Boland et al. (1988) and on the extensive geologic work of John Percival and coworkers (references cited below). The uplifted crustal cross section is separated from underthrusted lower crust by a SE-dipping wedge of lower crustal rocks (Fig. 1a). Large-scale imbrication and buckling of the lower crust isostatically balanced overthrusting of the Kapuskasing section, similar to the kinematic model of Percival et al. (1989) in which lower crustal material flows into a crustal root beneath the Kapuskasing section. Reflection seismic profiling indicates that the Ivanhoe Lake fault zone extends beneath the Kapuskasing section with a shallow (15 to 20°) dip to a detachment that originally lay at a depth of about 20 to 30 kilometers (Percival et al., 1989). Shortening above and below this intracrustal detachment is inferred to be equal (55 to 70 km; Percival et al., 1989) and occurred during the Early Proterozoic (Percival and Card, 1985; Percival and McGrath, 1986), possibly also during the Late Archean (Percival et al., 1989). The depth-temperature map in Figure 1b depicts the approximate geothermal conditions (thin lines) and P-T trajectories (arrows) of uplifted and underthrusted lower crustal rocks in the Kapuskasing region. Isotopic mineral ages indicate that the Kapuskasing section cooled slowly from regional metamorphic conditions at about 2.7 Ga. (box in Fig. 1b) to below the 300 °C blocking temperature of the Rb-Sr biotite system

490

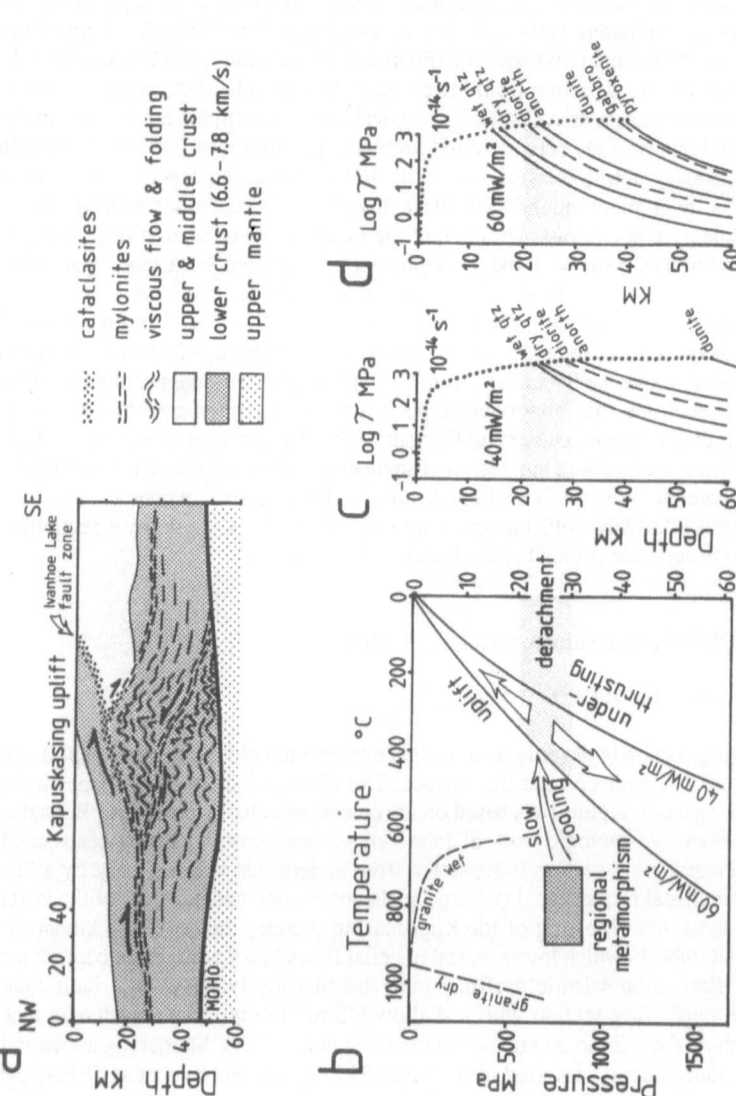

Figure 1. (a) Profile through the Kapuskasing uplift, Superior Province, Canada, modified from Percival et al. (1989); **(b)** Depth-temperature diagram showing conditions of regional metamorphism (Percival and Card, 1985) and predicted P-T paths of the Kapuskasing lower crust during uplift and underthrusting. Solidus curves for wet and dry granite from Johannes (1985), steady-state geotherms from Chapman (1986). **(c)** and **(d)** Log shear strength versus depth for steady-state geotherms in (b) at a regional shear strain-rate of 10^{-14} s^{-1}. Stippled area represents approximate depth of intracrustal detachment in (a) according to Percival et al. (1989). Dotted curves for frictional sliding (cataclasis) calculated with Byerlee's (1978) frictional coefficients. Solid curves are for viscous flow (dislocation creep) in wet and dry quartzite (Jaoul et al., 1984), anorthosite (Shelton and Tullis, 1981), and wet dunite (Chopra and Paterson, 1981). Dashed curves are for diorite and gabbro (see text).

some 2.2 to 1.9 Ga. ago (Fig. 15 in Percival, 1988). The age and blocking temperature of biotite corresponds with the retrograde greenschist facies metamorphic grade of mylonites and cataclasites that overprint granulites in the Ivanhoe Lake fault zone at the base of the cross section (Percival and Card, 1985). Therefore, the average geothermal gradient had dropped from 30 to 40 °C/km during regional metamorphism to about 10 to 20 °C/km by the time of final uplift. Steady-state geotherms in Figure 1b corresponding to surface heat flows of 40 to 60 mW/m^2 (Chapman, 1986) approximate the average geothermal gradient during uplift inferred from isotopic ages and syntectonic metamorphic conditions in the Ivanhoe Lake fault rocks.

Lower crustal rocks that were initially adjacent on either side of the intracrustal detachment experienced divergent thermo-barometric and rheologic evolutions as intracratonic shortening progressed (Figs. 1b to 1d). The P-T paths of lower crustal rocks in Figure 1b are inferred to have followed the steady-state geotherms because two important observations indicate that thermal and isostatic equilibrium prevailed throughout the shortening history: (1) a foreland basin is conspicuously absent from the Kapuskasing section (Percival and Green, 1988); (2) there is no marked structural and metamorphic asymmetry within the Ivanhoe Lake fault rocks at the base of the crustal section. Together, these observations suggest that tectonic and erosional denudation kept pace with uplift and that uplift was sufficiently slow to allow thermal equilibration across the boundary fault during thrusting.

The implications of this P-T path for the evolution of rheologic properties during intracratonic thrusting becomes apparent in the shear strength versus depth diagrams in Figures 1c and 1d. Such diagrams (e.g., Brace and Kohlstedt, 1980) conveniently depict depth-dependent variations in rock strength associated with frictional sliding (i.e., cataclasis, dotted curves) and viscous solid-state creep (solid and dashed curves in Figs. 1c and 1d). Although the solid curves in Figures 1c and 1d are strictly valid for monomineralic aggregates, they adequately describe the viscous creep strength of polymineralic rocks at high strains provided that (1) they represent the weakest mineral in the rock, and (2) this mineral forms interconnected (i.e., contiguous) domains and makes up at least 20% of the rock (Handy, 1990). A minimum estimate of the creep strength of rocks with 80% or more of a relatively strong mineral (i.e., gabbros, diorites; dashed lines in Figs. 1c and 1d) is obtained with the nonlinear strength-composition relation for porous sintered aggregates (Tharp,1983; see Handy, 1990). Due to the lack of reliable constraints on pore pressure, the effective pressure is simply assumed to equal the lithostatic pressure in Figures 1c and 1d, although it is recognized that pore pressure in hydrous shear zones can fluctuate extensively (Etheridge et al., 1984). Despite the problems of extrapolating laboratory flow laws to natural strain-rates (Paterson, 1987), the viscous strength relations predicted with the flow laws used to construct Figures 1c and 1d generally agree with the relative strength of lower crustal and upper mantle rocks inferred from strain measurements in the field (Handy and Zingg, in press) and so are a reliable qualitative guide to natural solid-state rheologies.

The intracrustal detachment at about 25 kilometers beneath the Kapuskasing section corresponds with two potential rheological discontinuities in Figures 1c and 1d: (1) the transition from frictional (i.e., brittle) to viscous deformation in quartz-bearing rock; (2) a change in the bulk composition of the lower crust from predominantly quartz- and feldspar-bearing paragneiss and diorite above the detachment (exposed in the Kapuskasing uplift) to quartz-free feldspathic, amphibolitic and gabbroic rocks below the detachment (in the thickened, underthrusted part of the crust in Fig. 1a). The latter interpretation is consistent with the velocity structure of the Kapuskasing region (Fig. 4 in Boland et al., 1988) indicating that lower crustal rocks with P-wave velocities greater than 7.0 km/s predominate below 25 to 30 kilometers depth.

Final uplift of the cold crustal cross section occurred largely in the frictional regime (Fig. 1c), as shown by the ubiquity of cataclastic fault rocks in the Kapuskasing section and along the

492

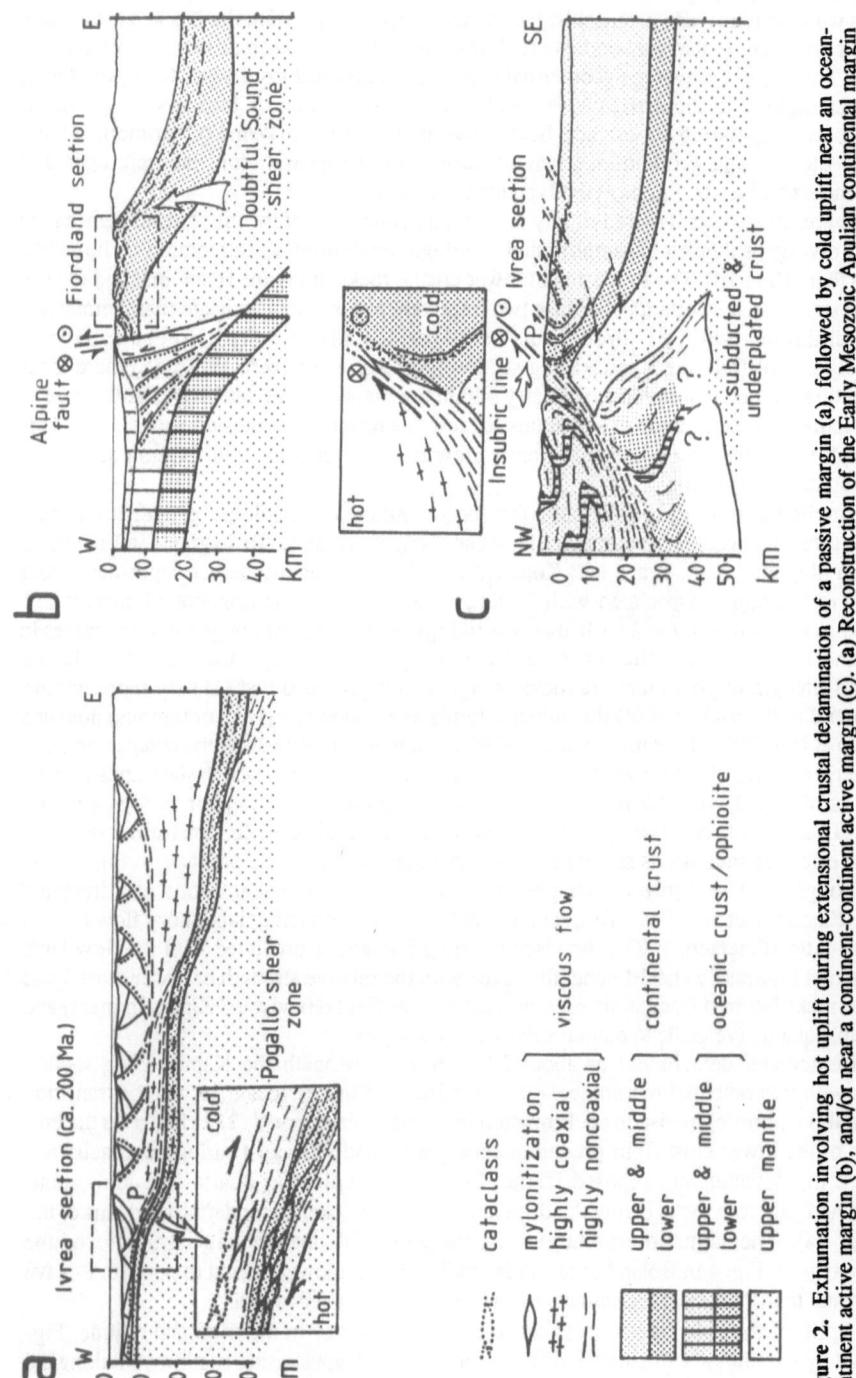

Figure 2. Exhumation involving hot uplift during extensional crustal delamination of a passive margin (a), followed by cold uplift near an ocean-continent active margin (b), and/or near a continent-continent active margin (c). (a) Reconstruction of the Early Mesozoic Apulian continental margin showing the approximate location of the Ivrea section in the dashed box (modified from Handy and Zingg, in press); (b) Present setting of the Fiordland section (dashed box), South Island, New Zealand (modified from Oliver and Coggon, 1979; Gibson et al., 1988); (c) Present setting of the Ivrea section south of the main part of the Alpine orogen (modified from Zingg et al., in press). Insets to (a) and (c) show distribution of fault rocks, respectively, in the Ivrea zone and at the Insubric line. P marks location of the Pogallo shear zone.

Ivanhoe Lake fault zone (Percival and Card, 1985; Percival and McGrath, 1986). Deformation within the underthrusted lower crust is predicted to have become localized within discrete mylonitic shear zones after relatively minor amounts of strain. Shearing of the basic granulite facies rocks under retrograde metamorphic conditions (300 - 600 °C according to Fig. 1b) would have facilitated the syntectonic growth of fine grained and/or weaker minerals within the shear zones. The viscous strength of such fine grained reaction products is much less than that of the original basic host rocks plotted in Figures 1c and 1d (e.g., Brodie and Rutter, 1985), so that narrow shear zones probably accommodated most of the strain in the underthrusted lower crust. In addition, the high viscosity contrasts associated with both localized grain size reduction in shear zones and the low homologous temperatures (T_h < 0.4) in the underthrusted lower crustal rocks may have lead to extensive buckling within the thickened lower crustal root underlying the Kapuskasing section (Fig. 1a). Temperatures in the underthrusted rocks increased somewhat during shortening (Fig. 1b), but this increase was probably not sufficiently large to induce significant changes in flow regime and structure.

An important feature of this model is that the brittly deformed and uplifted crustal cross section preserves its general structural integrity whereas the warmer underthrusted and wedged lower crust is predicted to lose its structural integrity through imbrication and viscous folding. This may explain the lack of strong reflectors in the crustal root underlying the Kapuskasing uplift (Fig. 2 in Percival et al., 1989).

3.2. EXTENSIONAL-CONVERGENT UPLIFT

The multistage uplift history and structural evolution of most crustal cross sections reflects a much more complicated interplay of kinematic, geothermal and rheologic factors than in the intracratonic Kapuskasing uplift. The Ivrea section (southern Alps of Europe) and the Fiordland section (Southern Alps, New Zealand) are two excellent examples of crustal cross sections that experienced early extensional uplift (Fig. 2a) followed by uplift either at an active ocean-continent margin (Fig. 2b) or near a continent-continent collisional boundary (Fig. 2c).

Extensional uplift of the crustal cross section to shallower depths (10-20 km) involved lithospheric shearing at very high transient geothermal gradients and broadly coincided with basin development and subsidence of the rifted upper crust (Fig. 2a). In the case of the Ivrea section, this extensional exhumation was probably polyphase (Handy and Zingg, in press). The metamorphic core complexes of the American Cordillera represent possible precursors to the passive margin stage depicted in Figure 2a. Subsequent exposure of the attenuated cross section at or near a convergent plate boundary occurred after the crustal section had cooled to temperatures similar to or less than that of the crust with which it was eventually juxtaposed at the surface. As with the extensional stage of exhumation, convergent uplift of the crustal section may be polyphase. The oblique convergent uplift of the Fiordland section shown schematically in Figure 2b is similar to the early (i.e., Eo-Alpine) episode of convergent uplift of the Ivrea section. At this time, the distal rim of the southern Alpine continental margin (Sesia zone) was partly subducted beneath the mantle and crust of the proximal part of the margin (i.e., the part containing the Ivrea section; Schmid et al., 1987), while the overriding part of the passive margin was uplifted. This early configuration may have resembled Figure 2b and preceded the final uplift configuration in Figure 2c. Thus, Figures 2a through 2c can be regarded as representing a complete evolutionary sequence of exhumation, beginning with a metamorphic core complex or a passive margin extensional stage, progressing through an ocean-continent collisional ("Fiordland") stage, and in some cases ending with a continent-continent collisional ("Ivrea") stage.

The faults at the base of the exposed crustal cross sections dip either toward the section (Fig. 2b; e.g., the Alpine fault west of the Fiordland Section; Oliver and Coggon, 1979) or away from the section (Fig. 2c; e.g., the Insubric line northwest of the Ivrea section; Schmid et al., 1987). The opposite fault dip directions probably reflect differences in the isostatic behaviour of continental and oceanic crust during different stages of plate collision. During early ocean-continent collision, oceanic crust subducts without substantial thickening along a master fault dipping beneath the crustal cross section. In contrast, continental crust is more buoyant and thickens during collision. If convergence and thickening rates are high compared to the rates of tectonic and erosional denudation, then continental crust eventually overthickens and escapes upwards and outwards. During the late stages of the Alpine orogeny, this escape was primarily in the direction of subduction and resulted in both south-vergent refolding ("backfolding") of the north-vergent Alpine basement nappes and rapid uplift concentrated along the Insubric line (Fig. 2c). The Insubric line formed at the northern rim of the Ivrea geophysical body because the cold, creep-resistant rocks of the previously uplifted upper mantle and lower crust acted like a rigid indenter within the thick pile of warm, relatively weak intermediate crust forming the core of the Alpine orogen. The Insubric line is therefore a comparatively young orogenic structure and is interpreted to flatten at depth within the crust, truncating the older SE-dipping master fault that was active during Eo-Alpine subduction (Fig. 2c).

The model of lithospheric attenuation and exhumation in Figures 2a and 3a involves stepwise crustal delamination along large zones of noncoaxial flow (like the Pogallo shear zone in the Ivrea section) that accommodate inhomogeneous pure shear extension of the entire lithosphere (e.g., Lister et al., 1986). The predicted pattern of shearing at the base of the extending crust is shown with the dashed (mylonitic) and dotted (cataclastic) zones in Figures 2a and 3a. Large extensional shear zones nucleate at major pre-existing compositional boundaries and then coalesce, stepping upwards towards the more thinned, distal part of the passive margin. The master fault is highly noncoaxial and dips towards the continent where it crosses compositional boundaries and drags the underlying lithosphere upwards and outwards. In the absence of well-constrained pressure - temperature - time (P-T-t) paths in deeply eroded extensional terrains, the extensional P-T paths for intermediate and lower crustal levels in Figure 3b (dotted curves) have been constructed conceptually for the lithospheric extensional model described above and depicted in Figure 3a. The corresponding particle paths for these crustal levels between times t_0 and t_2 within the evolving passive margin are indicated in Figure 3a. The predicted sinuosity of the extensional P-T paths reflects the passage of attenuating crustal material through highly noncoaxial, low-angle shear zones at crustal "necks" and through areas of more coaxial shear and passive uplift (Fig. 3a). The juxtaposition of disparate crustal levels along low-angle normal faults is associated with higher rates of thermal equilibration per amount of uplift than is homogeneous coaxial (pure) shear (Ruppel et al., 1988). Therefore, periods of highly noncoaxial shear in Figure 3a are depicted with the less arcuate and less steep parts of the P-T paths in Figure 3b than are periods of highly coaxial shear.

The overall concave upward arcuation of the P-T paths in Figure 3b reflects the fact that the rate of uplift is generally high with respect to the rate of thermal equilibration. Direct evidence in the field for tectonically induced thermal disequilibrium during uplift comes from the pronounced structural and metamorphic asymmetry of large shear zones in many exposed crustal cross sections (Tabel I). Cataclasite and low-grade mylonite at the cool contact between differentially uplifted crustal levels grade into high-grade mylonitic rock within the warmer uplifted footwall. Extensional shear zones that have exhumed hot lower crust are vertically zoned (inset to Fig. 2a; the Pogallo shear zone, Handy, 1987; the Doubtful Sound detachment, Gibson et al., 1988) whereas some steeply dipping, late-stage faults bounding several uplifted crustal cross sections are

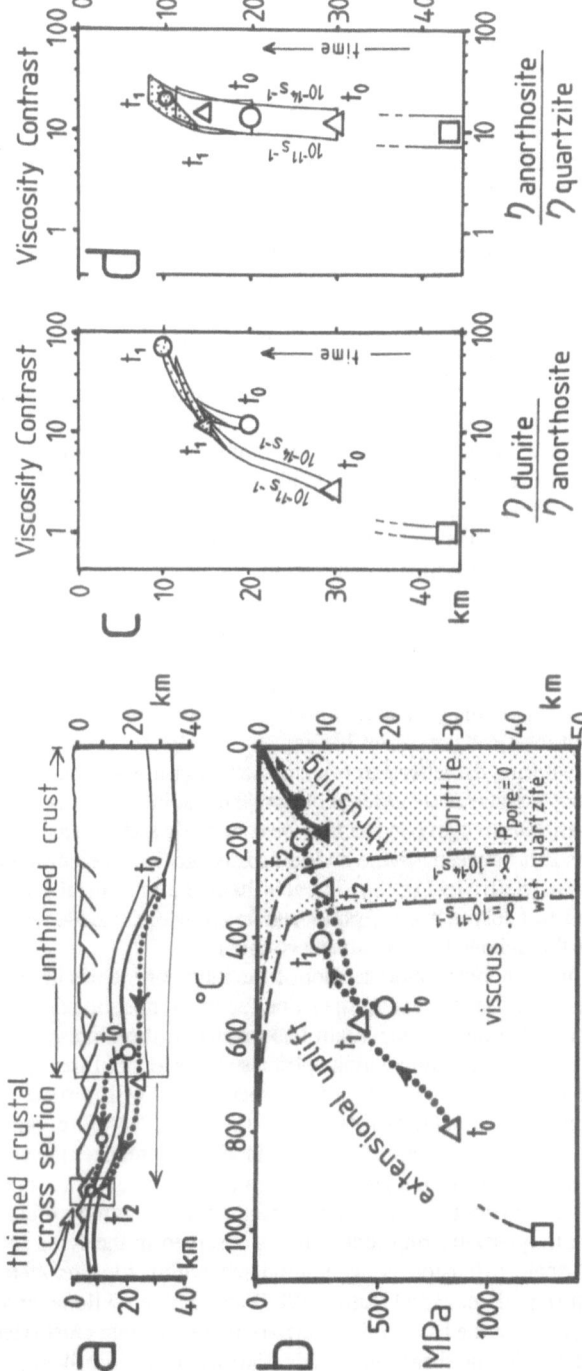

Figure 3. (a) Passive margin showing hypothetical particle paths during extensional uplift of lower and intermediate crustal levels (respectively, triangles and circles) between times t_0 at the onset of extension and the final position of the thinned crustal cross section at t_2. A major noncoaxial extensional shear zone separates the two crustal levels, as shown in Fig. 2a. Thin lines outline the crust prior to extension at t_0. (b) Depth-temperature diagram with hypothetical P-T paths for crustal levels shown in (a) during extension (dotted lines) and cold thrusting (solid lines). Dashed curves for transition from frictional (stippled region) to viscous flow in hydrous quartzite at shear strain-rates of 10^{-11} and 10^{-14} s^{-1} (taken from Handy, 1989). (c) and (d) Viscosity contrast versus depth diagrams, respectively, for dunite-anorthosite and anorthosite-quartzite during extensional uplift of the two crustal levels symbolized in (a) and (b). Squares in (b), (c), and (d) indicate conditions in hypothetically deeper crust (see text). Creep parameters as in Fig. 1 above.

subhorizontally zoned and reflect high lateral heat flux during rapid uplift and shearing (e.g., the Insubric line; inset to Fig. 2c).

An important consequence of inhomogeneous thinning is that different levels of the crust pass through highly noncoaxial shear zones at different times as the lithosphere attenuates. Uplift rates for different crustal levels vary, so pieces of crust that are vertically juxtaposed at time t_2 in the distal parts of the passive margin were not part of a vertically continuous crustal cross section at the onset of extension at time t_o (particle paths in Fig. 3a). The point emphasized here is that any extension involving a component of noncoaxial flow produces cross sections of attenuated crust whose constituent levels occupied laterally different positions prior to extension.

The long-term rheological characteristics of rocks in the extending lithosphere can be approximated by tracking the temperature and strain-rate dependent changes in the relative strength of rocks (Figs. 3c and 3d) between times t_o and t_2 on the P-T paths in Figure 3b. The curves in Figures 3c and 3d yield maximum estimates of strength contrast because the assumption of a constant regional strain-rate for all lithologies leads to an underestimation of strain-rate in the weak rocks and an overestimation of strain-rate in the strong rocks (see Handy, 1990). In addition, stress concentration at lithological contacts can cause a localized reduction in grain size and therefore also a strain-dependent decrease in the viscous strength of the relatively stronger rock with respect to that of the weaker rock (Rutter and Brodie, 1988). Nevertheless, Figures 3c and 3d demonstrate two important points: (1) decreasing temperature during uplift is associated with increased viscous strength contrast among the principle rock-forming minerals in the lithosphere; (2) strain localization at litho-rheological boundaries depends on the pre-extensional compositional and metamorphic characteristics of the lithosphere. For example, at an initial depth of 45 kilometers, an upper mantle rock comprising at least 20% olivine is predicted to have about the same viscous strength as a lower crustal rock comprising at least 20% feldpar (square symbol in Figs. 3b-3d). With decreasing temperature, olivine rock becomes progressively stronger with respect to feldspathic rock (Fig. 3c). This suggests that at high geothermal gradients in initially thick crust (> 45 km according to the flow laws used in Fig. 3c) the MOHO only becomes a significant rheological dicontinuity during the latter part of extensional uplift, whereas in thinner crust (< 45 km) the MOHO is already a potential rheological discontinuity at the onset of extension. Under the same conditions, the predicted strength contrast between quartzitic and feldspathic rocks is not as temperature dependent as that between olivine and feldspathic rocks. The considerable magnitude of this strength contrast (>10:1 in Fig. 3d) is such that the intracrustal boundary between quartz-bearing and quartz-free feldspathic compositional domains is predicted to remain an important rheological discontinuity throughout the extensional evolution.

Viscous instability leading to strain localization develops from inhomogeneities in the rock that either predate deformation (inherited structure and composition) or are deformationally induced (syntectonic reactions, thermal perturbations). Once strain has localized, the increased stress concentration associated with decreasing temperature during uplift also leads to grain size reduction within shear zones and so results in decreased viscosity of the shear zones with respect to the viscosity of the less sheared or unsheared country rock (Handy, 1989). Therefore, strain is progressively localized during uplift, because a localized decrease in viscosity reduces the volume of deforming rock needed to accommodate a given regional strain-rate. For a given strain-rate and grain size, strain localizes at higher temperatures (and therefore, at greater depths) in feldspar, amphibole, and olivine rocks than in quartz-bearing rock. This is observed in the Ivrea section (inset to Fig. 2a), where crustal-scale extension is accommodated within high-temperature mylonitic zones in the lower crust (e.g., Brodie and Rutter, 1987) and within a kilometer-wide zone of predominantly noncoaxial shear at the base of the quartz-rich intermediate crust (Handy, 1987). Anastomozing cataclastic and mylonitic shear zones grade downwards into broader zones of penetrative viscous shear at the originally deeper end of the crustal cross section (Handy and Zingg, in press).

Stress and strain partition not only on the lithospheric scale, but also on the outcrop and even on the microscopic scale. An important consequence of stress and strain partitioning during uplift is that large portions of the crust between shear zones remain undeformed or only weakly strained and therefore preserve the pre-extensional features of the rock. What makes extension (as opposed to collision) such an important exhumation mechanism is that intracrustal detachment along low-angle normal faults does not significantly change the depth-sequence of structural levels during uplift. While extensional detachments accommodate large lateral displacements of crust during thinning, a significant proportion of the vertical displacement results from passive uplift of all crustal levels as buoyancy forces act on the warm thinning continental lithosphere (Bott, 1981). However, it is important to note that extension on its own is not a viable mechanism for exposing large tracts of the deepest crust and upper mantle at the surface for long periods of time. Firstly, magmatic underplating related to decompression melting or incipient seafloor spreading can pre-empt large-scale uplift of the MOHO to the surface at very high extensional strains ($\beta > 3$ to 4; Le Pichon and Sibuet, 1981). Secondly, the tensional regional stress field and low crustal viscosities associated with lithospheric extension at high geothermal gradients are not suitable for supporting large volumes of dense, isostatically unstable material at the surface for long.

Final uplift and exposure of attenuated crustal cross sections is driven by compressional forces from a nearby convergent plate boundary (Figs. 2b and 2c). If the subducted crust is continental rather than oceanic, then the buoyancy of this crust exerts an additional upward force on the overlying column of rock, including the crustal cross section. In Figure 2c, the dipping sliver of mantle material responsible for the large positive Ivrea gravity anomaly was trapped between the overlying and subducted continental crust of the passive margin. Note that no large-scale obduction of mantle into the crust was necessary to generate this structure (compare with Laubscher and Bernoulli, 1982). At the surface, thrusts and folds associated with the latest stages of collision have the same vergence as subduction at the convergent boundary (Fig. 2c). Final emplacement of the crustal cross sections under low-temperature conditions is characterized by large overall crustal strengths. Shearing in most parts of the crustal section is extremely localized and involves cataclasis and brittle folding (solid P-T paths in the stippled frictional regime in Fig. 3b), as observed in the Ivrea zone (Zingg et al., in press).

4. Other Mechanisms for Exposing Sections of the Continental Crust

The crustal sections listed in Table II all come from the crystalline cores of deeply eroded orogens. The tectonic setting and history of these sections vary considerably, and in most cases, the details of their uplift history remain ambiguous pending further investigation. What distinguishes all of them from the crustal cross sections discussed so far is the high degree of internal deformation that accompanied exhumation. For this reason, none of these deep crustal exposures is coherent or complete in the sense defined in the Introduction.

4.1. HIGH-GRADE THRUSTING AND UPLIFT

Uplift under high-grade metamorphic conditions involved penetrative shearing and considerable imbrication in all structural levels of the deeply eroded Precambrian crustal sections listed in Table II. This internal shearing is temporally and kinematically related to high-grade thrusting at the base of the crustal sections which puts the higher grade lower crust above lower grade rocks of the

498

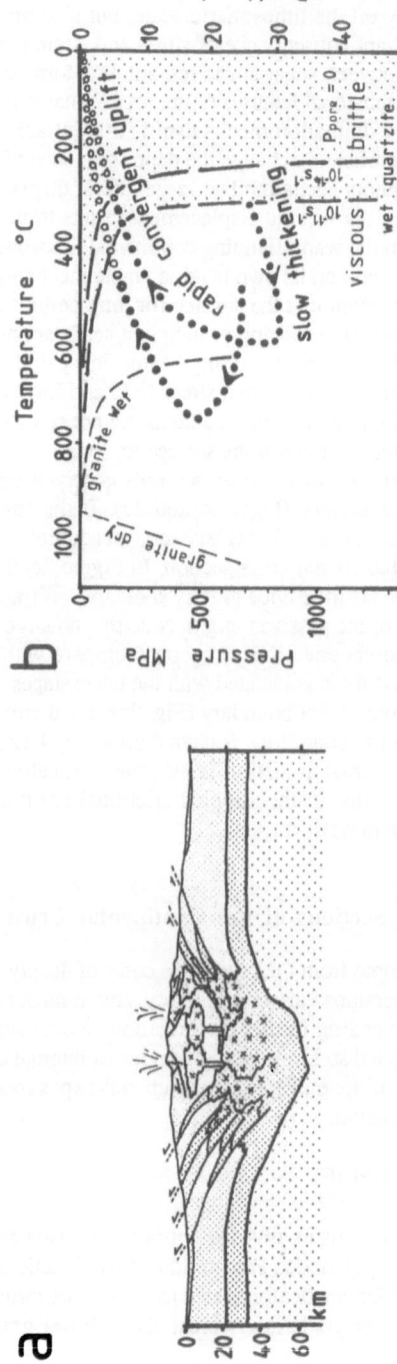

Figure 4. (a) Schematic profile through an oblique collisional magmatic arc based on Hollister and Crawford (1986); (b) Depth versus temperature diagram with P-T paths for two rapidly uplifted deep crustal segments in the Central Gneiss Complex, Coast Mountains, British Columbia (Hollister and Crawford, 1986). Open dots are extrapolated parts of P-T paths. Frictional to viscous transition (thick dashed lines) and granite solidus curves (thin dashed lines) as in Figs. 1b and 3b.

intermediate to upper crust. Thermal disequilibrium during thrusting is reflected in the asymmetrical zonation of fault rocks at the base of the cross section, with high-grade mylonite in the hot hanging wall grading to low-grade mylonite and cataclasite in the colder footwall (Table II). The sense of asymmetry of this zonation is therefore opposite to the asymmetry of fault rock zonation described above for extensional detachments. The observation of two or more generations of tectonites in some of these large shear zones (e.g., Redbank fault zone, Arunta Section; Obee and White, 1985) suggests that uplift of the crustal cross section involved several episodes of thrusting.

Reflection seismic profiling reveals that the high-grade thrusts bounding some of these crustal cross sections can be traced to the base of the crust (e.g., Grenville Front tectonic zone, Green et al., 1988) and are sometimes interpreted to cross, or even to displace the MOHO (e.g., Arunta Section, Goleby et al., 1989; for contrasting interpretation, see Teyssier, 1985). This raises the possibility that imbrication and uplift of the lower crust to the surface are directly connected to deep-seated shear zones within the mantle. An immediate implication of discrete faults that offset the MOHO is that the viscosity contrast between lower crustal and upper mantle rocks is much smaller than predicted from the extrapolation of laboratory creep laws for feldspar- and olivine aggregates to temperatures along typical orogenic geotherms (Carter and Tsenn, 1987).

4.2. MELT-ENHANCED THRUSTING AND UPLIFT:

Crustal cross sections from Cordilleran terrains of North America show substantial magmatism and crustal reconstitution associated with severe internal imbrication during oblique convergence and uplift. Hollister et al. (1989) have reconstructed their evolution from fragments of various crustal levels exposed in the Coast Mountains (western Canada; Table II) and the Idaho and Sierra Nevada batholiths (western United States). Figure 4a is a schematic profile through such an oblique convergent magmatic orogen, based on the work of Hollister and coworkers. Crustal shortening, metamorphism, magmatism, and uplift are broadly coeval and the interaction of these processes can be described in terms of a negative feedback cycle. The P-T paths of two deep crustal imbricates during two such cycles in the Central Gneiss Complex near Prince Rupert, British Columbia (Hollister and Crawford, 1986) are shown in Figure 4b. Collision leads to crustal shortening and thickening, burial metamorphism, and both mantle and crustal anatexis. These melts lubricate high- and low-angle faults that accommodate, respectively, rapid vertical and outward escape of deeper orogenic levels. Granitic intrusions in mid- to upper crustal levels may be associated with extension of the overthickened orogen (Hollister et al., 1989). Upward magma transport in such batholithic intrusions is also linked to localized viscous (mylonitic) downflow of the country rocks adjacent to the batholith (Saleeby, 1988). Decreasing temperatures during uplift freeze the magma, stop or slow the crustal uplift, and re-initiate a period of slower crustal thickening. Exhumation in these orogenic settings occurs during short periods or "surges" of very rapid uplift (1 to 2 mm/yr; Hollister, 1982) along melt-lubricated shear zones in the deep crust that grade upwards to mylonites at higher crustal levels (Hollister and Crawford, 1986). At the surface, sheets of higher grade rocks are emplaced over lower grade rocks (Fig. 4a).

This melt-enhanced deformational style contrasts with the solid-state creep deformation effecting exhumation of most other crustal cross sections. Partial melting is usually inferred to have a drastic weakening effect on the bulk strength of a rock (Arzi, 1978), but the magnitude of this strength drop as well as the long-term strength of partially melted rock depend strongly on the viscosity contrast between solid rock and melt, and on the amount and distribution of this melt within the deforming aggregate. If hot magma is injected into a significantly cooler rock, then the initial deformation may be brittle, while the magma intrudes rapidly along cracks and exerts a pore pressure that reduces the effective strength of the rock to levels favoring further fracture and

500

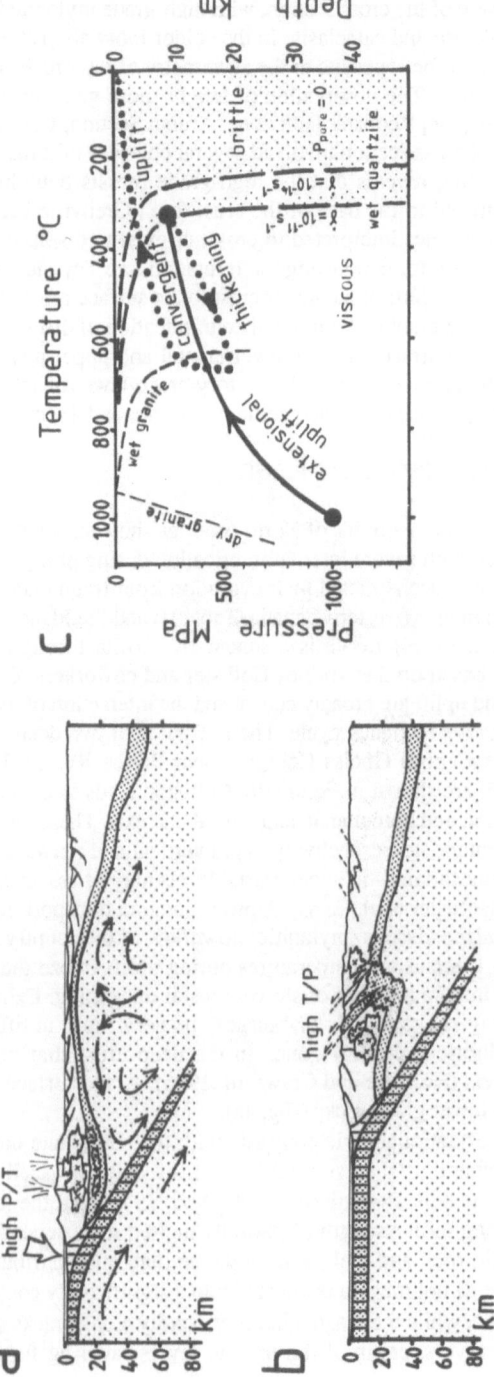

Figure 5. Extensional uplift of hot lower crust in a back-arc spreading environment (a) followed by imbrication and partial magmatic reconstitution during crustal thickening in (b). Hypothetical P-T evolution of the lower crust during extensional uplift (solid line) and shortening, thickening and uplift (dotted lines) appears in the depth-temperature diagram in (c). Brittle (i.e. frictional) to viscous transition in hydrous quartzite and solidus curves for granite as in previous figures.

dilation. After high strains or during in-situ synkinematic anatexis, however, the solid rock either comprises a viscously deforming framework containing lenticular layers of melt, or sufficient melt is present (> 20%; van der Molen and Paterson, 1979) to accommodate bulk strain within contiguous layers. These melt layers generally define a foliation parallel to the plane of shear and wrap about deformed lozenges of host rock (or restite; Fig. 1b, Hollister and Crawford, 1986). The mechanical role of the melt is not just primary (i.e., that of a weak, strain-accommodating phase), it is also secondary insomuch as the incipient melt forms thin films along grain boundaries that act like diffusive corridors and so enhances viscous grain boundary sliding in the host rock (Cooper and Kohlstedt, 1987). Thus, it is the combination of these effects that ultimately leads to a drop in flow strength coupled with an increase in strain-rate during syntectonic partial melting.

Generally, the volume proportion of sheared rock at a given strain-rate is inversely and nonlinearly proportional to the viscosity contrast between sheared (or melted) and unsheared rock. Shear zones are widest at deformational conditions above the solidus, where the viscosity contrasts both between rock and melt and among the constituent minerals in the rock are low. As partially melted middle and lower crustal rocks pass through the solidus on their way to the surface (Fig. 4b), shearing affects much of whatever original structure managed to survive magmatic reconstitution at depth. Therefore, rapid melt-enhanced uplift tends to preserve less of the pre-exhumational structure and metamorphism than do the other uplift mechanisms discussed above.

Other tectonic scenarios can be envisioned which exhume incomplete, dismembered sections of the continental crust in orogenic settings. In Figure 5a, for example, an early stage of subduction and crustal accretion is associated with granulite facies metamorphism and magmatism at the base of the continental crust. Localized instabilities in mantle convection lead to back-arc extension and roll-back of the oceanic subduction zone. Granulites at the bottom of the attenuating crust in the back-arc region are extensionally uplifted to midcrustal levels (Fig. 5a, solid P-T path in Fig. 5c). Exposure of the granulites at the surface results from thrusting and uplift while the thinned crust is still hot (dotted P-T path in Fig. 5c) or after this crust has cooled. Hot thrusting is associated with considerable internal imbrication and shearing of the uplifting crust under high-grade conditions (Fig. 5b), whereas cold thrusting and uplift might resemble the final emplacement stage depicted in Figure 2b. A similar exhumation mechanism (e.g., Fig. 2a in Fountain and Salisbury, 1981) may explain the pairing of metamorphic belts in many circum-Pacific oblique convergent margins, where the successive collision of island arcs resulted in the emplacement of rocks with high-T/P assemblages onto subducted and rapidly exhumed rocks with older high-P/T mineral assemblages (e.g., Hidaka crustal cross section, Hokkaido, North Japan; Komatsu et al., 1983).

5. The Exhumation of Crustal Cross Sections: A Selective Process

What part of the Earth's crust do we actually see in exposed crustal cross sections? A fundamental point to emerge in this paper is that crustal exhumation reveals certain types and features of the deep crust better than others. Moreover, different exhumation mechanisms preserve different stages of crustal evolution. This has implications for the question of how representative exposed crustal cross sections are of the unexposed continental crust. Intracratonic crustal sections emplaced as relatively cold wedges (the Kapuskasing section) best represent that part of the craton which lies above a deep intracrustal detachment. The unexposed crust below the detachment probably has different compositional and metamorphic characteristics. Crustal cross sections that were attenuated prior to final convergent emplacement are obviously good examples of thinned continental crust. However, the crustal levels that presently constitute these thinned crustal sections did not necessarily occupy the same vertical section of crust prior to the extensional stage of exhumation.

Finally, cross sections of crust that were uplifted in orogens at high geothermal gradients are rarely complete and are often internally imbricated and/or magmatically reconstituted under high-grade metamorphic conditions. Such exposures yield abundant information about crustal reconstitution and deformation in the core of an orogen, but preserve relatively little information about the crustal evolution prior to uplift.

What sets true crustal cross sections apart from most deeply eroded orogenic exposures is the high degree of structural and metamorphic coherence maintained during uplift and emplacement to the surface. Compressional and isostatic forces associated with the anomalous lithospheric structure beneath exposed crustal cross sections balance dense lower crust near the surface over long periods of time. It is not fortuitous that the bottom of all exposed crustal cross sections examined so far originally occupied depths no greater than about 45 kilometers. In some cases, this depth represents the thickness of the crust prior to extensional uplift, whereas in other cases it was the depth of basal intracrustal detachment and thrusting at the onset of convergent uplift. Original deep crustal features in crustal cross sections survived exhumation because decreasing temperatures during uplift led to increased viscous strength contrasts among different compositional domains and hence to progressive stress and strain localization within narrow shear zones.

In conclusion, the exhumation of coherent cross sections of the continental crust is a very selective, usually multistage process involving special kinematic and geothermal evolutions. The sampling of the deep crust exposed at the surface is therefore only as complete as the environments conducive to its uplift are varied.

Acknowledgements

I would like to thank D. Fountain, M. Salisbury, J. Barrett and the rest of the NATO-ASI staff for making the two weeks at Killarney, Ontario such an enjoyable experience. The ideas presented in this paper matured after many stimulating discussions at the conference and during subsequent communication with ASI-participants. G. Gibson, D. Hyndman, G. Oliver, V. Schenk, W. Weber, and H. Williams kindly provided reprints of their work. J. Percival's remarks on an early draft prompted a closer look at the Kapuskasing section. The reviews of C. Beaumont and G. Bassi, and the critical comments of A. Green and A. Zingg on the final manuscript lead to numerous substantive improvements. A. Pfiffner suggested several changes that clarified the text. Attendance at the NATO conference was made possible by a travel grant from the Karl Bretscher Stiftung of the University of Bern. The paper was researched and written with the support of the Swiss National Science Foundation (Project 2971-0.88).

References

Arzi, A.A., 1978, 'Critical Phenomena in the Rheology of Partially Melted Rocks', *Tectonophysics*, **44**, 173-184

Berckhemer, H., 1968, 'Topographie des "Ivrea Körpers" abgeleitet aus seismischen und gravimetrischen Daten', *Schweizerische Mineralogische und Petrographische Mitteilungen*, **48**, 235-246

Boland, A.V., Ellis, R.M., Northey, D.J., West, G.F., Green, A.G., Forsyth, D.A., Merev, R.F., Meyer, R.F., R.P., Morel-a-l'Huissier, P., Buchbinder, G.G.R., Asudeh, I., Haddon, R.A.W., 'Seismic delineation of upthrust Archean crust in Kapuskasing, Northern Ontario', *Nature*, **335**, 711-713

Boland, J.N. and Tullis, T.E., 1986, 'Deformation Behaviour of Wet and Dry Clinopyroxenite in the Brittle to Ductile Transition Region', in: *Mineral and Rock Deformation: Laboratory Studies, American Geophysical Union Monograph*, **36**, 35-50, edited by B.E. Hobbs and H.C. Heard, American Geophysical Union, Washington D.C.

Bott, M.H.P., 1981, 'Crustal doming and the mechanism of continental rifting', *Tectonophysics*, **73**, 1-8

Brace, W.F., and Kohlstedt, D.L., 1980, 'Limits on lithospheric stress imposed by laboratory experiments', *Journal of Geophysical Research*, **85**, 6248-6252

Brodie, K.H., and Rutter, E.H., 1985, 'On the relationship between deformation and metamorphism with special reference to the behaviour of basic rocks', in: *Kinematics, Textures, and Deformation, Advances in Physical Chemistry*, **4**, 138-179, edited by A.B. Thompson and D. Rubie, Springer Verlag, Heidelberg

Brodie, K.H. and Rutter,E.H., 1987, 'Deep Crustal Extensional Faulting in the Ivrea Zone of Northern Italy', *Tectonophysics*, **140**, 193-212

Byerlee, J., 1978, 'Friction of Rocks', *Pure and Applied Geophysics*, **116**, 615-626

Carter, N.L., and Tsenn, M.C., 1987, 'Flow properties of the continental lithosphere', *Tectonophysics*, **42**, 75-110

Chapman, D.S., 1986, 'Thermal gradients in the continental crust, in: *The Nature of the Lower Continental Crust, Geological Society Special Publication*, **24**, 51-62, edited by J.B. Dawson, D.A. Carswell, J. Hall, K.H. Wedepohl, Blackwell Scientific Publications, Oxford

Chopra, P.N., and Paterson, M.S., 1981, 'The experimental deformation of dunite', *Tectonophysics*, **42**, 75-110

Cooper, R.F., and Kohlstedt, D.L., 1987, 'Rheology and structure of olivine-basalt partial melt', *Journal of Geophysical Research*, **91**, 9315-9323

Davidson, A., 1988, 'Two sections across the Grenville Front, Killarney and Tyson Lake Areas, Ontario', *Field Guide, NATO Advanced Studies Institute Program*, Killarney, Ontario, Part I, 1-9

Ernst, W.G., 1988, 'Tectonic history of subduction zones inferred from retrograde blueschist P-T paths', *Geology*, **16**, 1081-1084

Etheridge, M.A., Wall, V.J., and Cox, S.F., 1984, 'High fluid pressures during regional metamorphism and deformation: Implications for mass transport and deformation mechanisms', *Journal of Geophysical Research*, **89**, 4344-4358

Fountain, D.M. and Salisbury, M.H., 1981, 'Exposed cross sections through the continental crust: implications for crustal structure, petrology, and evolution', *Earth and Planetary Science Letters*, **56**, 263-277

Gibson, G.M., McDougal, I., and Ireland, T.R., 1988, 'Age constraints on metamorphism and the development of a metamorphic core complex in Fiordland, southern New Zealand', *Geology*, **16**, 405-408

Goleby, B.R., Shaw, R.D., Wright, C., Kenneth, B.L., and Lambeck, K., 1989, 'Geophysical evidence for 'thick-skinned' crustal deformation in central Australia', *Nature*, **337**, 325-330

Green, A.G., Hajnal., Z., and Weber, W., 1985, 'An evolutionary model of the western Churchill Province and western margin of the Superior Province in Canada and the north-central United States', *Tectonophysics*, **116**, 281-322

Green, A.G., Milkereit, B., Davidson, A., Spencer, C., Hutchinson, D.R., Cannon, W.F., Lee, M.W., Agena, W.F., Behrendt, J.C., and Hinze, W.J., 1988, 'Crustal structure of the Grenville front and adjacent terranes', *Geology*, **16**, 788-792

Handy, M.R., 1987, 'The structure, age and kinematics of the Pogallo Fault Zone; southern Alps, northwestern Italy', *Eclogae Geologicae Helvetiae*, **80/3**, 593-632

Handy, M.R., 1989, 'Deformation regimes and the rheological evolution of fault zones in the lithosphere: The effects of pressure, temperature, grainsize and time', *Tectonophysics*, **163**, 119-152

Handy, M.R., 1990, 'The solid-state flow of polymineralic rocks', *Journal of Geophysical Research*

Handy, M.R. and Zingg, A., in press, 'The tectonic and rheological evolution of an exposed cross section through the continental crust (southern Alps, Europe)', *Bulletin of the Geological Society of America*

Hollister, L.S., 1982, 'Metamorphic evidence for rapid (2 mm/yr) uplift of a portion of the Central Gneiss Complex, Coast Mountains, British Columbia', *Canadian Mineralogist*, **20**, 319-332

Hollister, L.S., and Crawford, M.L., 1986, 'Melt-enhanced deformation: A major tectonic process', *Geology*, **14**, 558-561

Hollister, L.S., Hyndman, D.W., and Saleeby, J.B., 1989, 'A composite crustal section through mid- to late Cretaceous collisional magmatic arcs in western North America', *Abstract, Geological Society of America Meeting*, p. 94

Hubregtse, J.J.M.W., 1980, 'The Archean Pikwitonei Granulite Domain and its position at the margin of the northwestern Superior Province (Central Manitoba)', *Geological Survey of Canada, Mineral Resources Division*, **GP80-3**, 1-16

Hurford, A.J., 1986, 'Cooling and uplift patterns in the Lepontine Alps, South Central Switzerland and an age of vertical movement on the Insubric fault line', *Contributions to Mineralogy and Petrology*, **92**, 413-427

Jaoul, O., Tullis, J., and Kronenberg, A., 1984, 'The effect of varying water content on the behaviour of Heavitree quartzite', *Journal of Geophysical Research*, **89**, 4298-4312

Johannes, W., 1985, 'The significance of experimental studies for the formation of migmatites', in: *Migmatites*, edited by J.R. Ashworth, pp. 36-86, Blackie and Son Limited, Glasgow

Komatsu, M., Miyashita, S., Maeda, J., Osanai, Y., and Toyoshima, T., 1983, 'Dislosing of a Deepest Section of Continental-Type Crust Up-Thrust as the Final Event of Collision of Arcs in Hokkaido, North Japan', in: *Accretionary Tectonics in the Circum-Pacific Regions*, edited by M. Hashimoto and S. Uyeda, pp. 149-165, Terra Scientific Publishing Company, Tokyo

Laubscher, H.P., and Bernoulli, D., 1982, 'History and Deformation of the Alps', in: *Mountain Building Processes*, edited by K. Hsü, Chapter 2-3, pp. 169-180, Academic Press, London

Le Pichon, X., and Sibuet, J.C., 1981, 'Passive margins: A Model of Formation', *Journal of Geophysical Research*, **86**, 3708-3720

Lister, G.S., Etheridge, M.A., and Symonds, P.A., 1986, 'Detachment faulting and the evolution of passive continental margins', *Geology*, **14**, 246-250

Obee, H.K., and White, S.H., 1985, 'Faults and associated fault rocks of the southern Arunta block, Alice Springs, Central Australia', *Journal of Structural Geology*, **7/6**, 701-712

Oliver, G.J.H., and Coggon, J.H., 1979, 'Crustal structure of Fiordland, New Zealand', *Tectonophysics*, **54**, 253-292

Paterson, M.S., 1987, 'Problems in the extrapolation of laboratory rheological data', *Tectonophysics*, **133**, 33-43

Percival, J.A., 1988, 'The Kapuskasing Uplift: Cross section through the Archean Superior Province', *Field Guide, NATO Advanced Study Institutes Program*, Killarney, Ontario, Part I, pp. 1-31

Percival, J.A., and Card, K.D., 1985, 'Structure and Evolution of Archean crust in Central Superior Province, Canada', in: *Evolution of Archean Supracrustal Sequences, Geological Association of Canada Special Paper*, **28**, 179-189, edited by L.D. Ayres, P.C. Thurston, K.D. Card, and W. Weber

Percival, A.G., and Green, A.J., 1988, 'Towards a balanced crustal-scale cross section of the Kapuskasing uplift', *Abstract, KSZ Workshop II, Project Lithoprobe*, pp. 233-234

Percival, J.A., Green, A.G., Milkereit, B., Cook, F.A., Geis, W., and West, G.F., 1989, 'Seismic reflection profiles across deep continental crust exposed in the Kapuskasing uplift structure', *Nature*, 342, 416-420

Percival, J.A., and McGrath, P.H., 1986, 'Deep crustal structure and tectonic history of the northern Kapuskasing uplift of Ontario: An integrated petrological-geophysical study', *Tectonics*, 5/4, 553-572

Platt, J.P., 1986, 'Dynamics of orogenic wedges and the uplift of high-pressure metamorphic rocks', *Geological Society of America Bulletin*, **97**, 1037-1053

Ruppel, C., Royden, L., and Hodges, K.V., 1988, 'Thermal modelling of extensional tectonics: Application to pressure-temperature-time histories of metamorphic rocks', *Tectonics*, 7/5, 947-957

Rutter, E.H. and Brodie, K.H., 1988, 'The role of tectonic grain size reduction in the rheological stratification of the lithosphere', *Geologische Rundschau*, 77/1, 295-308

Saleeby, J., 1988, 'A down-plunge view of the Sierra Nevada batholith: structural and petrogenetic relations from resergent caldera to granulitic levels', *Abstract, Programme of NATO Advanced Studies Institute Course on Exposed Cross Sections of the Continental Crust*

Schmid, S.M., Zingg, A., and Handy, M.R., 1987, 'The kinematics of movements along the Insubric Line and the emplacement of the Ivrea Zone, *Tectonophysics*, **135**, 47-66

Shelton, G. and Tullis, J., 1981, 'Experimental flow laws for crustal rocks', *Abstract, Eos Transactions of the American Geophysical Union*, 62/17, p. 396

Teyssier, C., 1985, 'A crustal thrust system in an intracratonic tectonic environment', *Journal of Structural Geology*, 7/6, 689-700

Tharp, T.M., 1983, 'Analogies between the high-temperature deformation of polyphase rocks and the mechanical behaviour of porous powder metal', *Tectonophysics*, **96**, T1-T11

Van Breeman, O., and Davidson, A., 1988, 'Northeast extension of Proterozoic terranes of mid-continental North America', *Geological Society of America Bulletin*, **100**, 630-638

van der Molen, I., and Paterson, M.S., 1979, 'Experimental deformation of partially melted granite', *Contributions to Mineralogy and Petrology*, 70, 299-318

Weber, W., and Scoates, R.F.J., 1978, 'Archean and Proterozoic metamorphism in the northwestern Superior Province and along the Churchill-Superior Boundary, Manitoba', *Geological Survey of Canada Paper*, **78-10**, 5-16

Williams, H.R., 1988, 'The Archean Kasila Group of western Sierra Leone: Geology and Relations with adjacent granite-greenstone terranes', *Precambrian Research*, **38**, 201-213

Williams, H.R., and Culver, S.J., 1988, 'Structural terranes and their relationships in the Sierra Leone', *Journal of African Earth Sciences*, **7/2**, 473-477

Zingg, A., Handy, M.R., Hunziker, J.H., and Schmid, S.M., in press, 'Tectonometamorphic history of the Ivrea Zone and its relationship to the crustal evolution of the southern Alps', *Tectonophysics*

THE FLUID CRUSTAL LAYER AND ITS IMPLICATIONS FOR CONTINENTAL DYNAMICS

BRIAN WERNICKE
Department of Earth and Planetary Sciences
Harvard University
Cambridge, Massachusetts 02138

ABSTRACT. Extensional tectonism exposes relatively intact cross-sections of the pre-tectonic crust. The paleodepth of deep portions of the exposed crustal sections in strongly extended structural domains in the Basin and Range province constrains the amount of upper crustal thinning accommodated along extensional structures, which in many cases is in excess of 10 km. Regionally averaged topography in the province is generally lower in the strongly extended domains than in adjacent stable blocks in which the upper crust is not appreciably thinned. The difference in elevation between extended and unextended areas suggests that the differential thinning of the upper crust is probably not accommodated by inflow or outflow of asthenosphere, mantle lithosphere or mafic lower crust. Simple isostatic calculations suggest that the density of the compensating medium is probably within 100-200 kg/m^3 of the density of average upper continental crust, indicating that it lies within the crust and may be in large part quartzose. This conclusion is independently supported by laboratory experiments on the strengths of rocks, which suggest that quartzose rocks are substantially weaker than mafic and ultramafic rocks over a broad range of temperatures likely to exist in the deep crust, and with reflection seismograms in deformed regions which suggest that the Moho is subhorizontal beneath areas with large gradients in upper crustal vertical strain. It is suggested that the upper crust floats on a quartzose layer in the mid-crust, which under orogenic conditions appears to behave as a relatively inviscid fluid at geologic timescales (>10,000 a) and subcontinental lengthscales (100-1000 km). A fourfold division of the orogenic lithosphere is therefore proposed: (1) The upper crust, which at geologic timescales has the properties of a solid and is able to support shear stresses of 10's to 100's of MPa; (2) a fluid crustal layer (fluid in the sense of the asthenosphere), which flows so as to eliminate horizontal gradients in vertical stress on geologic timescales; (3) a solid lower crust, generally mafic, that may be substantially stronger than the overlying fluid layer; and (4) the mantle part of the lithosphere, which is much stronger than the fluid layer. The thickness of the fluid layer may range from 0 to over 30 km, but is generally in the 15-25 km range.

A model of continental deformation is thus proposed in which localized strain within the mantle lithosphere and solid lower crust is accompanied by regional failure of the fluid layer. Heterogeneously deforming solid upper crustal blocks ride buoyantly on the fluid layer, whose flow field is governed in large part by maintaining flotational equilibrium with solid upper crustal blocks, generally diverging from regions where deformation results in relative thickening of the upper crustal layer. The fluid layer may also accommodate uniform-sense simple shear, coupling the integrated upper crustal strain with discrete areas of increase or decrease of horizontal surface area at the top of the solid lower crust and mantle lithosphere. Intracrustal isostatic compensation during extensional or compressional orogenesis will be most likely to occur in areas of thick crust and high heat flow, as in the mid-Tertiary Basin and Range or the

M. H. Salisbury and D. M. Fountain (eds.), Exposed Cross-Sections of the Continental Crust, 509–544.
© 1990 *Kluwer Academic Publishers.*

modern-day Tibetan Plateau and Altiplano regions, resulting in broadly distributed crustal deformation and plateau-like topography. Areas of low geotherm or relatively thin crust may tend to accommodate strain in a more localized fashion with isostatic compensation occurring only by asthenospheric flow, accompanied by the development of high local topographic relief. A transition from one regime to the other may be exemplified by the contrast between regions such as the Basin and Range and Afar triangle regions (fluid layer present) and the northern Red Sea (fluid layer absent). Because vertical strains of the upper and fluid crustal layers are complementary, intense localized thickening of the upper crust may be accompanied by strong thinning of the fluid layer, depending on the initial ratio of their thicknesses. Similarly, areas of strongest upper crustal extension may result in overall thickening of the fluid layer beneath them. Outflow of fluid from beneath areas of crustal thickening may serve to thicken crust beneath some cratonic forelands (e.g., the western Great Plains), or conversely inflow toward areas of strong upper crustal thinning may attenuate crust on the margins of rifts (e.g., the western Colorado Plateau).

1. Introduction

The boundaries of continental lithospheric plates are generally diffusely strained. Central to plate theory is the idea that a solid lithosphere floats on a relatively inviscid, fluid asthenosphere. Broad aspects of the dynamics of orogenic belts can be successfully explained by treating the continental lithosphere as a thin viscous sheet in which a coupling exists between vertical and horizontal strain, and in which the overall rheology is governed principally by the strength of the mantle part of the lithosphere (e.g. England and McKenzie, 1982; Sonder et al., 1987). However, the thin sheet approximation does not account for large-scale decoupling of the crust from the mantle lithosphere, whereby strain may occur in portions of the crust and mantle that are horizontally far removed from one another. Deformation of the continental lithosphere may be strongly influenced by weak layers, particularly in the middle to lower crust (Armstrong and Dick, 1974; Brace and Kohlstedt, 1980). Models including multiple layers of varying strengths may simulate behavior not predicted by the thin sheet approximation, particularly at smaller scale (e.g. Zuber and Parmentier, 1986). Observations bearing on the degree to which rheological stratification influences continental deformation are thus central to understanding continental dynamics.

Studies of diffuse finite strain in the continental lithosphere have yielded insights independently of theoretical and experimental considerations via detailed field mapping and seismic imaging (e.g. Wernicke et al., 1988; Matthews and Cheadle, 1986). The results of this work suggest that in diffusely deformed zones such as the Basin and Range, the British Isles, the Grenville orogen and the New England Appalachians, the lower crust tends to be highly reflective (Allmendinger et al., 1987; Matthews and Cheadle, 1986; Green et al., 1988; Ando et al., 1984). There is currently no agreement as to the origin of this reflectivity, and explanations including pore fluid heterogeneity, basic sill complexes, pervasive mylonitization transposing preexisting layering into subhorizontality, metamorphic differentiation, and a number of other explanations have all been advanced. Experimental data on the strengths of rocks suggesting that the lower crust is weak have been used to attribute the reflectivity to low viscosity (Meissner and Wever, 1986), in that regions of low viscosity in the crust may be more subject to flow and therefore be more likely to develop subhorizontal fabric than stronger layers within the crust. In regions where upper crustal strain is strongly heterogeneous, the Moho appears to be relatively flat (e.g. Hauser et al., 1987a,b). This has led to the notion that the Moho may be a transient feature, somehow able to re-equilibrate or adjust itself in response to crustal thickening or thinning events (e.g. Allmendinger et al., 1987).

Geological mapping and geophysical data taken together define anomalous regions where variations in the amount of strain recorded in upper crustal rocks are not observed for the crust as a whole. For example, in the Basin and Range province, a number of areas have been identified where a discrepancy exists between large variations in upper crustal strain and little or no variation in crustal thickness. This observation has led to the hypothesis that lower crust in the Basin and Range may flow out from beneath unextended regions into extended ones, accommodated along simple shear zones between the undeformed upper crustal block and ductile lower crust (Wernicke, 1983, 1985, Fig. 11; Spencer and Reynolds, 1984; Gans, 1987; McKenzie, 1988). Smoothing of the Moho by flow and reinflation of extended crust with mantle-derived magma have been suggested as possible mechanisms by numerous workers (e.g. Thompson and McCarthy, 1986a; Allmendinger et al., 1987; Gans, 1987; Block and Royden, 1988).

In addition to the relatively flat Moho in the Basin and Range, the gravity and topographic expression of large-scale crustal strain is relatively subdued (Wernicke, 1983, 1985; Spencer and Reynolds, 1984, 1989; Thompson and McCarthy, 1986a,b; McKenzie, 1988; Block and Royden, 1988, 1990). According to Thompson and McCarthy (1986a), highly strained regions of the Basin and Range have no characteristic gravity or topographic signature, suggesting that some mechanism such as igneous addition from the mantle must be reinflating the extended regions to prevent them from becoming topographically low relative to surrounding regions. Block and Royden (1988; 1990) suggested that large-scale lateral inflow of lower crust of density very close to that of the upper crust could explain the lack of prominent topographic expression.

In this paper, these concepts are further developed by examining the relationship between topography and crustal strain in the extensional setting of the Basin and Range province, and in the contractional setting of the Wind River Range, a Laramide foreland uplift. By comparison of geological data on crustal sections with long-wavelength topography, we show that there is indeed a characteristic topographic signature to extended and shortened continental lithosphere, and that this signature constrains the density of the material that isostatically compensates thickening and thinning of the upper crust. These results indicate that the upper crust is probably afloat on a quartzose, fluid crustal layer. The layer is referred to as "fluid" in the sense of the asthenosphere: on geologic timescales (>10,000 a) and subcontinental lengthscales (100's to perhaps 1000's of km), the layer flows quickly enough to eliminate horizontal gradients in the vertical stress, thus maintaining hydrostatic equilibrium. In reality the rheology of the layer is probably quasi-plastic (conforming to flow laws of the type summarized by Brace and Kohlstedt, 1980), with yield stress and effective viscosity sufficiently low to allow the layer to act as a compensating medium for tectonically induced horizontal density contrasts in the upper crust.

The fluid layer concept differs from models advocating that the lower crust is simply weak or ductile because a weak, ductile layer need not flow quickly enough to attain hydrostatic equilibrium on geologic timescales (McKenzie, 1988). The purpose of this analysis is not to test the feasibility of intracrustal isostatic compensation by physically modeling the flow of materials conforming to experimentally determined flow laws. Rather, we use the lithosphere itself as an experimental apparatus to argue that such compensation occurs, in the hope that this may better focus modeling efforts. This result supports the conclusion that the orogenic lithosphere is characterized by decoupling of its upper and lower parts (e.g. Armstrong and Dick, 1974), perhaps with large-scale simple shear between them accommodated within the fluid layer.

2. Extended Lithosphere

Over the last decade, the amount and style of upper crustal extension has been increasingly well documented in a number of extended regions throughout the world. In particular, in a number of strongly extended regions in the Basin and Range, it has been shown that upper 15 or so km of the crust extended by stretching factors ranging from one hundred up to thousands of percent increase original width (Anderson, 1971; Guth, 1981; Miller et al., 1983; Wernicke et al., 1988; many others). Wernicke and Axen (1988) and Wernicke et al. (1988), based on field relations in the Basin and Range, argued that the extensions were so large that the upper crust can be thought of as being ruptured along zones where the initial widths of the extended terrains were in many cases only a few kilometers, much narrower than the relatively unextended blocks bounding them (Fig. 1). Between the separated blocks, the remainder of the lithosphere was considered to have rebounded isostatically to fill the void created between them. The minimum thickness of upper crust laterally removed from the strongly extended domains can be measured from the paleodepths of crustal sections exposed in steeply tilted, thin crustal slivers "stranded" atop the rebounded crust.

Figure 2 shows a compilation of such regions in the western United States, where large separation of relatively unstrained upper crustal blocks has resulted in a tectonic configuration more-or-less similar to that depicted in Figure 1. The minimum paleodepths of the deepest rocks exposed in the stranded crustal slabs (Table 1) range from as little as 5 km to as much as 18 km. In general, the extended regions correlate with areas that are topographically low relative to surrounding areas of little-extended crust (Fig. 3), suggesting a genetic relationship between large-scale tectonic denudation and the topographic depressions. The magnitude of topographic depression of the extended regions with respect to surrounding areas varies, ranging from as little as 200-300 m in the Okanagan example to as great as 1800 m in the Death Valley region (Fig. 4). There appears to be less relative topographic depression of extended domains adjacent to small unextended blocks (e.g. the Spring Mountains area of southern Nevada; Fig. 4) than there is adjacent to large ones (e.g. the Colorado Plateau or Sierra Nevada).

The correspondence between regional topographic depression and upper crustal denudation invites comparison to a simple model of isostatic rebound. In this model, we consider the lithosphere as in a problem of melting or adding glacial ice, that is removal from, or addition to, the upper lithosphere with compensating inflow at its base (Fig. 5). The only parameters that control the net topographic difference before and after denudation are the thickness of material removed, h_{crust} (or gained, as will be addressed below) and the ratio of the densities of the denuded layer and the compensating medium, ρ_{crust}/ρ_{comp} (Fig. 5). The densities of all intervening layers are the same before and after denudation and hence cancel each other out when the masses before and after denudation are equated above some depth of compensation in the fluid, giving

$$h_{comp} = (\rho_{crust}/\rho_{comp})h_{crust}, \tag{1}$$

where h_{comp} is the amount of isostatic rebound of the denuded region. The topographic

Table 1. Crustal sections in the Basin and Range province and environs. Letters keyed to locations of sections shown on Figs. 2 and 3.

Crustal Section	Uplift, km	Reference
A. Wasatch	11	Parry and Bruhn, 1987
B. Sevier Desert	8	Allmendinger et al., 1986
C. E.-central Nevada	10	Miller et al., 1983
D. Ruby Mountains	10	Snoke and Howard, 1983
E. Mormon Mountains	7	Axen and Wernicke, 1989
F. Gold Butte	10	Fryxell and Wernicke, in press
G. Death Valley	18	Wernicke et al., 1988
H. Yerington	7	Proffett, 1977
I. Mohave Mountains	10	Howard and John, 1987
J. Whipple Mountains	12	Davis, 1988
K. Newport	10	Harms and Price, 1983
L. Valhalla, Okanagan	15, 10	Parrish et al., 1988
M. Raft River	8	Compton et al., 1977

514

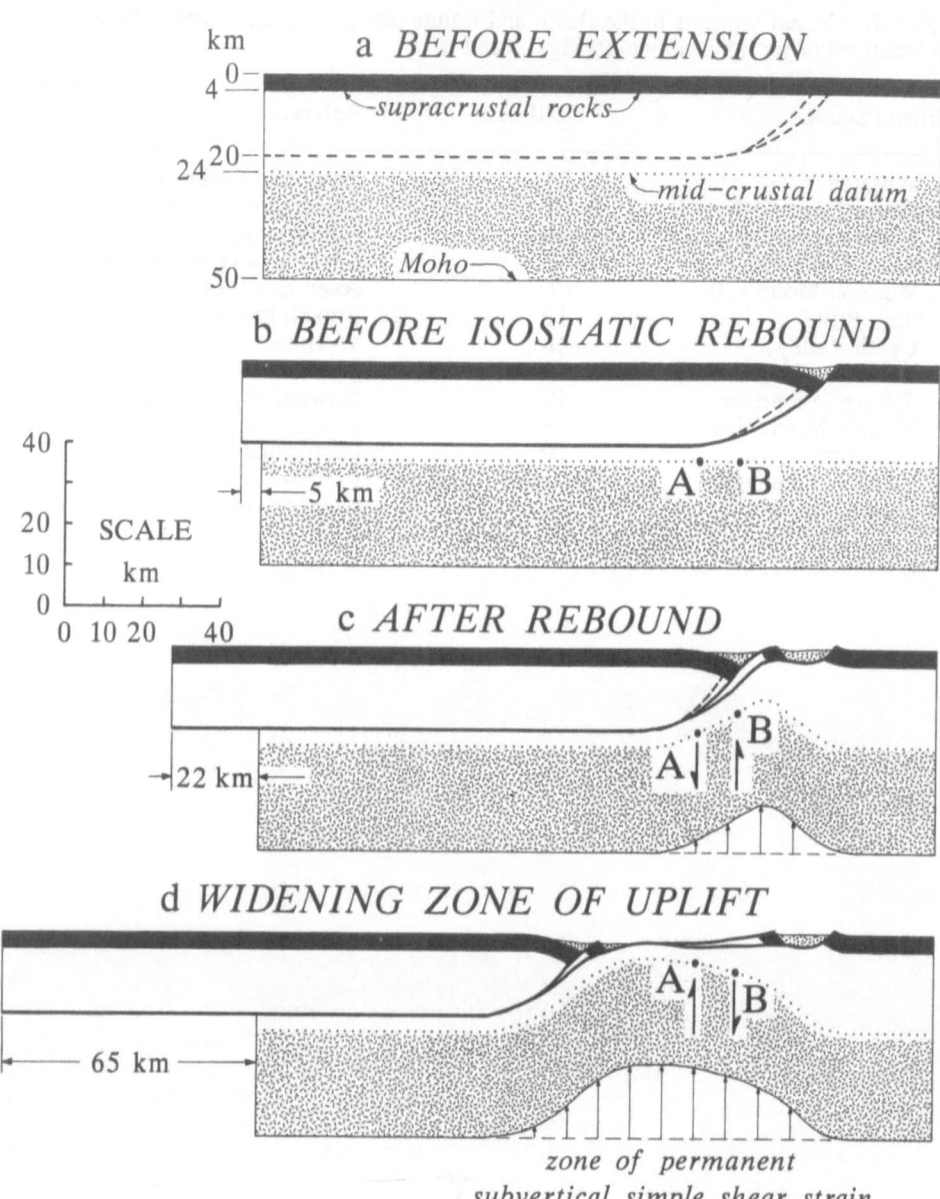

Figure 1. Kinematic model of the evolution of an extensional terrain, showing progressive upending of crustal sections detached from a migrating headwall, from Wernicke and Axen (1988; see also Hamilton, 1988). Isostatic compensation occurs by inflow of asthenosphere, causing deflection of the Moho (compare with Fig. 12).

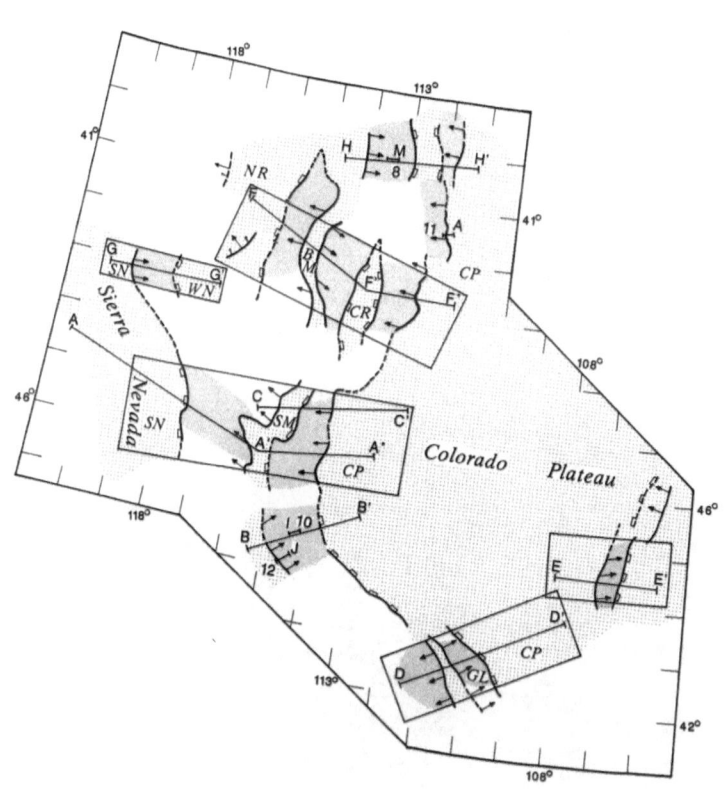

Figure 2. Map showing distribution of selected extended regions (darker shading), crustal sections and stable blocks (lighter shading) in the Basin and Range province and environs. Boxes show areas depicted on Fig. 3. Long lines correspond to topographic cross-sections shown on Fig. 4. Short lines show location of crustal sections (those within boxes are shown on Fig. 3), with single capital letters keyed to Table I. Numbers adjacent to crustal sections correspond to the minimum paleodepth of their deepest parts. Double capital letters indicate stable blocks: NR, Nevada rift block; BM, Butte Mountains block; CR, Confusion Range block; CP, Colorado Plateau block; PD, Paradise-Desatoya block; SN, Sierra Nevada block; SM, Spring Mountains block; GL, Galiuro block. Boundaries of extended terrains with arrows are proximal or "breakaway" (faults dip away from stable block), boundaries with square teeth are distal (faults dip beneath stable block).

516

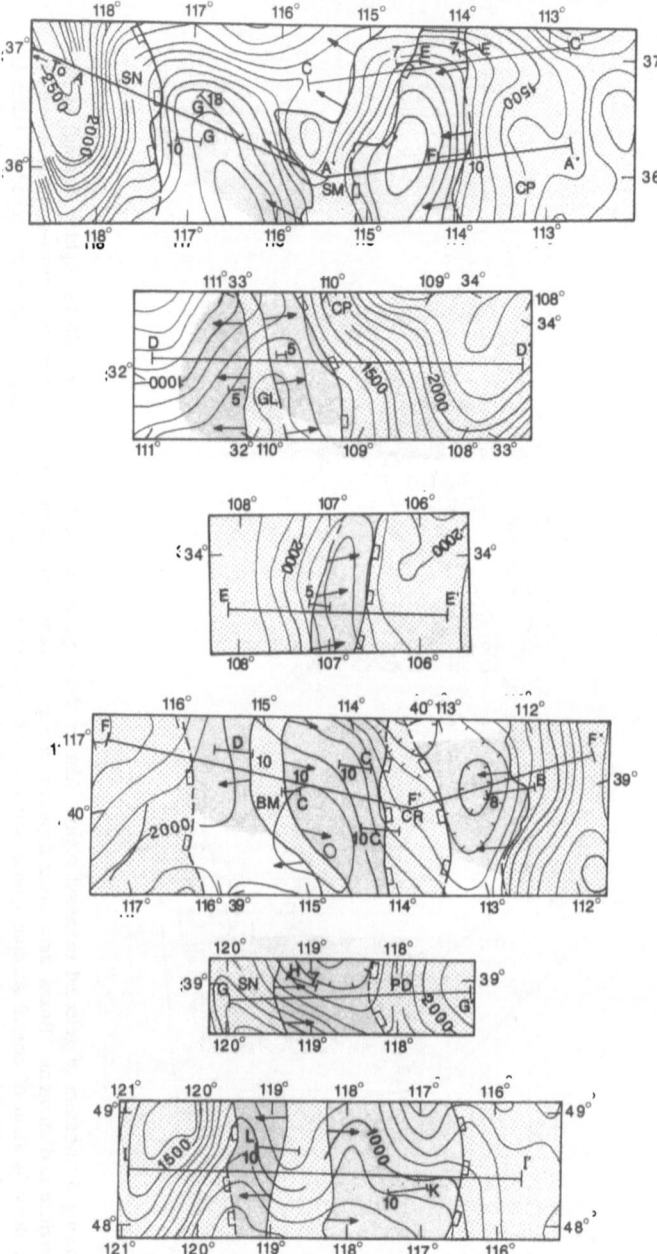

Figure 3. Regional elevation maps (in meters) of selected extended terrains in the Basin and Range shown on Fig. 2, except lowermost frame which is from extended terrain in northeastern Washington and northern Idaho. Symbols and patterns are the same as on Fig. 2. Regional elevation data from Diment and Urban (1981).

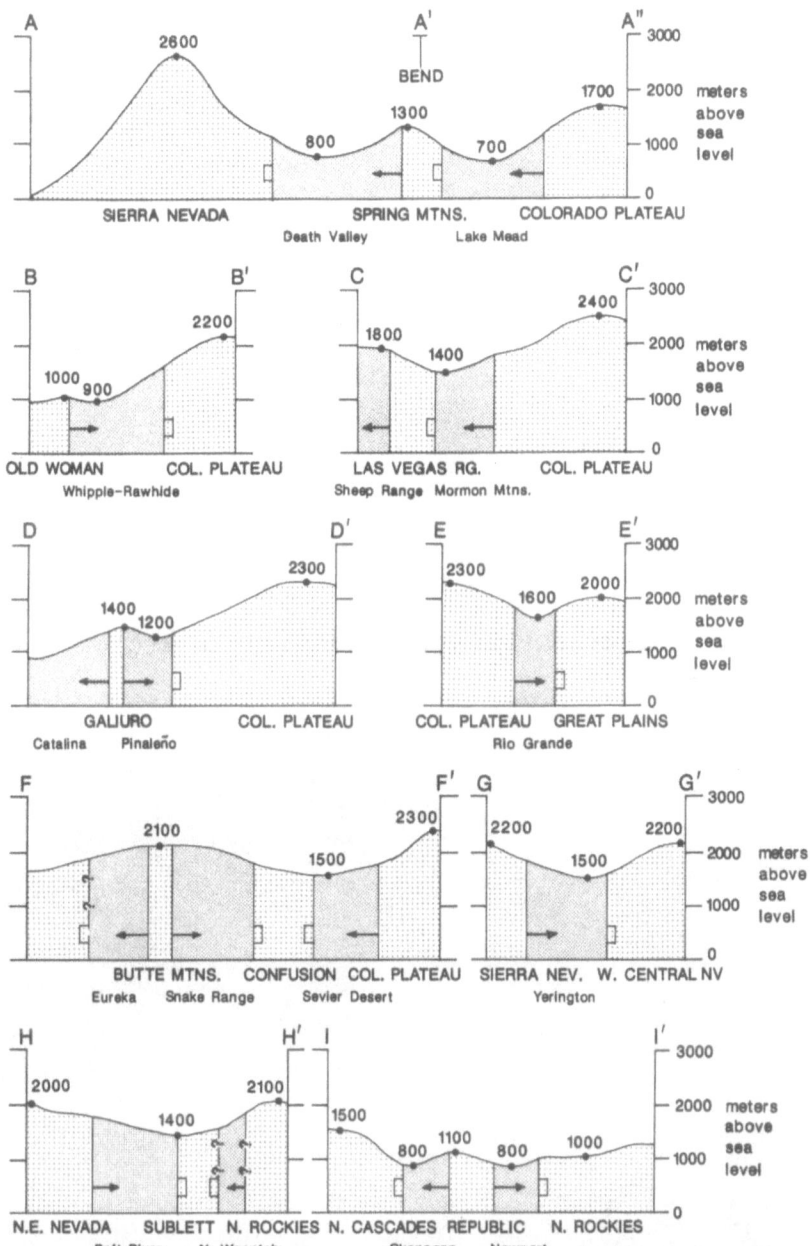

Figure 4. Cross-sections (locations on Figs. 2 and 3) of average elevation data from Fig. 3, showing distribution of extended domains and stable blocks. Stable blocks labeled in capital letters, extended domains in lower case. Numbered points show extremes in elevation used to establish maximum and minimum topographic depression plotted on Fig. 6 (see text for discussion). Symbols and patterns are the same as on Figs. 2 and 3.

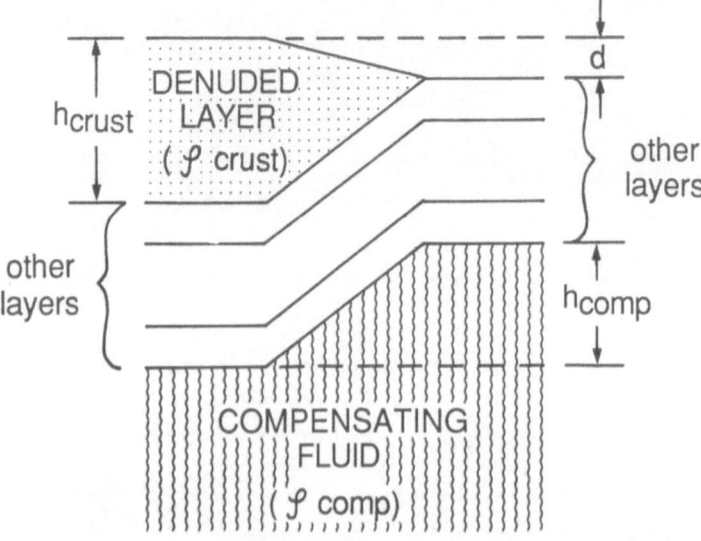

Figure 5. Simple denudation model showing variables in equations (1) and (2) and Fig. 6. See text for discussion.

depression d of a region of denudation where all other lithospheric layers are unmodified is the difference between the amount removed and the isostatic rebound, h_{crust}- h_{comp} (Fig. 5). Eliminating h_{comp} and solving for d gives

$$d = (1 - \rho_{crust}/\rho_{comp})h_{crust}. \qquad (2)$$

This relation allows a direct comparison of such a model with the topographic and geologic data. The depressions and amounts of denudation are difficult to constrain, and as such some criteria for assessing their magnitudes must be established. The parameter d plotted on Figure 5 represents the average of the extremes of elevation difference between the extended terrain and unextended blocks on either side of it (Fig. 4), showing error bars whose limits are the extreme values. The amount of denuded material h_{crust} is plotted as a minimum value cited in the literature (Table 1). In general, these values are probably too conservative. Not only are they the minimum of estimated paleodepth of exposed rock in the upended slabs, but as shown in Figure 1 the deepest rocks exposed in the slabs represents a minimum thickness of the block from which it was stripped. It is thus possible that the paleodepths underestimate the total denudation of the lithosphere in the extended regions by a large fraction of their values.

Assuming that the simple denudation model is approximately correct, the density of the compensating medium may be determined. Assuming an average density of denuded material of about 2700 kg/m³, the data would ideally fall along a line whose slope is proportional to the density of the compensating medium, shown as a family of lines in Figure 6. Because these parameters are poorly constrained, the range in density of compensating material spans the entire spectrum of densities from the upper crust to asthenosphere. However, even given the liberal upper bounds on topographic depression shown on Figure 6, if the paleodepths of upended slabs underestimate denudation by as little as 5 km, the bulk of the data would lie between 2800 and 2900 kg/m³. As will be argued below, a typical value for denudation is probably on the order of 15-20 km with a topographic depression of about 800 m. If typical, these values would indicate that the density of compensating material would on average lie within 100-200 kg/m³ of the average density of the upper crust, suggesting a maximum depth for the top of the compensating layer within the middle crust.

There are a number of difficulties with the assumptions of the simple denudation model, because the extending lithosphere may not have uniform density structure before or during denudation, due to the effects of (1) heterogeneous strain of the crust during extension, (2) differences in crustal thickness between extended and unextended areas prior to extension, and (3) differences in magmatic additions to the crust prior to and during extension. The model is applicable only to the extent that either the lithospheric layers between the denuded material and compensating medium were similar prior to extension and not affected by it, or to the extent that they were initially similar and uniformly modified by extension. Whether any of the stable blocks adjacent to the extended domains represent an unextended reference column, and whether the extended and stable domains have experienced contrasting strain histories in layers below the upper crust must be assessed before the above results can be considered further.

2.1. EXTENSIONAL ASYMMETRY, THE DEEP CRUST AND TOPOGRAPHY

While there is general agreement about how fault systems accommodate extension in the upper 15 km of crust, little is understood about the kinematics at depth. One of

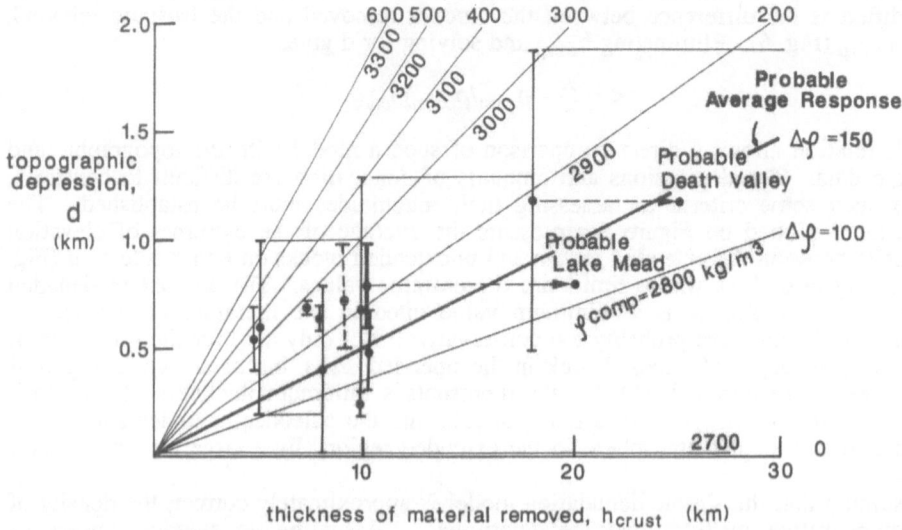

Figure 6. Plot of topographic depression vs. material denuded from various extended terrains (after equation 2). Slopes of reference lines correspond to varying the density of the compensating medium, based on the assumption that the average density of material denuded is 2700 kg/m³. Topographic depression taken from data on Figs. 4 and 9 (point with dashed uncertainty), and thicknesses of material removed are taken from Table I, except point with dashed uncertainty, which represents material added to Wind River Range area (Fig. 9). See text for discussion.

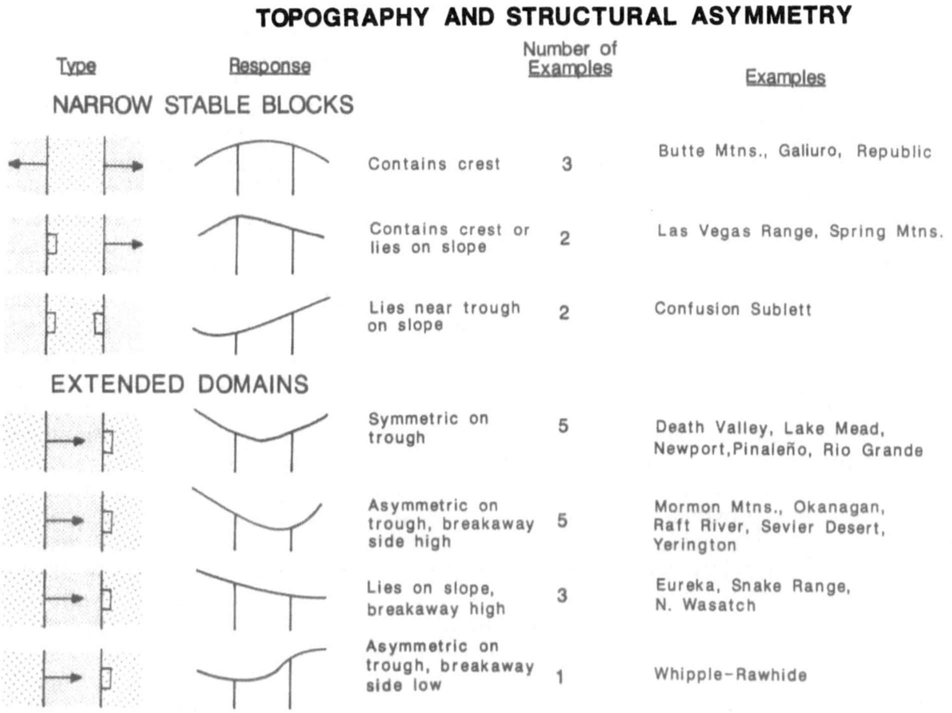

TOPOGRAPHY AND STRUCTURAL ASYMMETRY

Type	Response	Number of Examples	Examples
NARROW STABLE BLOCKS			
		Contains crest 3	Butte Mtns., Galiuro, Republic
		Contains crest or lies on slope 2	Las Vegas Range, Spring Mtns.
		Lies near trough on slope 2	Confusion Sublett
EXTENDED DOMAINS			
		Symmetric on trough 5	Death Valley, Lake Mead, Newport, Pinaleño, Rio Grande
		Asymmetric on trough, breakaway side high 5	Mormon Mtns., Okanagan, Raft River, Sevier Desert, Yerington
		Lies on slope, breakaway high 3	Eureka, Snake Range, N. Wasatch
		Asymmetric on trough, breakaway side low 1	Whipple-Rawhide

Figure 7.   Patterns of topography relative to structural asymmetry, from data in Figs. 2, 3 and 4. Symbols and patterns are the same as on Figs. 2, 3 and 4. See text for discussion.

the main characteristics of the upper crustal systems is their asymmetry, which complicates application of the simple rebound model discussed in the preceding section to areas such as the Basin and Range. In all of the examples of strongly extended domains shown on Figure 2, the normal faults in a given domain dip in the same direction. The asymmetry suggests distinction between boundaries between extended domains and stable blocks in which faults dip toward or away from the block (Figs. 2, 3 and 4), which will be referred to as the *proximal* and *distal* boundaries of the extended domain, respectively (Wernicke, 1985). The proximal boundaries are also commonly referred to as "breakaway zones" (e.g. Burchfiel et al., 1983; Howard and John, 1987). In order to understand possible differential modifications of unexposed lithospheric levels, we may first suppose that stable blocks with major fault systems projecting beneath them have a greater likelihood of being modified at deep levels than blocks where normal faults dip away from them (Wernicke, 1985). On the other hand, fault asymmetry may be restricted to the upper crust, such that deep levels are affected more uniformly, such that strain patterns there develop independently of upper crustal fault patterns (e.g. Gans, 1987; Klemperer, 1988). In the former case, we might expect systematic topographic patterns to exist with respect to fault asymmetry in the upper crust, whereas in the latter we might not expect to see a relationship.

Comparison of extensional asymmetry with topography does seem to reveal a pattern, wherein proximal boundaries tend to lie at higher elevation than distal boundaries, and wide stable blocks (e.g. Colorado Plateau and Sierra Nevada) tend to stand higher than adjacent extended terrains by a larger amount than narrow blocks (e.g., Spring Mountains block, AA'A", Fig. 4; Fig. 7). Narrow stable blocks with two proximal boundaries (three cases) contain topographic crests (DD', FF, and II', Fig. 4; Fig. 7). Stable blocks with one proximal and one distal boundary (two cases) lie on or near crests with the breakaway zone side resting higher (AA'A", BB', Fig. 4; Fig. 7). Blocks with two distal boundaries (two cases) lie on slopes near the bottoms of troughs (FF', HH', Fig. 4; Fig. 7). Of the extended domains, eleven of fourteen examples contain a topographic trough, with five of those being relatively symmetrical (Figs. 4 and 7). Where the boundaries of the domains lie at significantly different elevations, eight out of nine examples show higher topography on their proximal sides (Fig. 7). In the three examples of extended domains that lie well away from a trough, the block on the proximal side is higher (Fig. 7).

Large stable blocks blocks lie at higher elevations than extended domains regardless of asymmetry, especially at large distances from domain boundaries. In the one place a comparison can be made between two large blocks where one is bounded by a breakaway and the other by a distal boundary from a region of uniform fault dip (AA'A", Fig. 4), the breakaway zone side (Colorado Plateau) is low relative to the distal side (Sierra Nevada). This relation was explained by Wernicke (1985) and Jones (1987) by asymmetrical thinning of the mantle part of the lithosphere, with most of the thinning accommodated beneath the Sierran block.

2.1.1. *Differential Modification of Deep Crustal Layers During Extension.* These patterns suggest that heterogeneity in extending lithosphere is not restricted to the upper crust, because (1) unextended blocks immediately adjacent to the extended domains tend to be lower on their distal sides than on their proximal sides, (2) narrow stable blocks between two distal margins reside near troughs, and (3) narrow stable blocks between two proximal boundaries lie on topographic crests. Narrow stable blocks such as the Confusion Range block (FF', Fig. 4) where upper crustal shear zones project beneath on both sides are thus probably strongly modified at depth, such that lower crustal thinning balances or exceeds upper crustal denudation. Using differential topography to constrain topographic depression (parameter d, Fig. 6) is problematic in this case, leading to the conclusion that the compensating medium is less dense than the upper crust (equation 2).

A better case for comparison with Figure 5 might be that of opposing breakaways on either side of a stable block, because the asymmetry of upper crustal strain does not project beneath it. The position of these blocks astride topographic crests is consistent with the notion that there is less tendency for deep crustal strain beneath these blocks than beneath the extended domains or blocks with fault systems dipping beneath them. Therefore, these blocks, including the Galiuro, Butte and Republic blocks, more likely represent unextended reference columns for comparison with the extended domains.

Probably the most effective areas for comparison with Figure 5 from the point of view of differential deep crustal modification during extension are those in which the extended domains have breakaway zones against large stable blocks, such as the Lake Mead, Mormon Mountains, Yerington and Sevier Desert domains (Fig. 4). Of these, the Lake Mead and Mormon Mountains domains, adjacent to the Colorado Plateau stable block, are probably better for comparison than the Sevier Desert and Yerington examples because other systems with opposing fault dips are present at the latitude of the Sevier Desert and Yerington, potentially complicating the strain pattern at depth (Wernicke, 1985).

2.1.2. *Differences in Pre-Extensional Crustal Structure.* Further complicating the use of stable blocks for comparison with the extended domains is the possibility of major differences in pre-extensional elevation and density structure. For example, it is possible that the extended domains are systematically positioned above regions of greater Mesozoic crustal thickness (Coney and Harms, 1984) such that their current elevation contrast with surrounding stable blocks is substantially less than their subsidence during extension (parameter d in equation 2). For example, if the Sevier Desert region stood 1500 m higher than the adjacent Colorado Plateau prior to extension (as the Altiplano stands over 3000 m higher than the Brazilian foreland), then the actual subsidence would be nearly three times the maximum plotted on Figure 5.

However, the boundaries between extended domains and stable blocks are often abrupt, occurring over lengthscales of 10-20 km. The "patchwork" pattern of extended domains (Fig. 2), while broadly developed over crust shortened in Mesozoic time, does not correspond in detail to features in the surface geology that suggest greater shortening in extended domains relative to stable blocks, with the exception of the Colorado Plateau stable block, which displays little upper crustal shortening relative to areas to the west. Rather, the stable blocks reconstruct into a mosaic of upper crustal plates separated by comparatively narrow zones (say, 10-20 km vs. 50-100 km for stable blocks) of extensional failure (e.g. Wernicke et al., 1988, Fig. 6). Such a pattern would require improbably abrupt variations in deep crustal structure and topography, including major short wavelength relief on the Moho, in the pre-extension crust (e.g. Block and Royden, 1988, 1990).

Broad zones of intercontinental shortening comparable to the Cretaceous of the western U.S. (e.g. Tibet and the Altiplano) tend to lie at uniform regional elevation, suggesting that differences in crustal structure between extended domains and stable blocks were probably fairly minimal within the Basin and Range, although significant differences may have existed at long wavelength between the Colorado Plateau and the shortened crust of the Basin and Range. Thus, equation 2 may be somewhat compromised along the eastern side of the thrust belt, where extended domains may have developed along gradients in regional elevation and crustal thickness. However, extended domains and stable blocks developed within the region of Mesozoic shortening or to the east of it were probably of relatively uniform crustal structure at the onset of extension.

2.1.3. *Differential Magmatic Additions.* Large additions of mafic material from the mantle could occur differentially between the stable blocks and the extended domains. For example, major intermediate to silicic magmatic centers occur on the margins of the Colorado Plateau, and are centered upon regional topographic swells of about 500-1000 m (cf. Diment and Urban, 1981 and King and Beikman, 1974). If an extended domain experienced less crustal magmatism than adjacent stable blocks, the difference in topography between the two regions would be greater than that due to the effect of denudation alone, leading to overestimation of the density of the compensating fluid. This may be the case for the Mormon Mountains and the Sevier Desert domains, which are relatively amagmatic in comparison with portions of the bordering Plateau. Relatively greater magmatic addition in the extended domain would decrease the topographic contrast, resulting in underestimation of the density of the compensating medium (equation 2).

2.1.4. *Discussion.* The effect of differential topography or crustal structure due to Mesozoic shortening or Cenozoic magmatism is difficult to assess, particularly in the narrow stable blocks that lie within the province. Given all of the potential problems discussed above, it is concluded that the closest approximation to the simple denudation model of equation 2 is the Lake Mead region (Wernicke and Axen, 1988). The extended terrain is developed within cratonic North America, structurally within the foreland of the Mesozoic thrust belt, minimizing potential complications with differential initial crustal thickness. There is a major silicic magma center in the southern part of the extended domain, but the northern part is largely amagmatic. The Colorado Plateau contains a few Neogene basalt flows, as does the extended domain. The boundary between the extended and unextended crust is sharp, and the amount of denudation is at least 10 km (Table 1), and is probably about 15 km (Fryxell and Wernicke, in press).

Given the low probability of major topographic gradients between the extended and unextended domains in the Lake Mead region, the problem may be analyzed within the framework of magmatic and extensional effects (Fig. 8). In other words, the topographic contrast between the two regions is a good overall measure of differential modification of the crustal columns, including crustal thinning via upper crustal denudation (known to have occurred in the extended domain and not in the unextended domain), lower crustal thinning (not known to have occurred in either domain), and magmatic addition to the crust (probable in both areas but of unknown magnitude).

Several effects decrease estimates of the density of the compensating medium using equation 2, including crustal thinning below the extended domain other than that by simple denudation (Figs. 8a and 8b), and the possibility of greater addition of magma to the crust beneath the Colorado Plateau than to the extended domain. Two other possible effects, including greater deep crustal thinning beneath the Plateau relative to the extended domain (Fig. 8c) and greater addition of magma to the crust in the extended domain (Fig. 8d), would have the tendency to "smooth" the topography. Uniform thinning combined with flow of deep crust toward the extended domain has the tendency to smooth topography across the boundary relative to the case of uniform lower crustal thinning, but only near the boundary if the effect is fairly localized. Far from the boundary the net topographic difference should be the same. If the effect is regional (and that would mean in this case a region equal in size to a large fraction of the Colorado Plateau), then the only logical driving mechanism would be flotation of

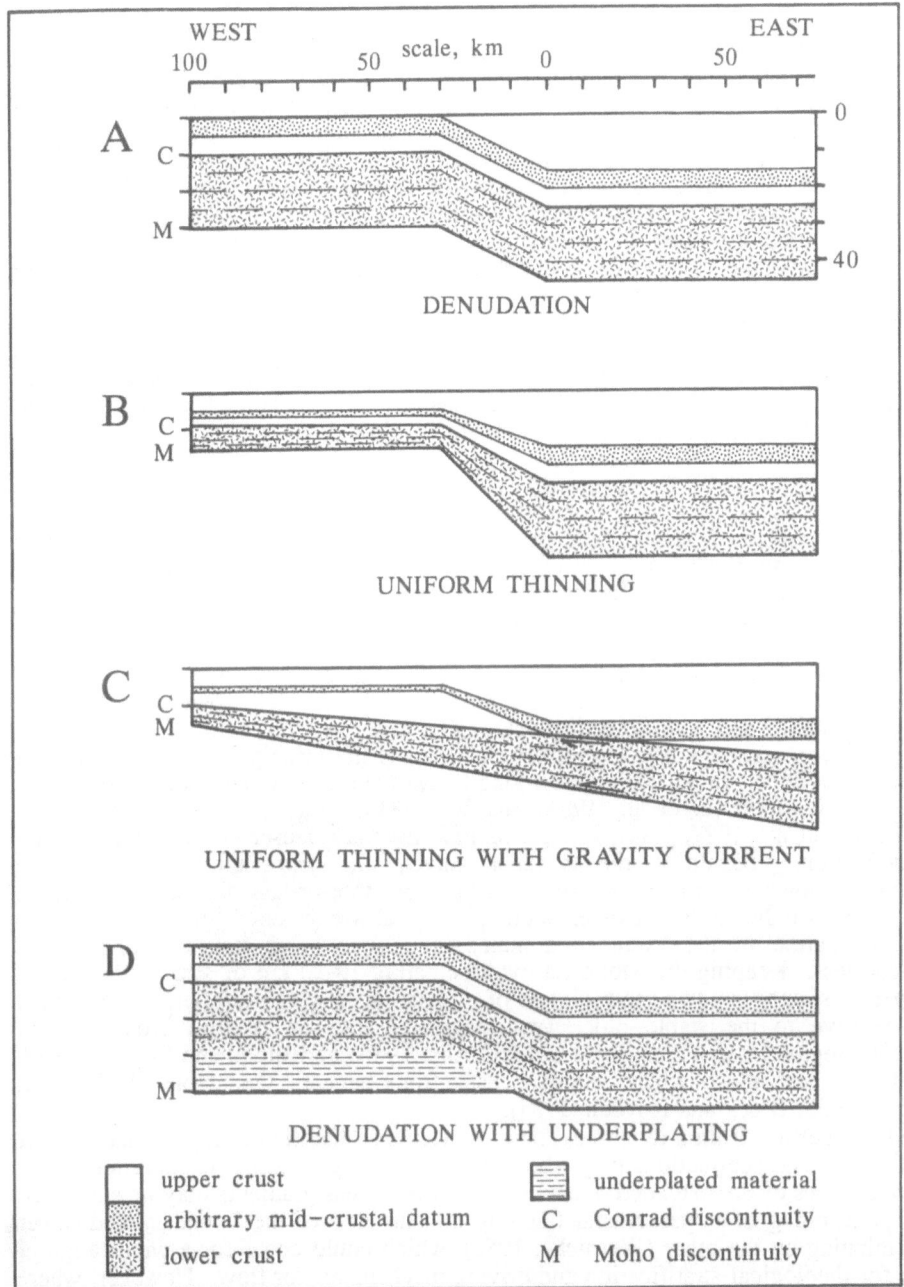

Figure 8. Non-exhaustive list of alternative kinematic models of how the lower crust might respond to an extreme gradient in upper crustal thinning or denudation. See text for discussion.

upper crust on the deep crust. In this case, the topographic contrast would directly measure the density of material flowing from beneath the Colorado Plateau (e.g. Block and Royden, 1988; 1990), and the density of deeper layers thus would not matter.

Although complications in applying equation 2 and Figure 6 are minimized in the Lake Mead region relative to others, the topographic pattern in the Lake Mead region is similar to that in other regions, suggesting that conclusions drawn from it may apply elsewhere. If we assume that differential modification of any two adjacent columns above the depth of compensation but below the denuded layer is minimal, and that initial crustal structures were not radically different, then Figure 5 suggests that the compensating medium must lie within the crust. Although poorly constrained in general, likely average values for topographic depression resulting from denudation are of the order 700-800 m for about 15 km of denudation, as shown by the position of points for the probable values of the Lake Mead and Death Valley extended domains. These values suggest a density contrast of between 100 and 200 kg/m^3 between the upper crust and the compensating medium. This conclusion is derived entirely from topography and paleobarometric data, independent of any consideration of actual crustal structure deduced from seismic and potential field data.

2.2. CRUSTAL STRUCTURE IN EXTENDED CRUST

The relatively subdued topographic response of large-scale upper crustal tectonic denudation suggests a shallow level of compensation, and therefore may indicate that the base of the uppermost rheologically strong layer of the earth lies within the crust. This has been suggested previously for the Basin and Range based on considerations of topography, heat flow and upper mantle P-wave velocities (Eaton et al., 1978). It is also supported by seismic reflection data across the Basin and Range, which indicate that the Moho is probably of relatively uniform depth across boundaries between extended domains and unextended blocks (Klemperer et al., 1986; Hauser et al., 1987a,b), if it is assumed that the reflection Moho is the principal density contrast at the base of the crust. In particular, the Moho across extended domain boundaries in the Pacific Northwest (Potter et al., 1986), the Butte Mountains area (Klemperer et al., 1986; Hauser et al., 1987a), and in west-central Arizona (Hauser et al., 1988b) shows little deflection. Similarly, the Bouguer gravity and topography (which at the wavelengths shown in Diment and Urban, 1981, are more-or-less directly proportional to one another) indicate that at other boundaries, local Moho relief equal in magnitude to the denudation in the extended domains is unlikely, and may be an order of magnitude less. Keeping the Moho flat while creating 10-20 km of structural relief in the upper crust suggests two main classes of effects: those which reinflate the extended region relative to the stable block (via magmatic addition or flow from beneath unextended regions, e.g. Thompson and McCarthy, 1986a and Block and Royden, 1988, respectively), and those which subtract mass from beneath the stable blocks (e.g. Wernicke, 1985; Block and Royden, 1988).

The key question from a dynamical point of view is what the driving mechanism for such additions and subtractions might be. In the case of smoothing Moho relief on the distal boundaries of the extended domains, crustal thickness gradients may be smoothed by the partitioning of upper crustal thinning on one side of the boundary and lower crustal thinning on the other (Wernicke, 1985), which could occur for a wide range of models for rheological stratification and driving mechanisms for flow. However, where upper crustal fault systems dip away from stable blocks, downward continuation of the asymmetry of faulting cannot be invoked to smooth Moho relief. If these boundaries show little or no Moho relief, lower crustal simple shear conjugate to upper crustal fault asymmetry must occur (Spencer and Reynolds, 1984; 1989). For narrow blocks in the middle of the province, this may be interpreted as increasing uniformity of extension with depth (e.g. Gans, 1987; Hamilton, 1988), but in and of itself does not necessarily indicate flotational equilibrium of upper crust on deep crust. For abrupt breakaway zones adjacent to large stable blocks, as in the Lake Mead region, a smooth Moho beneath the boundary would indicate a genetic relationship between upper crustal denudation and lower crustal strain beneath the Plateau, perhaps even more so than the case of narrow blocks located in the middle of the province. The various models in Figure 8 lead to contrasting crustal structures; however, the boundaries between extended domains and stable blocks do not have seismic data of sufficient detail to distinguish among these models. However, existing data combined with gravity and topography are more suggestive of the mechanisms shown in Figures 8c and 8d than in Figures 8a and 8b (Kruse et al., 1989).

3. Shortened Lithosphere

The concepts discussed above also apply to regions of crustal shortening. One of the best-characterized regions of shortening where the topographic response may be compared with upper crustal thickening is the Wind River uplift in Wyoming (e.g. Smithson et al., 1979). Here, the entire region lies within cratonic North America, and was near sea level just prior to tectonism in Late Cretaceous and early Tertiary time. Thus we can assume that the relative topographic difference between the shortened domain and surrounding regions indicates the topographic response to shortening. Although the Absaroka volcanic field lies immediately to the north of the region, it is largely post-tectonic. No magmatic centers are present astride the uplift itself, and there is no reason to suspect that the region of shortening was differentially affected by magmatic addition from the mantle compared to surrounding, unshortened regions.

Figure 9 shows the regional topography in relation to the zone of thrusting and upper crustal thickening. Based on seismic reflection data and gravity modeling, the center of the uplift has experienced a structural duplication by thrusting on the order of 13-14 km (Smithson et al., 1979). Considerable syn- and post-thrust erosion has occurred, affecting upper crustal thickness by redistributing mass into basins on the flanks of the uplift. The total thickening in the center of the uplift is the difference in thickness between post-Cretaceous strata away from the uplift and the remaining overthrust mass above the Cretaceous in the center of the uplift. According to the cross sections in Smithson et al. (1979), the magnitude of thickening is about 8-10 km, or roughly the total structural duplication less the amount of material eroded from the region of maximum duplication. The topographic uplift is well-centered on the region of thrusting, and has a relief of 500-1000 m, depending on precisely where the section is drawn. The system has an along-strike topographic gradient, with elevation increasing toward the Absaroka volcanic center, complicating assessment of the net effect related to the Wind River thrust system alone. Nonetheless, the magnitudes of upper crustal duplication and resulting topographic change in this intracratonic compressional system are comparable with those from the extensional examples discussed above (Fig. 6).

The validity of this result depends on whether the Wind River Range is locally or regionally compensated. If the overthrust mass, and hence the topography of the range were regionally compensated, the topographic deflection due to shortening would be larger than if compensated locally, resulting in overestimation of the density of the compensating medium below, according to equation 2. However, this effect is

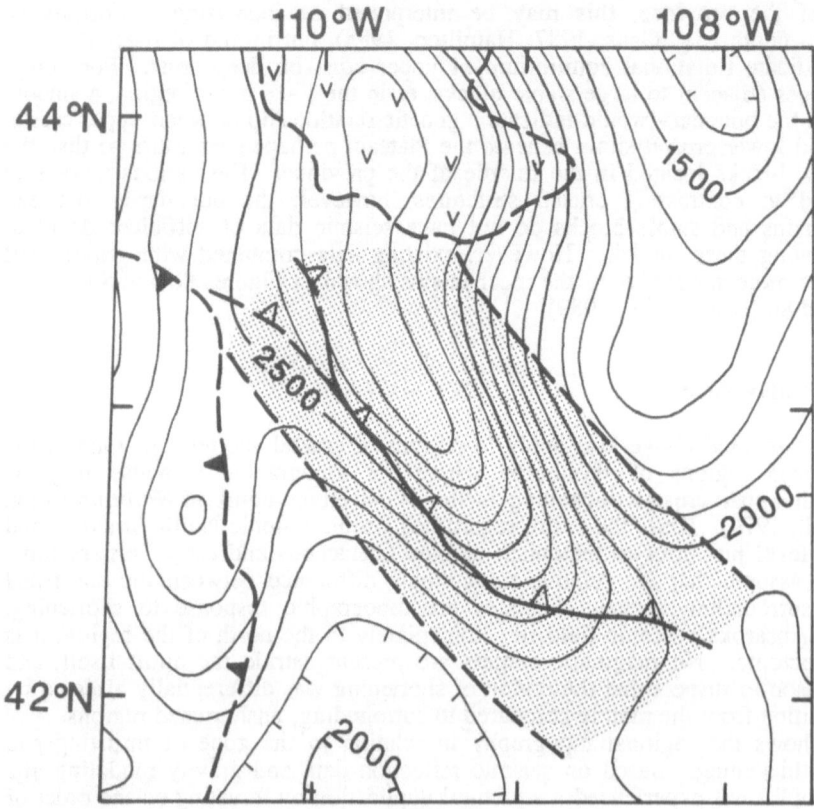

Figure 9. Map of Wind River Range and environs showing regional elevation (Diment and Urban, 1981), surface trace of the Wind River thrust (line with clear teeth), thrust front of Idaho-Wyoming thrust belt (line with black teeth), Absaroka volcanic field (v-pattern) and region of upper crustal thickening by the Wind River thrust system, including redistribution of thickened crust by sedimentation (shading). See text for discussion.

presumably minimized by using the regional topography of Diment and Urban (1981). In any event, this error has the same effect as that of underestimating the amount of tectonic denudation in the extensional case, so that the data provide an upper bound on the density of the compensating medium.

Another potential source of error is that much of the area of upper crustal thickening is sedimentary detritus eroded from the uplifted mass, and hence the average density of the load is lower than the density of the crystalline upper crust. If, for example, the average density of the added material were only 2600 kg/m^3, then the family of curves for density of the compensating medium on Figure 6 would shift upward and to the left, such that the line of zero slope would be 2600 kg/m^3 rather than 2700 kg/m^3, resulting in overestimation of the density of the compensating medium.

While the position of the Moho is not well known in detail beneath the Wind River uplift, as in the extensional cases, upper crustal strain must somehow also be absorbed in the lower crust. If it were absorbed directly beneath the uplift (analogous to Fig. 8b in the extensional case), then the topographic anomaly would be greater than that predicted by thickening of the upper crust only, which would alter the assumptions of equation 2 such that the density of the compensating medium would be overestimated for a given amount of topographic deflection. Sharry et al. (1986) and Thompson and Hill (1986) argued that the Wind River thrust probably flattens at depth, such that lower crust may be absorbing strain independently of the upper crust. To the extent that the assumptions of equation 2 are valid (the sources of error should all result in overestimation of the density of the compensating medium), it seems possible that in this case the differences in thickening of the upper crust are not compensated isostatically by outflow of mantle but instead by material in the deep crust not greatly different in density from the upper crust, probably in the range of 2800-2900 kg/m^3.

4. Discussion

If the high degree of rebound in response to denudation indicates compensation within the crust, it is not surprising that the Moho is flat beneath horizontal gradients in upper crustal thinning, assuming it was flat initially (Block and Royden, 1988, 1990). The principal question becomes the nature of interaction between the fluid crust and the remaining lithosphere below it. In particular, are there lower layers that are stronger than the proposed fluid layer, or is everything effectively inviscid below the solid upper crust?

Several lines of reasoning suggest that layers of density greater than about 2900 kg/m^3 are substantially stronger than the fluid crust. Although poorly constrained because the thermal structure and composition of the crust are not well known, quartz- and feldspar-rich lithologies (generally are less dense than 2900 kg/m^3 at crustal pressures) tend to be weaker than pyroxene or olivine-rich rocks (typically more dense than 2900 kg/m^3; e.g. summaries in Smith and Bruhn, 1984; Kuznir and Park, 1987). It is thus possible that the quartz-feldspar dominated middle crust (P-wave velocity of 6.0-6.6 km/s) is much less viscous than the lower crust and upper mantle for a variety of geotherms, although existing experimental data by no means demand this.

Seismic reflection profiling from rifted areas suggests highly reflective lower crust overlies transparent upper mantle. This has suggested to many workers that the lower crust deforms penetratively, while the upper crust and upper mantle deform as large, relatively coherent blocks bounded by relatively discrete shear zones (e.g. Matthews and Hirn, 1984; Wernicke, 1986; Reston, 1990; many others). Meissner and Wever (1986) concluded that the high reflectivity of the lower crust may in some way be related to its lower viscosity. The lack of earthquakes in the lower crust in diffusely strained continental lithosphere, and their presence in both the upper 15-25 km of the crust and in the upper mantle, suggest that the lower crust cannot maintain large (10's to 100's of MPa) shear stresses over geologic time, but that the upper mantle can (see, e.g. review of Molnar and Chen, 1983).

The topographic response data summarized in Figure 5 independently suggest that the weak material is probably not mafic lower crust, which on average would have a density of 3000 kg/m^3. In other words, these data suggest a limited role for any lower crust with densities appropriate to mafic (gabbroic) composition. Combined with the arguments above favoring a strong upper mantle during tectonism, and the reflectivity of mafic lower crust, the lower 5-10 km of the crust may represent a shear zone of finite strength that is in effect a transition zone between fluid crust above and relatively less deformed upper mantle below.

4.1. RELATIONSHIP TO PREVIOUS MODELS

If the flotation argument is correct, then several models proposed for the strain pattern in the lower crust during extension must be modified. Kinematic models of rifting proposed by McKenzie (1978), Wernicke (1985) and Wernicke and Axen (1988), for example, must be modified so as not to allow mantle to upwell differentially across zones of sharp gradients in upper crustal thinning. The lower crustal kinematics suggested by Wernicke and Axen (1988; Fig. 1), while suggesting negligible flexural rigidity of the lower crust, requires it to maintain constant thickness across the breakaway zone, transferring the difference in upper crustal thinning downward into Moho relief. Thus while presumably weak and ductile, such a layer would not be fluid. Simple shear models involving a single shear zone through two otherwise undeformed plates (e.g. Wernicke, 1985; Lister et al., 1986) must be modified to include complementary flow within the fluid layer.

Physical models such as the thin-sheet approximation of England and McKenzie (1982) are compromised if an effectively inviscid crustal layer generally intervenes between upper crust and upper mantle, because of the implied lack of coupling between upper crustal strain and strain in the upper mantle.

Models of rifting that call upon mafic material added from the mantle to systematically inflate the extended domains, thereby keeping them topographically high (e.g. Thompson and McCarthy, 1986a) may be inconsistent with Figure 5 to the extent that they imply the density of the compensating medium to be in the 3000 kg/m^3 range. According to Thompson and McCarthy (1986a), the strongly extended domains "have no prominent, characteristic gravity signature", and "are isostatically compensated by an active process of crustal intrusion, rather than a passive response to tectonic and erosional denudation." The compilation presented here shows that strongly extended domains do indeed have a signature in gravity and topography (albeit small, as indicated by Thompson and McCarthy's (1986a) suggestion that it is not present), and that its systematic development and magnitude seems better explained by "passive" lateral flow of a fluid crustal layer to smooth the gradient (Block and Royden, 1988, 1990) than by reinflation the crust beneath the extended areas with concentrations of basic intrusions. The fluid layer probably includes material added from the mantle, such as intermediate or silicic differentiates of mantle magmas, but if compensation were to occur primarily by a large subcrustal flux of magma beneath the extended domains, the compensating intrusions would have to be differentiated or mixed with crustal material to such a degree that their average compositions are silicic or intermediate. It is possible that the addition of basic material by underplating aids in the prevention of developing deep topographic troughs, but the areal distribution of such additions would have to be surprisingly systematic to mimic the effect of flotation on rock of mid-crustal density.

Gans (1987) proposed a stretching model similar to those of Eaton (1982) and Rehrig and Reynolds (1980), which "assumes that on a local and regional scale, the lower and middle crust extends in a 'pure shear' or taffylike fashion," and ascribed lateral inflow of lower crustal material into the extended domains as the result of "decoupling the heterogeneously deforming upper crust from the more uniformly deforming middle and lower crust..." As in the model of Wernicke and Axen (1988), the lower crust maintains relatively constant thickness across domain boundaries, such that lateral variations in upper crustal thickness would result in relief on the Moho of equal magnitude (Fig. 1), implying a limited role for lower crustal flow as a mechanism for smoothing the Moho (see also Fig. 3 in Gans, 1987). As in the model of Thompson and McCarthy (1986a), Gans' (1987) model also suggests underplating of mafic magma preferentially beneath the more strongly extended upper crust as a mechanism for smoothing the Moho (akin to Fig. 8d).

While broadly in agreement with this class of models, the fluid crust concept specifies that Moho smoothing be accomplished principally by complementary flow. While it can be shown that many magmatic centers developed astride strongly extended domains (e.g. Gans et al., 1989), a large proportion of broadly synextensional igneous centers in the Basin and Range, including some of the most voluminous, developed indiscriminately in space and time with respect to localized areas of large extension (e.g. Bartley et al., 1988; Taylor et al., 1989; Anderson, 1989; Best, 1990). If eruptive volume in the Basin and Range is at least a crude indicator of the subcrustal magmatic flux, it seems improbable that such a flux is coordinated in space and time with abrupt upper crustal strain gradients so as to precisely maintain topography and crustal thickness. If the fluid layer concept is correct, such coordination would be unlikely, in that the fluid layer eliminates lateral variations in vertical stress engendered by upper crustal extension, and thus the mantle would have no way of "knowing" the distribution of overlying extended domains, and vice-versa.

A subcrustal magmatic flux may weaken the crust, and may be required for (or at least contribute to) the development of a fluid layer. It seems likely that the fluid layer would be rich (10-50%?) in synrift magma. Thus mantle differentiates and remelted crust entrained in the fluid layer would be preferentially added to areas of strong upper crustal thinning, because the bulk of the fluid layer is "pumped" into these areas during extension (discussed further below). However, the fluid layer concept predicts that an initially flat Moho will remain flat after deformation so long as the layer is present, *regardless of the amount, timing and distribution of magma added from the mantle*. In other words, the lack of strong geophysical anomalies across strongly extended domains in the Basin and Range does not necessarily indicate that large volumes of magma have been emplaced into the crust, and thus may not be used to constrain the amount of added magma.

The fluid layer concept does not conflict with the notion that mafic magmatism plays an important role in shaping the overall topographic pattern, or with the important role played by mafic magmatism in some continental rifts, including those above mantle plumes (e.g. White and McKenzie, 1989) and those nearing the development of seafloor spreading (e.g. Lachenbruch et al., 1985). While an overall regime of "intrusion plus flow" advocated by many workers may be operative in the lower crust throughout much of the Basin and Range, the fluid layer concept suggests that hydraulic flow primarily accounts for the geophysical properties of strongly extended regions. The fluid layer concept explains the geophysical character of the province in conjunction with virtually any distribution, timing and amount of subcrustal magmatic flux, including no flux at all.

4.2. COMPLEMENTARY VERTICAL STRAINS OF FLUID AND UPPER CRUSTAL LAYERS

The flotation of blocks on a relatively inviscid crustal layer suggests that strain in the fluid layer is at least as heterogeneous as strain in the upper crust (Fig. 10). The net thinning of the fluid layer would approximately complement the thinning of the relatively solid upper crust (ignoring small differences due to density contrasts). If the layers are initially of uniform thickness, then

$$e_{zc}h_c = e_{zf}h_f + e_{zs}h_s \qquad (3)$$

where h_f, h_s and h_c are the initial thicknesses of the fluid, solid and fluid plus solid layers, respectively, and e_{zf}, e_{zs} and e_{zc} are their respective vertical finite strains (Fig. 10). If we consider the case where the initial thicknesses of the fluid and solid layers are approximately equal ($h_f = h_s$; Fig. 10), equation (3) becomes

$$e_{zf} = 2e_{zc} - e_{zs}, \qquad (4)$$

plotted in Figure 11. If thinning of the solid upper crustal layer generally corresponds to thinning of the crust as a whole (excluding considerations of crustal layers below the fluid layer), and similarly for thickening, then e_{zc} and e_{zs} will be of the same sign (negative for vertical thinning; shown as shaded domains in Fig. 11). However, for a variety of conditions of total crustal thinning and upper crustal thinning, the sign of e_{zf} may be opposite to that of e_{zs} and e_{zc}. For example, at point A on Figure 11, thinning of the combined fluid and solid layers e_{zc} is -0.3, but the thinning of the solid crust is -0.9, resulting in a net thickening of the fluid layer of +0.3 (Figs. 10 and 11). If the vertical strain of the solid crust were only -0.3 for the same amount of total strain of the two layers, then the fluid layer would thin, with a vertical strain of -0.3 (point B, Figs. 10 and 11).

Similar behavior might be expected for a fluid crustal layer in a shortened crust. For example, if the upper crust were doubled in thickness ($e_{zs} = 1$) but the total vertical strain were only 0.5, the fluid layer would neither thicken nor thin (point C, Fig. 11). Figure 11 illustrates that the fluid layer concept allows for greater heterogeneity of strain in the deep crust than in shallow levels. This is especially true near the line $e_{zs} = 0$ in Figure 11, where strain in the fluid layer exceeds that of the combined layers by a factor of two and that of the upper crust by an order of magnitude or more.

Equations 3 and 4 and Figures 10 and 11 represent only a specific case. For example, given a fluid layer that is thin in proportion to the solid upper crustal layer, small amounts of extension in the upper crust could lead to severe strains within the fluid layer, with extreme thinning beneath stable blocks and extreme thickening in areas where the upper crust is greatly thinned. Further, equation 3 does not address the relationship between the e_{zc} and e_{zs}. These variables are presumably related by conservation of mass over a domain D (assuming plane strain in cross section; Fig. 10) such that

$$\int_D e_{zs}(x)dx = e_{zc}, \qquad (5)$$

where x is the horizontal dimension, no horizontal flow occurs across the boundaries of

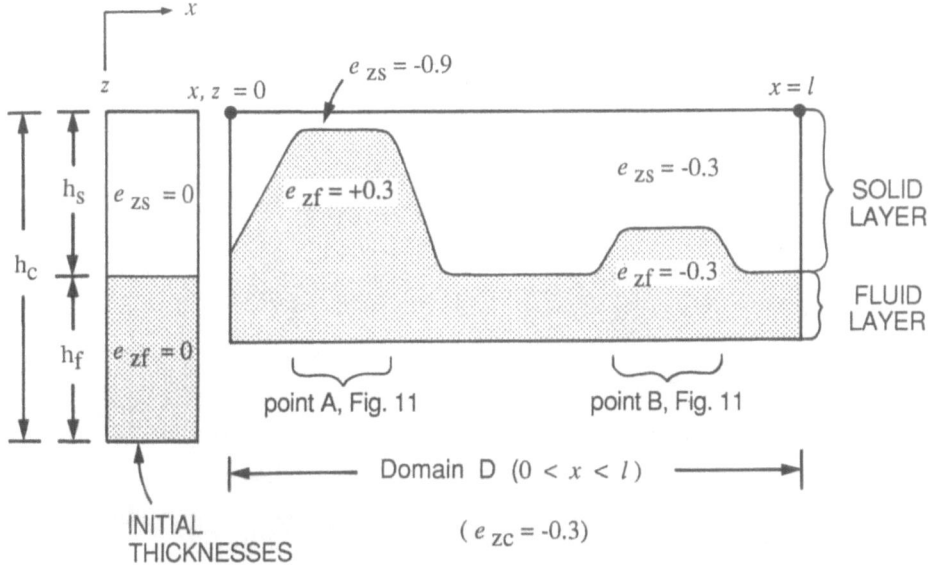

Figure 10. Hypothetical model depicting consequences of complementary vertical strain of fluid and solid crustal layers, where horizontal flow does not occur across domain boundaries x=0 and x=l. Variables correspond to equations (3), (4) and (5) and Fig. 11.

534

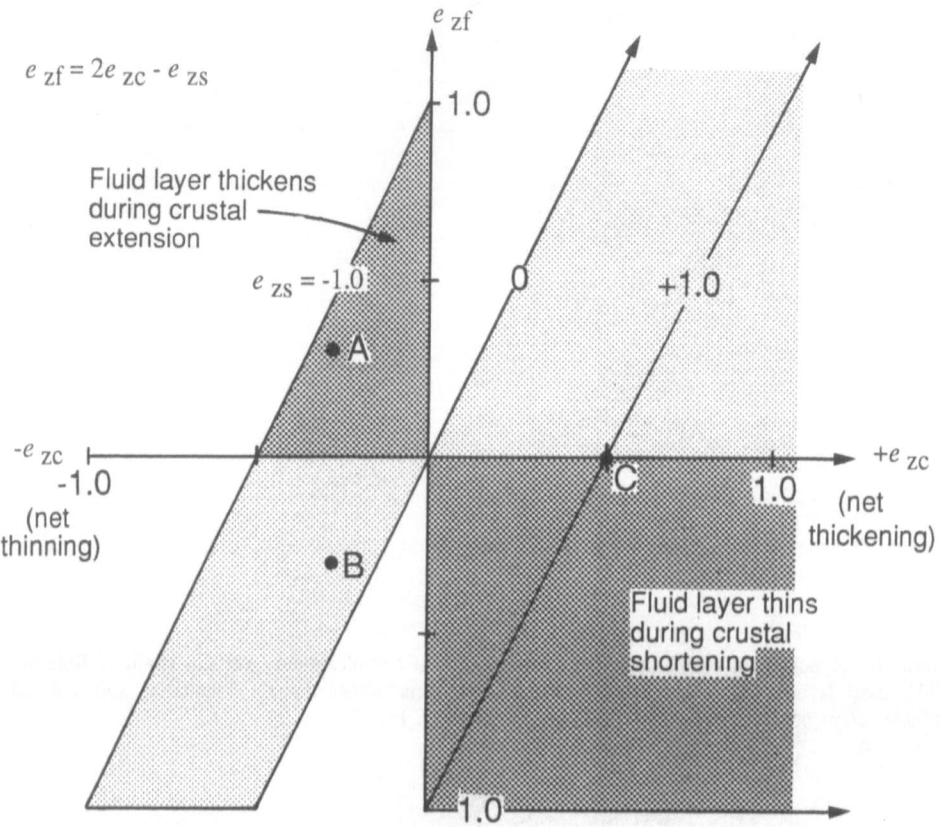

Figure 11. Graph of equation (4), showing vertical strain of the fluid layer as a function of vertical strain of the solid layer and combined solid and fluid layers. Darker shading shows domains in which vertical strain of fluid layer is opposite in sign to that of the solid and combined solid and fluid layers. Points A, B and C correspond to situations discussed in text.

the domain, and the total strain of both layers is uniform throughout the domain. If plane strain is not assumed, which is probably generally the case in the complex patchwork of extended domains in the Basin and Range, equation 5 would still hold (integrating over an area), but equation 3 would not be valid for a given vertical column.

Equation 5 does not account for possible additions of mass to the fluid layer during extension. As discussed above, voluminous magmatism apparent in extended areas such as the Basin and Range suggests volume-constant calculations such as these may be significantly affected by magmatic additions to the crust (e.g. McKenzie, 1984; Gans, 1987). A key, as yet unresolved problem alluded to above is the extent to which mantle-derived magmas make their way into the middle crust. If most of these magmas remain as mafic intrusions in the lower crust or near the Moho, such that magmas in the fluid layer are predominantly reworked crustal material, equation 5 may be a reasonable approximation. While isotopic data from intermediate synrift volcanic and intrusive rocks from the Basin and Range suggest some contain a significant component of mantle derived material (e.g. Gans et al., 1989; Asmerom et al., 1990), the relative volume of these magmas in the middle crust, and hence the volume of mantle-derived material, is highly uncertain.

5. Implications for Continental Dynamics

The arguments presented above suggest a fourfold division of diffusely deforming lithosphere: (1) an upper, solid crustal layer, (2) an underlying fluid crustal layer upon which the solid layer is in flotational equilibrium, (3) a solid lower crustal layer, much stronger than the fluid layer, and (4) the mantle part of the lithosphere, which is also much stronger than the fluid layer (Fig. 12). This framework resembles that proposed by Meissner and Wever (1986), except it suggests that the lower part of the crust is not a single weak layer, but may be divided into a strong, relatively mafic layer and a layer that is so weak that it behaves as a relatively inviscid fluid. This framework may provide fertile ground for testable new hypotheses, both in terms of kinematic interpretations of orogenic belts (based on geological and geophysical data) and in terms of thermal and dynamical modeling.

For example, if such a lithosphere were subjected to extension, one might expect that the mantle part of the lithosphere would deform along discrete shear zones, such as the Flannan shear zone beneath Scotland (Matthews and Hirn, 1984; Reston, 1990), which appears to extend at moderate dip through much of the thickness of the mantle part of the lithosphere (Warner and McGeary, 1987). If the crust contained a layer that could not support shear stress, then one would not in general expect to see mantle shear zones colinear with those in the crust (e.g. Wernicke, 1986). Using offshore Scotland as an example, the Flannan structure would represent a zone of normal shear of the mantle part of the lithosphere, increasing the surface area of the collective top of the solid lower crust and mantle lithosphere layers (shown schematically in Fig. 12). In this regard, it is interesting that the Flannan reflections appear to flatten upward into the lower crust at and above the Moho. The kinematics of extension of the lower crustal layer and mantle part of the lithosphere would thus be analogous to the beginning of a moving walkway, where the surface area beneath the fluid layer is progressively increased from the area where the shear zone flattens upward into the fluid layer (Fig. 12). In the case of the Flannan structure, new area would be created beneath the fluid crustal layer in the region west of where the structure flattens into the lower crust (left side of Fig. 12), while to the east the mantle would remain

536

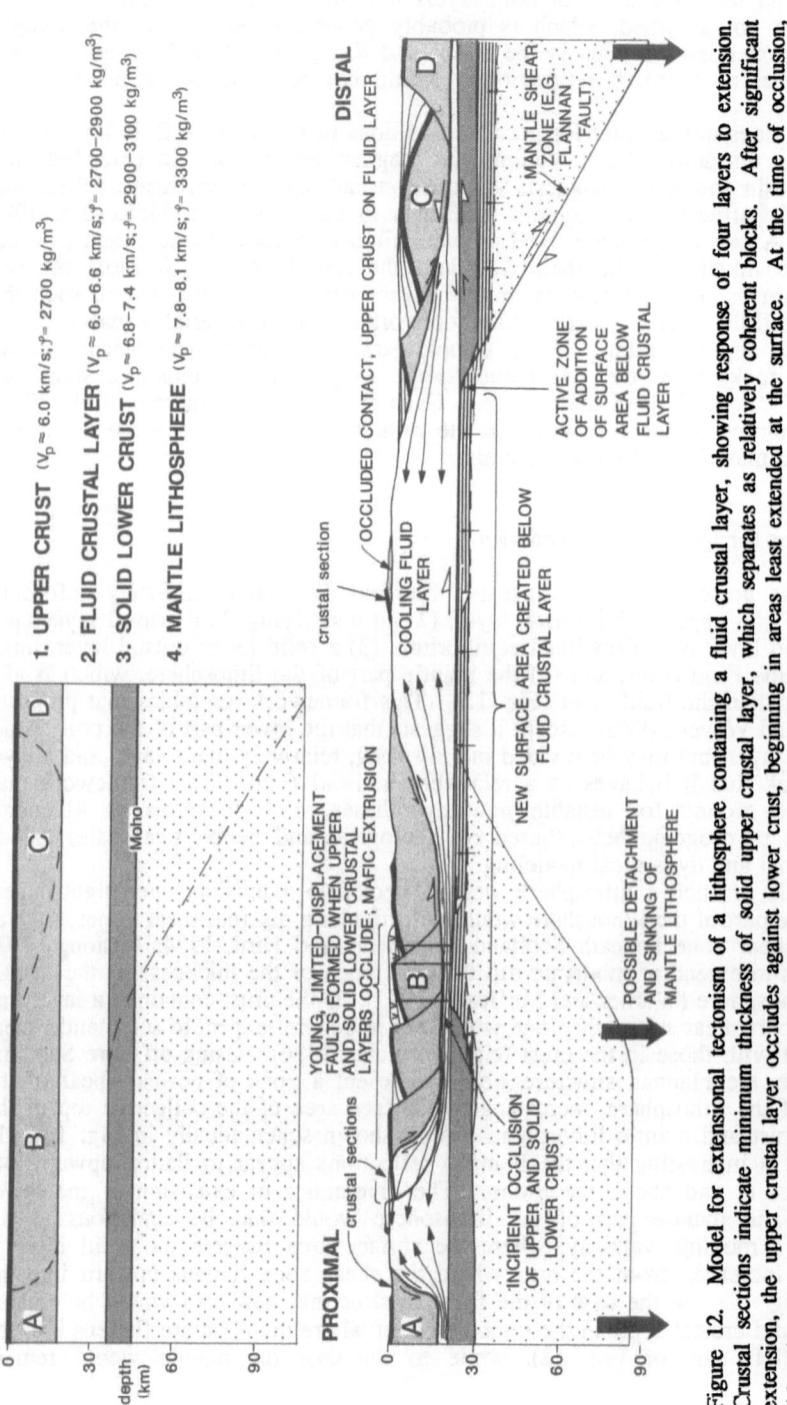

Figure 12. Model for extensional tectonism of a lithosphere containing a fluid crustal layer, showing response of four layers to extension. Crustal sections indicate minimum thickness of solid upper crustal layer, which separates as relatively coherent blocks. After significant extension, the upper crustal layer occludes against lower crust, beginning in areas least extended at the surface. At the time of occlusion, lithosphere may break into thick-skinned fault blocks accommodating limited extension, perhaps facilitating the eruption of mafic magmas underplated at the base of the crust. Lithosphere as a whole experiences uniform-sense simple shear, with upper crustal blocks B, C and D shearing top-to-the-right relative to mantle to the left of the principal mantle shear zone. Shear along the top of the asthenosphere may produce paired deep reflections such as seen in the North Sea (cf. model of Klemperer, 1988). Extreme weakness of fluid layer minimizes lateral variations in crustal thickness throughout extension. See text for discussion.

undeformed. This creates a situation where overlying upper crustal layers would shear over the solid lower crust with a sense toward the active zone of addition of surface area beneath the fluid crustal layer (Fig. 12), antithetically drawn in to the moving walkway (cf. Reston, 1990).

Depending upon the total thinning of the fluid layer, different senses of shear are predicted between the fluid layer and the solid lower crust and upper crust. For example, the right-hand upper crustal block on Figure 12 has an overall sense of shear top-to-the-left with respect to the solid lower crust, but the extension in the fluid layer is so intense that it has the opposite sense of shear with respect to the upper crustal block. Thus the lithosphere may shear in a uniform direction overall, with zones of antithetic shear common along the boundaries of the fluid layer (Fig. 12).

5.1. OCCLUSION AND COOLING OF THE FLUID LAYER

A consequence of extending the four-tiered lithosphere of Figure 12 is that after a large amount of strain, the solid layers occlude against one another, isolating cooling pockets of the fluid layer, such that in these areas isostatic compensation would then likely occur by asthenospheric inflow. Because any given crustal region uplifted to the surface tends to cool during extension, the fluid layer as a whole is likely to become more viscous during extension, depending on how fast mass transport occurs relative to thermal diffusion (Sonder and England, 1989). The upper part of the denuded fluid layer is in particular likely to strengthen during uplift (Block and Royden, 1988; 1990; Fig. 12). Thus, eventually the fluid crust will vanish; in the case of extension this should first occur beneath the least-strained upper crust (Fig. 12), and in the case of compression it should first occur beneath the most-strained upper crust. Subsequently, the crust may tend to deform in a more "thick skinned" fashion, with isostatic compensation occurring by asthenospheric inflow. Such a change may explain the observation that in many parts of the Basin and Range province, large-scale extension is overprinted by systems of steeply dipping younger faults that accommodate relatively little extension, but in some cases form deep local structural depressions (e.g. Owens Valley in eastern California, adjacent to the Death Valley extended terrain; Wernicke et al., 1988). Either freezing or squeezing all of the fluid layer out from beneath a stable block would increase tractions along its base, such that it may break up into smaller blocks as it is sheared across the lower crust and upper mantle. The predicted occlusion may also explain the evolution of magmatic style within some regions of the Basin and Range from intermediate-silicic to mafic, by promoting the injection and extrusion of mafic magma ponded at the base of the crust through whole-crust brittle cracks (Fig. 12).

These ideas can be tested with integrated geologic, potential field, and reflection and refraction data in key areas between the Basin and Range and Plateau interior, and between extended and unextended domains within the Basin and Range. The model in Figure 12 predicts that the transition from P-wave velocities of 6.0-6.6 km/s up to about 7.0 km/s is at relatively constant depth across the extended region, and that topographic variations principally result from horizontal density contrasts within the upper crust. This would suggest that strongly extended regions systematically have slightly higher upper crustal P-wave velocities than unextended blocks, and that the difference in density implied by the velocity contrast is sufficient to support topographic variations without the aid of deeper crustal layers. An exception to this pattern occurs in the case of narrow stable blocks where upper crustal faults dip beneath the block on both sides, such as the Confusion Range block, which is stable but lies near a topographic depression (FF', Figs. 3 and 4). The Confusion Range block may have been thinned by the shear zones projecting beneath them, such that there is more thinning of the upper crust compared with narrow blocks whose boundaries are both proximal, such as the Butte Mountains block (Figs. 4 and 7). Thus even though such blocks may be afloat, they would stand lower than the relatively unthinned upper crustal blocks.

Alternatively, cooling of the fluid layer with time may account for the observed differences in elevation between such blocks (Figs. 4 and 7). Cooling may result in a progressive increase in viscosity of the layer, such that in the latter stages of extension, the fluid layer is too viscous to flow out from beneath stable blocks hydraulically, and thus the effect of asymmetric simple shear draws relatively more deep crust out from beneath the blocks adjacent to distal boundaries, preferentially thinning deep crust there (cf. Wernicke, 1985). This could explain why proximal boundaries of either extended domains or stable blocks tend to be higher than distal boundaries.

5.2. VARIATIONS IN STRUCTURAL STYLE OF CONTINENTAL RIFTS

The presence or absence of a fluid crustal layer provides a plausible explanation for observed variations in rifts, from the diffuse rifts of the Basin and Range or Afar region to the narrow rifts such as the northern Red Sea. Relatively old, cold cratonic regions (such as the northern Red Sea prior to rifting) may contain weak or ductile layers, but these layers may not be sufficiently fluid to fill space between the separating blocks, allowing rapid breaching of the lithosphere (perhaps by a single shear zone) and upwelling of the asthenosphere to the surface. In other words, the deep crust is sufficiently strong that the asthenosphere "wins the race" to fill the void created as the upper crustal blocks separate. The magmatic character of the Basin and Range and Afar regions relative to the northern Red Sea supports this notion, assuming the flux of magma from the mantle during rifting is accompanied by an increase in heat flux sufficient to warm and weaken the middle part of the crust.

Another important control on asthenosphere upwelling into narrow voids created by upper crustal separation may be the rigidity of the upper mantle. If upper mantle failure is localized in the same place as upper crustal separation, the lithosphere may be breached relatively quickly. On the other hand, if a number of crustal blocks are separating across a wide zone of relatively strong upper mantle that initially fails only locally, the middle crust may win the race into the voids because the asthenosphere is "handicapped" by the flexural strength of the upper mantle.

5.3. PHYSICAL MODELING

It is stressed that these concepts are based purely on kinematic observations and interpretations, and may be tested by physical modeling. The existence of weak crust is favored by relatively thick, wet quartzose crust and a high geotherm (Kuznir and Park, 1987). An interesting question arising from the fluid layer concept is whether old shield areas contain such a layer. Are thick crust, an elevated geotherm, and a magmatic flux from the mantle prerequisites for intracrustal isostasy? This question may have important bearing on physical models relating glacial rebound to the rheology of the mantle. Does the rebound reflect the rheology of the mantle, or simply a fluid layer within the crust? The Wind River uplift formed in a cratonic area, but only shortly before voluminous magmatism in the Absaroka Range area just to the north, suggesting a higher geotherm than is typical of shield areas. It may be that fluid crustal layers typify only young continental margin orogenic belts with high heat flow and thick (50-70 km) continental crust, but even if this were true, most of the earth's orogenic record would be influenced by its behavior.

The lengthscale at which the fluid layer can respond to local thickness changes in the upper crust may be indicated by slopes on the Moho. For example, although there may not be any relief on the Moho across strain gradients between the Basin and Range and Colorado Plateau (lengthscale of about 100 km), the interior of the Plateau has crust approximately 40-50 km thick, compared with 25-30 km thick crust in the Basin and Range (e.g. Hauser and Lundy, 1989), suggesting that if fluid crust existed in the interior of the Plateau, it was not appreciably influenced by crustal thinning in the Basin and Range. Thus an upper bound on the lengthscale may be on the order of 1000 km. However, given the fourfold-division of the lithosphere suggested here, it is difficult to rule out "hydraulic coupling" over a region as large as the Plateau, because the solid lower crust could conceivably maintain the gradient on the Moho, while the base of the fluid layer remains at relatively constant depth across the region. Physical models of fluid flow in a narrow channel may serve to constrain deep crustal rheology (e.g. Kruse et al., 1989), which may be sensitive to the lengthscale of flow.

5.4. MAXIMUM DEPTH OF EXPOSURE OF CRUSTAL SECTIONS AND ATTENUATION OF METAMORPHIC ISOGRADS

Perhaps one of the broadest predictions of the lithospheric profile in Figure 12 lies in the transport of various crustal levels to the surface, or their emplacement relatively upward such that erosion may eventually expose them. In most tectonic regimes, it seems likely that rocks anywhere within the fluid layer may be tectonically transported upward. On the other hand, because the model implies decoupling of the mafic lower crust from the fluid and upper crustal layers and comparatively firm coupling between mafic lower crust and negatively buoyant mantle lithosphere, it would be difficult to tectonically emplace mafic lower crust at the surface or, in compressional regimes, move it to shallower levels where post-tectonic erosion may eventually lead to its exposure. Thus the model offers an explanation for the rarity of large tracts of mafic granulite and peridotite exposed within the continental lithosphere, and perhaps also explains the rarity of geobarometric measurements in excess of 1200 MPa on volumetrically important metamorphic assemblages exposed at the surface. Maximum pressures observed in orogenic belts are generally in the 1000-1500 MPa range (30-45 km depth), corresponding to the range of maximum depth of crust with velocities less than about 6.6 km/s observed in seismic refraction profiles.

In both extensional and compressional regimes, sites of relative thickening of the upper crust would be sites of extreme thinning of the fluid layer. The zones of occlusion shown on Figure 12, if ultimately exposed at the earth's surface, would show extreme attenuation of pre-tectonic isobaric surfaces. Attenuated metamorphic isograds in exposures of deep crust are often attributed to regimes of crustal extension, perhaps resulting from regional spreading of crust thickened in collision zones. If the fluid layer model is correct, it predicts that attenuation on the order of the thickness of the layer (10-25 km or so) may occur *during* major thickening of immediately overlying upper crust. If so, the attenuation of the isograds would occur in a regime of increasing pressure of structurally highest rocks. In the case of extensional tectonism, isogradic attenuation would occur beneath stable blocks (Fig. 12), in an overall environment of decreasing pressure of structurally deepest rocks.

At crustal scale, thickening of the crust may occur on the flanks of contractional orogens by outflow of fluid crust from areas of upper crustal thickening, similar to the withdrawal of fluid crust from large stable blocks during extension. For example, the western Great Plains, though unstrained at the surface, are underlain by crust about 50 km thick, which thins eastward toward the continental interior, where the crust averages only 40 km thick (e.g. Fig. 23b in Braile, 1989). This region may have experienced crustal thickening on the order of 10 km as a result of Creataceous to early Tertiary crustal shortening in the Cordillera to the west. Thickening may have occurred either during shortening or, if the viscosity of the fluid layer beneath the Plains was great enough, some time afterward.

6. Acknowledgements

The author is indebted to G.J. Axen, S. Kruse, P. Molnar, M.K. McNutt, R.J. O'Connell, B.M. Sheffels, J.K. Snow and G.A. Thompson for discussions which contributed substantially to views presented here. Constructive reviews by J.M. Stock and two anonymous referees were invaluable in clarifying the presentation. This research was supported by NSF Grants EAR-8805335 and EAR-8451181.

7. References

Allmendinger, R.W., Farmer, H., Hauser, E., Sharp, J., Von Tish, D., Oliver, J., and Kaufman, S., 1986, Phanerozoic tectonics of the Basin and Range - Colorado Plateau transition from COCORP data and geologic data: A review: in Barazangi, M. and Brown, L., eds., *Reflection Seismology: The Continental Crust*, Geodynamics Series, American Geophysical Union, Washington, D.C., 14, 257-267.

Allmendinger, R.W., Hauge, T.A., Hauser, E.C., Potter, C.J., and Oliver, J.E., 1987, Tectonic heredity and the layered lower crust in the Basin and Range province, western United States: in Coward, M.P., et al., eds., *Continental Extensional Tectonics*, Geological Society Special Publication, 28, 223-246.

Anderson, R.E., 1971, Thin-skin distension in Tertiary rocks of southeastern Nevada: *Geol. Soc. America Bull.*, 82, 43-58.

Anderson, R.E., 1989, Tectonic evolution of the intermontane system; Basin and Range Colorado Plateau, and High Lava Plains, in Pakiser, L.C. and Mooney, W.D., Geophysical framework of the continental United States: *Geol. Soc. America Memoir* 172, 163-177.

Ando, C.J., Chuchra, B.L., Klemperer, S.L., Brown, L.D., Cheadle, M.J., Cook, F.A., Oliver, J.E., Kaufman, S., Walsh, T., Thompson, J.B., Jr., Lyons, J.B., and Rosenfeld, J.L., 1984, Crustal profile of a mountain belt: COCORP deep seismic reflection profiling in New England Appalachians and implications for architecture of convergent mountain chains: *Am. Assoc. Petrol. Geol. Bull.*, 68, 819-837.

Armstrong, R.L. and Dick, H.J.B., 1974, A model for the development of thin overthrust sheets of crystalline rock: *Geology*, 2, 35-40.

Asmerom, Y., Snow, J.K., Holm, D.K., Jacobsen, S.B., Wernicke, B.P., and Lux, D.R., 1990, Rapid uplift and crustal growth in extensional environments: An isotopic study from the Death Valley region, California: *Geology*, 18, 223-226.

Axen, G.J. and Wernicke, B., 1989, Day 2. Superimposed thrust and normal faulting, Colorado Plateau to eastern Mormon Mountains: *Extensional Tectonics in the Basin and Range Province Between the Southern Sierra Nevada and the Colorado Plateau*, 28th International Geological Congress Field Trip Guidebook, American Geophysical Union, Washington, D.C., T138, 20-29.

Bartley, J.M., Axen, G.J., Taylor, W.J., and Fryxell, J.E., 1988, Cenozoic tectonics of a transect through eastern Nevada near 38°N. latitude, in Weide, D.L. and Faber, M.L., eds., *This Extended Land: Geological Journeys in the Southern Basin and*

Range: Field Trip Guidebook, Geological Society of America, Cordilleran Section, 1-20.

Best, M.G., and Christiansen, E.H., 1990, Limited extension during peak Tertiary volcanism in the southeastern Great Basin: *Geol. Soc. America Abs. Prgms.*, 22, 3, 7-8.

Block, L. and Royden, L., 1988, Core complex geometry implies flow in the lower crust: *Geol. Soc. America Abs. Prgms.*, 20, A64.

Block, L. and Royden, L., Core complex geometries and regional scale flow in the lower crust: *Tectonics* (in press).

Braile, L.W., 1989, Crustal structure of the continental interior, in Pakiser, L.C. and Mooney, W.D., Geophysical Framework of the Continental United States: *Geol. Soc. America Memoir* 172, 285-315.

Brace, W.F. and Kohlstedt, D.L., 1980, Limits on lithospheric strength imposed by laboratory experiments: *J. Geophys. Res.*, 85, 6248-6252.

Burchfiel, B.C., Walker, J.D., Davis, G.A. and Wernicke, B., 1983, Kingston Range and related detachment faults -- a major "breakaway" zone in the southern Great Basin: *Geol. Soc America Abs. Prgms.*, 15, 6, 536.

Compton, R.R., Todd, V.R., Zartman, R.E. and Naeser, C.W., 1977, Oligocene and Miocene metamorphism, folding and low-angle faulting in northwestern Utah: *Geol. Soc. America Bull.*, 88, 1237-1250.

Coney, P.J. and Harms, T.A., 1984, Cordilleran metamorphic core complexes -- Cenozoic extensional relics of Mesozoic compression: *Geology*, 12, 550-554.

Davis, G.A., 1988, Rapid upward transport of mid-crustal mylonitic gneisses in the footwall of a Miocene detachment fault, Whipple Mountains, southeastern California: *Geol. Rundsch.*, 77, 191-209.

Diment, W.H. and Urban, T.C., 1981, Average elevation map of the coterminous United States: U.S. Geological Survey Map GP-933.

Eaton, G.P., 1982, The Basin and Range province: Origin and tectonic significance: *Ann. Rev. Earth Planet Sci.*, 8, 409-440.

Eaton, G.P., Wahl, R.R., Prostka, H.J., Mabey, D.R., and Kleinkopf, M.D., 1978, Regional gravity and tectonic patterns: Their relation to Late Cenozoic epierogeny and lateral spreading in the western Cordillera: in Smith, R.B. and Eaton, G.P., eds., Cenozoic Tectonics and Regional Geophysics of the Western Cordillera: *Geol. Soc. America Memoir* 152, 51-92.

England, P.C. and McKenzie, D.P., 1982, A thin viscous sheet model for continental deformation: *Geophys. J. R. Astron. Soc.*, 70, 295-321.

Fryxell, J.E., and Wernicke, B.P., Gold Butte crustal section, South Virgin Mountains, Nevada: *Tectonics* (in press).

Gans, P.B., 1987, An open-system, two-layer crustal stretching model for the eastern Great Basin: *Tectonics*, 6, 1-12.

Gans, P.B., Mahood, G.A., and Schermer, E., 1989, Synextensional magmatism in the Basin and Range province; A case study from the eastern Great Basin: *Geol. Soc. America Spec. Paper* 233, 53 p.

Green, A.G., Milkereit, A., Davidson, A., Spencer, C., Hutchinson, D.R., Cannon, W.F., Lee, M.W., Agena, W.F., Behrendt, J.C., and Hinze, W.J., 1988, Crustal structure of the Grenville front and adjacent terranes: *Geology*, 16, 788-792.

Guth, P.L., 1981, Tertiary extension north of the Las Vegas Valley shear zone, Sheep and Desert Ranges, Clark County, Nevada: *Geol. Soc. America Bull.*, 92, 763-771.

Hamilton, W., 1988, Detachment in the Death Valley region, California and Nevada: *U.S. Geol. Survey Bull.*, 1790, 51-85.

Harms, T.A. and Price, R.A., 1983, The Newport fault: Eocene crustal stretching, necking and listric normal faulting in NE Washington and NW Idaho: *Geol. Soc. America Abs. Prgms.*, 15, 5, 309.

Hauser, E.C. and Lundy, J., 1989, COCORP deep reflections: Moho at 50 km (16 s) beneath the Colorado Plateau: *J. Geophys. Res.*, 94, 7071-7081.

Hauser, E.C., Potter, C., Hauge, T., Burgess, S., Burtch, S., Mutschler, J., Allmendinger, R., Brown, L., Kaufman, S., and Oliver, J., 1987a, Crustal structure of eastern Nevada from COCORP deep seismic reflection data: *Geol. Soc. America Bull.*, 99, 833-844.

Hauser, E.C., Gephart, J., Latham, T., Oliver, J., Kaufman, S., Brown, L., and Luchitta, I., 1987b, COCORP Arizona transect: strong crustal reflections and offset Moho beneath the transition zone: *Geology*, 15, 1103-1106.

Howard, K.A., and John B.E., 1987, Crustal extension along a rooted system of low-angle normal faults: Colorado River extensional corridor, California and Arizona, in Coward, M.P., Dewey, J.F., and Hancock, P.L., eds., *Continental Extensional Tectonics*: Geol. Soc. London Spec. Pub. 28, 299-312.

Jones, C.H., 1987, Is extension in Death Valley accommodated by thinning of mantle lithosphere beneath the Sierra Nevada, California?: *Tectonics*, 6, 449-473.

King, P.B. and Beikman, H.M., 1974, Geologic map of the United States, exclusive of Alaska and Hawaii (1:2,500,000): U.S. Geological Survey.

Klemperer, S.L., 1988, Crustal thinning and nature of extension in the northern North Sea from deep seismic reflection profiling: *Tectonics*, 7, 803-822.

Klemperer, S.L., Hauge, T.A., Hauser, E.C., Oliver, J.E., and Potter, C.J., 1986, The Moho in the northern Basin and Range province, Nevada, along the COCORP 40°N seismic-reflection transect: *Geol. Soc. America Bull.*, 97, 603-618.

Kruse, S., Sheffels, B., McNutt, M. and Wernicke, B., 1989, Gravitational constraints on the nature of lithospheric extension in the Lake Mead area, Nevada: *EOS Trans.*, 70, 465.

Kuznir, N.J. and Park, R.G., 1987, The extensional strength of the continental lithosphere: its dependence on geothermal gradient, and crustal composition and thickness: in Coward, M.P., et al., eds., *Continental Extensional Tectonics*, Geol. Soc. London Special Publication 28, 35-52.

Lachenbruch, A.H., Sass, J.H. and Galanis, S.P., 1985, Heat flow in southernmost California and the origin of the Salton trough: *J. Geophys. Res.*, 90, 6709-6736.

Lister, G.S., Etheridge, M.A., and Symonds, P.A., 1986, Detachment faulting and the evolution of passive continental margins: *Geology*, 14, 246-250.

Matthews, D. and Hirn, H., 1984, Crustal thickening in Himalayas and Caledonides: *Nature*, 308, 497-498.

Matthews, D.H. and Cheadle, M.J., 1986, Deep reflections from the Caledonides and Variscides west of Britain and comparison with the Himalayas: in Barazangi, M. and Brown, L., eds., *Reflection Seismology: A Global Perspective*, Geodynamics Series, American Geophysical Union, Washington, D.C., 13, 5-19.

McKenzie, D., 1978, Some remarks on the development of sedimentary basins: *Earth Planet. Sci. Lett.*, 40, 25-32.

McKenzie, D., 1984, A possible mechanism for epeirogenic uplift: *Nature*, 307, 616-618.

McKenzie, D., 1988, Gravity currents in the lower crust: *Geol. Soc. Newsletter* 17, p. 12.

Meissner, R. and Wever, T., 1986, Nature and development of the crust according to deep reflection data from the German Variscides: in Barazangi, M. and Brown, L., eds., *Reflection Seismology: A Global Perspective*, Geodynamics Series, American Geophysical Union, Washington, D.C., 13, 31-42.

Miller, E.L., Gans. P.B. and Garing, J.D., 1983, The Snake Range decollement: An exhumed mid-Tertiary ductile-brittle transition: *Tectonics*, 2, 239-263.

Molnar, P. and Chen, W.-P., 1983, Seismicity and mountain building: in Hsü, K.J, ed., *Mountain Building Processes*, Academic Press, London, 41-58.

Parrish, R.R., Carr, S.D. and Parkinson, D.L., 1988, Eocene extensional tectonics and geochronology of the southern Omineca belt, British Colombia and Washington: *Tectonics*, 7, 181-212.

Parry, W.T. and R.L. Bruhn, 1987, Fluid inclusion evidence for minimum 11 km vertical offset on the Wasatch fault, Utah: *Geology*, 15, 67-70.

Potter, C.J., Sanford, W.E., Yoos, T.R., Prussen, E.I., Keach, W.R., II, Oliver, J.E.. Kaufman, S., and Brown. L.D., 1986, COCORP deep seismic reflection traverse of the interior of the North American Cordillera, Washington and Idaho: *Tectonics*, 5, 1007-1027.

Proffett, J.M., Jr., 1977, Cenozoic geology of the Yerington District, Nevada, and implications for the nature and origin of Basin and Range faulting: *Geol. Soc. America Bull.*, 88, 247-266.

Rehrig, W.A. and Reynolds, S.J., 1980, Geologic and geochronologic reconnaissance of a northwest-trending zone of metamorphic core complexes, southern and western Arizona: in Crittenden, M.D., et al., eds., Cordilleran Metamorphic Core Complexes, *Geol. Soc. America Memoir* 153, 131-158.

Reston, T.J., 1990, Mantle shear zones and the origin of the northern North Sea Basin: *Geology*, 18, 272-275.

Sharry, J., Langan, R.T., Jovanovich, D.B., Jones, G.M., Hill, N.R., and Guidish, T.M., 1986, Enhanced imaging of the COCORP seismic line, Wind River Mountains, in Barazangi, M. and Brown, L., eds., *Reflection Seismology: A Global Perspective*, Geodynamics Series, American Geophysical Union, Washington, D.C., 13, 223-236.

Smith, R.B. and Bruhn, R.L., 1984, Intraplate extensional tectonics of the eastern Basin-Range: Inferences on the structural style from seismic reflection data, regional tectonics, and thermal-mechanical models of brittle-ductile deformation: *J. Geophys. Res.*, 89, 5733-5762.

Smithson, S.B., Brewer, J.A., Kaufman, S., Oliver, J.E., and Hurich, C.A., 1979, Structure of the Laramide Wind River uplift, Wyoming, from COCORP deep reflection data and from gravity data: *J. Geophys. Res.*, 84, 5955-5972.

Snoke, A.W. and Howard, K.A., 1984, Geology of the Ruby Mountains-East Humboldt Range, Nevada - A Cordilleran metamorphic core complex, in Lintz, J.P., ed., *Western Geological Excursions, Volume 4*, Mackay School of Mines, Reno, 260-303.

Sonder, L.J. and England, P.C., 1989, Effects of a temperature-dependent rheology on large-scale continental extension: *J. Geophys. Res.*, 94, 7603-7620.

Sonder, L.J., England, P.C., Wernicke, B.P., and Christiansen, R.L., 1987, A physical model for Cenozoic extension of western North America, in Coward, M.P., Dewey, J.F., and Hancock, P.L., eds., *Continental Extensional Tectonics*: Geol. Soc. London Spec. Pub. 28, 187-202.

Spencer, J.E. and Reynolds, S.J., 1984, Mid-Tertiary extension in Arizona: *Geol. Soc. America Abs. Prgms.*, 16, 6, 664.

Spencer, J.E., and Reynolds, S.J., 1989, Middle Tertiary tectonics of Arizona and adjacent areas, in Jenny, J.P. and Reynolds, S.J., eds., Geologic Evolution of Arizona: *Arizona Geol. Soc. Digest*, 17, 539-574.

Thompson, G.A. and Hill, J.L., 1986, The deep crust in convergent and divergent terranes: Laramide uplifts and Basin-Range rifts: in Barazangi, M. and Brown, L., eds., *Reflection Seismology: The Continental Crust*, Geodynamics Series, American

544

Geophysical Union, Washington, D.C., 14, 243-256.

Thompson, G.A. and McCarthy, J., 1986a, Geophysical evidence for igneous inflation of the crust in highly extended terranes: *EOS Trans.*, 67, 1184.

Thompson, G.A. and McCarthy, J., 1986b, A gravity constraint on the origin of highly extended terranes: *Geol. Soc. America Abs. Prgms.*, 18, 418.

Warner, M. and McGeary, S., 1987. Seismic reflection coefficients from mantle fault zones: *Geophys. J. R. Astr. Soc.*, 89, 223-230.

Wernicke, B., 1983, Evidence for large-scale simple-shear of the continental lithosphere during extension, Arizona and Utah: *Geol. Soc. America Abs. Prgms.*, 15, 5, 310-311.

Wernicke, B., 1985, Uniform-sense normal shear of the continental lithosphere: *Can. J. Earth Sci.*, 22, 108-125.

Wernicke, B., 1986, Whole-lithosphere normal simple shear: An interpretation of deep-reflection profiles in Great Britain: in Barazangi, M. and Brown, L., eds., *Reflection Seismology: The Continental Crust*, Geodynamics Series, American Geophysical Union, Washington, D.C., 14, 331-339.

Wernicke, B. and Axen, G.J., 1988, On the role of isostasy in the evolution of normal fault systems: *Geology*, 16, 848-851.

Wernicke, B., Axen, G.J. and Snow, J.K., 1988, Basin and Range extensional tectonics at the latitude of Las Vegas, Nevada: *Geol. Soc. America Bull.*, 100, 1738-1757.

White, R. and McKenzie, D., 1989, Magmatism at rift zones: The generation of volcanic continental margins and flood basalts: *J. Geophys. Res.*, 94, 7685-7729.

Zuber, M.T. and Parmentier, E.M. and Fletcher, R.C., 1986, Extension of the continental lithosphere: A model for two scales of Basin and Range deformation: *J. Geophys. Res.*, 91, 4826-4838.

GEOPHYSICAL INTERPRETATION OF ASTROGRAVIMETRIC DATA IN THE IVREA ZONE

B. Bürki
Institute of Geodesy and Photogrammetry
Swiss Federal Institute of Technology
ETH Hönggerberg
CH - 8093 Zürich
Switzerland

ABSTRACT. This paper deals with an interpretation of the gravity field caused by the prominent Ivrea-body in northern Italy. For this purpose a total of 117 astro-geodetic and of 500 gravimetric stations has been observed in the last decade in the framework of several international field campaigns, using special transportable zenith cameras and gravity meters. One of the main problems using such data is an appropriate reduction of topographic and geological effects as well as of the Moho- discontinuity. A new model for a combined adjustment procedure is presented which enables the use of both the anomalies of the gravity vector and its deflection from the vertical. Finally a new astro-gravimetric model of the Ivrea body is presented showing some new aspects concerning its geometry and structure: While up to now the extension of the geophysical Ivrea body ended some five kilometers east of Locarno, the new model indicates an eastern continuation of the structure with a reduced density contrast. The second aspect refers to the substructure of the Ivrea body on its northern part underneath the Insubric line. The formerly proposed steepness of the Ivrea body can not be confirmed by the astrogravimetric measurements. In contrary they indicate a mass surplus which reaches its maximum north of Locarno, where a big wedge reaching down to the Moho is needed to fit the observed gravity field. This result supports most recent tectonic interpretations of seismic data indicating combined plate tectonic processes of backthrusting and right lateral transpressive deformations.

1. Introduction

Geophysical methods are very powerful research tools for almost all earth sciences but mainly for mineral prospecting. The oil exploration in particular gained from continuous improvements and refinements of the geophysical methods.

Besides industrial applications the performance of these methods grew simultaneously in the field of applied sciences. This fact may be documented by numerous geo- oriented projects dealing with geophysical methods such as e. g. worldwide seismic reflection and refraction projects, the European Geo-Traverse Project (EGT) or the continental ultra- deep drilling project (KTB). In all those investigations, the earth's crust, its structure, properties and behaviour play a role of upmost importance.

Since geodesy is now in a position to determine not only intercontinental baselines but also their changes due to plate tectonic activities, geodesists also have become interested in the detailed structure of the earth's crust.

This paper reports on an investigation which was carried out in the vicinity of the Ivrea body in northern Italy and southern Switzerland. The observed Bouguer anomalies of this prominent intracrustal deposition of mantle material reach values of approximately 170 mgals. The main goal of the project was an accurate determination of the gravity field and beyond that to perform more data for an improved geoid computation (Bürki, 1988). To achieve this, the location, shape and density contrast of the northern part of

M. H. Salisbury and D. M. Fountain (eds.), Exposed Cross-Sections of the Continental Crust, 545–561.
© 1990 *Kluwer Academic Publishers.*

546

the intracrustal Ivrea body had to be interpreted. Therefore the gravity field has been observed in a widespread survey covering an area of about 100 km by 150 km as outlined in figure 1.

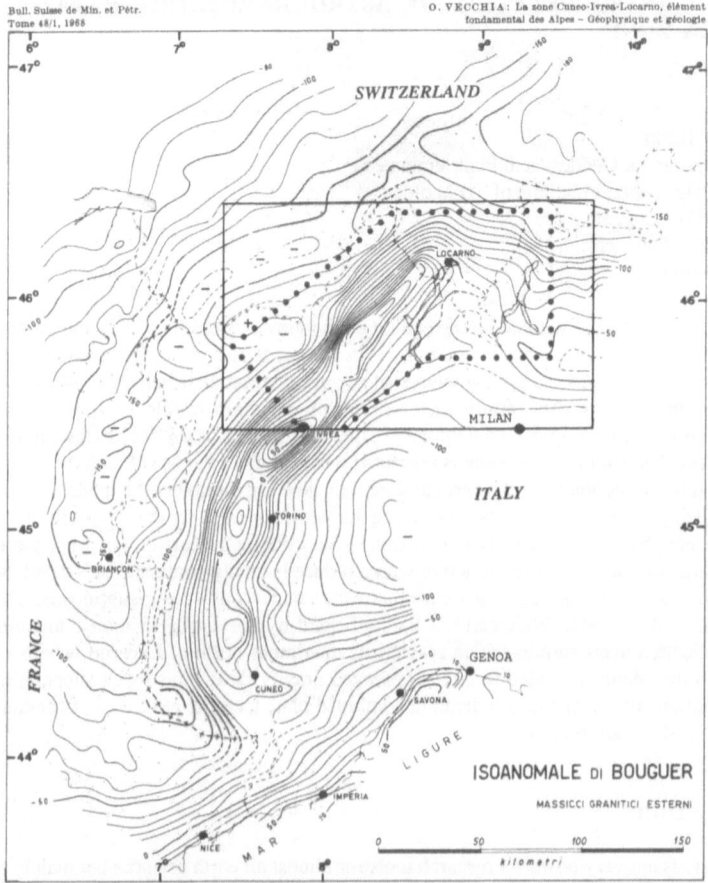

Bull. Suisse de Min. et Pétr.
Tome 48/1, 1968

O. VECCHIA: La zone Cuneo-Ivrea-Locarno, élément
fondamental des Alpes – Géophysique et géologie

Figure 1: Bouguer anomalies in the vicinity of the Ivrea- body in northern Italy and southern Switzerland based on (Vecchia, 1968). The isolines display equal anomalies in mgal with an equidistance of 10 mgal (1mgal = 10^{-5} ms^{-2}). The dotted lines show the area of investigation where the gravity field has been observed and interpreted as described in the following chapters. The rectangle indicates the range of figures 8, 9 and 10.

2 . Determination of the gravity field

The gravity field can be described with three parameters:

- two parameters describing the spatial direction of the gravity vector and
- one parameter defining its magnitude (the actual gravitational acceleration).

In geophysical prospecting and physical geodesy the gravity anomaly Δg is defined in the following way:

$$\Delta g = g_p - \gamma_Q \qquad (1)$$

with g_P = gravity value on the geoid and
               γ_Q = normal gravity value on the reference ellipsoid

The difference between the direction of the actual gravity vector **g** and the normal gravity vector **γ** defines the deflection of the vertical in two components (Heiskanen and Moritz, 1984):

$$\xi = \Phi - \varphi \qquad \text{north / south– component}$$

$$\eta = (\Lambda - \lambda) \cos \Phi \qquad \text{east / west– component}$$

of the deflection (2)

with
    Φ :   astronomical latitude
    Λ :   astronomical longitude
    φ :   ellipsoidal latitude
    λ :   ellipsoidal longitude

The ellipsoidal coordinates are determined either by classical geodetic surveys or by satellite methods such as the Global Positioning System (GPS).

While the gravity anomalies are measured with gravity meters, the astronomical position in terms of latitude Φ and longitude Λ (being identical with the direction of the physical plumb line) are determined with special zenith camera systems as shown in figure 2:

Figure 2: Transportable zenith camera system of the Institute of Geodesy and Photogrammetry at the ETH Zurich. The display box for the electronic levels is mounted on a small tripod at the left side. Below is the case with film cassettes of the size 6.5 x 9 cm. The monitor electronics is mounted in the assembly case at the right side. The direction of the local vertical, being identical with the optical axis of the device, can herewith be determined with respect to the fixed stars. While up to now this task was very time consuming using classical instruments, an occupation of up to 8 stations per night is possible with this automated equipment.

The principle of operation of the camera may be described briefly as follows: The optical axis of the camera is aligned towards the zenith by means of two electronic levels. They are mounted perpendicularly beside the objective which can be rotated by 180 degrees. The extended axis of rotation of the camera is identical with the direction of the plumb line in space. By taking a picture of the starry sky in the zenith of a station, one can determine the direction of the plumb line with respect to the system of the fixed stars.

Those coordinates are computed using appropriate star catalogues such as AGK3. Since the earth is rotating, the time of the exposures (epochs) must be fixed with suitable accuracy which in this case must be in the order of some milliseconds. To perform this task, a special microprocessor based time digitizing unit (TDU) was developed. It monitors automatically all important functions such as time keeping and measuring of the epochs, electronic shutter movements, level readings, data storing as well as the transfer of the data memory to the computer.

A complete film exposure consists of two sequences of exposures, taken in two opposite camera orientations. After the development, the films are evaluated on a x/y- comparator with computer aided positioning where the coordinates of the star traces are being measured with high accuracy. The further evaluation of the data on the mainframe computer includes automatic star recognition and the reduction of various effects, such as the stars' proper motion, precession, nutation, aberration, time- and polar variations, time signal- and shutter delays. The direction of the rotational axis is then computed by applying a transformation of the evaluated celestial coordinates onto an earth- fixed coordinate system. The resulting actual astronomical position in terms of latitude Φ and longitude Λ represents the physical direction of the local plumb line. This method enables the determination of the astronomical Parameters Φ and Λ with an accuracy of about 0.4 arc seconds for a single night observation. This accuracy may be improved by repeated measurements at different epochs.

Thanks to close international cooperation with European geodetic institutes, several joint field campaigns were made in which all available zenith camera systems in Europe were used. A total set of 117 astro-geodetic and gravimetric stations were finally determined. These measurements, completed by some 400 additional gravity measurements which were kindly made available from the Institute of Geophysics at the ETH, represent the base of this interpretation.

3. Reduction of the observed values

3.1 GENERAL REMARKS

In order to extract the effects of intracrustal inhomogeneities, various reductions of known mass disturbances have to be performed. The reductions within this project were applied to all three observed components of the gravity field: ξ, η and Δg. For this purpose we used appropriate formulae which describe the derivatives of the potential of suitable elementary structures such as prisms with either rectangular or triangular cross sections (Mader, 1951). Within the Ivrea project, all effects of known mass disturbances differing from the Ivrea body itself, have been modeled and removed (subtracted) from the measured values in both the gravity anomalies and the deflections of the vertical as well. The different reduction steps are described in the following.

3.2 EFFECT OF TOPOGRAPHY

Since the measured surface values are strongly dependant on the shape of the topography and its mean density, this reduction step is of upmost importance. To perform an accurate computation of this effect, a detailed digital terrain model is used. In general it may consist of either a regular array of geographical compartments or of an array of rectangular compartments outlined by the grid lines of a local cartesian coordinate system as is used in Switzerland. The compartment width depends on the ruggedness of the topography. For Switzerland a regular grid with a resolution of 250 by 250 m resp. 500 by 500 m is available covering the entire country and its close vicinity. Beside this relatively coarse grid, the topography has been digitized for every station of the Ivrea project in regular 50 by 50m compartments, covering an area of at least 750 m radius around the station. For the reductions a mean density of 2.67 gcm^{-3} for the crust material was applied using rigorous formulae for prisms as described in (Klingelé, 1980). The computations have shown values between -38.6" and +30.7" (arc seconds) in the deflection, and -59.5 to -7.8 mgal in gravity. These relatively high values demonstrate the importance of this reduction step which is often considered to be negligible.

3.3 EFFECT OF THE MOHO

The transition from the lower crust to the upper mantle is not only documented by increased seismic velocities but also by a density contrast of about +0.4 gcm^{-3} (Girdler, 1985). The effect of the varying Moho depth must be taken into account at least in regions where the Moho depth varies greatly as shown in figure 3:

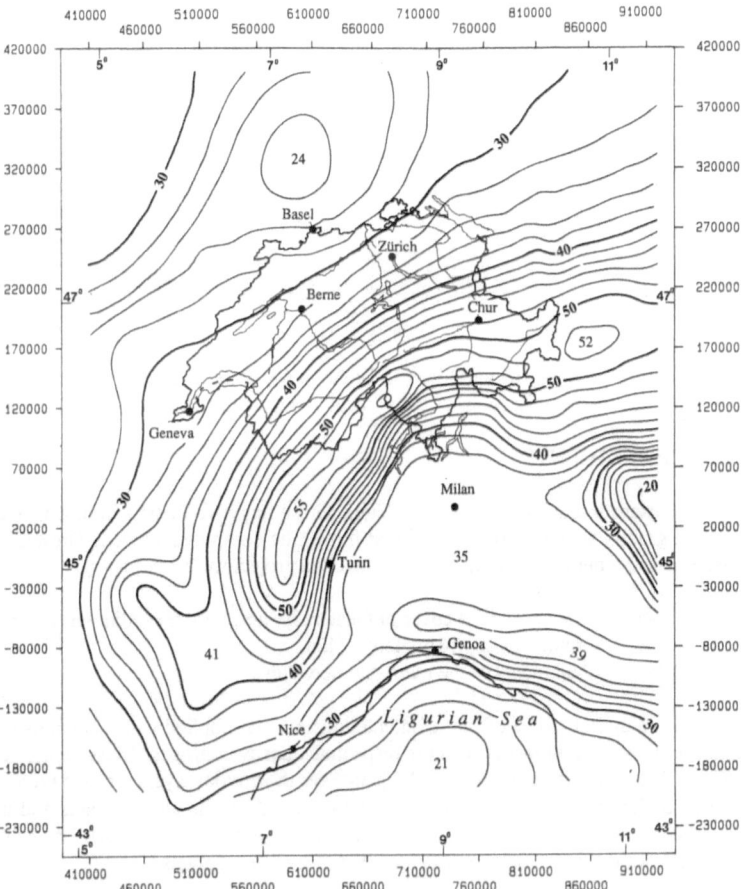

Figure 3: Contour lines of the Moho depth in Switzerland and its vicinity, after Mueller et al., (1980), Menard (1979) and Stein et al., (1978). Contour interval is 2 km.

To emphasize the significance of this reduction, which reaches amplitudes of up to 80 mgals, the computed gravity effect of the Moho is displayed in figure 4 (since this influence is only computed at the measurement stations, the extension of this map corresponds with the rectangle outlined in fig. 1):

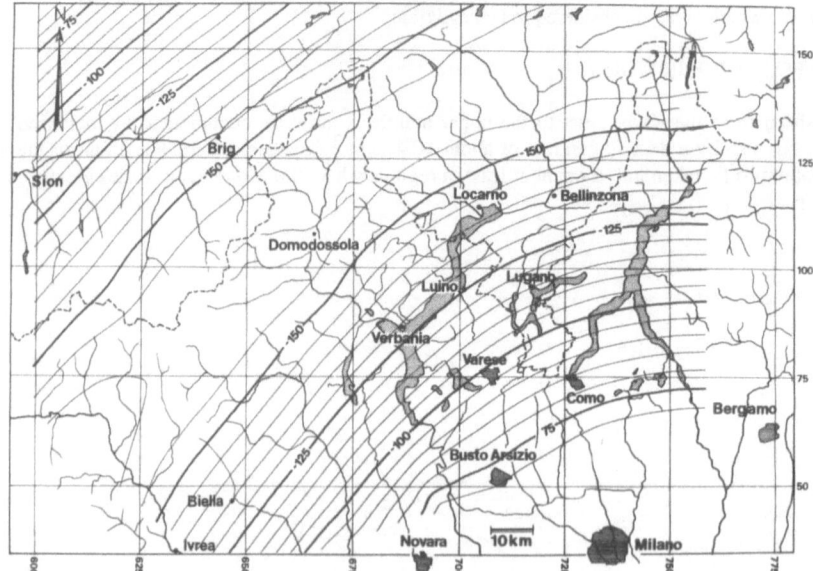

Figure 4: Gravity effect (in mgal) of variations in Moho depth as computed within the project area. The corresponding influence on the deflections of the vertical varies within a range of ± 15 arc seconds.

3.4   EFFECT OF SEDIMENTS

The sediments represent intracrustal inhomogeneities with a negative density contrast. A correction for their effect on the gravity field in terms of g, ξ and η must be computed as well. The sediment regions which were taken into account in the framework of the Ivrea project were:

- the Po plain, a cover of Quaternary and Tertiary sediments reaching maximum depths of approximately 12 km.
- sediments of the main Alpine valleys.

For these reductions a digital model of the sedimentation area was applied which contains information on the lateral extension of different layers with the respective density contrasts. These density contrasts vary between -0.8 gcm^{-3} for the near- surface and -0.2 gcm^{-3} for the bottom layers. For the project stations, the influence of the Po plain reaches magnitudes of some -70 mgals at the closest points and almost zero at remote points (the maximal effect due to the Po plain is in the order of 130 mgal). The magnitude of the respective reductions of the deflections varies in a range of ± 15 arc seconds.

3.5   EFFECT OF GEOLOGICAL FORMATIONS

The geological formation causing the largest influence on the measured values is the Ivrea zone itself. The observed densities of the near-surface rocks show a pronounced increase here, reaching values of up to 3.3 gcm^{-3}(Kissling, 1984). The masses between the stations and the reference level are subtracted from the measured values. The stations located directly above the Ivrea body were reduced mainly in gravity where the effect ranges from -2 to + 23 mgal. The resulting reductions in the deflection vary between -4.5 and +2.5 arc seconds which still is about ten times higher than the observation accuracy.

During the interpretation it became evident that both the deflections and the gravity residuals revealed systematic correlations with geological formations, such as gneisses or mesozoic ophiolites as outlined in the geological and tectonic maps of Switzerland, edited by the Swiss Geological Commission (1980).

3.6   EFFECT OF LAKES

The last reduction step concerns the water masses of the lakes. Since the digital terrain model refers to the bottom of the lakes, the respective effects of the water masses on the measurements would be neglected by applying the topographic correction only. Although this is a small effect, the computed magnitude reaches 6 mgals in gravity and 2 arc seconds in the deflection for some stations located directly at the shore of lake "Lago Maggiore", which begins near Locarno and extends towards south west (c. f. fig. 3, 4, 5). The other lakes of which the effects were computed are those in the vicinity of Lugano (lake Lugano), north of Como (lake Como) and lake Orta, located some 15 km south west of Verbania.

3.7   RESULTING RESIDUAL FIELD

After the subtraction of all computed effects from the measured gravity field, the remaining anomalies may be considered as being the isolated signal of the Ivrea body, superimposed on the regional effect. The figures 5 and 6 show two different components of the residual field representing the base of the interpretation.

3.8   ESTIMATION OF THE ACCURACY OF THE RESIDUAL FIELD

The obtained accuracy of the described reduction procedures, whose knowledge is very important with respect to the envisaged accuracy of the interpretation, depends on several parameters. The real existing mass and density distribution is unknown. Therefore a rigorous computation of true errors is not possible. The only way to perform this are numerical investigations with varying density contrasts and beyond that an estimation of the resulting error budget.

The final accuracy of the computed residual is mainly influenced by the following effects:

1.) Effect due to inadequately introduced density contrasts and/or mass distributions or structures.
2.) Effect of too large compartments widths used in the digital terrain or structure model.
3.) Effect of applied simplifications or approximations in terms of inadequate computational formulae.
4.) Effect of the neglection of unknown density disturbances.

An overview of the estimated rms (random mean square) errors is given in the following list:

effect and computed or estimated accuracy	$m(\xi)$ ["]	$m(\eta)$ ["]	$m(\Delta g)$ [mgal]
observed deflection and gravity anomaly	0.4	0.5	0.07
geodetic position	0.1	0.15	0.8
reduction of topographic effects incl. Geology	1.0	1.0	1.5
reduction of the Moho effects	1.0	1.0	3.5
reduction of sediments in the Po plain	0.3	0.3	0.8
Resulting accuracy of the residual field	**1.5 arc sec.**	**1.5 arc sec.**	**4.0 mgal**

These values represent the rms noise of the observed signals (ξ, η, Δg). Thereby they also represent the limit of attainable accuracy in any interpretation procedure. Thus it makes no sense to interpret too many details of a structure because it would only suggest an apparent accuracy beyond reality.

552

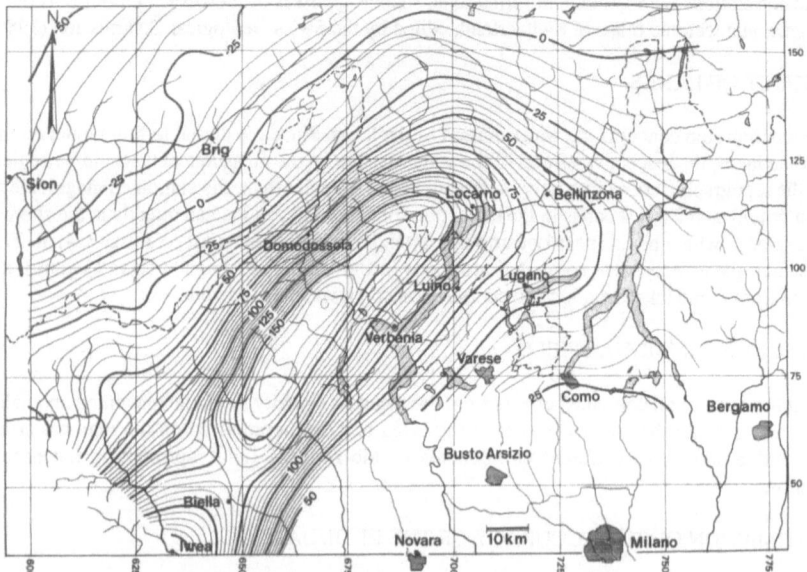

Figure 5: Isolated gravity anomalies in mgal caused by the Ivrea body after the subtraction of all computed effects due to topography, Moho- discontinuity, sediments, lakes and surrounding non- Ivrea geology.

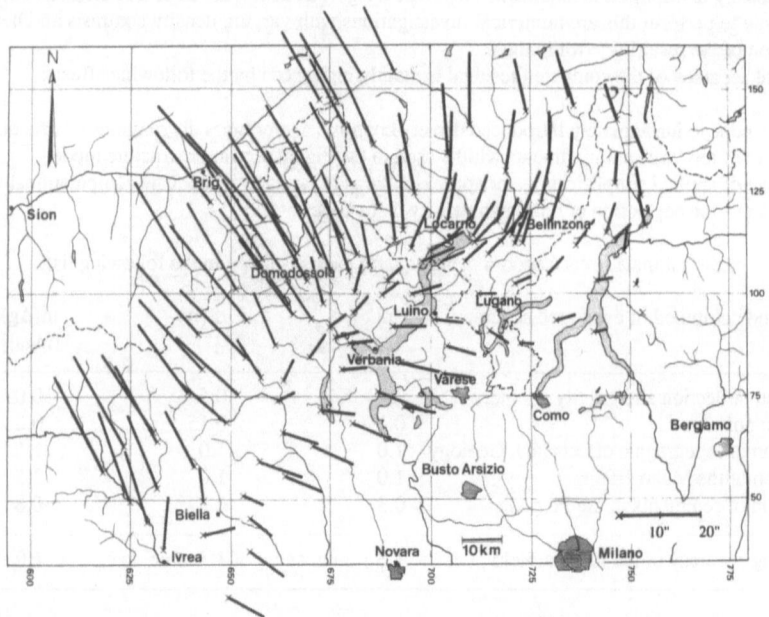

Figure 6: Isolated deflections of the vertical caused by the Ivrea body. The small cross signatures indicate the measurement locations while the thick lines indicate the direction and magnitude of the deflection of the vertical, being identical with the deflection of the astronomical zenith with respect to the normal of the reference ellipsoid (ellipsoidal zenith). The largest magnitudes, observed in the region of Brig - Domodossola are in the order of about 30 arc seconds ["].

4. Interpretation of the residual field

Prior to the interpretation of the residual field, the regional part of the gravity field has to be discussed. Numerical investigations have shown that the residual field may be approximated by a linear trend which forms an oblique plane coinciding with the strike of the Alps. The parameters of this trend which might be caused for example by a subduction zone, as proposed by Panza and Mueller (1978), are estimated simultaneously with the other unknown parameters within the adjustment procedure. Therefore the adjustment contains not only the the estimation of the parameters due to the introduced geometrical elements (prisms) for the interpretation, but also those of the regional field. This parameter set consists of three additional constants (offsets $\Delta \xi_0$, $\Delta \eta_0$ and Δg_0) and the slope of the trend plane in north- south and east- west direction.

4.1   MATHEMATICAL MODEL OF THE INTERPRETATION

Most of the interpretations published up to now dealt with gravity anomalies and two- dimensional structures only. While those simplifications are suitable for some simple structures in homogeneous areas, they are no longer valid in regions where the structure of the mass disturbance does not have a linear extension and varies in depth.

Since a detailed knowledge of the crustal structure is not only of interest for geophysical and geological purposes but also for geodetic applications, the interpretation to be applied should include and combine all available data performed by geodetic and geophysical methods as well.

For the interpretation of the Ivrea body, prisms with either rectangular or triangular cross sections whose spatial orientation may be defined arbitrarily, are used as structural elements for two reasons: the first reason is the possibility of a rigorous computation of the attraction effects in all three components. Secondly the construction of bodies with oblique planes requires much fewer elements.

To obtain not only the best possible estimation of the parameters of a structure but also to perform the proof of the achieved accuracies in terms of statistical quantities, a special software package has been designed. It performs a rigorous combination of all available gravity anomalies and deflections in the context of a common adjustment procedure. The design of the software enables a free combination of all parameters which are to be estimated whereby the estimation of the trend parameters for the regional field are included at every program run.

The parameters depend on the method actually applied. In one run the density contrast(s) of a particular structure may be kept fixed while the system gets solved to find out the best fitting geometry of the structure. In another run the geometry may remain fixed while the density contrast(s) are to be solved.

The mathematical model of such an adjustment consists of two parts: the functional model and the stochastic model. While the stochastic model defines all important relationships due to the observation accuracies to be introduced, the functional model describes the observation equations, connecting the observations (being represented by the vector $\underline{l}$), and the unknown parameters (vector $\underline{u}$):

$$\underline{l} = F(\underline{u}) \tag{3}$$

Depending on the available observations at the stations, the respective vector of observations $\underline{l}_i$ consists of the observed components of the residual gravity field (either completely or only partly filled):

$$\underline{l}_i^T = (\Delta g_{res}, \xi_{res}, \eta_{res}) \tag{4}$$

The vector of the unknowns may consist of the following components:

$$\underline{u}^T = (\Delta g_0 , \Delta \xi_0 , \Delta \eta_0 , \frac{\partial g_0}{\partial x}, \frac{\partial g_0}{\partial y}, \frac{\partial \xi_0}{\partial x}, \frac{\partial \xi_0}{\partial y}, \frac{\partial \eta_0}{\partial x}, \frac{\partial \eta_0}{\partial y}, \cdots$$

$$\cdots \Delta \rho_i , A_{x_i}, A_{y_i}, A_{z_i}, x_i , y_i , z_i , \alpha_i , \beta_i , \gamma_i) \tag{5}$$

where : $\Delta g_0 , \Delta \xi_0 , \Delta \eta_0$ are the offsets due to the regional field in three components
$\partial g_0/\partial x, \partial g_0/\partial y, \partial \xi_0/\partial x, \partial \xi_0/\partial y, \partial \eta_0/\partial x, \partial \eta_0/\partial y$ are the inclination of the trend planes.
(these 9 parameters are to remove the effect of the regional field).
$\Delta \rho_i$ is the density contrast of either a particular structure element i or a group of elements.
$A_{x,y,z}|_i$ are the geometrical dimensions of the prisms
$x, y, z|_i$ are the coordinates of the corresponding reference corners of the prisms
α, β, γ_i are the angles which define the orientation of the prisms in space

The system of the linearized observation equations is the base for the interpretation:

$$\underline{v}_1 = \underline{T} + \sum_{i=1}^{n} ef_{i_0} + \sum_{j=1}^{u} \frac{\partial ef_j}{\partial u_j} du_j - \underline{l} \tag{6}$$

$\underline{v}_1$ is the vector of the residuals between the observed and the adjusted values
$\underline{T}$ is the trend vector of the regional field
ef_i is the effect of the prism element i
$\underline{u}$ is the vector of the unknown parameters
$\underline{l}$ is the observation vector
n is the number of elementary prisms
u is the number of the unknown parameters

Numerical differentiation is applied to compute the derivatives representing the respective coefficients of the matrix of influence. By introducing the basic idea of a least squares adjustment which minimizes the sum of the weighted residuals:

$$v^T pv = min. \tag{7}$$

one can apply some well known matrix operations to perform the solution of the system.

This procedure is somehow similar to a trial-and-error-method but with much better support in terms of numerical residuals referring to all estimated parameters such as remaining density contrasts, geometrical improvements of the elements, their spatial orientations and so forth. To enable a simplification and to improve the performance of the method, the set(s) of unknown parameters may be chosen in completely independent assemblages.

4.2   RESULTS OF THE INTERPRETATION

The interpreted structure of the Ivrea body between Ivrea, Locarno and Lake Como reveals some new aspects (Bürki, 1988) with respect to former elaborations. Figure 7 displays the location and some details of the structure. A first detail is clearly visible north of Biella where the residual field (c. f. fig. 5) shows a saddle- like constriction which is interpreted as being the effect of an increased depth of the top and a reduced width of the Ivrea body. Another aspect is the eastward extension of the Ivrea structure toward the lake of Como (block G in figure 7). While here the model of Kissling (1984) ended some 5 kilometers east of Locarno (Kissling 1984) the new data used for this interpretation show clearly not only a northern but also an eastern continuatiuon of the structure whereby the found density contrast is only +0.1 gcm⁻³. The northern continuation underneath the Insubric line reaches its maximum in block F, located north-west of Locarno.

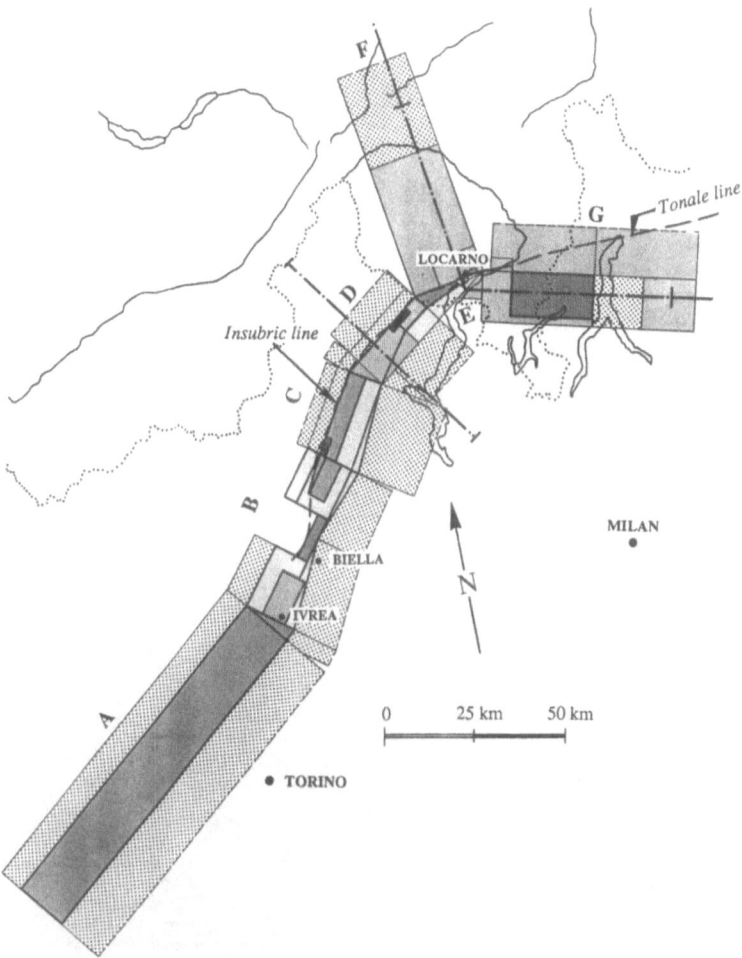

Figure 7: Location of the north eastern end of the Ivrea body along the Insubric line. The stipple- dashed lines crossing blocks D,F and G refer to the cross sections as outlined in fig. 8,9 and 10.

While the cross section of the models as proposed by Berckhemer (1968), Ansorge (1968), Giese (1968), Kaminski and Menzel (1968) and Kissling (1984) show a typical "overhanging" Ivrea structure which is interpreted as being caused by a pronounced seismic inversion zone, the analysis of the astro- gravimetric data could not confirm this "bird head" shape. On the contrary the data indicate the existence of a mass surplus underneath the Insubric Line. Therefore the gap had to be filled and even extended towards north to fit the data. Figure 8 shows a cross section of block D:

556

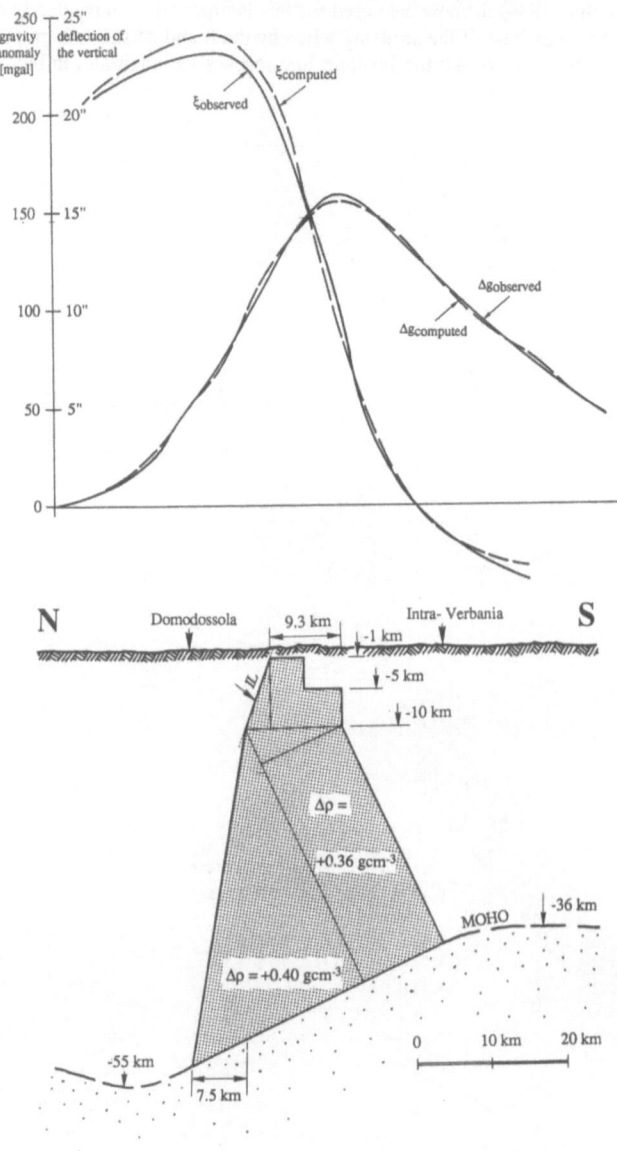

Figure 8 : Cross section of block B of the Ivrea body in the region Intra-Verbania - Domodossola. The major change compared to former models is the filled structure beneath the Insubric Line, indicated with *(IL)*. The upper profiles show the observed and the computed values for the gravity anomaly Δg and the north- south component of the deflection of the vertical.

The mass surplus reaches its maximum in the region of Locarno - Val Bedretto where a big wedge reaching down to the Moho is needed to satisfy the measurements (figure 9):

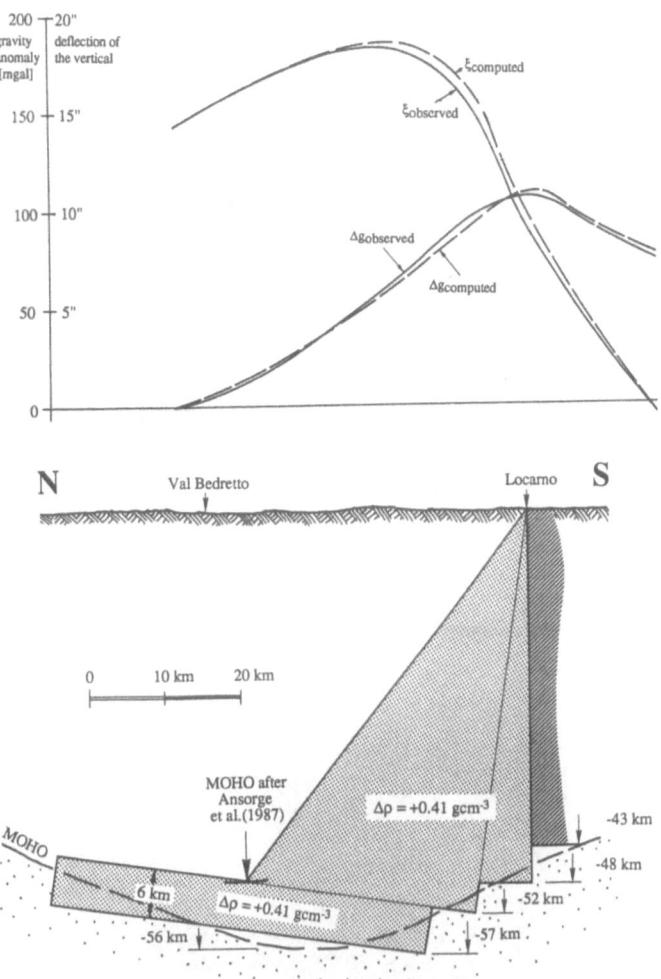

Figure 9: Cross section (blocks E- F) of the Ivrea structure in the region Locarno - Val Bedretto with the appropriate profiles of observed and computed values for Δg and the north- south component ξ of the deflection of the vertical. The astro- gravimetric interpretation indicates here the maximum of the mass surplus north of the Insubric line, where a big wedge with a density contrast of +0.41 gcm⁻³ is needed to satisfy the measurements. This interpretation of the Moho depth fits well with the re-evaluation of seismic refraction data from Ansorge et al., (1987) as marked with the arrow. The dashed bottom line indicates the Moho depth corresponding to figure 3.

In spite of the fact that the Insubric Line cuts this northern continuation, the found density contrast of +0.4 gcm⁻³ corresponds well with the density contrast encountered south of the tectonic lineament.

A further difference to geophysical models was found east of Locarno. While up to now the models of the Ivrea body ended beneath the Magadino plain some 5 km east of Locarno, the new interpretation has revealed an eastward continuation of the mass disturbance with an extension of approximately 60 km. The width of this structure is 12 km at 6 km depth and 27 km at the bottom (where it matches the Moho).

Two features worth to be mentioned were found in this part: The respective density contrast found here was only +0.1 gcm^{-3} and the azimuthal direction of this extension does not match the direction of the Insubric (Tonale) Line. Its axis is rotated by about 15 degrees towards south. In the longitudinal section, this mass extension is strongly undulated some ten kilometers east of Locarno and in the region of Lago di Como (lake of Como), where the mass subsides down to the Moho. 14 km towards the east it raises again by approximately 8 km (c. f. figure 10).

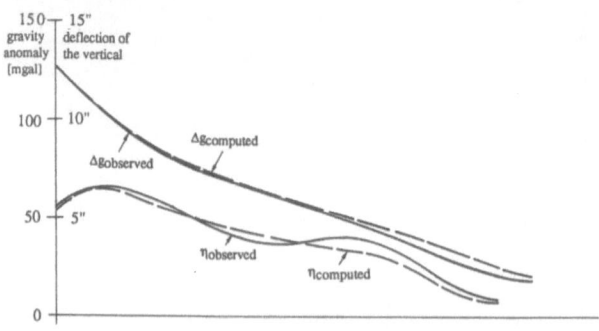

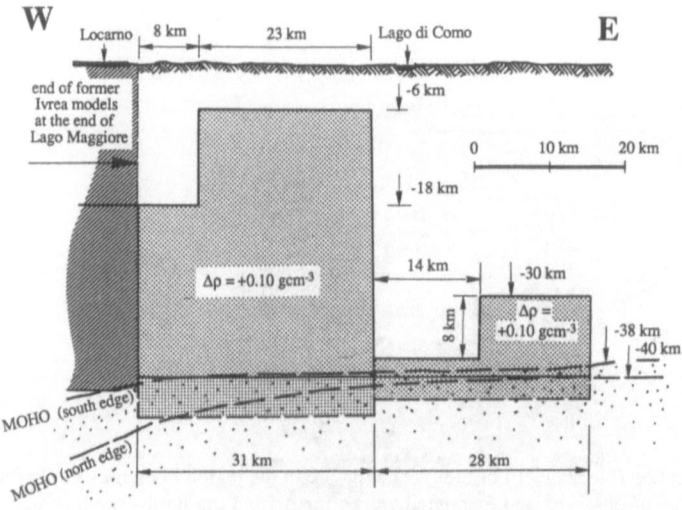

Figure 10: Longitudinal cross section through the continuation east of Locarno (block G). This part shows a density contrast of +0.1 gcm^{-3} and its axis does not correspond with the direction of the Tonale line.

4.3   REMARKS ON THE OBTAINED ACCURACY OF THE INTERPRETATION

As already mentioned earlier in this chapter, the applied software performs a quantitative justification of the actually achieved accuracy. This accuracy, which is deduced from differences between computed and observed signals, is describing to what extent the model is capable to reproduce the observed gravity field. The final adjustment run revealed the following values which agree well with the estimated accuracies of the residual field as explained in chapter 3.8:

rms error of the residual deflections (ξ and η) of the vertical : $m(\xi) = m(\eta) = 1.5"$
rms error of the residual gravity anomaly Δg: $m(\Delta g) = 4.5$ mgal

offsets of the regional field:

$\Delta\xi_0 = -4.5" \pm 0.2"$ $\Delta\eta_0 = +7.2" \pm 0.2"$ $\Delta g_0 = +7.1$ mgal" ± 1.3 mgal

trend parameters:

$\partial\xi/\partial x = -0.016 \pm 0.06$ ["/ 10 km] $\partial\xi/\partial y = +0.392 \pm 0.05$ ["/ 10 km]

$\partial\eta/\partial x = +0.289 \pm 0.06$ ["/ 10 km] $\partial\eta/\partial y = +0.293 \pm 0.06$ ["/ 10 km]

$\partial\Delta g/\partial x = +3.845 \pm 0.06$ ["/ 10 km] $\partial\Delta g/\partial y = -3.157 \pm 0.06$ ["/ 10 km]

beside these geometrical parameters, a group of some 15 differential density contrasts for small local geological units in the vicinity of the Ivrea body has been estimated, revealing values in the order of 0.1 ± 0.01 gcm^{-3}.

Instead of a list with a numerical description of all structural elements, a perspective illustration may be more suitable to illuminate the details of the new model of the Ivrea body:

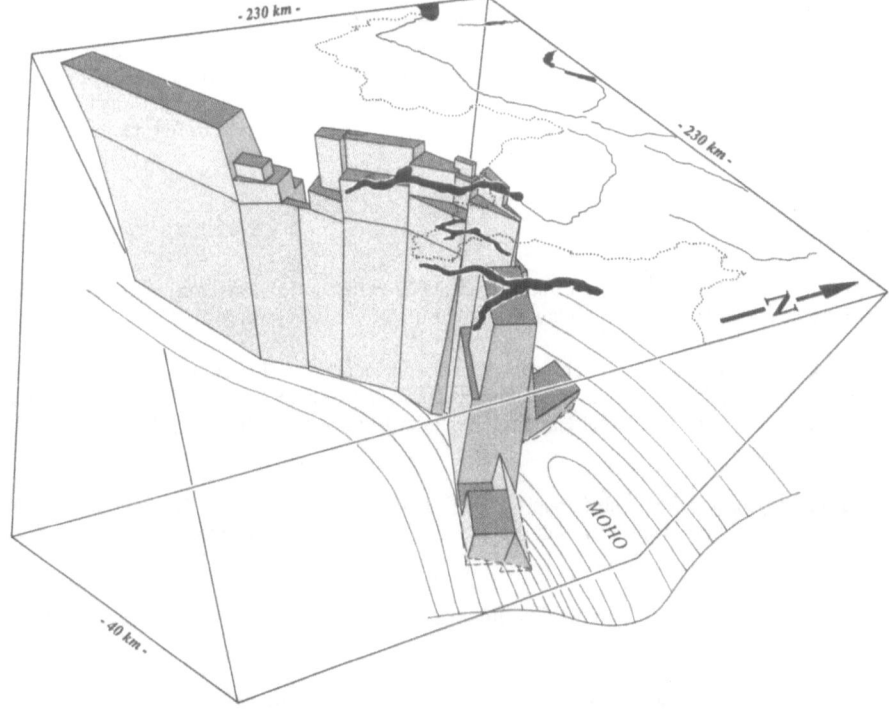

Figure 11: Perspective view of the interpreted Ivrea body as seen from east. In the middle at left, the striking constriction near Ivrea- Biella is seen. The eastern continuation, showing a density contrast of only $+0.1$ gcm^{-3} is visible in the foreground (underneath lake of Como) while the found wedge north of Locarno crossing the Moho trough is only partly visible at the right side.

5. Conclusions

Typical cross sections of geophysical models of the Ivrea body as proposed up to now indicate a seismic low velocity zone at the base of the body. This inversion zone was interpreted as being a density inversion zone as well.

The new interpretation of the gravity field in terms of astro- gravimetric observations has revealed some new aspects which cannot confirm this assumption. In contrary, it seems that the base of the Ivrea body is extending towards the alps, reaching its maximum north of Locarno.

Another aspect is the eastern continuation of the Ivrea body towards the lake of Como. Since the density contrast found for this part is only +0.1 gcm^{-3}, the related geological formation must not necessarily be a part of the geophysical Ivrea body which has a very pronounced mean density contrast of about +0.4 gcm^{-3}.

The distribution of excess masses proposed in this "geodetic" model seems to be supported by most recent interpretations of old (Giese, 1968) and new seismic data obtained in the framework of the Swiss National Research Program "Deep Crustal Structure in Switzerland" (NFP-20). The model can be understood tectonically by a combination of two plate tectonic processes such as Neo- Alpine backthrusting connected with right-lateral transpressive deformations (Heitzmann et al, 1989).

6. Acknowledgements

This contribution summarizes the results of several common field campaigns. We thank Prof. W. Torge (University of Hannover, FRG), Prof. K. Rinner (University of Graz, A), Prof. G. Birardi (University of Rome, I) and Gen. L. Zanetti from the Istituto Geografico Militare Italiano (IGMI) in Florence and all contributors from the Institute of Geodesy and Photogrammetry of the ETH Zurich for their excellent cooperation. We thank Prof. St. Mueller (Institute of Geophysics, ETH Zurich) for the gravimetric data he placed at our disposal.

7. References

Ansorge, J. (1968): Die Struktur der Erdkruste an der Westflanke der Zone von Ivrea. Schweizerische Mineralogische und Petrographische Mitteilungen (SMPM), Vol. 48/1, p. 247 - 254.

Ansorge, J., E. Kissling, N. Deichmann, H. Schwendener, E. Klingelé, St. Mueller (1987): Krustenmächtigkeit in der Schweiz aus Refraktionsseismik und Gravimetrie. 1. NFP 20- Symposium, Bad Ragaz. Bulletin Nr. 4 of the Swiss National Research Project NFP 20: "Deep Crustal Structure in Switzerland", p 12.

H. Berckhemer (German Research Group for Explosion Seismology, 1968): Topographie des "Ivrea-Körpers", abgeleitet aus seismischen und gravimetrischen Daten. Schweizerische Mineralogische und Petrographische Mitteilungen (SMPM), Vol. 48/1, p 235-246.

Bürki, B. (1988): Integrale Schwerefeldbestimmung in der Ivrea- Zone und deren geophysikalische Interpretation. Dissertation ETH Zürich, Nr. 8621, 186 p.

Giese, P. (1968): Die Struktur der Erdkruste im Bereich der Ivrea- Zone. Schweizerische Mineralogische und Petrographische Mitteilungen (SMPM), Vol. 48/1, p 261-284.

Girdler, R. W. (1985): Density contrast at the Lithosphere - Asthenosphere Boundary. In: Bericht Nr. 102 des Instituts für Geodäsie und Photogrammetrie, ETH Zürich.

Heiskanen, W. A. and Moritz, H. (1984): Physical Geodesy. Reprint Institute of Physical Geodesy. Technical University Graz, Austria.

Heitzmann, P., D. Bernoulli, G. Bertotti, W. Frei, P. Valasek and A. Zingg (1989): Geological Interpretation of the Deep Seismic Line NFP 20- South across the Central and Southern Alps (Switzerland). 3. NFP 20- Symposium, Lugano. Bulletin Nr. 8 of the Swiss National Research Project NFP 20: "Deep Crustal Structure in Switzerland", p 23.

Kaminski, W. and Menzel, H. (1968): Zur Deutung der Schwereanomalie des Ivrea- Körpers. Schweizerische Mineralogische und Petrographische Mitteilungen (SMPM), Vol. 48/1, 1968, p 255-260.

Kissling, E. (1984): Three-Dimensional Gravity Model of the Northern Ivrea-Verbano Zone. In: Geomagnetic and Gravimetric Studies of the Ivrea Zone. Beiträge zur Geologie der Schweiz, Nr. 21, p 53-61.

Klingelé, E. (1980): A new Method for Near- Topographic Correction in Gravity Surveys. Pageoph, Vol. 119 (1980/81).

Mader, K. (1951): Das Newtonsche Raumpotential prismatischer Körper und seineAbleitungen bis zur dritten Ordnung. Sonderheft 11 der Österreichischen Zeitschrift für Vermessungswesen. Wien 1951.

Menard, G. (1979): Relations entre Structures Profondes et Structures Superficielles dans le Sud- Est de la France. Institut de Recherches Interdisciplinaires de Géologie et de Mécanique, Universität Grenoble 1979, 178 p.

Mueller, St., Ansorge, J., Egloff, R. and Kissling, E. (1980) : A crustal cross section along the Swiss Geotraverse from the Rhinegraben to the Po plain. Eclogae Geol. Helv. Vol. 73, Nr. 2, 463-485.

Panza, G. F. and Mueller, St. (1978) : The plate boundary between Eurasia and Africa in the Alpine area. Mem. Sci. Geol. Univ. Padova, vol. 33.

Stein, A., Vecchia, O., Froehlich, R. (1978): A seismic Model of a Refraction Profile Across the Western Po Valley. In: Alps, Apennines and Hellenides; Closs, H., Roeder, D. and Schmidt, K. (eds.); E. Schweizerbart'sche Verlagsbuchhandlung, Stuttgart, BRD.

Swiss Geological Commission (1980): Geolocical and Tectonic maps of Switzerland, printed by the Federal Office of Topography.

Vecchia, O. (1968) : La Zone Cuneo - Ivrea - Locarno, élément fondamental des Alpes. Géophysique et géologie. Schweizerische Mineralogische und Petrographische Mitteilungen (SMPM), Vol. 48/, 215 - 226.

THE NATURE OF THE KAPUSKASING STRUCTURAL ZONE: RESULTS FROM THE 1984 SEISMIC REFRACTION EXPERIMENT

Jianjun Wu and Robert F. Mereu
Department of Geophysics, University of Western Ontario
London, Canada. N6A 5B7

ABSTRACT. A few years ago, the Canadian Consortium for Crustal Reconnaissance using Seismic Techniques (COCRUST) conducted a major long-range seismic refraction and wide-angle reflection experiment across the Kapuskasing structural zone (KSZ) in Northern Ontario. The main purpose of this experiment was to determine the structure of the crust below this zone and to help clarify geological theories on its origin. The interpretation of this data set made use of data processing which involved conventional travel-time procedures coupled with synthetic seismogram analysis using programs that were written to handle laterally heterogeneous structures. Two-dimensional P wave velocity models show that the velocities of the upper portion of the crust in the region varies from 5.9 to 6.5 km/s. The highest velocities were found to lie along the axis of the KSZ. Wide-angle reflection observations show that the Moho discontinuity in the survey area is not well defined and is transitional in nature. A combined P and S wave analysis shows that over most of the region Poisson's ratio does not differ very significantly from 0.25. There is some evidence that its value increases to 0.26-0.27 in the upper crust under the axis of the KSZ. A gravity inversion was also performed by using the seismic models as constraints. The calculations indicate that the largest densities lie along the axis of the KSZ and agree qualitatively well with the high 6.5 km/s P wave velocity found just below the surface along the axis. The results of our analyses give added support to the theory that the KSZ may contain rocks which were uplifted from the middle crust.

1. Introduction

A few years ago, the Canadian Consortium for Crustal Reconnaissance using Seismic Techniques (COCRUST) conducted a major long-range seismic refraction and wide-angle reflection experiment across the Kapuskasing structural zone (KSZ) in Northern Ontario (Northey and West, 1985). This experiment was composed of five seismic refraction lines of 350 to 430 km length (Figure 1). Lines BH, CJ, and DK trended NNE parallel to the KSZ whereas lines AE and HE crossed it obliquely. All lines were

563

M. H. Salisbury and D. M. Fountain (eds.), Exposed Cross-Sections of the Continental Crust, 563–586.
© 1990 *Kluwer Academic Publishers.* .

564

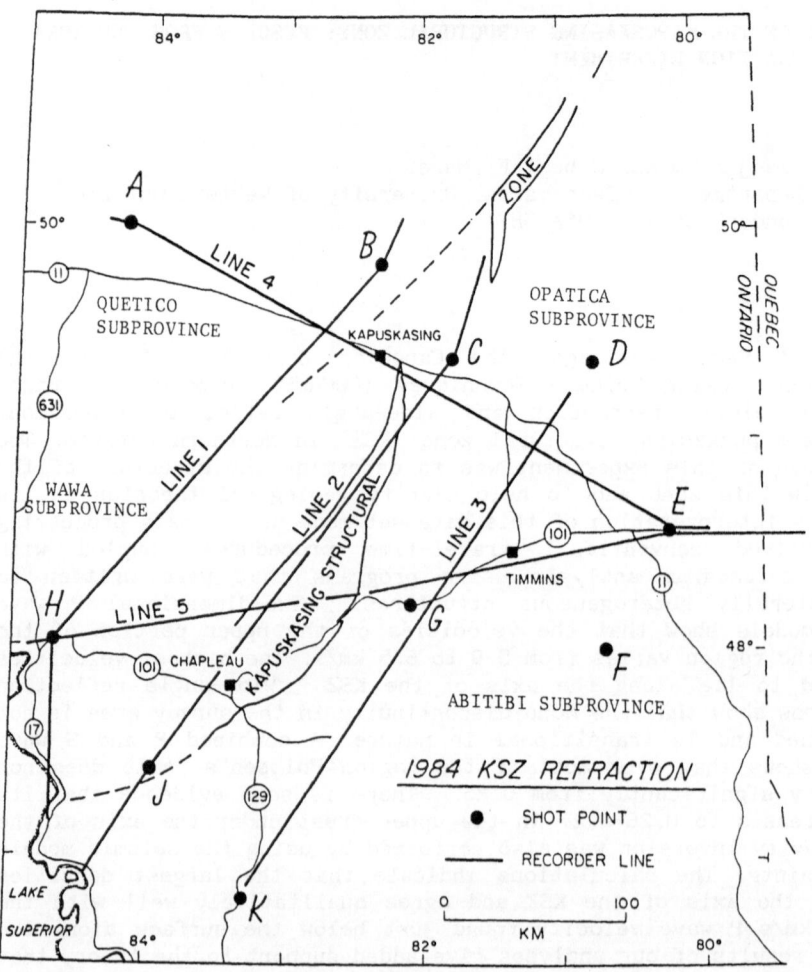

Figure 1. Kapuskasing Experiment seismic refraction lines shot
in 1984 (after Northey and West, 1985).

reverse shot and lines DK, AE and HE were also center shot. The lines CJ and HE were recorded in 2 sections to improve the spacing. In addition to the in-line profiles, two fan profiles were also obtained along line BH from shot F, and along line AE from shot J in order to increase the areal coverage. On the whole, the quality of the data collected from this experiment is good. Some examples of the observations are shown in Figures 2 and 3. The major objective of the experiment was to explore the nature of the Earth's crust under the KSZ and to help clarify the geological interpretation of its origin. Preliminary interpretations of the P wave data set were given by Wu (1987), Wu and Mereu (1988a, 1988b), Boland and Ellis (1988a, 1988b), Boland et al. (1988), and Boland and Ellis (1989). In this paper we present a combined P wave, S wave and gravity analysis of the data set.

The KSZ is characterized by the predominance of a north-northeast trending belt which extends from Lake Superior to the south of James Bay and cuts cross the main east-west structural trends of the adjacent Wawa, Abitibi, Quetico, and Opatica subprovinces (Figure 1), by an abrupt increase in the metamorphic grade, and by a broader positive gravity anomaly. Detailed geology and geophysics of the KSZ were given by Gaucher (1966), Gibb (1978), Watson (1980), Percival and Coe (1981), Percival and Card (1983) and Percival and Fountain (1986). The geological interpretation indicates that it may be an oblique cross-section through the upper two-thirds of the Superior Structural Province continental crust (Percival and Card, 1983). Thus the accompanying high-grade metamorphism and positive gravity anomaly may be due to the highly metamorphosed and high-density rocks found at the upper crust as a result of the uplift of the lower crust. Seismic reflection data (Cook, 1985, and Geis et al., 1988) give added support to this theory. Subsequent analyses based on the results of recent reflection survey have led Percival and Green (1988) to postulate that the uplift came from the middle crust rather than the lower crust.

2. Seismic Data Analysis Procedure

The P wave seismic refraction/wide-angle reflection data from each of the seismic lines was interpreted with an interactive 2-D ray-tracing method using the following steps:
(i) A simple starting model was first determined by combining the results from a least squares analysis of the first arrival refraction data for each of the shot points along each line.
(ii) In order to handle both vertically and laterally varying structures, the simple model was divided up into a set of triangular blocks. Each triangle had its own two-dimensional linear velocity function written in the form:

$$V(x, z) = ax + bz + c$$

where $V(x, z)$ is the seismic velocity at any point (x, z) in the model, and a, b, c are three constants which were determined from the given values of velocity at each vertex of each triangle. These three

566

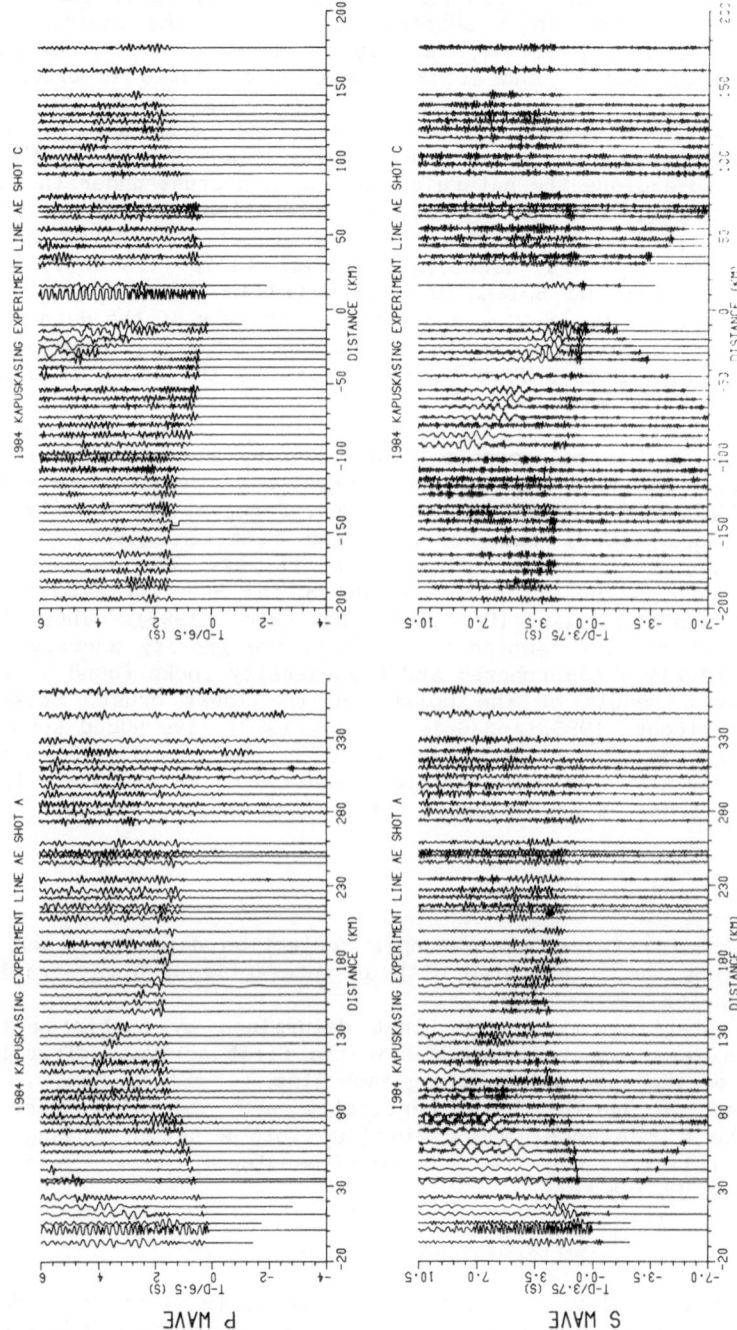

Figure 2. Comparison of observed P and S wave sections for line AE. Note the ratio of P wave reducing velocity to S wave reducing velocity is 6.5/3.75 = 1.73.

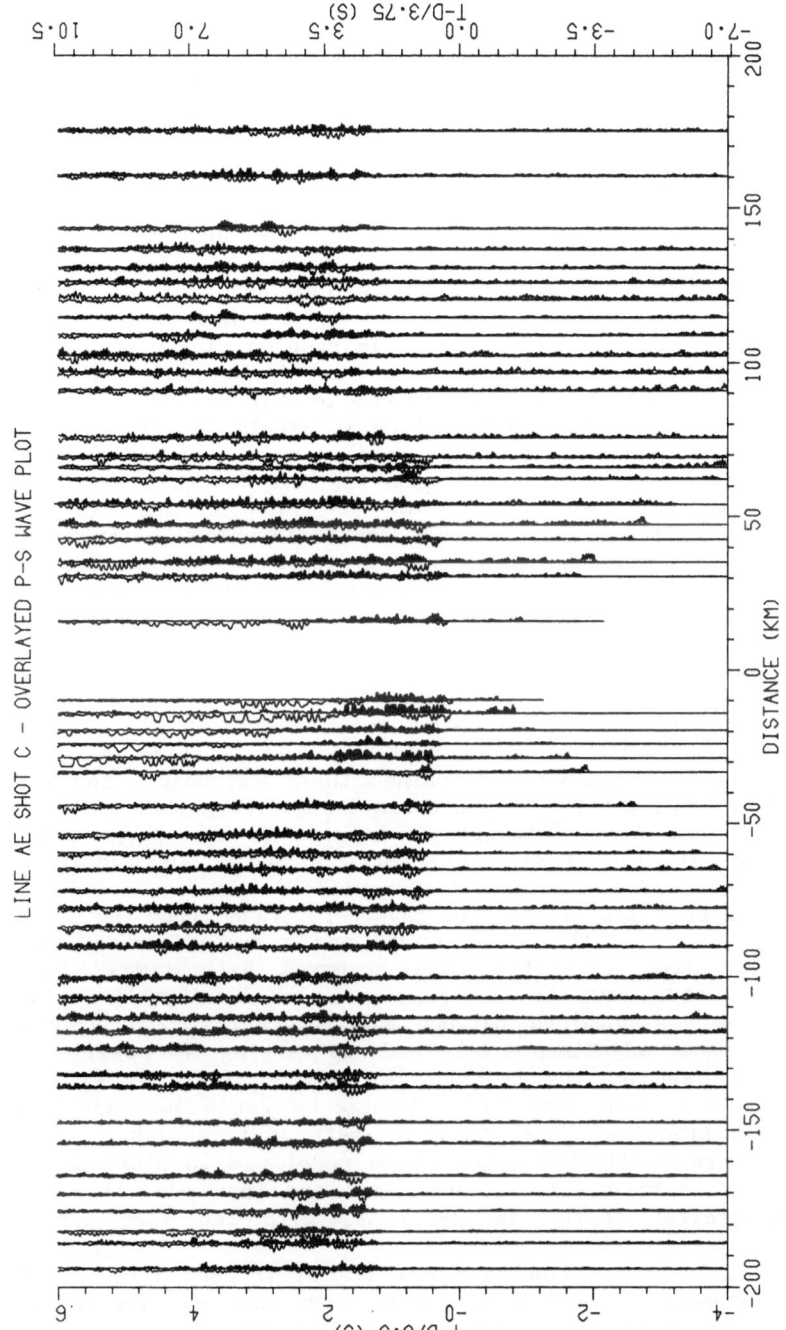

Figure 3. The composite P-S wave plot for line AE shot C with unshaded P waves plotted to the left and shaded S waves plotted to the right.

constants acquire different sets of values for each triangle.

The velocity gradient in each triangle is given by the gradient vector:

$$\text{Grad } \mathbf{V} = a\ \vec{i} + b\ \vec{k}$$

where $\vec{i}$ and $\vec{k}$ are unit vectors in the x and z directions respectively. The constant velocity gradient ensures that all ray paths within each triangle will form the arc of a circle thus enabling one to trace rays through the model very quickly. The tracing is done by solving ray arc equations with triangular boundary line equations and applying Snell's law at each intersection point. High speed is obtained if the actual ray paths are not plotted as then the whole problem can be handled by working with only the intersection points. The technique of dividing the model into triangular blocks enables one to work with a relatively small number of blocks. When the structure is simple the triangle may be large whereas when the structure is complex small triangles may be required. The presence of sloping boundaries and faults presents no inherent difficulty provided the sides of the triangles are chosen to lie along the geological boundaries. The manner in which the triangles is divided is not unique, however numerous tests have shown that in general the travel-time and amplitudes are not very sensitive to this non-unique problem. Synthetic seismograms were generated from transfer functions which were computed from the time delays, the free surface effects, the geometric spreading effects as well as all of the complex transmission and reflection coefficients which the ray encountered along its path. The computer program which does the ray tracing automatically searches the model for all major travel-time branches, numerically codes the branches, determines their end points and then ray traces through the model with appropriate angle ranges to determine arrival times and synthetic seismograms. Details of this program were first presented at the 1983 CCSS workshop held in Einsiedeln Switzerland (Mereu, 1983)

(iii) For a given modelling iteration, rays were traced from the source location in the model to the free surface and a comparison was made between the computed travel-time and observed data times. Small adjustments were then made to the gradients and boundary positions until a good fit was obtained between all theoretical and observed amplitude data. These adjustments were made by varying the position of the vertices as well as the seismic velocity at each vertex. The program was written such that the adjustments could be made very quickly in an interactive manner to whole sets of triangles at once. The whole procedure of iterating to a good model was to start at the top, fit the near source observations first and then gradually work one's way to greater depths in the model with the more distant observations.

Figure 4a illustrates how rays are traced through the line DK model in a forward direction to produce a set of theoretical travel-time curves which are superimposed on the DK record sections, and Figure 4b gives the corresponding synthetic seismogram. The final P

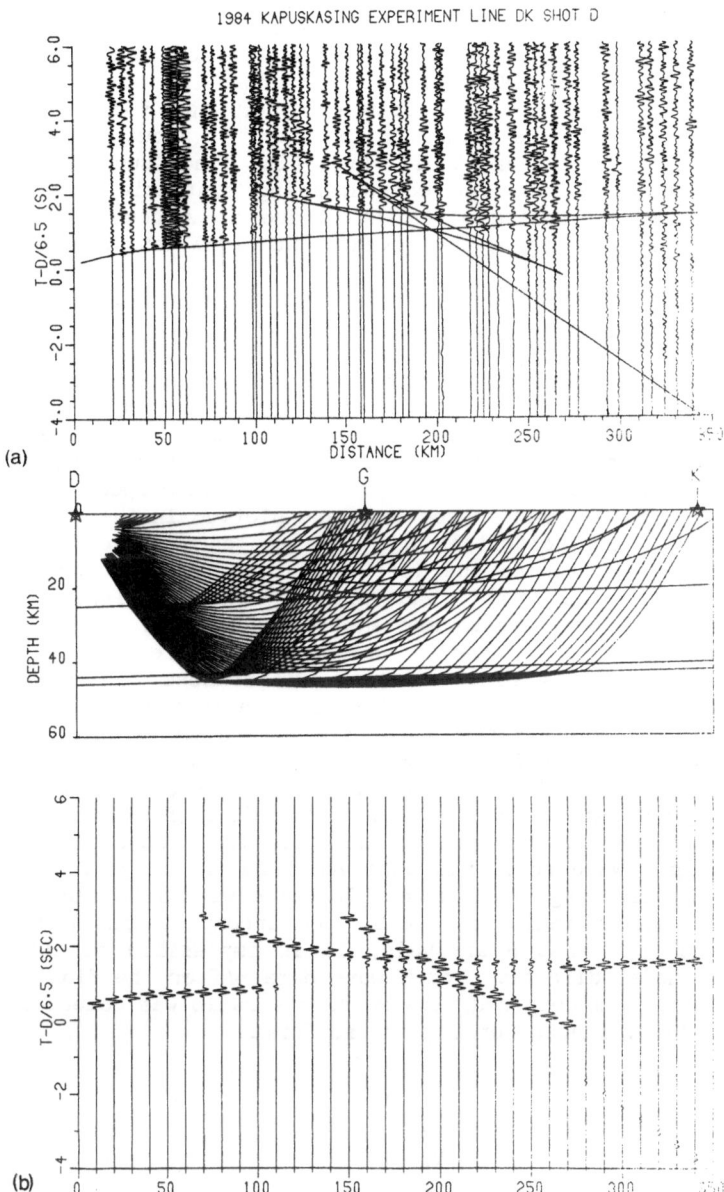

Figure 4 (a). Ray tracing diagram for model DK, with shot at D.
(b). Synthetic record section for line DK P wave velocity
model, with Shot at D.

wave models derived from all the lines are presented in Figure 5. The dotted lines in the diagrams are smoothed seismic velocity contours. Regions of high velocity gradient occur where the contours are close together as is seen in the Moho transition zones. By using a method which perturbs the model until misfits occur it was found that the precision of the contour line values in the model is about ± 0.05 km/s. Figures 6, 7, and 8 show theoretical travel-time lines superimposed on the observed record section. The corresponding synthetic seismograms were obtained from the models given in Figure 5. A complete set of these figures for all the lines was given by Wu (1987) and Wu and Mereu (1988a).

The S wave study was done by comparing reduced travel-time plots of S waves (reducing velocity 3.75 km/s) with those of P waves (reducing velocity 6.5 km/s) (see Figure 2). This choice of reducing velocities ensures that if Poisson's ratio is 0.25, then the scaling of the graphs will be such that the position of the travel-time lines on the P and S wave record sections will be the same (El-Isa et al., 1987). In order to facilitate the comparison of P and S waves, a new plotting method, called "composite plot", was developed. In this method, both P and S wave record sections are plotted in the same graph with different reducing velocities and time scales (Figure 3). All P wave peaks are plotted toward the left and unshaded and S wave peaks are plotted toward the right and shaded. The similarities and differences between P and S waves are easily seen in the composite P-S record section, especially when P and S waves are plotted with different colors, such as blue and red. Differences in the record sections can be interpreted as resulting from deviations of Poisson's ratio from 0.25. The subsurface location of these differences can be inferred by ray-tracing analysis. Two examples showing theoretical S wave travel-time lines superimposed on the observed S wave record sections are given in Figure 9. The procedure for the S wave interpretation is exactly the same as those for the P wave one, except that the starting S wave model was achieved from the final 2-D P wave velocity model by dividing the P wave velocities with 1.73 and keeping the same P wave discontinuities. Due to the incompleteness and large uncertainties in the first S arrivals, it is impossible to obtain a starting S wave model directly from S wave data. After the final 2-D S wave velocity model was obtained, the Poisson's ratio was computed from the $V_P(z)$ and $V_S(z)$ values according to the formula:

$$\sigma = \frac{1 - \dfrac{1}{2}(V_P/V_S)^2}{1 - (V_P/V_S)^2}$$

3. Discussion

An examination of all the P wave record sections indicated that the P waves are of high quality and first arrivals can be easily distinguished. In most of the in-line record sections, the first

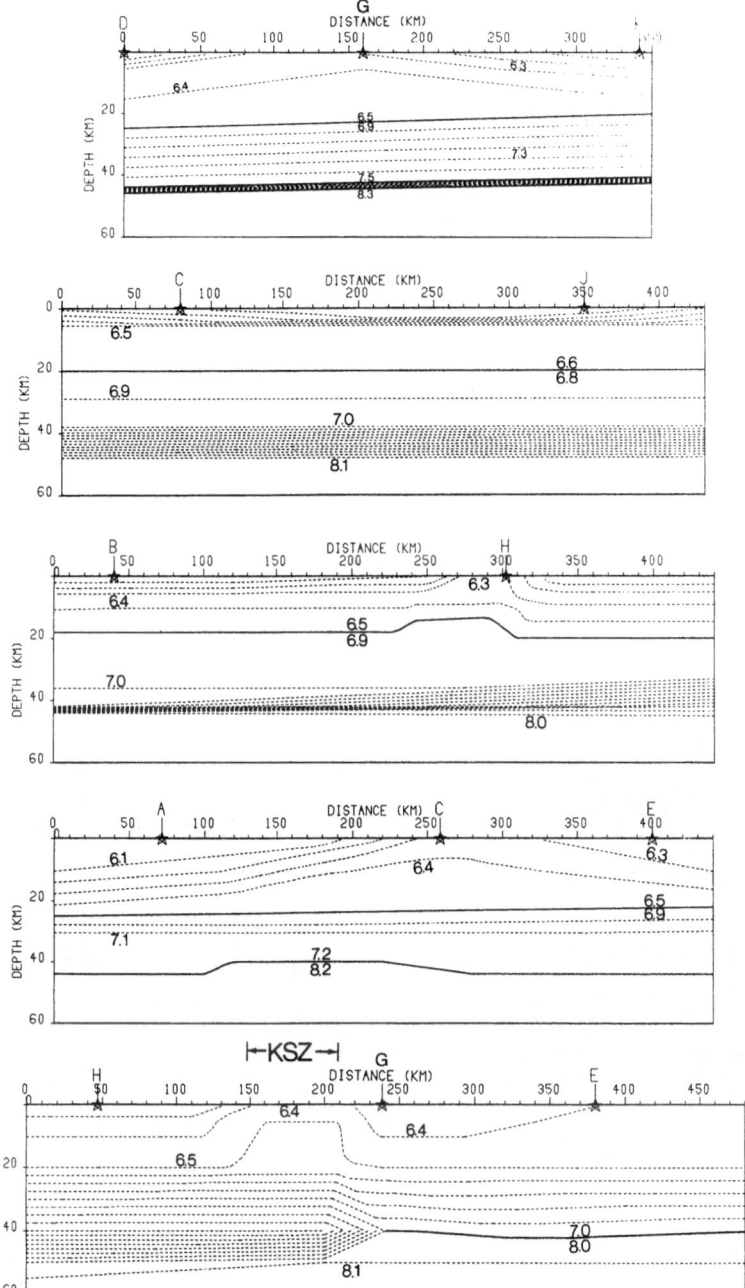

Figure 5. 2-D P wave velocity models for lines DK, CJ, BH, AE, and HE.

572

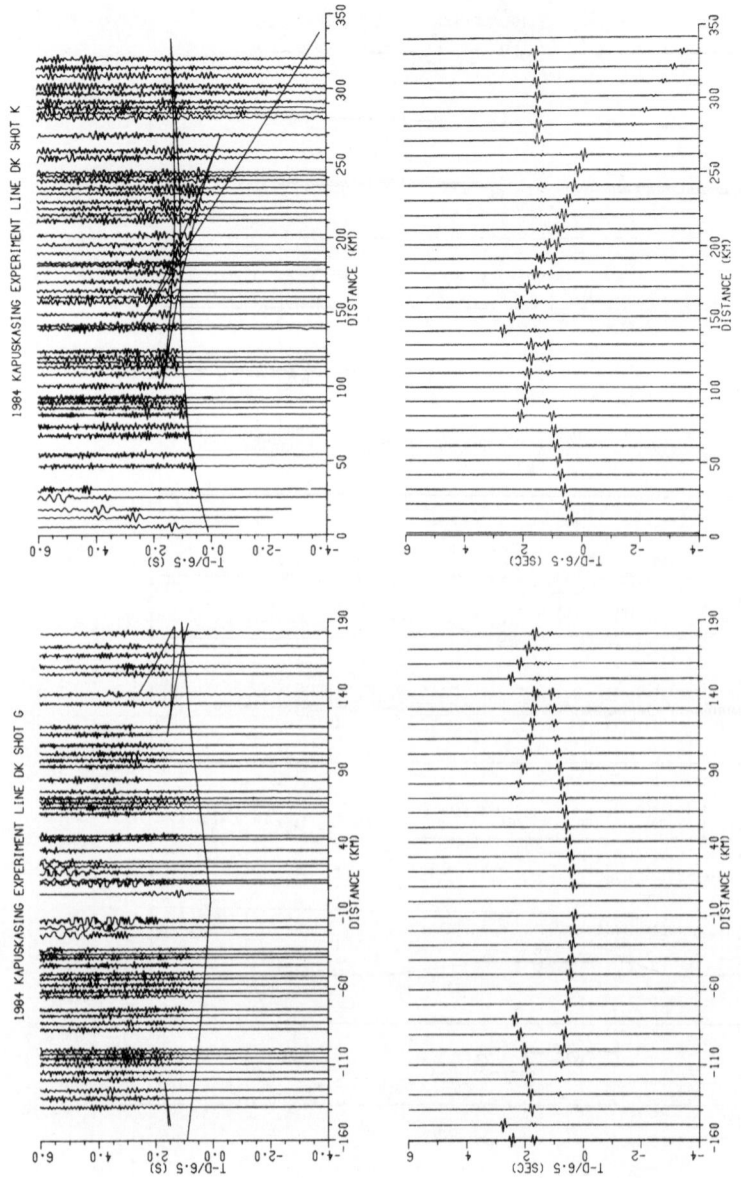

Figure 6. Theoretical travel-time curves and synthetic record sections for line DK.

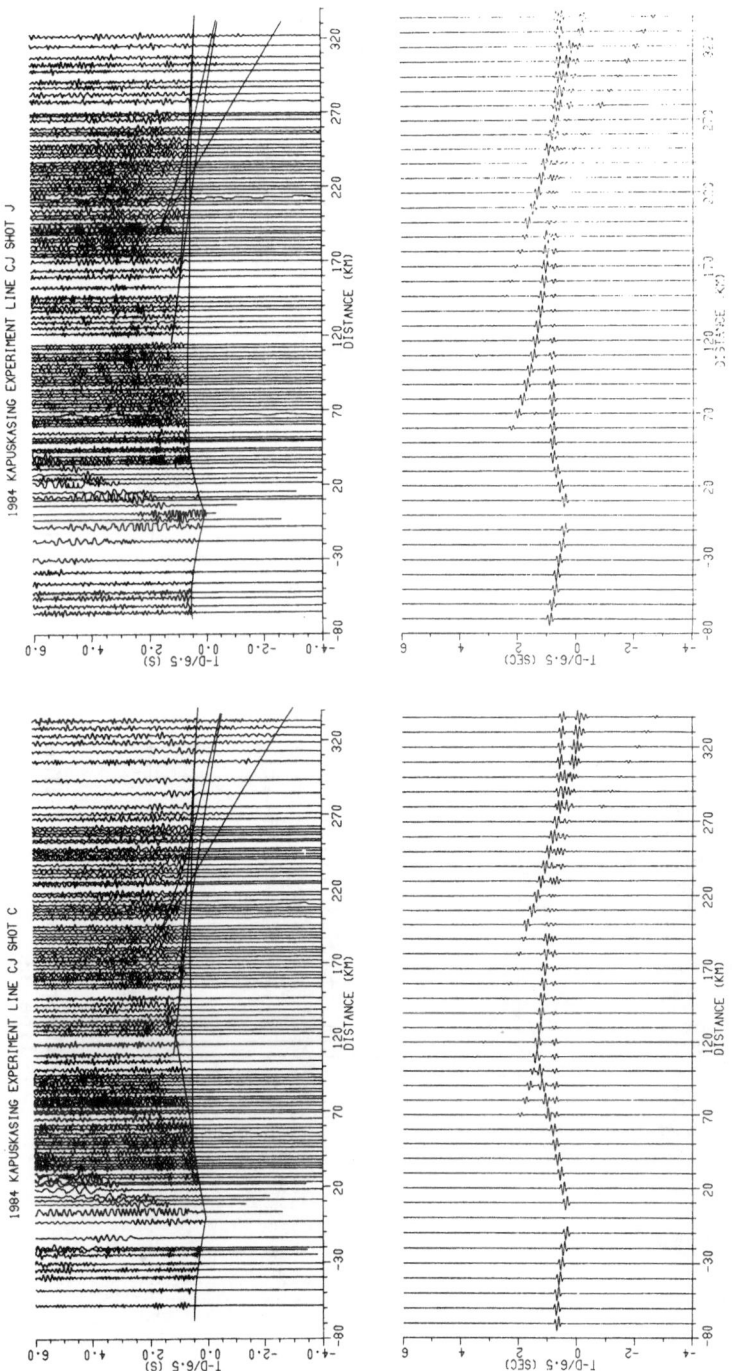

Figure 7. Theoretical travel-time curves and synthetic record sections for line CJ.

574

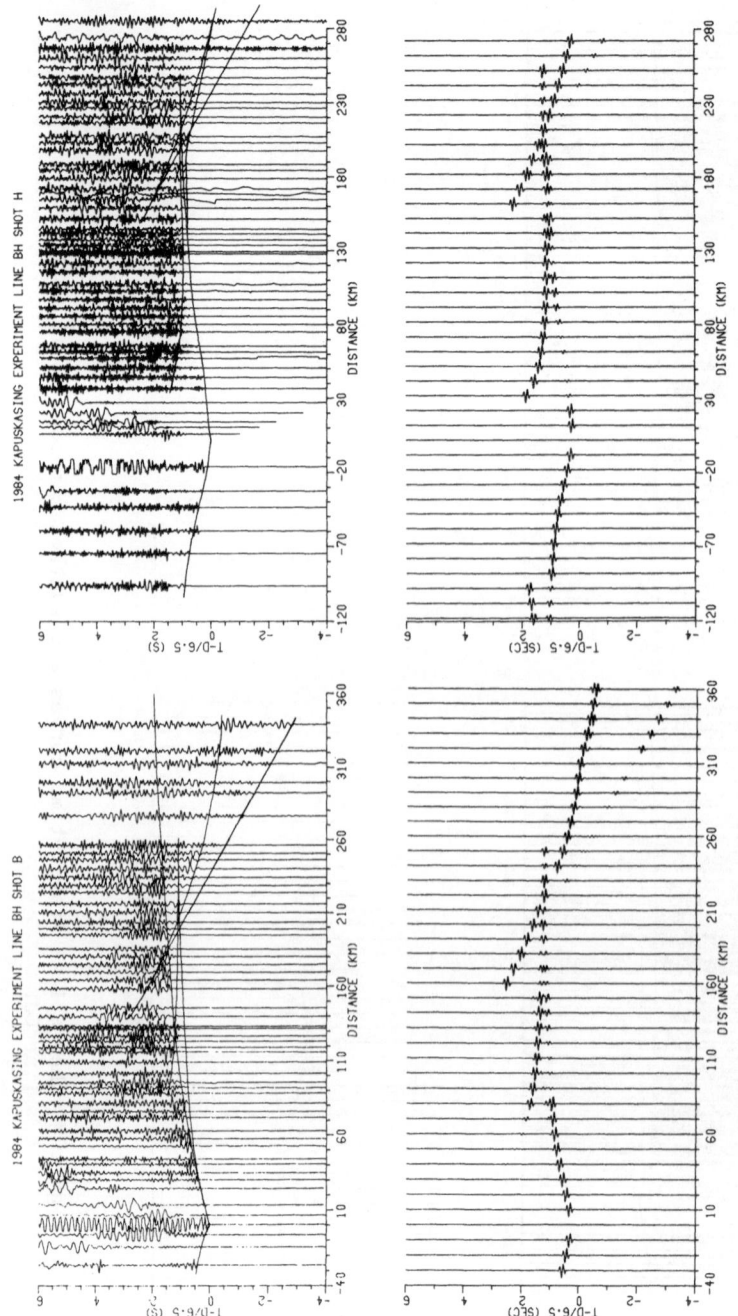

Figure 8. Theoretical travel-time curves and synthetic record sections for line BH.

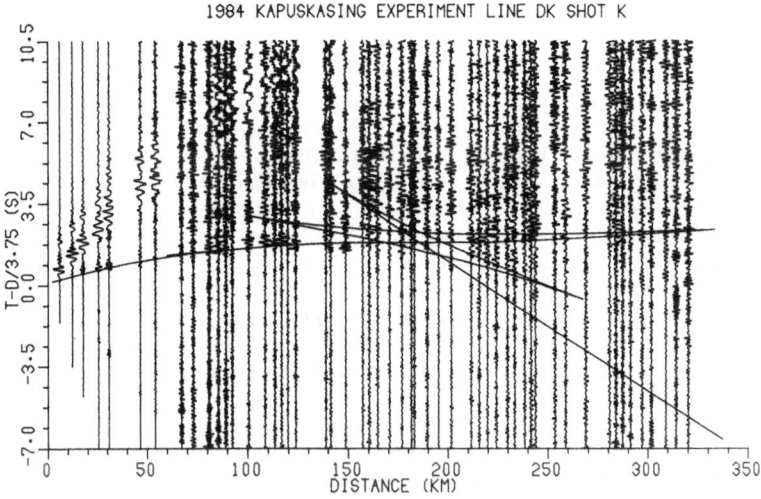

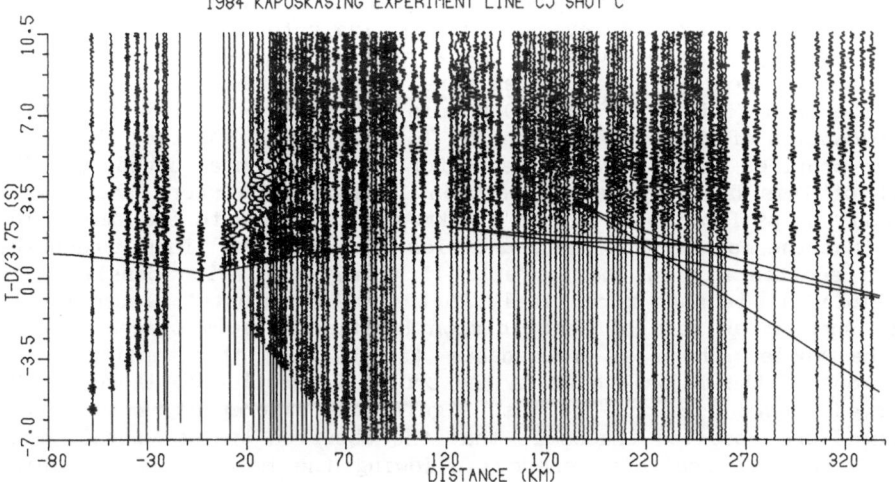

Figure 9. Theoretical S-wave travel-time curves superimposed on the S-wave record sections for line DK shot K and line CJ shot C.

arrivals show a clear curvature which implies that the velocity increases with depth. The apparent velocity analysis shows that a fairly large variation in the near-surface velocities exists across the region. The lowest value of about 5.9 km/s was found near shot point B (Figure 1). The largest values of approximately 6.5 km/s were found across the KSZ near shot point G.

A comparison of the three profiles which parallel the KSZ indicates that the coda energy following the main events was much more pronounced along the KSZ than across the Wawa and Abitibi greenstone belts, suggesting that rocks in the KSZ are more heterogeneous than those in the two belts. Mereu et al. (1989) measured the average signal complexities of sets of record sections from the COCRUST experiments conducted in Canada from 1979 to 1985. This study showed that the earth crust under the KSZ was much more complex than that under other terranes such as the Peach River Arch or Williston Basin in Canada. On the other hand, there is a strong similarity in the P wave travel-times for the forward and reversed record sections along the line CJ. Figure 10a shows the picked first arrivals from both forward and reversed shot points. It is surprising that most arrivals from shot C overlap with those from shot J. All these similarities imply that the KSZ is structurally more homogeneous on a large scale than the two adjacent belts. This can be seen from the 2-D seismic velocity model for line CJ (Figure 5). The model shows that from 5.5 km down to the upper mantle the KSZ is essentially homogeneous along its length. Figure 10b which displays the forward and reverse first arrivals for line AE shows that there are major differences in the 2 sets of travel-times. This can only be explained by large lateral structure as is shown in Figure 5.

The main feature of the 2-D models is the two-layered crust with the exception of model HE (Figure 5). This is due to the fact that most of the profiles showed some evidence for an intracrustal boundary. The depth of this boundary ranges between 13 and 25 km where the P wave velocity jumps from about 6.5 to 6.9 km/s or from 6.6 to 6.8 km/s. The fact that the models show the boundaries as continuous lines across the record section may not be correct. If a higher station density were available, it is much more likely that the intracrustal discontinuity would show up as sets of short discontinuities closely spaced layers. On the whole, there is reasonable fit between the amplitude and positions of the major events on the synthetic section to those of the observed record section. The most significant discrepancy between the computed and observed traces is that the observed traces contain much more incoherent scattered energy following the main arrivals, which were generated by small-scale lateral and vertical heterogeneities within the crust. It should be emphasized here that in order to obtain the simplest model consistent with the data, the effort to use as few layers as possible has been made, which means that strictly local features were not included in the 2-D model.

Poor observations of PmP along almost all of the profiles imply that the crust-mantle interface is not well defined because of its transitional nature. A least squares analysis was conducted on the Pn upper mantle arrivals to obtain apparent velocities of the various Pn data segment as well as the complete data set. The results indicate a

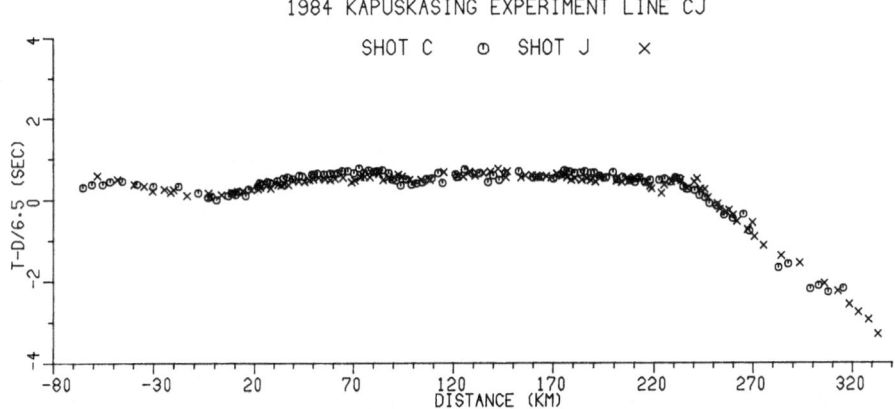

Figure 10 (a). Picked P wave first arrivals for line CJ. Note how the
arrivals from shot C overlap with those from shot J.

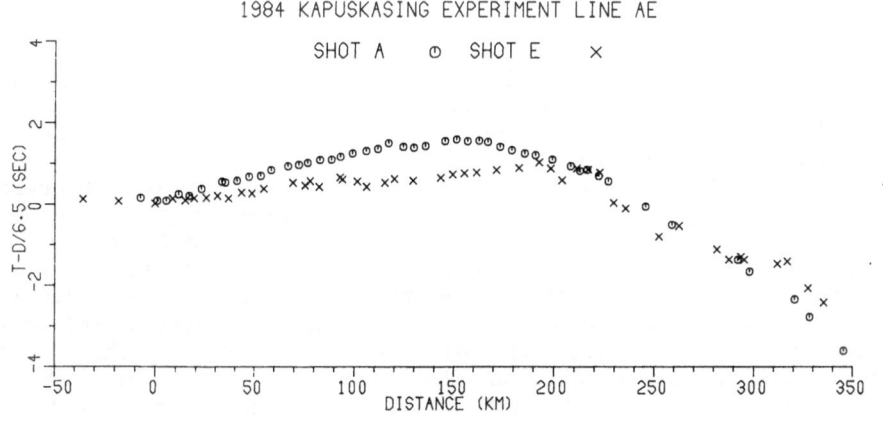

Figure 10 (b). Picked P wave first arrivals for line AE. Note how the
arrivals from shot E are advanced compared to those from
shot A.

very large variation in the apparent velocity values from 8.05 to 8.64 km/s, with the average of 8.14 ± 0.05 km/s. This may be caused by topography on the Moho surface; or significant variations of upper mantle velocity in the different regions of the survey area; or significant variations in crustal structure, or a combination of these. The average crust depth is about 40 to 45 km. There is a very good agreement between our P wave velocities and the laboratory measurement of rock velocities by Salisbury and Fountain (1988). The results of our P wave inversion analysis are also in good qualitative agreement with the findings of Boland and Ellis (1988a, 1988b).

Figure 3 shows examples of composite plots of line AE. Compared with P waves, S waves were of poor quality and often difficult to see clearly as they are always later arrivals which lie in the coda of the P waves. The identification of S waves was made based on travel times using the P waves as reference. On the whole, the S wave sections do show a great similarities with P wave sections. Usually Sg waves can be clearly seen near the shot point and then disappear. Reflected SmS waves are easily distinguished at the places where strong PmP waves are observed. No refracted Sn waves can be identified, even where Pn refractions have large amplitude, suggesting that S waves are strongly attenuated when they propagate through the Moho and the upper mantle. There exist some significant differences between the P and S wave sections. For line DK shot K, the early arrival Sg waves with large amplitudes are clearly seen in the range of 100 to 150 km, which could be explained with converted SP waves. The same phenomenon is also seen in the composite section of line AE shot A. However, the same thing is not observed from the reverse and center shots of line AE. This may indicate that there is a lower Poisson's ratio zone in the upper crust near shot point A. From line HE shot G, the Sg waves, coinciding with Pg waves but having large amplitudes, are recorded in the range of 0 to 60 km left. It is not clear why the Sg waves always have a relatively large amplitude in all these cases.

In general, the starting S wave models can fit the observed S waves well except for line CJ where the later large amplitude S arrivals in the range between 120-200 km seem to be delayed about 0.5 s. The excellent fit between the observed and calculated Sg waves indicates that rocks in the near-surface have a Poisson's ratio of 0.25. The good fit between the observed and calculated SmS waves, reflected waves from the Moho discontinuity, suggests that on the average the Poisson's ratio of the whole crust is about 0.25. The significant delay of the later S arrivals for line CJ probably implies a decrease in S wave velocity, therefore an increase in Poisson's ratio in the upper crust of the KSZ. The higher Poisson's ratio (about 0.26-0.27), if this is true, might indicate that rocks in the upper portion of the KSZ have been undergone different geological processes from those in the upper crust of the Wawa and Abitibi belts and have different physical properties.

Comparing our results with those of Assumpçâo and Bamford (1978) and Holbrook et al. (1988), there is a very good agreement in that the Poisson's ratio in the near-surface and the average Poisson's ratio of the whole crust are about 0.25. Our results differ from those of El-Isa et al. (1987) in that no high Poisson's ratio was found in the lower crust.

4. Gravity Analysis

One of the purposes of the experiment was to find an explanation for the high gravity anomaly along the KSZ. Hence, it is necessary to perform a gravity modelling after the seismic model has been obtained. The original Bouguer gravity data, which was obtained from the Gravity Division of the Geological Survey of Canada, consisted of 11,652 unequally spaced stations and covered an area between 80 to 85 degrees west longitude and 46 to 52 degrees north latitude. The gravity modelling was done using the following steps: (1) An interpolation program (Akima, 1978) was applied to this data to project the gravity data on an equally spaced grid. (2) The regional and residual components of the field were separated using 2-D low-pass Butterworth filters (Figure 11). (3) The inversion was performed by both 2-D and 3-D iterative computer programs which were based on the 2-D method of Bott (1960) and the 3-D method of Cordell and Henderson (1968).

In other gravity interpretations, see for example Percival and Card (1983), the constraints applied to the model were based on surface geological observations. Our interpretation differs from these studies in that rather than try to obtain a precise geometrical model of the subsurface structure of the KSZ we have set up the inversion process such that the solution would give us lateral variations in average crustal density. In order to constrain the models from an infinite number of possible choices, it is assumed that the residual gravity fields are entirely caused by lateral variations of the densities of rocks in the upper 20 km of the crust. This assumption was made due to the fact that the P wave velocity models show that the upper crustal rocks of the KSZ have higher velocity than those of the Wawa and Abitibi belts and it is known that gravity variations at the near-surface are much more sensitive to changes in the upper crustal rock density than changes in the lower crustal rock density. Based on this assumption, the upper crust (20 km deep) of the whole area is divided up into a set of rectangular blocks of equal dimensions with the gridding points at the center of each block and within each block the density is the same. Following a method similar to that of Bott (1960), the initial density of each block was obtained by treating each block as an infinite horizontal sheet that could generate the residual anomaly at the center of the block. Then the theoretical anomaly was computed by adding the effects of all blocks and differences between the theoretical and observed anomalies were obtained by subtracting them. Finally the new density was computed according to the difference and the anomaly was recalculated. The process is repeated until the best fit was achieved. If the density distribution in the survey area is known, this program can also be used to obtain the shape of the anomalous body automatically by changing depth in each iteration instead of changing the density. One example, showing the 2-D gravity modelling of a line perpendicular to the KSZ, is given in Figure 12.

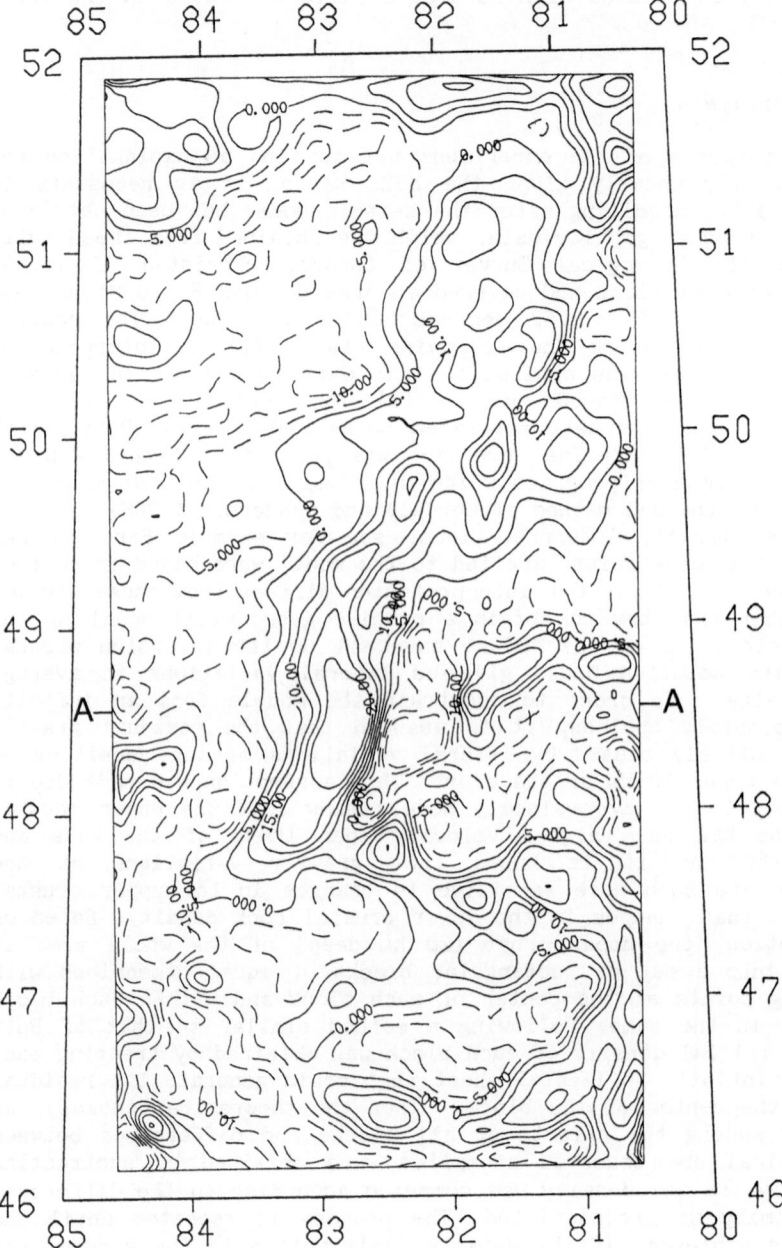

Figure 11. The residual component of the Bouguer gravity around the KSZ.

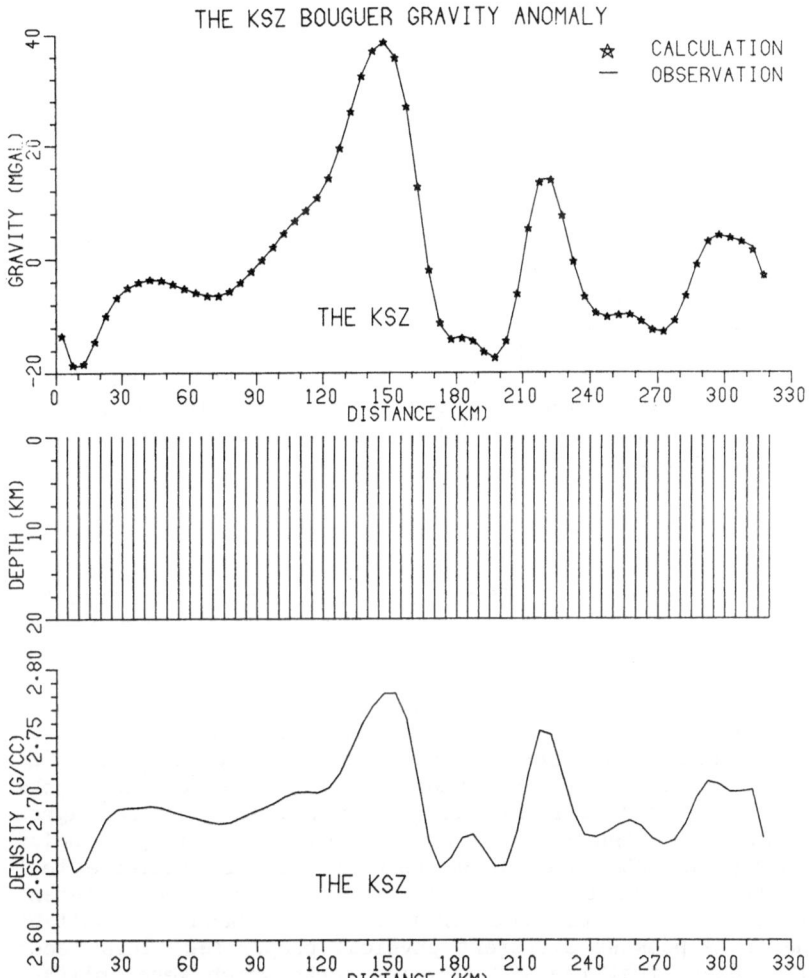

Figure 12. 2-D gravity modelling solution along line AA (see Figure 11).
Upper - Fit between the observations and calculations.
Middle- Crustal section used for the computation.
Lower - Computed densities (assuming the average density is
2.7 gm/cm^3).

The results of the complete 2-D and 3-D gravity modelling of the whole area indicate that the largest densities lie along the axis of the KSZ. In Figure 13, a comparison is made between 2-D and 3-D solutions and the main difference between them exists in density values. This is the expected result because in the 2-D gravity modelling only the W-E direction effect was considered, which is a very rough approximation. This suggests that the common 2-D gravity modelling is accurate in the pattern of density distribution but not in the absolute density values. However the 3-D gravity modelling is much more time-consuming and requires the use of the ETA Supercomputer due to a huge number of calculations.

It should be noted that the rock density in each block is assumed to be the average one of the whole upper crust. If we assume that the residual gravity field is entirely caused by lateral variations of the rock densities in the range of 5 to 20 km of the upper crust, the density contrast would be much larger. A more complicated gravity analysis could be obtained if the relationship between P wave velocity and rock density were known.

5. Conclusion

Our results can be summarized in Figure 14. The seismic data from the 1984 Kapuskasing experiment clearly indicates that the rocks in the upper crust of the KSZ have a higher P wave velocity than those in the Wawa and Abitibi belts. In the survey area, the near-surface rocks have a Poisson's ratio of 0.25 and the average Poisson's ratio of the whole crust does not differ significantly from 0.25. There was some evidence that its value increases to 0.26-0.27 in the upper crust under the KSZ. Both 2-D and 3-D gravity modellings show that the upper crustal rocks of the KSZ have a higher density and the high density values agree qualitatively with the high 6.5 km/s P wave velocity found just below the surface along the KSZ axis. These results probably indicate that the rocks in the upper part of the KSZ have been undergone different geological processes from the counterparts of the Wawa and Abitibi belts and therefore possess different physical properties. This adds support to the theory that the KSZ contains rocks which were uplifted from the middle crust.

Acknowledgments

We are indebted to all the participants of the 1984 Kapuskasing experiment for their efforts in making this experiment a success. Special thanks is extended to Dr. G. F. West and Dr. D. Northey of the University of Toronto who organized the experiment and made the complete data sets available to us. This research was supported by NSERC GRANT A-1793 and a scholarship from the Ministry of Education of China.

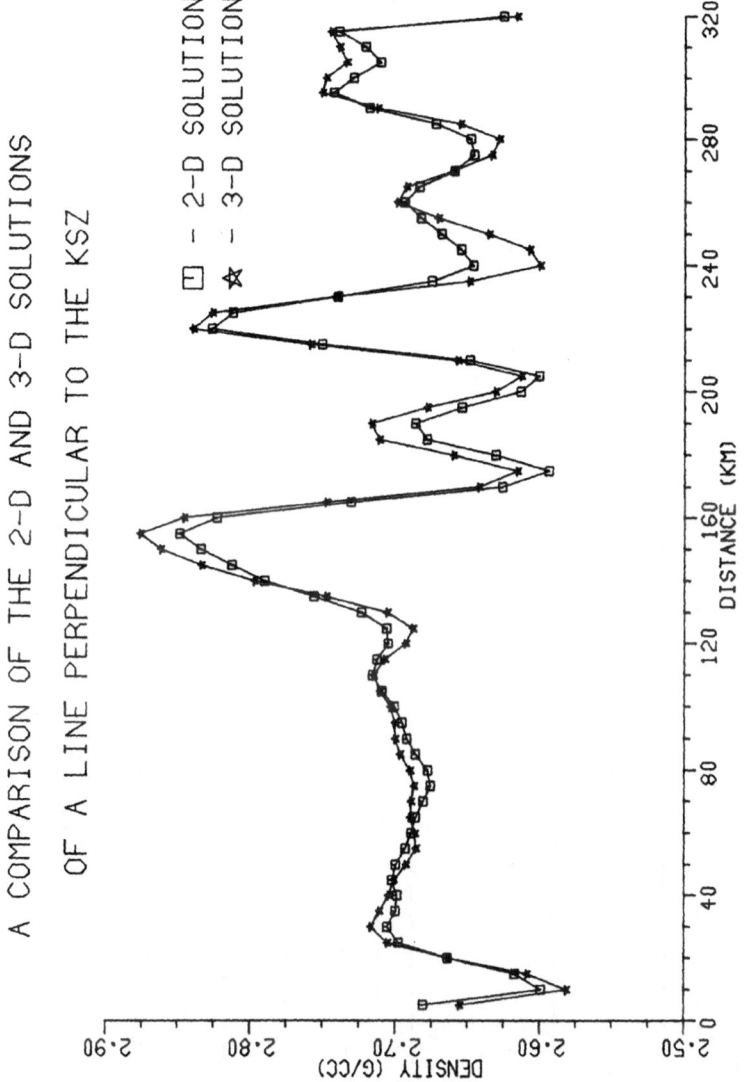

Figure 13. A comparison of the 2-D and 3-D solutions along line AA
(see Figure 11).

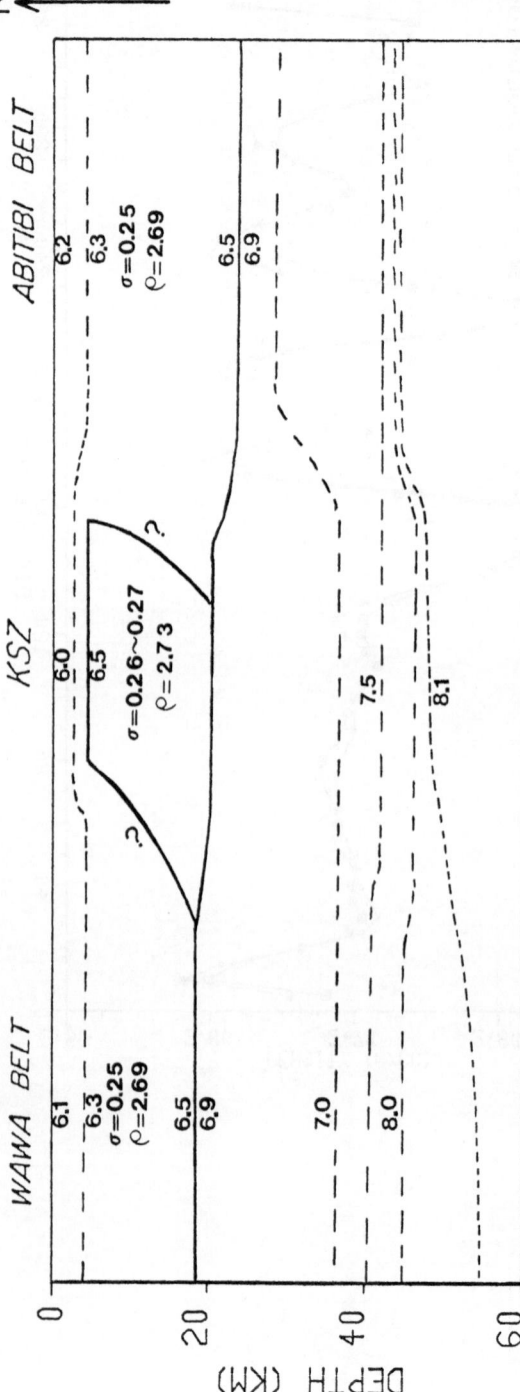

Figure 14. Composite sketch of a P wave velocity, Poisson's ratio and density cross-section of the KSZ along line AA (see Figure 11). Density values represent the average density of the whole upper crust (20 km deep).

References

Akima, H., 1978, 'A method of bivariate interpolation and smooth surface fitting for irregularly distributed data points', *ACM Transaction on Mathematical Software*, **4**, 148-159.

Assumpçâo, M., and Bamford, D., 1978, 'LISPB-V. Studies of crustal shear waves'. *Geophys. J. R. astr. Soc.*, **54**, 64-73.

Boland, A. V., and Ellis, R. M., 1988a, 'Velocity structure of the Kapuskasing Zone from seismic refraction studies'. In: Project Lithoprobe: *Kapuskasing Structural Zone Transect Lithoprobe Workshop Proceedings*. Ed. G. West, Toronto, February.

Boland, A. V., and Ellis, R. M., 1988b, 'Velocity structure of the Kapuskasing Zone from seismic refraction studies'. In: Project Lithoprobe: *Kapuskasing Structural Zone Transect Lithoprobe Workshop Proceedings*. Ed. G. West, Toronto, November.

Boland, A. V., Ellis, R. M., Northey, D. J., West, G. F., Green, A. G., Forsyth, D. A., Mereu, R. F., Meyer, R. P., Morel-à-l'Huissier, P., Buchbinder, G. G. R., Asudeh, I., and Haddon, R. A., 1988, 'Seismic delineation of upthrust Archaean crust in Kapuskasing, Northern Ontario'. *Nature*, **335**, 711-713.

Boland, A. V., and Ellis, R. M., 1989, 'Velocity Structure of the Kapuskasing Uplift, Northern Ontario, From Seismic Refraction Experiment Studies'. *J. Geophys. Res.*, **94**, 7189-7204.

Bott, M. H. P., 1960, 'The use of rapid digital computing methods for direct gravity interpretation of sedimentary basins'. *Roy.Geophys. Jour.*, **3**, 63-67.

Cook, F. A., 1985, 'Geometry of the Kapuskasing structure from a Lithoprobe pilot reflection survey'. *Geology*, **13**, 368-371.

Cordell, L., and Henderson, R. G., 1968, 'Iterative three-dimensional solution of gravity anomaly data using a digital computer'. *Geophysics*, **33**, 596-601.

El-Isa, Z., Mechie, J., and Prodehl, C., 1987, 'Shear velocity structure of Jordan from explosion seismic data'. *Geophys. J. R. astr. Soc.*, **90**, 265-281.

Gaucher, E. H., 1966, 'Elsas-Kapuskasing-Moosonee magnetic and gravity highs', in: *Report of activities.*, Geol. Surv. Can., **66-1**, 189-191.

Geis, W. T., Green, A. G., and Cook, F. A., 1988, 'Preliminary results from the high resolution seismic reflection profile: Chapleau Block, Kapuskasing Structural Zone.' In: Project Lithoprobe: *Kapuskasing Structural Zone Transect Lithoprobe Workshop Proceedings*. Ed. G. West, February.

Gibb, R. A., 1978, 'A gravity survey of James Bay and its bearing on the Kapuskasing Gneiss Belt, Ontario'. *Tectonophysics*, **45**, 7-13.

Holbrook, W. S., Gajewaski, D., Krammer, A., and Prodehl, C., 1988, 'An interpretation of wide-angle shear- and compressional wave data in southwest Germany: Poisson's ratio and petrological implications'. *J. Geophys. Res.*, **93**, 12,081-12,106.

Mereu, R. F., 1983, 'The generation of synthetic seismograms of two-dimensional laterally inhomogeneous media'. Conf. Proc. *Controlled Source Seismology Workshop on the interpretation of seismic wave propagation in laterally heterogeneous structures*, Einsiedeln, Switzerland, August.

Mereu, R. F., Baerg, J., and Wu, J., 1989, 'The complexity of the continental lower crust and Moho from PmP data: Results from COCRUST experiments'. In: Proceedings of the U7 interdisciplinary Symposium of the IUGG, Vancouver, 1987, *The Low Crust: Properties and Processes*. Eds: R.F. Mereu, St. Mueller, and D. Fountain, American Geophysical Union Geophysical Monograph **51**, IUGG Vol **6**, 103-119.

Northey, D. J., and West, G. F., 1985, 'The Kapuskasing Structural Zone Seismic Refraction Experiment-1984'. *A Phase I Lithoprobe Experiment,* Geological Survey of Canada Open File Report.

Percival, J. A., and Card, K. D., 1983, 'Archean crust as revealed in the Kapuskasing structural uplift, Superior Province, Canada'. *Geology*, **11**, 323-326.

Percival, J. A., and Coe, K., 1981, ' Parallel evolution of Archean low- and high-grade terrane: a view based on relationships between the Abitibi, Wawa and Kapuskasing belts'. *Precambrian Research*, **14**, 315-331.

Percival, J. A., and Fountain, D. M., 1986, 'Metamorphism and melting at an exposed example of the Conrad discontinuity, Kapuskasing uplift, Canada', in Fluid Movements: *Element Transport and the Composition of the Crust*. Ed. D. Bridgwater.

Percival, J. A., and Green, A. G., 1988, 'Towards a balanced crustal-scale cross section of the Kapuskasing uplift'. In: Project Lithoprobe: *Kapuskasing Structural Zone Transect Lithoprobe Workshop Proceedings*. Ed. G. West, Toronto, November.

Salisbury, M. H., and Fountain, D. M., 1988, 'Velocities of Rocks from the Kapuskasing Zone'. In: Project Lithoprobe: *Kapuskasing Structural Zone Transect Lithoprobe Workshop Proceedings*. Ed. G. West, Toronto, February.

Watson, J., 1980, 'The origin and history of the Kapuskasing structural zone, Ontario, Canada'. *Can. J. Earth Sci.*, **17**, 866-876.

Wu, J., 1987, 'The analysis of the data of the 1984 Kapuskasing seismic experiment'. *M.Sc thesis*, University of Western Ontario.

Wu, J., and Mereu, R. F., 1988a, 'Crustal models of the Kapuskasing structural zone: results from the 1984 seismic refraction experiment'. In: Project Lithoprobe: *Kapuskasing Structural Zone Transect Lithoprobe Workshop Proceedings*. Ed. G. West, Toronto, February.

Wu, J., and Mereu, R. F., 1988b, 'A P-S wave seismic-gravity analysis of the Kapuskasing structural zone'. In: Project Lithoprobe: *Kapuskasing Structural Zone Transect Lithoprobe Workshop Proceedings*. Ed. G. West, Toronto, November.

EXPOSED CONTINENTAL CRUST: SEISMIC RESULTS TO BE TESTED

Th. Wever
Institut für Geophysik, Universität
Leibniz Str. 15
2300 Kiel
West Germany

ABSTRACT.

Early results obtained from crustal refraction/wide-angle seismic investigations allowed a distinction of seperate 'velocity-layers' within the
continental crust. In some areas a pronounced velocity discontinuity
(the Conrad) has been found. Additionally, systematic age dependences of
average velocity or crustal thickness were proved.
    The near-vertical reflection seismic method generally cannot resol-
ve the velocity structure of the crust but is capable of imaging tectonic traces. The application of this method for crustal investigations in
different age provinces revealed also clear age (resp. heat flow) dependent trends of reflectivity distribution, reflection length and abundance of lower crustal lamellae.
    The results of both seismic methods were combined, especially the
depth of the Conrad and the top of the lower crust lamellae. A clear
difference is found implying that both phenomea are geologically unrelated. However, such a combination is problematic because of the rather
different physics underlying both methods.
    Questions arising from the restricted existence of the Conrad, the
variable velocity structure are combined with questions indicated by the
distribution of deep crustal reflections. An explanation of the obvious
age-dependence of several crustal (seismic-derived) properties is related to age and crustal development. Possibilities to test such ideas in
exposed cross-sections of continental crust are discussed.

INTRODUCTION

The refraction/wide-angle (RWA) and near-vertical reflection (NVR) seismic methods have been the most important tools for the investigation of
the deep continental crust. Most of our present knowledge about its physical properties originates from the these methods. While the RWA method
is capable of resolving the velocity structure (V-z functions), the NVR
method can directly image tectonic and other structural features. The
first interpretations of seismological data led to a simple two-layer

587

M. H. Salisbury and D. M. Fountain (eds.), Exposed Cross-Sections of the Continental Crust, 587–602.
© 1990 *Kluwer Academic Publishers.*

588

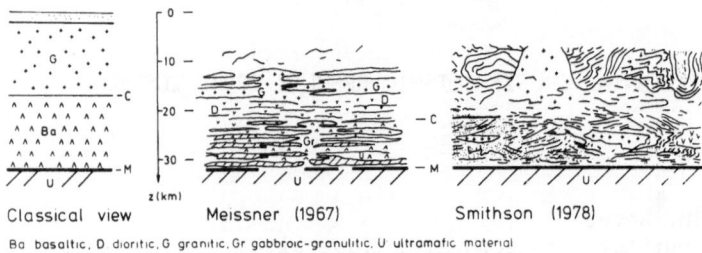

Ba basaltic, D dioritic, G granitic, Gr gabbroic-granulitic, U ultramafic material
C Conrad, M Moho

Figure 1: Development of ideas concerning continental crustal structure (a) classical view, (b) thin-layer model (Meissner, 1967, and (c) model of Smithson (1978).

crust, the granite-basalt model (Figure 1a). The results of an expanded spread profile (ESP, NVR to WA range) in southern Germany could only be interpreted by assuming a cyclic layering of high- and low-velocity lamellae causing strong reflections by interference (λ/4-concept of Meissner (1967), Figure 1b). Modern NVR profiling of e.g. COCORP resulted in similar models (Smithson, 1978, Figure 1c) showing a strong heterogeneity of the crust. One important implication of the interference concept is that laminated reflections do not represent single interfaces (Fuchs, 1969; Sandmeier et al., 1987).

There exist systematic variations of crustal properties determined with the seismic methods which are discussed below. Cross-sections of

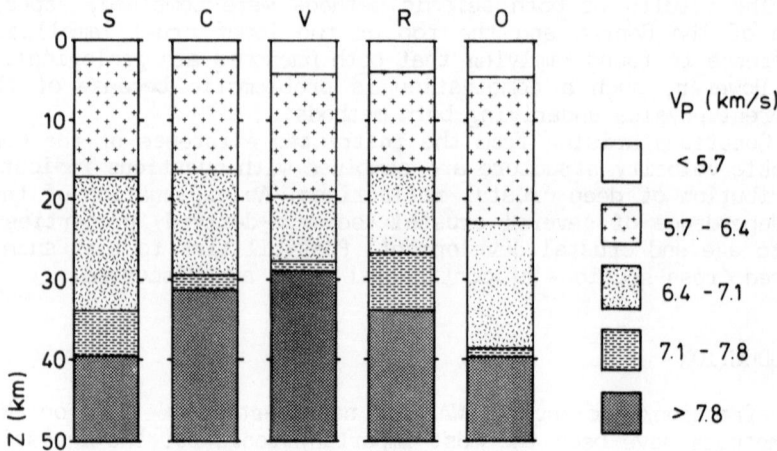

Figure 2: Standardized average V-z profiles for S: shield areas, C: Caledonian crusts (400-500 Ma), V: Variscan units (320-390 Ma), R: rifts, O: orogens. To allow an easy comparison, intervals of 0.7 km/s were chosen.

the continental crust offer the chance to test at least structural interpretations, whereas those depending on in situ conditions (e.g. on pore pressure) cannot be checked directly. Questions arising from seismic results are discussed.

RESULTS OF THE RWA SEISMIC METHOD

Long-range seismic RWA profiles have helped considerably to recognize fundamental differences between the various types of continental crust. There exists a large amount of RWA profiles all over the world, recently summarized by Prodehl (1984).

General Results

To allow an easy comparison of the data available, 135 RWA profiles from all over the world (Meissner, 1986) were used to derive the standardized average V-z profiles of Figure 2. From these profiles, two important facts can be deduced:
   (1) **The thickness of the continental crust (Moho depth, $V_p > 7.8$ km/s) increases with tectono-thermal age.** Tectonically active parts of the crust, zones of compression (orogens) and rifts have to be excluded from this comparison.
   (2) **The average crustal velocity is higher in older crusts.** This is mostly an effect of the high-velocity "layer" ($7.1 < V_p < 7.8$ km/s) above the Moho (Wever and Sadowiak, 1989). In most Phanerozoic crusts it has not been found. Only along profiles in the immediate neighbourhood of rifts (e.g. the Rhinegraben in Germany and France) such a layer was interpreted.

The Conrad

In some areas an intracrustal refractive boundary (the Conrad) has been detected. It seems to be best developed in younger crusts while older ones show a smooth, continuous increase of velocity.
   The reliability of RWA investigations has improved steadily under two aspects: on the acquisition side and on the interpretation side. While in former times geophones were separated by 10 km or more, modern profiles are shot with geophone distances of mostly less than 5 km. For the interpretation modern computers can be used integrating traveltime- and amplitude analyses.
   The Conrad was found in Europe. In Canada a similar boundary was named Riel implying that both possibly do not represent the same feature. Arguments may be the higher refraction velocities in Canada (ca. 7 km/s) and the greater depths.
   The existence of the Conrad has been put into question by results of Mereu and Ojo (1981). They used a continuous increase of velocity on which simple random velocity variations (< 0.1 km/s) were superimposed. Raytracing for this model reproduced traveltime branches of typical RWA field data (see their figures 1 and 2).

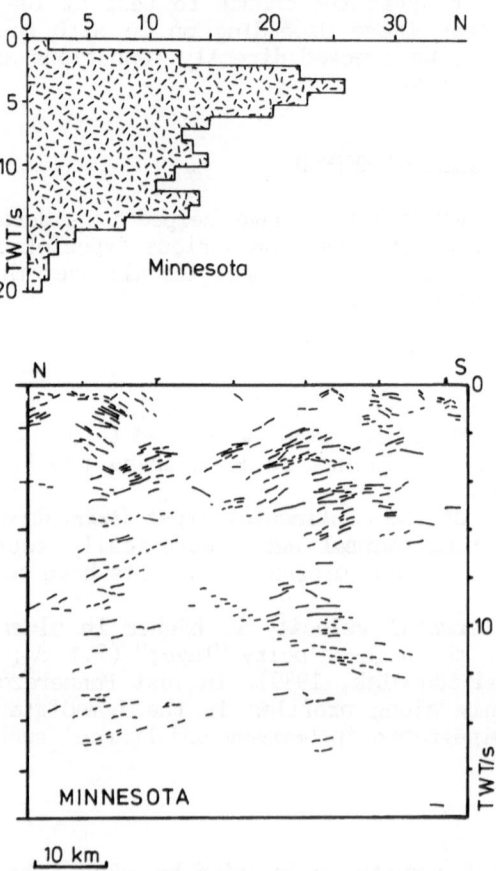

Figure 3: (a) Reflectivity distribution and (b) line drawing of a typical shield reflection line. Note the decrease of reflectivity with depth, the strong dips and the diffuse appearance of reflections.

RESULTS OF THE NVR SEISMIC METHOD

First used in the 1950s and 1960s for selected problems NVR profiling became an important tool for crustal studies after the first results of COCORP in the 1970s. This seismic method has also demonstrated an age-dependence of crustal properties. Among these properties are:

(1) **The distribution of reflectivity.** Figures 3b and 4b show typical line drawings from old and young crusts while in Figures 3a and 4a histograms quantitatively compare the reflectivity-depth distribution. Shield crusts show a decrease of reflectivity. Short reflections with strong dips are concentrateed in the upper crust (Figure 3a). The crust/

mantle boundary cannot simply be determined by an abrupt termination of reflections. In addition, Moho reflections are generally not observed.

In contrast, Phanerozoic crusts generally show an increase of reflectivity with depth as derived from Figure 4a. Their upper parts have been described as being 'transparent' (Meissner, 1986), underlain by a laminated lower crust. The laminated lower crust consists of flat lying reflections, often beginning and ending abruptly at midcrustal levels and at the Moho respectively.

(2) **The thickness of the reflective lower crust (RLC) varies systematically with surface heat flow (HF):** the higher the HF, the thinner is the RLC (Trappe et al., 1988). This strongly argues for a temperature-related model for the creation (and perhaps destruction) of seismic reflectors in a low-viscosity lower crust.

(3) **The length of reflections.** Seismic reflection lengths exhibit a strong tendency to decrease with increasing age. Although noise can mask reflections or bad statics can interrupt them, the deceasing length of true reflections could be verified. For describing the variable reflection length two parameters were used: L_{10}, the longest reflection length appearing at least ten times along a line, and L_{MAX}, the maximum length observed on a profile. L_{10} was chosen because of its poor sensitivity. Normally, it does not exceed 2.5 km but on marine profiles it is longer indicating the more constant source characteristics and less problems

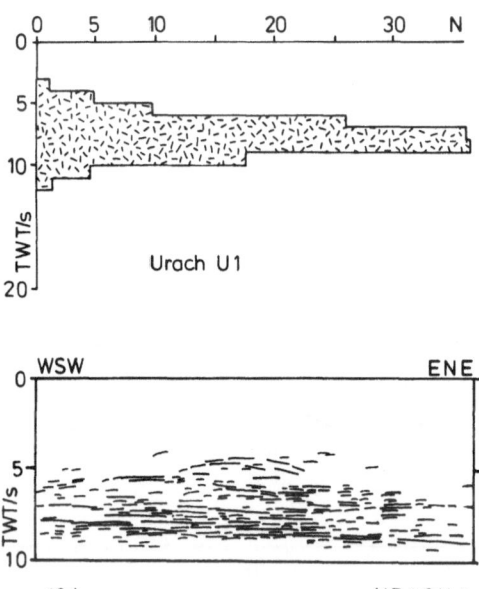

Figure 4: Same as Figure 3 but for a Phanerozoic crust. Strong reflections can be observed in the lower crust beneath a 'transparent' upper crust. The generally longer reflections are almost all subhorizontal.

and shows a stronger correlation to L_{MAX} than L_{10} does. L_{MAX} varies from less than 3 km (old crust) to 22 km (geothermal anomaly). In Figure 5, different crustal age fields can be distinguished in a L_{10} vs. L_{MAX} plot. The implications for crustal development is discussed later.

THE CONRAD AND THE LAMINATED LOWER CRUST

Both seismic methods reveal different parameters of the crust which show an age-dependence. A logical step was to combine the results and to look for correlations. The most obvious facts to be compared were the Conrad and the top of the reflective lower crust (TRLC). These have often been considered coincident features because both are found at roughly mid-crustal depths. A detailed analysis of 69 coincident RWA and NVR seismic profiles was made to prove this. Because this is a very critical comparison, only profiles were considered which intersected each other or were not seperated by more than 20 km (mostly less than 10 km). This strict limitation reduced the data but had to be choosen in order to avoid problems arising from dips of Conrad and/or TRLC. One shortcomming of the NVR method is the lack of velocity information. For converting measured traveltimes of reflections into depth, the velocities derived from the RWA data were used being the best estimates of velocity. The error introduced by this procedure has to be accepted. Assuming that crustal material is anisotropic, the horizontal velocity component (determined by the RWA method) shows higher values than the vertical one (affecting near-vertically reflected waves) (Meissner, 1986; Siegesmund, 1989). Therefore the depths determined by traveltime-conversion are systematically greater. This has to be kept in mind for the interpretation.

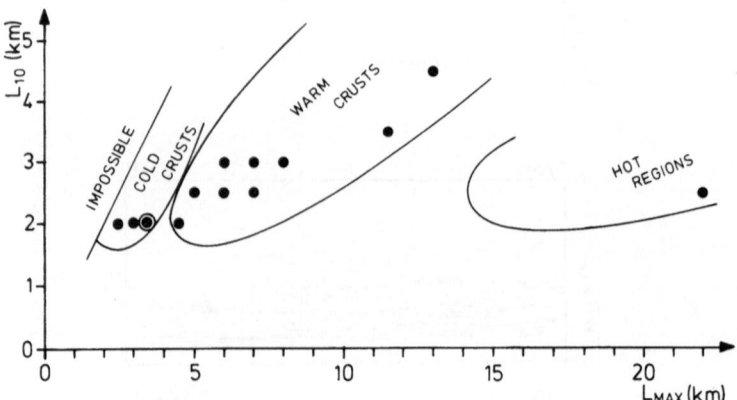

Figure 5: Plot of L_{10} versus L_{MAX}. Cold shield crusts are characterized by both short L_{MAX} and L_{10}, warmer Phanerozoic units by greater lengths of both and a geothermal anomaly by extremely high values of L_{MAX}.

Sometimes RWA and NVR data sets have been (re-)interpreted to yield identical depth of Conrad and TRLC (e.g. Mueller et al., 1987). Although this procedure is questionable such data were considered in the present comparison. However, 40 % of synoptically interpreted data sets showed identical depths whereas only 10 % of the data interpreted by different groups show this coincidence.

Examples of the combined data sets are shown in Figure 6. At first sight it is surprising that both Conrad and TRLC are not always found at the same depth. Table 1 gives a listing of the depths of Conrad (Z_C) and

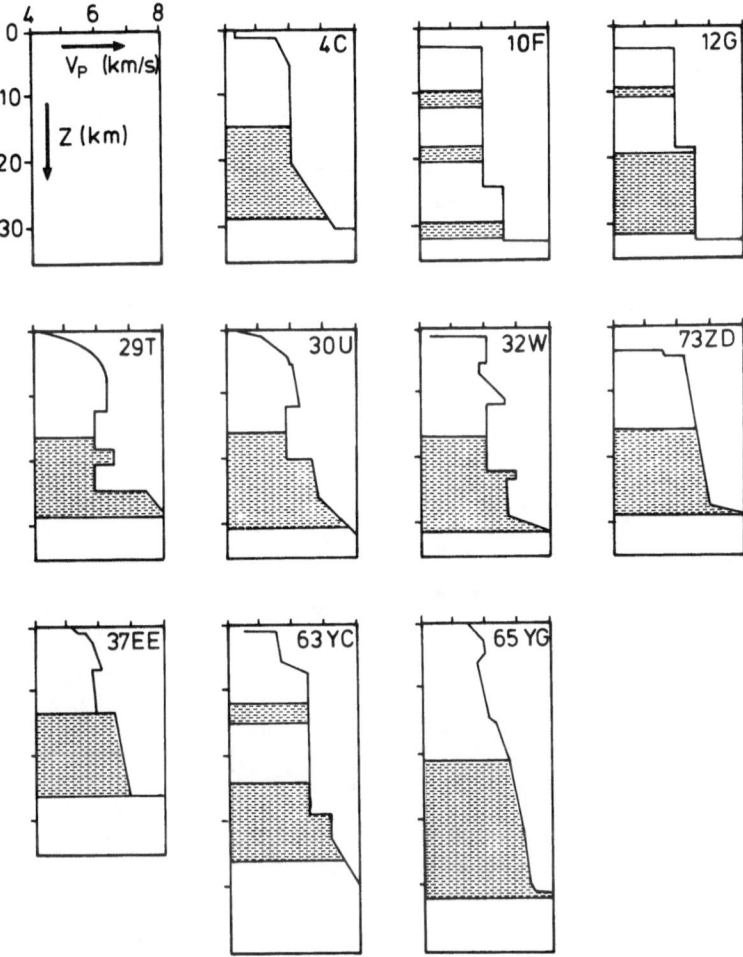

Figure 6: Depth relations of Conrad and deep crustal reflections. The curves represent the V-z function, the striped areas indicate the first occurence and depth extent of reflections. For a complete summary and reference see Wever (1989).

TRLC (Z_{TRLC}) and the depth pairs are plotted in Figure 7. All pairs of identical depth should lie on the line of equal depth. The points below/above the line represent a Conrad below/above the TRLC.

It can be easily derived from Figure 7 that the majority of lower crust reflections start above the Conrad. Accounting for the fact that the depths are overestimated because of the higher RWA-derived velocities, the difference between both is even greater than shown in Figure 7. This clear discrepancy of Conrad and RLC requires an explanation. However, we do not know whether the observations are real or if methodical differences (e.g. different wavelengths of both methods) are responsible. A reconstructed vertical cross-section through the continental crust of the Kapuskasing Uplift (Percival, 1986) may be cited to clarify the situation. This vertical section (Figure 8) clearly demonstrates that "domal" structures chararcterize the upper portion of the crust while layered contacts dominate in its lower part. Assuming that this model is typical for continental crust the conclusion can be drawn that different tectonic styles are encountered. On basis of their scale they affect the propagation of seismic waves depending on the wavelength

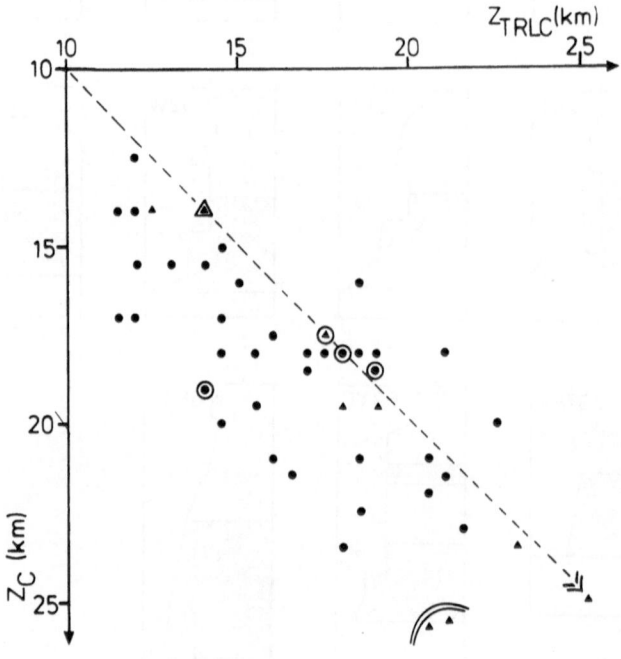

Figure 7: Plot of all comparisons showing a Conrad and lower crustal reflections, Z_C: depth to the Conrad and Z_{TRLC}: depth to the top of the reflective lower crust. Pairs of coincident depth must lie on the stippled line, triangles represent data from synoptical interpretations, points independent data evaluations. Lowermost triangles have greater depths than the scale shows.

TABLE 1

: Number (percentage) of comparisons	Observation	:
: 19 (28 %)	no Conrad but reflections	:
: 36 (52 %)	TRLC starts above Conrad	:
: 7 (10 %)	TRLC starts at Conrad	:
: 7 (10 %)	TRLC starts below Conrad	:
: 69 (100 %)		:

used: the example of Figure 8 allows to derive that short waves of the NVR method will already image lithological contacts found at depths of ca. 15 km whereas the 3...4 times longer waves of the RWA method will average over these structures and be refracted at depth exceeding 22 km thus showing the Conrad definitely below the uppermost reflections of the RLC.

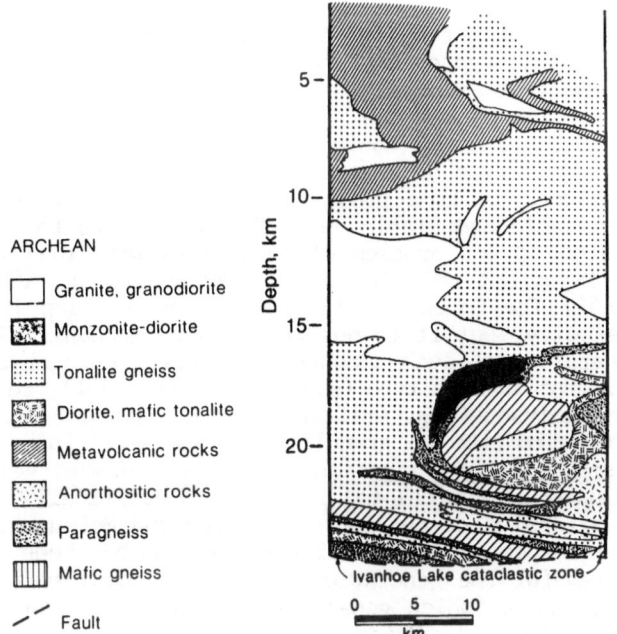

ARCHEAN

☐ Granite, granodiorite

▨ Monzonite-diorite

▨ Tonalite gneiss

▨ Diorite, mafic tonalite

▨ Metavolcanic rocks

▨ Anorthositic rocks

▨ Paragneiss

▨ Mafic gneiss

╱ Fault

Ivanhoe Lake cataclastic zone

0   5   10
km

Figure 8: Vertical cross-section through the continental crust reconstructed by Percival (1986) from mapping of surface geology of the Kapuskasing Uplift (Ontario, Canada). The transition to a layered appearance of lithological contacts is interpreted to represent the Conrad discontinuity.

Discussing Conrad and RLC as different phenomena is unevitable on basis of the conclusions drawn from the Kapuskasing example. At least it is not justified to interpret them as related phenomena found at the same depth. A strong supportive argument is the existence of reflections without the presence of a Conrad in 28 % of the compared data sets (Table 1). In addition, a certain systematic deviation is introduced by methodical difference.

HYPOTHESES TO BE TESTED

In this section actual models to explain the seismic results described above will be presented. Their uncertainties will be mentioned and opportunities to test the models in exposed cross-sections (Fountain and Salisbury, 1981; this volume) of the continental crust will be proposed. Questions arising from these seismic measurements concern both the nature and generation of the midcrustal refracting horizon, the Conrad, and the nature and creation of lower crustal reflectors.

The Conrad Problem

Important questions arising from the RWA results are: "Why are the old shield crusts thicker and generally show a continuous increase of velocity without prominent low-velocity layers?", "Is the Conrad real (see e.g. model of Mereu and Ojo, 1981)", if real: "What causes the intracrustal velocity discontinuity (Conrad): is it a change of metamorphic grade, a change of rock type or what else?", and "Why is the Conrad not a worldwide phenomenon: is it a product of crustal differentiation and thus missing in most old and stable units?". The answers to these questions may be related to the problem of crustal growth. Differentiation seems to play an important role (e.g. Meissner and Wever, 1986; Taylor and McLennan, 1985). The Conrad is interpreted as a sharp increase of density and/or velocity, but it is not known which mechanism is responsible for the generation of such a contrast between these physical properties. Both a change of rock type (granite-basalt model), recently revived by xenolith analyses (e.g. Griffin and O'Reilly, 1987) and of metamorphic grade (amphibolite to granulite facies, e.g. Fountain, 1976) are being discussed. Lab investigations (density, seismic velocity and anisotropy, chemistry) of typical materials could not finally solve this question (for an extensive discussion see Taylor and McLennan, 1985, or Kay and Kay, 1986). Repeated crustal differentiation has been considered a possible mechanism for the generation of a Conrad because it is missing in most shield areas (which generally show a transitional increase of velocity, discussed e.g. by Meissner, 1986).

Exposed cross-sections of continental crust thus offer an important chance to test these hypotheses. Sampling of rocks across a Conrad for laboratory measurements of their physical properties, mineral textures, and determination of the pressure/temperature history, combined with especially designed seismic field experiments, should provide an opportunity to improve our present knowledge. Most interesting would be a comparison of cross-sections of different age.

The Nature of Reflectors

Layered reflections from the crystalline crust are a fundamental observation in many areas. Questions concern the physical process (strain?,
heating?) which creates them, their geological nature (lithological contacts?, metamorphic contacts?, zones of localized strain?, fluid-filled
rocks?), and the geological process inducing the generation of reflectors (tectonic stress?, metamorphic events (partial melting)?, material
addition (intrusions) from the mantle?). A more detailed discussion is
given below.

The high energy of deep reflections excludes the assumption that
they are multiples from shallower reflectors (Liebscher, 1962). Based on
an especially designed wide-angle seismic experiment Meissner (1967) developed a first reasonable model. He explained the high-energy reflections from the "laminated" lower crust assuming constructive interference from thin $(\lambda/4)$ alternating high- and low-velocity layers. The layer
thickness was determined to be 80...140 m. Support for this theory was
provided by attempts of synthetic seismogram modelling (Fuchs, 1969):
the model of a layered lower crust resultet in the typical laminated appearance of deep crustal reflections. It is now widely accepted that a
RLC in seismograms is the product of interference effects having the important consequence that a lower crustal reflection does not represent a
single reflector but is the expression of a series of impedance $(\rho \cdot V)$
contrasts.

The Creation of Reflectors

The creation of subhorizontal reflecting interfaces in the deep crust is
also controversial:

(1) Based on NVR profiles e.g. in the northern Appalachians where
sedimentary rocks can be followed to the deep crust (Ando et al., 1984),
and lab investigations of metapelites from the Ivrea zone (Fountain,
1976), metamorphosed sediments have been proposed to be a major source
for the reflections.

(2) In order to explain thin layers in the lower crust Meissner
(1967) proposed mantle-derived mafic intrusions in the the lower crust.
These were assumed to spread horizontally thus forming thin lamellae.
Heating of host rock could lead to partial melts forming the granites of
the upper crust. No direct confirmation of this idea has been possible
up to now. However, the increasing percentage of mafic and ultramafic
material when approaching the Moho of the Ivrea zone may strengthen his
interpretation.

(3) Some attention has been paid to the idea that fluid-filled
cracks in the lower crust could be responsible for the layered reflections (Matthews, 1986). This idea has no observational support.

(4) Anisotropy as cause of the reflections has recently gained much
attention. However, no final decision can be made; Processing of NVR
seismic shear-wave experiments in the Black Forest (Germany) and northern Scotland (by BIRPS) has not yet been finished. Early results (Lüschen, priv. comm; Ward and Warner, 1989) indicate that shear wave energy has been reflected. Possibly observed shear-wave splitting is a

strong argument for anisotropy within the lower crust. For rocks from the Ivrea zone Siegesmund (1989) and Siegesmund et al. (1989) determined velocity anisotropies. The measured anisotropic velocities were the basis for synthetic modelling of the corresponding of the P-wave response which nicely duplicates the layered lower crust in NVR seismograms. In addition, shear wave splitting could be modelled using these anisotropic velocities (Siegesmund, 1989).

(5) The theory has gained greatest acceptance is that reflectors form in a low-viscosity lower crust. It is in good agreement with seismological observations which show that crustal seismicity is concentrated in the upper 10...15 km while generally no earthquakes occur in the lower ductile crust where lamellae are observed. In the low-viscosity lower crust stresses are assumed to induce movements along shear zones which reflect seismic energy. This model can also be combined with the intrusion model: low viscosity allows intrusions to spread easily and form thin layers. This rheological concept allows also to model different types of reflectivity distribution.

Exposed cross-sections of continental lower crust offer an unique possibility to study characteristics of each of the proposed mechanisms which generate reflectors. NVR seismic investigations of such zones may help to improve to understanding of the middle and lower crust. However, we are not dealing with plane waves. The same structure will not generate identical seismograms for different distances of the seismic source because the curvature of the incident wave may be completely different. In practice this means that an exposed cross-section will yield a seismogram which may not be compatible with the seismogram of an identical structure at lower crustal level. A second point to be taken in consideration is the tilted character of the exposed cross-section. To account for this, a well controlled migration of the data is unavoidable.

The Upward Migration And Disappearance Of Reflections

Questions also concern the different reflectivity distribution and reflection length in shield and young crusts. The first question to be answered is whether in shields lower crustal reflections are absent or whether they are not observed because seismic energy is too low or scattered too much. The application of marine NVR profiling of BIRPS has resultet in the best examples of deep reflection data. The extension of this method to the few water covered shields areas (e.g. the Baltic Shield) must be the first step to solve this basic question. If we don't find reflections with the marine technique, we have to find out if lamellae never existed in old crusts or whether they were destroyed with time. If they did exist, what is responsible for their destruction? Mechanical processes (e.g. earthquakes, folding, faulting) may have interrupted long reflectors. This can explain both the stronger dips of shield crust reflections and their shorter appearance at shallower levels which were reached by erosion and uplift. Chemical processes (e.g. long-time re-equilibration) may have been altered the material thus reducing or even erasing impedance contrasts. The decreasing lengths of reflections with increasing age can be taken as support for this destruction model.

If young shield crust had a laminated reflectivity then it must have lost it during slow uplift. Rapidly uplifted shield crust (e.g. in the Kapuskasing area or along the Canadian GLIMPCE in the Great Lakes, Behrendt et al., 1988) shows a strong reflectivity. The influence of mechanical disruption can be impressively demonstrated with reflection profiles across the Ries impact crater in S. Germany (Angenheister and Pohl, 1976) or the Siljan impact crater in Sweden (Juhlin and Pedersen, 1987). Although a very short event, the impact induced cracking which reduced reflector length and strength considerably.

In exposed cross-sections answers to the following questions can be sought: "Why are tectono-thermally old crusts characterized by a decrease of reflectivity?", "What causes lower crustal reflections in young areas, does the $\lambda/4$ concept adequately explain the generation of high-energy reflections?", and "Which mechanism is important for the generation of deep reflectors and why do different reflectivity types (Wever et al., 1987; Trappe et al., 1988) occur?"

The Relation Between Conrad And Lower Crustal Reflectivity

The combination of RWA and NVR results leads to further questions: "Why do we observe reflections where no Conrad is found and why don't they coincide where both are present". The first observation may be related to bad signal/noise ratios etc. The second one seems to be real although the uncertainties of the conversion of (NVR-) traveltime into depth exists. There are two arguments: 1. in average, TRLC starts 3.5 km above the Conrad which is a minimum estimate (see above chapters), and 2. there is a systematic trend of a Conrad below the TRLC. We have to check if the wavelengths of the methods in use are responsible or if both phenomena are geologically uncorrelated. Exposed cross-sections offer an opportunity to test these various ideas: detailed structural mapping and seismic modelling might answer if the wavelengths of both seismic methods "see" different boundaries, i.e. reflections occur at smaller structures at the beginning of a transition zone between upper and lower crust while refracted waves penetrate deep into the lower crust below this transition zone. This mapping should include studies to reveal the cause of the reflections and perhaps to explain the decreasing reflectivity in older areas. A combination of intensive seismic probing and lab investigations is an essential basis for decisions about the nature of deep reflections, possible mechanisms of their creation and disappearence.

CONCLUSIONS

Many questions arose from seismic studies of the continenal crust. Most of them cannot be answered without direct control. Drilling is not a suitable tool, the deepest borehole on the Kola Peninsula has just reached about 13,000 m which is 1/3 of the total crustal thickness. Technical limitations will not allow to reach the deep crust. Therefore, deep boreholes are no solution for answering the questions.

Mapping in shield areas where granulite terrains (representing lo-

600

wer crust) are exposed can only reveal characteristics of a limited depth level. Constructing a crustal model using depth levels from different areas can be misleading because of differences in composition and of evolutionary history. No unequivocal answers can be given from such studies.

The only alternative seems to be provided by continental cross-sections. They offer the best possibility to study a continuous vertical section of the continental crust. The opportunities encountered are multifold:

1. All properties influencing the propagation of seismic waves can be determined by sampling of rocks and investigation in laboratories. The results of these attempts may be used for synthetic modelling of reflection and refraction/wide-angle seismic data.

2. Structural mapping can be used for identifying different tectonic styles of upper and lower crust. Reconstructing vertical sections combined with velocity information from laboratory investigations may help to identify the basic cause of the midcrustal refraction boundary (Conrad) and its restricted appearance. The factor "seismic wavelength" may be investigated too. Synthetic modelling of the results may also help to identify the origin of deep crustal reflections.

3. The comparison both of relatively old (e.g. Kapuskasing) and young (e.g. Calabria) cross-sections can be a useful tool for investigating differences of tectonic and chemical evolution of the continental crust. The reason for age-dependent differences of reflectivity might also be detected by comparing the results. But also changes of crustal growth can possibly be recognized.

4. The creation of reflectors is not yet understood. A variety of mechanisms (all of them having their own attributes) has been proposed. The answers could be given by searching the expected individual characteristics.

5. Seismic field experiments can be a powerful tool especially when combined with the results of structural mapping and lab investigations. However, because of methodical pecularities caution is necessary as discussed above.

Only few of the problems and starting points for their solution have been mentioned so far. Comprehensive investigations are made possible in exposed cross-sections of the continental crust. They are cheaper to be performed than in deep boreholes. Additionally, they allow an areal investigation and control of changes whereas deep drilling samples only at one point (which at worst is an isolated anomaly without control).

ACKNOWLEDGEMENTS

Much of this work was made possible with the financial support of the Deutsche Forschungsgemeinschaft (DFG) under grant Me 335/77-3. Considerable improvements of the text were proposed by Matt Salisbury, Bob Mereu and an anonymous reviewer.

Contribution no. 383 from the Institut für Geophysik, Universität Kiel.

REFERENCES

Ando, C., Czuchra, B., Klemperer, S., Brown, L. D., Cheadle, M., Cook, F., Oliver, J., Kaufman, S., Walsh, T., Thompson, J., Lyons, J., Rosenfeld, J. (1984). Crustal profile of a mountain belt: COCORP deep seismic reflection profiling in New England Appalachians and implication for architecture of convergent mountain chains. Bull. AAPG **68**, 819-837.

Angenheister, G., Pohl, J. (1976). Results of seismic investigations in the Ries crater area (S. Germany), in "Exploration Seismology in Central Europe", P. Giese, C. Prodehl, A. Stein (eds.), Springer, Berlin, 290-302.

Fountain, D. (1976). The Ivrea-Verbano and Strona-Ceneri zones, northern Italy: A cross-section of the continental crust - new evidence from seismic velocities of rock samples. Tectonophysics **33**, 145-165.

Fountain, D., Salisbury, M. (1981). Exposed cross-sections through continental crust: implications for crustal structure, petrology and evolution. Earth Plan. Sci. Lett. **56**, 263-277.

Fuchs, K. (1969). On the properties of deep crustal reflectors. Z. Geophysik **35**, 133-149.

Griffin, W., O'Reilly, S. (1987). Is the continental Moho the crust-mantle boundary?, Geology **15**, 241-244.

Juhlin, C., Pedersen, L.B. (1987). Reflection seismic investigations of the Siljan impact structure, Sweden, J. Geophys. Res. **92**, 14113-14122.

Matthews, D.H. (1986). Seismic reflections from the lower crust around Britain, in "Nature of the Lower Continental Crust", J. Dawson, D. Carswell, J. Hall, K. Wedepohl (eds.), Blackwell, Oxford, 11-21.

Meissner, R. (1967). Zum Aufbau der Erdkruste: Ergebnisse der Weitwinkelmessungen im bayerischen Molassebecken. Gerl. Beitr. Geophysik **76**, 241-254, 295-314.

Meissner, R. (1986). The Continental Crust - A Geophysical Approach. Academic Press, Orlando, 426 pp.

Meissner, R., Wever, Th. (1986). Nature and development of the crust according to deep reflection data from the German Variscides, in "Reflection Seismology: A Global Perspective", M. Barazangi and L.D. Brown (eds.), Am. Geophys. Union, Washington, 31-42.

Mereu, R., Ojo, S. (1981). The scattering of seismic waves through a crust and upper mantle with random lateral and vertical inhomogeneities, Phys. Earth. Plan. Int. **26**, 233-240.

Mueller, S., Ansorge, J. Sierro, N. Finckh, P., Emter, D. (1987). Synoptic interpretation of seismic reflection and refraction data, Geophys. J. R. astr. Soc. **87**, 345-352.

Percival, J. (1986). A possible exposed Conrad discotinuity in the Kapuskasing Uplift, Ontario, in "Reflection Seismology: The Continental Crust", M. Barazangi and L.D. Brown (eds.), Am. Geophys. Union, Washington, 135-142.

Prodehl, C. (1984). Structure of the Earth's crust and upper mantle, in "Landolt-Börnstein: Geophysics of the Solid Earth, Moon and Planets", Group V, Vol. 2a, K. Fuchs, H. Soffel (eds.), Springer, Berlin, 97-206.

Sandmeier, K., Wälde, W., Wenzel, F. (1987). Physical properties and structure of the lower crust revealed by one- and two-dimensional modelling, Geophys. J. R. astr. Soc. **89**, 339-344.

Siegesmund, S. (1989). Texturelle und strukturelle Eigenschaften mylonitischer Gesteine der Insubrischen Linie (Ivrea-Zone, Italien) und ihr Einfluß auf die elastischen Gesteinseigenschaften. - Ein Beitrag zur Interpretation seismischer in-siu Messungen. Doktor Thesis, Universität Kiel, 178 pp.

Siegesmund, S., Takeshita, T., Kern, H. (1989). Anisotropy of V_P and V_S in an amphibolite of the deeper crust and its relationship to the mineralogical, microstructural and textural characteristics of the rock, Tectonophysics **157**, 25-38.

Smithson, S. (1978). Modelling continental crust: structural and chemical constraints. Geophys. Res. Lett. 5, 749-752.

Taylor, S., McLennan, S. (1985). The Continental Crust: its Composition and Evolution, Blackwell, Oxford, 312 pp.

Trappe, H., Wever, Th., Meissner, R. (1988). Crustal reflectivity pattern and its relation to geological provinces, Geophys. Prosp. **36**, 265-281.

Ward, G., Warner, M. (1989). S-wave images of the layered lower continental crust, Ann. Geophysicae 7 (Spec. Issue), 52.

Wever, Th. (1989). Conrad and top of the reflective lower crust - do they coincide? Tectonophysics **157**, 39-58.

Wever, Th., Sadowiak, P. (1989). A relation between crustal thickness and mean crustal velocity, Tectonophysics (in press).

Wever, Th., Trappe, H., Meissner, R. (1987). Possible relations between crustal reflectivity, crustal age, heat flow and the viscosity of the continents, Ann. Geophysicae **5B**, 255-266.

THE STRUCTURE OF THE CRUST AND UPPERMOST MANTLE OFFSHORE BRITAIN: DEEP SEISMIC REFLECTION PROFILING AND CRUSTAL CROSS-SECTIONS.

T. J. Reston
INSTOC,
Cornell University,
Ithaca, NY 14853.

Abstract: BIRPS deep seismic reflection profiles, recorded in the waters around the British Isles show dominantly extensional faults in the upper crust, a highly reflective lower crust and probable extensional shear zones in the mantle. As the reflective lower crust is best developed in regions of extension, where it seems to act as a divide between the extensional tectonics of the upper crust and the upper mantle, the reflectivity of the lower crust is probably related to extensional processes. Comparisons with exposed crustal cross-sections are used to provide constraints on the cause of lower crustal reflectivity, suggesting that the reflections could come from intrusions or from shear zones, but are less likely to be caused by fluids. The pattern of reflectivity within the lower crust is interpreted as evidence that at least some of the reflectivity of the lower crust comes from shear zones: large-lenticular transparent zones enclosed within highly reflective bands are interpreted as low-strain lozenges enclosed within high-strain zones, a style of deformation common in high-grade terranes.

The lower crust is the site of crust-mantle interaction. Fundamental to the typical BIRPS profile is the observation that the shear zones in the mantle are not collinear with the faults in the upper crust. Simple analysis is used to show that this requires the lower crust to be strongly sheared during extension, as it transfers deformation between the upper crust and the mantle. At least part of the reflectivity of the lower crust might be due to this shearing, a direct consequence of the role of the lower crust in the deformation of the lithosphere.

Introduction.

The nature of the deep continental crust can be addressed in a number of ways, all of which have their limitations. Direct study by deep drilling is prohibitively expensive and one-dimensional, whereas xenoliths, our only unequivocal samples of the lower crust, may be non-random samples (Griffin and O'Reilly, 1986) of atypical (i.e., volcanically active) crust. Consequently, most of our ideas about the deep structure of the crust are derived from high-grade terranes, some of which might be more or less complete sections through the crust, and from remote-sensing (geophysical) techniques (Dawson et al., 1986), such as deep seismic reflection profiling.

The seismic reflection method (Sheriff and Geldart, 1982 and 1983) is based on the principle that a seismic wave is partially reflected at an interface across which there is a change in seismic velocity, density or both. As such interfaces often correspond to geological boundaries, the method provides an acoustic image of the subsurface geology. Unfortunately a seismic section differs from a geological section in a number of ways. For example, a dipping reflection will not be in its true location on a seismic section, unless restored there by a process called migration. Furthermore, the resolution of the reflection method, although better on a crustal-scale than any other geophysical method, is limited in proportion to the seismic wavelength. For the frequencies used in deep seismic profiling, the vertical resolution is limited to about 100 metres (Blundell and Raynaud, 1986), whereas the limit of lateral resolution in the deep crust is about 4 km, the diameter of the first Fresnel zone (Sheriff and Geldart, 1982), but can be improved by migration. However, migration only works well if the velocity structure is well

603

M. H. Salisbury and D. M. Fountain (eds.), Exposed Cross-Sections of the Continental Crust, 603–621.
© 1990 *Kluwer Academic Publishers.*

known, if the structure is within the plane of section, and if noise levels are low. Unfortunately, these criteria are rarely met in deep seismic reflection profiling, so an alternative approach is sometimes adopted (e.g. Hurich and Smithson, 1987), namely the use of seismic modelling to convert a geological model into an unmigrated seismic section, which can then be compared with the seismic data.

Deep seismic reflection profiling thus provides a section that, although based on geology, is fundamentally different from a true geological section: the technique provides a low resolution and distorted image of the *in situ* crust, although both the resolution and accuracy of the image can be improved by migration. In contrast, field mapping of exposed high-grade terranes reveals a high resolution, but commonly incomplete view of surface geology, which can, in some cases, be projected into a crustal cross-section. The two approaches are thus both radically different and somewhat complementary: seismic sections can help constrain the general applicability of exposed crustal cross-sections, and exposed high-grade terranes can help us understand the significance of seismic reflections. This paper attempts to do both of these, integrating results from deep seismic profiling with seismic modelling of high grade terranes to aid understanding of the structure of the upper lithosphere around Britain.

Lithospheric Reflectivity Offshore Britain

Since 1981, the British Institutions Reflection Profiling Syndicate (BIRPS) have acquired over 10,000 km of deep seismic data in the waters around the British Isles (Figure 1). As the profiles provide a complete, if low-resolution and acoustic, image of the upper 50 km of the continental lithosphere over the length of these profiles, the north-west European continental shelf has become the most intensively studied piece of lithosphere in the world.

Although the BIRPS profiles are located in a geographically small area, the geological evolution of the region is quite complex. For instance, the Phanerozoic geology of NW Europe can be divided into four major episodes. The continent-continent collision that followed the closure of the Iapetus ocean resulted in the Silurian development of a major orogenic belt, the Caledonides, trending SW-NE through Ireland, Wales, northern England, and Scotland (Dewey, 1982). This was followed, during the Devonian, by post-orogenic extensional collapse, and continued regional extension during the Carboniferous. Both were largely controlled by SW-NE trending faults, sub-parallel to the Caledonian grain (e.g. Ziegler, 1982; McClay et al., 1986).

The Carboniferous also saw the development of the Variscan orogeny in the southwest of Britain and Ireland, and in much of Europe (Ziegler, 1982). In Britain and Ireland, the Variscides strike WNW-ESE, cutting across the earlier Caledonian structures. This was followed, throughout most of NW Europe, by renewed rifting, starting in the Permian and continuing intermittently through the Mesozoic, eventually culminating in the opening of the northernmost Atlantic Ocean near the end of the Cretaceous (Glennie, 1984).

The BIRPS profiles (Figure 1) cross the trend of all four of these major tectonic units, and also the Precambrian foreland to the Caledonian orogen (the Lewisian in NW Scotland). However, despite the different geological provinces traversed, the reflectivity of much of the upper lithosphere around Britain (although not necessarily elsewhere) can be summarised by a "typical BIRP" (Matthews and Cheadle, 1986). This is a cartoon (Figure 2) showing a sedimentary basin developed in the hanging wall of a dipping, extensional fault that passes through an otherwise transparent upper crust (i.e. approximately 0-18 km depth) and soles out at the top of a highly reflective lower crust (approximately between 18 and 30 km). When coincident high quality refraction data are available, the base of the lower crustal reflective zone always corresponds, within error, to the refraction Moho (e.g. Matthews, 1986; Powell and

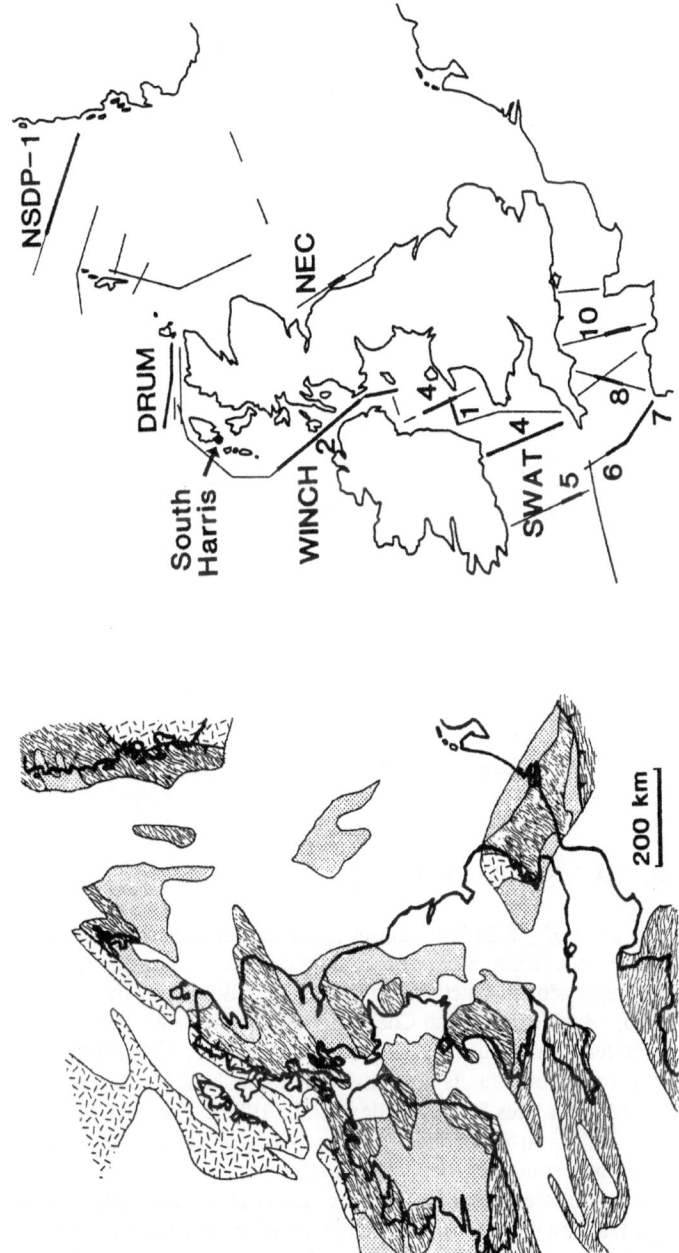

Figure 1 A Pre-Tertiary geological map of the British Isles (derived from Ziegler, 1982 and Cheadle et al., 1987). Accompanying map shows the location the BIRPS profiles discussed here (thick lines are portions of data shown). The geology is divided into 5 units: Pre-Caledonian basement (random dash), the Caledonides (SW-NE trending fabric), Devono-Carboniferous basins (mainly extensional - shown stippled), the Variscides (E-W trending fabric), and Permo-Mesozoic rift basins (unornamented). Note that extensional tectonics dominate, particularly offshore.

Sinha, 1987). Beneath the Moho, the upper mantle appears largely unreflective, apart from occasional dipping bands of reflections, generally interpreted as mantle shear zones (Matthews et al., 1987; Warner and McGeary, 1987).

In the following sections, I shall use some of the BIRPS dataset (Figures 1 and 3) to look at the different parts of the typical BIRP, in order to outline the features' significance and interpretation, making reference to structures found in high-grade terranes, including possible crustal sections.

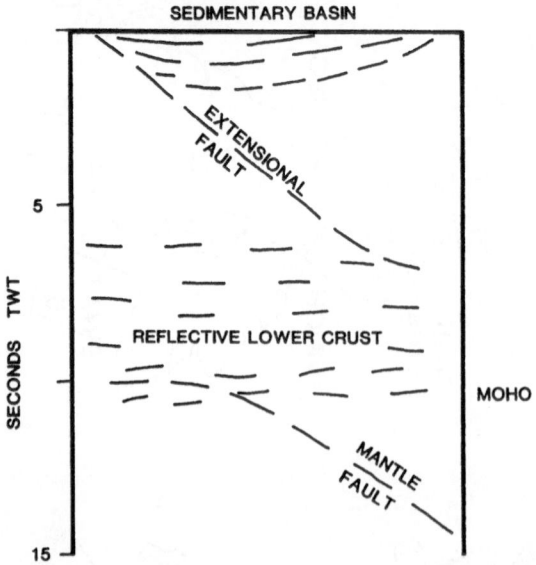

Figure 2 The "typical BIRP" (modified after Matthews and Cheadle, 1986) - a cartoon of lithospheric reflectivity around Britain.

EXTENSIONAL FEATURES IN THE CRUST

One remarkable feature of the BIRPS dataset is the dominance of demonstrably extensional structures within the upper crust. Most of the faults that are imaged, including those that have been mapped as compressional structures on land, have sedimentary basins developed in their hanging-wall (Figure 3 - Matthews and Cheadle, 1986): both Caledonian and Variscan thrusts appear to have been reactivated during extension (Cheadle et al., 1987). Other structures that are imaged offshore as extensional faults may have had a long history of movement back into the Proterozoic (e.g. the Outer Isles Fault - Lailey et al., 1989). The DRUM profile (Figure 3), recorded off the north coast of Scotland, and across the Caledonian trend, illustrates the dominance of extensional structures particularly well.

Not all the structures seen on the profiles are clearly extensional, and some compressional features are imaged, although these are the exception rather than the rule. A few south-dipping structures (e.g. at the southern end of SWAT 4 - Figure 3), interpreted as Variscan thrusts, do not have sedimentary basins developed in their hanging-wall, and so were probably not active during Permo-Mesozoic extension. Furthermore, Freeman et al. (1988) trace structures within the lower crust along the strike of the Caledonian orogen, and interpret them as the Iapetus suture zone (Figure 4a), pointing out that the features occur in a region of relative stability during Permo-Mesozoic extension, and at a high angle to the Permo-Mesozoic extension direction. However, the crust in this region is only about 30 km thick, rather less than might be expected for the heart of an orogen before extension, supporting the idea that the area underwent considerable extension (with a Caledonian strike) during the Devonian-Carboniferous (Ziegler, 1982). If so, the reflective features might have formed, or at least been modified, during these phases of post-Caledonian extension.

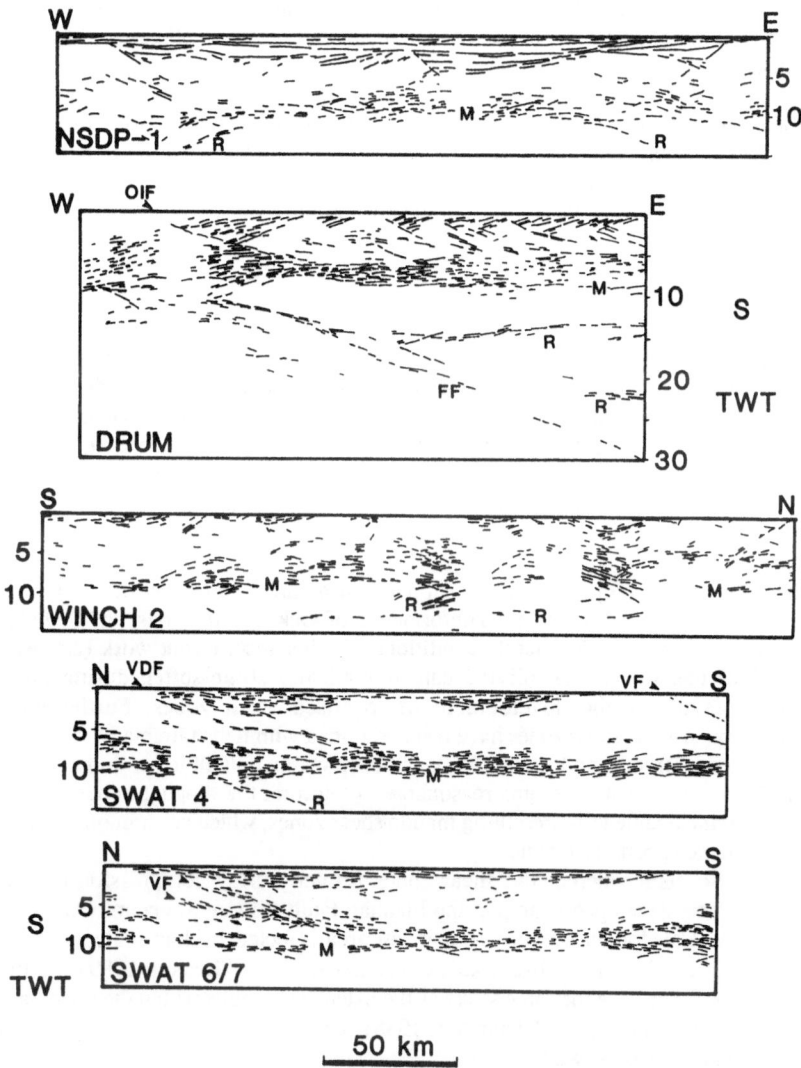

Figure 3 Line-drawings of deep seismic profiles (after Klemperer, 1988; McGeary and Warner, 1985; Matthews and Cheadle, 1986 and Cheadle et al., 1987), showing numerous sedimentary basins, dominantly extensional faults in the upper crust, a reflective lower crust, whose base corresponds to the Moho (M), and reflections (R) in the mantle, interpreted as shear zones. VDF is the Variscan Deformation Front, VF are Variscan faults, OIF is the Outer Isles Fault, and FF is the so-called Flannan Fault. Note that the vertical scale is in seconds: multiply by three for approximate depth conversion.

None of the profiles show the thickened crust associated with modern mountain belts, such as the Alps (Meissner et al., 1987). Using both reflection and refraction profiles to map the depth to the Moho, Meissner et al. (1987), demonstrated that both the Caledonian and Variscan belts lack mountain roots. A combination of erosion, post-orogenic extension and ductile creep within the lower crust appears to have reduced the thickness of the crust to 30 km or less (Meissner et al., 1987; Meissner and Kusznir, 1987) over most of the area covered. Given the extent and duration of post-orogenic extension, particularly offshore (Figure 1), this is perhaps not surprising.

The BIRPS data thus provide an acoustic image of the upper lithosphere of a dominantly extensional province. Consequently, it is best compared to crustal sections that show deep crustal extensional processes, although perhaps uplifted during compression. Such crustal sections include the Ivrea and Kapuskasing zones (Brodie and Rutter, 1987; Percival, this volume): the applicability of these zones to the crust around Britain is a central theme of this paper.

MANTLE REFLECTIONS.

The uppermost mantle appears to be largely nonreflective on BIRPS profiles, apart from occasional bands of reflections (Figure 3) which remain in the mantle after migration. Some are quite spectacular: one mantle reflection, the so-called Flannan Fault (McGeary and Warner, 1985) can be traced to 30s TWT, or about 80 km after depth migration (the DRUM line - Figure 3). Because many of these bands, when mapped in three-dimensions, are sub-planar, dipping at about 30-45°, and are too numerous and too widespread to all be subduction zones, it is generally accepted that these features are shear zones (e.g. Klemperer, 1988; Warner and McGeary, 1987).

The notion that deformation in the uppermost mantle is localised into shear zones is supported by experimental work on the deformation of rocks and minerals, and by field studies of peridotites deformed under mantle conditions. For instance, recent work (e.g. Rutter and Brodie, 1988) has shown that olivine can dramatically strain-soften during progressive deformation, favouring the development of localised shear zones. Furthermore, high temperature, extensional shear zones have been described within the ultramafic rocks at the base of the Ivrea Zone (Brodie and Rutter, 1987), and deep within ophiolitic peridotites (Norrell and Harper, 1988). As a result, it seems reasonable to interpret the dipping bands of reflections imaged within the mantle as representing mantle shear zones, which accommodate much of the deformation in the uppermost mantle.

It is not so clear whether the shear zones formed during compression or extension. However, the best developed example, the Flannan Fault (Figure 3) occurs well beneath the foreland to the Caledonian orogeny, but in an area affected by Late Paleozoic-Mesozoic extension. As a result, it is probably an extensional structure (Reston, 1989). Furthermore, those in the North Sea have the same strike as the extensional faults within the crust, rather than that of the Caledonian orogeny (Klemperer, 1988), and so are interpreted as extensional shear zones. The others may be as well.

THE REFLECTIVE LOWER CRUST

The third, and possibly the most controversial, aspect of the typical BIRP is the reflective lower crust. This is a zone, approximately between 6 and 10s TWT (~18-30 km depth) that is dominated by numerous, bright, sub-horizontal reflections (Figure 4). These are often described as layered, implying that the reflecting features are extensive, planar, parallel and sub-horizontal.

However, this is a misleading: on many profiles the lower crust is dominated by convex-up, cross-cutting, segmented reflections (e.g. NEC - Figure 4a), observations that indicate that much of the reflecting structure is not fully resolved (Hurich and Smithson, 1987; Reston, 1988) and may be quite complex. Furthermore, some profiles image large-scale features within the lower crust. For instance, the same portion of the NEC line (Figure 4a) images a dipping feature, probably a shear zone, and perhaps part of the Iapetus suture zone (Freeman et al., 1988). Furthermore, the reflectivity of the lower crust on SWAT 10 (Figure 4b) is concentrated into distinct bands, separating zones of poor reflectivity. The interrelationship of these features is not clear on the unmigrated data, so that the overall structure of the lower crust is not resolved. Elsewhere, bands of reflections within the lower crust outline large lenticular zones of poor reflectivity, described here as transparent zones (Figures 4c and 5). The significance of these features is discussed later in this paper, but it is immediately clear that the reflective lower crust is not uniformly layered, but instead appears to have some internal structure.

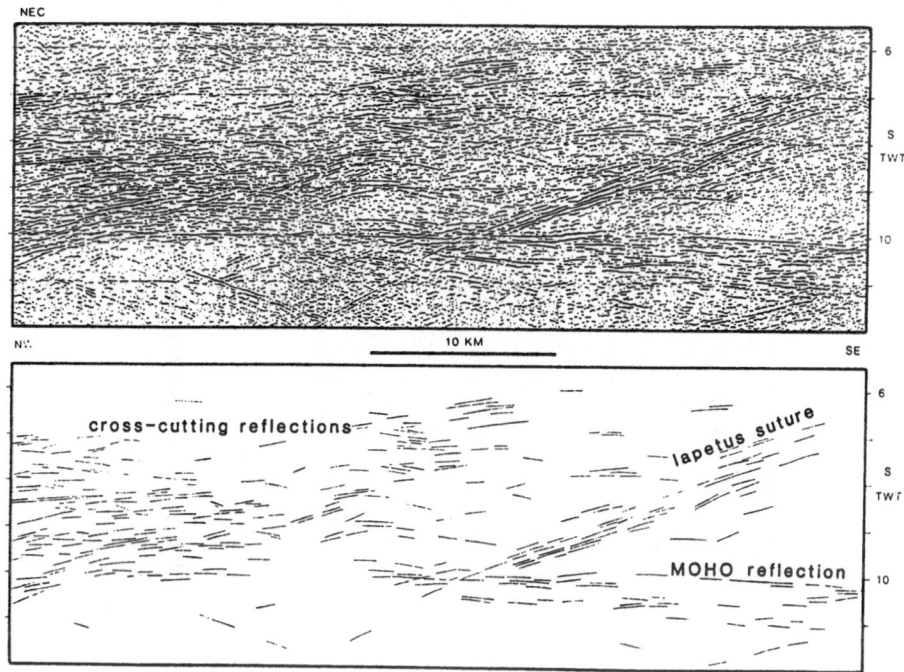

Figure 4 Portions of BIRPS data showing some of the variability within the reflective lower crust. A: detail from the NEC line, shows a dipping shear zone within the lower crust (perhaps the Iapetus suture), and convex-up, cross-cutting, segmented reflections.

610

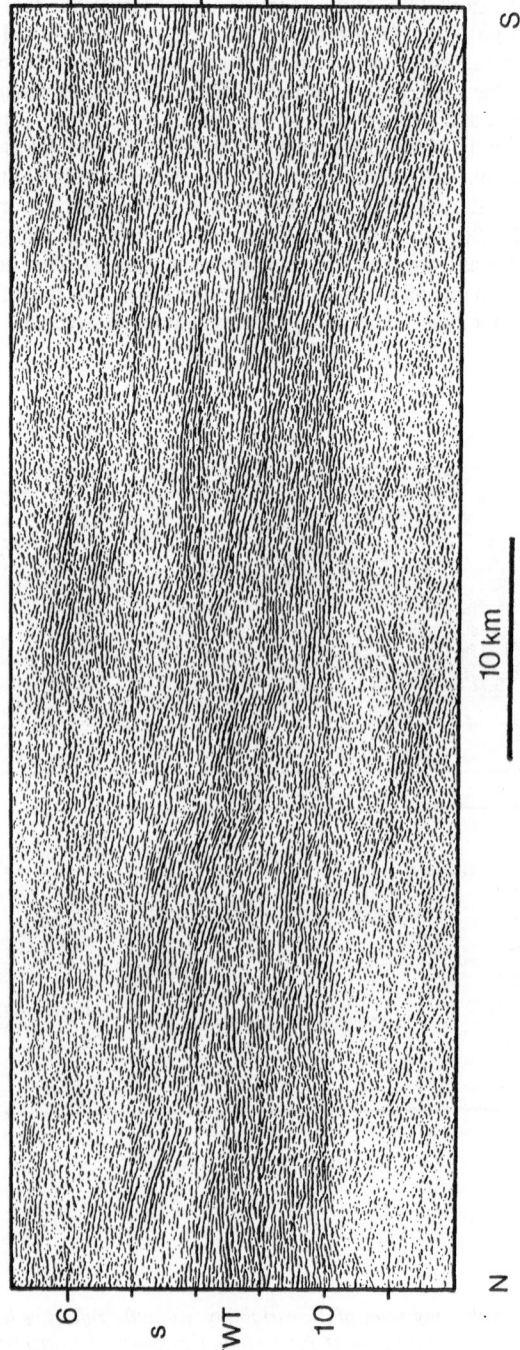

Figure 4b: SWAT 10: the reflectivity of the lower crust is concentrated into distinct south-dipping bands.

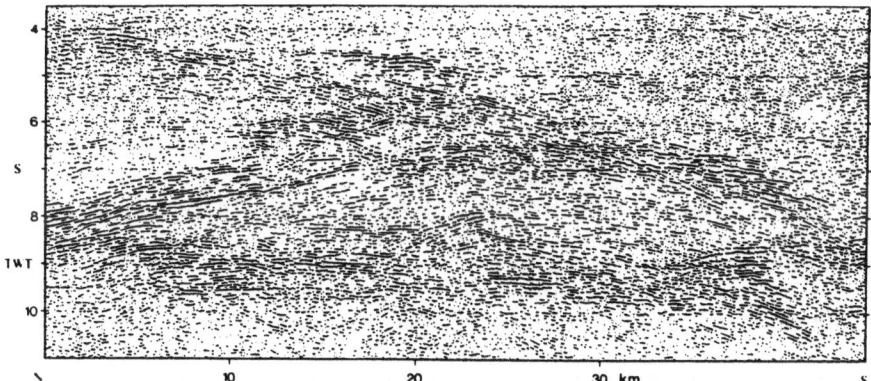

Figure 4c: SWAT 5 - the reflectivity of the lower crust is concentrated into a few bands that clearly outline a large, lenticular transparent zone. These features are probably structural in nature.

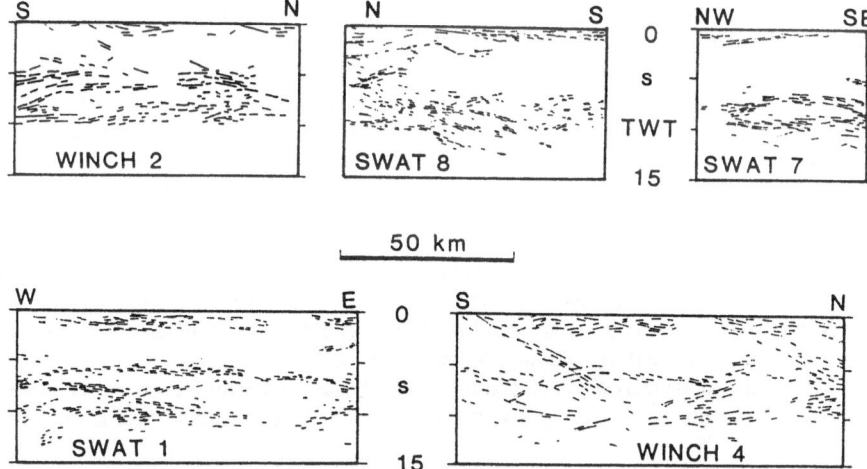

Figure 5   Line-drawings of BIRPS data showing other examples of reflective bands and lenticular transparent zones within the lower crust. A structural origin for these features seems most reasonable. Also note that extensional structures dominate the upper crust (cf the "typical BIRP").

Lower crustal reflections have been variously attributed to fluids, intrusions, and deformation fabrics (e.g., Matthews and Cheadle, 1986). One of the key observations in favour of the fluid hypothesis was the observation that many high-grade rocks are not reflective when uplifted into the upper crust (Matthews and Cheadle, 1986). Klemperer (1987) points out that the Lewisian complex, among others, formed under lower crustal conditions, but is not reflective where exposed at the surface. Klemperer argued that the poor reflectivity of granulite terranes indicated that crustal reflectivity is a mobile or transient feature, related to the current temperature, and that consequently only the *in situ* lower crust was reflective. He suggested that the apparent mobility of the reflective lower crust might be related to the mobility of fluids. However, recent results have shown that old, cold granulites can be reflective: the reflection profiles recorded over both the Kapuskasing zone and the Grenville Front (Green et al., 1989) image bands of sub-parallel bright reflections similar to those seen in the lower crust around Britain, but within the upper crust of cold, old, high-grade terranes. In both cases, the reflectivity is probably due to plastic shearing (perhaps transposing pre-existing lithological units into a strong, sub-planar fabric) and so a relationship between temperature and the *formation* of a reflective fabric is possible (Wever et al., 1987). The poor reflectivity of the Lewisian complex may instead arise because its principal structures are all sub-vertical (see later), and unlikely to be well-imaged by standard seismic techniques.

This does not mean that fluids play no part in deep crustal processes, as in many places the lower crust shows anomalously high electrical conductivity (Jones, 1987). However, the notion that lower crustal reflectivity is caused by a property unique to the *in situ* lower crust (and thus could not be explained by studies of exposed crustal sections) has generally been abandoned.

Attempts have been made to calculate the amplitude of lower crustal reflections in order to provide constraints on the nature of the reflecting boundaries within the lower crust. Using reverberations within the water column as calibration, Warner (1989) has calculated that some reflections within the lower crust require a reflection coefficient of at least 0.1. This is approximately that of a contact between a granite and a gabbro, or between a gabbro and a peridotite (e.g. Hurich and Smithson, 1987). Warner further argues that a reflection coefficient of this magnitude cannot easily be explained just by the presence of free fluids, nor just by the development of varying degrees of anisotropy. Instead, the amplitude of the observed reflections is best explained by contacts between different rock-types (e.g., Fountain et al., 1987), suggesting that the lower crust is highly heterogeneous, a result that will not surprise those geologists who work on high-grade terranes.

Unfortunately, amplitude studies cannot distinguish between sheared and unsheared lithological contacts: the reflection coefficient between a granite and a gabbro is much the same before and after shearing (Christensen, 1982; Fountain et al., 1987). Strong reflections are likely both from unsheared intrusions, and from transposed lithological units within high-strain zones, which may also exhibit varying degrees of anisotropy (e.g. Fountain et al., 1987). Amplitude studies alone can only constrain the physical cause of the reflectivity (i.e., lithological contacts), and may not indicate the geological significance of the reflections (i.e., shear zones or intrusions).

It is clearly important to distinguish between unsheared intrusions and transposed units within shear zones as each has a bearing on the nature of continental extension. The former would call for significant synextensional intrusion of the lower crust, whereas the latter requires considerable sub-horizontal deformation within the lower crust during extension. Both sub-horizontal deformation fabrics and igneous intrusions are observed in proposed crustal cross-sections. For instance, the Ivrea zone contains numerous sub-horizontal mafic intrusions and localised high temperature shear zones that both might be suitable analogs for the reflectors seen in the lower crust (Brodie and Rutter, 1987; Hale and Thompson, 1982). To investigate this possibility, computer modelling (see Reston, 1987, 1988 for technical details) has been used to generate the synthetic two-dimensional seismic response of the Ivrea Zone. Previous seismic modelling has either been one-dimensional (Hale and Thompson, 1982), or limited by the use of an unrestored section at a few kilometres depth (Hurich and Smithson, 1987). Here, however, the detailed restored section of Brodie and Rutter (1987) is modelled at lower crustal depths (Figure 6), thus providing a more legitimate comparison with the deep seismic data. The principal difference between the input model and the restored section of Brodie and Rutter is that the input model has been extended laterally by attaching a mirror image of their section to one edge. The result is a symmetric input model and synthetic section (only one half of each is shown in Figure 6). This step is necessary to capture reflections from the entire model; otherwise, reflections from dipping structures such as the high temperature shear zones would not have been synthesised.

The velocities and densities used are from Fountain (1976), except for the values adopted for the high temperature shear zones, for which no data were available. These were assigned a velocity and density that was the mean for the rocks in the section, but with a 15% velocity anisotropy. It turns out that they are too thin to be resolved on the section, so the values used are somewhat irrelevant.

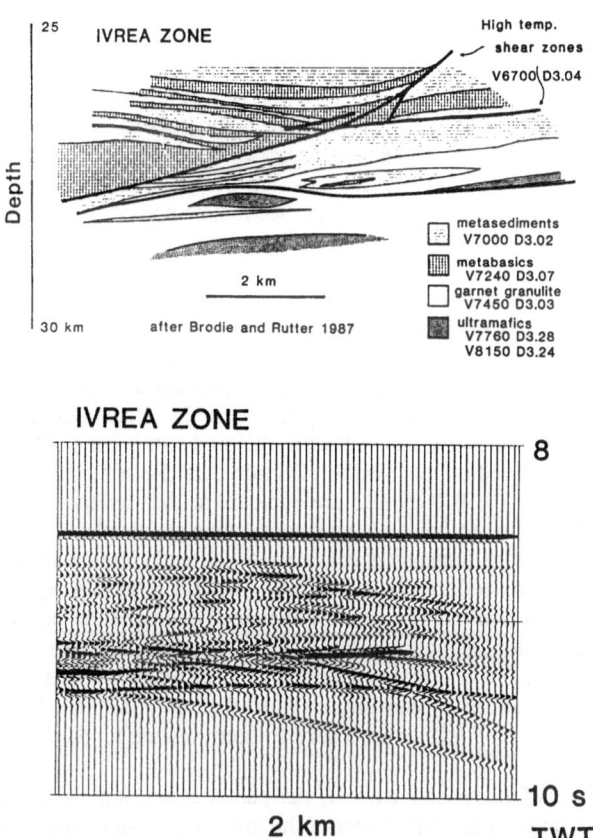

Figure 6 A model for the Ivrea Zone (based on a restored section by Brodie and Rutter, 1987, and on velocities from Fountain, 1976) has been used to generate a synthetic seismic section. Note that the seismic section cannot resolve the details of the geological model, but does resemble parts of the deep seismic data.

The flat reflection at the top of the resulting synthetic section (Figure 6) is a reference reflection, correponding to a reflection coefficient of 0.1, and is included to calibrate the other reflections. The brightest of these are from the base of the section, and correspond to contacts between the garnet granulites and either metasediments or ultramafics. The high temperature shear zones are too thin (20 m. thick in the model) to be clearly imaged. Overall, the synthetic section is a complex interference pattern (Hurich and Smithson, 1987; Reston, 1987) bearing little resemblance to the geological section, but showing many of the characteristics of real seismic data: the synthetic is dominated by convex-up (diffractive), cross-cutting and segmented reflections. The fairly good match, at least in overall appearance, between this synthetic section and the reflectivity of the lower crust around Britain suggests that the Ivrea zone may be a suitable analog for the reflectivity of the lower crust around Britain. However, as is discussed later in the paper, in other ways the Ivrea section does not match the lower crust beneath Britain.

The Kapuskasing zone (Percival, this volume) is another possible crustal section dominated by extensional structures. Part of a restored section (after Percival, 1986) through the Kapuskasing zone in shown in Figure 7. Towards the base of the section, the lithological units form a strong subhorizontal fabric composed of different gneisses, apart from a large anorthosite body, the Shawmere anorthosite. As this has strongly tectonised margins, yet retains early corona structures in its core, it seems likely that this anorthosite is a low-strain lozenge within more highly strained rocks (the surrounding gneisses) and that the dominant subhorizontal fabric of the deep section is the result of the transposition of preexisting lithological units during extensional shearing (Percival, this volume). This section, combined with appropriate velocities and the densities given by Percival (1986), has been used as an input model to generate a synthetic section (Figure 7). Note that the upper crust is poorly reflective in the synthetic section: the reflection coefficients between tonalites and granites are unlikely to produce strong reflections, and there is no strong lithological fabric. The BIRPS data similarly display a poorly reflective upper crust, probably also because of the lack of a suitably oriented and reflective fabric. The Kapuskasing section therefore appears to provide a possible explanation for the reflectivity of the crust around Britain. Also note that the Shawmere anorthosite and an associated diorite together form a transparent zone enclosed within bands of reflections: similar structures have been observed on the BIRPS reflection data (Reston, 1988 and Figure 5). The transparent zones seen on the BIRPS data are enclosed within reflective bands, they cannot represent areas of poor signal-to-noise but must represent a genuine lack of a suitably-oriented reflective fabric. Furthermore, transparent zones are consistently lenticular in shape and appear to be intimately related to the reflectivity of the lower crust: the reflections appear to wrap around the zones (Figure 5). As a result, they cannot be explained as anomalous portions of a generally reflective lower crust: instead any explanation of lower crustal reflectivity needs to explain the transparent zones as well. One possible explanation of these features is that they represent large low-strain lozenges surrounded by high strain zones (Reston, 1988), much as observed in the Kapuskasing zone. Hamilton (1982) made a similar interpretation for the midcrust of the Basin and Range.

The interpretation of the reflective bands/transparent zones as high- and low-strain zones has implications not only for the reflectivity of the lower crust (suggesting that it is related to shearing), but also for the way it deforms: deformation might be concentrated into high-strain zones which wrap around and enclose relatively undeformed bodies. This style of deformation is common in high-grade terranes (Ramsay, 1980), such as the Lewisian of NW Scotland: the South Harris Shear Zone (Figure 8) shows deformation concentrated into the shear zones of the Leverburgh and Langavat Belts, enclosing low-strain lozenges of norite, diorite/tonalite, and anorthosite, which preserve original igneous features (Graham, 1980). In contrast, pre-existing lithological units within the shear belts have been drawn into a strong fabric, which would be

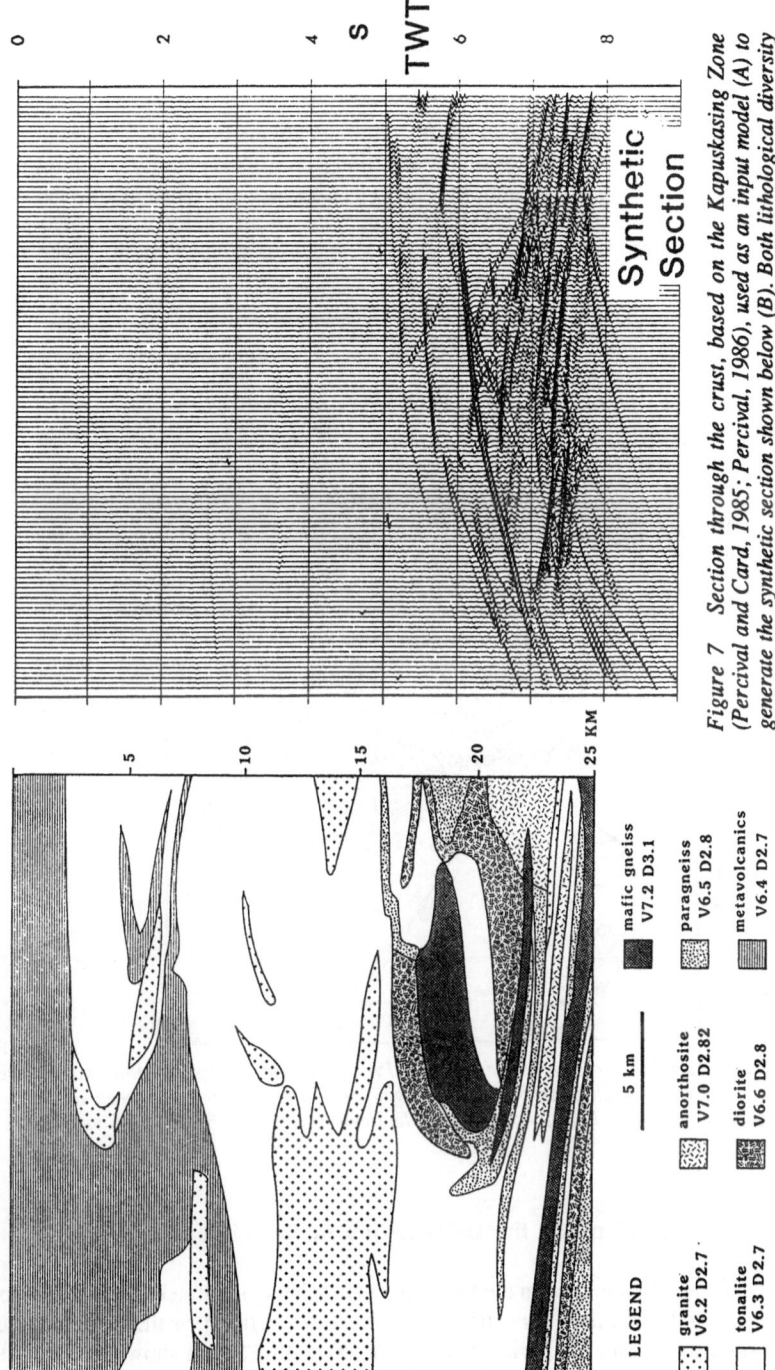

Figure 7 Section through the crust, based on the Kapuskasing Zone (Percival and Card, 1985; Percival, 1986), used as an input model (A) to generate the synthetic section shown below (B). Both lithological diversity and the shear fabric increase downwards, apart from the Shawmere Anorthosite, which has remained internally undeformed. The strong fabric makes the lower part of the section very reflective, in contrast to the upper part of the section, and to the Shawmere Anorthosite which appears as part of a transparent zone, similar to those in Figures 4c and 5.

616

highly reflective if suitably oriented. Unfortunately, the South Harris Shear Zone, along with the other major shear zones within the Lewisian complex (Coward and Park, 1987), is nearly vertical at present, and so not imaged on the seismic reflection data.

Of course, large lenticular transparent zones are not imaged on every profile, and in places the lower crust is a layer of uniform reflectivity (Figure 4a). However, this may be a function of resolution: low-strain lozenges smaller than about 5 km in length might simply contribute to the reflectivity as part of a complex, poorly resolved reflection pattern. It is possible that transparent zones, and cross-cutting, segmented reflections (Figure 4) are scale-dependent manifestations of similar structures. Certainly, low-strain lozenges are seen on all scales within high-grade terranes, which may provide good analogies for the style of deformation within the lower crust.

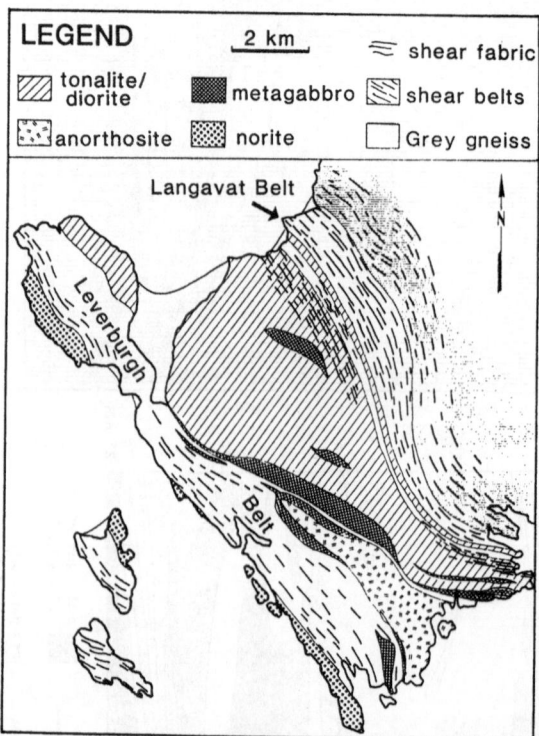

Figure 8 The South Harris Shear Zone (after Graham, 1980). Deformation is concentrated into shear belts that wrap around and enclose low-strain lozenges. Note the similarity between this style of deformation and the pattern of reflectivity seen within the lower crust of the BIRPS profiles (Figures 4c and 5).

The Role of the Lower Crust in the Deformation of the Lithosphere.

So far I have considered deformation in the lower crust completely separately from that in both the upper crust and in the mantle. This is however not strictly justified, as the different levels within the lithosphere must interact during deformation. The BIRPS data show one form that this interaction can take.

Fundamental to the "typical BIRP" is the observation of numerous extensional faults in the upper crust, and widely-spaced shear zones in the mantle (Figures 2 and 3). These do not appear to be collinear with each other, but instead both disappear into the reflective lower crust. This observation implies that the lower crust decouples the upper crust from the mantle (Matthews et al., 1987; Kusznir and Matthews, 1988), a model that has been likened to a "jelly sandwich" (Matthews and Cheadle, 1986).

The hypothesis that the lower crust decouples the upper crust from the mantle is compatible with rheological models for the lithosphere, particularly if the lower crust contains sufficient quartz to control the rheology (Kusznir and Park, 1987). As a result, the rheological stratification of the lithosphere may depend on the presence or absence of quartz, and hence on the bulk composition and the compositional heterogeneity of the lower crust.

Seismic refraction provides information about the bulk velocity of the *in situ* lower crust. As velocity is principally dependent on pressure, temperature, and mineralogy (Christensen, 1982), refraction results can be used to constrain the bulk composition of the lower crust (e.g. Smithson, 1978). Modern interpretations of refraction and wide-angle reflection experiments around the British Isles generally (Table I) show that the Moho represents a jump from around 6.7-6.8 km/s in the lowermost crust to 8 km/s or more in the mantle. These values contrast with the high velocity (>7.2 km/s) zone identified at the base of the crust in some other regions of continental extension (e.g. White et al., 1987): such high velocity zones are generally interpreted as pillows of mafic underplate, which appear to be absent around Britain (White et al., 1987). As gabbroic rocks should have a seismic velocity in excess of 7.0 km/s under appropriate conditions of temperature and pressure (Christensen, 1982; Furlong and Fountain, 1986) a 6.7-6.8 km/s lower crust is unlikely to be dominantly mafic, but instead is likely to be intermediate in composition between quartz-rich and mafic lithologies. In this respect, it might be similar to the Kapuskasing section, which also has velocities of 6.6-7.0 km/s when in place 25 km below the surface (Percival this volume). In contrast, the lowermost crust beneath the Po Basin, interpreted as corresponding to the quartz-poor granulites at the base of the Ivrea Zone, has a velocity of 7.4 km/s (Fountain, 1976), significantly higher than anywhere beneath Britain.

TABLE I: Refraction velocity of the lower crust around the British Isles.

Authors	Location (BIRPS profile)	Lower crustal velocity in km/second.
Edwards and Blundell, 1984	South-west Approaches (SWAT)	6.7
Barton et al. 1984	The Central North Sea (SALT)	6.5-7.0
Jones et al. 1984	North of Scotland (DRUM)	6.6
Ginzburg eta l. 1985	Western Approaches, Atlantic Margin (WAM)	6.5
Jacob et al., 1985	Irish Sea (WINCH)	6.8
Bott et al. 1985	North Sea (NEC)	6.8
Powell and Sinha, 1987	West of Outer Hebrides (WINCH)	6.7
Powell, pers. comm. 1988	West of Outer Hebrides (WINCH)	6.7-7.0

Although refraction studies of the lower crust around Britain constrain its bulk composition they give no indication of its compositional heterogeneity. Many high grade terranes (e.g., the Lewisian granulites of NW Scotland, and the deeper parts of the Kapuskasing section - Percival, 1986) have an intermediate bulk composition, but are dominantly bimodal in composition, containing quartzo-feldspathic and mafic gneisses. Furthermore, it has already been noted that the extreme reflectivity of the lower crust can best be explained if it is compositionally heterogeneous. It therefore seems reasonable to suggest that the lower crust around Britain is a complex mixture of quartzo-feldspathic and mafic gneisses. As rheology is dominated by the weakest major constituent (Kirby, 1985), quartz may be a major influence on lower crustal strength. The velocity structure thus provides theoretical support for the interpretation, based on the "typical BIRP", that the lower crust does decouple the upper crust from the mantle.

Of course, the lower crust cannot completely decouple the upper crust from the mantle, it must also transfer deformation between the two, assuming that all lithospheric levels undergo the same total deformation. This can only result in considerable simple shear strain within the lower crust. To illustrate the need for this simple shear, it is assumed that in some places there is no such simple shear, i.e. that the upper crust is pinned to the mantle (Figure 9a). The model is based on the typical BIRP and shows the upper crust and the upper mantle extending along a non-colinear fault and shear zone pair, both of which flatten into the lower crust. Because the upper crust is pinned to the mantle at the ends of the model, all the extension along the fault has to be transferred by top-to-the-left simple shear within the lower crust so it can be accommodated along the mantle shear zone. Removing these pins would allow the extension to be transferred out of the model by simple shear strain within the lower crust, perhaps shifting the locus of simple shear strain in the lower crust, but not removing the need for this simple shear. If extension in the upper crust is laterally displaced from that in the uppermost mantle, as the BIRPS data would seem to suggest, the lower crust is likely to be strongly sheared, both as it is stretched and because of its role in the deformation of the lithosphere.

A similar effect might occur during compression. Again, if deformation in the upper crust is not colinear with that in the mantle, the lower crust must transfer deformation between the two, and be strongly sheared in the process (Figure 9b). However, the typical BIRP geometry has only been observed in extensional settings, and might not apply during compression.

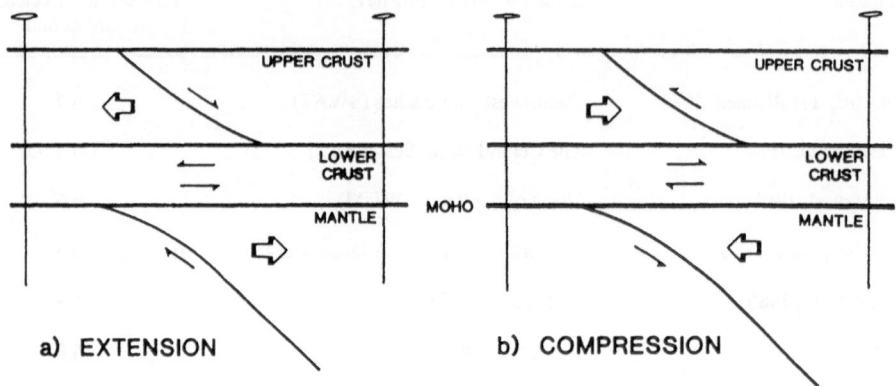

Figure 9 Lithospheric models, showing how the lower crust can be strongly sheared as it transfers deformation between the upper crust and the mantle, during both extension (A) and compression (B). The models are based on the typical BIRP.

Unfortunately, where the lower crust decouples the upper crust from the mantle, the faults and shear zones, including those that are instrumental in bringing crustal slices to the surface, are likely to detach at the top of the lower crust decoupling zone (cf the "typical BIRP"). As a result, these faults are unlikely to bring true exposures of the lower crust to the surface, but instead uplift only the upper and middle crust (e.g. Kapuskasing - Green et al., 1989). Conversely, true slices of the lower crust are most likely to be exposed when the lower crust does not decouple the upper crust from mantle (e.g. the Ivrea Zone - Brodie and Rutter, 1987). Either way, we are unlikely to see, exposed at the surface, sections of the lower crust that have been sheared in the manner required by the typical BIRP pattern of crustal faults and mantle shear zones.

Conclusions

This study has attempted to build up a picture of lithospheric reflectivity and structure around Britain, by drawing analogies between the results of the reflection profiles and exposed crustal sections. The basic conclusions are:

1. The crust around Britain is dominated by post-orogenic extension, represented by sedimentary basins, thin crust and demonstrably extensional faults.

2. Reflections in the mantle almost certainly represent mantle shear zones, accommodating much of the deformation in the uppermost mantle. At least some of these features are probably extensional in nature; the others may be as well.

3. The lower crust is generally quite reflective, and appears to act as a divide between localised deformation in both the upper crust and in the mantle. As the upper crust and upper mantle are both dominated by post-orogenic extensional tectonics, it seems likely that the reflectivity of the lower crust is also related to some form of extension. It is however not a uniformly reflective layer, but contains numerous bands of reflections and other less-reflective (transparent) zones. These are interpreted as high strain zones, enclosing low-strain lozenges, a style of deformation observed in the Kapuskasing zone. This crustal section appears in some ways analogous to the structure of the crust offshore Britain.

4. The lower crust around Britain is likely to contain significant amounts of quartz, and so may partly decouple the upper crust from the mantle. This is consistent with the pattern of lithospheric reflectivity present on the BIRPS data. However, the lower crust must also transfer deformation (strain) between the upper crust and mantle. As a result it is likely to be strongly sheared, along sub-horizontal structures, during both extension and compression. This shearing may enhance its reflectivity: the reflectivity of the lower crust around Britain may best be explained by shearing of a lithologically heterogeneous lower crust.

5. If the lower crust does decouple the upper crust from the mantle, it seems likely that the lowermost part of the crust will not be exposed at the surface. Conversely, the most complete sections through the crust likely arise where the lower crust does not decouple the upper crust from the mantle. Either way, the complete typical BIRP is unlikely to be represented as an exposed crustal cross-section.

Acknowledgements. Computer modelling made use of the AIMS package, donated to Cornell by GeoQuest International Inc. I am indebted to various individuals, too numerous to list, from the University of London, BIRPS and COCORP for lively discussion about the nature of the reflective lower crust. Finally, I would like to thank the NATO conference organisers and participants for a most enjoyable meeting. A NERC NATO Research Fellowship is gratefully acknowledged. INSTOC contribution number 112.

620

REFERENCES:

Allmendinger, R., D. Nelson, C. Potter, M. Barazangi, L.Brown and J. Oliver, 1987. Deep seismic characteristics of the continental crust. *Geology*, **15**, 304-310.

Barton, P., D. Matthews, J. Hall and M. Warner, 1984. Moho beneath the North Sea compared on normal incidence and wide-angle seismic records. *Nature*, **323**, 53-55.

Bott, M, R. Long, A. Green, A. Lewis, M. Sinha, D. Stevenson, 1985. Crustal structure south of the Iapetus suture beneath northern England, *Nature*, **314**, 274-727.

Brodie, K. and E. Rutter, 1987, Deep crustal extensional faulting in the Ivrea zone of Northern Italy, *Tectonophysics*, **140**, 193-212.

Burke, M. and D. Fountain, 1987. Seismic character of continental crust based on laboratory seismic velocity data from the Ivrea and Strona-Ceneri zones, northern Italy. *EOS*, **68**, 1481.

Cheadle, M., S. McGeary, M. Warner, and D. Matthews, 1987, Extensional structures on the western United Kingdom continental shelf: A review of evidence from deep seismic reflection profiling. *Spec. Publ. geol. Soc. London*, **28**, 445-466.

Christensen, N., 1982. Seismic velocities. In: *Handbook of Physical Properties of Rocks*, vol. II (ed: R.S. Carmichael), CRC Press, pp 1-228.

Coward, M. and R. G. Park, 1987. The role of mid-crustal shear zones in the early Proterozoic evolution of the Lewisian. *Spec. Publ. geol. Soc. London*, **27**, 127-138.

Dewey, J., 1982, Plate tectonics and the evolution of the British Isles, *J. geol. Soc. London*, **139**, 371-412.

Edwards, J., and D. Blundell, 1984, Summary of seismic refraction experiments in the English Channel, Celtic Sea and St. Georges Channel, Rep. Inst. Geol. Sci. U.K., **144**, 14 pp.

Fountain, D., 1976. The Ivrea-Verbano and Strona-Ceneri zones, northern Italy: a cross-section of the continental crust - new evidence from seismic velocities of rock samples. *Tectonophysics*, **33**, 145-165.

Fountain, D., McDonough, D., and Gorham, J., 1987, Seismic reflection models of the continental crust based on metamorphic terrains. *Geophys. J. Roy. astron. Soc.*, **89**, 61-66.

Freeman, B., S. Klemperer, and R. Hobbs, 1988. The deep structure of northern England and the Iapetus suture from BIRPS deep seismic reflection profiles. *J. geol. Soc. London*, **145**, 727-740.

Furlong, K., and D. Fountain, 1986, Continental crustal underplating: thermal considerations and seismic-petrologic consequences, *J. Geophys. Res.*, **91**, 8285-8294.

Ginzburg, A., R. Whitmarsh, D. Roberts, L. Montadert, A. Camus and F. Avedik, 1985. The deep seismic structure of the northern continental margin of the Bay of Biscay. *Ann. Geophys.*, **3**, 499-510.

Glennie, K., 1984, The structural framework and pre-Permian history of the North Sea area, In: *Introduction to the petroleum geology of the North Sea*, Blackwell, Oxford, 17-40.

Graham, R., 1980, The role of shear belts in the structural evolution of the South Harris igneous complex, *J. Struct. Geol.*, **2**, 29-37.

Green, A., A. Davidson, J. Percival, R. Milkereit, R. Parrish, F. Cook, W. Cannon, D. Hutchinson, G. West, and R. Clowes, 1989, Orgin of deep crustal reflections: results from seismic profiling across high grade metamorphic terranes in Canada. *Tectonophysics*, in press.

Griffin, W. and S. O'Reilly, 1986, The lower crust in eastern Australia: xenolith evidence. *Spec. Publ. geol. Soc. London*, **24**, 363-374.

Hale, L., and Thompson, G., 1982, The seismic reflection character of the continental Mohorovicic discontinuity, *J. Geophys. Res.*, **87**, 4625-4635.

Hamilton, W., 1987, Crustal extension in the Basin and Range Province, southwestern United States, *Spec. Publ. geol. soc. London*, **28**, 155-175.

Hurich, C., and Smithson, S., 1987, Compositional variations and the origin of deep crustal reflections. *Earth Planet. Sci. Letts.*, **8**, 416-426.

Jacob, A., W. Kaminski, T. Murphy, W. Phillips and C. Prodehl, 1985. A crustal model for a NE-SW profile through Ireland. *Tectonophysics*, **113**, 75-103.

Jones, A., 1987. MT and reflection: an essential combination. *Geophys. J. R. astr. Soc.*, **89**, 7-18.

Jones, E., R. White, V. Hughes, D. Matthews and B. Clayton, 1984. Crustal structure of the continental shelf off NW Britain from two ship experiments. *Geophysics*, **49**, 1605-1621.

Kirby, S., 1985, Rock mechanics observations pertinent to the rheology of the continental lithosphere and the localization of strain along shear zones., *Tectonophysics*, **119**, 1-27.

Klemperer, S., 1987. A relation between continental heat flow and the seismic reflectivity of the lower crust. *J. Geophysics*, **61**, 1-11.

Klemperer, S., 1988, Crustal thinning and nature of extension in the northern North Sea from deep seismic reflection profiling, *Tectonics*, **7**, 803-821.

Kusznir, N., and R. Park, 1987, The extensional strength of the continental lithosphere: Its dependence on geothermal gradient, and crustal composition and thickness, *Spec. Publ. Geol. Soc. London*, **28**, 35-52.

Kusznir, N. and D. Matthews, 1988. Deep seismic reflections and the deformational mechanics of the continental lithosphere, *Journal of Petrology*, Special Lithosphere Issue, 63-87.

Lailey, M., Stein, A. and Reston, T., 1988. The Outer Hebrides Fault: a major Proterozoic structure in NW Scotland. *J. geol. Soc. London*, **146**, 253-259.

Matthews, D., 1986, Seismic reflections from the lower crust around Britain, *Spec. Publ. Geol. Soc. London*, **24**, 11-21.

Matthews, D., and M. Cheadle,1986, Deep reflections from the Caledonides and Variscides west of Britain, and comparisons with the Himalayas, in *Reflection Seismology: A Global Perspective*, Geodyn. Ser.,vol. 13, edited by M. Barazangi and L. Brown, pp. 5-19, AGU, Washington, D.C.

Matthews, D., and the BIRPS group, 1987, Some unresolved BIRPS problems, *Geophys. J. R. Astron. Soc.*, **89**, 209-216.

McClay, K., M. Norton, P. Coney and G. Davis, 1986. Collapse of the Caledonian orogeny and the Old Red Sandstone. *Nature*, **323**, 147-149.

McGeary, S. and M. Warner, 1985. Seismic profiling the lower continental lithosphere, *Nature*, **317**, 795-797.

Meissner, R., D. Matthews, and T. Wever, 1986. The Moho in and around Britain. *Ann. Geophys.*, **4B**, 659-666.

Meissner, R., and N. Kusznir, 1987, Crustal viscosity and the reflectivity of the lower crust, *Ann. Geophys.* **5B**, 365-373.

Norrell, G. and Harper, G., 1988. Detachment faulting and amagmatic extension at mid-ocean ridges: the Josephine ophiolite as an example. *Geology*, **16**, 827-830.

Percival, J., 1986. A possible exposed Conrad Discontinuity in the Kapuskasing Uplift, Ontario. *AGU Geodynamics Series*, **14**, 135-141.

Powell, C. and Sinha, M., 1987. The PUMA experiment west of Lewis, UK, *Geophys. J. R. astr. Soc.*, **89**, 259-264.

Ramsay, J., 1980, Shear zone geometry: A review, *J. Struct. Geol.*, **2**, 83-99.

Reston, T., 1987, Spatial interference, reflection character and the structure of the lower crust under extension - Results from 2-D seismic modelling, *Ann. Geophys.*, **5B**, 339-348.

Reston, T., 1988. Evidence for shear zones in the lower crust offshore Britain, *Tectonics*, **7**, 929-945.

Reston, T., 1989, Shear in the lower crust during extension: not so pure and simple. *Tectonophysics*, in press.

Rutter, E., and K. Brodie, 1988. The role of tectonic grainsize reduction in the rheological stratification of the lithosphere, *Geol. Rundsch.*, **77**, 295-308.

Sheriff, R., and L. Geldart, 1982, *Exploration Seismology*, vol. 1, 253 pp.,Cambridge University Press, New York.

Sheriff, R., and L. Geldart, 1983, *Exploration Seismology*, vol. 2, 221 pp.,Cambridge University Press, New York.

Smithson, S., 1978, Modeling continental crust: structural and chemical constraints, *Geophys. Res. Letts.*, **5**, 749-752.

Warner, M., 1989. Seismic reflections from the lower continental crust: free fluids. *J. Geophys. Res.*, in press.

Warner, M., and S. McGeary, 1987, Seismic reflection coefficients from mantle fault zones. *Geophys. J. R. Astron. Soc.*, **89**, 223-230.

Wever, T., H. Trappe, and R. Meissner, 1987. Possible relations between crustal reflectivity, crustal age, heat flow and viscosity of the continents, *Ann. Geophys.*, **5B**, 255-266.

White, R., G. Westbrook, S. Fowler, G. Spence, P. Barton, M. Joppen, J. Morgan, A. Bowen, C. Prestcott, and M. Bott, Hatton Bank (northwest U.K.) continental margin structure. *Geophys. J. R. Astron. Soc.*, **89**, 265-272.

Ziegler, P., 1982. *Geological Atlas of Western and Central Europe*. Shell International Petroleum, Maatshappij B.V., 130 pp.

INTRACRUSTAL DETACHMENT AND WEDGING ALONG A DETAILED CROSS SECTION IN CENTRAL EUROPE

St. Mueller
Institute of Geophysics
ETH-Hönggerberg
CH-8093 Zürich
Switzerland

ABSTRACT.   Detailed features of the crustal structure beneath the Black Forest basement complex (in southwestern Germany), the adjacent Jura mountains and Molasse basin (in northern Switzerland) combined with the eastern and southern geotraverses across the Swiss Alps have recently been surveyed by various geophysical methods. An interpretation of the near-vertical seismic reflection data obtained in conjunction with seismic refraction and wide-angle reflection measurements along the same profiles has made it possible to image the fine structure of the crust along a 420 km long cross section through Central Europe. Based on both reliable travel-time and amplitude information this simultaneous inversion procedure led to the unambiguous identification of low-velocity zones and velocity-depth gradients. The depth ranges of lowered velocity are associated with increased electrical conductivity and appear as zones of weakness which favor detachment at levels in the upper and middle crust. In particular, during the continuing collision process which has formed the Alps and their northern foreland the southward dipping crust is detached at one or two levels and deformed by northward protruding wedges of presumably lower crustal material. This mechanism can explain both the observed earthquake focal depths and the recent uplift pattern.

INTRODUCTION

The application of both deep seismic reflection and refraction sounding on a regional or local scale permits a synoptic view of crustal properties based on two different imaging techniques (see e.g. Mueller et al. 1987). Near-vertically incident seismic waves illuminate the detailed vertical structure in terms of their reflectivity pattern with a resolution which is determined by the dominant wavelength of the incident signal, i.e. a few hundred meters. Wide-angle reflection and refraction ray paths are sensitive to large-scale vertical and horizontal velocity gradients. The obvious question arises if the reflecting structures detected in near-vertical sounding experiments at a chosen surface location are compatible - or even identical - with first- and second-order discontinuities deduced from wide-angle studies.

        In order to answer this question a detailed crustal cross section has been constructed in Central Europe based on combined seismic reflection and refraction observations. The cross section extends from the northern Black Forest near Karlsruhe in southwestern Germany across the Jura mountains, the Molasse basin and the Central Alps in Switzerland to the northern limit of the Po valley near Como in Italy (Fig. 1). A synoptic interpretation of the various data sets has led to a better understanding of the regional crustal structure and has provided some clues to the physical mechanisms governing the intracrustal tectonic processes.

M. H. Salisbury and D. M. Fountain (eds.), Exposed Cross-Sections of the Continental Crust, 623–643.
© 1990 *Kluwer Academic Publishers.*

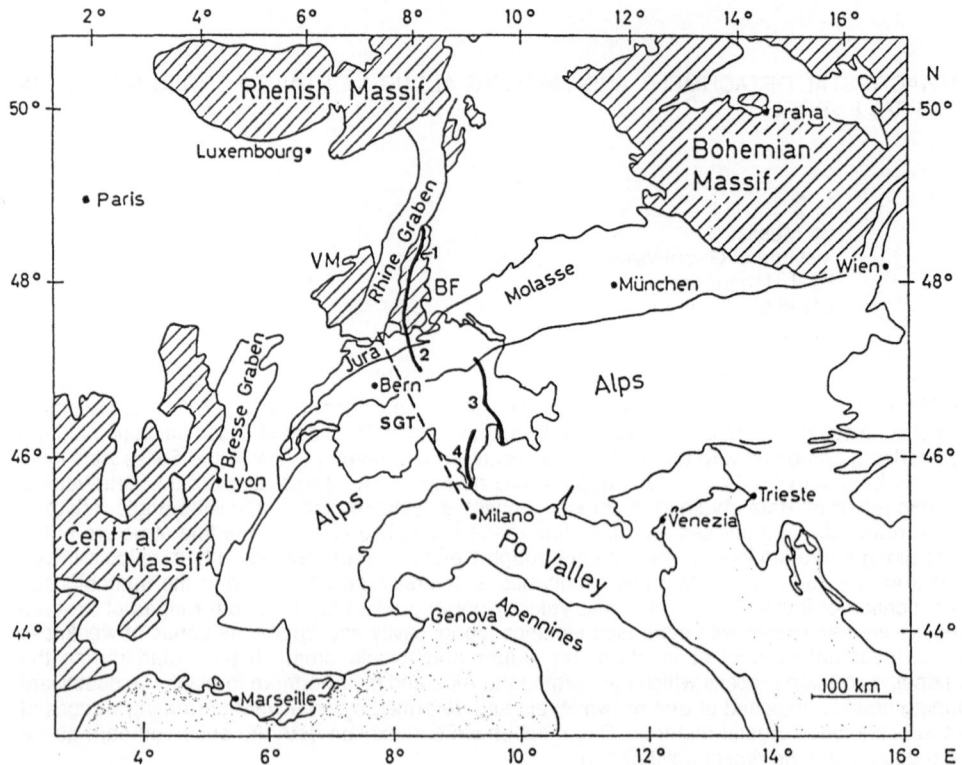

Figure 1.   Regional sketch map showing the position of the seismic profiles in Central Europe discussed in this paper. Profile segments: 1 = Black Forest massif; 2 = Jura mountains and Molasse basin; 3 = Central Alps (Eastern Traverse); 4 = Southern Alps (Southern Traverse). SGT = Swiss Geotraverse (from Basel to Milano). Crystalline massifs (hachured), where BF = Black Forest; VM = Vosges Mountains.

THE BLACK FOREST MASSIF

The Black Forest (BF) is situated at the eastern flank of the Tertiary Rhine Graben in Central Europe (Fig. 1); it is one of several extended areas where rocks of the Variscan (Hercynian) crystalline basement complex are exposed at the surface. Its position is about half way between the French Central Massif in the southwest and the Bohemian Massif to the northeast. The metamorphic and plutonic rocks of the northern Black Forest are separated from the granites in that region by the southward dipping Saxothuringian-Moldanubian suture of the Variscan system (Fig. 2). Gneisses and migmatites are the prevailing rock types of the central Black Forest. In the west the Black Forest is bounded by the master fault of the Rhine Graben. Along the eastern margin granites are exposed, and to the south the central gneiss complex is bordered by the Paleozoic lineament of the Badenweiler-Lenzkirch thrust zone (exposed near Todtnau, cf. Figs. 2 and 3a) which comprises Lower Carboniferous, Upper Devonian and Paleozoic rock sequences. The southern Black Forest consists mainly of Variscan granites and gneisses. To the north and east the crystalline basement disappears underneath the gently dipping Permian and

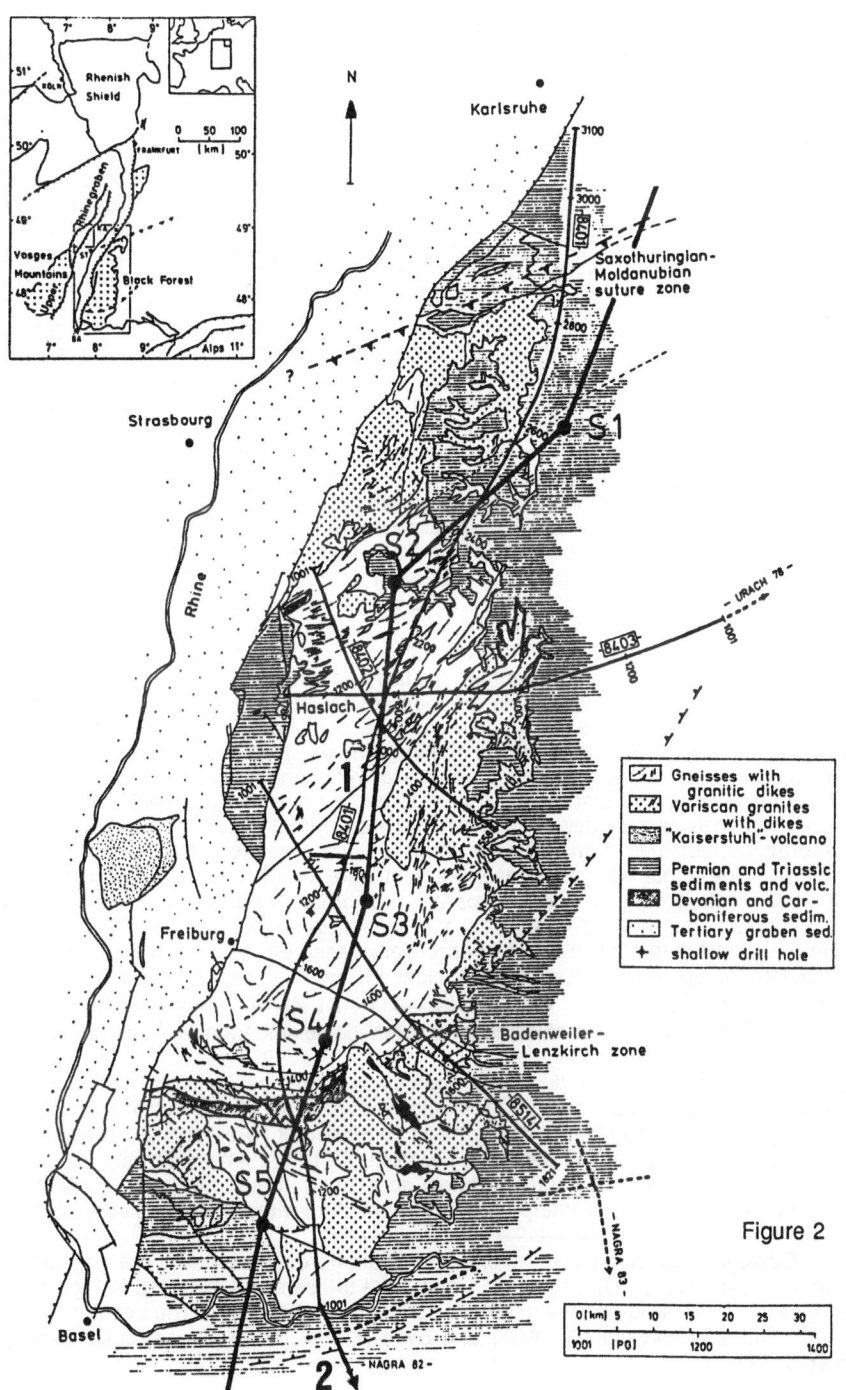

Figure 2

Triassic sedimentary cover (cf. Fig. 2). Further to the east Upper Jurassic rocks are exposed in the Swabian Jura; identical rock sequences are found in the tabular Jura and the folded Jura mountains south of the Black Forest in northern Switzerland (Fig. 5). Beneath the thickening Quaternary and Tertiary sediments of the Molasse basin (cf. Fig. 1) the Variscan basement dips southeastward towards the Alpine orogen.

Up to 1984 the crustal structure of the Black Forest proper had only sporadically been studied (see e.g. Rothé and Peterschmitt 1950, Emter 1971). A few regional seismic surveys in or close to the Black Forest were conducted in connection with the geophysical investigations of the Rhine Graben (Mueller et al. 1969, 1973; Edel et al. 1975; Prodehl et al. 1976; Zucca 1984). During the past few years some results of regional studies to the east and south of the Black Forest have been published (Deichmann and Ansorge 1983; Sierro et al. 1983; Finckh et al. 1984, 1986; Gajewski and Prodehl 1985; Gajewski et al. 1987). The recent unified seismic exploration program (cf. Fig. 2) consisting of a 240 km long seismic refraction profile (Gajewski and Prodehl 1987), 345 km of deep reflection profiling, one expanding-spread profile and near-surface high-resolution measurements (Lüschen et al. 1987), have revealed a strongly differentiated crust beneath the Black Forest which contains several unusual features. A carefully designed expanding-spread experiment in the central part of the Black Forest (near Haslach, see Fig. 2), utilizing high-frequency signals (20 to 30 Hz) and a detector spacing of 80 m, produced an excellent data set which allowed continuous phase correlation. Inversion and modelling of these data have demonstrated beyond doubt that the major crustal discontinuities located in this area by the two seismic sounding techniques are indeed coincident (Lüschen et al. 1987).

As can be seen in Fig. 2 the seismic reflection network was planned to obtain a structural image of the Black Forest crystalline massif from a north-south profile, 170 km in length, supplemented by three shorter intersecting profiles in the region of Haslach and further south near shotpoint S 3. The north-south profile follows roughly the morphological axis of the Black Forest (Figs. 1 and 2) traversing the extended central gneiss complex, the adjacent thrust zones and the granitic outcrops in the south. At the southern end the north-south reflection profile is directly connected with the network in northern Switzerland (Finckh et al. 1984, 1986). In Fig. 3a a line drawing is shown of the migrated north-south reflection section in the Black Forest. The crustal image is characterized by a highly reflective lower crust between 5 and 9 seconds (s) two-way time (TWT), probably due to a sequence of thin alternating high- and low-velocity layers, and an upper crust with a significantly lower density of reflecting elements. Dominant dipping reflectors, corresponding to thrust faults, appear to die out at about 5 s TWT. Although near-vertical reflection profiling provides a detailed image of crustal structure, the reliability of the velocity information obtained from relatively short detector spreads is rather poor. Therefore, supplementary wide-angle reflection measurements are needed for a realistic time-depth conversion and a subsequent synoptic interpretation. A gross two-dimensional P-wave velocity distribution (Fig. 3b) has been derived from the seismic refraction profile whose position is depicted in Fig. 2 (Lüschen et al. 1987). It is essentially based on the correlation of the Pg wave which penetrates the uppermost crust, the reflections PiP from the top of the lower crust and the wide-angle reflections P_MP from the crust-mantle boundary ("MOHO"). The velocity-depth model was successively improved by the calculation of synthetic seismogram sections based on two-dimensional asymptotic ray theory.

It is important to note that an additional signal phase could be identified which corresponds to reflections from the upper boundary of a low-velocity zone. This zone which lies at a depth of 7 to 14 km is very accentuated under the northern and central Black Forest and is characterized by a velocity decrease of about 9 to 10 per cent, i.e. from 5.9 to 5.4 km/s (Fig. 3b). This contrast diminishes gradually towards the south. A "bright spot" seen in the reflections at about 9.5 km depth beneath Haslach (Fig. 3a) is interpreted as fine structure within the low-velocity zone. The

---

Figure 2. Geological-tectonic sketch map of the Black Forest massif with the adjacent Rhine Graben (to the west). The profile segments 1 and 2 are the same as indicated in Figure 1. Shot points of the seismic refraction profile (thick line) are labelled S1 ..... S5. The position of the four seismic reflection lines 8401, 8402, 8403 and 8414 is depicted by thinner solid lines. (After Lüschen et al. 1987).

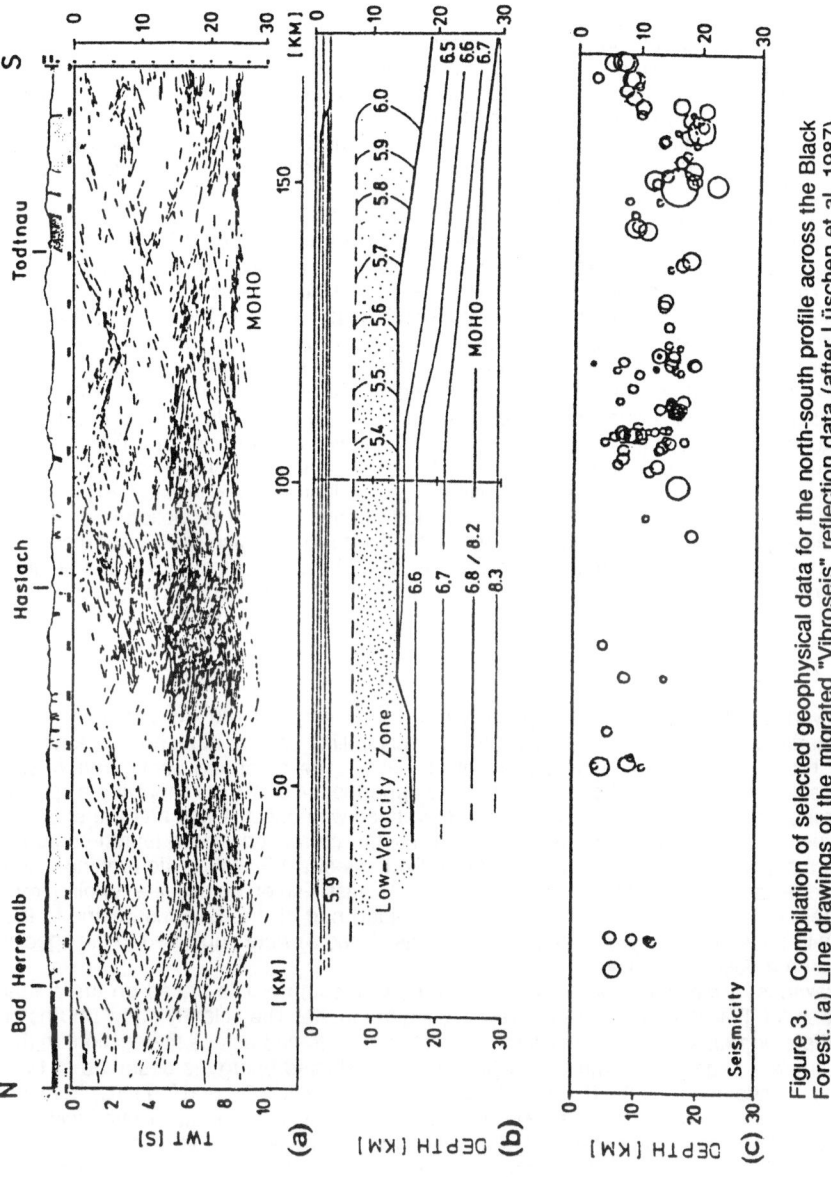

Figure 3. Compilation of selected geophysical data for the north-south profile across the Black Forest. (a) Line drawings of the migrated "Vibroseis" reflection data (after Lüschen et al. 1987) along the profile segment 1 (N-S line 8401 in Figure 2); (b) P-wave velocity (km/s) model (after Gajewski and Prodehl 1987) from the reversed N-S seismic refraction profile (thick solid line in Figure 2); (c) Distribution of earthquake foci (magnitudes: 1<M <5). The events located within a swath of + 10 km in width were projected into the cross section. (After Bonjer et al. 1984).

lower crust whose top coincides with the lower boundary of the low-velocity zone and which is normally called the "Conrad (C) discontinuity", shows a very complex response in the seismic record sections (Gajewski and Prodehl 1987). Based on a combined interpretation of the near-vertical and wide-angle reflection data the lower crust appears as a laminated zone consisting of thin layers with alternating high and low velocities whose average velocity is 6.7 km/s (Sandmeier and Wenzel 1986). In regions where the highest density of reflecting elements is detected, i.e. in the northern part of the Black Forest, only sporadic seismicity (Fig. 3c) is observed. The Mohorovičić (M) boundary is an almost horizontal first-order discontinuity at a relatively shallow depth of 25 to 27 km above a transparent upper mantle (Fig. 3a).

The lower part of the upper crust, i.e. the depth range 7 - 14 km, which is characterized by a P-wave low-velocity zone (Fig. 3b), is also a zone of relative transparency (Fig. 3a) and, in its upper part coincides with a depth range of increased electrical conductivity (LOTEM Working Group 1986). This zone widens somewhat towards the south. Under the central gneiss complex a thin highly conductive layer has been mapped at a depth of 12 km, i.e. in the lowermost part of the low-velocity zone (Tezkan 1988). Earthquake foci appear to be concentrated close to the upper and lower bounds of the low-velocity zone (Fig. 3c); they split up into two groups at the southern end of the crustal section (cf. Figs. 4c and 4d). Observations of the same type have earlier been associated with a quasi-continuous laccolithic zone of granitic intrusions (Mueller 1977) which are subject to dilatancy-induced cracking thus providing pathways for enhanced fluid circulation at elevated pore pressures (Brace 1972). Dehydration reactions which lead to the release of free water into existing and newly forming cracks, will provide a consistent explanation for the lowered seismic velocities, the increase in signal attenuation, the reduction in cohesive strength (eventually leading to detachment and/or brittle fracture) and also the higher electrical conductivity. The closer the pore pressure approaches the lithostatic pressure, i.e. the smaller the effective pressure, the more dominant will these interconnected phenomena become.

THE JURA MOUNTAINS AND MOLASSE BASIN

In 1982 a network of "Vibroseis" reflection profiles with a total length of 180 km was surveyed in northern Switzerland to investigate the suitability of the crystalline basement as host rock for the deposition of highly radioactive waste (Finckh et al. 1986). The recording configuration was designed to resolve in great detail the structures in the uppermost 4 s TWT. By applying a special processing procedure in which the field data were correlated with only the first half (10 s) of the up-sweep covering a frequency range from 11 to 36 Hz the correlated record length could be extended to a maximum of 14 s (Finckh et al. 1984). Portions of a NE-SW profile running along the strike of the Swiss Jura mountains (Fig. 1) from west to east, and of a cross line running from the Black Forest in the north across the tabular and folded part of the Jura mountains to the Molasse basin in the south (see profile segment 2 in Figs. 1 and 5) were subjected to the special processing scheme described above.

Line drawings of the reflecting elements for the north-south profile based on the actual seismic sections (cf. Finckh et al. 1986) are reproduced in Fig. 4a. They clearly show returns of energy from the entire crust with major reflectors which can be traced across the section. In the central and southern segments of the profile pronounced reflected energy is observed at two-way travel times between 2.0 - 3.0 s (G) from the upper crust, between 6.0 and 7.0 s (C) from the middle crust and between 8.0 and 9.0 s (M). The latter reflections must be associated with the crust-mantle transition zone (Mueller et al. 1980), while the sloping reflector (1.5 to 5.0 s) at the northern end of the N-S section must be ascribed to a prominent northward directed thrust fault which extends into the southernmost Black Forest crystalline massif (cf. Fig. 3a).

In the same area of northern Switzerland detailed seismic refraction and wide-angle reflection measurements were carried out with station spacings of less than one kilometer as part of the national geophysical survey (Sierro et al. 1983; Sierro 1988). As an example, part of the record section for a north-south refraction profile extending from the Black Forest massif to the Molasse basin is shown (Fig. 4b) whose position coincides with the N-S reflection line across the Jura mountains (Fig. 4a). The clearly delineated signal phases permit an indisputable correlation

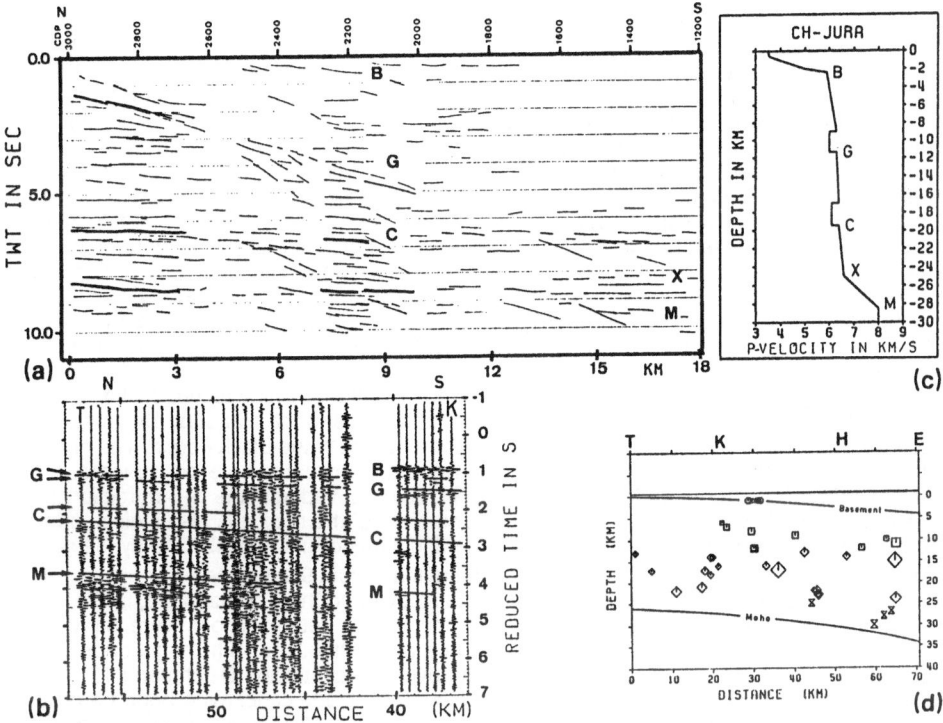

Figure 4. Seismic data for the Swiss Jura mountains and Molasse basin (cf. line segment 2 in Figs. 2 and 5). (a) Line drawings of the reflection data along the profile segment 2 (TWT = two-way time in seconds). (b) Part of the refraction record section with correlations (reduced time tR = t - d/6, where d = distance in km) for a reversed N-S profile from the Black Forest (T) across the Jura mountains (K) to the Molasse basin (H). (c) P-wave velocity-depth structure based on the synoptic interpretation of reflection and refraction data. B = top of crystalline basement, G = upper crustal low-velocity zone, C = mid-crustal low-velocity zone, X = sharp gradient zone in the lower crust, M = crust-mantle transition zone (after Mueller et al. 1987 and Sierro 1988). (d) Distribution of earthquake foci projected into a cross section roughly perpendicular to the Moho depth contours (Mueller et al. 1980). The different symbols correspond to different depth intervals and their size is proportional to magnitude. T = Todtmoos (Black Forest), K = Kaisten (Jura), H = Lake of Hallwil (Molasse), E = Emmen (Folded Molasse). (After Deichmann 1987).

of first and later arrivals as indicated. Both surveys are detailed enough for a synoptic interpretation of the reflection and refraction data.

The inversion leads to a consistent P-wave velocity-depth section in that area (Fig. 4c) which, except for the absolute depth, resembles the velocity-depth section proposed by Emter (1971) for the western Molasse basin (cf. Mueller et al. 1987). The velocity structure derived from the refraction and wide-angle reflection data contains relatively abrupt interfaces which cause normal-incidence reflections from the top and bottom of low-velocity zones in the upper and middle crust. It is worth noting that the wide-angle reflections from the top of the velocity reversals are less pronounced if compared to the corresponding reflections from the bottom (G and C) of those structures. Both low-velocity zones exhibit significantly smaller velocity contrasts than the crustal structures proposed for the Rhine Graben area by Mueller et al. (1969, 1973) and for the Black Forest by Gajewski and Prodehl (1987), which are both characterized by a single well-developed low-velocity "channel" at mid-crustal depths. As in the Black Forest a thin extraordinarily well-conducting layer has been detected beneath the Molasse basin (Schnegg and Fischer 1989) which coincides with the bottom of the low-velocity zone (C) in the middle crust (cf. Fig. 4c).

The conspicuous reverberations between C and M in the refraction record section (Fig. 4b) are most likely due to lamination of the lower crust analogous to the model by Deichmann and Ansorge (1983) for a nearby profile close to the eastern margin of the Black Forest. The gradients in the schematic velocity-depth function for the lower crust presented here are concordant with the more detailed model of Sandmeier and Wenzel (1986) derived from the N-S refraction profile through the Black Forest (Fig. 3b). A lamination at the top of the mantle is also indicated by the reverberations following the onset of the P_MP wide-angle reflections (Fig. 4b) and by the normal-incidence reflections with two-way travel times of more than about 9 s (Fig. 4a).

The results of an ongoing detailed seismicity study in northern Switzerland, in the course of which the accuracy and reliability of earthquake focal depth determinations were improved by several different methods, revealed the occurrence of earthquakes with magnitudes $\leqslant$ 4.2 throughout the crust all the way down to the Mohorovičić discontinuity (Deichmann 1987). This important observation raises a number of fundamental questions concerning both the physical properties of the crust and the mechanism of earthquakes, in particular of those occurring within the lower crust which conventionally is assumed to be "ductile" and, therefore, essentially aseismic.

In order to ensure uniform data quality, only recordings for the time span from 1984 to 1986 were used. The results for a cross section which as closely as possible coincides with the profile segment 2 in Figs. 1 and 5, and which extends from Todtmoos in the southern Black Forest to Emmen in the Swiss Molasse basin, are shown in Fig. 4d. It is obvious that earthquakes in northern Switzerland appear to be distributed from close to the surface all the way down to the crust-mantle boundary (Moho). Upon closer inspection it can, however, be seen that southward from Kaisten (K) - except for a few very shallow events - the earthquake foci are found in two "bands" whose depths under the Jura mountains are either above or below the two low-velocity zones, i.e. depth ranges of lowered strength, in the upper and middle crust, respectively (Fig. 4c). As mentioned earlier, the same pattern has also been observed at the southern end of the Black Forest crustal section (cf. Fig. 3c). It should be noted that the two "bands" of hypocenters separate more and more as the Alps are approached indicating increasing crustal detachment and wedging which seems to reach as far north as the belt of the folded Molasse.

THE CENTRAL AND SOUTHERN ALPS

Deep seismic sounding experiments have been carried out in the Alps since 1956 and are still going on. A number of important results have been obtained which can be summarized as follows (Miller et al. 1982): The Alpine crust is characterized by an asymmetric crust-mantle boundary which reaches its greatest depth of some 55 to 60 km south of the central region and apparently rises rather steeply towards the inner arc side of the Alps. The internal structure of the crust differs markedly between the eastern and the western Alps with more pronounced lateral velocity variations existing in the central and western Alps, which might be an indication that intracrustal detachment and wedging were the dominant processes forming the present-day crust, since the deformation of the lithosphere was greater in the central and western Alps than in the eastern Alps.

A comprehensive study of the deep structure and recent dynamics in the central portion of the Alps in Switzerland was carried out between 1974 and 1980 as part of the International Geodynamics Project. The geophysical results obtained were presented in the form of a crust-mantle cross section for the "Swiss Geotraverse" (SGT in Figs. 1 and 5) which transects the main tectonic and geologic units of the northern Alpine foreland and the Alps (Rybach et al. 1980). As sketched in Fig. 5 it extends from the southern end of the Rhine Graben across the tabular and folded Jura mountains through the Molasse basin, the Helvetic zone of the northern Alps, the Aar and Gotthard (external) crystalline massifs, the Penninic zone and the Southern Alps to the Tertiary sediments of the Po Valley (for details of the geology see Trümpy 1980).

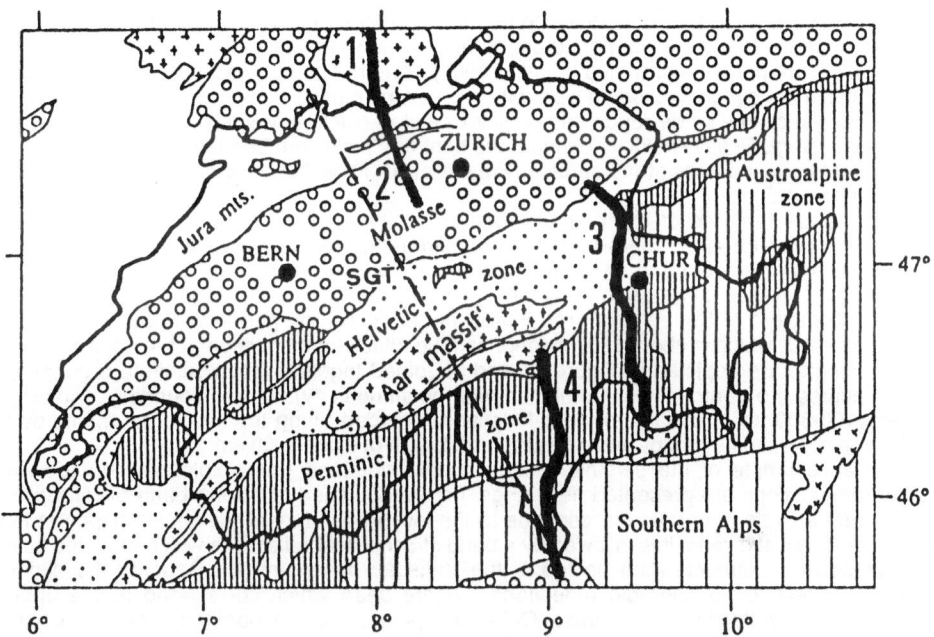

Figure 5.   Schematic geologic map of Switzerland (after Pfiffner et al. 1988) with the traces of the four seismic profile segments (heavy solid lines) described in this paper (cf. Figure 1) SGT = Swiss Geotraverse (from Basel to Milano). Vertical crosses = Variscan basement; diagonal crosses = Tertiary intrusives; open circles = Tertiary sediments of the Molasse basin and the Rhine Graben; dots = Helvetic zone; vertical shading = Penninic and Austroalpine zones and Southern Alps, respectively; white areas = Mesozoic of Jura mountains and tabular Jura. The Insubric Line (I.L. in Figure 10) separates the Penninic zone from the Southern Alps.

By combining all available seismic refraction and reflection data along the "Swiss Geotraverse" a representative crustal cross section could be constructed (Mueller et al. 1980). The crustal structure derived appears quite complex, but is well documented for the northern foreland of the Alps with the Helvetic zone, the Penninic zone and the Southern Alps, while the details of the contacts between these latter three major structural units, i.e. the extensions to depth of the Rhine-Rhône Line (R.R.L. in Fig. 8) and of the Insubric Line (I.L. in Fig. 10), have only recently been investigated more thoroughly (Finckh et al. 1987; Heitzmann 1987).

Proceeding from the southern end of the Rhine Graben in the northwest towards the southeast the entire crust submerges beneath the Jura mountains and the thickening wedge of the Molasse basin. Only minor internal variations of seismic velocities and thicknesses of the crustal layers have been mapped, with the lower crustal layers and the M discontinuity running roughly parallel to the contact between the Mesozoic formations and the Variscan basement (Mueller et al. 1980). A low-velocity zone - with a velocity reduction of 3 to 10 per cent (cf. Figs. 4c and 6b) - separates the upper part of the crystalline basement (with P-wave velocities of about 6 km/s) from the middle crust under the northern Alpine foreland (cf. Mueller 1977, 1982). An interpretation of the following segment further to the southeast, i.e. the crustal structure beneath the Helvetic nappes and the Aar massif, appears to be rather straightforward, but is not yet fully understood (cf. Figs. 9a & b).

In contrast to this relatively simple tectonic style in the north a rather different crustal structure has been mapped in the adjoining Penninic domain (Figs. 8, 9a and 10). There is now mounting evidence that the lowermost crust of the northern foreland extends all the way under the Penninic block in the central Alps (Mueller 1989). Whether in this regions two slabs of lower crustal material are superimposed on each other or several crustal wedges are shoved into each other cannot yet clearly be decided. This complex crustal structure which is characterized by an overall low average P-wave velocity beneath the central Alps is corroborated by the presence of highly metamorphosed units at the surface (Frey et al. 1980) which must have been uplifted by about 15 to 20 km.

The gross features of the crustal structure of the Alps as described for the "Swiss Geotraverse" (Mueller et al. 1980) have been confirmed by recent results of continuous deep seismic profiling in eastern and southern Switzerland obtained between 1986 and 1988. This project is part of Research Program 20 of the Swiss National Science Foundation entitled "Deep Geological Structure of Switzerland" (ref. Finckh et al. 1987). The position of both the eastern profile segment 3, which is about 120 km long, and the offset southern segment 4, approximately 70 km in length, are depicted in Figures 1 and 5. Joined together they constitute a complete transect across the central Alps. Combined with continuous "Vibroseis" sounding, explosive charges (100-300 kg of dynamite) were fired and recorded along the profile segments by a geophone array with 240 groups and a total spread length of close to 20 km.

As an example of the seismic reflection data obtained from explosive sources two seismogram sections are presented here (Figs. 6a and 7). The first strong band of reflections after the first arrivals in Fig. 6a corresponds to the lower boundary of the Säntis nappe in the northernmost Helvetic zone. It is followed by a band of consistent reflections which mark the base of the Mesozoic sediments (Mz) on top of the Variscan basement (B). Beneath the almost transparent upper crust sporadic reflections (C) are seen which correspond to the upper boundary of the reflective lower crust. Quite conspicuous is the dominant reflection band between 12 and 13 seconds two-way time (TWT). It images the crust-mantle transition zone (M) which dips towards the center of the Swiss Alps and can be traced continuously all the way into the Penninic domain (cf. Figs. 8, 9 and 10) as postulated earlier (Mueller 1989).

To illuminate details of the crust-mantle transition band in Fig. 6a an enlargement of single traces for the two-way time interval of 5 to 14 seconds is displayed in Fig. 7. Clear intracrustal reflections can be traced across the spread at about 9.2 seconds two-way time; these are the C reflections of Fig. 6a which must be associated with the classical "Conrad discontinuity". Furthermore two discrete reflection bands between 12 and 13 seconds TWT, about 0.8 seconds apart, can be discerned which were not so clearly resolved in Fig. 6a. They are termed X (after Klemperer et al. 1986) and M (for Mohorovičić discontinuity). The compound (X-M) character of the "Moho" reflections is most likely due to a lamination, i.e. an alternating sequence of thin high- and low-velocity layers, superimposed on a positive velocity gradient, similar to the model proposed by Deichmann and Ansorge (1983) for the lowermost crust at the eastern margin of the Black Forest.

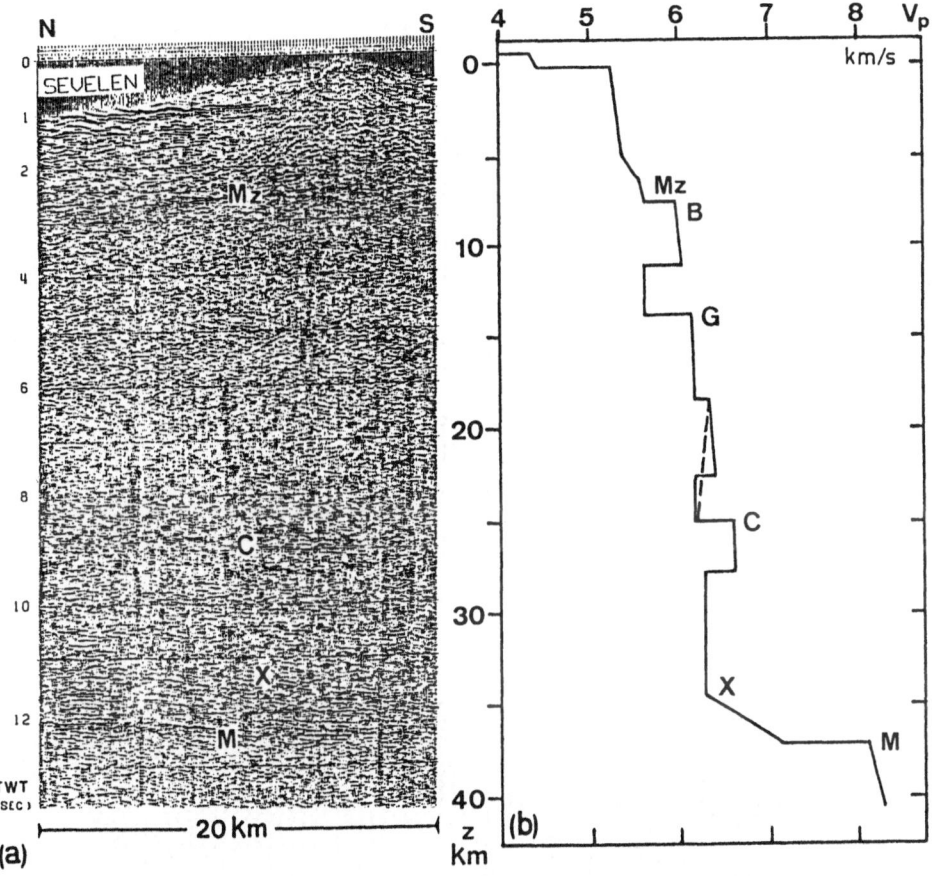

Figure 6. Seismic data for the northern part of profile segment 3 across the Central Alps in eastern Switzerland. (a) Reflection seismogram section recorded near Sevelen (shotpoint 4 in Figure 8), in the northern part of the Helvetic zone (after Schweizerische Arbeitsgruppe für Reflexionsseismik 1988). Four reflection bands (Mz, C, X, M) are marked. For a discussion see text. TWT = two-way time in seconds. (b) P-wave velocity-depth structure (z = depth) obtained near the shotpoint Sevelen from a reversed seismic refraction profile along strike in the northern border zone of the Swiss Alps. Mz = base of the Mesozoic sediments, B = top of crystalline basement, G = upper crustal low-velocity zone, C = mid-crustal low-velocity zone (dashed line = alternate solution), X = sharp gradient zone in the lower crust, M = crust-mantle transition zone (after Maurer 1989).

634

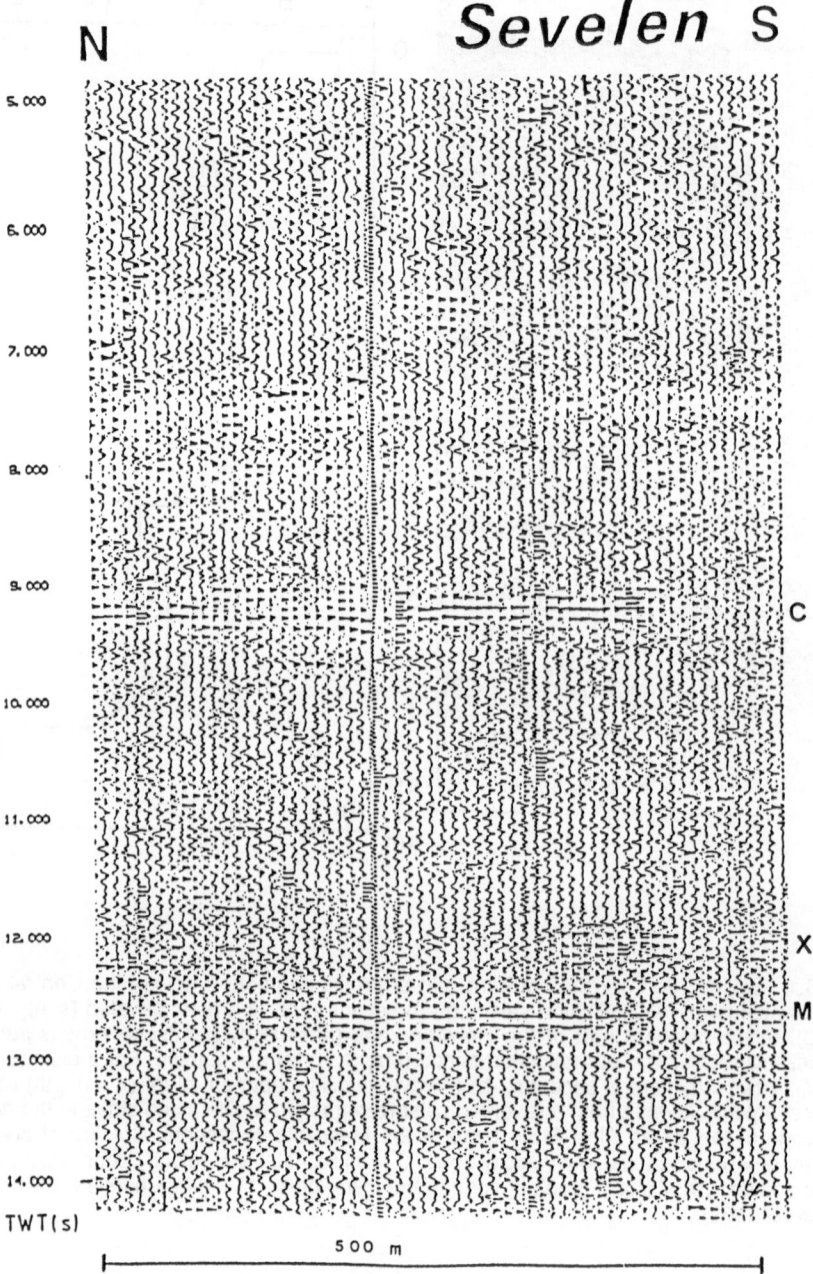

Figure 7. Enlarged reflection seismogram section (single traces only, 80 m apart) for a shot near Sevelen (shotpoint 4 in Figure 8). The C, X and M reflection bands can be clearly seen (cf. Figure 6a). For a detailed discussion see text (cf. Mueller 1989).

In September 1987 a seismic refraction experiment was carried out along the strike of the northern border zone of the Swiss Alps (Maurer 1989). The northeastern shotpoint of a 200 km long reversed profile was located in the Säntis nappe of the Helvetic domain. A P-wave velocity-depth function has been derived for the vicinity of the Säntis shotpoint (Fig. 6b) to be compared with the seismic reflection results nearby (Fig. 6a). The top of the basement (B) was found at a depth of 7 to 8 km with an average velocity of 6.0 km/s (Fig. 6b). An upper crustal low-velocity zone (G) with a velocity decrease of 6 per cent has to be postulated in the depth range from 11 to 14 km to satisfy the data. In the middle crust a second velocity inversion exists, which is less pronounced, but is of the same type as in the northern Alpine foreland (cf. Fig. 4c and Emter 1971). The "Conrad" transition zone (C) has an average velocity of 6.6 km/s, while the lowermost crust (between X and M) is characterized by a strong positive velocity gradient (6.3 to 7.2 km/s). A first-order discontinuity at a depth of 37 km corresponds to the crust-mantle boundary (M). The entire velocity-depth pattern as found in the Jura-Molasse region is translated to greater depths in the Alps with all the major features still preserved.

All the explosion-seismic reflection data for the traverse through eastern Switzerland (with 17 shotpoints) are presented in the preliminary form of hand-migrated line drawings in Fig. 8. In the northern part of the section the Triassic marker can clearly be seen. The middle and lower crustal section with sporadic "Conrad" reflections and the two "Moho" reflection bands is detached from the upper crust in the depth range between 12 and 22 km (cf. Fig. 6b). From the line-up of the reflections it is obvious that the delaminated lower crust dips towards the south and seems to have moved intact down into the deep-reaching subduction zone beneath the Alps (Mueller and Panza 1986; Mueller 1989). The upper boundary (C) of the lower crustal slab can be traced beyond the Rhine-Rhône-Line (R.R.L.) all the way beneath the Penninic nappes near Thusis (Fig. 8).

The same deep-reflection pattern that has been identified in the Helvetic and northern Penninic domains in eastern Switzerland (Fig. 8) has also been found in the region of the western transect through the Swiss Alps (cf. Schweizerische Arbeitsgruppe für Reflexionsseismik 1988). A corresponding signal pattern has also been shown for the French-Italian ECORS/CROP profile which roughly runs from Geneva in southwestern Switzerland to Torino in northern Italy (Bayer et al. 1987). There the "chaînes subalpines" and the Belledonne massif play the same role as the Helvetic nappes and the Aar massif (Fig. 9), and the Vanoise region is the exact counterpart of the northern Penninic domain in Switzerland.

When approaching the southern part of the Penninic domain some steeply dipping reflectors can be seen in Fig. 8. They should be compared to the reflecting elements at the northern end of the southern traverse (Fig. 10) where they apparently outline a wedge of lower crustal (or partly mantle?) material which penetrates into the lowermost Penninic crust. The flat reflectors in the upper crust must be attributed to the well-known pile of crystalline nappes (Suretta, Tambo, Adula, Simano and Leventina) in the central part of the Swiss Alps. The northern front of these Penninic nappes (P.F. in Fig. 8) can be traced close to the surface at the Rhine-Rhône-Line (R.R.L. in Fig. 9b).

Results obtained from commercial seismic reflection measurements exhibit a much more complex structure for the Mesozoic and Tertiary sediments in the Molasse basin and their interaction with the northern front of the Aar massif (cf. Figs. 8 and 9) than observed in previously published crustal cross sections. The pattern of deep reflections indicates that beneath the Helvetic nappes the upper crust has been sheared off near or at the base of the first ("sialic") low-velocity zone (at a two-way time of about 6 s as sketched in Fig. 9b). The compressional shortening of the lithosphere combined with dilatancy effects and the involvement of fluids with elevated pore pressures in the low-velocity zone have led to a detachment or "decoupling" of the upper crustal layers in such a way that by northward directed wedging they could be pushed upward in a flake-type manner (cf. Figs. 8 and 9b) as suggested earlier by Mueller et al. (1976, 1980).

It should be noted that under the central part of the Aar massif the northward tilted geologic units observed near the surface seem to be restricted only to the upper portion of an otherwise relatively "homogeneous" crust without pronounced velocity variations down to depths of approximately 15 to 20 km. The grade of metamorphism determined in that region of the Alps

636

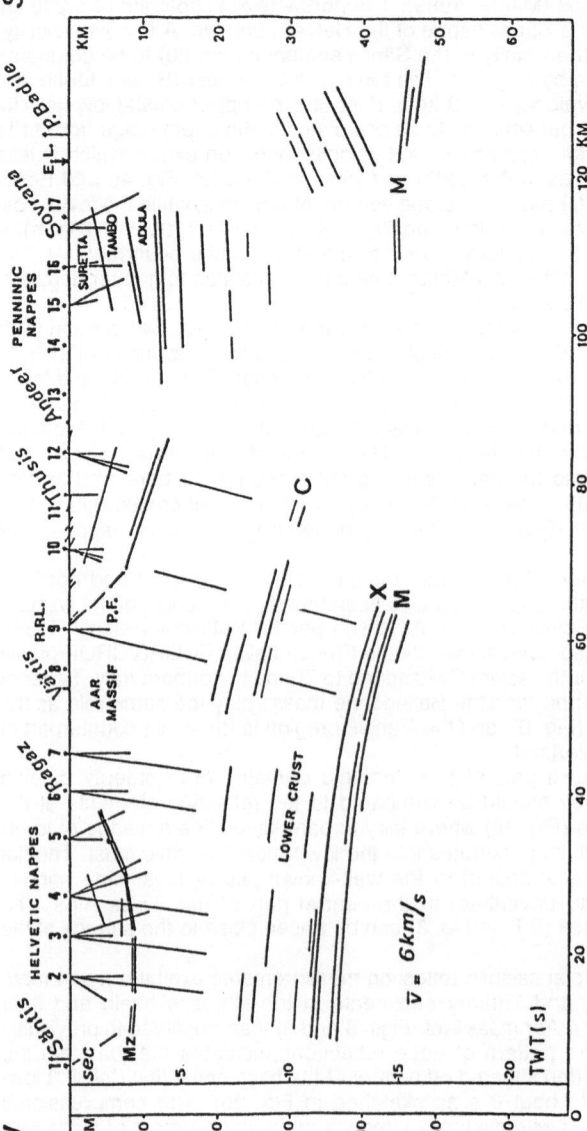

Figure 8. Simplified interpretation of the explosion-seismic reflection data for the profile segment 3 (cf. Figure 5) in eastern Switzerland. The most conspicuous reflection bands are represented by hand-migrated line drawings (after Finckh et al. 1987). Shotpoint 4 is near Sevelen (cf. Figures 6a and 7). In the north the middle and lower crust with the C, X and M reflectors appear to be "decoupled" from the upper crust and seem to move intact down into the steeply southward dipping subduction zone (cf. Mueller and Panza 1986). In the upper crust the Aar massif (cf. Figure 5) which is exposed in a "window" near Vättis separates the Helvetic domain with its distinct Triassic marker (Mz in Figure 6a) from the Penninic domain with the Suretta, Tambo and Adula nappes. The "Penninic Front" (P.F.) outcrops at the Rhine-Rhône Line (R.R.L.). The steeply northward dipping reflectors in the southernmost part of the section have probably not correctly been migrated. There is a distinct gap in the C, X and M reflection bands beneath the Penninic nappes, but when approaching the Engadine Line (E.L.) the M discontinuity appears again. For a discussion see text.

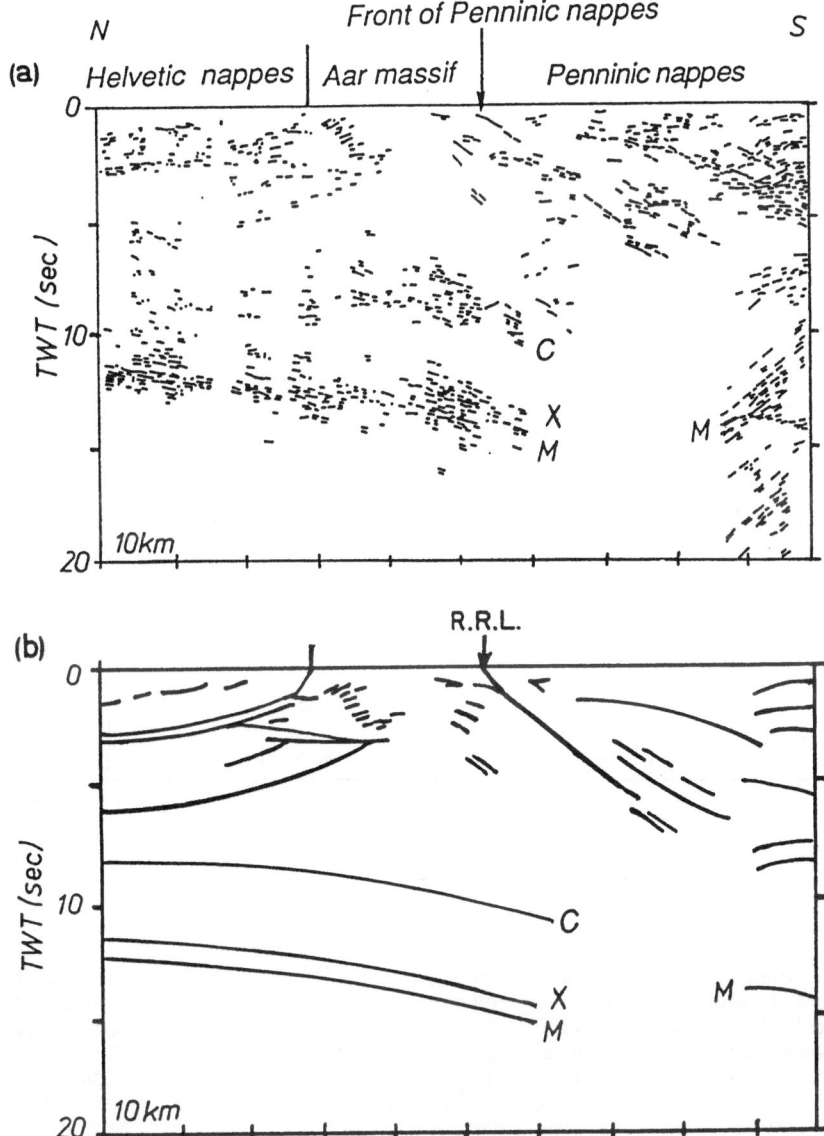

Figure 9. Explosion-seismic reflection data for profile segment 3 across the Central Alps in eastern Switzerland (cf. Figure 5). (a) Composite unmigrated line drawings derived from individual shot gathers of the reflection line NFP 20-East (after Pfiffner et al. 1988). C = Top of lower crust, X = sharp gradient zone in the lowermost crust, M = crust-mantle transition zone (cf. Figures 4c and 6b). (b) Preliminary schematic interpretation illustrating the "anticlinal flexure" in the Aar massif and the northward thrust of its northern front. The "Penninic Front" outcrops at the Rhine-Rhône Line (R.R.L.) near Tamins (shotpoint 9 in Figure 8). For a discussion see text.

638

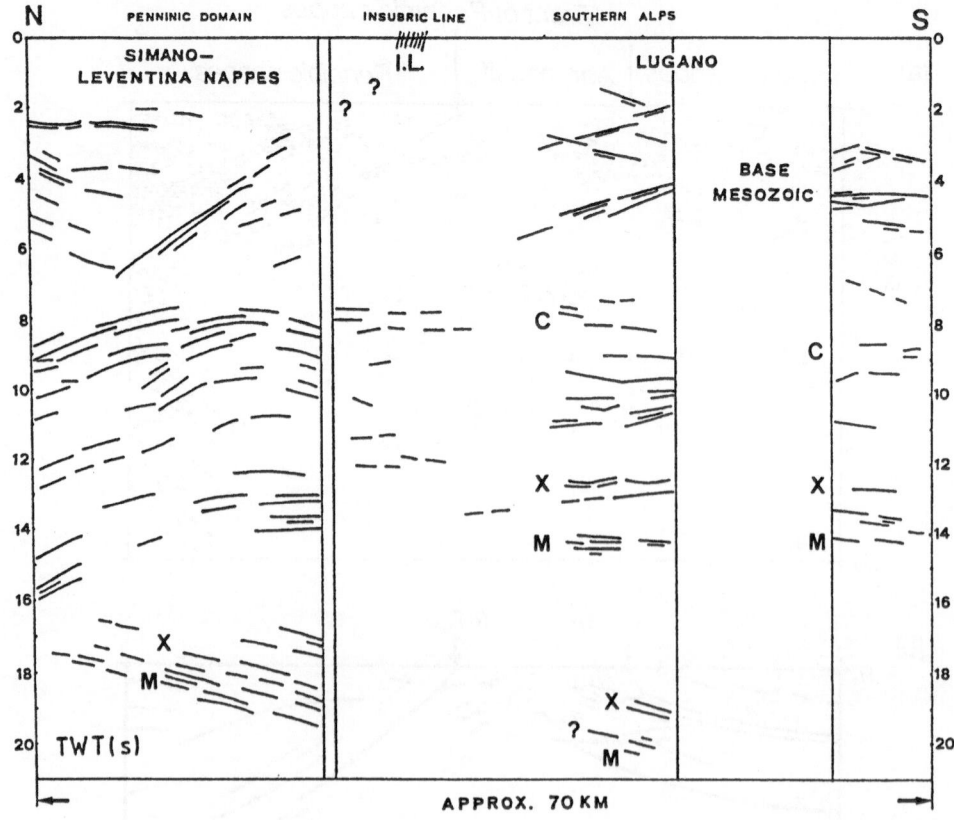

Figure 10.   Preliminary interpretation of the explosion-seismic reflection data for the profile segment 4 (cf. Figure 1) in southern Switzerland. The most conspicuous reflection bands are represented by line drawings (after Frei et al. 1989). In the north the uplifted and highly metamorphosed Simano and Leventina nappes of the Penninic domain are now exposed at the surface and separated from the Southern Alps (cf. Figure 5) by the northward dipping Insubric Line (I.L.). Below 8 seconds TWT it appears that the lower crust of the Adriatic promontory of the African plate protrudes wedge-like northward into the midlle crust of the Penninic domain (cf. also Figure 8). This indentation is most likely responsible for the axial culmination and the accompanying uplift of the Alps in this region (cf. Mueller 1982).

(Frey et al. 1980) suggests that the basement rocks of the exposed Aar massif - which incidentally are of the same type and age (Variscan) as in the Black Forest (cf. Fig. 2) - must once have resided at depths of about 15 km and have subsequently been uplifted by mid-crustal wedging. If the crustal "flexure" sketched in Fig. 9b is restored into its presumed original position the rock suites would fall into the appropriate depth range. Variants of the same type of structure with crustal wedges protruding both to the north and south are found beneath the Southern Alps, i.e. in the upper crust between the Insubric Line and the region of Lugano (cf. Fig. 10). Again the upper crust has been detached from the underlying middle and lower crust. Delaminated crustal slices wedging into the adjacent stack of crustal layers have very descriptively been termed "flake" tectonics (Oxburgh 1972) or "crocodile" tectonics.

Both the newly obtained data from the seismic reflection measurements in eastern Switzerland and new seismic refraction observations in the Aar massif are in agreement with the schematic model in Fig. 9b. This "delamination" model implies that the northwestern part of the Aar massif - consisting of the gneissic envelope and the granitic masses in the axial core - have been thrust northward thereby strongly deforming the Mesozoic autochthonous sediments on top. Additional evidence of the thrust character at the northern boundary of the Aar massif is also provided by the focal mechanisms of earthquakes in that region which contain a clear component of thrust motion.

The southern traverse (profile segment 4 in Figs. 1 and 5) crosses the Penninic zone, the Insubric Line and the Southern Alps in a north-south direction somewhat displaced to the east of the axial culmination between the eastern and western Alps. At this stage only the first preliminary results of the explosion seismic data of the southern traverse are available (Frei et al. 1989). The structural units exposed at the surface in the Penninic domain (Fig. 10), such as the Simano-Leventina gneiss nappes, lie at a depth of about 8 to 10 km and reach a depth of approximately 15 km under segment 3 to the east (cf. Fig. 8).

The structural configuration of the southern traverse is dominated by a major crustal indentation (Frei et al. 1989). The lower crust of the Adriatic promontory of the African plate protrudes wedge-like northward into the middle and lower crust of the Penninic domain (Fig. 10).The resulting vertical "escape" of the upper crust is concordant with the results of repeated first-order precise levellings for different periods of time (cf. Mueller 1982). The observed uplift during the past 120 years shows a distinct maximum in the region of the Penninic nappes north of the Insubric Line.

Beneath the Penninic nappes the "Moho" zone (X-M) of the descending European plate is found at 16 to 19 s TWT, i.e. at depths between 48 and 58 km (Fig. 10). The depth of the crust-mantle boundary (M) below the Penninic domain and at the southern end of the section is in very good agreement with earlier seismic refraction work (Mueller et al. 1980). A rather complex structure has been found for the upper crust of the Southern Alps south of the Insubric Line (I.L.) which appears to be detached at a depth of about 20 to 25 km (7 to 8 s TWT). There basement slices are thrust southward into an approximately 10 km thick wedge of Mesozoic sediments whose base dips northward towards the Insubric fault zone (cf. Laubscher 1985; Heitzmann 1987). In this model the pile of basement thrust slices is resting on top of the apparently undisturbed Adriatic basement beneath the Southern Alps.

CONCLUSIONS

The seismic reflection and refraction data presented in this paper and their synoptic interpretation emphasize the significant role of intracrustal low-velocity zones as depth ranges of lowered strength which - in compressional regimes -facilitates detachment followed by wedging of more competent layers into these zones of weakness. Large protruding crustal wedges apparently control the uplift of individual "finger-like" regions in the collision zones and thus may provide an explanation for the axial culminations in the Alps. The observed focal depths of earthquakes suggest that the main seismic activity is concentrated close to the bounds of low-velocity zones, i.e. above or below depth intervals of lowered mechanical strength, as is to be suspected.

REFERENCES

Bayer, R., Cazes, M., Dal Piaz, G.V., Damotte, B., Elter, G., Gosso, G., Hirn, A., Lanza, R., Lombardo, B., Mugnier, J.-L., Nicolas, A., Nicolich, R., Polino, R., Roure, F., Sacchi, R., Scarascia, S., Tabacco, I., Tapponier, P., Tardy, M., Taylor, M., Thouvenot, F., Toreilles, G. and Villien, A., 1987. Premiers résultats de la traversée des Alpes occidentales par sismique reflexion verticale (Programme ECORS-CROP). Compte Rendu de l'Academie des Siences de Paris, 305, II: 1461-1470.

Bonjer, K.-P., Gelbke, C., Gilg, B., Rouland, D., Mayer-Rosa, D. and Massinon, B., 1984. Seismicity and dynamics of the Upper Rhinegraben. J. Geophys., 55: 1-12.

Brace, W.F., 1972. Pore pressure in geophysics. In: Heard, H.C., Borg, I.Y., Carter, N.L. and Raleigh, C.B. (eds.), Flow and Fracture of Rocks, Amer. Geophys. Union (Washington), Geophys. Monograph, 16: 265-273.

Deichmann, N., 1987. Focal depths of earthquakes in northern Switzerland. Annales Geophysicae, 5B: 395-402.

Deichmann, N. and Ansorge, J., 1983. Evidence for lamination in the lower continental crust beneath the Black Forest (Southwestern Germany). J. Geophys., 52: 109-118.

Edel, J.B., Fuchs, K., Gelbke, C. and Prodehl, C., 1975. Deep structure of the southern Rhinegraben area from seismic refraction investigations. J. Geophys., 41: 333-356.

Emter, D., 1971. Ergebnisse seismischer Untersuchungen der Erdkruste und des obersten Erdmantels in Südwestdeutschland. Ph.D. Thesis, University of Stuttgart (F.R. Germany), 108 pp.

Finckh, P., Ansorge, J., Mueller, St. and Sprecher, Ch., 1984. Deep crustal reflections from a Vibroseis survey in northern Switzerland. Tectonophysics, 109: 1-14.

Finckh, P., Frei, W., Fuller, W., Johnson, R., Mueller, St., Smithson, S. and Sprecher, Ch., 1986. Detailed crustal structure from a seismic reflection survey in northern Switzerland. In: Barazangi, M. and Brown, L. (eds.), Reflection Seismology: A Global Perspective, Amer. Geophys. Union (Washington), Geodyn. Ser., 13: 43-54.

Finckh, P., Frei, W., Freeman, R., Heitzmann, P., Lehner, P., Mueller, St., Pfiffner, A. and Valasek, P., 1987. Nationales Forschungsprogramm 20 "Geologische Tiefenstruktur der Schweiz" - Problemstellung und erste Resultate. Bull. Ver. schweiz. Petroleum-Geol. u. -Ing., 53: 59-74.

Frei, W., Heitzmann, P., Lehner, P., Mueller, St., Olivier, R., Pfiffner, A., Steck, A. and Valasek, P., 1989. Geotraverses across the Swiss Alps. Nature, 340: 544-548.

Frey, M., Bucher, K., Frank, E. and Mullis, J., 1980. Alpine metamorphism along the Geotraverse Basel-Chiasso: a review. *Eclogae geol. Helv.*, **73**: 527-546.

Gajewski, D. and Prodehl, C., 1985. Crustal structure beneath the Swabian Jura, SW-Germany, from seismic refraction investigations. *J. Geophys.*, **56**: 69-80.

Gajewski, D. and Prodehl, C., 1987. Seismic refraction investigation of the Black Forest. *Tectonophysics*, **142**: 27-48.

Gajewski, D., Holbrook, W.S. and Prodehl, C., 1987. A three-dimensional crustal model of southwest Germany derived from seismic refraction data. *Tectonophysics*, **142**: 49-70.

Heitzmann, P., 1987. Evidence of late Oligocene/Early Miocene backthrusting in the central Alpine "root zone". *Geodinamica Acta* (Paris), **1**: 183-192.

Klemperer, S.L., Hauge, T.A., Hauser, E.C., Oliver, J.E. and Potter, C.J., 1986. The Moho in the northern Basin and Range province, Nevada, along the COCORP 40° N seismic-reflection transect. *Geol. Soc. Amer. Bull.*, **97**: 603-618.

Laubscher, H., 1985. Large-scale, thin-skinned thrusting in the Southern Alps: Kinematic models. *Geol. Soc. Amer. Bull.*, **96**: 710-718.

LOTEM Working Group: Schwarzwald, LOTEM-Tiefensondierung. Abstract of Poster Program, 2nd KTB-Colloquium, 19-21 September 1986, Seeheim (F.R.G.), p. 48.

Lüschen, E., Wenzel, F., Sandmeier, K.-J., Menges, D., Rühl, Th., Stiller, M., Janoth, W., Keller, F., Söllner, W., Thomas, R., Krohe, A., Stenger, R., Fuchs, K., Wilhelm, H. and Eisbacher, G., 1987. Near-vertical and wide-angle seismic surveys in the Black Forest, SW Germany. *J. Geophys.*, **62**: 1-30.

Maurer, H.-R., 1989. Die Struktur der Erdkruste unter dem schweizerischen Alpennordrand. Diploma Thesis, Institute of Geophysics, ETH Zürich (Switzerland), 108 pp. (with additional appendices).

Miller, H., Mueller, St. and Perrier, G., 1982. Structure and dynamics of the Alps - a geophysicsl inventory. In: Berckhemer, H. and Hsü, K.J. (eds.), *Alpine-Mediterranean Geodynamics*, Amer. Geophys. Union (Washington), Geodyn. Ser., **7**: 175-203.

Mueller, St., 1977. A new model of the continental crust. In: Heacock, J.G. (ed.), *The Earth's Crust*, Amer. Geophys. Union (Washington), Geophys. Monograph, **20**: 289-317.

Mueller, St., 1982. Deep structure and recent dynamics in the Alps. In: Hsü, K.J. (ed.), *Mountain Building Processes*, Academic Press (London), 189-199.

Mueller, St., 1989. Deep-reaching geodynamic processes in the Alps. In: Coward, M.P., Dietrich, D. and Park, R.G. (eds.), *Alpine Tectonics*, Geol. Soc. (London), Spec. Publ., **45**: 303-328.

Mueller, St. and Panza, G.F., 1986. Evidence of a deep-reaching lithospheric root under the Alpine arc. In: Wezel, F.-C. (ed.), *The Origin of Arcs*, Developments in Geotectonics (Elsevier, Amsterdam), **21**: 93-113.

Mueller, St., Peterschmitt, E., Fuchs, K. and Ansorge, J., 1969. Crustal structure beneath the Rhinegraben from seismic refraction and reflection measurements. *Tectonophysics*, **8**: 529-542.

Mueller, St., Peterschmitt, E., Fuchs, K., Emter, D. and Ansorge, J., 1973. Crustal structure of the Rhinegraben area. *Tectonophysics*, **20**: 381-392.

Mueller, St., Egloff, R. and Ansorge, J., 1976. Struktur des tieferen Untergrundes entlang der Schweizer Geotraverse. *Schweiz. mineral. petrogr. Mitt.*, **56**: 685-692.

Mueller, St., Ansorge, J., Egloff, R. and Kissling, E., 1980. A crustal cross section along the Swiss Geotraverse from the Rhinegraben to the Po plain. *Eclogae geol. Helv.*, **73**: 463-485.

Mueller, St., Ansorge, J., Sierro, N., Finckh, P. and Emter, D., 1987. Synoptic interpretation of seismic reflection and refraction data. *Geophys. J.R. astr. Soc.*, **89**: 345-352.

Oxburgh, E.R., 1972. Flake tectonics and continental collision. *Nature*, **239**: 202-204.

Pfiffner, O.A., Frei, W., Finckh, P. and Valasek, P., 1988. Deep seismic reflection profiling in the Swiss Alps: Explosion seismology results for line NFP 20-East. *Geology*, **16**: 987-990.

Prodehl, C., Ansorge, J., Edel, J.B., Emter, D., Fuchs, K., Mueller, St. and Peterschmitt, E., 1976. Explosion seismology research in the central and southern Rhinegraben - a case history. In: Giese, P., Prodehl, C. and Stein, A. (eds.), *Explosion Seismology in Central Europe - Data and Results*. Springer-Verlag (Berlin-Heidelberg-New York), 313-328.

Rothé, J.-P. and Peterschmitt, E., 1950. Etude séismique des explosions d'Haslach. *Annales de l'Institut de Physique du Globe de Strasbourg*, **5** (3me partie: Géophysique): 3-28.

Rybach, L., Mueller, St., Milnes, A., Ansorge, J., Bernoulli, D. and Frey, M., 1980. The Swiss Geotraverse Basel-Chiasso: a review. *Eclogae geol. Helv.*, **73**: 437-462.

Sandmeier, K.-J. and Wenzel, F., 1986. Synthetic seismograms for a complex model. *Geophys. Res. Lett.*, **13**: 22-25.

Schnegg, P.-A. and Fischer, G., 1989. Magnetotelluric soundings in western Switzerland. In: Freeman, R. and Mueller, St. (eds.), *Proc. 6th Workshop on the European Geotraverse Project (EGT)*, European Science Foundation (Strasbourg), 323-332.

Schweizerische Arbeitsgruppe für Reflexionsseismik, 1988. Erste Ergebnisse der Alpentraversen von NFP-20 "Geologische Tiefenstruktur der Schweiz". *Vierteljahrsschrift der Naturforschenden Gesellschaft in Zürich*, **133**: 61-98.

Sierro, N.A.M.J., 1988. Regionale Struktur der Erdkruste in der Nordschweiz. Ph.D. Thesis No. 8712, ETH Zürich (Switzerland), 205 pp. (with additional appendices).

Sierro, N., Bindschädler, A., Ansorge, J. and Mueller, St., 1983. Geophysikalisches Untersuchungsprogramm Nordschweiz: Regionale refraktionsseismische Messungen 81/82. NAGRA (Nationale Genossenschaft für die Lagerung radioaktiver Abfälle), Baden (Switzerland), Technischer Bericht 83-21, 59 pp.

Tezkan, B., 1988. Electromagnetic sounding experiments in the Schwarzwald central gneiss massiv. *J. Geophys.*, **62**: 109-118.

Trümpy, R., 1980. Geology of Switzerland - a guide book. Part A: An outline of the ‚geology of Switzerland. Wepf & Co., Publishers (Basel-New York), 104 pp.

Zucca, J.J., 1984. The crustal structure of the southern Rhinegraben from re-interpretation of seismic refraction data. *J. Geophys.*, **55**: 13-22.

STRATEGY FOR EXPLORATION OF THE BURIED CONTINENTAL CRUST

Jack Oliver
Institute for the Study of the Continents
Cornell University
Ithaca, NY 14853

1. EXPLORING THE THIRD DIMENSION OF CONTINENTAL GEOLOGY

In earth science, we are currently in the era of exploration of the
third dimension of continental geology -- from the surface of the earth
to the very base of the crust at depths of some 40 km. This part of
the earth is a prime frontier of modern earth science for many reasons,
including the following:

(1) The continental crust holds the only record of most
of earth history.

(2) The nature and complexity of rocks at and near the
surface show unambiguously that the continental
crust has had a rich and varied history.

(3) Understanding of the geology of the continents has
practical implications of enormous importance to
society.

(4) The great success of earth science in developing
the concept of plate tectonics through study of the
previously poorly-known ocean basins suggests that
improved understanding of the entire continental
crust will provide yet another major advance in
earth science.

(5) Powerful and proven methods and tools for exploring
the continental crust exist yet must still be
applied comprehensively. We need to observe and
study the entire earth, not just some supposedly
"typical" parts of it.

(6) From the history of earth science and the history
of human exploration of the earth it seems obvious
that the continental crust is next in line as the
leading frontier of earth exploration.

To state the above for those who work in this field is to state
the obvious. The setting seems appropriate, however, for the question
that is at the heart of this paper, that is "How can we learn about the
third dimension of continental crustal geology in the most effective

645

M. H. Salisbury and D. M. Fountain (eds.), Exposed Cross-Sections of the Continental Crust, 645–651.
© 1990 Kluwer Academic Publishers.

manner?"   Some likely answers to that question are the core of what
follows.

First, note that "in the most effective manner" is a key phrase in
the question.   Times are changing in science.   Transportation has
improved to the point where almost all parts of the earth's surface are
open to direct scientific observation.   A striking array of instruments
and techniques is available for making observations of outstanding
quality, large quantity and great diversity.   Methods of analyzing and
handling large quantities of observations are growing very rapidly.

In other words, science has expanded its capability and potential
enormously.   In fact, it has already come to the point where it is no
longer possible to hope that all "good" science that is proposed can be
funded.   Instead, there must be some selection among the good projects,
selection because of the potential effectiveness of a particular
project for advancing the science and for providing practical benefit
to society.   Science of the modern era, in other words, must be done in
the most effective manner.

The third dimension of geology has long drawn the attention of
earth scientists.   Road cuts, topographic relief, tunnels, structural
relief and shallow drilling have provided some information on the third
dimension of rocks at shallow depths.   Geophysical methods - for
example, gravitational, magnetic, electrical and certain kind of
seismological methods, all generally of modest resolution - as well as
various geochemical methods have provided valuable information.   And
there is much yet to be gained through such efforts.   It is only within
the last year, for example, that we have had readily available gravity
and magnetic maps of the entire North American continent and even these
need refinement and augmentation.

The last decade and a half has seen the coming of age of two
techniques, both already well tested at shallow depth, for study of the
deeper crust.   They are (1) seismic reflection profiling and (2) deep
drilling for scientific purposes.   Coupled closely to the deep drilling
story is the subject of exposed high grade terranes, the focus of this
conference.   Most of the remainder of this article is devoted to these
two key subjects.

2.   STUDYING THE DEEP CRUST BY SEISMIC REFLECTION PROFILING

Let us begin with deep seismic reflection profiling.   Although a number
of spot samples of deep crustal reflectivity were obtained earlier in
the U.S., Canada, Europe and elsewhere, it might be said that the
present era of deep seismic reflection profiling of the crust began
with the COCORP survey in Hardeman County, Texas in 1975.   The longest
line of that survey was 17 km, minuscule by today's standards, but the
style quickly changed.   Soon COCORP was running profiles that sampled
the entire crust and were hundreds of kilometers in length.   The
effectiveness of the method was demonstrated readily as buried major
features of the crust were revealed and major advances in geological
understanding was achieved.   Others quickly joined in, and surveyed new
and different geologic features.   Activity snowballed.   At present,

over 20 countries have done some deep seismic profiling of the
continental crust and the surveying of over 30,000 km of seismic line
has been completed. Soon the equivalent of one circumference of the
earth will have been surveyed. That is impressive, particularly if one
remembers that much of this work was done using the Vibroseis technique
and hence that trucks have injected energy into the ground at points
spaced every few meters along this very large distance. The most
common alternative source is the air gun which may be used where
shallow seas overly continental crust. In some cases, explosives
provide the source of seismic energy. In any case, the effort is a
large one for basic earth science, though not nearly so large as
commercial exploration for hydrocarbons.

The reason for the rapidly expanding scientific effort is simple
enough. It is that a new, detailed, important kind of basic
information on the earth is obtained as a consequence of application of
the method that was originally developed by industry for hydrocarbon
exploration in sedimentary basins to study of the entire crust.

3. COMBINING SEISMIC DATA WITH SURFACE GEOLOGY

An important point is that the new kind of information, although
obtained by a geophysical technique, is much more geological in nature
than is the product of most subsurface exploration techniques.
Reflecting horizons are found which can sometimes be traced for long
distances, sometimes from the surface to great depths. Sometimes the
horizons are configured to indicate tectonic deformation, sometimes
they truncate older horizons to indicate relative ages. In short, an
entire spectrum of geologic reasoning can be applied to the
observations of seismic reflection profiling and the seismic
information can readily be coupled to surface geology as well. One
might say that a principal consequence of seismic profiling is the
extension of not only surface geology, but also surface geological
reasoning to great depths in the crust.

The results of such surveying are so numerous and varied that only
some can be mentioned here. They include demonstration that the
present configuration of the Moho is younger than past orogenic
activity, the demonstration of thin skinned thrusting of large scale in
orogenic belts and the discovery of buried Precambrian basins and
orogenic belts. They reveal great crust-penetrating thrusts, rifts and
rift basins, peculiar isolated strong reflectors that are possibly mid-
crustal fluid bodies, and a host of other features.

A new perspective of the deep crust is evolving from a combination
of surface geological information and seismic reflection profiling. It
is a crust decidedly different from the simple layered models of the
crust that were derived from early studies of earthquake-generated
seismic waves. It is a crust made of igneous, sedimentary and
metamorphic rocks of great variety, substantial deformation, and
complex history, all bounded below by an enigmatic, dynamic,
oversimplified feature named the Moho.

4.   LUCKY BREAKS, SURPRISING RESULTS AND GREAT POTENTIAL

But amazingly and to our good fortune, although the crust is complex it is not so hopelessly complex that we cannot unravel its history. One could easily imagine a fictitious earth 4 1/2 billion years old that has undergone deformation so frequently that it would be impossible to reconstruct its history. It has not turned out that way. In the case of the real earth, as geologists learned from surface studies, the number of orogenies affecting a given part of the crust is commonly more than one, but not usually more than a few.

A similar kind of good fortune seems to have struck in the case of seismic reflection studies of the deep crust. And this is an important point that rarely receives explicit attention. When deep profiling began, many felt intuitively that the basement rocks of the continents were far too complex to produce seismic information of much value in geological interpretation. Seismic wave propagation in a contorted structure can be hopelessly complex and it might have turned out that all information on the basement obtained by seismic reflection profiling was beyond geologic interpretation.

As this subject developed, however, it become clear that an important fraction of the information obtained is, in fact, surprisingly simple, comprehensible, and revealing of earth history. A part of the information continues to be befuddling, of course, but a growing and very substantial portion of it is coming under control. Just within the last few years it has become obvious that deep seismic profiling is revealing simple basic features of the crust, not just within a single profile but on many profiles, in such a way that features of large areal extent, i.e. regional scale, can be recognized and the crust better understood as a consequence of these data.

Let us note a few of these features as examples. At present, there are at least a dozen deep seismic reflection profiles that cross the Appalachian orogen in North America or its Caledondian or Hercynian extensions across the Atlantic. Each profile is distinctive in some way of course, but many of these profiles show common features, such as the decollement, the crust-penetrating zone of deformation, post-orogenic extensional basins and spatial consistency in Moho reflections. I do not wish to minimize the differences from one profile to the next for they occur and are surely informative, but those differences should not obscure a degree of consistency among adjoining profiles and hence a certain simplicity to the continent at this scale. This simplicity not only facilitates study of this orogen but also bodes well for similar seismic studies of other as-yet-unsurveyed major tectonic features.

A particularly encouraging and exciting result was recently revealed through COCORP surveying along two long traverses in the eastern half of the United States and the GLIMPCE surveying in the Great Lakes near the U.S.-Canadian border. From these data it seems very likely that major features such as the east-dipping Grenville Front and a major west-dipping crust-penetrating feature to the east of the Front can be correlated from one seismic profile to the next over distances of at least 1000 km. If this interpretation is correct, it means that deep seismic profiling will reveal the third dimension of the basic building blocks of the continents, and will provide information or how those blocks came to their present status and configuration. A modest increment of seismic profiling will verify and supplement this conclusion. The story that evolves will surely be an important fundamental part of continental tectonics and continental geology, an exciting prospect.

A third example of seismic information of relative simplicity and concerning rather large features of the crust has to do with the so-called "layered lower crust". Seismic data in certain areas shows a pattern of layered reflections in the lower half to one-third of the crust, often with low reflectivity above and below. Some of the earliest reflection studies revealed the layered lower crust and for a time some felt it was a universal characteristic of the lower continental crust. That assumption turned out to be false. The layered lower crust is not everywhere but it does occur commonly on a regional basis. Most of the area surveyed by BIRPS around Great Britain displays this feature, for example.

Similar layering is also observed in parts of Europe, in parts of Australia, and in the Basin and Range Province of the U.S. The interpretation of the layering is controversial but its existence in volumes of regional scale seems certain.

A fourth example of a feature of large areal extent concerns the reflective character of the Moho. When only a small quantity of deep seismic reflection data was available, the occurrence or nonoccurrence of a Moho reflection in the data was often noticed, but the observation lacked impact because of multiple possible explanations. For example, was absence of a Moho reflection evidence for absence of a Moho reflector, or merely the consequence of weak signal penetration in that area? Now, with much greater data coverage, it is becoming obvious that, although other factors still play a role in specific cases, spatial variation in Moho reflectivity is decidely real. In some areas there is a strong reflector at Moho depths and in others there is not. Furthermore, the spatial pattern of occurrences or reflectors is emerging, and certain relationships of the pattern to tectonic history are being proposed. Some claim, for example, that strong Moho reflections are observed in areas where extension is the most recent mode of deformation.

Although other examples might be cited, these four should suffice to make the case that deep seismic reflection data are beginning to reveal some relatively simple features of the crust of regional scale, in addition, of course, to other more detailed information.

This conclusion is one of great promise and it virtually guarantees that truly major discovery about the geology of the continents will result as deep seismic reflection profiling of the continents proceeds.

5.   SOME COMMENTS ON DRILLING AND CHOOSING DRILLING TARGETS

This conclusion also has relevance to the subject of deep drilling for
scientific purposes.   Drilling is a scientific tool with unique
potential; it surely must be an important component of exploration of
the buried continental crust.   Drilling, like the surface exposures of
portions of formerly deep crust, makes possible the actual sampling of
rocks that are only sensed by other methods.

       Drilling, however, has two major disadvantages: (1) it is very
expensive and (2) it samples only a very limited portion of the earth.
It is, therefore, absolutely essential that deep drilling sites be
chosen carefully and at places where information of great significance
to crustal studies can be obtained from the small sample.   Just how to
judge the significance of a particular site or target is a difficult
matter, however.   Controversy often arises.

       In my opinion, there is one almost fool proof way of recognizing
that a proposed target is significant.   It has to do with the volume of
material from which the sample is to be obtained.   In the volume of
unknown rocks is huge, say of crustal, or better yet, regional scale,
and the entire volume of those rocks can be identified for the first
time by the drill, then major significance to the science is
guaranteed.

       For this reason, I think the highest priority drilling targets in
a program of deep drilling for scientific purposes should be the great
unknown features of large volume revealed by reflection profiling.   The
layered lower crust is the prime example.   It is so large and forms so
great a fraction of the continent that it must be of fundamental
significance.   Drilling it at one place might provide identification of
rocks at all places where such reflections are observed.

       There are some difficulties with this suggestion.   One is that, in
most places, the layered lower crust is too deep to drill.   In order
for this feature to be a target, it would be necessary to find, by
seismic profiling, a place where the layered, normally-lower, crust
unequivocally rises sufficiently near to the surface so that it can be
be reached by the drill.   Futhermore an optimum place, i.e. where a
particular feature is the shallowest, should be sought in order to
minimize drilling costs.   Therefore, it is essential that, not merely a
profile, but a reasonably comprehensive seismic reflection survey be
carried out prior to selection of a site and prior to commencement of
drilling.   To identify the rocks of the layered lower crust a
reasonable understanding of the spatial extent of such rocks must be
achieved.   The seismic survey should produce a comprehensive
determination of the configuration of the volume of layered lower crust
rocks.

       A second possible difficulty is that such seismic profiling, and
related geologic studies, may reveal that the layered lower crust
outcrops at places such as those discussed at this conference.   If so,
deep drilling for identification would be largely unnecessary and
perhaps wasteful.   The surface, in fact, could be thought of as the
ultimate drill hole.

And of course, it is possible to define other drilling targets based on some of the other examples given above. In all cases, however, sound decisions on drilling sites would require more seismic profiling than currently exists.

Finally, it is important to note and remember that the goal of our collective endeavors is to explore and understand the earth, not to apply this technique or that technique to study of the earth. So long as we keep that prime goal before us, decisions as to what we do next will be easier and our activities far more effective than will be the case if we see our efforts only through the eyes of the specialist in a particular technique.

EXPOSED CROSS SECTIONS OF THE CONTINENTAL CRUST - SYNOPSIS

David M. Fountain
Department of Geology and Geophysics, University of Wyoming
Laramie, Wyoming 82071 U.S.A.

John Percival
Geological Survey of Canada, 588 Booth Street
Ottawa, Ontario K1A OE4 Canada

Matthew H. Salisbury
Atlantic Geoscience Centre, Geological Survey of Canada
Bedford Institute of Oceanography,
Dartmouth, Nova Scotia B2Y 4A2 Canada

ABSTRACT. Exposed cross sections of the continental crust are recognized on the basis of geophysical and geological data. In some cases, high-grade metamorphic portions of the sections can be traced into contemporary lower crust by geophysical methods. Geobarometric, structural and geophysical data, however, indicate that the deepest crustal levels are not exposed. Exposed cross sections can be emplaced by thrust faults in collisional or intracratonic settings, in transpressional uplifts, extensional regimes and impactogens. Continental crustal evolution, as portrayed by these sections, involves addition of magmas in continental magmatic arcs or island arcs and incorporation of supracrustal rocks into the lower crust. The role of magmatic underplating in crustal growth is not resolved by these sections. Geophysical models of the crust based on measurement of physical properties of rocks from exposed cross sections provide important constraints on the interpretation of crustal geophysical data.

1. Introduction

Important constraints on the nature of the continental crust come from investigation of terrains scattered around the world regarded as intact or nearly intact cross sections of the crust (Fountain and Salisbury, 1981). Recently, advances in disciplines such as crustal seismology, isotope geology and geobarometry have provided many new and important insights into these sections, and numerous other terrains have been discovered. In 1988, a NATO Advanced Studies Institute focused on these developments with the goals of evaluating the various proposed crustal cross sections, reviewing the insights these provide concerning crustal structure, composition and evolution, and exploring the advantages and limitations of the study of exposed cross sections.

M. H. Salisbury and D. M. Fountain (eds.), Exposed Cross-Sections of the Continental Crust, 653–662.

2. Historical Background

Just a little more than 20 years ago, a collection of papers was published in *Schweizerische Mineralogische und Petrographische Mitteilungen* (1968, v. 48, no. 1) in which various geophysical and geological data were marshalled to advance the hypothesis that the crust-mantle boundary was uplifted to the surface and exposed in the Ivrea-Verbano zone (IVZ) in northern Italy. Central to the argument were gravity and seismological data that were interpreted to show that material with high density and high seismic velocity at or near the surface was connected with the high velocity, high density lower crust under the Po Basin (e.g., Giese, 1968; Berckhemer, 1969). Laboratory investigations subsequently demonstrated that metamorphic rocks exposed at the surface in the Ivrea zone equilibrated at lower crustal depths and that these rocks had physical properties similar to the deep crust under the Po Basin (Schmid and Wood, 1976; Fountain, 1976). This confluence of geophysical and geological data led to the notion that a cross section of the continental crust was exposed in the South Alpine region; granulite facies rocks of the Ivrea zone represent the base and adjacent lower grade rocks in the Strona-Ceneri zone to the east represent higher levels.

For several years the Ivrea zone stood as the lone example of an exposed cross section of the continental crust. But the similarity of the shape and magnitude of the pronounced Ivrea zone Bouguer gravity anomaly to other Bouguer anomaly patterns around the world suggested that other such terrains could be found. For instance, Brooks (1970) compared the Ivrea anomaly to a similar anomaly over the Lofoten-Vesteralen region in Norway and speculated that this region also exposed the lower portions of the continental crust. Forman and Shaw (1973), in a perceptive but neglected monograph, interpreted the large gravity anomalies and pattern of metamorphism in the Musgrave and Arunta blocks in central Australia in much the same way. Gibb and Thomas (1976) realized that continental sutures were commonly characterized by paired high and low Bouguer gravity anomalies similar to the Ivrea anomaly.

Salisbury and Fountain (1974) brought these observations into a cohesive picture that relied on a variety of geological and geophysical data, including gravity, to prospect for exposed cross sections of the continental crust. In their expanded and amplified work (Fountain and Salisbury, 1981) they identified five terrains (Ivrea zone, Fraser Range, Musgrave Range, Pikwitonei domain, Kasila series) that satisfied these criteria and used these exposed cross sections to explore the broader issues of the composition, structure and evolution of the continental crust.

Since 1981, many other terrains have been added to the list of possible cross sections of the continental crust including the Kapuskasing uplift (KU), Doubtful Sound, Calabria and Prince Rupert. Modern geobarometric techniques have refined estimates of the depth levels from which these terrains were exhumed (e.g., Percival, 1983; Mezger et al., 1989) and seismic methods have been successful, in some cases, in linking the surface exposures to deeper crustal levels (e.g., Boland et al., 1988; Boland and Ellis, 1989; Percival et al., 1989).

3. Recognition of exposed cross sections of the continental crust

Fountain and Salisbury (1981) emphasized investigation of exposed sections where various geophysical surveys confirm the physical connection of high-grade rocks near or at the surface with deeper crustal levels (Fig. 1). As indicated previously, geophysical data gathered in the 1960s provided good evidence that the deeper crust of the Po Basin was elevated to near-surface levels under the Ivrea zone. The recent ECORS-CROP reflection data (Bayer et al., 1987), however, failed to confirm this geometry. Refraction and reflection data over the Kapuskasing structural zone provide compelling evidence that high-grade rocks of the KU can be traced to depths of about 20 to 25 km beneath the Michipicoten greenstone belt to the west. Geophysical data collected over the Arunta block in central Australia (e.g., Goleby et al., 1989; Glikson et al., this volume) and Doubtful Sound in New Zealand (Oliver and Coggon, 1979; Priestley and Davey, 1983; Oliver, this volume) suggest

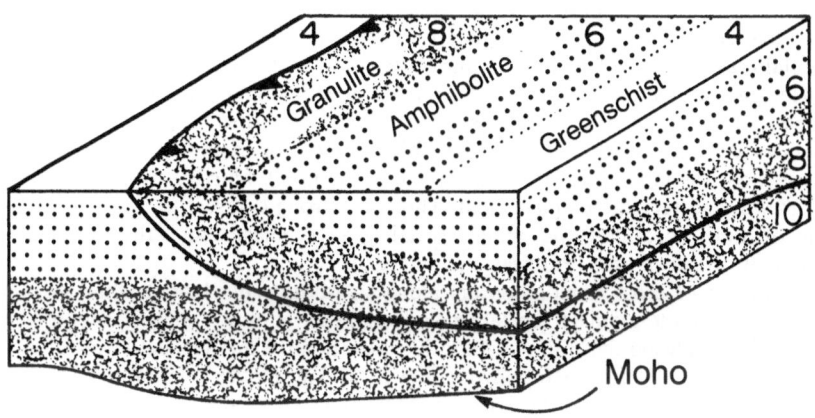

Figure 1. Characteristics of exposed crustal cross sections. Depth to Moho is about 30-50 km; lateral extent of oblique cross section is 20-150 km. Right-hand and top scales are in paleopressure (in kilobars). (from Percival, 1988).

that this type of geometry may exist in these situations as well. Ideally, the deepest level rocks at the surface should project down-dip into the crust-mantle boundary but this is rarely the case. If the physical properties of the exposed rocks match those at the correlative levels in the modern crust (e.g., Fountain, 1976; Fountain et al., in press), it is then possible to interpret surface rocks as samples of the contemporary deep crust.

Alternatively, cross sections of the continental crust may be reconstructed from various surface exposures strictly on the basis of geological lines of evidence (e.g., down-plunge projection, geobarometry, etc.). In these cases the appropriate geophysical surveys have not yet been conducted that would reveal the direct connection of the exposed rocks with the contemporary deeper crustal levels. These reconstructions, however, are extremely useful as they represent the crust as it appeared at some time in the past. For instance, Saleeby (this volume) used detailed geochronology, geobarometry, geothermometry and structural analysis to develop a cross section of the Sierra Nevada crust as it existed 100 million years ago. Similarly Kriens and Wernicke (this volume) and, recently, Whitney and McGroder (1989) presented Cretaceous reconstructions of the crust of the North Cascades, Washington.

4. Emplacement of Cross Sections

Early ideas concerning tectonic emplacement of cross sections (Fountain and Salisbury, 1981) stressed that cross sections of the continental crust are emplaced along crustal scale thrust faults in collisional orogenic belts. Recent research, however, demonstrates that these terrains can be emplaced in a variety of tectonic environments. Several cross sections were emplaced along major intracratonic thrust systems (e.g., Kapuskasing, Arunta, Musgrave) whereas others were emplaced in transpressional settings (e.g., Doubtful Sound) and in impactogens (e.g., Slawson, 1976). Detailed structural and geochronological analyses of a few cross sections revealed that several stages of uplift may be necessary to elevate the deepest levels of the continental crust to the surface. For example, the Ivrea zone was apparently uplifted both in a post-Hercynian extensional event and a later Alpine shortening episode (Zingg, this volume).

In many cases, uplift of a section is closely linked to the same processes that form and modify the section. Hollister and Crawford (this volume), for example, showed that the Prince Rupert section was inflated by arc magmas during its uplift, blurring the distinction between crustal growth and tectonic emplacement. This is in strong contrast to the KU where a large amount of time separated the main crustal growth phases and the ultimate tectonic emplacement of the terrain (Percival, this volume).

Emplacement of cross sections occurs along significant detachments that, in most cases, only slice down to intermediate crustal levels (e.g., Percival et al., 1989). Geobarometric data, where available, indicate that the metamorphic conditions of the deepest levels of the terrains rarely exceed 800 to 900 MPa (approximately 24-27 km depth) equilibration pressures and, with the possible exception of the Ivrea zone, do not expose the continental Moho. Bohlen and Mezger (1989), among others, recently articulated this point for granulite facies terrains in general and further argued that granulite facies xenoliths that record much higher pressures are likely representative of deeper crustal levels. These sections may only represent 15 to 25 km thick windows into the crust. An excellent illustration is provided by seismic data for the KU (Boland et al., 1988; Boland and Ellis, 1989; Percival et al., 1989) that, when coupled with geological data, indicate that the detachment soles between 20 and 25 km, leaving the lowermost 15 to 20 km of the crust unsampled. Analysis of laboratory-determined seismic velocities of rocks from the KU (Fountain et al., in press) also demonstrates their depth of origin at only 20 to 25 km through comparison with the seismic refraction section.

5. Are cross sections complete?

Recognition that the main detachment responsible for emplacement of the KU cuts down only to about 20 km suggests that the KU and, possibly, other cross sections are not complete samples of the continental crust. Fountain (1987) pointed out that cross sections can only account for a portion of the total contemporary crustal thickness for the region of interest and emphasized that a significant portion of the lower crust is not exposed in the sections (Fig. 2).

With the possible exception of the ultramafic rocks exposed at the base of the Ivrea zone, the continental Moho is not exposed in cross sections. Even in the Ivrea zone, the mylonitic contacts between granulite facies rocks and the ultramafic bodies suggested to Brodie and Rutter (1987) that an unknown amount of lower crustal section may have been deleted. If this boundary represents the Moho, it must be viewed as a tectonically modified Moho or the type of Moho expected in extensional terrains (Fountain, 1989).

Other portions of cross sections are often missing, especially the top. The upper portions of most of the sections studied to date are within the greenschist facies suggesting erosional removal of about 3 to 9 km of crust. The severity of this problem depends upon one's point of view. For instance, Archean granite-greenstone belts are exposed over much of the Superior province and constitute the top of the KU and Pikwitonei sections (Percival and Card, 1985; Weber and Mezger, this volume). Thus the top *is not* missing when these sections are regarded as cross sections of the contemporary crust but the top *is* missing if they are reconstructed to their crustal position during the Archean (Fig. 2). Several sections do preserve the uppermost levels; volcanic rocks constitute the top of the southern Sierra section (Saleeby, this volume) and a Mesozoic carbonate platform sequence caps the Ivrea section.

Pre-emplacement faults replicate or delete section within some cross sections and provide critical evidence concerning the tectonic processes that shaped that particular crustal column. A possible example is the Pogallo fault zone in the Ivrea zone, a postulated low-angle normal fault that presumably extended the Ivrea section before emplacement into its present near-vertical position (e.g., Hodges and Fountain, 1984; Handy, 1987). This interpretation of the Pogallo fault, however, remains controversial (see Zingg, this volume, for review).

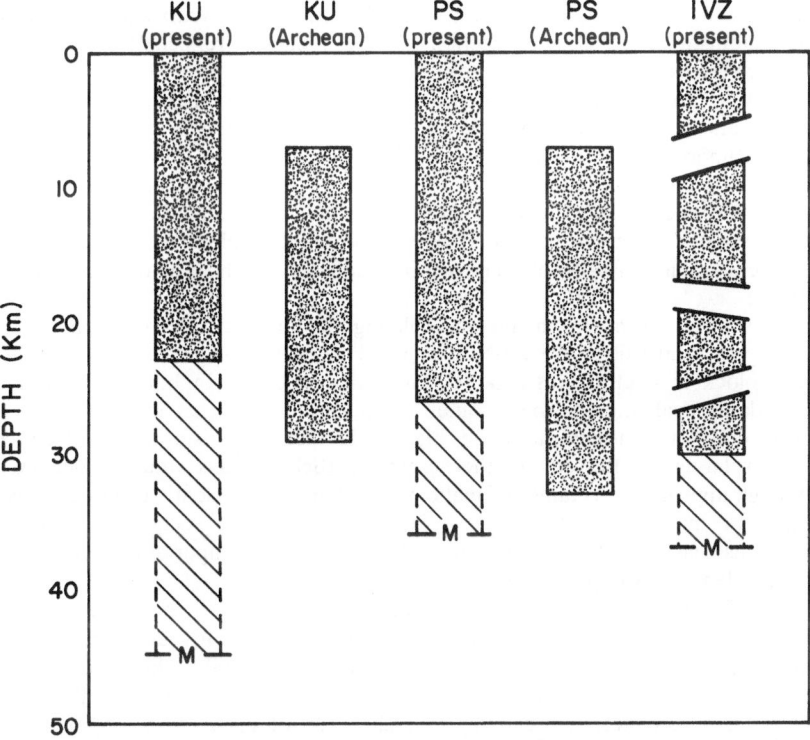

Figure 2. Estimated crustal positions (shaded) of the Kapuskasing uplift (KU), Pikwitonei-Sachigo (PS), and Ivrea-Verbano (IVZ) cross sections for the present day configuration. Ruled area provides an estimate of the amount of lower continental crust not exposed in these cross sections. Also shown are the estimated crustal positions of the KU and PS sections during the late Archean. Tectonically excised sections are represented by gaps in the IVZ section.

6. What have we learned?

What can be learned through the study of cross sections? Although incomplete cross sections, these terrains commonly provide views of half to two-thirds of the contemporary or fossil crust. Importantly, cross sections allow investigation of spatial and temporal relationships between rock units, a serious weakness of xenolith investigations. Another advantage to the study of sections is that deep crustal levels retain linkages with the upper crust. Thus cryptic events or processes at deep levels can be deciphered through analysis of the upper crustal levels where the geology may be easier to understand. Vertical crustal linkages have been particularly informative in the IVZ, where Late Paleozoic-Early Mesozoic rift basin development, volcanism and shallow-level plutonism have been related to extensional shear zone movement and mafic plutonism at mid- and lower crustal levels (Handy, 1987; Schmid et al., 1987; Zingg, this volume). Similarly, deep-level tectonic denudation of the high-grade rocks in Doubtful Sound may have occurred during Cretaceous rifting of New Zealand and Australia (Gibson et al., 1988; Gibson, this volume). Saleeby (this volume), with the aid of geochronological data, postulated that steep mid-crustal structures in the Sierra Nevada region developed in response to downward return flow of supracrustal rocks during magma diapirism (or magmatoconvective recycling) and caldera formation 100 million years ago. Hypabyssal S-type granite plutonism in the Calabria section may be a shallow reflection of depletion of high-grade, lower-crustal metapelites (Schenk, 1984).

Despite the wide range of age and geological disparities of cross sections, almost all contain significant magmatic components. The crust of many of the younger examples (e.g., the Sierras, Prince Rupert, Skagit, Idaho batholith, Doubtful Sound and the Ruby Mountains) show clear signs of growth in continental magmatic arc or island arc environments (Hamilton, 1981). Archean cross sections, such as the Pikwitonei-Sachigo terrane in Manitoba and the KU are similarly dominated by magmatic components, indicating that older crust may differ from young magmatic arcs only in rates of magma production and width of active magmatic belts (Percival, 1988).

Evidence for lower crustal magmatic underplating can only be inferred in a few cross sections. Large mafic intrusions are exposed in the Ivrea zone (Rivalenti et al., 1984) and the Musgrave block in central Australia (Glikson, this volume). In both cases, however, intrusion post-dates peak metamorphic conditions suggesting, at least in these cases, that mafic magmas did not add heat to drive high-grade metamorphism in the deep crust. Magmatic underplating is inferred in other examples (Schenk, this volume; Saleeby, this volume) although appropriate volumes of mafic material are not presently exposed. Although clear evidence for magmatic underplating is lacking in the cross sections, high pressure granulite xenoliths may provide the key evidence that this is an important mode of crustal growth (e.g., Bohlen and Mezger, 1989).

Another significant, although volumetrically minor, component of many deep levels of cross sections is metamorphosed sedimentary and volcanic rocks. Their widespread occurrence implies that crustal formation involves early transport of supracrustal rocks to deep levels by various mechanisms such as tectonic underplating, magmatic overplating (e.g., Percival and Card, 1985) or magmatoconvective recycling (Saleeby, this volume).

The common geologic history revealed in cross sections from single geological provinces indicates that crustal growth and modification processes can occur simultaneously within large regions. For example, peak metamorphic conditions were attained at similar times in the late Archean in the KU and Pikwitonei granulite domain separated by 1400 km across strike in the Superior province. In addition to strong lithological similarities, the Ivrea and Calabria sections, 1200 km apart, both experienced high-grade metamorphic conditions during the Hercynian (Schenk, 1984; Zingg, this volume).

Although crustal cross sections appear not to tap the crust to full Moho depths, they still provide important clues, albeit indirect, about the nature of the lowermost continental crust. The Kapuskasing zone provides an excellent example. Seismic refraction and reflection data (Boland et al., 1988; Boland and Ellis, 1989; Percival et al., 1989) indicate that the Kapuskasing structure only represents about 20 to 25 km of the crust of that region. Typical regional Moho depths are about 40 to 45 km, suggesting that about 20 km of crust are not sampled in the KU. The refraction models show that the lower 20 km of crust is characterized by compressional wave velocities greater than 7 km/s. The weighted mean velocity of the Kapuskasing granulite facies rocks, based on laboratory measurements (Fountain et al., in press), is only about 6.8 to 6.9 km/s suggesting that the Kapuskasing granulites, in their exposed proportions, do not constitute the deeper crustal levels of the Superior province in this region. Some of the garnet-rich mafic rocks, however, have appropriate velocities suggesting that these, or kindred rocks, could be present in greater abundances with depth.

Exposed cross sections of the continental crust are also useful in understanding the variation of physical properties with depth in the crust. Several groups have determined heat production (Galson, 1983; Ashwal et al., 1987; Fountain et al., 1987), seismic velocities (Fountain, 1976; Kern and Schenk, 1985, 1988; Chroston and Simmons, 1989; Burke and Fountain, in press; Fountain et al., in press) and magnetic susceptibility (Wasilewski and Fountain, 1982; Williams et al., 1985; Shive and Fountain, 1988) for suites of rocks collected from these terrains. These data, coupled with the geometry of rock units in the exposed sections, form the basis of geophysical models of the crust that can be compared to field geophysical data to assess various theories concerning the properties of the crust. For example, seismic models based on exposed cross sections (Hale and Thompson, 1982; Fountain, 1986; Burke and Fountain, in press) indicate that large variations of seismic velocity associated with lithologic layers of variable composition could be responsible for the pronounced seismic reflectivity observed in the lower continental crust in some regions. This explanation is but only one of many and is based on models derived from exposed cross sections that have long since terminated their evolution in the deep crust. Rutter and Brodie (in press) emphasize that the physical properties of the exposed cross sections were likely substantially different when these rocks occupied deeper levels because of the activity of fluids and the effect of temperature.

7. Future directions

The study of cross sections of the continental crust will continue to be part of a broader strategy for exploration of the continental crust that includes geophysical methods, deep drilling, and xenolith investigations. Future research will likely take three paths. First, exposed cross sections will be most valuable if they can be traced into the contemporary crust. To date this has only been accomplished to any convincing degree for the Ivrea zone and Kapuskasing zone. Geophysical surveys designed to test such connections for other examples must be mounted. Second, the exposed cross sections provide excellent arenas to calibrate geophysical data. A drill hole through reflections arising from gneisses within the KU, for example, could address the question of the causes of reflections in the deeper crystalline crust at a fraction of the cost of penetrating such structures *in situ*. Finally, more geological, geochemical and geochronological work is required to unravel the complex history of many of the terrains, thereby providing key information about processes of crustal evolution. The Ivrea zone is perhaps the best studied of all the sections, yet many aspects of its evolution remain enigmatic or are hotly debated. Many other examples have seen much less investigation and we can only surmise what interesting processes of crustal evolution might be revealed through further in-depth study. Despite the numerous limitations of exposed cross sections of the continental crust, they remain an extremely valuable guide to the processes that shape much of the continental crust.

References

Ashwal, L.D., Morgan, P., Kelley, S.A., and Percival, J.A., 1987. 'Heat production in an Archean crustal profile and implications for heat flow and mobilization of heat-producing elements' *Earth Planet. Sci. Lett.* **85**, 439-450.

Bayer, R., Cazes, M., Dal Piaz, G.V., Damotte, B., Elter, G., Gosso, G., Hirn, A., Lanza, R., Lombardo, B., Mugnier, J.-L., Nicolas, A., Nicolich, R., Polino, R., Roure, F., Sacchi, R., Scarascia, S., Tabacco, I., Taponnier, P., Tardy, M., Taylor, M., Thouvenot, F., Torreilles, G., and Villien, A., 1987. 'Premiers résultats de la traversée des Alpes occidentales par sismique reflexion verticale (Programme ECORS-CROP)' *C. R. Acad. Sci. Paris* **305**, Serie II, 1461-1470.

Berckhemer, H., 1969. 'Direct evidence for the composition of the lower crust and the Moho' *Tectonophysics* **8**, 97-105.

Bohlen, S.R., and Mezger, K., 1989. 'Origin of granulite terranes and the formation of the lowermost continental crust' *Science* **244**, 326-329.

Boland, A.V., and Ellis, R.M., 1989. 'Velocity structure of the Kapuskasing uplift, northern Ontario, from seismic refraction studies' *J. Geophys. Res.* **94**, 7189-7204.

Boland, A.V., Ellis, R.M., Northey, D.J., West, G.F., Green, A.G., Forsyth, D.A. Mereu, R.F., Meyer, R.P., Morel-à-l'Huissier, P.A., Buchbinder, G.G.R., Asudeh, I., and Haddon, R.A.W., 1988. 'Seismic delineation of upthrust Archaean crust in Kapuskasing, northern Ontario' *Nature* **335**, 711-713.

Brodie, K.H., and Rutter, E.H., 1987. 'Deep crustal extensional faulting in the Ivrea Zone of northern Italy' *Tectonophysics* **140**, 193-212.

Brooks, M., 1970. 'Positive gravity anomalies in some orogenic belts' *Geol. Mag.* **111**, 399-400.

Burke, M., and Fountain, D.M., in press. 'Seismic properties of rocks from extended continental crust - new laboratory measurements from the Ivrea zone' *Tectonophysics*.

Chroston, P.N., and G. Simmons, 1989. 'Seismic velocities from the Kohistan Arc, northern Pakistan' *J. Geol. Soc.* **146**, 971-979.

Forman, D.J., and Shaw, R.D., 1973. 'Deformation of the crust and mantle in central Australia' *Aust. Bur. Miner. Resour., Geol. Geophys. Bull.* **144**, 20p.

Fountain, D.M., 1976. 'The Ivrea-Verbano and Strona-Ceneri zones, northern Italy, a cross-section of the continental crust-new evidence from seismic velocities' *Tectonophysics* **33**, 145-166.

Fountain, D.M., 1986. 'Implications of deep crustal evolution for seismic reflection seismology' In M. Barazangi and L. Brown, eds, *Reflection Seismology: The Continental Crust. Am. Geophys. Un. Geodyn. Ser.* **14**, 1-7.

Fountain, D.M., 1987. 'Geological and geophysical nature of the lower continental crust as revealed by exposed cross sections of the continental crust' In J.S. Noller, S.H. Kirby and

J.E. Nielsen-Pike, eds, *Geophysics and Petrology of the Deep Crust and Upper Mantle. U.S. Geol. Surv. Circular,* **0956**, 25-26.

Fountain, D.M., 1989. 'Growth and modification of lower continental crust in extended terrains: the role of extension and magmatic underplating' In R.F. Mereu, S. Mueller and D.M. Fountain, eds, *Lower Crust: Properties and Processes. Am. Geophys. Monogr.* **51**, 287-299.

Fountain, D.M., and Salisbury, M.H., 1981. 'Exposed cross sections through the continental crust: Implications for crustal structure, petrology and evolution' *Earth Planet. Sci. Lett.* **5**, 263-277.

Fountain, D.M., Salisbury, M.H., and Furlong, K.P., 1987. 'Heat production and thermal conductivity of rocks from the Pikwitonei-Sachigo continental cross section, central Manitoba: implications for the thermal structure of Archean crust' *Can. J. Earth Sci.* **24**, 1583-1594.

Fountain, D.M., Salisbury, M.H., and Percival, J.A., in press. 'Seismic properties of the Superior Province crust based on seismic velocity measurements on rocks from the Michipicoten, Wawa and Kapuskasing terranes, Ontario' *J. Geophys. Res.*.

Galson, D.A., 1983. 'Heat production and temperature of the Alpine crust' Ph.D. Diss., Cambridge, 171p.

Gibb, R.A., and Thomas, M.D., 1976. 'Gravity signature of fossil plate boundaries in the Canadian shield' *Nature* **262**, 199-200.

Gibson, G.M., McDougall, I., and Ireland, T.R., 1988. 'Age constraints on metamorphism and development of a metamorphic core complex in Fiordland, southern New Zealand' *Geology* **16**, 405-408.

Giese, P., 1968. 'Die Struktur der Erdkruste im Bereich der Ivrea-Zone. Ein Vergleich verschiedener, seismischer Interpretationen und der Versuch einer petrographisch-geologischen Deutung' *Schweiz. Mineral. Petrogr. Mitt.* **48**, 261-284.

Goleby, B.R., Shaw, R.D., Wright, C., Kennett, B.L.N., and Lameck, K., 1989. 'Geophysical evidence for 'thick-skinned' crustal deformation in central Australia' *Nature* **337**, 325-330.

Hale, L.D., and Thompson, G.A., 1982. 'The seismic reflection character of the continental Mohorovicic discontinuity' *J. Geophys. Res.* **87**, 4625-4635.

Hamilton, W., 1981. 'Crustal evolution by arc magmatism' *Phil. Trans. R. Soc. Lond.* **A301**, 279-291.

Handy, M.R., 1987. 'The structure, age and kinematics of the Pogallo fault zone; Southern Alps, northwestern Italy' *Eclogae Helv.* **80**, 593-632.

Hodges, K.V., and Fountain, D.M., 1984. 'Pogallo Line, South Alps, northern Italy: An intermediate crustal level, low-angle normal fault?' *Geology* **12**, 151-155.

Kern, H., and Schenk, V., 1985. 'Elastic wave velocities in rocks from a lower crustal section in southern Calabria (Italy)' *Earth Planet. Sci. Lett.* **40**, 147-160.

Kern, H., and Schenk, V., 1988. 'A model of velocity structure beneath Calabria southern Italy, based on laboratory data' *Earth Planet. Sci. Lett.* **87**, 325-337.

Mezger, K., Hanson, G.N., and Bohlen, S.R., 1989. 'U-Pb systematics of garnet: dating the growth of garnet in the Late Archean Pikwitonei granulite domain at Cauchon and Natawahunan Lakes, Manitoba, Canada' *Contrib. Mineral. Petrol.* **101**, 136-148.

Oliver, G.J.H., and Coggon, J.H., 1979. 'Crustal structure of Fiordland, New Zealand' *Tectonophysics* **54**, 253-292.

Percival, J.A., 1983. 'High-grade metamorphism in the Chapleau-Foleyet area, Ontario' *Am. Mineral.* **68**, 667-686.

Percival, J.A., 1988. 'Deep geology out in the open' *Nature* **335**, 671.

Percival, J.A., and Card, K.D., 1983. 'Archean crust as revealed in the Kapuskasing uplift, Superior Province, Canada' *Geology* **11**, 323-326.

Percival, J.A., and Card, K.D., 1985. 'Structure and evolution of Archean crust in central Superior Province, Canada' In L.D. Ayres, P.C. Thurston, K.D. Card and W. Weber, eds, *Evolution of Archean Supracrustal Sequences. Geol. Assoc. Can. Spec. Pap.*, **28**, 179-192.

Percival, J.A., Green, A.G., Milkereit, B., Cook, F.A., Geis, W., and West, G.F., 1989. 'Seismic reflection profiles across deep continental crust exposed in the Kapuskasing uplift structure' *Nature* **342**, 416-420.

Priestley, K., and Davey, F., 1983. 'Crustal structure of Fiordland, New Zealand, from seismic refraction measurements' *Geology* **11**, 660-663.

Rivalenti, G., Rossi, A., Siena F., and Sinigoi, S., 1984. 'The layered series of the Ivrea Verbano igneous complex, Western Alps, Italy' *Tschermaks Mineral. Petrogr. Mitt.* **33**, 77-99.

Rutter, E.H., and Brodie, K.H. in press. 'Some geophysical implications of the deformation and metamorphism of the Ivrea zone, northern Italy' *Tectonophysics*.

Salisbury, M.H., and Fountain, D.M. 1974. 'Continent/continent obduction zones: geological and geophysical case studies' *EOS Trans. Am. Geophys. Un.* **57**, 335.

Schenk, V., 1984. 'Petrology of felsic granulites, metapelites, metabasics, ultramafics and metacarbonates from southern Calabria (Italy): Prograde metamorphism, uplift and cooling of a former lower crust' *J. Petrol.* **25**, 255-298.

Schmid, R., and Wood, B.J., 1976. 'Phase relationships in granulitic metapelites from the Ivrea-Verbano Zone (Northern Italy)' *Contrib. Mineral. Petrol.* **54**, 255-279.

Schmid, S.M., Zingg, A., and Handy, M., 1987. 'The kinematics of movements along the Insubric Line and the emplacement of the Ivrea zone' *Tectonophysics* **135**, 47-66.

Shive, P.N., and Fountain, D.M., 1988. 'Magnetic mineralogy in an Archean crustal cross section: Implications for crustal magnetization' *J. Geophys. Res.* **93**, 12,177-12,186.

Slawson, W.F., 1976. 'Vredefort dome: A cross-section of the upper crust' *Geochim. Cosmochim. Acta* **40**, 117-121.

Wasilewski, P., and Fountain, D.M., 1982. 'The Ivrea zone as a model for the distribution of magnetization in the continental crust' *Geophys. Res. Lett.* **9**, 333-336.

Whitney, D.L., and McGroder, M.F., 1989. 'Cretaceous crustal section through the proposed Insular-Intermontane suture, North Cascades, Washington' *Geology* **17**, 555-558.

Williams, M.C., Shive, P.N., Fountain, D.M. and Frost, B.R., 1985. 'Magnetic properties of exposed deep crustal rocks from the Superior Province of Manitoba' *Earth Planet. Sci. Lett.* **76**, 176-184.